W9-CMU-458

Metals & Alloys in the
UNIFIED NUMBERING SYSTEM

Fourth Edition

WITH A DESCRIPTION OF THE SYSTEM AND A CROSS INDEX OF CHEMICALLY SIMILAR SPECIFICATIONS

SAE HSJ 1086 APR86
ASTM DS-56 C

A Joint Publication of the
SOCIETY OF AUTOMOTIVE ENGINEERS, INC.
AMERICAN SOCIETY FOR TESTING AND MATERIALS

Permission to photocopy for internal or personal use, or the internal or personal use of specific clients, is granted by SAE for libraries and other users registered with the Copyright Clearance Center (CCC), provided that the base fee of $3.00 per copy is paid directly to CCC, 21 Congress St., Salem, MA 01970. Special requests should be addressed to the SAE Publications Division.

No part of this publication may be reproduced in any form, in an electronic retrieval system or otherwise, without the prior written permission of the publisher.

Published by:
Society of Automotive Engineers, Inc.
400 Commonwealth Drive
Warrendale, PA 15096

ISBN 0-89883-419-8
Library of Congress Catalog Card Number: 77-89064
Copyright 1986 Society of Automotive Engineers, Inc.

PREFACE TO THE FOURTH EDITION

This Fourth Edition contains in excess of 3000 designations, an increase of approximately 200 from the number which appeared in the Third Edition, as well as descriptions and cross-reference specifications related to these designations.

All descriptions and references have been refined and expanded, including many new American Society of Mechanical Engineers (ASME), American Society for Testing and Materials (ASTM), SAE, Federal, Military, American Iron and Steel Institute (AISI), and American Welding Society (AWS) specification references.

Included are cross references by specifications to applicable UNS designations, as well as a reference list by some of the common trade designations.

Information regarding the UNS organization and the manner in which designations are issued, is contained in the combined document SAE J1086 and ASTM E527 which is included in the Appendix.

The latest application form for UNS number assignment (and for input of the new data to the UNS computer system) has been included.

Alvin G. Cook, Chairman
UNS Advisory Board

TABLE OF CONTENTS

INTRODUCTION TO THE UNIFIED NUMBERING SYSTEM

The Unified Numbering System (UNS) provides a means of correlating many nationally used metal and alloy numbering systems currently administered by societies, trade associations, and individual users and producers of metals and alloys, thereby avoiding confusion caused by use of more than one identification number for the same metal or alloy, and by the opposite situation of having the same number assigned to two or more different metals or alloys.

It provides the uniformity necessary for efficient indexing, record keeping, data storage and retrieval, and cross-referencing.

A UNS designation is not in itself a specification, since it establishes no requirements for form, condition, properties, or quality. It is a unified identifier of a metal or an alloy for which controlling limits have been established in specifications published elsewhere.

The UNS establishes 18 series of designations for metals and alloys. Each UNS designation consists of a single-letter prefix followed by five digits. In most cases, the letter is suggestive of the family of metals identified, for example A for aluminum, P for precious metals, S for stainless steels.

Although some of the digits in certain UNS designation groups have special assigned meanings, each series is independent of the others in regard to significance of digits, thus permitting greater flexibility and avoiding complicated and lengthy UNS designations.

Wherever feasible, and for user convenience, identification "numbers" from existing systems are incorporated into the UNS designations. For example, carbon steel presently identified by American Iron and Steel Institute "AISI 1020" is covered by "UNS G10200."

The UNS designation assignments for certain metals and alloys are established by the relevant trade associations which in the past have administered their own numbering systems; for other metals and alloys, UNS designation assignments are processed by the SAE or the American Society for Testing and Materials (ASTM). Each of these assigners has the responsibility for administering a specific UNS series of designations. Each considers requests for the assignment of new UNS designations, and informs applicants of the action taken. UNS designation assigners report immediately to both SAE and ASTM details of each assignment for inclusion into the system.

For additional details on the UNS System, see ASTM E527 and SAE J1086, "Recommended Practice for Numbering Metals and Alloys," shown in the Appendix.

The cross-referenced specifications are representative only and are not necessarily a complete list of specifications applicable to a particular UNS designation.

LISTING OF CHEMICAL ELEMENTS

Actinium	Ac	Gallium	Ga
Aluminum	Al	Germanium	Ge
Americium	Am	Gold, Aurum	Au
Antimony, Stibium	Sb	Hafnium	Hf
Argon	A	Helium	He
Arsenic	As	Holmium	Ho
Astatine	At	Hydrogen	H
Barium	Ba	Indium	In
Berkelium	Bk	Iodine	I
Beryllium, Glucinum	Be	Iridium	Ir
Bismuth	Bi	Iron, Ferrum	Fe
Boron	B	Krypton	Kr
Bromine	Br	Lanthanum	La
Cadmium	Cd	Lawrencium	Lr
Calcium	Ca	Lead, Plumbum	Pb
Californium	Cf	Lithium	Li
Carbon	C	Lutetium, Cassiopeium	Lu
Cassiopeium, Lutetium	Lu	Magnesium	Mg
Cerium	Ce	Manganese	Mn
Cesium	Cs	Mendelevium	Md
Chlorine	Cl	Mercury, Hydrargyrum	Hg
Chromium	Cr	Molybdenum	Mo
Cobalt	Co	Neodymium	Nd
Columbium, Niobium	Cb[1]	Neon	Ne
Copper, Cuprum	Cu	Neptunium	Np
Curium	Cm	Nickel	Ni
Dysprosium	Dy	Niobium, Columbium	Nb[1]
Einsteinium	Es	Nitrogen	N
Erbium	Er	Nobelium	No
Europium	Eu	Osmium	Os
Fermium	Fm	Oxygen	O
Fluorine	F	Palladium	Pd
Francium	Fr	Phosphorus	P
Gadolinium	Gd	Platinum	Pt

[1] The UNS uses either Columbium (Cb) or Niobium (Nb) in its listings depending upon the prevailing practice associated with that particular alloy system.

UNS NUMBERS ASSIGNED TO DATE

With Description of Each Material Covered and References To Documents In Which The Same or Similar Materials are Described

Axxxxx Number Series
Aluminum and Aluminum Alloys

ALUMINUM AND ALUMINUM ALLOYS

UNIFIED NUMBER	DESCRIPTION	CHEMICAL COMPOSITION	CROSS REFERENCE SPECIFICATIONS
A01001	Aluminum Foundry Alloy, Ingot	**Al** 99.00 min **Cu** 0.10 max **Fe** 0.6-0.8 **Si** 0.15 max **Zn** 0.05 max **Other** each 0.03 max, total 0.10 max, Mn+Cr+Ti+V 0.025 max	**AA** 100.1 **ASTM** B179 (100.1)
A01301	Aluminum Foundry Alloy, Ingot	**Al** 99.30 min **Cu** 0.10 **Zn** 0.05 max **Other** each 0.03 max, total 0.10 max, Fe/Si ratio 2.5 min, Mn+Cr+Ti+V 0.025 max	**AA** 130.1 **ASTM** B179 (130.1)
A01501	Aluminum Foundry Alloy, Ingot	**Al** 99.50 min **Cu** 0.05 max **Zn** 0.05 max **Other** each 0.03 max, total 0.10 max, Fe/Si ratio 2.0 min, Mn+Cr+Ti+V 0.025 max	**AA** 150.1 **ASTM** B179 (150.1)
A01601	Aluminum Foundry Alloy, Ingot	**Al** rem **Fe** 0.25 max **Si** 0.10 max **Zn** 0.05 **Other** each 0.03 max, total 0.10 max, Fe/Si ratio 2.0 min, Mn+Cr+Ti+V 0.025 max	**AA** 160.1
A01701	Aluminum Foundry Alloy, Ingot	**Al** 99.70 min **Zn** 0.05 max **Other** each 0.03 max, total 0.10 max, Fe/Si ratio 1.5 min, Mn+Cr+Ti+V 0.025 max	**AA** 170.1 **ASTM** B179 (170.1)
A02010	Aluminum Foundry Alloy, Casting	**Ag** 0.40-1.2 **Al** rem **Cu** 4.0-5.2 **Fe** 0.15 max **Mg** 0.15-0.55 **Mn** 0.20-0.50 **Si** 0.10 max **Ti** 0.15-0.35 **Other** each 0.05 max, total 0.10 max	**AA** 201.0 **AMS** 4188/1 **ASTM** B26 (201.0) **SAE** J452 (382); J453 (382)
A02012	Aluminum Foundry Alloy, Ingot	**Ag** 0.40-1.2 **Al** rem **Cu** 4.0-5.2 **Fe** 0.10 max **Mg** 0.20-0.55 **Mn** 0.20-0.50 **Si** 0.10 max **Ti** 0.15-0.35 **Other** each 0.05 max, total 0.10 max	**AA** 201.2 **ASTM** B179 (201.2) **SAE** J453 (382)
A02020	Aluminum Foundry Alloy, Casting	**Ag** 0.40-1.0 **Al** rem **Cr** 0.20-0.6 **Cu** 4.0-5.2 **Fe** 0.15 max **Mg** 0.15-0.55 **Mn** 0.20-0.8 **Si** 0.10 max **Ti** 0.15-0.35 **Other** each 0.05 max, total 0.10 max	**AA** 202.0
A02022	Aluminum Foundry Alloy, Ingot	**Ag** 0.40-1.0 **Al** rem **Cr** 0.20-0.6 **Cu** 4.0-5.2 **Fe** 0.10 max **Mg** 0.20-0.55 **Mn** 0.20-0.8 **Si** 0.10 max **Ti** 0.15-0.35 **Other** each 0.05 max, total 0.10 max	**AA** 202.2
A02030	Aluminum Foundry Alloy, Casting (Hiduminium)	**Al** rem **Co** 0.20-0.30 **Cu** 4.5-5.5 **Fe** 0.50 max **Mg** 0.10 max **Mn** 0.20-0.30 **Ni** 1.3-1.7 **Sb** 0.20-0.30 **Si** 0.30 max **Ti** 0.15-0.25 **Zn** 0.10 max **Zr** 0.10-0.30 **Other** each 0.05 max, total 0.20 max, plus Ti+Zr 0.50 max	**AA** 203.0 **AMS** 4225
A02032	Aluminum Foundry Alloy, Ingot	**Al** rem **Co** 0.20-0.30 **Cu** 4.8-5.2 **Fe** 0.35 max **Mg** 0.10 max **Mn** 0.20-0.30 **Ni** 1.3-1.7 **Sb** 0.20-0.30 **Si** 0.20 max **Ti** 0.15-0.25 **Zn** 0.10 max **Zr** 0.10-0.30 **Other** each 0.05 max, total 0.20 max, plus Ti+Zr 0.50 max	**AA** 203.2
A02040	Aluminum Foundry Alloy, Casting	**Al** rem **Cu** 4.2-5.0 **Fe** 0.35 max **Mg** 0.15-0.35 **Mn** 0.10 max **Ni** 0.05 max **Si** 0.20 max **Sn** 0.05 max **Ti** 0.15-0.30 **Zn** 0.10 max **Other** each 0.05 max, total 0.15 max	**AA** 204.0 **ASTM** B26 (204.0); B108 (204.0)
A02042	Aluminum Foundry Alloy, Ingot	**Al** rem **Cu** 4.2-4.9 **Fe** 0.10-0.20 **Mg** 0.20-0.35 **Mn** 0.05 max **Ni** 0.03 max **Si** 0.15 max **Sn** 0.05 max **Ti** 0.15-0.25 **Zn** 0.05 max **Other** each 0.05 max, total 0.15 max	**AA** 204.2 **ASME** SFA5.10 (R242.0) **ASTM** B179 (204.2) **AWS** A5.10 (R242.0)
A02060	Aluminum Foundry Alloy, Casting	**Al** rem **Cu** 4.2-5.0 **Fe** 0.15 max **Mg** 0.15-0.35 **Mn** 0.20-0.50 **Ni** 0.05 max **Si** 0.10 max **Sn** 0.05 max **Ti** 0.15-0.30 **Zn** 0.10 max **Other** each 0.05 max, total 0.15 max	**AA** 206.0 **AMS** 4188/2; 4237 **ASME** SFA5.10 (R206.0) **AWS** A5.10 (R206.0)

The chemical compositions listed are for identification purposes and should not be used in lieu of the cross referenced specifications.

ALUMINUM AND ALUMINUM ALLOYS

UNIFIED NUMBER	DESCRIPTION	CHEMICAL COMPOSITION	CROSS REFERENCE SPECIFICATIONS
A02062	Aluminum Foundry Alloy, Ingot	Al rem Cu 4.2-5.0 Fe 0.10 max Mg 0.20-0.35 Mn 0.20-0.50 Ni 0.03 max Si 0.10 max Sn 0.05 max Ti 0.15-0.25 Zn 0.05 max Other each 0.05 max, total 0.15 max	AA 206.2
A02080	Aluminum Foundry Alloy, Casting	Al rem Cu 3.5-4.5 Fe 1.2 max Mg 0.10 max Mn 0.50 max Ni 0.35 max Si 2.5-3.5 Ti 0.25 max Zn 1.0 max Other total 0.50 max	AA 208.0 ASTM B26 (208.0); B108 (208.0) FED QQ-A-601 (208.0) MIL SPEC MIL-F-3922; MIL-F-39000
A02081	Aluminum Foundry Alloy, Ingot	Al rem Cu 3.5-4.5 Fe 0.9 max Mg 0.10 max Mn 0.50 max Ni 0.35 max Si 2.5-3.5 Ti 0.25 max Zn 1.0 max Other total 0.50 max	AA 208.1 ASTM B179 (208.1) FED QQ-A-371 (208.1)
A02082	Aluminum Foundry Alloy, Ingot	Al rem Cu 3.5-4.5 Fe 0.8 max Mg 0.03 max Mn 0.30 max Si 2.5-3.5 Ti 0.20 max Zn 0.20 max Other total 0.30 max	AA 208.2 ASTM B179 (208.2) FED QQ-A-371 (208.2)
A02130	Aluminum Foundry Alloy, Casting	Al rem Cu 6.0-8.0 Fe 1.2 max Mg 0.10 max Mn 0.6 max Ni 0.35 max Si 1.0-3.0 Ti 0.25 max Zn 2.5 max Other total 0.50 max	AA 213.0 FED QQ-A-596 (213)
A02131	Aluminum Foundry Alloy, Ingot	Al rem Cu 6.0-8.0 Fe 0.9 max Mg 0.10 max Mn 0.6 max Ni 0.35 max Si 1.0-3.0 Ti 0.25 max Zn 2.5 max Other total 0.50 max	AA 213.1 FED QQ-A-371 (213.1)
A02220	Aluminum Foundry Alloy, Casting	Al rem Cu 9.2-10.7 Fe 1.5 max Mg 0.15-0.35 Mn 0.50 max Ni 0.50 max Si 2.0 max Ti 0.25 max Zn 0.8 max Other total 0.35 max	AA 222.0 ASTM B26 (222.0); B108 (222.0) FED QQ-A-596 (222); QQ-A-601 (222) SAE J452 (34); J453 (34)
A02221	Aluminum Foundry Alloy, Ingot	Al rem Cu 9.2-10.7 Fe 1.2 max Mg 0.20-0.35 Mn 0.50 max Ni 0.50 max Si 2.0 max Ti 0.25 max Zn 0.8 max Other total 0.35 max	AA 222.1 ASTM B179 (222.1) FED QQ-A-371 (222.1) SAE J453 (34)
A02240	Aluminum Foundry Alloy, Casting	Al rem Cu 4.5-5.5 Fe 0.10 max Mn 0.20-0.50 Si 0.06 max Ti 0.35 max V 0.05-0.15 Zr 0.10-0.25 Other each 0.03 max, total 0.10 max	AA 224.0 AMS 4226
A02242	Aluminum Foundry Alloy, Ingot	Al rem Cu 4.5-5.5 Fe 0.04 max Mn 0.20-0.50 Si 0.02 max Ti 0.25 max V 0.05-0.15 Zr 0.10-0.25 Other each 0.03 max, total 0.10 max	AA 224.2
A02380	Aluminum Foundry Alloy, Casting	Al rem Cu 9.0-11.0 Fe 1.5 max Mg 0.15-0.35 Mn 0.6 max Ni 1.0 max Si 3.5-4.5 Ti 0.25 max Zn 1.5 max Other total 0.50 max	AA 238.0 ASTM B108 (238.0)
A02381	Aluminum Foundry Alloy, Ingot	Al rem Cu 9.0-11.0 Fe 1.2 max Mg 0.20-0.35 Mn 0.6 max Ni 1.0 max Si 3.5-4.5 Ti 0.25 max Zn 1.5 max Other total 0.50 max	AA 238.1 FED QQ-A-371 (238.1)
A02382	Aluminum Foundry Alloy, Ingot	Al rem Cu 9.5-10.5 Fe 1.2 max Mg 0.20-0.35 Mn 0.50 max Ni 0.50 max Si 3.5-4.5 Ti 0.20 max Zn 0.50 max Other total 0.50 max	AA 238.2 FED QQ-A-371 (238.2)
A02400	Aluminum Foundry Alloy, Casting	Al rem Cu 7.0-9.0 Fe 0.50 max Mg 5.5-6.5 Mn 0.3-0.7 Ni 0.30-0.7 Si 0.50 max Ti 0.20 max Zn 0.10 max Other each 0.05 max, total 0.15 max	AA 240.0 AMS 4227
A02401	Aluminum Foundry Alloy, Ingot	Al rem Cu 7.0-9.0 Fe 0.40 max Mg 5.6-6.5 Mn 0.30-0.7 Ni 0.30-0.7 Si 0.50 max Ti 0.20 max Zn 0.10 max Other each 0.05 max, total 0.15 max	AA 240.1 FED QQ-A-371 (A240.1)

The chemical compositions listed are for identification purposes and should not be used in lieu of the cross referenced specifications.

UNIFIED NUMBER	DESCRIPTION	CHEMICAL COMPOSITION	CROSS REFERENCE SPECIFICATIONS
A02420	Aluminum Foundry Alloy, Casting	Al rem Cr 0.25 max Cu 3.5-4.5 Fe 1.0 max Mg 1.2-1.8 Mn 0.35 max Ni 1.7-2.3 Si 0.7 max Ti 0.25 max Zn 0.35 max Other each 0.05 max, total 0.15 max	AA 242.0 AMS 4222 ASME SFA5.10 (R242.0) ASTM B26 (242.0); B108 (242.0) AWS A5.10 (R242.0) FED QQ-A-596 (242); QQ-A-601 (242.0) SAE J452 (39); J453 (39)
A02421	Aluminum Foundry Alloy, Ingot	Al rem Cr 0.25 max Cu 3.5-4.5 Fe 0.8 max Mg 1.3-1.8 Mn 0.35 max Ni 1.7-2.3 Si 0.7 max Ti 0.25 max Zn 0.35 max Other each 0.05 max, total 0.15 max	AA 242.1 ASTM B179 (242.1) FED QQ-A-371 (242.1)
A02422	Aluminum Foundry Alloy, Ingot	Al rem Cu 3.5-4.5 Fe 0.6 max Mg 1.3-1.8 Mn 0.10 max Ni 1.7-2.3 Si 0.6 max Ti 0.20 max Zn 0.10 max Other each 0.05 max, total 0.15 max	AA 242.2 ASTM B179 (242.2) FED QQ-A-371 (242.2)
A02430	Aluminum Foundry Alloy, Casting	Al rem Cr 0.20-0.40 Cu 3.5-4.5 Fe 0.40 max Mg 1.8-2.3 Mn 0.15-0.45 Ni 1.9-2.3 Si 0.35 max Ti 0.06-0.20 V 0.06-0.20 Zn 0.05 max Other each 0.05 max, total 0.15 max	AA 243.0 AMS 4224
A02431	Aluminum Foundry Alloy, Ingot	Al rem Cr 0.20-0.40 Cu 3.5-4.5 Fe 0.30 max Mg 1.9-2.3 Mn 0.15-0.45 Ni 1.9-2.3 Si 0.35 max Ti 0.06-0.20 Zn 0.05 max Other each 0.05 max, total 0.15 max	AA 243.1
A02490	Aluminum Foundry Alloy, Casting	Al rem Cu 3.8-4.6 Fe 0.10 max Mg 0.25-0.50 Mn 0.25-0.50 Si 0.05 max Ti 0.02-0.35 Zn 2.5-3.5 Other each 0.03 max, total 0.10 max	AA 249.0
A02492	Aluminum Foundry Alloy, Ingot	Al rem Cu 3.8-4.6 Fe 0.07 max Mg 0.30-0.50 Mn 0.25-0.50 Si 0.05 max Ti 0.02-0.12 Zn 2.5-3.5 Other each 0.03 max, total 0.10 max	AA 249.2
A02950	Aluminum Foundry Alloy, Casting	Al rem Cu 4.0-5.0 Fe 1.0 max Mg 0.03 max Mn 0.35 max Si 0.7-1.5 Ti 0.25 max Zn 0.35 max Other each 0.05 max, total 0.15 max	AA 295.0 AMS 4231 ASME SFA5.10 (R295.0) ASTM B26 (295.0) AWS A5.10 (R295.0) FED QQ-A-601 (295.0) MIL SPEC MIL-F-3922; MIL-C-11866 (195); MIL-F-39000 SAE J452 (38); J453 (38)
A02951	Aluminum Foundry Alloy, Ingot	Al rem Cu 4.0-5.0 Fe 0.8 max Mg 0.03 max Mn 0.35 max Si 0.7-1.5 Ti 0.25 max Zn 0.35 max Other each 0.05 max, total 0.15 max	AA 295.1 ASTM B179 (295.1) FED QQ-A-371 (295.1)
A02952	Aluminum Foundry Alloy, Ingot	Al rem Cu 4.0-5.0 Fe 0.8 max Mg 0.03 max Mn 0.30 max Si 0.7-1.2 Ti 0.20 max Zn 0.30 max Other each 0.05 max, total 0.15 max	AA 295.2 ASTM B179 (295.2) FED QQ-A-371 (295.2) SAE J453 (38)
A02960	Aluminum Foundry Alloy, Casting	Al rem Cu 4.0-5.0 Fe 1.2 max Mg 0.05 max Mn 0.35 max Ni 0.35 max Si 2.0-3.0 Ti 0.25 max Zn 0.50 max Other total 0.35 max	AA 296.0 FED QQ-A-596 (296) SAE J452 (380)
A02961	Aluminum Foundry Alloy, Ingot	Al rem Cu 4.0-5.0 Fe 0.9 max Mg 0.05 max Mn 0.35 max Ni 0.35 max Si 2.0-3.0 Ti 0.25 max Zn 0.50 max Other total 0.35 max	AA 296.1 AMS 4282 FED QQ-A-371 (B295.1)
A02962	Aluminum Foundry Alloy, Ingot	Al rem Cu 4.0-5.0 Fe 0.8 max Mg 0.03 max Mn 0.30 max Si 2.0-3.0 Ti 0.20 max Zn 0.30 max Other each 0.05 max, total 0.15 max	AA 296.2 FED QQ-A-371 (B295.2)
A03050	Aluminum Foundry Alloy, Casting	Al rem Cr 0.25 max Cu 1.0-1.5 Fe 0.6 max Mg 0.10 max Mn 0.50 max Si 4.5-5.5 Ti 0.25 max Zn 0.35 max Other each 0.05 max, total 0.15 max	AA 305.0

The chemical compositions listed are for identification purposes and should not be used in lieu of the cross referenced specifications.

UNIFIED NUMBER	DESCRIPTION	CHEMICAL COMPOSITION	CROSS REFERENCE SPECIFICATIONS
A03052	Aluminum Foundry Alloy, Ingot	**Al** rem **Cu** 1.0-1.5 **Fe** 0.14-0.25 **Mn** 0.05 max **Si** 4.5-5.5 **Ti** 0.20 max **Zn** 0.05 max **Other** each 0.05 max, total 0.15 max	**AA** 305.2
A03080	Aluminum Foundry Alloy, Casting	**Al** rem **Cu** 4.0-5.0 **Fe** 1.0 max **Mg** 0.10 max **Mn** 0.50 max **Si** 5.0-6.0 **Ti** 0.25 max **Zn** 1.0 max **Other** total 0.50 max	**AA** 308.0 **FED** QQ-A-596 (308)
A03081	Aluminum Foundry Alloy, Ingot	**Al** rem **Cu** 4.0-5.0 **Fe** 0.8 max **Mg** 0.10 max **Mn** 0.50 max **Si** 5.0-6.0 **Ti** 0.25 max **Zn** 1.0 max **Other** total 0.50 max	**AA** 308.1 **FED** QQ-A-371 (308.1)
A03082	Aluminum Foundry Alloy, Ingot	**Al** rem **Cu** 4.0-5.0 **Fe** 0.8 max **Mg** 0.10 max **Mn** 0.30 max **Si** 5.0-6.0 **Ti** 0.20 max **Zn** 0.50 max **Other** total 0.50 max	**AA** 308.2 **FED** QQ-A-371 (308.2)
A03190	Aluminum Foundry Alloy, Casting	**Al** rem **Cu** 3.0-4.0 **Fe** 1.0 max **Mg** 0.10 max **Mn** 0.50 max **Ni** 0.35 max **Si** 5.5-6.5 **Ti** 0.25 max **Zn** 1.0 max **Other** total 0.50 max	**AA** 319.0 **ASTM** B26 (319.0); B108 (319.0) **FED** QQ-A-596 (319); QQ-A-601 (319.0) **SAE** J452 (326); J453 (326)
A03191	Aluminum Foundry Alloy, Ingot	**Al** rem **Cu** 3.0-4.0 **Fe** 0.8 max **Mg** 0.10 max **Mn** 0.50 max **Ni** 0.35 max **Si** 5.5-6.5 **Ti** 0.25 max **Zn** 1.0 max **Other** total 0.50 max	**AA** 319.1 **ASTM** B179 (319.1) **FED** QQ-A-371 (319.1) **SAE** J453 (326)
A03192	Aluminum Foundry Alloy, Ingot	**Al** rem **Cu** 3.0-4.0 **Fe** 0.6 max **Mg** 0.10 max **Mn** 0.10 max **Ni** 0.10 max **Si** 5.5-6.5 **Ti** 0.20 max **Zn** 0.10 max **Other** total 0.20 max	**AA** 319.2 **ASTM** B179 (319.2) **FED** QQ-A-371 (319.2)
A03200	Aluminum Foundry Alloy, Casting	**Al** rem **Cu** 2.0-4.0 **Fe** 1.2 **Mg** 0.05-0.6 **Mn** 0.8 **Ni** 0.35 **Si** 5.0-8.0 **Ti** 0.25 **Zn** 3.0 **Other** total 0.50 max	**AA** 320.0
A03201	Aluminum Foundry Alloy, Ingot	**Al** rem **Cu** 2.0-4.0 **Fe** 0.9 **Mg** 0.10-0.6 **Mn** 0.8 **Ni** 0.35 **Si** 5.0-8.0 **Ti** 0.25 **Zn** 3.0 **Other** total 0.50 max	**AA** 320.1
A03240	Aluminum Foundry Alloy, Casting	**Al** rem **Cu** 0.40-0.6 **Fe** 1.2 max **Mg** 0.40-0.7 **Mn** 0.50 max **Ni** 0.30 max **Si** 7.0-8.0 **Ti** 0.20 max **Zn** 1.0 max **Other** each 0.15 max, total 0.20 max	**AA** 324.0
A03241	Aluminum Foundry Alloy, Ingot	**Al** rem **Cu** 0.40-0.6 **Fe** 0.9 max **Mg** 0.45-0.7 **Mn** 0.50 max **Ni** 0.30 max **Si** 7.0-8.0 **Ti** 0.20 max **Zn** 1.0 max **Other** each 0.15 max, total 0.20 max	**AA** 324.1 **FED** QQ-A-371 (324.1)
A03242	Aluminum Foundry Alloy, Ingot	**Al** rem **Cu** 0.40-0.6 **Fe** 0.6 max **Mg** 0.45-0.7 **Mn** 0.10 max **Ni** 0.10 max **Si** 7.0-8.0 **Ti** 0.20 max **Zn** 0.10 max **Other** each 0.05 max, total 0.15 max	**AA** 324.2
A03280	Aluminum Foundry Alloy, Casting	**Al** rem **Cr** 0.35 max **Cu** 1.0-2.0 **Fe** 1.0 max **Mg** 0.20-0.6 **Mn** 0.20-0.6 **Ni** 0.25 max **Si** 7.5-8.5 **Ti** 0.25 max **Zn** 1.5 max **Other** total 0.50 max	**AA** 328.0 **ASTM** B26 (328.0) **FED** QQ-A-601 (328.0) **SAE** J452 (327); J453 (327)
A03281	Aluminum Foundry Alloy, Ingot	**Al** rem **Cr** 0.35 max **Cu** 1.0-2.0 **Fe** 0.8 max **Mg** 0.25-0.6 **Mn** 0.20-0.6 **Ni** 0.25 max **Si** 7.5-8.5 **Ti** 0.25 max **Zn** 1.5 max **Other** total 0.50 max	**AA** 328.1 **ASTM** B179 (328.1) **FED** QQ-A-371 (328.1) **SAE** J453 (327)
A03320	Aluminum Foundry Alloy, Casting	**Al** rem **Cu** 2.0-4.0 **Fe** 1.2 max **Mg** 0.50-1.5 **Mn** 0.50 max **Ni** 0.50 max **Si** 8.5-10.5 **Ti** 0.25 max **Zn** 1.0 max **Other** total 0.05 max	**AA** 332.0 **ASTM** B108 (F332.0) **FED** QQ-A-596 (332) **SAE** J452 (332); J453 (332)
A03321	Aluminum Foundry Alloy, Ingot	**Al** rem **Cu** 2.0-4.0 **Fe** 0.9 max **Mg** 0.6-1.5 **Mn** 0.50 max **Ni** 0.50 max **Si** 8.5-10.5 **Ti** 0.25 max **Zn** 1.0 max **Other** total 0.50 max	**AA** 332.1 **ASTM** B179 (F332.1) **FED** QQ-A-371 (F332.1) **SAE** J453 (332)

The chemical compositions listed are for identification purposes and should not be used in lieu of the cross referenced specifications.

UNIFIED NUMBER	DESCRIPTION	CHEMICAL COMPOSITION	CROSS REFERENCE SPECIFICATIONS
A03322	Aluminum Foundry Alloy, Ingot	**Al** rem **Cu** 2.0-4.0 **Fe** 0.6 max **Mg** 0.9-1.3 **Mn** 0.10 max **Ni** 0.10 max **Si** 8.5-10.0 **Ti** 0.20 max **Zn** 0.10 max **Other** total 0.30 max	**AA** 332.2 **ASTM** B179 (F332.2) **FED** QQ-A-371 (F332.2)
A03330	Aluminum Foundry Alloy, Casting	**Al** rem **Cu** 3.0-4.0 **Fe** 1.0 max **Mg** 0.05-0.50 **Mn** 0.50 max **Ni** 0.50 max **Si** 8.0-10.0 **Ti** 0.25 max **Zn** 1.0 max **Other** total 0.50 max	**AA** 333.0 **ASTM** B108 (333.0) **FED** QQ-A-596 (333) **SAE** J452 (331); J453 (331)
A03331	Aluminum Foundry Alloy, Ingot	**Al** rem **Cu** 3.0-4.0 **Fe** 0.8 max **Mg** 0.10-0.50 **Mn** 0.50 max **Ni** 0.50 max **Si** 8.0-10.0 **Ti** 0.25 max **Zn** 1.0 max **Other** total 0.50 max	**AA** 333.1 **ASTM** B179 (333.1) **FED** QQ-A-371 (333.1)
A03360	Aluminum Foundry Alloy, Casting	**Al** rem **Cu** 0.5-1.5 **Fe** 1.2 max **Mg** 0.7-1.3 **Mn** 0.35 max **Ni** 2.0-3.0 **Si** 11.0-13.0 **Ti** 0.25 max **Zn** 0.35 max **Other** each 0.05 max	**AA** 336.0 **ASTM** B108 **FED** QQ-A-596 (336) **SAE** J452 (321); J453 (321)
A03361	Aluminum Foundry Alloy, Ingot	**Al** rem **Cu** 0.50-1.5 **Fe** 0.9 max **Mg** 0.8-1.3 **Mn** 0.35 max **Ni** 2.0-3.0 **Si** 11.0-13.0 **Ti** 0.25 max **Zn** 0.35 max **Other** each 0.05 max	**AA** 336.1 **ASTM** B179 **FED** QQ-A-371 (A332.1)
A03362	Aluminum Foundry Alloy, Ingot	**Al** rem **Cu** 0.50-1.5 **Fe** 0.9 max **Mg** 0.9-1.3 **Mn** 0.10 max **N** 2.0-3.0 **Si** 11.0-13.0 **Ti** 0.20 max **Zn** 0.10 max **Other** each 0.05 max, total 0.15 max	**AA** 336.2 **ASTM** B179 **FED** QQ-A-371 (A332.2)
A03390	Aluminum Foundry Alloy, Casting	**Al** rem **Cu** 1.5-3.0 **Fe** 1.2 max **Mg** 0.50-1.5 **Mn** 0.50 max **Ni** 0.50-1.5 **Si** 11.0-13.0 **Ti** 0.25 max **Zn** 1.0 max **Other** total 0.50 max	**AA** 339.0
A03391	Aluminum Foundry Alloy, Ingot	**Al** rem **Cu** 1.5-3.0 **Fe** 0.9 max **Mg** 0.6-1.5 **Mn** 0.50 max **Ni** 0.50-1.5 **Si** 11.0-13.0 **Ti** 0.25 max **Zn** 1.0 max **Other** total 0.50 max	**AA** 339.1
A03430	Aluminum Foundry Alloy, Casting	**Al** rem **Cr** 0.10 max **Cu** 0.50-0.9 **Fe** 1.2 max **Mg** 0.10 max **Mn** 0.50 max **Si** 6.7-7.7 **Sn** 0.05 max **Zn** 1.2-2.0 **Other** each 0.10 max, total 0.35 max	**AA** 343.0
A03431	Aluminum Foundry Alloy, Ingot	**Al** rem **Cr** 0.10 max **Cu** 0.50-0.9 **Fe** 0.50-0.9 **Mg** 0.10 max **Mn** 0.50 max **Si** 6.7-7.7 **Sn** 0.05 max **Zn** 1.2-1.9 **Other** each 0.10 max, total 0.35 max	**AA** 343.1
A03540	Aluminum Foundry Alloy, Casting	**Al** rem **Cu** 1.6-2.0 **Fe** 0.20 max **Mg** 0.40-0.6 **Mn** 0.10 max **Si** 8.6-9.4 **Ti** 0.20 max **Zn** 0.10 max **Other** each 0.05 max, total 0.15 max	**AA** 354.0 **ASTM** B108 (354.0) **MIL SPEC** MIL-A-21180 (354.0)
A03541	Aluminum Foundry Alloy, Ingot	**Al** rem **Cu** 1.6-2.0 **Fe** 0.15 max **Mg** 0.45-0.6 **Mn** 0.10 max **Si** 8.6-9.4 **Ti** 0.20 max **Zn** 0.10 max **Other** each 0.05 max, total 0.15 max	**AA** 354.1 **ASTM** B179 (354.1) **FED** QQ-A-371 (354.1)
A03550	Aluminum Foundry Alloy, Casting	**Al** rem **Cr** 0.25 max **Cu** 1.0-1.5 **Fe** 0.6 max **Mg** 0.40-0.6 **Mn** 0.50 max **Si** 4.5-5.5 **Ti** 0.25 max **Zn** 0.35 max **Other** each 0.05 max, total 0.15 max. Note: If Fe exceeds 0.45, Mn shall not be less than 0.5 × Fe	**AA** 355.0 **AMS** 4188/3; 4210; 4212; 4214; 4280; 4281 **ASME** SFA5.10 (R355.0) **ASTM** B26 (355.0); B108 (355.0) **AWS** A5.10 (R355.0) **FED** QQ-A-596 (355); QQ-A-601 (355.0) **MIL SPEC** MIL-F-3922; MIL-C-11866 (355); MIL-F-39000 **SAE** J452 (322)
A03551	Aluminum Foundry Alloy, Ingot	**Al** rem **Cr** 0.25 max **Cu** 1.0-1.5 **Fe** 0.50 max **Mg** 0.45-0.6 **Mn** 0.50 max **Si** 4.5-5.5 **Ti** 0.25 max **Zn** 0.35 max **Other** each 0.05 max, total 0.15 max. Note: If Fe exceeds 0.45, Mn shall not be less than 0.5 × Fe	**AA** 355.1 **ASTM** B179 (355.1) **FED** QQ-A-371 (355.1)

The chemical compositions listed are for identification purposes and should not be used in lieu of the cross referenced specifications.

UNIFIED NUMBER	DESCRIPTION	CHEMICAL COMPOSITION	CROSS REFERENCE SPECIFICATIONS
A03552	Aluminum Foundry Alloy, Ingot	Al rem Cu 1.0-1.5 Fe 0.14-0.25 Mg 0.50-0.6 Mn 0.05 max Si 4.5-5.5 Ti 0.20 max Zn 0.05 max Other each 0.05 max, total 0.15 max	AA 355.2 ASTM B179 (355.2) FED QQ-A-371 (355.2)
A03560	Aluminum Foundry Alloy, Casting	Al rem Cu 0.25 max Fe 0.6 max Mg 0.20-0.45 Mn 0.35 max Si 6.5-7.5 Ti 0.25 max Zn 0.35 max Other each 0.05 max, total 0.15 max	AA 356.0 AMS 4188/4; 4217; 4260; 4261; 4284; 4285; 4286 ASME SFA5.10 (R356.0) ASTM B26 (356.0); B108 (356.0) AWS A5.10 (R356.0) FED QQ-A-596 (356); QQ-A-601 (356.0) MIL SPEC MIL-F-3922; MIL-C-10387; MIL-C-11866; MIL-F-39000 SAE J452 (323); J453 (323)
A03561	Aluminum Foundry Alloy, Ingot	Al rem Cu 0.25 max Fe 0.50 max Mg 0.25-0.45 Mn 0.35 max Si 6.5-7.5 Ti 0.25 max Zn 0.35 max Other each 0.05 max, total 0.15 max	AA 356.1 ASTM B179 (356.1) FED QQ-A-371 (356.1) SAE J453 (323)
A03562	Aluminum Foundry Alloy, Ingot	Al rem Cu 0.10 max Fe 0.12-0.25 Mg 0.30-0.45 Mn 0.05 max Si 6.5-7.5 Ti 0.20 max Zn 0.05 max Other each 0.05 max, total 0.15 max	AA 356.2 ASTM B179 (356.2) FED QQ-A-371 (356.2)
A03570	Aluminum Foundry Alloy, Casting	Al rem Cu 0.05 max Fe 0.15 max Mg 0.45-0.6 Mn 0.03 max Si 6.5-7.5 Ti 0.20 max Zn 0.05 max Other each 0.05 max, total 0.15 max	AA 357.0 AMS 4188/5; 4238; 4239 FED QQ-A-596 (357)
A03571	Aluminum Foundry Alloy, Ingot	Al rem Cu 0.05 max Fe 0.12 max Mg 0.45-0.6 Mn 0.03 max Si 6.5-7.5 Ti 0.20 max Zn 0.05 max Other each 0.05 max, total 0.15 max	AA 357.1 FED QQ-A-371 (357.1)
A03580	Aluminum Foundry Alloy	Al rem Be 0.10-0.30 Cr 0.20 max Cu 0.20 max Fe 0.30 max Mg 0.40-0.6 Mn 0.20 max Si 7.6-8.6 Ti 0.10-0.20 Zn 0.20 max Other each 0.05 max, total 0.15 max	AA 358.0
A03582	Aluminum Foundry Alloy, Ingot	Al rem Be 0.15-0.30 Cr 0.05 max Cu 0.10 max Fe 0.20 max Mg 0.45-0.6 Mn 0.10 max Si 7.6-8.6 Ti 0.12-0.20 Zn 0.10 max Other each 0.05 max, total 0.15 max	AA 358.2 FED QQ-A-371 (B358.2)
A03590	Aluminum Foundry Alloy, Casting	Al rem Cu 0.20 max Fe 0.20 max Mg 0.50-0.7 Mn 0.10 max Si 8.5-9.5 Ti 0.20 max Zn 0.10 max Other each 0.05 max, total 0.15 max	AA 359.0 ASTM B108 (359.0) MIL SPEC MIL-A-21180 (359.0)
A03592	Aluminum Foundry Alloy, Ingot	Al rem Cu 0.10 max Fe 0.12 max Mg 0.55-0.7 Mn 0.10 max Si 8.5-9.5 Ti 0.20 max Zn 0.10 max Other each 0.05 max, total 0.15 max	AA 359.2 ASTM B179 (359.2)
A03600	Aluminum Foundry Alloy, Casting	Al rem Cu 0.6 max Fe 2.0 max Mg 0.40-0.6 Mn 0.35 max Ni 0.50 max Si 9.0-10.0 Sn 0.15 max Zn 0.50 max Other total 0.25 max	AA 360.0 AMS 4290 ASTM B85 (360.0) FED QQ-A-591 (360.0)
A03602	Aluminum Foundry Alloy, Ingot	Al rem Cu 0.10 max Fe 0.7-1.1 Mg 0.45-0.6 Mn 0.10 max Ni 0.10 max Si 9.0-10.0 Sn 0.10 max Zn 0.10 max Other total 0.20 max	AA 360.2 ASTM B179 (360.2) FED QQ-A-371 (360.2)
A03610	Aluminum Foundry Alloy, Casting	Al rem Cr 0.20-0.30 Cu 0.50 max Fe 1.1 max Mg 0.40-0.6 Mn 0.25 max Ni 0.20-0.30 Si 9.5-10.5 Sn 0.10 max Ti 0.20 max Zn 0.50 max Other each 0.05 max, total 0.15 max	AA 361.0
A03611	Aluminum Foundry Alloy, Ingot	Al rem Cr 0.20-0.30 Cu 0.50 max Fe 0.8 max Mg 0.45-0.6 Mn 0.25 max Ni 0.20-0.30 Si 9.5-10.5 Sn 0.10 max Ti 0.20 max Zn 0.40 max Other each 0.05 max, total 0.15 max	AA 361.1

The chemical compositions listed are for identification purposes and should not be used in lieu of the cross referenced specifications.

UNIFIED NUMBER	DESCRIPTION	CHEMICAL COMPOSITION	CROSS REFERENCE SPECIFICATIONS
A03630	Aluminum Foundry Alloy, Casting	Al rem Cu 2.5-3.5 Fe 1.1 max Mg 0.15-0.40 Ni 0.25 max Si 4.5-6.0 Sn 0.25 max Ti 0.20 max Zn 3.0-4.5 Other total 0.30 max, plus Mn+Cr 0.8 max	AA 363.0
A03631	Aluminum Foundry Alloy, Ingot	Al rem Cu 2.5-3.5 Fe 0.8 max Mg 0.20-0.40 Ni 0.25 max Si 4.5-6.0 Sn 0.25 max Ti 0.20 max Zn 3.0-4.5 Other total 0.30 max, plus Mn+Cr 0.8 max	AA 363.1
A03640	Aluminum Foundry Alloy, Casting	Al rem Be 0.02-0.04 Cr 0.25-0.50 Cu 0.20 max Fe 1.5 max Mg 0.20-0.40 Mn 0.10 max Ni 0.15 max Si 7.5-9.5 Sn 0.15 max Zn 0.15 max Other each 0.05 max, total 0.15 max	AA 364.0
A03642	Aluminum Foundry Alloy, Ingot	Al rem Be 0.02-0.04 Cr 0.25-0.50 Cu 0.20 max Fe 0.7-1.1 Mg 0.25-0.40 Mn 0.10 max Ni 0.15 max Si 7.5-9.5 Sn 0.15 max Zn 0.15 max Other each 0.05 max, total 0.15 max	AA 364.2 FED QQ-A-371 (364.2)
A03690	Aluminum Foundry Alloy, Casting	Al rem Cr 0.30-0.40 Cu 0.50 max Fe 1.3 max Mg 0.25-0.45 Mn 0.35 max Ni 0.05 max Si 11.0-12.0 Sn 0.10 max Zn 1.0 max Other each 0.05 max, total 0.15 max	AA 369.0
A03691	Aluminum Foundry Alloy, Ingot	Al rem Cr 0.30-0.40 Cu 0.50 max Fe 1.0 max Mg 0.30-0.45 Mn 0.35 max Ni 0.05 max Si 11.0-12.0 Sn 0.10 max Zn 0.9 max Other each 0.05 max, total 0.15 max	AA 369.1
A03800	Aluminum Foundry Alloy, Casting	Al rem Cu 3.0-4.0 Fe 2.0 max Mg 0.10 max Mn 0.50 max Ni 0.50 max Si 7.5-9.5 Sn 0.35 max Zn 3.0 max Other total 0.50 max	AA 380.0 ASTM B85 (380.0) FED QQ-A-591 (380) SAE J452 (308); J453 (308)
A03802	Aluminum Foundry Alloy, Ingot	Al rem Cu 3.0-4.0 Fe 0.7-1.1 Mg 0.10 max Mn 0.10 max Ni 0.10 max Si 7.5-9.5 Sn 0.10 max Zn 0.10 max Other total 0.20 max	AA 380.2 ASTM B179 (380.2) FED QQ-A-371 (380.2)
A03830	Aluminum Foundry Alloy, Casting	Al rem Cu 2.0-3.0 Fe 1.3 max Mg 0.10 max Mn 0.50 max Ni 0.30 max Si 9.5-11.5 Sn 0.15 max Zn 3.0 max Other total 0.50 max	AA 383.0 ASTM B85 (383.0) FED QQ-A-591 (383.0) SAE J452 (383); J453 (383)
A03831	Aluminum Foundry Alloy, Ingot	Al rem Cu 2.0-3.0 Fe 0.6-1.0 Mg 0.10 max Mn 0.50 max Ni 0.30 max Si 9.5-11.5 Sn 0.15 max Zn 2.9 max Other total 0.50 max	AA 383.1 ASTM B179 (383.1) SAE J453 (383)
A03832	Aluminum Foundry Alloy, Ingot	Al rem Cu 2.0-3.0 Fe 0.6-1.0 Mg 0.10 max Mn 0.10 max Ni 0.10 max Si 9.5-11.5 Sn 0.10 max Zn 0.10 max Other total 0.20 max	AA 383.2 ASTM B179 (383.2)
A03840	Aluminum Foundry Alloy, Casting	Al rem Cu 3.0-4.5 Fe 1.3 max Mg 0.10 max Mn 0.50 max Ni 0.50 max Si 10.5-12.0 Sn 0.35 max Zn 3.0 max Other total 0.50 max	AA 384.0 ASTM B85 (384.0) FED QQ-A-591 (384.0) SAE J452 (303); J453 (303)
A03841	Aluminum Foundry Alloy, Ingot	Al rem Cu 3.0-4.5 Fe 1.0 max Mg 0.10 max Mn 0.50 max Ni 0.50 max Si 10.5-12.0 Sn 0.35 max Zn 2.9 max Other total 0.50 max	AA 384.1 ASTM B179 (384.1) FED QQ-A-371 (384.1) SAE J453 (303)
A03842	Aluminum Foundry Alloy, Ingot	Al rem Cu 3.0-4.5 Fe 0.6-1.0 Mg 0.10 max Mn 0.10 max Ni 0.10 max Si 10.5-12.0 Sn 0.10 max Zn 0.10 max Other total 0.20 max	AA 384.2 ASTM B179 (384.2) FED QQ-A-371 (384.2)

The chemical compositions listed are for identification purposes and should not be used in lieu of the cross referenced specifications.

UNIFIED NUMBER	DESCRIPTION	CHEMICAL COMPOSITION	CROSS REFERENCE SPECIFICATIONS
A03850	Aluminum Foundry Alloy, Casting	Al rem Cu 2.0-4.0 Fe 2.0 max Mg 0.30 max Mn 0.50 max Ni 0.50 max Si 11.0-13.0 Sn 0.30 max Zn 3.0 max Other total 0.50 max	AA 385.0
A03851	Aluminum Foundry Alloy, Ingot	Al rem Cu 2.0-4.0 Fe 1.1 max Mg 0.30 max Mn 0.50 max Ni 0.50 max Si 11.0-13.0 Sn 0.30 max Zn 2.9 max Other total 0.50 max	AA 385.1
A03900	Aluminum Foundry Alloy, Casting	Al rem Cu 4.0-5.0 Fe 1.3 max Mg 0.45-0.65 Mn 0.10 max Si 16.0-18.0 Ti 0.20 max Zn 0.10 max Other each 0.10 max, total 0.20 max	AA 390.0
A03902	Aluminum Foundry Alloy, Ingot	Al rem Cu 4.0-5.0 Fe 0.6-1.0 Mg 0.50-0.65 Mn 0.10 max Si 16.0-18.0 Ti 0.20 max Zn 0.10 max Other each 0.10 max, total 0.20 max	AA 390.2 FED QQ-A-371 (390.2)
A03920	Aluminum Foundry Alloy, Casting	Al rem Cu 0.40-0.8 Fe 1.5 max Mg 0.8-1.2 Mn 0.20-0.6 Ni 0.50 max Si 18.0-20.0 Sn 0.30 max Ti 0.20 max Zn 0.50 max Other each 0.15 max, total 0.50 max	AA 392.0
A03921	Aluminum Foundry Alloy, Ingot	Al rem Cu 0.40-0.8 Fe 1.1 max Mg 0.9-1.2 Mn 0.20-0.6 Ni 0.50 max Si 18.0-20.0 Sn 0.30 max Ti 0.20 max Zn 0.40 max Other each 0.15 max, total 0.50 max	AA 392.1
A03930	Aluminum Foundry Alloy, Casting	Al rem Be 0.15-0.30 Cu 0.7-1.1 Fe 1.3 max Mg 0.7-1.3 Mn 0.10 max Ni 2.0-2.5 Si 21.0-23.0 Ti 0.10-0.20 Zn 0.10 max Other each 0.05 max, total 0.15 max	AA 393.0
A03931	Aluminum Foundry Alloy, Ingot	Al rem Cu 0.7-1.1 Fe 1.0 max Mg 0.8-1.3 Mn 0.10 max Ni 2.0-2.5 Si 21.0-23.0 Ti 0.10-0.20 V 0.08-0.15 Zn 0.10 max Other each 0.05 max, total 0.15 max	AA 393.1
A03932	Aluminum Foundry Alloy, Ingot	Al rem Cu 0.7-1.1 Fe 0.8 max Mg 0.8-1.3 Mn 0.10 max Ni 2.0-2.5 Si 21.0-23.0 Ti 0.10-0.20 V 0.08-0.15 Zn 0.10 max Other each 0.05 max, total 0.15 max	AA 393.2
A04082	Aluminum Foundry Alloy, Ingot (Steel Coating Alloy)	Al rem Cu 0.10 max Fe 0.6-1.3 Mn 0.10 max Si 8.5-9.5 Zn 0.10 max Other each 0.10 max, total 0.20 max	AA 408.2
A04092	Aluminum Foundry Alloy, Ingot (Steel Coating Alloy)	Al rem Cu 0.10 max Fe 0.6-1.3 Mn 0.10 max Si 9.0-10.0 Zn 0.10 max Other each 0.10 max, total 0.20 max	AA 409.2
A04112	Aluminum Foundry Alloy, Ingot (Steel Coating Alloy)	Al rem Cu 0.20 max Fe 0.6-1.3 Mn 0.10 max Si 10.0-12.0 Zn 0.10 max Other each 0.10 max, total 0.20 max	AA 411.2
A04130	Aluminum Foundry Alloy, Casting	Al rem Cu 0.6 max Fe 2.0 max Mg 0.10 max Mn 0.35 max Ni 0.50 max Si 11.0-13.0 Sn 0.15 max Zn 0.50 max Other total 0.25 max	AA 413.0 ASTM B85 (413.0) FED QQ-A-591 (413.0)
A04132	Aluminum Foundry Alloy, Ingot	Al rem Cu 0.10 max Fe 0.7-1.1 Mg 0.07 max Mn 0.10 max Ni 0.10 max Si 11.0-13.0 Sn 0.10 max Zn 0.10 max Other total 0.20 max	AA 413.2 FED QQ-A-371 (413.2)
A04352	Aluminum Foundry Alloy, Ingot	Al rem Cu 0.05 Fe 0.40 Mg 0.05 Mn 0.05 Si 3.3-3.9 Zn 0.10 Other each 0.05, total 0.20	AA 435.2

The chemical compositions listed are for identification purposes and should not be used in lieu of the cross referenced specifications.

UNIFIED NUMBER	DESCRIPTION	CHEMICAL COMPOSITION	CROSS REFERENCE SPECIFICATIONS
A04430	Aluminum Foundry Alloy, Casting	**Al** rem **Cr** 0.25 max **Cu** 0.6 max **Fe** 0.8 max **Mg** 0.05 max **Mn** 0.50 max **Si** 4.5-6.0 **Ti** 0.25 max **Zn** 0.50 max **Other** total 0.35 max	**AA** 443.0 **ASTM** B26 (443.0); B108 (443.0) **FED** QQ-A-591 (443.0) **SAE** J452 (35); J453 (35)
A04431	Aluminum Foundry Alloy, Ingot	**Al** rem **Cr** 0.25 max **Cu** 0.6 max **Fe** 0.6 max **Mg** 0.05 max **Mn** 0.50 max **Si** 4.5-6.0 **Ti** 0.25 max **Zn** 0.50 max **Other** total 0.35 max	**AA** 443.1 **ASTM** B179 (443.1) **FED** QQ-A-371 (443.1) **SAE** J453 (35)
A04432	Aluminum Foundry Alloy, Ingot	**Al** rem **Cu** 0.10 max **Fe** 0.6 max **Mg** 0.05 max **Mn** 0.10 max **Si** 4.5-6.0 **Ti** 0.20 max **Zn** 0.10 max **Other** each 0.05 max, total 0.15 max	**AA** 443.2 **ASTM** B179 (443.2) **FED** QQ-A-371 (443.2)
A04440	Aluminum Foundry Alloy, Casting	**Al** rem **Cu** 0.25 max **Fe** 0.6 max **Mg** 0.10 max **Mn** 0.35 max **Si** 6.5-7.5 **Ti** 0.25 max **Zn** 0.35 max **Other** each 0.05 max, total 0.15 max	**AA** 444.0
A04442	Aluminum Foundry Alloy, ingot	**Al** rem **Cu** 0.10 max **Fe** 0.13-0.25 **Mg** 0.05 max **Mn** 0.05 max **Si** 6.5-7.5 **Ti** 0.20 max **Zn** 0.05 max **Other** each 0.05 max, total 0.15 max	**AA** 444.2
A04452	Aluminum Foundry Alloy, Ingot	**Al** rem **Cu** 0.10 max **Fe** 0.6-1.3 **Mn** 0.10 max **Si** 6.5-7.5 **Zn** 0.10 max **Other** each 0.10 max, total 0.20 max	**AA** 445.2
A05110	Aluminum Foundry Alloy, Casting	**Al** rem **Cu** 0.15 max **Fe** 0.50 max **Mg** 3.5-4.5 **Mn** 0.35 max **Si** 0.30-0.7 **Ti** 0.25 max **Zn** 0.15 max **Other** each 0.05 max, total 0.15 max	**AA** 511.0
A05111	Aluminum Foundry Alloy, Ingot	**Al** rem **Cu** 0.15 max **Fe** 0.40 max **Mg** 3.6-4.5 **Mn** 0.35 max **Si** 0.30-0.7 **Ti** 0.25 max **Zn** 0.15 max **Other** each 0.05 max, total 0.15	**AA** 511.1 **FED** QQ-A-371 (F514.1)
A05112	Aluminum Foundry Alloy, Ingot	**Al** rem **Cu** 0.10 max **Fe** 0.30 max **Mg** 3.6-4.5 **Mn** 0.10 max **Si** 0.30-0.7 **Ti** 0.20 max **Zn** 0.10 max **Other** each 0.05 max, total 0.15 max	**AA** 511.2 **FED** QQ-A-371 (F514.2)
* A05120	Aluminum Foundry Alloy, Casting	**Al** rem **Cr** 0.25 max **Cu** 0.35 max **Fe** 0.6 max **Mg** 3.5-4.5 **Mn** 0.8 max **Si** 1.4-2.2 **Ti** 0.25 max **Zn** 0.35 max **Other** each 0.05 max, total 0.15 max	**AA** 512.0 **ASTM** B26; B108 (B514.0) **FED** QQ-A-601 (512.0)
A05122	Aluminum Foundry Alloy, Ingot	**Al** rem **Cu** 0.10 max **Fe** 0.30 max **Mg** 3.6-4.5 **Mn** 0.10 max **Si** 1.4-2.2 **Ti** 0.20 max **Zn** 0.10 max **Other** each 0.05 max, total 0.15 max	**AA** 512.2 **ASTM** B179 **FED** QQ-A-371 (B514.2)
A05130	Aluminum Foundry Alloy, Casting	**Al** rem **Cu** 0.10 max **Fe** 0.40 max **Mg** 3.5-4.5 **Mn** 0.30 max **Si** 0.30 max **Ti** 0.20 max **Zn** 1.4-2.2 **Other** each 0.05 max, total 0.15 max	**AA** 513.0 **ASTM** B108 **FED** QQ-A-596 (513)
A05132	Aluminum Foundry Alloy, Ingot	**Al** rem **Cu** 0.10 max **Fe** 0.30 max **Mg** 3.6-4.5 **Mn** 0.10 max **Si** 0.30 max **Ti** 0.20 max **Zn** 1.4-2.2 **Other** each 0.05 max, total 0.15 max	**AA** 513.2 **ASTM** B179 **FED** QQ-A-371 (A514.2)
A05140	Aluminum Foundry Alloy, Casting	**Al** rem **Cu** 0.15 max **Fe** 0.50 max **Mg** 3.5-4.5 **Mn** 0.35 max **Si** 0.35 max **Ti** 0.25 max **Zn** 0.15 max **Other** each 0.05 max, total 0.15 max	**AA** 514.0 **ASTM** B26 (514.0) **FED** QQ-A-601 (514.0) **MIL SPEC** MIL-C-11866 (214) **SAE** J452 (320); J453 (320)

*Boxed entries are no longer active and are retained for reference purposes only.

The chemical compositions listed are for identification purposes and should not be used in lieu of the cross referenced specifications.

UNIFIED NUMBER	DESCRIPTION	CHEMICAL COMPOSITION	CROSS REFERENCE SPECIFICATIONS
A05141	Aluminum Foundry Alloy, Ingot	Al rem **Cu** 0.15 max **Fe** 0.40 max **Mg** 3.6-4.5 **Mn** 0.35 max **Si** 0.35 max **Ti** 0.25 max **Zn** 0.15 max **Other** each 0.05 max, total 0.15 max	**AA** 514.1 **ASTM** B179 (514.1) **FED** QQ-A-371 (514.1) **SAE** J453 (320)
A05142	Aluminum Foundry Alloy, Ingot	Al rem **Cu** 0.10 max **Fe** 0.30 max **Mg** 3.6-4.5 **Mn** 0.10 max **Si** 0.30 max **Ti** 0.20 max **Zn** 0.10 max **Other** each 0.05 max, total 0.15 max	**AA** 514.2 **ASTM** B179 (514.2) **FED** QQ-A-371 (514.2)
A05150	Aluminum Foundry Alloy, Casting	Al rem **Cu** 0.20 max **Fe** 1.3 max **Mg** 2.5-4.0 **Mn** 0.40-0.6 **Si** 0.5-1.0 **Zn** 0.10 max **Other** each 0.05 max, total 0.15 max	**AA** 515.0
A05152	Aluminum Foundry Alloy, Ingot	Al rem **Cu** 0.10 max **Fe** 0.6-1.0 **Mg** 2.7-4.0 **Mn** 0.40-0.6 **Si** 0.50-1.0 **Zn** 0.05 max **Other** each 0.05 max, total 0.15 max	**AA** 515.2 **FED** QQ-A-371 (L514.2)
A05160	Aluminum Foundry Alloy, Casting	Al rem **Cu** 0.30 **Fe** 0.35-1.0 **Mg** 2.5-4.5 **Mn** 0.15-0.40 **Ni** 0.25-0.40 **Pb** 0.10 max **Si** 0.30-1.5 **Sn** 0.10 max **Ti** 0.10-0.20 **Zn** 0.20 max **Other** each 0.05 max	**AA** 516.0
A05161	Aluminum Foundry Alloy, Ingot	Al rem **Cu** 0.30 max **Fe** 0.35-0.7 **Mg** 2.6-4.5 **Mn** 0.15-0.40 **Ni** 0.25-0.40 **Pb** 0.10 max **Si** 0.30-1.5 **Sn** 0.10 max **Ti** 0.10-0.20 **Zn** 0.20 max **Other** each 0.05 max	**AA** 516.1
A05180	Aluminum Foundry Alloy, Die Casting	Al rem **Cu** 0.25 max **Fe** 1.8 max **Mg** 7.5-8.5 **Mn** 0.35 max **Ni** 0.15 max **Si** 0.35 max **Sn** 0.15 max **Zn** 0.15 max **Other** total 0.25 max	**AA** 518.0 **ASTM** B85 (68A) **FED** QQ-A-591 (518.0) **MIL SPEC** MIL-F-3922; MIL-F-39000
A05181	Aluminum Foundry Alloy, Ingot	Al rem **Cu** 0.25 max **Fe** 1.0 max **Mg** 7.6-8.5 **Mn** 0.35 max **Ni** 0.15 max **Si** 0.35 max **Sn** 0.15 max **Zn** 0.15 max **Other** total 0.25 max	**AA** 518.1 **FED** QQ-A-371 (518.1)
A05182	Aluminum Foundry Alloy, Ingot	Al rem **Cu** 0.10 max **Fe** 0.7 max **Mg** 7.6-8.5 **Mn** 0.10 max **Ni** 0.05 max **Si** 0.25 max **Sn** 0.05 max **Other** total 0.10 max	**AA** 518.2 **FED** QQ-A-371 (518.2)
A05200	Aluminum Foundry Alloy, Casting	Al rem **Cu** 0.25 max **Fe** 0.30 max **Mg** 9.5-10.6 **Mn** 0.15 max **Si** 0.25 max **Ti** 0.25 max **Zn** 0.15 max **Other** each 0.05 max, total 0.15 max	**AA** 520.0 **AMS** 4240 **ASTM** B26 (520.0) **FED** QQ-A-601 (520.0) **SAE** J452 (324); J453 (324)
A05202	Aluminum Foundry Alloy, Ingot	Al rem **Cu** 0.20 max **Fe** 0.20 max **Mg** 9.6-10.6 **Mn** 0.10 max **Si** 0.15 max **Ti** 0.20 max **Zn** 0.10 max **Other** each 0.05 max, total 0.15 max	**AA** 520.2 **ASTM** B179 (520.2) **FED** QQ-A-371 (520.2)
A05350	Aluminum Foundry Alloy, Casting	Al rem **B** 0.005 max **Be** 0.003-0.007 **Cu** 0.05 max **Fe** 0.15 max **Mg** 6.2-7.5 **Mn** 0.10-0.25 **Si** 0.15 max **Ti** 0.10-0.25 **Other** each 0.05 max, total 0.15 max	**AA** 535.0 **AMS** 4238; 4239 **ASTM** B26 (535.0); B108 (535.0) **FED** QQ-A-601 (535.0) **MIL SPEC** MIL-C-11866 (Almag 35)
A05352	Aluminum Foundry Alloy, Ingot	Al rem **B** 0.002 max **Be** 0.003-0.007 **Cu** 0.05 max **Fe** 0.10 max **Mg** 6.6-7.5 **Mn** 0.10-0.25 **Si** 0.10 max **Ti** 0.10-0.25 **Other** each 0.05 max, total 0.15 max	**AA** 535.2 **ASTM** B179 (535.2) **FED** QQ-A-371 (535.2)
A07050	Aluminum Foundry Alloy, Casting	Al rem **Cr** 0.20-0.40 **Cu** 0.20 max **Fe** 0.8 max **Mg** 1.4-1.8 **Mn** 0.40-0.6 **Si** 0.20 max **Ti** 0.25 max **Zn** 2.7-3.3 **Other** each 0.05 max, total 0.15 max	**AA** 705.0 **ASTM** B26 (705.0); B108 (705.0) **FED** QQ-A-596 (705); QQ-A-601 (705.0) **SAE** J452 (311); J453 (311)

The chemical compositions listed are for identification purposes and should not be used in lieu of the cross referenced specifications.

UNIFIED NUMBER	DESCRIPTION	CHEMICAL COMPOSITION	CROSS REFERENCE SPECIFICATIONS
A07051	Aluminum Foundry Alloy, Ingot	Al rem Cr 0.20-0.40 Cu 0.20 max Fe 0.6 max Mg 1.5-1.8 Mn 0.40-0.6 Si 0.20 max Ti 0.25 max Zn 2.7-3.3 Other each 0.05 max, total 0.15 max	AA 705.1 ASTM B179 (705.1) FED QQ-A-371 (705.1) SAE J453 (311)
A07070	Aluminum Foundry Alloy, Casting	Al rem Cr 0.20-0.40 Cu 0.20 max Fe 0.8 max Mg 1.8-2.4 Mn 0.40-0.6 Si 0.20 max Ti 0.25 max Zn 4.0-4.5 Other 0.05 max, total 0.15 max	AA 707.0 ASTM B26 (707.0); B108 (707.0) FED QQ-A-596 (707); QQ-A-601 (707.0) SAE J452 (312); J453 (312)
A07071	Aluminum Foundry Alloy, Ingot	Al rem Cr 0.20-0.40 Cu 0.20 max Fe 0.6 max Mg 1.9-2.4 Mn 0.40-0.6 Si 0.20 max Ti 0.25 max Zn 4.0-4.5 Other each 0.05 max, total 0.15 max	AA 707.1 ASTM B179 (707.1) FED QQ-A-371 (707.1) SAE J453 (312)
A07100	Aluminum Foundry Alloy, Casting	Al rem Cu 0.35-0.65 Fe 0.50 max Mg 0.6-0.8 Mn 0.05 max Si 0.15 max Ti 0.25 max Zn 6.0-7.0 Other each 0.05 max, total 0.15 max	AA 710.0 ASTM B26 FED QQ-A-601 (710.0) SAE J452 (313); J453 (313)
A07101	Aluminum Foundry Alloy, Ingot	Al rem Cu 0.35-0.65 Fe 0.40 max Mg 0.65-0.8 Mn 0.05 max Si 0.15 max Ti 0.25 max Zn 6.0-7.0 Other each 0.05 max, total 0.15 max	AA 710.1 ASTM B179 FED QQ-A-371 (A712.1)
A07110	Aluminum Foundry Alloy, Casting	Al rem Cu 0.35-0.65 Fe 0.7-1.4 Mg 0.25-0.45 Mn 0.5 max Si 0.30 max Ti 0.20 max Zn 6.0-7.0 Other each 0.05 max, total 0.15 max	AA 711.0 ASTM B108 (C112.0) MIL SPEC Mil-F-3922 (C712.0); Mil-F-39000 (C712.0) SAE J452 (314); J453 (314)
A07111	Aluminum Foundry Alloy, Ingot	Al rem Cu 0.35-0.65 Fe 0.7-1.1 Mg 0.30-0.45 Mn 0.05 max Si 0.30 max Ti 0.20 max Zn 6.0-7.0 Other each 0.05 max, total 0.15 max	AA 711.1 ASTM B179 (C712.1) FED QQ-A-371 (C712.1)
A07120	Aluminum Foundry Alloy, Casting	Al rem Cr 0.40-0.6 Cu 0.25 max Fe 0.50 max Mg 0.50-0.65 Mn 0.10 max Si 0.30 max Ti 0.15-0.25 Zn 5.0-6.5 Other each 0.05 max, total 0.20 max	AA 712.0 ASTM B26 (D712.0) FED QQ-A-601 (712.0); QQ-R-566 MIL SPEC Mil-F-3922 (D712.0); Mil-C-11866 (40E); Mil-F-39000 (D712.0) SAE J452 (310); J453 (310)
A07122	Aluminum Foundry Alloy, Ingot	Al rem Cr 0.40-0.6 Cu 0.25 max Fe 0.40 max Mg 0.50-0.65 Mn 0.10 max Si 0.15 max Ti 0.15-0.25 Zn 5.0-6.5 Other each 0.05 max, total 0.20 max	AA 712.2 ASTM B179 (712.2) FED QQ-A-371 (D712.2)
A07130	Aluminum Foundry Alloy, Casting	Al rem Cr 0.35 max Cu 0.40-1.0 Fe 1.1 max Mg 0.20-0.50 Mn 0.6 max Ni 0.15 max Si 0.25 max Ti 0.25 max Zn 7.0-8.0 Other each 0.10 max, total 0.25 max	AA 713.0 ASTM B26 (713.0); B108 (713.0) FED QQ-A-596 (713-Tenzaloy); QQ-A-601 (713.0) SAE J452 (315); J453 (315)
A07131	Aluminum Foundry Alloy, Ingot	Al rem Cr 0.35 max Cu 0.40-1.0 Fe 0.8 max Mg 0.25-0.50 Mn 0.6 max Ni 0.15 max Si 0.25 max Ti 0.25 max Zn 7.0-8.0 Other each 0.10 max, total 0.25 max	AA 713.1 ASTM B179 (713.1) FED QQ-A-371 (713.1) SAE J453 (315)
A07710	Aluminum Foundry Alloy, Casting	Al rem Cr 0.06-0.20 Cu 0.10 max Fe 0.15 max Mg 0.8-1.0 Mn 0.10 max Si 0.15 max Ti 0.10-0.20 Zn 6.5-7.5 Other each 0.05 max, total 0.15 max	AA 771.0 ASTM B26 (771.0) FED QQ-A-601 (771.0)
A07712	Aluminum Foundry Alloy, Ingot	Al rem Cr 0.06-0.20 Cu 0.10 max Fe 0.10 max Mg 0.85-1.0 Mn 0.10 max Si 0.10 max Ti 0.10-0.20 Zn 6.5-7.5 Other each 0.05 max, total 0.15 max	AA 771.2 ASTM B179 (771.2) FED QQ-A-371 (771.2)
A07720	Aluminum Foundry Alloy, Casting	Al rem Cr 0.06-0.20 Cu 0.10 max Fe 0.15 max Mg 0.6-0.8 Mn 0.10 max Si 0.15 max Ti 0.10-0.20 Zn 6.0-7.0 Other each 0.05 max, total 0.15 max	AA 772.0

The chemical compositions listed are for identification purposes and should not be used in lieu of the cross referenced specifications.

UNIFIED NUMBER	DESCRIPTION	CHEMICAL COMPOSITION	CROSS REFERENCE SPECIFICATIONS
A07722	Aluminum Foundry Alloy, Ingot	**Al** rem **Cr** 0.06-0.20 **Cu** 0.10 max **Fe** 0.10 max **Mg** 0.65-0.8 **Mn** 0.10 max **Si** 0.10 max **Ti** 0.10-0.20 **Zn** 6.0-7.0 **Other** each 0.05 max, total 0.15 max	**AA** 772.2
A08500	Aluminum Foundry Alloy, Casting	**Al** rem **Cu** 0.7-1.3 **Fe** 0.7 max **Mg** 0.10 max **Mn** 0.10 max **Ni** 0.7-1.3 **Si** 0.7 max **Sn** 5.5-7.0 **Ti** 0.20 max **Other** total 0.30 max	**AA** 850.0 **AMS** 4275 **ASTM** B26 (771.0); B108 (850.0) **FED** QQ-A-596 (850); QQ-A-601 (850.0)
A08501	Aluminum Foundry Alloy, Ingot	**Al** rem **Cu** 0.7-1.3 **Fe** 0.50 max **Mg** 0.10 max **Mn** 0.10 max **Ni** 0.7-1.3 **Si** 0.7 max **Sn** 5.5-7.0 **Ti** 0.20 max **Other** total 0.30 max	**AA** 850.1 **ASTM** B179 (850.1) **FED** QQ-A-371 (850.1)
A08510	Aluminum Foundry Alloy, Casting	**Al** rem **Cu** 0.7-1.3 **Fe** 0.7 max **Mg** 0.10 max **Mn** 0.10 max **Ni** 0.30-0.7 **Si** 2.0-3.0 **Sn** 5.5-7.0 **Ti** 0.20 max **Other** total 0.30 max	**AA** 851.0 **ASTM** B26; B108 **FED** QQ-A-596 (851); QQ-A-601 (851)
A08511	Aluminum Foundry Alloy, Ingot	**Al** rem **Cu** 0.7-1.3 **Fe** 0.50 max **Mg** 0.10 max **Mn** 0.10 max **Ni** 0.30-0.7 **Si** 2.0-3.0 **Sn** 5.5-7.0 **Ti** 0.20 max **Other** total 0.30 max	**AA** 851.1 **ASTM** B179 **FED** QQ-A-371 (A850.1)
A08520	Aluminum Foundry Alloy, Casting	**Al** rem **Cu** 1.7-2.3 **Fe** 0.7 max **Mg** 0.6-0.9 **Mn** 0.10 max **Ni** 0.9-1.5 **Si** 0.40 max **Sn** 5.5-7.0 **Ti** 0.20 max **Other** total 0.30 max	**AA** 852.0 **ASTM** B26 (B850.0); B108 (B850.0) **FED** QQ-A-596 (852); QQ-A-601 (852.0)
A08521	Aluminum Foundry Alloy, Ingot	**Al** rem **Cu** 1.7-2.3 **Fe** 0.50 max **Mg** 0.7-0.9 **Mn** 0.10 max **Ni** 0.9-1.5 **Si** 0.40 max **Sn** 5.5-7.0 **Ti** 0.20 max **Other** total 0.30 max	**AA** 852.1 **ASTM** B179 (B850.1) **FED** QQ-A-371 (B850.1)
A08530	Aluminum Foundry Alloy, Casting	**Al** rem **Cu** 3.0-4.0 **Fe** 0.7 max **Mn** 0.05 max **Si** 5.5-6.5 **Sn** 5.5-7.0 **Ti** 0.20 max **Other** total 0.30 max	**AA** 853.0
A08532	Aluminum Foundry Alloy, Ingot	**Al** rem **Cu** 3.0-4.0 **Fe** 0.50 max **Mn** 0.10 max **Si** 5.5-6.5 **Sn** 5.5-7.0 **Ti** 0.20 max **Other** total 0.30 max	**AA** 853.2
A12010	Aluminum Foundry Alloy, Casting	**Ag** 0.40-1.0 **Al** rem **Cu** 4.0-5.0 **Fe** 0.10 max **Mg** 0.15-0.35 **Mn** 0.20-0.40 **Si** 0.05 max **Ti** 0.15-0.35 **Other** each 0.03 max, total 0.10 max	**AA** A201.0 **AMS** 4223; 4228; 4229 **MIL SPEC** MIL-A-21180 (A210.0)
A12011	Aluminum Foundry Alloy, Ingot	**Ag** 0.40-1.0 **Al** rem **Cu** 4.0-5.0 **Fe** 0.07 max **Mg** 0.20-0.35 **Mn** 0.20-0.40 **Si** 0.05 max **Ti** 0.15-0.35 **Other** each 0.03 max, total 0.10 max	**AA** A201.1
A12012	Aluminum Foundry Alloy, Ingot	**Ag** 0.40-1.0 **Al** rem **Cu** 4.0-5.0 **Fe** 0.07 max **Mg** 0.20-0.35 **Mn** 0.20-0.40 **Si** 0.05 max **Ti** 0.15-0.35 **Other** each 0.03 max, total 0.10 max	**AA** A201.2
A12060	Aluminum Foundry Alloy, Casting	**Al** rem **Cu** 4.2-5.0 **Fe** 0.10 max **Mg** 0.15-0.35 **Mn** 0.20-0.50 **Ni** 0.05 max **Si** 0.05 max **Sn** 0.05 max **Ti** 0.15-0.30 **Zn** 0.10 max **Other** each 0.05 max, total 0.15 max	**AA** A206.0 **AMS** 4235; 4236
A12062	Aluminum Foundry Alloy, Ingot	**Al** rem **Cu** 4.2-5.0 **Fe** 0.07 max **Mg** 0.20-0.35 **Mn** 0.20-0.50 **Ni** 0.03 max **Si** 0.05 max **Sn** 0.05 max **Ti** 0.15-0.25 **Zn** 0.05 max **Other** each 0.05 max, total 0.15 max	**AA** A206.2
A12400	Aluminum Foundry Alloy, Casting replaced by A02400		
A12401	Aluminum Foundry Alloy, Ingot replaced by A02401		

The chemical compositions listed are for identification purposes and should not be used in lieu of the cross referenced specifications.

UNIFIED NUMBER	DESCRIPTION	CHEMICAL COMPOSITION	CROSS REFERENCE SPECIFICATIONS
A12420	Aluminum Foundry Alloy, Casting	**Al** rem **Cr** 0.15-0.25 **Cu** 3.7-4.5 **Fe** 0.8 max **Mg** 1.2-1.7 **Mn** 0.10 max **Ni** 1.8-2.3 **Si** 0.6 max **Ti** 0.07-0.20 **Zn** 0.10 max **Other** each 0.05 max, total 0.15 max	**AA** A242.0 **AMS** 4222
A12421	Aluminum Foundry Alloy, Ingot	**Al** rem **Cr** 0.15-0.25 **Cu** 3.7-4.5 **Fe** 0.6 max **Mg** 1.3-1.7 **Mn** 0.10 max **Ni** 1.8-2.3 **Si** 0.6 max **Ti** 0.07-0.20 **Zn** 0.10 max **Other** each 0.05 max, total 0.15 max	**AA** A242.1 **FED** QQ-A-371 (A242.1)
A12422	Aluminum Foundry Alloy, Ingot	**Al** rem **Cr** 0.15-0.25 **Cu** 3.7-4.5 **Fe** 0.6 max **Mg** 1.3-1.7 **Mn** 0.10 max **Ni** 1.8-2.3 **Si** 0.35 max **Ti** 0.07-0.20 **Zn** 0.10 max **Other** each 0.05 max, total 0.15 max	**AA** A242.2 **FED** QQ-A-371 (A242.2)
A13050	Aluminum Foundry Alloy, Casting	**Al** rem **Cu** 1.0-1.5 **Fe** 0.20 max **Mg** 0.10 max **Mn** 0.10 max **Si** 4.5-5.5 **Ti** 0.20 max **Zn** 0.10 max **Other** each 0.05 max, total 0.15 max	**AA** A305.0
A13051	Aluminum Foundry Alloy, Ingot	**Ag** 4.5-5.5 **Al** rem **Cu** 1.0-1.5 **Fe** 0.15 max **Mn** 0.05 max **Ti** 0.20 max **Zn** 0.05 max **Other** each 0.05 max, total 0.15 max	**AA** A305.1
A13052	Aluminum Foundry Alloy, Ingot	**Al** rem **Cu** 1.0-1.5 **Fe** 0.13 max **Mn** 0.05 max **Si** 4.5-5.5 **Ti** 0.20 max **Zn** 0.05 max **Other** each 0.05 max, total 0.15 max	**AA** A305.2
A13190	Aluminum Foundry Alloy, Casting	**Al** rem **Cu** 3.0-4.0 **Fe** 1.0 max **Mg** 0.10 max **Mn** 0.50 max **Ni** 0.35 max **Si** 5.5-6.5 **Ti** 0.25 max **Zn** 3.0 max **Other** total 0.50 max	**AA** A319.0
A13191	Aluminum Foundry Alloy, Ingot	**Al** rem **Cu** 3.0-4.0 **Fe** 0.8 max **Mg** 0.10 max **Mn** 0.50 max **Ni** 0.35 max **Si** 5.5-6.5 **Ti** 0.25 max **Zn** 3.0 max **Other** total 0.50 max	**AA** A319.1
A13320	Aluminum Foundry Alloy, Casting replaced by A03360		
A13321	Aluminum Foundry Alloy, Ingot replaced by A03361		
A13322	Aluminum Foundry Alloy, Ingot replaced by A03362		
A13330	Aluminum Foundry Alloy, Casting	**Al** rem **Cu** 3.0-4.0 **Fe** 1.0 max **Mg** 0.05-0.50 **Mn** 0.50 max **Ni** 0.50 max **Si** 8.0-10.0 **Ti** 0.25 max **Zn** 3.0 max **Other** total 0.50 max	**AA** A333.0
A13331	Aluminum Foundry Alloy, Ingot	**Al** rem **Cu** 3.0-4.0 **Fe** 0.8 max **Mg** 0.10-0.50 **Mn** 0.50 max **Ni** 0.50 max **Si** 8.0-10.0 **Ti** 0.25 max **Zn** 3.0 max **Other** total 0.50 max	**AA** A333.1
A13550	Aluminum Foundry Alloy, Casting	**Al** rem **Cu** 1.0-1.5 **Fe** 0.09 **Mg** 0.45-0.6 **Mn** 0.05 **Si** 4.5-5.5 **Ti** 0.04-0.20 **Zn** 0.05 **Other** each 0.05, total 0.15	**AA** A355.0
A13552	Aluminum Foundry Alloy, Ingot	**Al** rem **Cu** 1.0-1.5 **Fe** 0.06 **Mg** 0.50-0.6 **Mn** 0.03 **Si** 4.5-5.5 **Ti** 0.04-0.20 **Zn** 0.03 **Other** each 0.05, total 0.15	**AA** A355.2
A13560	Aluminum Foundry Alloy, Casting	**Al** rem **Cu** 0.20 max **Fe** 0.20 max **Mg** 0.20-0.40 **Mn** 0.10 max **Si** 6.5-7.5 **Ti** 0.20 max **Zn** 0.10 max **Other** each 0.05 max, total 0.15 max	**AA** A356.0 **AMS** 4218 **ASTM** B26 (A356.0); B108 (A356.0) **FED** QQ-A-596 (A356); QQ-A-601 (A356.0) **MIL SPEC** MIL-A-21180 (356.0) **SAE** J452 (336); J453 (336)

The chemical compositions listed are for identification purposes and should not be used in lieu of the cross referenced specifications.

UNIFIED NUMBER	DESCRIPTION	CHEMICAL COMPOSITION	CROSS REFERENCE SPECIFICATIONS
A13561	Aluminum Foundry Alloy, Ingot	**Al** rem **Cu** 0.25 max **Fe** 0.50 max **Mg** 0.25-0.40 **Mn** 0.35 max **Si** 6.5-7.5 **Ti** 0.25 max **Zn** 0.35 max **Other** each 0.05 max, total 0.15 max	**AA** A356.1
A13562	Aluminum Foundry Alloy, Ingot	**Al** rem **Cu** 0.10 max **Fe** 0.12 max **Mg** 0.30-0.40 **Mn** 0.05 max **Si** 6.5-7.5 **Ti** 0.20 max **Zn** 0.05 max **Other** each 0.05 max, total 0.15 max	**AA** A356.2 **ASTM** B179 (A356.2) **FED** QQ-A-371 (A356.2)
A13570	Aluminum Foundry Alloy, Casting	**Al** rem **Be** 0.04-0.07 **Cu** 0.20 max **Fe** 0.20 max **Mg** 0.40-0.7 **Mn** 0.10 max **Si** 6.5-7.5 **Ti** 0.10-0.20 **Zn** 0.10 max **Other** each 0.05 max, total 0.15 max	**AA** A357.0 **AMS** 4219 **ASTM** B108 (A357.0) **MIL SPEC** MIL-A-21180 (A357.0)
A13572	Aluminum Foundry Alloy, Ingot	**Al** rem **Be** 0.04-0.07 **Cu** 0.10 max **Fe** 0.12 max **Mg** 0.45-0.7 **Mn** 0.05 max **Si** 6.5-7.5 **Ti** 0.10-0.20 **Zn** 0.05 max **Other** each 0.03 max, total 0.10 max	**AA** A357.2 **FED** QQ-A-371 (A357.2)
A13600	Aluminum Foundry Alloy, Casting	**Al** rem **Cu** 0.6 max **Fe** 1.3 max **Mg** 0.40-0.6 **Mn** 0.35 max **Ni** 0.50 max **Si** 9.0-10.0 **Sn** 0.15 max **Zn** 0.50 max **Other** total 0.25 max	**AA** A360.0 **AMS** 4290 **ASTM** B85 (A360.0) **FED** QQ-A-591 (A360.0) **MIL SPEC** MIL-F-3922; MIL-F-39000 **SAE** J452 (309); J453 (309)
A13601	Aluminum Foundry Alloy, Ingot	**Al** rem **Cu** 0.6 max **Fe** 1.0 max **Mg** 0.45-0.6 **Mn** 0.35 max **Ni** 0.50 max **Si** 9.0-10.0 **Sn** 0.15 max **Zn** 0.40 max **Other** total 0.25 max	**AA** A360.1 **ASTM** B179 (A360.1) **FED** QQ-A-371 (A360.1) **SAE** J453 (309)
A13602	Aluminum Foundry Alloy, Ingot	**Al** rem **Cu** 0.10 max **Fe** 0.6 max **Mg** 0.45-0.6 **Mn** 0.05 max **Si** 9.0-10.0 **Zn** 0.05 max **Other** each 0.05 max, total 0.15 max	**AA** A360.2 **FED** QQ-A-371 (A360.2)
A13800	Aluminum Foundry Alloy, Casting	**Al** rem **Cu** 3.0-4.0 **Fe** 1.3 max **Mg** 0.10 max **Mn** 0.50 max **Ni** 0.50 max **Si** 7.5-9.5 **Sn** 0.35 max **Zn** 3.0 max **Other** total 0.50 max	**AA** A380.0 **AMS** 4291 **ASTM** B85 (A380.0) **FED** QQ-A-591 (A380.0) **SAE** J452 (306); J453 (306)
A13801	Aluminum Foundry Alloy, Ingot	**Al** rem **Cu** 3.0-4.0 **Fe** 1.0 max **Mg** 0.10 max **Mn** 0.50 max **Ni** 0.50 max **Si** 7.5-9.5 **Sn** 0.35 max **Zn** 2.9 max **Other** total 0.50 max	**AA** A380.1 **ASTM** B179 (A380.1) **FED** QQ-A-371 (A380.1) **SAE** J453 (306)
A13802	Aluminum Foundry Alloy, Ingot	**Al** rem **Cu** 3.0-4.0 **Fe** 0.6 max **Mg** 0.10 max **Mn** 0.10 max **Ni** 0.10 max **Si** 7.5-9.5 **Zn** 0.10 max **Other** each 0.05 max, total 0.15 max	**AA** A380.2 **FED** QQ-A-371 (A380.2)
A13840	Aluminum Foundry Alloy, Casting	**Al** rem **Cu** 3.0-4.5 **Fe** 1.3 max **Mg** 0.10 max **Mn** 0.50 max **Ni** 0.50 max **Si** 10.5-12.0 **Sn** 0.35 max **Zn** 1.0 max **Other** total 0.50 max	**AA** A384.0
A13841	Aluminum Foundry Alloy, Ingot	**Al** rem **Cu** 3.0-4.5 **Fe** 1.0 max **Mg** 0.10 max **Mn** 0.50 max **Ni** 0.50 max **Si** 10.5-12.0 **Sn** 0.35 max **Zn** 0.9 max **Other** total 0.50 max	**AA** A384.1
A13900	Aluminum Foundry Alloy, Casting	**Al** rem **Cu** 4.0-5.0 **Fe** 0.50 max **Mg** 0.45-0.65 **Mn** 0.10 max **Si** 16.0-18.0 **Ti** 0.20 max **Zn** 0.10 max **Other** each 0.10 max, total 0.20 max	**AA** A390.0
A13901	Aluminum Foundry Alloy, Ingot	**Al** rem **Cu** 4.0-5.0 **Fe** 0.40 max **Mg** 0.50-0.65 **Mn** 0.10 max **Si** 16.0-18.0 **Ti** 0.20 max **Zn** 0.10 max **Other** each 0.10 max, total 0.20 max	**AA** A390.1 **FED** QQ-A-371 (A390.1)
A14130	Aluminum Foundry Alloy, Casting	**Al** rem **Cu** 0.6 max **Fe** 1.3 max **Mg** 0.10 max **Mn** 0.35 max **Ni** 0.50 max **Si** 11.0-13.0 **Sn** 0.15 max **Zn** 0.50 max **Other** total 0.25 max	**AA** A413.0 **ASTM** B85 (A413.0) **FED** QQ-A-591 (A413.0) **SAE** J452 (305); J453 (305)

The chemical compositions listed are for identification purposes and should not be used in lieu of the cross referenced specifications.

UNIFIED NUMBER	DESCRIPTION	CHEMICAL COMPOSITION	CROSS REFERENCE SPECIFICATIONS
A14131	Aluminum Foundry Alloy, Ingot	**Al** rem **Cu** 0.6 max **Fe** 1.0 max **Mg** 0.10 max **Mn** 0.35 max **Ni** 0.50 max **Si** 11.0-13.0 **Sn** 0.15 max **Zn** 0.40 max **Other** total 0.25 max	**AA** A413.1 **FED** QQ-A-371 (A413.1) **SAE** J453 (305)
A14132	Aluminum Foundry Alloy, Ingot	**Al** rem **Cu** 0.10 max **Fe** 0.6 max **Mg** 0.05 max **Mn** 0.05 max **Ni** 0.05 max **Si** 11.0-13.0 **Sn** 0.05 max **Zn** 0.05 max **Other** total 0.10 max	**AA** A413.2 **FED** QQ-A-371 (A413.2)
A14430	Aluminum Foundry Alloy, Casting	**Al** rem **Cr** 0.25 max **Cu** 0.30 max **Fe** 0.8 max **Mg** 0.05 max **Mn** 0.50 max **Si** 4.5-6.0 **Ti** 0.25 max **Zn** 0.50 max **Other** total 0.35 max	**AA** A443.0 **ASTM** B26 (A443.0); B108 (A443.0)
A14431	Aluminum Foundry Alloy, Ingot	**Al** rem **Cr** 0.25 max **Cu** 0.30 max **Fe** 0.6 max **Mg** 0.05 max **Mn** 0.50 max **Si** 4.5-6.0 **Ti** 0.25 max **Zn** 0.50 max **Other** total 0.35 max	**AA** A443.1
A14440	Aluminum Foundry Alloy, Casting	**Al** rem **Cu** 0.10 max **Fe** 0.20 max **Mg** 0.05 max **Mn** 0.10 max **Si** 6.5-7.5 **Ti** 0.20 max **Zn** 0.10 max **Other** each 0.05 max, total 0.15 max	**AA** A444.0 **ASTM** B108 (A444.0)
A14441	Aluminum Foundry Alloy, Ingot	**Al** rem **Cu** 0.10 max **Fe** 0.15 max **Mg** 0.05 max **Mn** 0.10 max **Si** 6.5-7.5 **Ti** 0.20 max **Zn** 0.10 max **Other** each 0.05 max, total 0.15 max	**AA** A444.1
A14442	Aluminum Foundry Alloy, Ingot	**Al** rem **Cu** 0.05 max **Fe** 0.12 max **Mg** 0.05 max **Mn** 0.05 max **Si** 6.5-7.5 **Ti** 0.20 max **Zn** 0.05 max **Other** each 0.05 max, total 0.15 max	**AA** A444.2 **ASTM** B179 (A444.2) **FED** QQ-A-371 (A444.2)
A15140	Aluminum Foundry Alloy, Casting replaced by A05130		
A15142	Aluminum Foundry Alloy, Ingot replaced by A05132		
A15350	Aluminum Foundry Alloy, Sand Castings	**Al** rem **Cu** 0.10 max **Fe** 0.20 max **Mg** 6.5-7.5 **Mn** 0.10-0.25 **Si** 0.20 max **Ti** 0.25 max **Other** each 0.05 max, total 0.15 max	**AA** A535.0
A15351	Aluminum Foundry Alloy, Ingot	**Al** rem **Cu** 0.10 max **Fe** 0.15 max **Mg** 6.6-7.5 **Mn** 0.10-0.25 **Si** 0.20 max **Ti** 0.25 max **Other** each 0.05 max, total 0.15 max	**AA** A535.1 **FED** QQ-A-371 (A535.1)
A17120	Aluminum Foundry Alloy, Casting replaced by A07100		
A17121	Aluminum Foundry Alloy, Ingot replaced by A07101		
A18500	Aluminum Foundry Alloy, Casting replaced by A08510		
A18501	Aluminum Foundry Alloy, Ingot replaced by A08511		
A22010	Aluminum Foundry Alloy, Casting	**Al** rem **Cu** 4.5-5.0 **Fe** 0.05 max **Mg** 0.25-0.35 **Mn** 0.20-0.50 **Si** 0.05 max **Ti** 0.15-0.35 **Other** each 0.05 max, total 0.15 max	**AA** B201.0
A22950	Aluminum Foundry Alloy, Casting replaced by A02960		
A22951	Aluminum Foundry Alloy, Ingot replaced by A02961		

The chemical compositions listed are for identification purposes and should not be used in lieu of the cross referenced specifications.

UNIFIED NUMBER	DESCRIPTION	CHEMICAL COMPOSITION	CROSS REFERENCE SPECIFICATIONS
A22952	Aluminum Foundry Alloy, Ingot replaced by A02962		
A23190	Aluminum Foundry Alloy, Casting	Al rem Cu 3.0-4.0 Fe 1.2 Mg 0.10-0.50 Mn 0.8 Ni 0.50 Si 5.5-6.5 Ti 0.25 Zn 1.0 Other total 0.50	AA B319.0 SAE J452 (329); J453 (329)
A23191	Aluminum Foundry Alloy, Ingot	Al rem Cu 3.0-4.0 Fe 0.9 max Mg 0.15-0.50 Mn 0.8 Ni 0.50 Si 5.5-6.5 Ti 0.25 Zn 1.0 Other total 0.50	AA B319.1 SAE J453 (329)
A23560	Aluminum Foundry Alloy, Casting	Al rem Cu 0.05 Fe 0.09 Mg 0.25-0.45 Mn 0.05 Si 6.5-7.5 Ti 0.04-0.20 Zn 0.05 Other each 0.05, total 0.15	AA B356.0
A23562	Aluminum Foundry Alloy, Ingot	Al rem Cu 0.03 Fe 0.06 Mg 0.30-0.45 Mn 0.03 Si 6.5-7.5 Ti 0.04-0.20 Zn 0.03 Other each 0.03, total 0.10	AA B356.2
A23570	Aluminum Foundry Alloy, Casting	Al rem Cu 0.05 Fe 0.09 Mg 0.40-0.6 Mn 0.05 Si 6.5-7.5 Ti 0.04-0.20 Zn 0.05 Other each 0.5, total 0.15	AA B237.0
A23572	Aluminum Foundry Alloy, Ingot	Al rem Cu 0.03 Fe 0.06 Mg 0.45-0.6 Mn 0.03 Si 6.5-7.5 Ti 0.04-0.20 Zn 0.03 Other each 0.03, total 0.10	AA B357.2
A23580	Aluminum Foundry Alloy, Casting replaced by A03580		
A23582	Aluminum Foundry Alloy, Ingot replaced by A03582		
A23800	Aluminum Foundry Alloy, Casting	Al rem Cu 3.0-4.0 Fe 1.3 max Mg 0.10 max Mn 0.50 max Ni 0.50 max Si 7.5-9.5 Sn 0.35 max Zn 1.0 max Other total 0.50 max	AA B380.0
A23801	Aluminum Foundry Alloy, Ingot	Al rem Cu 3.0-4.0 Fe 1.0 max Mg 0.10 max Mn 0.50 max Ni 0.50 max Si 7.5-9.5 Sn 0.35 max Zn 0.9 max Other total 0.50 max	AA B380.1
A23840	Aluminum Foundry Alloy, Casting replaced by A03850		
A23841	Aluminum Foundry Alloy, Ingot replaced by A03851		
A23900	Aluminum Foundry Alloy, Casting	Al rem Cu 4.0-5.0 Fe 0.50 max Mg 0.45-0.65 Mn 0.50 max Ni 0.10 max Si 16.0-18.0 Ti 0.20 max Zn 1.5 max Other each 0.10 max, total 0.20 max	AA B390.0
A23901	Aluminum Foundry Alloy, Ingot	Al rem Cu 4.0-5.0 Fe 1.0 max Mg 0.50-0.65 Mn 0.50 max Ni 0.10 max Si 16.0-18.0 Ti 0.20 max Zn 1.4 max Other each 0.10 max, total 0.20 max	AA B390.1
A24130	Aluminum Foundry Alloy, Casting	Al rem Cu 0.10 max Fe 0.50 max Mg 0.05 max Mn 0.35 max Ni 0.05 max Si 11.0-13.0 Ti 0.25 max Zn 0.10 max Other each 0.05 max, total 0.20 max	AA B413.0
A24131	Aluminum Foundry Alloy, Ingot	Al rem Cu 0.10 max Fe 0.40 max Mg 0.05 max Mn 0.35 max Ni 0.05 max Si 11.0-13.0 Ti 0.25 max Zn 0.10 max Other each 0.05 max, total 0.20 max	AA B413.1
A24430	Aluminum Foundry Alloy, Casting	Al rem Cu 0.15 max Fe 0.8 max Mg 0.05 max Mn 0.35 max Si 4.5-6.0 Ti 0.25 max Zn 0.35 max Other each 0.05 max, total 0.15 max	AA B443.0 ASTM B26 (B443.0); B108 (B443.0) FED QQ-A-596 (B443.0); QQ-A-601 (B443.0) MIL SPEC MIL-C-11866

The chemical compositions listed are for identification purposes and should not be used in lieu of the cross referenced specifications.

UNIFIED NUMBER	DESCRIPTION	CHEMICAL COMPOSITION	CROSS REFERENCE SPECIFICATIONS
A24431	Aluminum Foundry Alloy, Ingot	**Al** rem **Cu** 0.15 max **Fe** 0.6 max **Mg** 0.05 max **Mn** 0.35 max **Si** 4.5-6.0 **Ti** 0.25 max **Zn** 0.35 max **Other** each 0.05 max, total 0.15 max	**AA** B443.1 **ASTM** B179 (B443.1)
A24442	Aluminum Foundry Alloy, Ingot (Steel Coating Alloy) replaced by A04452		
A25140	Aluminum Foundry Alloy, Casting replaced by A05120		
A25142	Aluminum Foundry Alloy, Ingot replaced by A05122		
A25350	Aluminum Foundry Alloy, Casting	**Al** rem **Cu** 0.10 max **Fe** 0.15 max **Mg** 6.5-7.5 **Mn** 0.05 max **Si** 0.15 max **Ti** 0.10-0.25 **Other** each 0.05 max, total 0.15 max	**AA** B535.0
A25352	Aluminum Foundry Alloy, Ingot	**Al** rem **Cu** 0.05 max **Fe** 0.12 max **Mg** 6.6-7.5 **Mn** 0.05 max **Si** 0.10 max **Ti** 0.10-0.25 **Other** each 0.05 max, total 0.15 max	**AA** B535.2
A27710	Aluminum Foundry Alloy, Casting replaced by A07720		
A27712	Aluminum Foundry Alloy, Ingot replaced by A07722		
A28500	Aluminum Foundry Alloy, Casting replaced by A08520		
A28501	Aluminum Foundry Alloy, Ingot replaced by A08521		
A33550	Aluminum Foundry Alloy, Casting	**Al** rem **Cu** 1.0-1.5 **Fe** 0.20 max **Mg** 0.40-0.6 **Mn** 0.10 max **Si** 4.5-5.5 **Ti** 0.20 max **Zn** 0.10 max **Other** each 0.05 max, total 0.15 max	**AA** C355.0 **AMS** 4215 **ASTM** B26 (C355.0); B108 (C355.0) **FED** QQ-A-596 (C355); QQ-A-601 (C355.0) **MIL SPEC** MIL-A-21180 (C355.0) **SAE** J452 (335); J453 (335)
A33551	Aluminum Foundry Alloy, Ingot	**Al** rem **Cr** 0.25 max **Cu** 1.0-1.5 **Fe** 0.50 max **Mg** 0.45-0.6 **Mn** 0.50 max **Si** 4.5-5.5 **Ti** 0.25 max **Zn** 0.35 max **Other** each 0.05 max, total 0.15 max. Note: If Fe exceeds 0.45, Mn shall be not less than ½ Fe.	**AA** C355.1
A33552	Aluminum Foundry Alloy, Ingot	**Al** rem **Cu** 1.0-1.5 **Fe** 0.13 max **Mg** 0.45-0.6 **Mn** 0.05 max **Si** 4.5-5.5 **Ti** 0.20 max **Zn** 0.05 max **Other** each 0.05 max, total 0.15 max	**AA** C355.2 **ASTM** B179 (C355.2) **FED** QQ-A-371 (C355.2)
A33560	Aluminum Foundry Alloy, Casting	**Al** rem **Cu** 0.05 max **Fe** 0.07 max **Mg** 0.25-0.45 **Mn** 0.05 max **Si** 6.5-7.5 **Ti** 0.04-0.20 **Zn** 0.05 max **Other** each 0.05 max, total 0.15 max	**AA** C356.0
A33562	Aluminum Foundry Alloy, Ingot	**Al** rem **Cu** 0.05 max **Fe** 0.04 max **Mg** 0.30-0.45 **Mn** 0.03 max **Si** 6.5-7.5 **Ti** 0.04-0.20 **Zn** 0.03 max **Other** each 0.05 max, total 0.10 max	**AA** C356.2
A33570	Aluminum Foundry Alloy, Casting	**Al** rem **Be** 0.04-0.07 **Cu** 0.05 **Fe** 0.09 **Mg** 0.45-0.7 **Mn** 0.05 **Si** 6.5-7.5 **Ti** 0.04-0.20 **Zn** 0.05 **Other** each 0.05, total 0.15	**AA** C357.0
A33572	Aluminum Foundry Alloy, Ingot	**Al** rem **Be** 0.04-0.07 **Cu** 0.03 **Fe** 0.06 **Mg** 0.50-0.7 **Mn** 0.03 **Si** 6.5-7.5 **Ti** 0.04-0.20 **Zn** 0.03 **Other** each 0.03, total 0.10	**AA** C357.2

The chemical compositions listed are for identification purposes and should not be used in lieu of the cross referenced specifications.

UNIFIED NUMBER	DESCRIPTION	CHEMICAL COMPOSITION	CROSS REFERENCE SPECIFICATIONS
A34430	Aluminum Foundry Alloy, Casting	Al rem Cu 0.6 max Fe 2.0 max Mg 0.10 max Mn 0.35 max Ni 0.50 max Si 4.5-6.0 Sn 0.15 max Zn 0.50 max Other total 0.25 max	AA C443.0 ASTM B85 (C443.0) SAE J452 (304); J453 (304)
A34431	Aluminum Foundry Alloy, Ingot	Al rem Cu 0.6 max Fe 1.0 max Mg 0.10 max Mn 0.35 max Ni 0.50 max Si 4.5-6.0 Sn 0.15 max Zn 0.40 max Other total 0.25 max	AA C443.1 ASTM B179 (C443.1) FED QQ-A-371 (C443.1)
A34432	Aluminum Foundry Alloy, Ingot	Al rem Cu 0.10 max Fe 0.7-1.1 Mg 0.05 max Mn 0.10 max Si 4.5-6.0 Zn 0.10 max Other each 0.05 max, total 0.15 max	AA C443.2 ASTM B179 (C443.2) FED QQ-A-371 (C443.2)
A37120	Aluminum Foundry Alloy, Casting replaced by A07110		
A37121	Aluminum Foundry Alloy, Ingot replaced by A07111		
A43570	Aluminum Foundry Alloy	Al rem Be 0.04-0.07 Fe 0.20 max Mg 0.55-0.6 Mn 0.10 max Si 6.5-7.5 Ti 0.10-0.20 Other each 0.05 max, total 0.15 max	AA D357.0
A47120	Aluminum Foundry Alloy, Casting replaced by A07120		
A47122	Aluminum Foundry Alloy, Ingot replaced by A07122		
A63320	Aluminum Foundry Alloy, Casting replaced by A03320		
A63321	Aluminum Foundry Alloy, Ingot replaced by A03321		
A63322	Aluminum Foundry Alloy, Ingot replaced by A03322		
A63560	Aluminum Foundry Alloy, Casting	Al rem Cu 0.20 max Fe 0.20 max Mg 0.17-0.25 Mn 0.10 max Si 6.5-7.5 Ti 0.20 max Zn 0.10 max Other each 0.05 max, total 0.15 max	AA F356.0
A63562	Aluminum Foundry Alloy, Ingot	Al rem Cu 0.10 max Fe 0.12 max Mg 0.17-0.25 Mn 0.05 max Si 6.5-7.5 Ti 0.20 max Zn 0.05 max Other each 0.05 max, total 0.15 max	AA F356.2
A65140	Aluminum Foundry Alloy, Casting replaced by A05110		
A65141	Aluminum Foundry Alloy, Ingot replaced by A05111		
A65142	Aluminum Foundry Alloy, Ingot replaced by A05112		
A91030	Wrought Aluminum Alloy, Non-Heat Treatable	Al 99.30 min Cu 0.10 max Fe 0.6 max Mg 0.05 max Mn 0.05 max Si 0.35 max Ti 0.03 max V 0.05 max Zn 0.10 max Other each 0.03 max	AA 1030
A91035	Wrought Aluminum Alloy, Non-Heat Treatable	Al 99.35 min Cu 0.10 max Fe 0.6 max Mg 0.05 max Mn 0.05 max Si 0.35 max Ti 0.03 max V 0.05 max Zn 0.10 max Other each 0.03 max	AA 1035
A91040	Wrought Aluminum Alloy, Non-Heat Treatable	Al 99.40 min Cu 0.10 max Fe 0.50 max Mg 0.05 max Mn 0.05 max Si 0.30 max Ti 0.03 max V 0.05 max Zn 0.10 max Other each 0.03 max	AA 1040

The chemical compositions listed are for identification purposes and should not be used in lieu of the cross referenced specifications.

UNIFIED NUMBER	DESCRIPTION	CHEMICAL COMPOSITION	CROSS REFERENCE SPECIFICATIONS
A91045	Wrought Aluminum Alloy, Non-Heat Treatable	**Al** 99.45 min **Cu** 0.10 max **Fe** 0.45 max **Mg** 0.05 max **Mn** 0.05 max **Si** 0.30 max **Ti** 0.03 max **V** 0.05 max **Zn** 0.05 max **Other** each 0.03 max	**AA** 1045
A91050	Wrought Aluminum Alloy, Non-Heat Treatable	**Al** 99.50 min **Cu** 0.05 max **Fe** 0.40 max **Mg** 0.05 max **Mn** 0.05 max **Si** 0.25 max **Ti** 0.03 max **V** 0.05 max **Zn** 0.05 max **Other** each 0.03 max	**AA** 1050 **ASTM** B491 (1050)
A91055	Wrought Aluminum Alloy, Non-Heat Treatable	**Al** 99.55 min **Cu** 0.05 max **Fe** 0.40 max **Mg** 0.05 max **Mn** 0.05 max **Si** 0.25 max **Ti** 0.03 max **V** 0.05 max **Zn** 0.05 max **Other** each 0.03 max	**AA** 1055
A91060	Wrought Aluminum Alloy, Non-Heat Treatable	**Al** 99.60 min **Cu** 0.05 max **Fe** 0.35 max **Mg** 0.03 max **Mn** 0.03 max **Si** 0.25 max **Ti** 0.03 max **V** 0.05 max **Zn** 0.05 max **Other** each 0.03 max (Be 0.0008 max for welding electrode and filler metal only)	**AA** 1060 **AMS** 4000 **ASTM** B209 (1060); B210 (1060); B211 (1060); B221 (1060); B234 (1060); B241 (1060); B345 (1060); B361 (1060); B404 (1060); B483 (1060); B548 (1060) **SAE** J454 (1060)
A91065	Wrought Aluminum Alloy, Non-Heat Treatable	**Al** 99.65 min **Cu** 0.05 max **Fe** 0.30 max **Mg** 0.03 max **Mn** 0.03 max **Si** 0.25 max **Ti** 0.03 max **V** 0.05 max **Zn** 0.05 max **Other** each 0.03 max	**AA** 1065
A91070	Wrought Aluminum Alloy, Non-Heat Treatable	**Al** 99.70 min **Cu** 0.04 max **Fe** 0.25 max **Mg** 0.03 max **Mn** 0.03 max **Si** 0.20 max **Ti** 0.03 max **V** 0.05 max **Zn** 0.04 max **Other** each 0.03 max	**AA** 1070
A91075	Wrought Aluminum Alloy, Non-Heat Treatable	**Al** 99.75 min **Cu** 0.04 max **Fe** 0.20 max **Mg** 0.03 max **Mn** 0.03 max **Si** 0.20 max **Ti** 0.03 max **V** 0.05 max **Zn** 0.04 max **Other** each 0.03 max	**AA** 1075
A91080	Wrought Aluminum Alloy, Non-Heat Treatable	**Al** 99.80 min **Cu** 0.03 max **Fe** 0.15 max **Ga** 0.03 max **Mg** 0.02 max **Mn** 0.02 max **Si** 0.15 max **Ti** 0.03 max **V** 0.03 max **Zn** 0.03 max **Other** each 0.02 max	**AA** 1080
A91085	Wrought Aluminum Alloy, Non-Heat Treatable	**Al** 99.85 min **Cu** 0.03 max **Fe** 0.12 max **Ga** 0.03 max **Mg** 0.02 max **Mn** 0.02 max **Si** 0.10 max **Ti** 0.02 max **V** 0.03 max **Zn** 0.03 max **Other** each 0.01 max	**AA** 1085
A91090	Wrought Aluminum Alloy, Non-Heat Treatable	**Al** 99.90 min **Cu** 0.02 max **Fe** 0.07 max **Ga** 0.03 max **Mg** 0.01 max **Mn** 0.01 max **Si** 0.07 max **Ti** 0.01 max **V** 0.03 max **Zn** 0.03 max **Other** each 0.01 max	**AA** 1090
A91095	Wrought Aluminum Alloy, Non-Heat Treatable	**Al** 99.95 min **Cu** 0.010 max **Fe** 0.040 max **Mg** 0.010 max **Mn** 0.010 max **Si** 0.030 max **Ti** 0.005 max **Zn** 0.010 max **Other** each 0.005 max	**AA** 1095
A91100	Wrought Aluminum Alloy, Non-Heat Treatable	**Al** 99.00 min **Cu** 0.05-0.20 **Mn** 0.05 max **Zn** 0.10 max **Other** each 0.05 max (Be 0.0008 max for welding electrode and filler wire only), total 0.15 max, Si+ Fe 0.95 max	**AA** 1100 **AMS** 4001; 4002; 4062; 4102; 4180; 7220 **ASME** SFA5.10 (ER1100) **ASTM** B209 (1100); B210 (1100); B211 (1100); B221 (1100); B241 (1100); B247 (1100); B313 (1100); B316 (1100); B361 (1100); B483 (1100); B491 (1100); B547 (1100) **AWS** A5.10 (ER1100) **FED** QQ-A-225/1; QQ-A-250/1; QQ-A-430; QQ-A-1876; QQ-R-566; WW-T-700/1 **MIL SPEC** MIL-W-85; MIL-R-5674; MIL-W-6712; MIL-A-12545; MIL-E-15597; MIL-S-24149/5; MIL-C-26094; MIL-A-52174; MIL-A-52177 **SAE** J454 (1100)
A91135	Wrought Aluminum Alloy, Non-Heat Treatable	**Al** 99.35 min **Cu** 0.05-0.20 **Mg** 0.05 max **Mn** 0.04 max **Ti** 0.03 max **Zn** 0.10 max **Other** each 0.03 max, Si+Fe 0.60 max	**AA** 1135

The chemical compositions listed are for identification purposes and should not be used in lieu of the cross referenced specifications.

ALUMINUM AND ALUMINUM ALLOYS

UNIFIED NUMBER	DESCRIPTION	CHEMICAL COMPOSITION	CROSS REFERENCE SPECIFICATIONS
A91145	Wrought Aluminum Alloy, Non-Heat Treatable	Al 99.45 min Cu 0.05 max Mg 0.05 max Mn 0.05 max Ti 0.03 max Zn 0.05 max Other each 0.03 max, Si+Fe 0.55 max	AA 1145 AMS 4011 ASTM B373 (1145) FED QQ-A-1876
A91170	Wrought Aluminum Alloy, Non-Heat Treatable	Al 99.70 min Cr 0.03 max Cu 0.03 max Mg 0.02 max Mn 0.03 max Ti 0.03 max Zn 0.03 max Other each 0.03 max, Si+Fe 0.30 max	AA 1170
A91175	Wrought Aluminum Alloy, Non-Heat Treatable	Al 99.75 min Cu 0.10 max Ga 0.03 max Mg 0.02 max Mn 0.02 max Ti 0.02 max V 0.03 max Zn 0.03 max Other each 0.02 max, Si+Fe 0.15 max	AA 1175
A91180	Wrought Aluminum Alloy, Non-Heat Treatable	Al 99.80 min Cu 0.01 max Fe 0.09 max Ga 0.03 max Mg 0.02 max Mn 0.02 max Si 0.09 max Ti 0.02 max V 0.03 max Zn 0.03 max Other each 0.02 max	AA 1180
A91185	Wrought Aluminum Alloy, Non-Heat Treatable	Al 99.85 min Cu 0.01 max Mg 0.02 max Mn 0.02 max V 0.03 max Zn 0.03 max Other each 0.03 max, Si+Fe 0.15 max	AA 1185
A91188	Wrought Aluminum Alloy, Non-Heat Treatable	Al 99.88 min Cu 0.005 max Fe 0.06 max Ga 0.03 max Mg 0.01 max Mn 0.01 max Si 0.06 max Ti 0.01 max V 0.03 max Zn 0.03 max Other each 0.01 max	AA 1188 ASME SFA5.10 (ER1188) AWS A5.10 (ER1188)
* A91193	Wrought Aluminum Alloy, Non-Heat Treatable	Al 99.93 min Cu 0.006 max Fe 0.04 max Ga 0.03 max Si 0.04 max Ti 0.01 max V 0.03 max Zn 0.03 max Other each 0.01 max	AA 1193
A91199	Wrought Aluminum Alloy, Non-Heat Treatable	Al 99.99 min Cu 0.006 max Fe 0.006 max Ga 0.005 max Mg 0.006 max Mn 0.002 max Si 0.006 max Ti 0.002 max V 0.005 max Zn 0.006 max Other each 0.002 max	AA 1199
A91200	Wrought Aluminum Alloy, Non-Heat Treatable	Al 99.00 min Cu 0.05 max Mn 0.05 max Zn 0.10 max Other each 0.05 max, total 0.15 max, Si+Fe 1.0 max	AA 1200 ASTM B491 (1200)
A91230	Wrought Aluminum Alloy, Non-Heat Treatable	Al 99.30 min Cu 0.10 max Mn 0.05 max Zn 0.10 max Other each 0.05 max, plus Si+Fe 0.7 max	AA 1230 ASTM B209 (1230)
A91235	Wrought Aluminum Alloy, Non-Heat Treatable	Al 99.35 min Cu 0.05 max Mg 0.05 max Mn 0.05 max Ti 0.03 max Zn 0.010 max Other each 0.03 max, Si+Fe 0.65 max	AA 1235 ASTM B373 (1235); B491 (1235) FED QQ-A-1876
A91250	Wrought Aluminum Alloy, Non-Heat Treatable	Al 99.50 min Cr 0.01 max Cu 0.10 max Fe 0.40 max Mg 0.01 max Mn 0.01 max Si 0.20 max Zn 0.05 max Other each 0.03 max, Ti+V 0.02 max	AA 1250
* A91260	Wrought Aluminum Alloy, Non-Heat Treatable	Al 99.60 min Cu 0.04 max Mg 0.03 max Mn 0.01 max Ti 0.03 max Zn 0.03 max Other each 0.03 max (Be 0.0008 max for welding electrode and filler wire only), Si+Fe 0.40 max	AA 1260

*Boxed entries are no longer active and are retained for reference purposes only.

The chemical compositions listed are for identification purposes and should not be used in lieu of the cross referenced specifications.

UNIFIED NUMBER	DESCRIPTION	CHEMICAL COMPOSITION	CROSS REFERENCE SPECIFICATIONS
A91285	Wrought Aluminum Alloy, Non-Heat Treatable	**Al** 99.85 min **Cu** 0.02 max **Fe** 0.08 max **Ga** 0.03 max **Si** 0.08 max **Ti** 0.02 max **V** 0.03 max **Zn** 0.03 max **Other** each 0.01 max, Si+Fe 0.14 max	**AA** 1285
A91345	Wrought Aluminum Alloy, Non-Heat Treatable	**Al** 99.45 min **Cu** 0.10 max **Fe** 0.40 max **Mg** 0.05 max **Mn** 0.05 max **Si** 0.30 max **Ti** 0.03 max **Zn** 0.05 max **Other** each 0.03 max	**AA** 1345
A91350	Wrought Aluminum Alloy, Non-Heat Treatable	**Al** 99.50 min **B** 0.05 max **Cr** 0.01 max **Cu** 0.05 max **Fe** 0.40 max **Mn** 0.01 max **Si** 0.10 max **Zn** 0.05 max **Other** Gallium 0.03 max; Vanadium + Titanium 0.02 max; other unspecified elements each 0.03 max, total 0.10 max	**AA** 1350 **ASTM** B230; B231; B233; B236; B324; B401; B524; B544; B549; B609 **SAE** J454 (EC-O)
A91435	Wrought Aluminum Alloy, Non-Heat Treatable	**Al** 99.35 min **Cu** 0.02 max **Fe** 0.30-0.50 **Mg** 0.05 max **Mn** 0.05 max **Si** 0.15 max **Ti** 0.03 max **Zn** 0.10 max **Other** each 0.03 max	**AA** 1435 **ASTM** B483 (1435)
A92011	Wrought Aluminum Alloy, Heat Treatable	**Al** rem **Bi** 0.20-0.6 **Cu** 5.0-6.0 **Fe** 0.7 max **Pb** 0.20-0.6 **Si** 0.40 max **Zn** 0.30 max **Other** each 0.05 max, total 0.15 max	**AA** 2011 **ASTM** B210 (2011); B211 (2011) **FED** QQ-A-225/3 **SAE** J454 (2011)
A92014	Wrought Aluminum Alloy, Heat Treatable	**Al** rem **Cr** 0.10 max **Cu** 3.9-5.0 **Fe** 0.7 max **Mg** 0.20-0.8 **Mn** 0.40-1.2 **Si** 0.50-1.2 **Ti** 0.15 max **Zn** 0.25 max **Other** each 0.05 max (Be 0.0008 max for welding electrode and filler metal only), total 0.15 max, Ti + Zn 0.20 max	**AA** 2014 **AMS** 4028; 4029; 4121; 4133; 4134; 4135; 4153; 4314 **ASTM** B209 (2014); B210 (2014); B211 (2014); B221 (2014); B241 (2014); B247 (2014) **FED** QQ-A-200/2; QQ-A-225/4; QQ-A-367 **MIL SPEC** MIL-F-5509; MIL-A-12545; MIL-T-15089; MIL-F-18280; MIL-A-22771 **SAE** J454 (2014)
A92017	Wrought Aluminum Alloy, Heat Treatable	**Al** rem **Cr** 0.10 max **Cu** 3.5-4.5 **Fe** 0.7 max **Mg** 0.40-0.8 **Mn** 0.40-1.0 **Si** 0.8 max **Zn** 0.25 max **Other** each 0.05 max, total 0.15 max, Ti + Zn 0.20 max	**AA** 2017 **AMS** 4118 **ASTM** B211 (2017); B316 (2017) **FED** QQ-A-225/5; QQ-A-430 **MIL SPEC** MIL-R-5674 **SAE** J454 (2017)
A92018	Wrought Aluminum Alloy, Heat Treatable	**Al** rem **Cr** 0.10 max **Cu** 3.5-4.5 **Fe** 1.0 max **Mg** 0.45-0.9 **Mn** 0.20 max **Ni** 1.7-2.3 **Si** 0.9 max **Zn** 0.25 max **Other** each 0.05 max, total 0.15 max	**AA** 2018 **AMS** 4140 **ASTM** B247 (2018) **FED** QQ-A-367 **SAE** J454 (2018)
* A92020	Wrought Aluminum Alloy, Heat Treatable	**Al** rem **Cd** 0.10-0.35 **Cu** 4.0-5.0 **Fe** 0.40 max **Li** 0.9-1.7 **Mg** 0.03 max **Mn** 0.30-0.8 **Si** 0.40 max **Ti** 0.10 max **Zn** 0.25 max **Other** each 0.05 max, total 0.15 max	**AA** 2020
* A92021	Wrought Aluminum Alloy, Heat Treatable	**Al** rem **Cd** 0.05-0.20 **Cu** 5.8-6.8 **Fe** 0.30 max **Mg** 0.02 max **Mn** 0.20-0.40 **Si** 0.20 max **Sn** 0.03-0.08 **Ti** 0.02-0.10 **V** 0.05-0.15 **Zn** 0.10 max **Zr** 0.10-0.25 **Other** each 0.05 max, total 0.15 max	**AA** 2021
A92024	Wrought Aluminum Alloy, Heat Treatable	**Al** rem **Cr** 0.10 max **Cu** 3.8-4.9 **Fe** 0.50 max **Mg** 1.2-1.8 **Mn** 0.30-0.9 **Si** 0.50 max **Zn** 0.25 max **Other** each 0.05 max, total 0.15 max	**AA** 2024 **AMS** 4007; 4035; 4037; 4086; 4087; 4088; 4112; 4119; 4120; 4152; 4164; 4165; 4192; 4193; 7223 **ASTM** B209 (2024); B210 (2024); B211 (2024); B221 (2024); B241 (2024); B316 (2024); F467; F468 **FED** QQ-A-200/3; QQ-A-225/6; QQ-A-250/4; QQ-A-430; WW-T-700/3 **MIL SPEC** MIL-F-5509; MIL-R-5674; MIL-B-6812; MIL-T-15089; MIL-F-18280; MIL-T-50777; MIL-A-81596 **SAE** J454 (2024)
A92025	Wrought Aluminum Alloy, Heat Treatable	**Al** rem **Cr** 0.10 max **Cu** 3.9-5.0 **Fe** 1.0 max **Mg** 0.05 max **Mn** 0.40-1.2 **Si** 0.50-1.2 **Ti** 0.15 max **Zn** 0.25 max **Other** each 0.05 max, total 0.15 max	**AA** 2025 **AMS** 4130 **ASTM** B247 **FED** QQ-A-367 **SAE** J454 (2025)

*Boxed entries are no longer active and are retained for reference purposes only.

The chemical compositions listed are for identification purposes and should not be used in lieu of the cross referenced specifications.

UNIFIED NUMBER	DESCRIPTION	CHEMICAL COMPOSITION	CROSS REFERENCE SPECIFICATIONS
A92034	Wrought Aluminum Alloy, Heat Treatable	Al rem Cr 0.05 Cu 4.2-4.8 Fe 0.12 Mg 1.3-1.9 Mn 0.8-1.3 Si 0.10 Ti 0.15 Zn 0.20 Zr 0.08-0.15 Other each 0.15, total 0.15	AA 203A
A92036	Wrought Aluminum Alloy, Heat Treatable	Al rem Cr 0.10 max Cu 2.2-3.0 Fe 0.50 max Mg 0.30-0.6 Mn 0.10-0.40 Si 0.50 max Ti 0.10 max Zn 0.25 max Other each 0.05, total 0.15 max	AA 2036
A92037	Wrought Aluminum Alloy, Heat Treatable	Al rem Cr 0.10 max Cu 1.4-2.2 Fe 0.50 max Mg 0.30-0.8 Mn 0.10-0.40 Si 0.50 max Ti 0.15 max V 0.05 max Zn 0.25 max Other each 0.05 max, total 0.15 max	AA 2037
A92038	Wrought Aluminum Alloy, Heat Treatable	Al rem Cr 0.20 Cu 0.8-1.8 Fe 0.6 Ga 0.05 Mg 0.40-1.0 Mn 0.10-0.40 Si 0.50-1.3 Ti 0.15 V 0.05 Zn 0.50 Other each 0.05, total 0.15	AA 2038
A92048	Wrought Aluminum Alloy, Heat Treatable	Al rem Cu 2.8-3.8 Fe 0.20 max Mg 1.2-1.8 Mn 0.20-0.6 Si 0.15 max Ti 0.10 max Zn 0.25 max Other each 0.05 max, total 0.15 max	AA 2048
A92090	Wrought Aluminum Alloy, Heat Treatable	Al rem Cr 0.05 max Cu 2.4-3.0 Fe 0.12 max Li 1.9-2.6 Mg 0.25 max Mn 0.05 max Si 0.10 max Ti 0.15 max Zn 0.10 max Zr 0.08-0.15 Other each 0.05 max, total 0.15 max	AA 2090
A92117	Wrought Aluminum Alloy, Heat Treatable	Al rem Cr 0.10 max Cu 2.2-3.0 Fe 0.7 max Mg 0.20-0.50 Mn 0.20 max Si 0.8 max Zn 0.25 max Other each 0.05 max, total 0.15 max	AA 2117 AMS 7222 ASTM B316 (2117) FED QQ-A-430 MIL SPEC MIL-R-5674; MIL-R-8814 SAE J454 (2117)
A92124	Wrought Aluminum Alloy, Heat Treatable	Al rem Cr 0.10 max Cu 3.8-4.9 Fe 0.30 max Mg 1.2-1.8 Mn 0.30-0.90 Si 0.20 max Ti 0.15 max Zn 0.25 max Other each 0.05 max, total 0.15 max, plus Ti + Zn 0.20 max	AA 2124 AMS 4101 ASTM B209 (2124) FED QQ-A-250/9; QQ-A-250/29
A92214	Wrought Aluminum Alloy, Heat Treatable	Al rem Cr 0.10 max Cu 3.9-5.0 Fe 0.30 max Mg 0.20-0.8 Mn 0.40-1.2 Si 0.50-1.2 Ti 0.15 max Zn 0.25 max Other each 0.05 max, total 0.15 max, plus Ti + Zn 0.20 max	AA 2214
A92218	Wrought Aluminum Alloy, Heat Treatable	Al rem Cr 0.10 max Cu 3.5-4.5 Fe 1.0 max Mg 1.2-1.8 Mn 0.20 max Ni 1.7-2.3 Si 0.9 max Zn 0.25 max Other each 0.05 max, total 0.15 max	AA 2218 AMS 4142 ASTM B247 (2218) FED QQ-A-367 SAE J454 (2218)
A92219	Wrought Aluminum Alloy, Heat Treatable	Al rem Cu 5.8-6.8 Fe 0.30 max Mg 0.02 max Mn 0.20-0.40 Si 0.20 max Ti 0.02-0.10 V 0.05-0.15 Zn 0.10 max Zr 0.10-0.25 Other each 0.05 max, total 0.15 max	AA 2219 AMS 4031; 4066; 4068; 4143; 4144; 4162; 4163; 4313 ASTM B209 (2219); B211 (2219); B221 (2219); B241 (2219); B247 (2219); B316 (2219) FED QQ-A-250/30; QQ-A-367; QQ-A-430 MIL SPEC MIL-A-22771; MIL-A-46118; MIL-A-46808 SAE J454 (2219)
A92224	Wrought Aluminum Alloy, Heat Treatable	Al rem Cr 0.10 max Cu 3.8-4.4 Fe 0.15 max Mg 1.2-1.8 Mn 0.30-0.9 Si 0.12 max Ti 0.15 max Zr 0.25 max Other each 0.05 max, total 0.15 max	AA 2224
A92319	Wrought Aluminum Alloy, Heat Treatable (Filler Alloy)	Al rem Be 0.0008 max Cu 5.8-6.8 Fe 0.30 max Mg 0.02 max Mn 0.20-0.40 Si 0.20 max Ti 0.10-0.20 V 0.05-0.15 Zn 0.10 max Zr 0.10-0.25 Other each 0.05 max (Be 0.0008 max for welding electrode and filler wire only), total 0.15 max	AA 2319 AMS 4191 ASME SFA5.10 (ER2319) AWS A5.10 (ER2319) FED QQ-R-566

The chemical compositions listed are for identification purposes and should not be used in lieu of the cross referenced specifications.

UNIFIED NUMBER	DESCRIPTION	CHEMICAL COMPOSITION	CROSS REFERENCE SPECIFICATIONS
A92324	Wrought Aluminum Alloy	**Al** balance **Cr** 0.10 max **Cu** 3.8-4.4 **Fe** 0.12 max **Mg** 1.2-1.8 **Mn** 0.30-0.9 **Si** 0.10 max **Ti** 0.15 max **Zn** 0.25 max **Other** each 0.05 max, total 0.15 max	**AA** 2324
A92419	Wrought Aluminum Alloy, Heat Treatable	**Al** rem **Cu** 5.8-6.8 **Fe** 0.18 max **Mg** 0.02 max **Mn** 0.20-0.40 **Si** 0.15 max **Ti** 0.02-0.10 **V** 0.05-0.15 **Zn** 0.10 max **Zr** 0.10-0.25 **Other** each 0.05 max, total 0.15 max	**AA** 2419
A92519	Wrought Aluminum Alloy, Heat Treatable	**Al** rem **Cu** 5.3-6.4 **Fe** 0.30 max **Mg** 0.05-0.40 **Mn** 0.10-0.50 **Si** 0.25 max **Ti** 0.02-0.10 **V** 0.05-0.15 **Zn** 0.10 max **Zr** 0.10-0.25 **Other** each 0.05 max, total 0.15 max NOTE: Si + Fe 0.40 max	**AA** 2519
A92618	Wrought Aluminum Alloy, Heat Treatable	**Al** rem **Cu** 1.9-2.7 **Fe** 0.9-1.3 **Mg** 1.3-1.8 **Ni** 0.9-1.2 **Si** 0.25 max **Ti** 0.04-0.10 **Other** each 0.05 max, total 0.15 max	**AA** 2618 **AMS** 4132 **ASTM** B247 (2618); B248 (2618) **FED** QQ-A-367 **MIL SPEC** MIL-A-22771 (2618) **SAE** J454 (2618)
A93002	Wrought Aluminum Alloy, Non-Heat Treatable	**Al** rem **Cu** 0.15 max **Ir** 0.10 max **Mg** 0.05-0.20 **Mn** 0.10-0.25 **Si** 0.08 max **Ti** 0.03 max **Zn** 0.03 max **Other** each 0.03 max, total 0.10 max	**AA** 3002
A93003	Wrought Aluminum Alloy, Non-Heat Treatable	**Al** rem **Cu** 0.05-0.20 **Fe** 0.7 max **Mn** 1.0-1.5 **Si** 0.6 max **Zn** 0.10 max **Other** each 0.05 max (Be 0.0008 max for welding electrode and filler wire only), total 0.15 max	**AA** 3003 **AMS** 4006; 4008; 4010; 4065; 4067. **ASTM** B209 (3003); B210 (3003); B211 (3003); B221 (3003); B234 (3003); B241 (3003); B247 (3003); B313 (3003); B316 (3003); B345 (3003); B404 (3003); B483 (3003); B491 (3003); B547 (3003) **FED** QQ-A-200/1; QQ-A-225/2; QQ-A-250/2; QQ-A-430; WW-T-700/2 **MIL SPEC** MIL-S-12875; MIL-E-15597; MIL-P-25995; MIL-A-52174; MIL-A-81596 **SAE** J454 (3003)
A93004	Wrought Aluminum Alloy, Non-Heat Treatable	**Al** rem **Cu** 0.25 max **Fe** 0.7 max **Mg** 0.8-1.3 **Mn** 1.0-1.5 **Si** 0.30 max **Zn** 0.25 max **Other** each 0.05 max (Be 0.0008 max for welding electrode and filler wire only), total 0.15 max	**AA** 3004 **ASTM** B209 (3004); B221 (3004); B313 (3004); B547 (3004); B548 (3004) **SAE** J454 (3004)
A93005	Wrought Aluminum Alloy, Non-Heat Treatable	**Al** rem **Cr** 0.10 max **Cu** 0.30 max **Fe** 0.7 max **Mg** 0.20-0.6 **Mn** 1.0-1.5 **Si** 0.6 max **Ti** 0.10 max **Zn** 0.25 max **Other** each 0.05 max, total 0.15 max	**AA** 3005 **ASTM** B209 (3005)
A93006	Wrought Aluminum Alloy, Non-Heat Treatable	**Al** rem **Cr** 0.20 max **Cu** 0.10-0.30 **Fe** 0.7 max **Mg** 0.30-0.6 **Mn** 0.50-0.8 **Si** 0.50 max **Ti** 0.10 max **Zn** 0.15-0.40 **Other** each 0.05 max, total 0.15 max	**AA** 3006
A93007	Wrought Aluminum Alloy, Non-Heat Treatable	**Al** rem **Cr** 0.20 max **Cu** 0.10-0.30 **Fe** 0.7 max **Mg** 0.6 max **Mn** 0.30-0.8 **Si** 0.50 max **Ti** 0.10 max **Zn** 0.40 max **Other** each 0.05 max, total 0.15 max	**AA** 3007
A93009	Wrought Aluminum Alloy, Non-Heat Treatable	**Al** rem **Cr** 0.05 max **Cu** 0.10 max **Fe** 0.7 max **Mg** 0.10 max **Mn** 1.2-1.8 **Ni** 0.05 max **Si** 1.0-1.8 **Ti** 0.10 max **Zn** 0.05 max **Zr** 0.10 max **Other** each 0.05 max, total 0.15 max	**AA** 3009
A93010	Wrought Aluminum Alloy, Non-Heat Treatable	**Al** rem **Cr** 0.05-0.40 **Cu** 0.03 max **Fe** 0.20 max **Mn** 0.20-0.9 **Si** 0.10 max **Ti** 0.05 max **V** 0.05 max **Zn** 0.05 max **Other** each 0.03 max, total 0.10 max	**AA** 3010
A93011	Wrought Aluminum Alloy, Non-Heat Treatable	**Al** rem **Cr** 0.10-0.40 **Cu** 0.05-0.20 **Fe** 0.7 max **Mn** 0.8-1.2 **Si** 0.40 max **Ti** 0.10 max **Zn** 0.10 max **Zr** 0.10-0.30 **Other** each 0.05 max, total 0.15 max	**AA** 3011

The chemical compositions listed are for identification purposes and should not be used in lieu of the cross referenced specifications.

UNIFIED NUMBER	DESCRIPTION	CHEMICAL COMPOSITION	CROSS REFERENCE SPECIFICATIONS
A93102	Wrought Aluminum Alloy, Non-Heat Treatable	Al rem Cu 0.10 max Fe 0.7 max Mn 0.05-0.40 Si 0.40 max Ti 0.10 max Zn 0.30 max Other each 0.05 max, total 0.15 max	AA 3102 ASTM B210 (3102)
A93104	Wrought Aluminum Alloy, Non-Heat Treatable	Al rem Cu 0.05-0.25 Fe 0.8 max Ga 0.05 max Mg 0.8-1.3 Mn 0.8-1.4 Si 0.6 max Ti 0.10 max V 0.05 max Zn 0.25 max Other each 0.05 max, total 0.15 max	AA 3104
A93105	Wrought Aluminum Alloy, Non-Heat Treatable	Al rem Cr 0.20 max Cu 0.30 max Fe 0.7 max Mg 0.20-0.8 Mn 0.30-0.8 Si 0.6 max Ti 0.10 max Zn 0.40 max Other each 0.05 max, total 0.15 max	AA 3105 ASTM B209 (3105)
A93107	Wrought Aluminum Alloy, Non-Heat Treatable	Al rem Cu 0.05-0.15 Fe 0.7 max Mn 0.40-0.9 Si 0.6 max Ti 0.10 max Zn 0.20 max Other each 0.05 max, total 0.15 max	AA 3107
A93303	Wrought Aluminum Alloy, Non-Heat Treatable	Al rem Cu 0.05-0.20 Fe 0.7 max Mn 1.0-1.5 Si 0.6 max Zn 0.30 max Other each 0.05 max, total 0.15 max	AA 3303 ASTM B210 (3303)
* A94002	Wrought Aluminum Alloy, Non-Heat Treatable	Al rem Cd 0.8-1.4 Cu 0.05-0.15 Fe 0.35 max Mg 0.05-0.15 Mn 0.03 max Si 3.5-4.5 Ti 0.02 max Zn 0.15 max Other each 0.05 max, total 0.15 max	AA 4002
A94004	Wrought Aluminum Alloy, Non-Heat Treatable (Brazing Alloy)	Al rem Bi 0.02-0.20 Cu 0.25 max Fe 0.8 max Mg 1.0-2.0 Mn 0.10 max Si 9.0-10.5 Zn 0.20 max Other each 0.05 max, total 0.15 max	AA 4004 ASME SFA5.8 (BAlSi-7) AWS A5.8 (BAlSi-7)
A94008	Wrought Aluminum Alloy, Non-Heat Treatable	Al rem Be 0.0008 max Cu 0.05 max Fe 0.09 max Mg 0.30-0.45 Mn 0.05 max Si 6.5-7.5 Ti 0.04-0.15 Zn 0.05 max Other each 0.05 max, total 0.15 max	AA 4008
A94032	Wrought Aluminum Alloy, Heat Treatable	Al rem Cr 0.10 max Cu 0.50-1.3 Fe 1.0 max Mg 0.8-1.3 Ni 0.50-1.3 Si 11.0-13.5 Zn 0.25 max Other each 0.05 max, total 0.15 max	AA 4032 AMS 4145 ASTM B247 (4032) FED QQ-A-367 SAE J454 (4032)
A94043	Wrought Aluminum Alloy, Non-Heat Treatable	Al rem Cu 0.30 max Fe 0.8 max Mg 0.05 max Mn 0.05 max Si 4.5-6.0 Ti 0.20 max Zn 0.10 max Other each 0.05 max (Be 0.0008 max for welding electrode and filler wire only), total 0.15 max	AA 4043 AMS 4190 ASME SFA5.10 (ER4043) AWS A5.10 (ER4043) FED QQ-B-655 MIL SPEC MIL-W-6712; MIL-E-15597 SAE J454 (4043)
A94044	Wrought Aluminum Alloy, Non-Heat Treatable (Brazing Alloy)	Al rem Cu 0.25 max Fe 0.8 max Mn 0.10 max Si 7.8-9.2 Zn 0.20 max Other each 0.05 max, total 0.15 max	AA 4044
A94045	Wrought Aluminum Alloy, Non-Heat Treatable	Al rem Cu 0.30 max Fe 0.8 max Mg 0.05 max Mn 0.05 max Si 9.0-11.0 Ti 0.20 max Zn 0.10 max Other each 0.05 max, total 0.15 max	AA 4045 ASME SFA5.8 (BAlSi-5) AWS A5.8 (BAlSi-5) FED QQ-B-655
A94047	Wrought Aluminum Alloy, Non-Heat Treatable (Brazing Alloy)	Al rem Cu 0.30 max Fe 0.8 max Mg 0.10 max Mn 0.15 max Si 11.0-13.0 Zn 0.20 max Other each 0.05 max (Be 0.0008 max for welding electrode and filler wire only), total 0.15 max	AA 4047 AMS 4185 ASME SFA5.8 (BAlSi-4); SFA5.10 (ER4047) AWS A5.8 (BAlSi-4); A5.10 (ER4047) FED QQ-B-655; QQ-R-566 MIL SPEC MIL-B-20148
A94104	Wrought Aluminum Alloy, Non-Heat Treatable	Al rem Bi 0.02-0.20 Cu 0.25 max Fe 0.8 max Mg 1.0-2.0 Mn 0.10 max Si 9.5-10.5 Zn 0.20 max Other each 0.05 max, total 0.15 max	AA 4104 ASME SFA5.8 (BAlSi-11) AWS A5.8 (BAlSi-11)

*Boxed entries are no longer active and are retained for reference purposes only.

The chemical compositions listed are for identification purposes and should not be used in lieu of the cross referenced specifications.

UNIFIED NUMBER	DESCRIPTION	CHEMICAL COMPOSITION	CROSS REFERENCE SPECIFICATIONS
A94145	Wrought Aluminum Alloy, Non-Heat Treatable (brazing and Welding Alloy)	**Al** rem **Cr** 0.15 max **Cu** 3.3-4.7 **Fe** 0.8 max **Mg** 0.15 max **Mn** 0.15 max **Si** 9.3-10.7 **Zn** 0.20 max **Other** each 0.05 max (Be 0.0008 max for welding electrode and filler wire only), total 0.15 max	**AA** 4145 **AMS** 4184 **ASME** SFA5.8 (BAlSi-3); SFA5.10 (ER4145) **AWS** A5.8 (BAlSi-3); A5.10 (ER4145) **FED** QQ-B-655; QQ-R-566
A94343	Wrought Aluminum Alloy, Non-Heat Treatable (Brazing Alloy)	**Al** rem **Cu** 0.25 max **Fe** 0.8 max **Mn** 0.10 max **Si** 6.8-8.2 **Zn** 0.20 max **Other** each 0.05 max, total 0.15 max	**AA** 4343 **ASME** SFA5.8 (BAlSi-2) **AWS** A5.8 (BAlSi-2) **FED** QQ-B-655 **MIL SPEC** MIL-B-20148
A94543	Wrought Aluminum Alloy, Non-Heat Treatable	**Al** rem **Cr** 0.05 max **Cu** 0.10 max **Fe** 0.50 max **Mg** 0.10-0.40 **Mn** 0.05 max **Si** 5.0-7.0 **Ti** 0.10 max **Zn** 0.10 max **Other** each 0.05 max, total 0.15 max	**AA** 4543
A94643	Wrought Aluminum Alloy, Non-Heat Treatable (Welding Alloy)	**Al** rem **Be** 0.0008 max **Cu** 0.10 max **Fe** 0.8 max **Mg** 0.10-0.30 **Mn** 0.05 max **Si** 3.6-4.6 **Ti** 0.15 max **Zn** 0.10 max **Other** each 0.05 max, total 0.15 max	**AA** 4643 **AMS** 4189 **ASME** SFA5.10 (ER4643) **AWS** A5.10 (ER4643)
A95005	Wrought Aluminum Alloy, Non-Heat Treatable	**Al** rem **Cr** 0.10 max **Cu** 0.20 max **Fe** 0.7 max **Mg** 0.50-1.1 **Mn** 0.20 max **Si** 0.40 max **Zn** 0.25 max **Other** each 0.05 max, total 0.15 max	**AA** 5005 **ASTM** B209 (5005); B210 (5005); B316 (5005); B396 (5005); B397 (5005); B483 (5005); B531 (5005) **FED** QQ-A-430 (5005) **MIL SPEC** MIL-C-26094 **SAE** J454 (5005)
A95006	Wrought Aluminum Alloy, Non-Heat Treatable	**Al** rem **Cr** 0.10 max **Cu** 0.10 max **Fe** 0.8 max **Mg** 0.8-1.3 **Mn** 0.40-0.8 **Si** 0.40 max **Ti** 0.10 max **Zn** 0.25 max **Other** each 0.05 max, total 0.15 max	**AA** 5006
A95010	Wrought Aluminum Alloy, Non-Heat Treatable	**Al** rem **Cr** 0.15 max **Cu** 0.25 max **Fe** 0.7 max **Mg** 0.20-0.6 **Mn** 0.10-0.30 **Si** 0.40 max **Ti** 0.10 max **Zn** 0.30 max **Other** each 0.05 max, total 0.15 max	**AA** 5010
A95016	Wrought Aluminum Alloy, Non-Heat Treatable	**Al** rem **Cr** 0.10 max **Cu** 0.20 max **Fe** 0.6 max **Mg** 1.4-1.9 **Mn** 0.40-0.7 **Si** 0.25 max **Ti** 0.05 max **Zn** 0.15 max **Other** each 0.05 max, total 0.15 max	**AA** 5016
* A95034	Wrought Aluminum Alloy	**Al** rem **Cr** 0.10 max **Cu** 0.40 max **Fe** 0.40-0.9 **Mg** 0.6-1.1 **Mn** 0.20-0.50 **Si** 0.40 max **Ti** 0.10 max **Zn** 0.40 max **Other** each 0.05 max, total 0.15 max	**AA** 5034
* A95039	Wrought Aluminum Alloy, Non-Heat Treatable (Welding Electrode)	**Al** rem **Be** 0.0008 max **Cr** 0.10-0.20 **Cu** 0.03 max **Fe** 0.40 max **Mg** 3.3-4.3 **Mn** 0.30-0.50 **Si** 0.10 max **Ti** 0.10 max **Zn** 2.4-3.2 **Other** each 0.05 max, total 0.10 max	**AA** 5039 **FED** QQ-R-566
A95040	Wrought Aluminum Alloy, Non-Heat Treatable	**Al** rem **Cr** 0.10-0.30 **Cu** 0.25 max **Fe** 0.7 max **Mg** 1.0-1.5 **Mn** 0.9-1.4 **Si** 0.30 max **Zn** 0.25 max **Other** each 0.05 max, total 0.15 max	**AA** 5040
A95042	Wrought Aluminum Alloy, Non-Heat Treatable	**Al** rem **Cr** 0.10 max **Cu** 0.15 max **Fe** 0.35 max **Mg** 3.0-4.0 **Mn** 0.20-0.50 **Si** 0.20 max **Sn** 0.10 max **Zn** 0.25 max **Other** each 0.05 max, total 0.15 max	**AA** 5042
A95043	Wrought Aluminum Alloy, Non-Heat Treatable	**Al** rem **Cr** 0.05 max **Cu** 0.05-0.35 **Fe** 0.7 max **Ga** 0.05 max **Mg** 0.7-1.3 **Mn** 0.7-1.2 **Si** 0.40 max **Ti** 0.10 max **V** 0.05 max **Zn** 0.25 max **Other** each 0.05 max, total 0.15 max	**AA** 5043

*Boxed entries are no longer active and are retained for reference purposes only.

The chemical compositions listed are for identification purposes and should not be used in lieu of the cross referenced specifications.

ALUMINUM AND ALUMINUM ALLOYS

UNIFIED NUMBER	DESCRIPTION	CHEMICAL COMPOSITION	CROSS REFERENCE SPECIFICATIONS
A95050	Wrought Aluminum Alloy, Non-Heat Treatable	**Al** rem **Cr** 0.10 max **Cu** 0.20 max **Fe** 0.7 max **Mg** 1.1-1.8 **Mn** 0.10 max **Si** 0.40 max **Zn** 0.25 max **Other** each 0.05 max (Be 0.0008 max for welding electrode and filler wire only), total 0.15 max	**AA** 5050 **ASTM** B209 (5050); B210 (5050); B313 (5050); B483 (5050); B547 (5050); B548 (5050) **SAE** J454 (5050)
A95051	Wrought Aluminum Alloy, Non-Heat Treatable	**Al** rem **Cr** 0.10 max **Cu** 0.25 max **Fe** 0.7 max **Mg** 1.7-2.2 **Mn** 0.20 max **Si** 0.40 max **Ti** 0.10 max **Zn** 0.25 max **Other** each 0.05 max, total 0.15 max	**AA** 5051
A95052	Wrought Aluminum Alloy, Non-Heat Treatable	**Al** rem **Cr** 0.15-0.35 **Cu** 0.10 max **Mg** 2.2-2.8 **Mn** 0.10 max **Zn** 0.10 max **Other** each 0.05 max (Be 0.0008 max for welding electrode and filler wire only), total 0.15 max, plus Si+Fe 0.45 max	**AA** 5052 **AMS** 4004; 4015; 4016; 4017; 4069; 4070; 4071; 4114; 4175; 4178 **ASTM** B209 (5052); B210 (5052); B211 (5052); B221 (5052); B234 (5052); B241 (5052); B313 (5052); B316 (5052); B404 (5052); B483 (5052); B547 (5052) **FED** QQ-A-225/7; QQ-A-250/8; QQ-A-430; WW-T-700/4 **MIL SPEC** MIL-S-12875; MIL-G-18014; MIL-G-18015; MIL-C-26094; MIL-A-81596
A95056	Wrought Aluminum Alloy, Non-Heat Treatable	**Al** rem **Cr** 0.05-0.20 **Cu** 0.10 max **Fe** 0.40 max **Mg** 4.5-5.6 **Mn** 0.05-0.20 **Si** 0.30 max **Zn** 0.10 max **Other** each 0.05 max (Be 0.0008 max for welding electrode and filler wire only), total 0.15 max	**AA** 5056 **AMS** 4005; 4176; 4177; 4182; 4349 **ASTM** B211 (5056); B316 (5056) **FED** QQ-A-430 **MIL SPEC** MIL-R-5674; MIL-R-8814; MIL-A-81596 **SAE** J454 (5056)
A95082	Wrought Aluminum Alloy, Non-Heat Treatable	**Al** rem **Cr** 0.15 max **Cu** 0.15 max **Fe** 0.35 max **Mg** 4.0-5.0 **Mn** 0.15 max **Si** 0.20 max **Ti** 0.10 max **Zn** 0.25 max **Other** each 0.05 max, total 0.15 max	**AA** 5082
A95083	Wrought Aluminum Alloy, Non-Heat Treatable	**Al** rem **Cr** 0.05-0.25 **Cu** 0.10 max **Fe** 0.40 max **Mg** 4.0-4.9 **Mn** 0.40-1.0 **Si** 0.40 max **Ti** 0.15 max **Zn** 0.25 max **Other** each 0.05 max, total 0.15 max	**AA** 5083 **AMS** 4056; 4057; 4058; 4059 **ASTM** B209 (5083); B210 (5083); B221 (5083); B241 (5083); B247 (5083); B345 (5083); B361 (5083); B547 (5083); B548 (5083) **FED** QQ-A-200/4; QQ-A-250/6; QQ-A-367 **MIL SPEC** MIL-A-45225; MIL-A-46027; MIL-A-46083; MIL-G--S-24149/2 **SAE** J454 (5083)
A95086	Wrought Aluminum Alloy, Non-Heat Treatable	**Al** rem **Cr** 0.05-0.25 **Cu** 0.10 max **Fe** 0.50 max **Mg** 3.5-4.5 **Mn** 0.20-0.7 **Si** 0.40 max **Ti** 0.15 max **Zn** 0.25 max **Other** 0.05 max, total 0.15 max	**AA** 5086 **ASTM** B209 (5086); B210 (5086); B221 (5086); B241 (5086); B313 (5086); B345 (5086); B361 (5086); B547 (5086); B548 (5086) **FED** QQ-A-200/5; QQ-A-250/7; WW-T-700/5 **MIL SPEC** MIL-G-18014; MIL-S-24149/2; MIL-C-26094 **SAE** J454 (5086)
A95151	Wrought Aluminum Alloy, Non-Heat Treatable	**Al** rem **Cr** 0.10 max **Cu** 0.15 max **Fe** 0.35 max **Mg** 1.5-2.1 **Mn** 0.10 max **Si** 0.20 max **Ti** 0.10 max **Zn** 0.15 max **Other** each 0.05 max, total 0.15 max	**AA** 5151
A95154	Wrought Aluminum Alloy, Non-Heat Treatable	**Al** rem **Cr** 0.15-0.35 **Cu** 0.10 max **Fe** 0.40 max **Mg** 3.1-3.9 **Mn** 0.10 max **Si** 0.25 max **Ti** 0.20 max **Zn** 0.20 max **Other** each 0.05 max (Be 0.0008 max for welding electrode and filler wire only), total 0.15 max	**AA** 5154 **AMS** 4018; 4019 **ASTM** B209 (5154); B210 (5154); B211 (5154); B221 (5154); B313 (5154); B361 (5154); B547 (5154); B548 (5154) **MIL SPEC** MIL-C-26094 **SAE** J454 (5154)
A95182	Wrought Aluminum Alloy, Non-Heat Treatable	**Al** rem **Cr** 0.10 max **Cu** 0.15 max **Fe** 0.35 max **Mg** 4.0-5.0 **Mn** 0.20-0.50 **Si** 0.20 max **Ti** 0.10 max **Zn** 0.25 max **Other** 0.05 max, total 0.15 max	**AA** 5182
A95183	Wrought Aluminum Alloy, Non-Heat Treatable (Welding Electrode)	**Al** rem **Be** 0.0008 max **Cr** 0.05-0.25 **Cu** 0.10 max **Fe** 0.40 max **Mg** 4.3-5.2 **Mn** 0.50-1.0 **Si** 0.40 max **Ti** 0.15 max **Zn** 0.25 max **Other** each 0.05 max, total 0.l5 max	**AA** 5183 **ASME** SFA5.10 (ER5183) **AWS** A5.10 (ER5183) **FED** QQ-R-566
A95205	Wrought Aluminum Alloy, Non-Heat Treatable	**Al** rem **Cr** 0.10 max **Cu** 0.03-0.10 **Fe** 0.7 max **Mg** 0.6-1.0 **Mn** 0.10 max **Si** 0.15 max **Zn** 0.05 max **Other** each 0.05 max, total 0.15 max	**AA** 5205

The chemical compositions listed are for identification purposes and should not be used in lieu of the cross referenced specifications.

UNIFIED NUMBER	DESCRIPTION	CHEMICAL COMPOSITION	CROSS REFERENCE SPECIFICATIONS
A95250	Wrought Aluminum Alloy, Non-Heat Treatable	**Al** rem **Cu** 0.10 **Fe** 0.10 **Ga** 0.03 **Mg** 1.3-1.8 **Mn** 0.05-0.15 **Si** 0.08 **V** 0.05 **Zn** 0.05 **Other** each 0.03, total 0.10	**AA** 5250
A95252	Wrought Aluminum Alloy, Non-Heat Treatable	**Al** rem **Cu** 0.10 max **Fe** 0.10 max **Mg** 2.2-2.8 **Mn** 0.10 max **Si** 0.08 max **Other** each 0.03 max, total 0.10 max	**AA** 5252 **ASTM** B209 (5252) **SAE** J454 (5252)
A95254	Wrought Aluminum Alloy, Non-Heat Treatable	**Al** rem **Cr** 0.15-0.35 **Cu** 0.05 max **Fe** 0.40 max **Mg** 3.1-3.9 **Mn** 0.01 max **Si** 0.25 max **Ti** 0.05 max **Zn** 0.20 max **Other** each 0.05 max (Be 0.0008 max for welding electrode and filler wire only), total 0.15 max	**AA** 5254 **ASTM** B209 (5254); B241 (5254); B548 (5254) **SAE** J454 (5254)
A95351	Wrought Aluminum Alloy	**Al** rem **Cu** 0.10 max **Fe** 0.10 max **Mg** 1.6-2.2 **Mn** 0.10 max **Si** 0.08 max **V** 0.05 max **Zn** 0.05 max **Other** each 0.03 max, total 0.10 max	**AA** 5351
A95352	Wrought Aluminum Alloy, Non-Heat Treatable	**Al** rem **Cr** 0.10 max **Cu** 0.10 max **Mg** 2.2-2.8 **Mn** 0.10 max **Ti** 0.10 max **Zn** 0.10 max **Other** each 0.05 max, total 0.15 max, plus Si+Fe 0.45 max	**AA** 5352
A95356	Wrought Aluminum Alloy, Non-Heat Treatable	**Al** rem **Cr** 0.05-0.20 **Cu** 0.10 max **Fe** 0.40 max **Mg** 4.5-5.5 **Mn** 0.05-0.20 **Si** 0.25 max **Ti** 0.06-0.20 **Zn** 0.10 max **Other** each 0.05 max (Be 0.0008 max for welding electrode and filler wire only), total 0.15 max	**AA** 5356 **ASME** SFA5.10 (ER5356) **AWS** A5.10 (ER5356) **FED** QQ-R-566 **MIL SPEC** MIL-S-24149/2
A95357	Wrought Aluminum Alloy, Non-Heat Treatable	**Al** rem **Cu** 0.20 max **Fe** 0.17 max **Mg** 0.8-1.2 **Mn** 0.15-0.45 **Si** 0.12 max **Other** each 0.05 max, total 0.15 max	**AA** 5357
A95451	Wrought Aluminum Alloy, Non-Heat Treatable	**Al** rem **Cr** 0.15-0.35 **Cu** 0.10 **Fe** 0.40 **Mg** 1.8-2.4 **Mn** 0.10 **Ni** 0.05 **Si** 0.25 **Ti** 0.05 **Zn** 0.10 **Other** each 0.05, total 0.15	**AA** 5451
A95454	Wrought Aluminum Alloy, Non-Heat Treatable	**Al** rem **Cr** 0.05-0.20 **Cu** 0.10 max **Mg** 2.4-3.0 **Mn** 0.50-1.0 **Ti** 0.20 max **Zn** 0.25 max **Other** each 0.05 max, total 0.15 max, Si+Fe 0.40 max	**AA** 5451 **ASTM** B209 (5454); B221 (5454); B234 (5454); B241 (5454); B404 (5454); B547 (5454); B548 (5454) **FED** QQ-A-200/6; QQ-A-250/10 **SAE** J454 (5454)
A95456	Wrought Aluminum Alloy, Non-Heat Treatable	**Al** rem **Cr** 0.05-0.20 **Cu** 0.10 max **Fe** 0.40 max **Mg** 4.7-5.5 **Mn** 0.50-1.0 **Si** 0.25 max **Ti** 0.20 max **Zn** 0.25 max **Other** each 0.05 max, total 0.15 max	**AA** 5456 **ASTM** B209 (5456); B210 (5456); B221 (5456); B241 (5456); B247 (5456); B548 (5456) **FED** QQ-A-200/7; QQ-A-250/9 **MIL SPEC** MIL-G-18014; MIL-S-24149/2; MIL-A-45225; MIL-A-46027; MIL-A-46083 **SAE** J454 (5456)
A95457	Wrought Aluminum Alloy, Non-Heat Treatable	**Al** rem **Cu** 0.20 max **Fe** 0.10 max **Mg** 0.8-1.2 **Mn** 0.15-0.45 **Si** 0.08 max **Zn** 0.03 max **Other** each 0.03 max, total 0.10 max	**AA** 5457 **ASTM** B209 (5457) **SAE** J454 (5457)
A95554	Wrought Aluminum Alloy, Non-Heat Treatable (Welding Electrode)	**Al** rem **Be** 0.0008 max **Cr** 0.05-0.20 **Cu** 0.10 max **Mg** 2.4-3.0 **Mn** 0.50-1.0 **Ti** 0.05-0.20 **Zn** 0.25 max **Other** each 0.05 max, total 0.15 max, Si+Fe 0.40 max	**AA** 5554 **ASME** SFA5.10 (ER5554) **AWS** A5.10 (ER5554) **FED** QQ-R-566 **MIL SPEC** MIL-G-18014
A95556	Wrought Aluminum Alloy, Non-Heat Treatable (Welding Electrode)	**Al** rem **Be** 0.0008 max **Cr** 0.05-0.20 **Cu** 0.10 max **Fe** 0.40 max **Mg** 4.7-5.5 **Mn** 0.50-1.0 **Si** 0.25 max **Ti** 0.05-0.20 **Zn** 0.25 max **Other** each 0.05 max, total 0.15 max	**AA** 5556 **ASME** SFA5.10 (ER5556) **AWS** A5.10 (ER5556) **FED** QQ-R-566
A95557	Wrought Aluminum Alloy, Non-Heat Treatable	**Al** rem **Cu** 0.15 max **Fe** 0.12 max **Mg** 0.40-0.8 **Mn** 0.10-0.40 **Si** 0.10 max **V** 0.05 max **Other** each 0.03 max, total 0.10 max	**AA** 5557

The chemical compositions listed are for identification purposes and should not be used in lieu of the cross referenced specifications.

UNIFIED NUMBER	DESCRIPTION	CHEMICAL COMPOSITION	CROSS REFERENCE SPECIFICATIONS
A95652	Wrought Aluminum Alloy, Non-Heat Treatable	Al rem Cr 0.15-0.35 Cu 0.04 max Mg 2.2-2.8 Mn 0.01 max Zn 0.10 max Other each 0.05 max (Be 0.0008 max for welding electrode and filler wire only), total 0.15 max, Si+Fe 0.40 max	AA 5652 ASTM B209 (5652); B241 (5652); B548 (5652) SAE J454 (5652)
A95654	Wrought Aluminum Alloy, Non-Heat Treatable (Welding Electrode)	Al rem Be 0.0008 max Cr 0.15-0.35 Cu 0.05 max Mg 3.1-3.9 Mn 0.01 max Ti 0.05-0.15 Zn 0.20 max Other each 0.05 max, total 0.15 max, Si+Fe 0.45 max	AA 5654 ASME SFA5.10 (ER5654) AWS A5.10 (ER5654) FED QQ-R-566
A95657	Wrought Aluminum Alloy, Non-Heat Treatable	Al rem Cu 0.10 max Fe 0.10 max Ga 0.03 max Mg 0.6-1.0 Mn 0.03 max Si 0.08 max Zn 0.03 max Other each 0.02 max, total 0.05 max	AA 5657 ASTM B209 (5657) SAE J454 (5657)
A96003	Wrought Aluminum Alloy, Heat Treatable (Cladding Alloy)	Al rem Cr 0.35 max Cu 0.10 max Fe 0.6 max Mg 0.8-1.5 Mn 0.8 max Si 0.35-1.0 Ti 0.10 max Zn 0.20 max	AA 6003
A96004	Wrought Aluminum Alloy, Heat Treatable	Al rem Cu 0.10 max Fe 0.10-0.30 Mg 0.40-0.7 Mn 0.20-0.6 Si 0.30-0.6 Zn 0.05 max Other each 0.05 max, total 0.15 max	AA 6004
A96005	Wrought Aluminum Alloy, Heat Treatable	Al rem Cr 0.10 max Cu 0.10 max Fe 0.35 max Mg 0.40-0.6 Mn 0.10 max Si 0.6-0.9 Ti 0.10 max Zn 0.10 max Other each 0.05 max, total 0.15 max	AA 6005; 6005A ASTM B221 (6005)
A96006	Wrought Aluminum Alloy, Heat Treatable	Al rem Cr 0.10 max Cu 0.15-0.30 Fe 0.35 max Mg 0.45-0.9 Mn 0.05-0.20 Si 0.20-0.6 Ti 0.10 max Zn 0.10 max Other each 0.05 max, total 0.15 max	AA 6006
A96007	Wrought Aluminum Alloy, Heat Treatable	Al rem Cr 0.05-0.25 Cu 0.20 max Fe 0.7 max Mg 0.6-0.9 Mn 0.05-0.25 Si 0.9-1.4 Ti 0.15 max Zn 0.25 max Zr 0.05-0.20 Other each 0.05 max, total 0.15 max	AA 6007
A96009	Wrought Aluminum Alloy, Heat Treatable	Al rem Cr 0.10 max Cu 0.15-0.6 Fe 0.50 max Mg 0.40-0.8 Mn 0.20-0.8 Si 0.6-1.0 Ti 0.10 max Zn 0.25 max Other each 0.05 max, total 0.15 max	AA 6009
A96010	Wrought Aluminum Alloy, Heat Treatable	Al rem Cr 0.10 max Cu 0.15-0.6 Fe 0.5 max Mg 0.6-1.0 Mn 0.20-0.8 Si 0.8-1.2 Ti 0.10 max Zn 0.25 max Other each 0.05 max, total 0.15 max	AA 6010
A96011	Wrought Aluminum Alloy, Heat Treatable	Al rem Cr 0.30 max Cu 0.40-0.9 Fe 1.0 max Mg 0.6-1.2 Mn 0.8 max Ni 0.20 max Si 0.6-1.2 Ti 0.20 max Zn 1.5 max Other each 0.05 max, total 0.15 max	AA 6011
A96017	Wrought Aluminum Alloy, Heat Treatable	Al rem Cr 0.10 max Cu 0.05-0.20 Fe 0.15-0.30 Mg 0.45-0.6 Mn 0.10 max Si 0.55-0.7 Ti 0.04 max Zn 0.05 max Other each 0.05 max, total 0.15 max	AA 6017
A96053	Wrought Aluminum Alloy, Heat Treatable	Al rem Cr 0.15-0.35 Cu 0.10 max Fe 0.35 max Mg 1.1-1.4 Si 45.0-65.0 of Mg Zn 0.10 max Other each 0.05 max, total 0.15 max	AA 6053 ASTM B316 (6053) FED QQ-A-430 SAE J454 (6053)
A96060	Wrought Aluminum Alloy, Heat Treatable	Al rem Cr 0.05 max Cu 0.10 max Fe 0.10-0.30 Mg 0.35-0.6 Mn 0.10 max Si 0.30-0.6 Ti 0.10 max Zn 0.15 max Other each 0.05 max, total 0.15 max	AA 6060

The chemical compositions listed are for identification purposes and should not be used in lieu of the cross referenced specifications.

UNIFIED NUMBER	DESCRIPTION	CHEMICAL COMPOSITION	CROSS REFERENCE SPECIFICATIONS
A96061	Wrought Aluminum Alloy, Heat Treatable	**Al** rem **Cr** 0.04-0.35 **Cu** 0.15-0.40 **Fe** 0.7 max **Mg** 0.8-1.2 **Mn** 0.15 max **Si** 0.40-0.8 **Ti** 0.15 max **Zn** 0.25 max **Other** each 0.05 max, total 0.15 max	**AA** 6061 **AMS** 4009; 4025; 4026; 4027; 4079; 4080; 4081; 4082; 4083; 4113; 4115; 4116; 4117; 4127; 4128; 4146; 4150; 4160; 4161; 4172; 4173; 4312 **ASTM** B209 (6061); B210 (6061); B211 (6061); B221 (6061); B234 (6061); B241 (6061); B247 (6061); B308 (6061); B313 (6061); B316 (6061); B345 (6061); B361; B404 (6061); B429 (6061); F467; F468; B483 (6061); B547 (6061); B548; B632 **FED** QQ-A-200/8; QQ-A-200/16; QQ-A-225/8; QQ-A-250/11; QQ-A-367; QQ-A-430; WW-T-700/6 **MIL SPEC** MIL-W-85; MIL-F-3922; MIL-T-7081; MIL-T-10794; MIL-A-12545; MIL-G-18014; MIL-G-18015; MIL-F-18280; MIL-A-22771; MIL-W-23351; MIL-P-25995; MIL-F-39000 **SAE** J454 (6061)
A96063	Wrought Aluminum Alloy, Heat Treatable	**Al** rem **Cr** 0.10 max **Cu** 0.10 max **Fe** 0.35 max **Mg** 0.45-0.9 **Mn** 0.10 max **Si** 0.20-0.6 **Ti** 0.10 max **Zn** 0.10 max **Other** each 0.05 max, total 0.15 max	**AA** 6063 **AMS** 4156 **ASTM** B210 (6063); B221 (6063); B241 (6063); B345 (6063); B361; B429 (6063); B483 (6063); B491 (6063) **FED** QQ-A-200/9 **MIL SPEC** MIL-W-85; MIL-G-18014; MIL-G-18015; MIL-P-25995 **SAE** J454 (6063)
A96066	Wrought Aluminum Alloy, Heat Treatable	**Al** rem **Cr** 0.40 max **Cu** 0.7-1.2 **Mg** 0.8-1.4 **Mn** 0.6-1.1 **Pb** 0.50 max **Si** 0.9-1.8 **Ti** 0.20 max **Zn** 0.25 max **Other** each 0.05 max, total 0.15 max	**AA** 6066 **ASTM** B221 (6066); B247 (6066) **FED** QQ-A-200/10; QQ-A-367 **SAE** J454 (6066)
A96070	Wrought Aluminum Alloy, Heat Treatable	**Al** rem **Cr** 0.10 max **Cu** 0.15-0.40 **Fe** 0.50 max **Mg** 0.50-1.2 **Mn** 0.40-1.0 **Si** 1.0-1.7 **Ti** 0.15 max **Zn** 0.25 max **Other** each 0.05 max, total 0.15 max	**AA** 6070 **ASTM** B345 (6070) **MIL SPEC** MIL-A-12545; MIL-A-46104 **SAE** J454 (6070)
A96101	Wrought Aluminum Alloy, Heat Treatable	**Al** rem **B** 0.06 max **Cr** 0.03 max **Cu** 0.10 max **Fe** 0.50 max **Mg** 0.35-0.8 **Mn** 0.03 max **Si** 0.30-0.7 **Zn** 0.10 max **Other** each 0.03 max, total 0.10 max	**AA** 6101 **ASTM** B317 (6101) **SAE** J454 (6101)
A96105	Wrought Aluminum Alloy, Heat Treatable	**Al** rem **Cr** 0.10 max **Cu** 0.10 max **Fe** 0.35 max **Mg** 0.45-0.8 **Mn** 0.10 max **Si** 0.6-1.0 **Ti** 0.10 max **Zn** 0.10 max **Other** each 0.05 max, total 0.15 max	**AA** 6105
A96110	Wrought Aluminum Alloy, Heat Treatable	**Al** rem **Cr** 0.04-0.25 **Cu** 0.20-0.7 **Fe** 0.8 max **Mg** 0.50-1.1 **Mn** 0.20-0.7 **Si** 0.7-1.5 **Ti** 0.15 max **Zn** 0.30 max **Other** each 0.05 max, total 0.15 max	**AA** 6110
A96111	Wrought Aluminum Alloy, Heat Treatable	**Al** rem **Cr** 0.10 **Fe** 0.40 **Mg** 0.50-1.0 **Mn** 0.15-0.45 **Si** 0.7-1.1 **Ti** 0.10 **U** 0.50-0.9 **Zn** 0.15 **Other** each 0.05, total 0.15	**AA** 6111
A96151	Wrought Aluminum Alloy, Heat Treatable	**Al** rem **Cr** 0.15-0.35 **Cu** 0.35 max **Fe** 1.0 max **Mg** 0.45-0.8 **Mn** 0.20 max **Si** 0.6-1.2 **Ti** 0.15 max **Zn** 0.25 max **Other** each 0.05 max, total 0.15 max	**AA** 6151 **AMS** 4125 **ASTM** B247 **FED** QQ-A-367 **MIL SPEC** MIL-C-10387; MIL-A-22771 **SAE** J454 (6151)
A96162	Wrought Aluminum Alloy, Heat Treatable	**Al** rem **Cr** 0.10 max **Cu** 0.20 max **Fe** 0.50 max **Mg** 0.7-1.1 **Mn** 0.10 max **Si** 0.40-0.8 **Ti** 0.10 max **Zn** 0.25 max **Other** each 0.05 max, total 0.15 max	**AA** 6162 **FED** QQ-A-200/17
A96201	Wrought Aluminum Alloy, Heat Treatable	**Al** rem **B** 0.06 max **Cr** 0.03 max **Cu** 0.10 max **Fe** 0.50 max **Mg** 0.6-0.9 **Mn** 0.03 max **Si** 0.50-0.9 **Zn** 0.10 max **Other** each 0.03 max, total 0.10 max	**AA** 6201 **ASTM** B398 (6201); B399 (6201); B524 (6201) **SAE** J454 (6201)
A96205	Wrought Aluminum Alloy, Heat Treatable	**Al** rem **Cr** 0.05-0.15 **Cu** 0.20 max **Fe** 0.7 max **Mg** 0.40-0.6 **Mn** 0.05-0.15 **Si** 0.6-0.9 **Ti** 0.15 max **Zn** 0.25 max **Zr** 0.05-0.15 **Other** each 0.05 max, total 0.15 max	**AA** 6205
A96253	Wrought Aluminum Alloy, Heat Treatable (Cladding Alloy)	**Al** rem **Cr** 0.15-0.35 **Cu** 0.10 max **Fe** 0.50 max **Mg** 1.0-1.5 **Si** 45.0-65.0 of Mg **Zn** 1.6-2.4 **Other** each 0.05 max, total 0.15 max	**AA** 6253 **ASTM** B211 (6253)

The chemical compositions listed are for identification purposes and should not be used in lieu of the cross referenced specifications.

UNIFIED NUMBER	DESCRIPTION	CHEMICAL COMPOSITION	CROSS REFERENCE SPECIFICATIONS
A96261	Wrought Aluminum Alloy, Heat Treatable	**Al** rem **Cr** 0.10 max **Cu** 0.15-0.40 **Fe** 0.40 max **Mg** 0.7-1.0 **Mn** 0.20-0.35 **Si** 0.40-0.7 **Ti** 0.10 max **Zn** 0.20 max **Other** each 0.05 max, total 0.15 max	**AA** 6261
A96262	Wrought Aluminum Alloy, Heat Treatable, Free Machining	**Al** rem **Bi** 0.40-0.7 **Cr** 0.04-0.14 **Cu** 0.15-0.40 **Fe** 0.7 max **Mg** 0.8-1.2 **Mn** 0.15 max **Pb** 0.40-0.7 **Si** 0.40-0.8 **Ti** 0.15 max **Zn** 0.25 max **Other** each 0.05 max, total 0.15 max	**AA** 6262 **ASTM** B210 (6262); B211 (6262); B221 (6262); B467; B483 (6262) **FED** QQ-A-225/10 **SAE** J454 (6262)
A96301	Wrought Aluminum Alloy, Heat Treatable	**Al** rem **Cr** 0.10 max **Cu** 0.10 max **Fe** 0.7 max **Mg** 0.6-0.9 **Mn** 0.15 max **Si** 0.50-0.9 **Ti** 0.15 max **Zn** 0.25 max **Other** each 0.05 max, total 0.15 max	**AA** 6301
A96351	Wrought Aluminum Alloy, Heat Treatable	**Al** rem **Cu** 0.10 max **Fe** 0.50 max **Mg** 0.40-0.8 **Mn** 0.40-0.8 **Si** 0.7-1.3 **Ti** 0.20 max **Zn** 0.20 max **Other** each 0.05 max, total 0.15 max	**AA** 6351 **ASTM** B221 (6351); B345 (6351)
A96463	Wrought Aluminum Alloy, Heat Treatable	**Al** rem **Cu** 0.20 max **Fe** 0.15 max **Mg** 0.45-0.9 **Mn** 0.05 max **Si** 0.20-0.6 **Other** each 0.05 max, total 0.15 max	**AA** 6463 **ASTM** B221 (6463) **SAE** J454 (6463)
A96763	Wrought Aluminum Alloy, Heat Treatable	**Al** rem **Cu** 0.04-0.16 **Fe** 0.08 max **Mg** 0.45-0.9 **Mn** 0.03 max **Si** 0.20-0.6 **V** 0.05 max **Zn** 0.03 max **Other** each 0.03 max, total 0.10 max	**AA** 6763
A96951	Wrought Aluminum Alloy, Heat Treatable	**Al** rem **Cu** 0.15-0.40 **Fe** 0.8 max **Mg** 0.40-0.8 **Mn** 0.10 max **Si** 0.20-0.50 **Zn** 0.20 max **Other** each 0.05 max, total 0.15 max	**AA** 6951
A97001	Wrought Aluminum Alloy, Heat Treatable	**Al** rem **Cr** 0.18-0.35 **Cu** 1.6-2.6 **Fe** 0.40 max **Mg** 2.6-3.4 **Mn** 0.20 max **Si** 0.35 max **Ti** 0.20 max **Zn** 6.8-8.0 **Other** each 0.05 max, total 0.15 max	**AA** 7001 **MIL SPEC** MIL-A-52242 (7001) **SAE** J454 (7001)
A97004	Wrought Aluminum Alloy, Heat Treatable	**Al** rem **Cr** 0.05 max **Cu** 0.05 max **Fe** 0.35 max **Mg** 1.0-2.0 **Mn** 0.20-0.7 **Si** 0.25 max **Ti** 0.05 max **Zn** 3.8-4.6 **Zr** 0.10-0.20 **Other** each 0.05 max, total 0.15 max	**AA** 7004
A97005	Wrought Aluminum Alloy, Heat Treatable	**Al** rem **Cr** 0.06-0.20 **Cu** 0.10 max **Fe** 0.40 max **Mg** 1.0-1.8 **Mn** 0.20-0.7 **Si** 0.35 max **Ti** 0.01-0.06 **Zn** 4.0-5.0 **Zr** 0.08-0.20 **Other** each 0.05 max, total 0.15 max	**AA** 7005 **ASTM** B221 (7005)
A97008	Wrought Aluminum Alloy, Heat Treatable (Cladding Alloy)	**Al** rem **Cr** 0.12-0.25 **Cu** 0.05 max **Fe** 0.10 max **Mg** 0.7-1.4 **Mn** 0.05 max **Si** 0.10 max **Ti** 0.05 max **Zn** 4.5-5.5 **Other** each 0.05 max, total 0.10 max	**AA** 7008 **ASTM** B209 (7008)
* A97011	Wrought Aluminum Alloy, Heat Treatable (Cladding Alloy)	**Al** rem **Cr** 0.05-0.20 **Cu** 0.05 max **Fe** 0.20 max **Mg** 1.0-1.6 **Mn** 0.10-0.30 **Si** 0.15 max **Ti** 0.05 max **Zn** 4.0-5.5 **Other** each 0.05 max, total 0.15 max	**AA** 7011 **ASTM** B209 (7011)
A97013	Wrought Aluminum Alloy, Heat Treatable (Cladding Alloy)	**Al** rem **Cu** 0.10 max **Fe** 0.7 max **Mn** 1.0-1.5 **Si** 0.6 max **Zn** 1.5-2.0 **Other** each 0.05 max, total 0.15 max	**AA** 7013
A97016	Wrought Aluminum Alloy, Heat Treatable	**Al** rem **Cu** 0.45-1.0 **Fe** 0.12 max **Mg** 0.8-1.4 **Mn** 0.03 max **Si** 0.10 max **Ti** 0.03 **V** 0.05 max **Zr** 4.0-5.0 **Other** each 0.03 max, total 0.10 max	**AA** 7016

*Boxed entries are no longer active and are retained for reference purposes only.

The chemical compositions listed are for identification purposes and should not be used in lieu of the cross referenced specifications.

UNIFIED NUMBER	DESCRIPTION	CHEMICAL COMPOSITION	CROSS REFERENCE SPECIFICATIONS
A97021	Wrought Aluminum Alloy, Heat Treatable	**Al** rem **Cr** 0.05 max **Cu** 0.25 max **Fe** 0.40 max **Mg** 1.2-1.8 **Mn** 0.10 max **Si** 0.25 max **Ti** 0.10 max **Zn** 5.0-6.0 **Zr** 0.08-0.18 **Other** each 0.05 max, total 0.15 max	**AA** 7021
A97029	Wrought Aluminum Alloy, Heat Treatable	**Al** rem **Cu** 0.50-0.9 **Fe** 0.12 max **Mg** 1.3-2.0 **Mn** 0.3 max **Si** 0.10 max **Ti** 0.05 max **V** 0.05 max **Zn** 4.2-5.2 **Other** each 0.03 max, total 0.10 max	**AA** 7029
A97039	Wrought Aluminum Alloy, Heat Treatable	**Al** rem **Cr** 0.15-0.25 **Cu** 0.10 max **Fe** 0.40 max **Mg** 2.3-3.3 **Mn** 0.10-0.40 **Si** 0.30 max **Ti** 0.10 max **Zn** 3.5-4.5 **Other** each 0.05 max, total 0.15 max	**AA** 7039 **MIL SPEC** MIL-A-22771; MIL-A-45225; MIL-A-46063; MIL-A-46083
A97046	Wrought Aluminum Alloy, Heat Treatable	**Al** rem **Fe** 0.40 max **Mg** 1.0-1.6 **Si** 0.20 max **Ti** 0.06 max **Zn** 6.6-7.6 **Zr** 0.10-0.18 **Other** each 0.05 max, total 0.15 max	**AA** 7046
A97049	Wrought Aluminum Alloy, Heat Treatable	**Al** rem **Cr** 0.10-0.22 **Cu** 1.2-1.9 **Fe** 0.35 max **Mg** 2.0-2.9 **Mn** 0.20 max **Si** 0.25 max **Ti** 0.10 max **Zn** 7.2-8.2 **Other** each 0.05 max, total 0.15 max	**AA** 7049 **AMS** 4111; 4157; 4159; 4200 **ASTM** B247 (7049) **FED** QQ-A-367
A97050	Wrought Aluminum Alloy, Heat Treatable	**Al** rem **Cr** 0.04 max **Cu** 2.0-2.6 **Fe** 0.15 max **Mg** 1.9-2.6 **Mn** 0.10 max **Si** 0.12 max **Ti** 0.06 max **Zn** 5.7-6.7 **Zr** 0.08-0.15 **Other** each 0.05 max, total 0.15 max	**AA** 7050 **AMS** 4050; 4107; 4108; 4201; 4340; 4341; 4342 **ASTM** B247 (7050) **FED** QQ-A-430
A97070	Wrought Aluminum Alloy, Heat Treatable	**Al** rem **Cu** 0.05 max **Fe** 0.25 max **Si** 0.15 max **Zn** 1.3-1.8 **Other** each 0.05 max, total 0.15 max	**AA** 7070
A97072	Wrought Aluminum Alloy, Non-Heat Treatable	**Al** rem **Cu** 0.10 max **Mg** 0.10 max **Mn** 0.10 max **Zn** 0.8-1.3 **Other** each 0.05 max, total 0.15 max, Si+Fe 0.7 max	**AA** 7072 **ASTM** B209 (7072); B221 (7072); B234 (7072); B241 (7072); B313 (7072); B345 (7072); B404 (7072); B547 (7072)
A97075	Wrought Aluminum Alloy, Heat Treatable	**Al** rem **Cr** 0.18-0.28 **Cu** 1.2-2.0 **Fe** 0.50 max **Mg** 2.1-2.9 **Mn** 0.30 max **Si** 0.40 max **Ti** 0.20 max **Zn** 5.1-6.1 **Other** each 0.05 max, total 0.15 max	**AA** 7075 **AMS** 4044; 4045; 4078; 4122; 4123; 4124; 4126; 4131; 4141; 4147; 4154; 4166; 4167; 4168; 4169; 4174; 4186; 4187; 4310; 4311 **ASTM** B209 (7075); B210 (7075); B211 (7075); B221 (7075); B241 (7075); B247 (7075); B316 (7075) **FED** QQ-A-200/11; QQ-A-200/15; QQ-A-225/9; QQ-A-250/13; QQ-A-250/24; QQ-A-367; QQ-A-430; WW-T-700/7 **MIL SPEC** MIL-F-5509; MIL-A-12545; MIL-F-18280; MIL-A-22771 **SAE** J454 (7075)
A97076	Wrought Aluminum Alloy, Heat Treatable	**Al** rem **Cu** 0.30-1.0 **Fe** 0.6 max **Mg** 1.2-2.0 **Mn** 0.30-0.8 **Si** 0.40 max **Ti** 0.20 max **Zn** 7.0-8.0 **Other** each 0.05 max, total 0.15 max	**AA** 7076 **ASTM** B247 (7076) **FED** QQ-A-367
A97079	Wrought Aluminum Alloy, Heat Treatable	**Al** rem **Cr** 0.10-0.25 **Cu** 0.40-0.8 **Fe** 0.40 max **Mg** 2.9-3.7 **Mn** 0.10-0.30 **Si** 0.30 max **Ti** 0.10 max **Zn** 3.8-4.8 **Other** each 0.05 max, total 0.15 max	**AA** 7079 **AMS** 4024; 4136; 4138; 4139 **ASTM** B247 (7079) **FED** QQ-A-367 **SAE** J454 (7079)
A97090	Wrought Aluminum Alloy, Heat Treatable	**Al** rem **Co** 1.0-1.9 **Cu** 0.6-1.3 **Fe** 0.15 **Mg** 2.0-3.0 **O** 0.20-0.50 **Si** 0.12 **Zn** 7.3-8.7 **Other** each 0.05	**AA** 7090
A97091	Wrought Aluminum Alloy, Heat Treatable	**Al** rem **Co** 0.20-0.6 **Cu** 1.1-1.8 **Fe** 0.15 **Mg** 2.0-3.0 **O** 0.20-0.50 **Si** 0.12 **Zn** 5.8-7.1 **Other** each 0.05, total 0.15	**AA** 7091
A97104	Wrought Aluminum Alloy, Heat Treatable	**Al** rem **Cu** 0.03 max **Fe** 0.40 max **Mg** 0.50-0.9 **Si** 0.25 max **Ti** 0.10 max **Zn** 3.6-4.4 **Other** each 0.05 max, total 0.15 max	**AA** 7104

The chemical compositions listed are for identification purposes and should not be used in lieu of the cross referenced specifications.

UNIFIED NUMBER	DESCRIPTION	CHEMICAL COMPOSITION	CROSS REFERENCE SPECIFICATIONS
A97108	Wrought Aluminum Alloy, Heat Treatable (Cladding Alloy)	Al rem Cu 0.05 max Fe 0.10 max Mg 0.7-1.4 Mn 0.05 max Si 0.10 max Ti 0.05 max Zn 4.5-5.5 Zr 0.12-0.25 Other each 0.05 max, total 0.15 max	AA 7108
A97116	Wrought Aluminum Alloy, Heat Treatable	Al rem Cu 0.50-1.1 Fe 0.30 max Ga 0.03 max Mg 0.8-1.4 Mn 0.05 max Si 0.15 max Ti 0.05 max V 0.05 max Zn 4.2-5.2 Other each 0.05 max, total 0.15 max	AA 7116
A97129	Wrought Aluminum Alloy, Heat Treatable	Al rem Cr 0.10 max Cu 0.50-0.9 Fe 0.30 max Ga 0.03 max Mg 1.3-2.0 Mn 0.10 max Si 0.15 max Ti 0.05 max V 0.05 max Zn 4.2-5.2 Other each 0.05 max, total 0.15 max	AA 7129
A97146	Wrought Aluminum Alloy, Heat Treatable	Al rem Cr 0.10-0.22 Cu 1.2-1.9 Fe 0.20 max Mg 2.0-2.9 Mn 0.20 max Si 0.15 max Ti 0.10 max Zn 7.2-8.2 Other each 0.05 max, total 0.15 max	AA 7146
A97149	Wrought Aluminum Alloy, Heat Treatable	Al rem Cr 0.10-0.22 Cu 1.2-1.9 Fe 0.20 max Mg 2.0-2.9 Mn 0.20 max Si 0.15 max Ti 0.10 max Zn 7.2-8.2 Other each 0.05 max, total 0.15 max	AA 7149 AMS 4320; 4343
A97150	Wrought Aluminum Alloy, Heat Treatable	Al rem Cr 0.04 max Cu 1.9-2.5 Fe 0.15 max Mg 2.0-2.7 Mn 0.10 max Si 0.12 max Ti 0.06 max Zn 5.9-6.9 Zr 0.08-0.15 Other each 0.05 max, total 0.15 max	AA 7150
A97175	Wrought Aluminum Alloy, Heat Treatable	Al rem Cr 0.18-0.28 Cu 1.2-2.0 Fe 0.20 max Mg 2.1-2.9 Mn 0.10 max Si 0.15 max Ti 0.10 max Zn 5.1-6.1 Other each 0.05 max, total 0.15 max	AA 7175 AMS 4147; 4148; 4149; 4179; 4344 ASTM B247 (7175)
A97178	Wrought Aluminum Alloy, Heat Treatable	Al rem Cr 0.18-0.35 Cu 1.6-2.4 Fe 0.50 max Mg 2.4-3.1 Mn 0.30 max Si 0.40 max Ti 0.20 max Zn 6.3-7.3 Other each 0.05 max, total 0.15 max	AA 7178 ASTM B209 (7178); B221 (7178); B241 (7178); B316 (7178) FED QQ-A-200/13; QQ-A-200/14; QQ-A-250/14; QQ-A-250/21; QQ-A-250/28; QQ-A-430 SAE J454 (7178)
A97179	Wrought Aluminum Alloy, Heat Treatable	Al rem Cr 0.10-0.25 Cu 0.40-0.8 Fe 0.20 max Mg 2.9-3.7 Mn 0.10-0.30 Si 0.15 max Ti 0.10 max Zn 3.8-4.8 Other each 0.05 max, total 0.15 max	AA 7179
A97277	Wrought Aluminum Alloy, Heat Treatable	Al rem Cr 0.18-0.35 Cu 0.8-1.7 Fe 0.7 max Mg 1.7-2.3 Si 0.50 max Ti 0.10 max Zn 3.7-4.3 Other each 0.05 max, total 0.15 max	AA 7277 MIL SPEC MIL-R-12221 (7277)
A97472	Wrought Aluminum Alloy, Heat Treatable	Al rem Cu 0.05 max Fe 0.6 max Mg 0.9-1.5 Mn 0.05 max Si 0.25 max Zn 1.3-1.9 Other each 0.05 max, total 0.15 max	AA 7472
A97475	Wrought Aluminum Alloy, Heat Treatable	Al rem Cr 0.18-0.25 Cu 1.2-1.9 Fe 0.12 max Mg 1.9-2.6 Mn 0.06 max Si 0.10 max Ti 0.06 max Zn 5.2-6.2 Other each 0.05 max, total 0.15	AA 7475 AMS 4084; 4085; 4089; 4090; 4100; 4202
A98001	Wrought Aluminum Alloy, Non-Heat Treatable	Al rem B 0.001 max Cd 0.003 max Co 0.001 max Cu 0.15 max Fe 0.45-0.7 Li 0.008 max Ni 0.9-1.3 Si 0.17 max Other each 0.05 max, total 0.15 max	AA 8001
A98006	Wrought Aluminum Alloy, Non-Heat Treatable	Al rem Cu 0.30 Fe 1.2-2.0 Mg 0.10 Mn 0.30-1.0 Si 0.40 Zn 0.10 Other each 0.05, total 0.15	AA 8006
A98007	Wrought Aluminum Alloy, Non-Heat Treatable	Al rem Cu 0.10 Fe 1.2-2.0 Mg 0.10 Mn 0.30-1.0 Si 0.40 Zn 0.8-1.8 Other each 0.05, total 0.15	AA 8007

The chemical compositions listed are for identification purposes and should not be used in lieu of the cross referenced specifications.

UNIFIED NUMBER	DESCRIPTION	CHEMICAL COMPOSITION	CROSS REFERENCE SPECIFICATIONS
* A98013	Wrought Aluminum Alloy, Non-Heat Treatable	**Al** rem **Cr** 0.20-0.50 **Ti** 0.10 max **Other** each 0.03 max, total 0.10 max, plus Si+Fe 0.25 max	**AA** 8013
A98014	Wrought Aluminum Alloy, Non-Heat Treatable	**Al** rem **Cu** 0.20 max **Fe** 1.2-1.6 **Mg** 0.10 max **Mn** 0.20-0.6 **Si** 0.30 max **Ti** 0.10 max **Zn** 0.10 max **Other** each 0.05 max, total 0.15 max	**AA** 8014
A98017	Wrought Aluminum Alloy, Non-Heat Treatable	**Al** rem **B** 0.04 max **Cu** 0.10-0.20 **Fe** 0.55-0.8 **Li** 0.003 max **Mg** 0.01-0.05 **Si** 0.10 max **Zn** 0.05 max **Other** each 0.03 max, total 0.10 max	**AA** 8017
A98020	Wrought Aluminum Alloy, Non-Heat Treatable	**Al** rem **Bi** 0.10-0.50 **Cu** 0.005 max **Fe** 0.10 max **Mn** 0.005 max **Si** 0.10 max **Sn** 0.10-0.25 **V** 0.05 max **Zn** 0.005 max **Other** each 0.03 max, total 0.10 max	**AA** 8020
A98030	Wrought Aluminum Alloy, Non-Heat Treatable	**Al** balance **B** 0.001-0.04 **Cu** 0.15-0.30 **Fe** 0.30-0.8 **Mg** 0.05 max **Si** 0.10 max **Zn** 0.05 max **Other** each 0.03 max, total 0.10 max	**AA** 8030
A98040	Wrought Aluminum Alloy, Non-Heat Treatable	**Al** rem **Cu** 0.20 max **Mn** 0.05 max **Zn** 0.20 max **Zr** 0.10-0.30 **Other** each 0.05 max, total 0.15 max, plus Si+Fe 1.0 max	**AA** 8040
A98076	Wrought Aluminum Alloy, Non-Heat Treatable	**Al** rem **B** 0.04 max **Cu** 0.04 max **Fe** 0.6-0.9 **Mg** 0.08-0.22 **Si** 0.10 max **Zn** 0.05 max **Other** each 0.03 max, total 0.10 max	**AA** 8076
A98077	Wrought Aluminum Alloy, Non-Heat Treatable	**Al** rem **B** 0.05 max **Cu** 0.05 max **Fe** 0.10-0.40 **Mg** 0.10-0.30 **Si** 0.10 max **Zn** 0.05 max **Zr** 0.02-0.08 **Other** each 0.03 max, total 0.10 max	**AA** 8077
A98079	Wrought Aluminum Alloy, Non-Heat Treatable	**Al** rem **Cu** 0.05 max **Fe** 0.7-1.3 **Si** 0.05-0.30 **Zn** 0.10 max **Other** each 0.05 max, total 0.15 max	**AA** 8079
A98081	Wrought Aluminum Alloy, Heat Treatable	**Al** rem **Cu** 0.7-1.3 **Fe** 0.7 max **Mn** 0.10 max **Si** 0.7 max **Sn** 18.0-22.0 **Ti** 0.10 max **Other** each 0.05 max, total 0.15 max	**AA** 8081
A98111	Wrought Aluminum Alloy, Non-Heat Treatable	**Al** rem **Cr** 0.05 **Cu** 0.10 **Fe** 0.40-1.0 **Mg** 0.05 **Mn** 0.10 **Si** 0.30-1.1 **Ti** 0.08 **Zn** 0.10 **Other** each 0.05, total 0.15	**AA** 8111
A98112	Wrought Aluminum Alloy, Non-Heat Treatable	**Al** rem **Cr** 0.20 max **Cu** 0.40 max **Fe** 1.0 max **Mg** 0.7 max **Mn** 0.6 max **Si** 1.0 max **Ti** 0.20 max **Zn** 1.0 max **Other** each 0.05 max, total 0.15 max	**AA** 8112
A98130	Wrought Aluminum Alloy, Non-Heat Treatable	**Al** rem **Cu** 0.05-0.15 **Fe** 0.40-1.0 **Si** 0.15 max **Zn** 0.10 max **Other** each 0.03 max, total 0.10 max	**AA** 8130
A98176	Wrought Aluminum Alloy, Non-Heat Treatable	**Al** rem **Fe** 0.40-1.0 **Ga** 0.03 max **Si** 0.03-0.15 **Zn** 0.10 max **Other** each 0.05 max, total 0.15 max	**AA** 8176
A98177	Wrought Aluminum Alloy, Non-Heat Treatable	**Al** rem **B** 0.04 **Cu** 0.04 **Fe** 0.25-0.45 **Mg** 0.04-0.12 **Si** 0.10 **Zn** 0.05 **Other** each 0.03, total 0.10	**AA** 8177

*Boxed entries are no longer active and are retained for reference purposes only.

The chemical compositions listed are for identification purposes and should not be used in lieu of the cross referenced specifications.

UNIFIED NUMBER	DESCRIPTION	CHEMICAL COMPOSITION	CROSS REFERENCE SPECIFICATIONS
A98280	Wrought Aluminum Alloy, Heat Treatable	**Al** rem **Cu** 0.7-1.3 **Fe** 0.7 max **Ni** 0.20-0.7 **Si** 1.0-2.0 **Sn** 5.5-7.0 **Ti** 0.10 max **Other** each 0.05 max, total 0.15 max	**AA** 8280 **MIL SPEC** MIL-A-11267 (8280)

The chemical compositions listed are for identification purposes and should not be used in lieu of the cross referenced specifications.

UNS NUMBERS ASSIGNED TO DATE

With Description of Each Material Covered and References To Documents In Which The Same or Similar Materials are Described

Cxxxxx Number Series
Copper and Copper Alloys

COPPER AND COPPER ALLOYS

UNIFIED NUMBER	DESCRIPTION	CHEMICAL COMPOSITION	CROSS REFERENCE SPECIFICATIONS
C10100	Oxygen-Free Electronic Copper (OFE)	**Bi** 0.0010 max **Cd** 0.0001 max **Cu** 99.99 min **Hg** 0.0001 **O** 0.0010 max **P** 0.0003 max **Pb** 0.0010 max **S** 0.0018 max **Se** 0.0010 max **Te** 0.0010 max **Zn** 0.0001 max **Other** total As, Bi, Mn, Sb, Se, Sn, and Te 0.0040 max	**ASTM** B1; B2; B3; B33; B48; F68; B75 (101); B133 (101); B152 (101); B187 (101); B189; B246; B272; B298; B355; B432 (101); B451 **FED** QQ-C-502; QQ-C-576; QQ-W-343 **MIL SPEC** MIL-W-85; MIL-W-3318; MIL-B-18907; MIL-W-23068
C10200	Oxygen-Free Copper (OF)	**Cu** 99.95 min **Other** Ag included in Cu	**AMS** 4501; 4602; 4701 **ASME** SB12; SB42; SB75; SB111; SB152; SB359; SB395 **ASTM** B1; B2; B3; F9; B12 (102); B33; B42 (102); B48; B49; B68 (102); B75 (102); B88 (102); B111 (102); B133 (102); B152 (102); B170 (102); B187 (102); B188 (102); B189; B246; B272; B280 (102); B298; B355; B359 (102); B372 (102); B395 (102); B432 (102); B447 (102); B451; B506 (102); B566 **FED** QQ-C-502; QQ-C-576; QQ-R-571 (RCu-1; QQ-W-343 **MIL SPEC** MIL-W-85; MIL-T-3235; MIL-W-3318; MIL-W-23068 **SAE** J461 (CA102); J463 (CA102)
C10300	Oxygen-Free Extra Low Phosphorus Copper (OFXLP)	**Cu** 99.95 min **P** 0.001-0.005 **Other** Ag and P included in Cu	**ASTM** B12 (103); B42 (103); B68 (103); B75 (103); B88 (103); B111 (103); B152 (103); B187 (103); B188 (103); B280 (103); B302 (103); B306 (103); B359 (103); B372 (103); B379 (103); B395 (103); B432 (103); B447 (103) **MIL SPEC** MIL-W-23068
C10400	Oxygen-Free Copper with Silver (OFS)	**Ag** 0.027 min **Cu** 99.95 min **Other** Ag included in Cu	**ASME** SB152 **ASTM** B1; B2; B3; B48; B49; B133 (104); B152 (104); B187 (104); B188 (104); B189; B246; B272; B298; B334; B355; B506 (104) **FED** QQ-C-502; QQ-C-576; QQ-W-343 **MIL SPEC** MIL-W-3318
C10500	Oxygen-Free Copper with Silver (OFS)	**Ag** 0.034 min **Cu** 99.95 min **Other** Ag included in Cu	**ASME** SB152 **ASTM** B1; B2; B3; B12 (104); B49; B133 (104); B152 (105); B187 (105); B188 (105); B189; B246; B272; B298; B334; B355; B506 **FED** QQ-C-502; QQ-C-576; QQ-W-343 **MIL SPEC** MIL-W-3318
C10700	Oxygen-Free Copper with Silver (OFS)	**Ag** 0.085 min **Cu** 99.95 min **Other** Ag included in Cu	**ASME** SB152 **ASTM** B1; B2; B3; B49; B133 (107); B152 (107); B187 (107); B188 (107); B189; B246; B272; B298; B334; B355; B506 (107) **FED** QQ-C-502; QQ-C-576; QQ-W-343 **MIL SPEC** MIL-W-3318; MIL-B-19231
C10800	Oxygen-Free Low Phosphorus Copper (OFLP)	**Cu** 99.95 min **P** 0.005-0.012 **Other** Ag and P included in Cu	**ASTM** B12 (108); B42 (108); B68 (108); B75 (108); B88 (108); B111 (108); B152 (108); B280 (108); B302 (108); B306 (108); B359 (108); B360; B379 (108); B395 (108); B432 (108); B447 (108); B543 (108)
C10920	Copper (Controlled Oxygen)	**Cu** 99.90 min **O** 0.02 max **Other** Includes Ag	
C10930	Copper (Controlled Oxygen)	**Ag** 0.044 min **Cu** 99.90 min **O** 0.02 max **Other** Includes Ag	
C10940	Copper (Controlled Oxygen)	**Ag** 0.085 min **Cu** 99.90 min **O** 0.02 max	
C11000	Electrolytic Tough Pitch Copper (ETP)	**Cu** 99.90 min **Other** Ag included in Cu	**AMS** 4500 **ASME** SB11; SB12 **ASTM** B1; B2; B3; B8; B11 (110); B12 (110); B47; B48; B49; B116; B124 (110); B133 (110); B152 (110); B172; B173; B174; B187 (110); B188 (110); B189; B226; B228; B229; B246; B272; B283 (110); B286; B298; B334; B355; B370; B447 (110); B451; B506 (110); B566 **FED** QQ-C-502; QQ-C-576; QQ-W-343 **MIL SPEC** MIL-W-3318; MIL-W-6712; MIL-C-12166 **SAE** J461 (CA110); J463 (CA110)
C11010	Remelted High Conductivity Tough Pitch Copper	**Cu** 99.90 min **Other** Ag included in Cu; unspecified oxygen and trace elements	**FED** QQ-B-650
C11020	Fine Refined High Conductivity Tough Pitch Copper	**Cu** 99.90 min **Other** Ag included in Cu; unspecified oxygen and trace elements	**FED** QQ-B-650
C11030	Chemically Refined Tough Pitch Copper	**Cu** 99.90 min **Other** Ag included in Cu; unspecified oxygen and trace elements	**FED** QQ-B-650
C11100	Electrolytic Tough Pitch Anneal Resistant	**Cu** 99.90 min **Other** Includes Ag	**ASTM** B1; B2; B3; B8; B33; B47; B49; B116; B133; B172; B173; B174; B189; B226; B228; B229; B246; B272; B286; B293; B355; B470; B496 **CDA** C11100 **MIL SPEC** MIL-W-3318; MIL-W-8777; MIL-C-12166; MIL-B-20292; MIL-W-22759; MIL-W-81044; MIL-W-81381

The chemical compositions listed are for identification purposes and should not be used in lieu of the cross referenced specifications.

UNIFIED NUMBER	DESCRIPTION	CHEMICAL COMPOSITION	CROSS REFERENCE SPECIFICATIONS
C11300	Tough Pitch Copper with Silver (STP)	**Ag** 0.027 min **Cu** 99.90 min **Other** Ag included in Cu	**ASME** SB152 **ASTM** B1; B2; B3; B8; B47; B49; B116; B152 (113); B172; B173; B174; B187 (113); B188 (113); B189; B226; B228; B229; B246; B272; B286; B298; B334; B355; B506 (113) **FED** QQ-C-502; QQ-C-576; QQ-W-343 **MIL SPEC** MIL-W-3318 **SAE** J461 (CA113); J463 (CA113)
C11400	Tough Pitch Copper with Silver (STP)	**Ag** 0.034 min **Cu** 99.90 min **Other** Ag included in Cu	**ASTM** B1; B2; B3; B8; B47; B48; B49; B116; B122 (114); B152 (114); B172; B173; B174; B187 (114); B188 (114); B189; B226; B228; B229; B246; B272; B286; B298; B334; B355; B506 (114) **FED** QQ-C-502; QQ-C-576; QQ-W-343 **MIL SPEC** MIL-W-3318 **SAE** J461 (CA114); J463 (CA114)
C11500	Tough Pitch Copper with Silver (STP)	**Ag** 0.054 min **Cu** 99.90 min **Other** Ag included in Cu	**ASTM** B1; B2; B3; B8; B47; B49; B116; B122 (115); B146; B172; B173; B174; B189; B226; B228; B229; B246; B272; B286; B298; B334; B355 **SAE** J461 (CA115); J463 (CA115)
C11600	Tough Pitch Copper with Silver (STP)	**Ag** 0.085 min **Cu** 99.90 min **Other** Ag included in Cu	**ASTM** B1; B2; B3; B8; B47; B48; B49; B116; B152 (116); B172; B173; B174; B187 (116); B188 (116); B189; B226; B228; B229; B246; B272; B286; B298; B334; B355; B506 (116) **FED** QQ-C-502; QQ-C-576; QQ-W-343 **MIL SPEC** MIL-W-3318; MIL-B-19231 **SAE** J461 (CA116); J463 (CA116)
C11700	Boron Deoxidized Copper	**B** 0.004-0.02 **Cu** 99.9 min **P** 0.04 max **Other** Ag + P + B included in Cu	
C11904	Tough Pitch Copper with Silver	**Ag** 0.027 min **Cu** 99.9 min **Other** Ag included in Cu. For oxygen-free and deoxidized grades, B, Li, P etc limits shall be as agreed upon.	
C11905	Tough Pitch Copper with Silver	**Ag** 0.034 min **Cu** 99.9 min **Other** Ag included in Cu. For oxygen-free and deoxidized grades, B, Li, P, etc limits shall be as agreed upon.	
C11906	Tough Pitch Copper with Silver	**Ag** 0.054 min **Cu** 99.9 min **Other** Ag included in Cu. For oxygen-free and deoxidized grades, B, Li, P, etc limits shall be as agreed upon.	
C11907	Tough Pitch Copper with Silver	**Ag** 0.085 min **Cu** 99.9 min **Other** Ag included in Cu. For oxygen-free and deoxidized grades, B, Li, P, etc limits shall be as agreed upon.	
C12000	Phosphorus Deoxidized, Low Residual Phosphorus Copper (DLP)	**Cu** 99.90 min **P** 0.004-0.012 **Other** Ag included in Cu	**ASME** SB12; SB42; SB75; SB111; SB359; SB395 **ASTM** B12 (120); B42 (120); B68 (120); B75 (120); B88 (120); B111 (120); B152 (120); B187 (120); B188 (120); B272; B280 (120); B302 (120); B306 (120); B359 (120); B372 (120); B395 (120); B447 (120); B451; B506 (120) **FED** QQ-C-502; QQ-C-576 **MIL SPEC** MIL-W-85; MIL-T-3235; MIL-W-3318; MIL-B-18907; MIL-W-23068; MIL-T-24107 **SAE** J461 (CA120); J463 (CA120)
C12100	Silver-Bearing, Low Residual Phosphorus Copper (DLP)	**Ag** 0.014 min **Cu** 99.90 min **P** 0.005-0.012 **Other** Ag included in Cu	**ASME** SB12 **ASTM** B12 (121); B133 (121) **FED** QQ-C-502; QQ-C-576 **MIL SPEC** MIL-W-3318
C12200	Phosphorus Deoxidized, High Residual Phosphorus Copper (DHP)	**Cu** 99.9 min **P** 0.015-0.040 **Other** Ag included in Cu	**ASME** SB11; SB12; SB42; SB75; SB111; SB152; SB359; SB395; SB543 **ASTM** B11 (122); B12 (122); B42 (122); B68 (122); B75 (122); B88 (122); B111 (122); B133 (122); B152 (122); B272; B280 (122); B302 (122); B306 (122); B359 (122); B360 (122); B370; B395 (122); B432 (122); B447 (122); B506 (122); B543 (122) **FED** QQ-C-502; QQ-C-576 **MIL SPEC** MIL-T-3235; MIL-W-3318; MIL-B-18907; MIL-T-24107 **SAE** J461 (CA122); J463 (CA122)
C12210	Phosphorus Deoxidized Copper	**Cu** 99.90 **P** 0.015-0.025 **Other** Ag included in Cu	
C12220	Phosphorus Deoxidized Copper	**Cu** 99.9 min **P** 0.040-0.065 **Other** Ag included in Cu	
C12300	Silver-Bearing, High Residual Phosphorus Copper (DHP)	**Ag** 0.014 min **Cu** 99.90 min **P** 0.015-0.040 **Other** Ag included in Cu	**ASME** SB12; SB152 **ASTM** B12 (123); B133 (123); B152 (123); B506 (123) **FED** QQ-C-502; QQ-C-576 **MIL SPEC** MIL-W-3318

The chemical compositions listed are for identification purposes and should not be used in lieu of the cross referenced specifications.

UNIFIED NUMBER	DESCRIPTION	CHEMICAL COMPOSITION	CROSS REFERENCE SPECIFICATIONS
C12500	Fire-Refined Tough Pitch Copper (FRTP)	**As** 0.012 max **Bi** 0.003 max **Cu** 99.88 min **Ni** 0.050 max **Pb** 0.004 max **Sb** 0.003 max **Te** 0.025 max **Other** Ag included in Cu, Se included in Te	**ASME** SB11; SB12 **ASTM** B11 (125); B12 (125); B133 (125); B152 (125); B370; B506 (125) **FED** QQ-C-502; QQ-C-576 **MIL SPEC** MIL-W-3318
C12700	Fire-Refined Tough Pitch Copper with Silver (FRSTP)	**Ag** 0.027 min **As** 0.012 max **Bi** 0.003 max **Cu** 99.88 min **Ni** 0.050 max **Pb** 0.004 max **Sb** 0.003 max **Te** 0.025 max **Other** Ag included in Cu, Se included in Te	**ASME** SB12 **ASTM** B12 (127); B133 (127) **FED** QQ-C-502; QQ-C-576 **MIL SPEC** MIL-W-3318
C12800	Fire-Refined Tough Pitch Copper with Silver (FRSTP)	**Ag** 0.034 min **As** 0.012 max **Bi** 0.003 max **Cu** 99.88 min **Ni** 0.050 max **Pb** 0.004 max **Sb** 0.003 max **Te** 0.025 max **Other** Ag included in Cu, Se included in Te	**ASME** SB12 **ASTM** B12 (128); B133 (128) **FED** QQ-C-502; QQ-C-576 **MIL SPEC** MIL-W-3318
C12900	Fire-Refined Tough Pitch Copper with Silver (FRSTP)	**Ag** 0.054 min **As** 0.012 max **Bi** 0.003 max **Cu** 99.88 min **Ni** 0.050 max **Pb** 0.004 max **Sb** 0.003 max **Te** 0.025 max **Other** Ag included in Cu, Se included in Te	
C13000	Fire-Refined Tough Pitch Copper with Silver (FRSTP)	**Ag** 0.085 min **As** 0.012 max **Bi** 0.003 max **Cu** 99.88 min **Ni** 0.050 max **Pb** 0.004 max **Sb** 0.003 max **Te** 0.025 max **Other** Ag included in Cu, Se included in Te	**ASME** SB12 **ASTM** B12 (130); B133 (130) **FED** QQ-C-502; QQ-C-576 **MIL SPEC** MIL-W-3318; MIL-B-20292
* C14100	Arsenical, Tough Pitch Copper (ATP)	**As** 0.15-0.50 **Cu** 99.40 min **Other** Ag included in Cu	**ASME** SB11; SB12 **ASTM** B11 (141); B12 (141); B152 (141)
C14180	Copper Brazing Filler Metal	**Al** 0.01 max **Cu** 99.90 min **P** 0.075 max **Pb** 0.02 max **Other** Ag included in Cu	**AMS** 4700 **ASME** SFA5.8 (BCu-1) **AWS** A5.8 (BCu-1) **FED** QQ-C-576
C14181	Copper Vacuum Grade Brazing Filler Metal	**C** 0.005 max **Cd** 0.002 max **Cu** 99.99 min **P** 0.002 max **Pb** 0.002 max **Zn** 0.002 **Other** Ag included in Cu	**ASME** SFA5.8 (BVCu-1x) **AWS** A5.8 (BVCu-1x)
C14200	Phosphorus Deoxidized Copper, Arsenical (DPA)	**As** 0.15-0.50 **Cu** 99.4 min **P** 0.015-0.040 **Other** Ag included in Cu	**ASME** SB11; SB12; SB75; SB111; SB359; SB395 **ASTM** B11 (142); B12 (142); B75 (142); B111 (142); B359 (142); B395 (142); B447 (142)
C14210	Phosphorus Deoxidized Arsenical Copper	**As** 0.30-0.50 **Cu** 99.20 min **P** 0.013-0.050 **Other** Cu 99.20 min (Includes Ag)	
C14300	Cadmium Copper, Deoxidized	**Cd** 0.05-0.15 **Cu** 99.90 min **Other** Ag + Cd included in Cu, Li by agreement	
C14310	Cadmium Copper, Deoxidized	**Cd** 0.10-0.30 **Cu** 99.90 min **Other** Cd included in Cu	
C14400	Tin Copper Radiator Strip	**Cu** 99.90 min **Fe** 0.03 max **Ni** 0.05 max **P** 0.013-0.025 **Sb** 0.003 max **Se** 0.02 max **Sn** 0.10-0.20 **Te** 0.02 max **Zn** 0.05 max **Other** Cu 99.90 min (Includes Cu+Sn+P)	
C14410	Copper-Tin-Phosphorus Alloy	**Cu** 99.90 **Fe** 0.05 max **P** 0.005-.020 **Pb** 0.05 max **Sn** 0.10-0.20 **Other** Includes Ag and Sn	
C14420	Copper-Tellurium-Tin Alloy	**Cu** 99.90 min **Sn** 0.05-0.15 **Te** 0.02-0.05 **NOTE:** Cu includes Cu + Ag + Sn + Te	

*Boxed entries are no longer active and are retained for reference purposes only.

The chemical compositions listed are for identification purposes and should not be used in lieu of the cross referenced specifications.

41

UNIFIED NUMBER	DESCRIPTION	CHEMICAL COMPOSITION	CROSS REFERENCE SPECIFICATIONS
C14500	Phosphorus Deoxidized Copper, Tellurium Bearing (DPTE) Includes oxygen-free or deoxidized grades with deoxidizers (such as phosphorus, boron, lithium, or other) in an amount agreed upon.	**Cu** 99.90 min **P** 0.004-0.012 **Te** 0.40-0.6 **Other** Ag + Te included in Cu, other deoxidizers may be used instead of P	**ASTM** B124 (145); B283 (145); B301 (145) **SAE** J462 (CA145); J463 (CA145)
C14510	Copper-Tellurium Alloy Includes oxygen-free or deoxidized grades with deoxidizers (such as phosphorus, boron, lithium, or other) in an amount agreed upon.	**Cu** 99.90 **Other** Includes Tellurium	**ASTM** B124; B283; B301
C14520	Copper-Tellurium	**Cu** 99.90 **P** 0.004-0.020 **Te** 0.40-0.7 **Other** Includes Ag and Te	**ASTM** B124; B283; B301
C14700	Sulfur-Bearing Copper	**Cu** 99.90 min **S** 0.20-0.50 **Other** Ag + S included in Cu. For oxygen-free and deoxidized grades: O, P, B, Li, etc. limits shall be as agreed upon.	**ASTM** B301 (147) **SAE** J461 (CA147); J463 (CA147)
C14710	Sulfur Bearing Copper	**Cu** 99.90 min **P** 0.010-0.030 **Pb** 0.05 max **S** 0.05-0.15 **Other** Ag+P+S+Pb included in Cu. For oxygen-free and deoxidized grades: O, P, B, Li, etc. limits shall be as agreed upon.	**ASTM** B301
C14720	Sulfur Copper	**Cu** 99.50 min **P** 0.010-0.030 **Pb** 0.10 max **S** 0.20-0.50 **Other** Ag+S+P+Pb included in Cu. (Includes oxygen-free or deoxidized grades with deoxidizers (such as phosphorus, boron, lithium, or other) in an amount agreed upon.)	**ASTM** B301
* C14730	Sulfur Bearing Copper	**Cu** 99.80 **Other** (Includes silver, sulfur, phosphorus, and lead) (Includes oxygen-free or deoxidized grades with deoxidizers (such as phosphorous, boron, lithium, or other) in an amount agreed upon.)	
C15000	Zirconium Copper	**Cu** 99.80 min **Zr** 0.10-0.20 **Other** Ag included in Cu	**SAE** J461 (CA150); J463 (CA150)
C15100	Zirconium Copper	**Al** 0.005 max **Cu** 99.82 min **Mn** 0.005 max **Zr** 0.05-0.15 **Other** Ag included in Cu	
C15500	Silver-Bearing Copper	**Ag** 0.027-0.10 **Cu** 99.75 min **Mg** 0.08-0.13 **P** 0.040-0.080 **Other** Ag included in Cu	
C15600	Copper Cobalt	**Co** 0.20-0.30 **Cu** 99.6 min **Mg** 0.02 max **P** 0.06-0.09 **Other** Ag included in Cu	
C15710	Dispersion Strengthened Copper	**Al** 0.08-0.12 **Cu** + Ag 99.71 min **Fe** 0.01 max **O$_2$** 0.07-0.15 **P** 0.01 max **Pb** 0.01 max **NOTE:** All Al present as Al$_2$O$_3$; 0.04% of O$_2$ as Cu$_2$O with a negligible amount in solid solution with Cu.	

*Boxed entries are no longer active and are retained for reference purposes only.

The chemical compositions listed are for identification purposes and should not be used in lieu of the cross referenced specifications.

UNIFIED NUMBER	DESCRIPTION	CHEMICAL COMPOSITION	CROSS REFERENCE SPECIFICATIONS
C15715	Dispersion Strengthened Copper	**Al** 0.13-0.17 **Cu** 99.62 min **Fe** 0.01 max **O₂** 0.12-0.19 **P** 0.01 max **Pb** 0.01 max **NOTE:** Cu content includes Ag, and all Al present as Al₂O₃; 0.04% O₂ present as Cu₂O with a negligible amount in solid solution with Cu.	
C15720	Dispersion Strengthened Copper	**Al** 0.18-0.22 **Cu** + Ag 99.52 min **Fe** 0.01 max **O₂** 0.16-0.24 **P** 0.01 max **Pb** 0.01 max **NOTE:** All Al present as Al₂O₃; 0.04% O₂ present as Cu₂O with a negligible amount in solid solution with Cu.	
C15735	Dispersion Strengthened Copper	**Al** 0.33-0.37 **Cu** + Ag 99.24 min **Fe** 0.01 max **O₂** 0.29-0.37 **P** 0.01 max **Pb** 0.01 max **NOTE:** All Al present as Al₂O₃; 0.04% O₂ present as Cu₂O with a negligible amount in solid solution with Cu.	
C15760	Dispersion Strengthened Copper	**Al** 0.58-0.62 **Cu** + Ag 98.77 min **Fe** 0.01 max **O₂** 0.52-0.59 **P** 0.01 max **Pb** 0.01 max **NOTE:** All Al present as Al₂O₃; 0.04% O₂ present as Cu₂O with a negligible amount in solid solution with Cu.	
C16200	Cadmium Copper	**Cd** 0.7-1.2 **Cu** 99.8 min **Fe** 0.02 max **Other** Ag + Cd + Fe included in Cu	**ASTM** B9; B105 **MIL SPEC** MIL-W-82598 **SAE** J461 (CA162); J463 (CA162)
C16210	Cadmium Copper	**Cd** 0.50-1.2 **Cu** 99.95 min	
* C16400	Cadmium Copper Alloy	**Cd** 0.6-0.9 **Cu** 99.8 min **Fe** 0.02 max **Sn** 0.20-0.40 **Other** Ag + Cd + Fe + Sn included in Cu	**ASTM** B105
C16500	Cadmium Copper Alloy	**Cd** 0.6-1.0 **Cu** 99.8 min **Fe** 0.02 max **Sn** 0.50-0.7 **Other** Ag + Cd + Fe + Sn included in Cu	**ASTM** B9; B105
C17000	Beryllium Copper	**Be** 1.60-1.79 **Cu** 99.5 min **Other** Be + Co + Fe + Ni included in Cu, Co + Ni 0.20 min, Co + Fe + Ni 0.6 max	**ASTM** B194 (170); B196 (170); B570 (170) **FED** QQ-C-533 **SAE** J461 (CA170); J463 (CA170)
C17200	Beryllium Copper	**Be** 1.80-2.00 **Cu** 99.5 min **Other** Be + Co + Fe + Ni included in Cu, Co + Ni 0.20 min, Co + Fe + Ni 0.6 max	**AMS** 4530; 4532; 4650; 4651; 4725 **ASTM** B194 (172); B196 (172); B197 (172); B570 (172) **FED** QQ-C-530; QQ-C-533 **MIL SPEC** MIL-C-21657 **SAE** J461 (CA172); J463 (CA172)
C17300	Beryllium Copper with Lead	**Be** 1.80-2.00 **Cu** 99.5 min **Pb** 0.20-0.6 **Other** Be + Co + Fe + Ni + Pb included in Cu, Co + Ni 0.20 min, Co + Fe + Ni 0.6 max	**ASTM** B196 (173) **FED** QQ-C-530
C17400	Copper-Beryllium Alloy (Beryllium-Copper)	**Al** 0.20 max **Be** 0.15-0.50 **Co** 0.15-0.35 **Cu** rem **Fe** 0.20 max **Si** 0.20 max	
C17410	Copper-Beryllium Alloy (Beryllium-Copper)	**Al** 0.20 max **Be** 0.15-0.50 **Co** 0.35-0.6 **Cu** rem **Fe** 0.20 max **Si** 0.20 max	
C17420	Copper-Beryllium Alloy (Beryllium-Copper)	**Al** 0.20 max **Be** 0.05-0.15 **Co** 0.05-0.6 **Cu** rem **Fe** 0.20 max **Si** 0.20 max	
C17500	Beryllium Copper	**Be** 0.40-0.7 **Co** 2.4-2.7 **Cu** 99.5 min **Fe** 0.10 max **Other** Be + Co + Fe included in Cu	**ASTM** B441 (175); B534 (175) **MIL SPEC** MIL-C-46087; MIL-C-81021 **SAE** J461 (CA175); J463 (CA175)
C17510	Copper Beryllium Alloy	**Al** 0.20 max **Be** 0.20-0.6 **Co** 0.30 max **Cu** rem **Fe** 0.10 max **Mg** 0.03-0.06 **Ni** 0.50-1.5 **Zr** 0.10-0.30 **Other** Ag included in Cu	

*Boxed entries are no longer active and are retained for reference purposes only.

The chemical compositions listed are for identification purposes and should not be used in lieu of the cross referenced specifications.

UNIFIED NUMBER	DESCRIPTION	CHEMICAL COMPOSITION	CROSS REFERENCE SPECIFICATIONS
C17520	Copper Beryllium	**Be** 0.10-0.30 **Cu** rem **Mg** 0.03-0.06 **Ni** 0.50-1.5 **Zr** 0.10-0.30 **Other** Includes Ag	
C17600	Beryllium Copper	**Ag** 0.9-1.1 **Be** 0.25-0.50 **Co** 1.4-1.7 **Cu** 99.5 min **Fe** 0.10 max **Other** Be + Co + Fe included in Cu	**SAE** J461 (CA176); J463 (CA176)
C17700	Beryllium Copper	**Be** 0.40-0.7 **Co** 2.4-2.7 **Cu** 99.5 min **Fe** 0.10 max **Te** 0.40-0.6 **Other** Be + Co + Fe + Te included in Cu	
C18000	Chromium Silicon Copper	**Cr** 0.10-0.6 **Cu** rem **Fe** 0.15 max **Ni** 2.0-3.0 **Si** 0.40-0.8 **Other** Ag included in Cu; Ni includes Co	
C18090	Copper Alloy	**Cr** 0.30-1.0 **Cu** 96.0 min **Ni** 0.7-1.2 **Sn** 0.7-1.2 **Ti** 0.40-0.8 **Other** Ag included in Cu; Cu + all named elements = 99.85 min	
C18100	Chromium Magnesium Copper Alloy	**Cr** 0.40-10.0 **Cu** 98.7 min **Mg** 0.03-0.06 **Zr** 0.08-0.20	
C18135	Cadmium Chromium Copper	**Cd** 0.20-0.6 **Cr** 0.20-0.6 **Cu** rem **Other** Ag included in Cu	
C18200	Chromium Copper	**Cr** 0.6-1.2 **Cu** 99.5 min **Fe** 0.10 max **Pb** 0.05 max **Si** 0.10 max **Other** Ag + Cr + Fe + Pb + Si included in Cu	**ASTM** F9 **MIL SPEC** MIL-C-19311
C18400	Chromium Copper	**As** 0.005 max **Ca** 0.005 max **Cr** 0.40-1.2 **Cu** 99.8 min **Fe** 0.15 max **Li** 0.05 max **P** 0.05 max **Si** 0.10 max **Zn** 0.7 max **Other** Ag + all named elements included in Cu	**MIL SPEC** MIL-C-19311 **SAE** J461 (CA184); J463 (CA184)
C18500	Chromium Copper	**Ag** 0.08-0.12 **Cr** 0.40-1.0 **Cu** 99.8 min **P** 0.04 max **Pb** 0.015 max **Other** Ag + Cr + P + Pb included in Cu	**MIL SPEC** MIL-C-19311
C18700	Leaded Copper	**Cu** 99.9 min **Pb** 0.8-1.5 **Other** Ag + Pb included in Cu	**ASTM** B301 (187) **SAE** J461 (CA187); J463 (CA187)
C18900	High Copper Alloy	**Al** 0.01 max **Cu** 99.9 min **Mn** 0.10-0.30 **P** 0.05 max **Pb** 0.02 max **Si** 0.15-0.40 **Sn** 0.6-0.9 **Zn** 0.10 max **Other** Ag + all named elements included in Cu	**FED** QQ-R-571
C18980	Copper Welding Filler Metal	**Al** 0.01 max **Cu** 98.0 min **Mn** 0.50 max **P** 0.15 max **Pb** 0.02 max **Si** 0.50 max **Sn** 1.0 max	**ASME** SFA5.7 (ERCu); SFA5.27 (ERCu) **AWS** A5.7 (ERCu); A5.27 (ERCu) **FED** QQ-R-571c
C19000	High Copper Alloy	**Cu** 99.5 min **Fe** 0.10 max **Ni** 0.9-1.3 **P** 0.15-0.35 **Pb** 0.05 max **Zn** 0.8 max **Other** Ag + Fe + P + Pb + Sn + Zn included in Cu	
C19010	Copper Alloy	**Cu** rem **Ni** 0.8-1.8 **P** 0.01-0.05 **Si** 0.15-0.35	
C19100	High Copper Alloy	**Cu** 99.5 min **Fe** 0.20 max **Ni** 0.9-1.3 **P** 0.15-0.35 **Pb** 0.10 max **Te** 0.35-0.6 **Zn** 0.50 max **Other** Ag + all named elements included in Cu	
C19200	High Copper Alloy	**Cu** 98.7 min **Fe** 0.8-1.2 **P** 0.01-0.04	**ASME** SB111; SB359; SB395 **ASTM** B111 (192); B359 (192); B395 (192); B469; B585 (192) **SAE** J461 (CA192); J463 (CA192)
C19210	Copper-Iron Alloy	**Cu** rem **Fe** 0.05-0.15 **P** 0.025-0.040	
C19220	Copper-Iron Alloy	**B** 0.005-0.015 **Cu** rem **Fe** 0.10-0.30 **Ni** 0.10-0.25 **P** 0.03-0.07 **Sn** 0.05-0.10	

The chemical compositions listed are for identification purposes and should not be used in lieu of the cross referenced specifications.

UNIFIED NUMBER	DESCRIPTION	CHEMICAL COMPOSITION	CROSS REFERENCE SPECIFICATIONS
C19250	Copper-Iron Alloy	**B** 0.005-0.025 **Cu** rem **Fe** 0.05-0.40 **Ni** 0.05-0.40 **P** 0.03-0.10 **Sn** 0.05-0.20	
C19280	Copper-Iron Alloy	**Cu** rem **Fe** 0.50-1.5 **P** 0.005-0.015 **Sn** 0.30-0.7 **Zn** 0.30-0.7	
C19400	High Copper Alloy	**Cu** 97.0 min **Fe** 2.1-2.6 **P** 0.015-0.15 **Pb** 0.03 max **Zn** 0.05-0.20	**ASME** SB543 **ASTM** B465; B543 (194); B586 (194)
C19500	High Copper Alloy	**Al** 0.02 max **Co** 0.30-1.3 **Cu** 96.0 min **Fe** 1.0-2.0 **P** 0.01-0.35 **Pb** 0.02 max **Sn** 0.10-1.0 **Zn** 0.20 max	
C19520	Copper-Iron-Tin Alloy	**Cu** 96.6 min **Fe** 0.50-1.5 **P** 0.01-0.35 **Sn** 0.50-1.5	
C19600	High Copper Alloy	**Fe** 0.9-1.2 **P** 0.25-0.35 **Zn** 0.35 **Other** Ag included in Cu; Cu (including Ag) + all named elements = 99.7 min	
C19700	Copper Alloy	**Co** 0.05 max **Cu** rem **Fe** 0.30-1.2 **Mg** 0.01-0.20 **Mn** 0.05 max **Ni** 0.05 max **P** 0.10-0.40 **Pb** 0.05 max **Sn** 0.02 max **Zn** 0.20 max **Other** Cu + all named elements /me 99.8 min	
C19750	Copper Alloy	**Co** 0.05 max **Cu** rem **Fe** 0.35-1.2 **Mg** 0.01-0.20 **Mn** 0.05 max **Ni** 0.05 max **P** 0.10-0.40 **Pb** 0.05 max **Sn** 0.05-0.40 **Zn** 0.20 max **Other** Cu + all named elements = 99.0 min	
C20500	Brass	**Cu** 97.0-98.0 **Fe** 0.05 max **Pb** 0.02 max **Zn** rem	
C21000	Gilding, 95% (Copper/Zinc)	**Cu** 94.0-96.0 **Fe** 0.05 max **Pb** 0.03 max **Zn** rem	**ASTM** B36 (210); B134 (210) **FED** QQ-W-321 **MIL SPEC** MIL-C-21768 **SAE** J461 (CA210); J463 (CA210)
C22000	Commercial Bronze, 90%	**Cu** 89.0-91.0 **Fe** 0.05 max **Pb** 0.05 max **Zn** rem	**ASTM** B36 (220); B130 (220); B131 (220); B134 (220); B135 (220); B372 (220); B587 (220) **FED** QQ-W-321 **MIL SPEC** MIL-W-85; MIL-C-3383; MIL-W-6712; MIL-B-18907; MIL-B-20292; MIL-C-21768; MIL-W-23068 **SAE** J461 (CA220); J463 (CA220)
C22600	Jewelry Bronze, 87-1/2%	**Cu** 86.0-89.0 **Fe** 0.05 max **Pb** 0.05 max **Zn** rem	
C23000	Red Brass, 85%	**Cu** 84.0-86.0 **Fe** 0.05 max **Pb** 0.06 max **Zn** rem	**ASME** SB43; SB111; SB359; SB395; SB543 **ASTM** B36 (230); B43 (230); B111 (230); B134 (230); B135 (230); B359 (230); B395 (230); B543 (230); B587 (230) **FED** QQ-B-613; QQ-B-626; QQ-W-321 **MIL SPEC** MIL-T-20168 **SAE** J461 (CA230); J463 (CA230)
C23030	Red Brass	**Cu** 83.5-85.5 **Fe** 0.05 max **Pb** 0.05 max **Si** 0.20-0.40 **Zn** rem	
C23400	Brass	**Cu** 81.0-84.0 **Fe** 0.05 max **Pb** 0.05 max **Zn** rem	
C24000	Low Brass, 80%	**Cu** 78.5-81.5 **Fe** 0.05 max **Pb** 0.05 max **Zn** rem	**ASTM** B36 (240); B134 (240) **FED** QQ-B-613; QQ-B-626; QQ-B-650; QQ-W-321 **SAE** J461 (CA240); J463 (CA240)
C24080	Low Brass	**Al** 0.10 max **Cu** 78.0-82.0 **Pb** 0.20 max **Zn** rem	**ASME** SFA5.8 (BCuZn-H) **AWS** A5.8 (BCuZn-H)
C25000	Brass	**Cu** 74.0-76.0 **Fe** 0.05 max **Pb** 0.05 max **Zn** rem	
C26000	Cartridge Brass, 70%	**Cu** 68.5-71.5 **Fe** 0.05 max **Pb** 0.07 max **Zn** rem	**AMS** 4505; 4507; 4555 **ASME** SFA5.8 (BCuZn-H) **ASTM** B19 (260); B36 (260); B129 (260); B134 (260); B135 (260); B569 (260); B587 (260) **AWS** A5.8 (BCuZn-H) **FED** QQ-B-613; QQ-B-626; QQ-B-650; QQ-W-321 **MIL SPEC** MIL-C-50; MIL-C-10375; MIL-T-20219; MIL-S-22499 **SAE** J461 (CA260); J463 (CA260)

The chemical compositions listed are for identification purposes and should not be used in lieu of the cross referenced specifications.

UNIFIED NUMBER	DESCRIPTION	CHEMICAL COMPOSITION	CROSS REFERENCE SPECIFICATIONS
C26100	Brass	Cu 68.5-71.5 Fe 0.05 max P 0.02-0.05 Pb 0.05 max Zn rem	ASTM B129 (261) MIL SPEC MIL-C-10375; MIL-T-20219; MIL-S-22499
C26130	Arsenic Inhibited Cartridge Brass	As 0.02-0.08 Cu 68.5-71.5 Fe 0.05 max Pb 0.05 max Zn rem	
C26200	Brass	Cu 67.0-70.0 Fe 0.05 max Pb 0.07 max Zn rem	
C26380	Brass Brazing Rod	Al 0.10 max Cr 68.0-72.0 Pb 0.30 max Zn rem	ASME SFA5.8 (BCuZn-G) AWS A5.8 (BCuZn-G)
C26800	Yellow Brass, 66%	Cu 64.0-68.5 Fe 0.05 max Pb 0.15 max Zn rem	ASTM B36 (268); B587 (268) FED QQ-B-613; QQ-B-626 MIL SPEC MIL-W-6712 SAE J461 (CA268); J463 (CA268)
C27000	Yellow Brass, 65%	Cu 63.0-68.5 Fe 0.07 max Pb 0.10 max Zn rem	AMS 4710; 4712; 4713 ASTM B134 (270); B135 (270); B587 (270) FED QQ-W-321 SAE J461 (CA270); J463 (CA270)
C27200	Brass	Cu 62.0-65.0 Fe 0.07 max Pb 0.07 max Zn rem	ASTM B36 (272); B135 (272); B587 (272)
C27400	Yellow Brass, 63%	Cu 61.0-64.0 Fe 0.05 max Pb 0.10 max Zn rem	ASTM B134 (274) FED QQ-W-321
C28000	Muntz Metal, 60%	Cu 59.0-63.0 Fe 0.07 max Pb 0.30 max Zn rem	ASME SB111 ASTM B111 (280); B135 (280) FED WW-T-791
* C28200	Brass	Al 0.005 max Cu 58.0-61.0 Fe 0.05 max P 0.12-0.22 Pb 0.03 max Si 0.005 max Sn 0.05 max Zn rem	
C28580	Brass Brazing Rod	Al 0.10 max Cu 49.0-52.0 Fe 0.10 max Pb 0.50 max Zn rem	ASME SFA5.8 (BCuZn-E) AWS A5.8 (BCuZn-E)
* C29800	Brazing Alloy	Al 0.10 max Cu 49.0-52.0 Fe 0.10 max Pb 0.50 max Zn rem	FED QQ-B-650
* C31000	Leaded Commercial Bronze (Low Lead)	Cu 89.0-91.0 Fe 0.10 max Pb 0.30-0.7 Zn rem	ASTM B121 (310)
C31200	Medium Leaded Commercial Bronze	Cu 87.5-90.5 Fe 0.10 max Ni 0.25 max Pb 0.7-1.2 Zn rem	
C31400	Leaded Commercial Bronze	Cu 87.5-90.5 Fe 0.10 max Ni 0.7 max Pb 1.3-2.5 Zn rem	ASTM B140 (314)
C31600	Leaded Commercial Bronze (Nickel Bearing)	Cu 87.5-90.5 Fe 0.10 max Ni 0.7-1.2 P 0.04-0.10 Pb 1.3-1.2 Zn rem	ASTM B140 (316)
C32000	Leaded Red Brass	Cu 83.5-86.5 Fe 0.10 max Ni 0.25 max Pb 1.5-2.2 Zn rem	ASTM B140 (320)
* C32500	Leaded Brass	Cu 72.0-74.5 Fe 0.10 max Pb 2.5-3.0 Zn rem	
C32510	Leaded 70/30 Arsenical Brass	As 0.20-0.06 Cu 69.0-72.0 Pb 0.30-0.7	
C33000	Low Leaded Brass (Tube)	Cu 65.0-68.0 Fe 0.07 max Pb 0.25-0.7 Zn rem	AMS 4555 ASTM B135 (330) MIL SPEC MIL-T-46072 SAE J461 (CA330); J463 (CA330)
C33100	Leaded Brass	Cu 65.0-68.0 Fe 0.06 max Pb 0.8-1.5 Zn rem	MIL SPEC MIL-T-46072 SAE J461 (CA331); J463 (CA331)
C33200	High Leaded Brass (Tube)	Cu 65.0-68.0 Fe 0.07 max Pb 1.5-2.5 Zn rem	AMS 4558 ASTM B135 (332) MIL SPEC MIL-T-46072

*Boxed entries are no longer active and are retained for reference purposes only.

The chemical compositions listed are for identification purposes and should not be used in lieu of the cross referenced specifications.

UNIFIED NUMBER	DESCRIPTION	CHEMICAL COMPOSITION	CROSS REFERENCE SPECIFICATIONS
C33500	Low Leaded Brass	**Cu** 62.0-65.0 **Fe** 0.10 max **Pb** 0.25-0.7 **Zn** rem	**ASTM** B121 (335); B453 (335)
C33530	Leaded 65/35 Arsenical Brass	**As** 0.02-0.06 **Cu** 62.5-66.5 **Fe** 0.10 max **Pb** 0.30-0.8 **Zn** rem	
C34000	Medium Leaded Brass, 64-1/2%	**Cu** 62.0-65.0 **Fe** 0.10 max **Pb** 0.8-1.5 **Zn** rem	**ASTM** B121 (340); B453 (340)
C34200	High Leaded Brass, 64-1/2%	**Cu** 62.0-65.0 **Fe** 0.10 max **Pb** 1.5-2.5 **Zn** rem	**ASTM** B121 (342) **FED** QQ-B-613; QQ-B-626 **SAE** J461 (CA342); J463 (CA342)
* C34400	Leaded Brass	**Cu** 62.0-66.0 **Fe** 0.10 max **Pb** 0.50-1.0 **Zn** rem	
C34500	Leaded Brass	**Cu** 62.0-65.0 **Fe** 0.15 max **Pb** 1.5-2.5 **Zn** rem	**ASTM** B453 (345) **SAE** J461 (CA345); J463 (CA345)
* C34700	Leaded Brass	**Cu** 62.5-64.5 **Fe** 0.10 max **Pb** 1.0-1.8 **Zn** rem	
* C34800	Leaded Brass	**Cu** 61.5-63.5 **Fe** 0.10 max **Pb** 0.40-0.8 **Zn** rem	
* C34900	Leaded Brass	**Cu** 61.0-64.0 **Fe** 0.10 max **Pb** 0.10-0.50 **Zn** rem	
C35000	Medium Leaded Brass, 62%	**Cu** 60.0-63.0 **Fe** 0.10 max **Pb** 0.8-2.0 **Zn** rem **Other** For rod, Cu 61.0 min	**ASTM** B121 (350); B453 (350) **SAE** J461 (CA350); J463 (CA350)
C35300	High Leaded Brass, 62%	**Cu** 60.0-63.0 **Fe** 0.10 max **Pb** 1.5-2.5 **Zn** rem **Other** For rod, Cu 61.0 min	**ASTM** B121 (353); B453 (353) **FED** QQ-B-613; QQ-B-626
C35330	63/37 Leaded Brass	**As** 0.02-0.25 **Cu** 60.5-64.0 **Pb** 1.5-2.5 **Zn** rem **Other** Lead content may be reduced to 1% by agreement between purchaser and supplier.	**ASTM** B121
C35340	Leaded Brass	**Cu** 60.0-63.0 **Fe** 0.10-0.30 **Pb** 1.5-2.5 **Zn** rem	
C35600	Leaded Brass Alloy	**Cu** 60.0-63.0 **Fe** 0.10 max **Pb** 2.0-3.0 **Zn** rem	**ASTM** B121 (356)
C36000	Free Cutting Brass	**Cu** 60.0-63.0 **Fe** 0.35 max **Pb** 2.5-3.7 **Zn** rem	**AMS** 4610 **ASTM** B16 (360) **FED** QQ-B-626 **SAE** J461 (CA360); J463 (CA360)
C36200	Leaded Brass	**Cu** 60.0-63.0 **Fe** 0.15 max **Pb** 3.5-4.5 **Zn** rem	
C36500	Leaded Muntz Metal, Uninhibited (Cu-Zn-Pb)	**Cu** 58.0-61.0 **Fe** 0.15 max **Pb** 0.25-0.7 **Sn** 0.25 max **Zn** rem	**ASME** SB171 **ASTM** B171 (365); B432 (365)
C36600	Leaded Muntz Metal, Arsenical (Cu-Zn-Pb)	**As** 0.02-0.10 **Cu** 58.0-61.0 **Fe** 0.15 max **Pb** 0.25-0.7 **Sn** 0.25 max **Zn** rem	**ASME** SB171 **ASTM** B432 (366)
C36700	Leaded Muntz Metal, Antimonial (Cu-Zn-Pb)	**Cu** 58.0-61.0 **Fe** 0.15 max **Pb** 0.25-0.7 **Sb** 0.02-0.10 **Sn** 0.25 max **Zn** rem	**ASME** SB171 **ASTM** B432 (367)
C36800	Leaded Muntz Metal, Phosphorized (Cu-Zn-Pb)	**Cu** 58.0-61.0 **Fe** 0.15 max **P** 0.02-0.10 **Pb** 0.25-0.7 **Sn** 0.25 max **Zn** rem	**ASME** SB171 **ASTM** B432 (368)
C37000	Free Cutting Muntz Metal (Cu-Zn-Pb)	**Cu** 58.0-61.0 **Fe** 0.15 max **Pb** 0.8-1.5 **Zn** rem	**ASTM** B135 (370) **MIL SPEC** MIL-T-46072

*Boxed entries are no longer active and are retained for reference purposes only.

The chemical compositions listed are for identification purposes and should not be used in lieu of the cross referenced specifications.

COPPER AND COPPER ALLOYS

UNIFIED NUMBER	DESCRIPTION	CHEMICAL COMPOSITION	CROSS REFERENCE SPECIFICATIONS
C37100	Leaded Brass	**Cu** 58.0-62.0 **Fe** 0.15 max **Pb** 0.6-1.2 **Zn** rem	
C37700	Forging Brass	**Cu** 58.0-61.0 **Fe** 0.30 max **Pb** 1.5-2.5 **Zn** rem	**AMS** 4614 **ASME** SB283 **ASTM** B124 (377); B283 (377) **FED** QQ-B-626 **MIL SPEC** MIL-C-13351 **SAE** J461 (CA377); J463 (CA377)
C37710	2% Leaded 60/40 (Forging) Brass	**Cu** 56.5-60.0 **Fe** 0.30 max **Pb** 1.0-2.5	
C37800	Leaded Brass	**Cu** 56.0-59.0 **Fe** 0.30 max **Pb** 1.5-2.5 **Zn** Rem	
C38000	Leaded Brass	**Al** 0.50 max **Cu** 55.0-60.0 **Fe** 0.35 max **Pb** 1.5-2.5 **Sn** 0.30 max **Zn** rem	**ASTM** B455 (380)
C38010	Section Brass	**Al** 0.10-0.6 **Cu** 55.0-60.0 **Fe** 0.30 max **Pb** 1.5-3.0	
C38500	Architectural Bronze	**Cu** 56.0-59.0 **Fe** 0.35 max **Pb** 2.5-3.5 **Zn** rem	**ASTM** B455 (385)
C38510	Free Cutting Brass	**Cu** 56.0-60.0 **Pb** 2.5-4.5 **Zn** rem	
C38590	Architectural Brass	**Cu** 56.5-60.0 **Fe** 0.35 max **Pb** 2.0-3.5 **Zn** Rem	
* C38600	Copper	**Cu** 56.0-59.0 **Fe** 0.35 max **Pb** 2.5-3.5 **Sb** 0.02 max **Zn** rem	
C40400	Tin Brass	**Cu** rem **Sn** 0.35-0.7 **Zn** 2.0-3.0	
C40500	Tin Brass	**Cu** 94.0-96.0 **Fe** 0.05 max **Pb** 0.05 max **Sn** 0.7-1.3 **Zn** rem	**ASTM** B591 (405)
C40800	Tin Brass	**Cu** 94.0-96.0 **Fe** 0.05 max **Pb** 0.05 max **Sn** 1.8-2.2 **Zn** rem	**ASTM** B591 (408)
* C40900	Tin Brass	**Cu** 92.0-94.0 **Fe** 0.05 max **Pb** 0.05 max **Sn** 0.50-0.8 **Zn** rem	
C41000	Tin Brass	**Cu** 91.0-93.0 **Fe** 0.05 max **Pb** 0.05 max **Sn** 2.0-2.8 **Zn** rem	
C41100	Tin Brass	**Cu** 89.0-92.0 **Fe** 0.05 max **Pb** 0.10 max **Sn** 0.30-0.7 **Zn** rem	**ASTM** B508 (411); B591 (411) **MIL SPEC** MIL-B-13501 **SAE** J459 (795); J460 (795)
C41300	Tin Brass	**Cu** 89.0-93.0 **Fe** 0.05 max **Pb** 0.10 max **Sn** 0.7-1.3 **Zn** rem	**ASTM** B591 (413)
C41500	Tin Brass	**Cu** 89.0-93.0 **Fe** 0.05 max **Pb** 0.10 max **Sn** 1.5-2.2 **Zn** rem	**ASTM** B591 (415)
* C41900	Tin Brass	**Cu** 89.0-92.0 **Fe** 0.05 max **Pb** 0.10 max **Sn** 4.8-5.5 **Zn** rem	
C42000	Tin Brass	**Cu** 88.0-91.0 **P** 0.25 max **Sn** 1.5-2.0 **Zn** rem	
C42100	Tin Brass	**Cu** 87.5-89.0 **Fe** 0.05 max **Mn** 0.15-0.35 **P** 0.35 max **Pb** 0.05 max **Sn** 2.2-3.0 **Zn** rem	
C42200	Tin Brass	**Cu** 86.0-89.0 **Fe** 0.05 max **P** 0.35 max **Pb** 0.05 max **Sn** 0.8-1.4 **Zn** rem	**ASTM** B591 (422)
C42500	Tin Brass	**Cu** 87.0-90.0 **Fe** 0.05 max **P** 0.35 max **Pb** 0.05 max **Sn** 1.5-3.0 **Zn** rem	**ASTM** B591 (425)

*Boxed entries are no longer active and are retained for reference purposes only.

The chemical compositions listed are for identification purposes and should not be used in lieu of the cross referenced specifications.

UNIFIED NUMBER	DESCRIPTION	CHEMICAL COMPOSITION	CROSS REFERENCE SPECIFICATIONS
C43000	Tin Brass	**Cu** 84.0-87.0 **Fe** 0.05 max **Pb** 0.10 max **Sn** 1.7-2.7 **Zn** rem	**ASTM** B591 (430)
C43200	Tin Brass	**Cu** 85.0-88.0 **Fe** 0.05 max **P** 0.35 max **Pb** 0.05 max **Sn** 0.40-0.6 **Zn** rem	
C43400	Tin Brass	**Cu** 84.0-87.0 **Fe** 0.05 max **Pb** 0.05 max **Sn** 0.40-1.0 **Zn** rem	**ASTM** B591 (434)
C43500	Tin Brass	**Cu** 79.0-83.0 **Fe** 0.05 max **Pb** 0.10 max **Sn** 0.6-1.2 **Zn** rem	
C43600	Tin Brass	**Cu** 80.0-83.0 **Fe** 0.05 max **Pb** 0.05 max **Sn** 0.20-0.50 **Zn** rem	
* C43800	Tin Brass	**Cu** 79.0-82.0 **Fe** 0.05 max **Pb** 0.05 max **Sn** 1.0-1.5 **Zn** rem	
C44300	Admiralty, Arsenical (Cu-Zn-Sn)	**As** 0.02-0.06 **Cu** 70.0-73.0 **Fe** 0.06 max **Pb** 0.07 max **Sn** 0.8-1.2 **Zn** rem **Other** For tubular products, Sn 0.9 min	**ASME** SB111; SB171; SB359; SB395; SB543 **ASTM** B111 (443); B171 (443); B359 (443); B395 (443); B432 (443); B543 (443)
C44400	Admiralty, Antimonial (Cu-Zn-Sn)	**Cu** 70.0-73.0 **Fe** 0.06 max **Pb** 0.07 max **Sb** 0.02-0.10 **Sn** 0.8-1.2 **Zn** rem **Other** For tubular products, Sn 0.9 min	**ASME** SB111; SB171; SB359; SB395; SB543 **ASTM** B111 (444); B171 (444); B359 (444); B395 (444); B432 (444); B543 (443)
C44500	Admiralty, Phosphorized (Cu-Zn-Sn)	**Cu** 70.0-73.0 **Fe** 0.06 max **P** 0.02-0.10 **Pb** 0.07 max **Sn** 0.8-1.2 **Zn** rem **Other** For tubular products, Sn 0.9 min	**ASME** SB111; SB171; SB359; SB395; SB543 **ASTM** B111 (445); B171 (445); B359 (445); B395 (445); B432 (445); B543 (445)
C46200	Naval Brass, 63-1/2%	**Cu** 62.0-65.0 **Fe** 0.10 max **Pb** 0.20 max **Sn** 0.50-1.0 **Zn** rem	**ASTM** B21 (462) **FED** QQ-B-637; QQ-B-639
C46210	Copper	**Al** 0.03 max **Cu** 61.0-64.0 **Pb** 0.05 max **Si** 0.50 max **Sn** 1.0 max **Zn** rem	
C46400	Naval Brass, Uninhibited	**Cu** 59.0-62.0 **Fe** 0.10 max **Pb** 0.20 max **Sn** 0.50-1.0 **Zn** rem	**AMS** 4611; 4612 **ASME** SB171 **ASTM** B21 (464); B124 (464); B171 (464); B283 (464); B432 (464) **FED** QQ-B-637; QQ-B-639 **SAE** J461 (CA464); J463 (CA464)
C46420	Naval Brass	**Cu** 61.0-63.5 **Fe** 0.20 max **Pb** 0.20 max **Sn** 1.0-1.4 **Zn** rem	
C46500	Naval Brass, Arsenical	**As** 0.02-0.06 **Cu** 59.0-62.0 **Fe** 0.10 max **Pb** 0.20 max **Sn** 0.50-1.0 **Zn** rem	**ASME** SB171 **ASTM** B432 (465) **SAE** J461 (CA465); J463 (CA465)
C46600	Naval Brass, Antimonial	**Cu** 59.0-62.0 **Fe** 0.10 max **Pb** 0.20 max **Sb** 0.02-0.10 **Sn** 0.50-1.0 **Zn** rem	**ASME** SB171 **ASTM** B432 (465) **SAE** J461 (CA466); J463 (CA466)
C46700	Naval Brass, Phosphorized	**Cu** 59.0-62.0 **Fe** 0.10 max **P** 0.02-0.10 **Pb** 0.20 max **Sn** 0.50-1.0 **Zn** rem	**ASME** SB171 **ASTM** B432 (467) **FED** QQ-B-626 **SAE** J461 (CA467); J463 (CA467)
C47000	Naval Brass, Welding and Brazing Rod	**Al** 0.01 max **Cu** 57.0-61.0 **Pb** 0.05 max **Sn** 0.25-1.0 **Zn** rem	**ASME** SFA5.8 (RBCuZn-A); SFA5.27 (RBCuZn-A); **AWS** A5.8 (RBCuZn-A); A5.27 (RBCuZn-A) **FED** QQ-R-571 (RBCuZn-A)
* C47200	Brazing Alloy (Cu-Zn-Sn)	**Cu** 49.0-52.0 **Fe** 0.10 max **Pb** 0.50 max **Sn** 3.0-4.0 **Zn** rem	**FED** QQ-B-650
C47600	Tin Brass, Leaded	**Cu** 86.0-88.0 **Fe** 0.05 max **Mn** 0.05-0.15 **P** 0.03-0.07 **Pb** 1.8-2.2 **Sn** 1.8-2.2 **Zn** rem	
C47940	Brass Alloy	**Cu** 63.0-66.0 **Fe** 0.10-1.0 **Ni** 0.10-0.50 **Pb** 1.0-2.0 **Sn** 1.2-2.0 **Zn** rem **Other** Includes Co	**ASTM** B21
C48200	Naval Brass, Medium Leaded	**Cu** 59.0-62.0 **Fe** 0.10 max **Pb** 0.40-1.0 **Sn** 0.50-1.0 **Zn** rem	**ASTM** B21 (482) **FED** QQ-B-637; QQ-B-639

*Boxed entries are no longer active and are retained for reference purposes only.

The chemical compositions listed are for identification purposes and should not be used in lieu of the cross referenced specifications.

COPPER AND COPPER ALLOYS

UNIFIED NUMBER	DESCRIPTION	CHEMICAL COMPOSITION	CROSS REFERENCE SPECIFICATIONS
C48500	Naval Brass, High Leaded	**Cu** 59.0-62.0 **Fe** 0.10 max **Pb** 1.3-2.2 **Sn** 0.50-1.0 **Zn** rem	**ASTM** B21 (485); B124 (485); B283 (485) **FED** QQ-B-637; QQ-B-639
*C48510	Leaded Naval Brass, Arsenical	**As** 0.02-0.25 **Cu** 59.0-62.0 **Pb** 1.0-2.5 **Sn** 0.7-1.5 **Zn** rem	
C48600	Naval Brass, Arsenical, Leaded	**As** 0.02-0.25 **Cu** 59.0-62.0 **Pb** 1.0-2.5 **Sn** 0.8-1.5 **Zn** rem	
C48650	Replaced by C48600		
C49080	Brass Brazing Filler Metal	**Al** 0.10 max **Cu** 49.0-52.0 **Pb** 0.50 max **Sn** 3.0-4.0 **Zn** rem	**ASME** SFA5.8 (BCuZn-F) **AWS** A5.8 (BCuZn-F)
C50100	Phosphor Bronze	**Cu** 99.5 min **Fe** 0.05 max **P** 0.01-0.05 **Pb** 0.05 max **Sn** 0.50-0.8 **Other** P + Sn included in Cu	
C50200	Phosphor Bronze	**Cu** 99.5 min **Fe** 0.10 max **P** 0.04 max **Pb** 0.05 max **Sn** 1.0-1.5 **Other** P + Sn included in Cu	**ASTM** B105 (502)
C50500	Phosphor Bronze, 1.25% E	**Cu** 99.5 min **Fe** 0.10 max **P** 0.03-0.35 **Pb** 0.05 max **Sn** 1.0-1.7 **Zn** 0.30 max **Other** P + Sn included in Cu	**ASTM** B9; B105; B508 (505)
C50700	Phosphor Bronze	**Cu** 99.5 min **Fe** 0.10 max **P** 0.30 max **Pb** 0.05 max **Sn** 1.5-2.0 **Other** P + Sn included in Cu	**ASTM** B105 (507)
C50710	Phosphor Bronze Alloy (MF 202)	**Cu** rem **Ni** 0.10-0.40 **P** 0.15 max **Sn** 1.7-2.3	
C50715	Phosphor Bronze Alloy	**Cu** rem **Fe** 0.05-0.15 **P** 0.025-0.04 **Pb** 0.02 max **Sn** 1.7-2.3 **Other** Cu + Sn + Pb + P = 99.5 min.	
C50800	Phosphor Bronze	**Cu** 99.5 min **Fe** 0.10 max **P** 0.01-0.07 **Pb** 0.05 max **Sn** 2.6-3.4 **Other** P + Sn included in Cu	**ASTM** B105 (508)
C50900	Phosphor Bronze	**Cu** 99.5 min **Fe** 0.10 max **P** 0.03-0.30 **Pb** 0.05 max **Sn** 2.5-3.8 **Zn** 0.30 max **Other** P + Sn included in Cu	
C51000	Phosphor Bronze, 5% A	**Cu** 99.5 min **Fe** 0.10 max **P** 0.03-0.35 **Pb** 0.05 max **Sn** 4.2-5.8 **Zn** 0.30 max **Other** P + Sn included in Cu	**AMS** 4510; 4625; 4720 **ASTM** B100 (510); B103 (510); B139 (510); B159 (510) **FED** QQ-B-750; QQ-W-321 **MIL SPEC** MIL-T-3595; MIL-W-6712; MIL-B-13501 **SAE** J461 (CA510); J463 (CA510)
C51100	Phosphor Bronze	**Cu** 99.5 min **Fe** 0.10 max **P** 0.03-0.35 **Pb** 0.05 max **Sn** 3.5-4.9 **Zn** 0.30 max **Other** P + Sn included in Cu	**ASTM** B100 (511); B103 (511) **SAE** J461 (CA511); J463 (CA511)
C51800	Phosphor Bronze	**Al** 0.01 max **Cu** 99.5 min **P** 0.10-0.35 **Pb** 0.02 max **Sn** 4.0-6.0 **Other** P + Sn included in Cu	**ASME** SFA5.7 (ERCuSn-A) **AWS** A5.7 (ERCuSn-A); A5.13 (ERCuSn-A) **FED** QQ-R-571
C51900	Phosphor Bronze	**Cu** 99.5 min **Fe** 0.10 max **P** 0.03-0.35 **Pb** 0.05 max **Sn** 5.0-7.0 **Zn** 0.30 max **Other** P + Sn included in Cu	
C52100	Phosphor Bronze, 8% C	**Cu** 99.5 min **Fe** 0.10 max **P** 0.03-0.35 **Pb** 0.05 max **Sn** 7.0-9.0 **Zn** 0.20 max **Other** P + Sn included in Cu	**ASTM** B103 (521); B139 (521); B159 (521) **FED** QQ-R-571 (RCuSn-C) **MIL SPEC** MIL-E-21659 (MIL-CuSn-C); MIL-E-23765 (MIL-CuSn-C) **SAE** J461 (CA521); J463 (CA521)
C52400	Phosphor Bronze, 10% D	**Cu** 99.5 min **Fe** 0.10 max **P** 0.03-0.35 **Pb** 0.05 max **Sn** 9.0-11.0 **Zn** 0.20 max **Other** P + Sn included in Cu	**ASTM** B103 (524); B139 (524); B159 (524) **AWS** A5.13 (RCuSn-D) **FED** QQ-B-750 **SAE** J461 (CA524); J463 (CA524)

*Boxed entries are no longer active and are retained for reference purposes only.

The chemical compositions listed are for identification purposes and should not be used in lieu of the cross referenced specifications.

UNIFIED NUMBER	DESCRIPTION	CHEMICAL COMPOSITION	CROSS REFERENCE SPECIFICATIONS
* C52600	Phosphor Bronze	**Cu** 99.5 min **Fe** 0.10 max **Mn** 1.0-2.0 **P** 0.03-0.35 **Pb** 0.05 max **Sn** 2.2-3.3 **Zn** 0.20 max **Other** Mn + P + Sn included in Cu	
* C52900	Phosphor Bronze	**Cu** 99.5 min **Fe** 0.10 max **Mn** 1.0-2.0 **P** 0.03-0.35 **Pb** 0.05 max **Sn** 7.0-9.0 **Zn** 0.20 max **Other** Mn + P + Sn included in Cu	
* C53200	Phosphor Bronze B	**Cu** 99.5 min **Fe** 0.10 max **P** 0.03-0.35 **Pb** 2.5-4.0 **Sn** 4.0-5.5 **Zn** 0.20 max **Other** P + Pb + Sn included with Cu	**ASTM** B103 (532)
C53400	Phosphor Bronze B-1	**Cu** 99.5 min **Fe** 0.10 max **P** 0.03-0.35 **Pb** 0.8-1.2 **Sn** 3.5-5.8 **Zn** 0.30 max **Other** P + Pb + Sn included in Cu	**ASTM** B103 (534); B139 (534)
C54400	Phosphor Bronze B-2	**Cu** 99.5 min **Fe** 0.10 max **P** 0.01-0.50 **Pb** 3.5-4.5 **Sn** 3.5-4.5 **Zn** 1.5-4.5 **Other** P + Pb + Sn + Zn included in Cu	**AMS** 4520 **ASTM** B103 (544); B139 (544) **FED** QQ-B-750 **SAE** J459 (791); J460 (791); J461 (CA544); J463 (CA544)
* C54600	Phosphor Bronze B-2 (P 0.50 max)	**Cu** 99.5 min **Fe** 0.10 max **P** 0.50 max **Pb** 3.5-4.5 **Sn** 3.5-4.5 **Zn** 1.5-4.5 **Other** P + Pb + Sn + Zn included in Cu	**MIL SPEC** MIL-B-13501
C54800	Leaded Phosphor Bronze	**Cu** 99.5 min **Fe** 0.10 max **P** 0.03-0.35 **Pb** 4.0-6.0 **Sn** 4.0-6.0 **Zn** 0.30 max **Other** P + Pb + Sn + Zn included in Cu	
C55180	Copper-Phosphorus Brazing Filler Metal	**Cu** rem **P** 4.8-5.2 **Other** Cu + P = 99.85 min	**ASME** SFA5.8 (BCuP-1) **AWS** A5.8 (BCuP-1) **FED** QQ-B-650 (FS-BCuP-1)
C55181	Copper-Phosphorus Brazing Filler Metal	**Cu** rem **P** 7.0-7.5 **Other** Cu + P = 99.85 min	**ASME** SFA5.8 (BCuP-2) **AWS** A5.8 (BCuP-2) **FED** QQ-B-650 (FS-BCuP-2)
C55280	Cu-Ag-P Brazing Filler Metal	**Ag** 1.8-2.2 **Cu** rem **P** 6.8-7.2 **Other** Cu + all named elements = 99.85 min	**ASME** SFA5.8 (BCuP-6) **AWS** A5.8 (BCuP-6)
C55281	Cu-Ag-P Brazing Filler Metal	**Ag** 4.8-5.2 **Cu** rem **P** 5.8-6.2 **Other** Cu + all named elements = 99.85 min	**ASME** SFA5.8 (BCuP-3) **AWS** A5.8 (BCuP-3) **FED** QQ-B-650 (FS-BCuP-3) **MIL SPEC** MIL-B-7883 (BCuP-3)
C55282	Cu-Ag-P Brazing Filler Metal	**Ag** 4.8-5.2 **Cu** rem **P** 6.5-7.0 **Other** Cu + all named elements = 99.85 min	**ASME** SFA5.8 (BCuP-7) **AWS** A5.8 (BCuP-7)
C55283	Cu-Ag-P Brazing Filler Metal	**Ag** 5.8-6.2 **Cu** rem **P** 7.0-7.5 **Other** Cu + all named elements = 99.85 min	**ASME** SFA5.8 (BCuP-4) **AWS** A5.8 (BCuP-4) **FED** QQ-B-650 (FS-BCuP-4)
C55284	Cu-Ag-P Brazing Filler Metal	**Ag** 14.5-15.5 **Cu** rem **P** 4.8-5.2 **Other** Cu + all named elements = 99.85 min	**ASME** SFA5.8 (BCuP-5) **AWS** A5.8 (BCuP-5) **FED** QQ-B650 (FS-BCuP-5); QQ-B-654 (III); QQ-S-561 (Class 3) **MIL SPEC** MIL-B-7883 (BCuP-5)
C60600	Aluminum Bronze	**Al** 4.0-7.0 **Fe** 0.50 max	**ASTM** B169 (606) **FED** QQ-C-450; QQ-C-465
C60700	Aluminum Bronze	**Al** 2.3-2.9 **Pb** 0.01 max **Sn** 1.7-2.0	**ASTM** B105 (607)
C60800	Aluminum Bronze	**Al** 5.0-6.5 **As** 0.2-0.35 **Fe** 0.10 max **Pb** 0.10 max	**ASME** SB111; SB359; SB395 **ASTM** B111 (608); B359 (608); B395 (608) **SAE** J461 (CA608); J463 (CA608)
C61000	Aluminum Bronze	**Al** 6.0-8.5 **Fe** 0.50 max **Pb** 0.02 max **Si** 0.10 max **Zn** 0.20 max	**ASME** SB169; SFA5.7 (ERCuAl-A1) **ASTM** B169 (610) **AWS** A5.7 (ERCuAl-A1) **FED** QQ-C-450 **MIL SPEC** MIL-E-21659 (CuAl-A1)

*Boxed entries are no longer active and are retained for reference purposes only.

The chemical compositions listed are for identification purposes and should not be used in lieu of the cross referenced specifications.

UNIFIED NUMBER	DESCRIPTION	CHEMICAL COMPOSITION	CROSS REFERENCE SPECIFICATIONS
C61300	Aluminum Bronze	**Al** 6.0-7.5 **Fe** 2.0-3.0 **Mn** 0.20 max **Ni** 0.15 max **P** 0.015 max **Pb** 0.01 max **Si** 0.10 max **Sn** 0.20-0.50 **Zn** 0.10 max **NOTE:** When the product is for subsequent welding applications and is so specified, the Cr, Cd, Zr, and Zn limits shall be 0.05 max each.	**CDA** 613 **FED** QQ-C-450
C61400	Aluminum Bronze D	**Al** 6.0-8.0 **Fe** 1.5-3.5 **Mn** 1.0 max **P** 0.015 max **Pb** 0.01 max **Zn** 0.20 max	**ASME** SB150; SB169; SB171 **ASTM** B150 (614); B169 (614); B171 (614); B432 (614) **FED** QQ-C-450; QQ-C-465 **SAE** J461 (CA614); J463 (CA614)
C61470	Aluminum Bronze Welding Electrode replaced by W60614		
C61500	Aluminum Bronze	**Al** 7.7-8.3 **Ni** 1.8-2.2 **Pb** 0.015 max	
C61550	Aluminum Bronze	**Al** 5.5-6.5 **Cu** rem **Fe** 0.20 max **Mn** 1.0 max **Ni** 1.5-2.5 **Pb** 0.05 max **Sn** 0.05 max **Zn** 0.8 max **Other** Cu + all named elements = 99.5 min	
C61800	Aluminum Bronze	**Al** 8.5-11.0 **Fe** 0.50-1.5 **Pb** 0.02 max **Si** 0.10 max **Zn** 0.02 max	**ASME** SFA5.7 (ERCuAl-A2) **AWS** A5.7 (ERCuAl-A2) **FED** QQ-R-571 **MIL SPEC** MIL-E-21659 (MIL-CuAl-A2); MIL-E-23765 (MIL-CuAl-A2) **SAE** J461 (CA618); J463 (CA618)
C61810	Aluminum Bronze	**Al** 8.5-11.0 **Fe** 0.50-1.5 **Pb** 0.02 max **Si** 0.10 max **Zn** 0.02 max	
C61900	Aluminum Bronze	**Al** 8.5-10.0 **Fe** 3.0-4.5 **Pb** 0.02 max **Sn** 0.6 max **Zn** 0.8 max	**ASTM** B129 (619); B150 (619); B283 (619) **MIL SPEC** MIL-E-23765 (MIL-CuAl)
C62200	Aluminum Bronze	**Al** 11.0-12.0 **Fe** 3.0-4.2 **Pb** 0.02 max **Si** 0.10 max **Zn** 0.02 max	**FED** QQ-R-571 (RCuAl-B) **MIL SPEC** MIL-E-21659 (MIL-CuAl-B)
C62300	Aluminum Bronze	**Al** 8.5-10.0 **Fe** 2.0-4.0 **Mn** 0.50 max **Ni** 1.0 max **Si** 0.25 max **Sn** 0.6 max	**AMS** 4635 **ASME** SB150 **ASTM** B124 (623); B150 (623); B283 (623) **MIL SPEC** MIL-B-16166 **SAE** J461 (CA623); J463 (CA623)
C62400	Aluminum Bronze	**Al** 10.0-11.5 **Fe** 2.0-4.5 **Mn** 0.30 max **Si** 0.25 max **Sn** 0.20 max	**ASME** SFA5.7 (ERCuAl-A3) **AWS** A5.7 (ERCuAl-A3) **SAE** J461 (CA624); J463 (CA624)
C62500	Aluminum Bronze	**Al** 12.5-13.5 **Fe** 3.5-5.0 **Mn** 2.0 max	
C62580	Aluminum Bronze Welding Filler Metal	**Al** 12.0-13.0 **Fe** 3.0-5.0 **Pb** 0.02 max **Si** 0.04 max **Zn** 0.02 max	**ASME** SFA5.13 (RCuAl-C) **AWS** A5.13 (RCuAl-C)
C62581	Aluminum Bronze Welding Filler Metal	**Al** 13.0-14.0 **Fe** 3.0-5.0 **Pb** 0.02 max **Si** 0.04 max **Zn** 0.02 max	**ASME** SFA5.13 (RCuAl-D) **AWS** A5.13 (RCuAl-D)
C62582	Aluminum Bronze Welding Filler Metal	**Al** 14.0-15.0 **Fe** 3.0-5.0 **Pb** 0.02 max **Si** 0.04 max **Zn** 0.02 max	**ASME** SFA5.13 (RCuAl-E) **AWS** A5.13 (RCuAl-E)
C62730	Aluminum Bronze	**Al** 8.5-11.0 **Fe** 4.0-6.0 **Mg** 0.05 max **Mn** 0.50 max **Ni** 4.0-6.0 **Pb** 0.05 max **Si** 0.10 max **Sn** 0.10 max **Zn** 0.40 max	
C63000	Aluminum Bronze	**Al** 9.0-11.0 **Fe** 2.0-4.0 **Mn** 1.5 max **Ni** 4.0-5.5 **Si** 0.25 max **Sn** 0.20 max **Zn** 0.30 max	**AMS** 4640 **ASME** SB150; SB171 **ASTM** B124 (630); B150 (630); B171 (630); B283 (630) **FED** QQ-C-450; QQ-C-465 **MIL SPEC** MIL-B-16166 **SAE** J461 (CA630); J463 (CA630)
C63010	Aluminum Bronze	**Al** 9.7-10.9 **Cu** 78.0 min **Fe** 2.0-3.5 **Mn** 1.5 max **Ni** 4.5-5.5 **Sn** 0.20 max **Zn** 0.30 max **NOTE:** Cu content includes Ag, Ni content includes Co, also Cu + named elements = 99.8 min.	
C63200	Aluminum Bronze	**Al** 8.5-9.5 **Fe** 3.0-5.0 **Mn** 3.5 max **Ni** 4.0-5.5 **Pb** 0.02 max **Si** 0.10 max	
C63230	Aluminum Bronze	**Al** 8.5-9.5 **Cu** 75.9-84.5 **Fe** 3.0-5.0 **Mn** 3.5 max **Ni** 4.0-5.5 **Pb** 0.02 max **Si** 0.10 max **Other** Fe content not to exceed Ni content	

The chemical compositions listed are for identification purposes and should not be used in lieu of the cross referenced specifications.

UNIFIED NUMBER	DESCRIPTION	CHEMICAL COMPOSITION	CROSS REFERENCE SPECIFICATIONS
C63280	Nickel Aluminum Bronze Welding Filler Metal	**Al** 8.5-9.5 **Fe** 3.0-5.0 **Mn** 0.6-3.5 **Ni** 4.0-5.5 **Pb** 0.02 max **Si** 0.10 max **Zn** 0.10 max	**ASME** SFA5.7 (ERCuNiAl) **AWS** A5.7 (ERCuNiAl) **MIL SPEC** MIL-E-23765/3, CuNiAl
C63380	Manganese Aluminum Bronze Welding Filler Metal	**Al** 7.0-8.5 **Fe** 2.0-4.0 **Mn** 11.0-14.0 **Ni** 1.5-3.0 **Pb** 0.02 max **Si** 0.10 max **Zn** 0.15 max	**ASME** SFA5.7 (ERCuMnNiAl) **AWS** A5.7 (ERCuMnNiAl) **MIL SPEC** MIL-E-23765/3 (CuMnNiAl)
C63400	Aluminum Bronze	**Al** 2.6-3.2 **As** 0.15 max **Fe** 0.15 max **Ni** 0.15 max **Pb** 0.05 max **Si** 0.25-0.45 **Sn** 0.20 max **Zn** 0.50 max	
C63600	Aluminum Bronze	**Al** 3.0-4.0 **As** 0.15 max **Fe** 0.15 max **Ni** 0.15 max **Pb** 0.05 max **Si** 0.7-1.3 **Sn** 0.20 max **Zn** 0.50 max	
C63800	Aluminum Bronze	**Al** 2.5-3.1 **Co** 0.25-0.55 **Fe** 0.20 max **Mn** 0.10 max **Ni** 0.20 max **Pb** 0.05 max **Si** 1.5-2.1 **Zn** 0.80 max	
C64110	Aluminum Bronze, Free Machining	**Al** 8.0-11.0 **Mn** 0.50 max **Pb** 1.0-2.0	
C64200	Aluminum Bronze	**Al** 6.3-7.6 **As** 0.15 max **Fe** 0.30 max **Mn** 0.10 max **Ni** 0.25 max **Pb** 0.05 max **Si** 1.5-2.2 **Sn** 0.20 max **Zn** 0.50 max	**ASME** SB150 **ASTM** B124 (642); B150 (642); B283 (642) **FED** QQ-C-465 **SAE** J461 (CA462); J463 (CA642)
C64210	Aluminum Silicon Bronze	**Al** 6.3-7.0 **As** 0.15 max **Fe** 0.30 max **Mn** 0.10 max **Ni** 0.25 max **Pb** 0.05 max **Si** 1.5-2.0 **Sn** 0.20 max **Zn** 0.50 max	
C64250	Aluminum Silicon Bronze	**Al** 5.5-7.5 **Fe** 1.0 max **Mn** 0.50 max **Si** 1.5-3.0	
C64400	Aluminum Silicon Bronze	**Al** 3.5-4.5 **Cu** rem **Fe** 0.05 max **Ni** 4.2-5.0 **Pb** 0.03 max **Si** 0.8-1.3 **Sn** 0.10 max **Zn** 0.20 max **Other** includes Ag, includes Co	
C64700	Silicon Bronze	**Cu** rem **Fe** 0.10 max **Ni** 1.6-2.2 **Pb** 0.110 max **Si** 0.40-0.8 **Zn** 0.50 max **Other** Ag included in Cu	**ASTM** B411 (647); B412 (647); B422 (647) **FED** QQ-C-591
C64710		**Cu** 95.0 min **Mn** 0.10 max **Ni** 2.9-3.5 **Si** 0.50-0.9 **Zn** 0.20-0.50 **Other** Ag included in Cu, Co included in Ni, total of specified elements = 99.5 min.	
C64900	Silicon Bronze	**Al** 0.10 max **Cu** rem **Fe** 0.10 max **Ni** 0.10 max **Pb** 0.05 max **Si** 0.8-1.2 **Sn** 1.2-1.6 **Zn** 0.20 max **Other** Ag included in Cu	
C65100	Low Silicon Bronze B	**Cu** rem **Fe** 0.8 max **Mn** 0.7 max **Pb** 0.05 max **Si** 0.8-2.0 **Zn** 1.5 max **Other** Ag included in Cu	**ASME** SB98; SB315 **ASTM** B97 (651); B98 (651); B99 (651); B105 (651); B315 (651); B432 (651) **FED** QQ-C-591
C65300	Silicon Bronze	**Cu** rem **Fe** 0.8 max **Mn** 0.7 max **Pb** 0.05 max **Si** 2.0-2.6 **Other** Ag included in Cu	
C65400	Copper Silicon	**Cr** 0.01-0.12 **Cu** rem **Pb** 0.05 max **Si** 2.7-3.4 **Sn** 1.2-1.9 **Zn** 0.50 max **Other** Ag included in Cu	
C65500	High Silicon Bronze A	**Cu** rem **Fe** 0.8 max **Mn** 0.50-1.3 **Ni** 0.6 max **Pb** 0.05 max **Si** 2.8-3.8 **Zn** 1.5 max **Other** Ag included in Cu	**AMS** 4615; 4665 **ASME** SB96; SB98; SB315 **ASTM** B96 (655); B97 (655); B98 (655); B99 (655); B100 (655); B105 (655); B124 (655); B283 (655); B315 (655); B432 (655) **FED** QQ-C-591 **MIL SPEC** MIL-T-8231; MIL-E-23765 **SAE** J461 (CA655); J463 (CA655)

The chemical compositions listed are for identification purposes and should not be used in lieu of the cross referenced specifications.

UNIFIED NUMBER	DESCRIPTION	CHEMICAL COMPOSITION	CROSS REFERENCE SPECIFICATIONS
C65600	Silicon Bronze	Al 0.01 max Cu rem Fe 0.50 max Mn 1.5 max Pb 0.02 max Si 2.8-4.0 Sn 1.5 max Zn 1.5 max Other Ag included in Cu	ASME SFA5.7 (ERCuSi-A); SFA5.13 (ERCuSi-A); SFA5.27 (ERCuSi-A) AWS A5.7 (ERCuSi-A); A5.13 (ERCuSi-A); A5.27 (ERCuSi-A) FED QQ-R-571 MIL SPEC MIL-E-21659 (MIL-CuSi); MIL-E-23765 (MIL-CuSi)
C65800	Silicon Bronze	Al 0.01 max Cu rem Fe 0.25 max Mn 0.50-1.3 Ni 0.6 max Pb 0.05 max Si 2.8-3.8 Other Ag included in Cu	ASME SB315 ASTM B315 (658)
C66100	Silicon Bronze	Cu rem Fe 0.25 max Mn 1.5 max Pb 0.20-0.8 Si 2.8-3.5 Zn 1.5 max Other Ag included in Cu	ASME SB98 ASTM B98 (661) FED QQ-C-591
C66400	Copper Alloy	Ag 0.05 max Al 0.05 max Co 0.30-0.7 Cu rem Fe 1.3-1.7 Mn 0.05 max Ni 0.05 max P 0.02 max Pb 0.015 max Si 0.05 max Sn 0.05 max Zn 11.0-12.0 Other Co + Fe 1.8-2.3	
C66410	Misc. Copper Zinc Alloy	Al 0.05 max As 0.05 max Cu rem Fe 1.8-2.3 Mn 0.05 max Ni 0.05 max P 0.02 max Pb 0.015 max Si 0.05 max Sn 0.05 max Zn 11.0-12.0 Other Incl Ag	
C66700	Manganese Brass	Cu 68.5-71.5 Fe 0.10 max Mn 0.8-1.5 Pb 0.07 max Zn rem	ASTM B291 (667)
C66800	Manganese Bronze	Al 0.25 max Cu 60.0-63.0 Fe 0.35 max Mn 2.0-3.5 Ni 0.25 max Pb 0.50 max Si 0.50-1.5 Sn 0.30 max Zn rem	
C66900	Manganese Bronze	Cu 62.5-64.5 Fe 0.25 max Mn 11.5-12.5 Pb 0.05 max Zn rem	
C67000	Manganese Bronze B	Al 3.0-6.0 Cu 63.0-68.0 Fe 2.0-4.0 Mn 2.5-5.0 Pb 0.20 max Sn 0.50 max Zn rem	ASTM B138 (670) FED QQ-B-728 SAE J461 (CA670); J463 (CA670)
C67130	Leaded Manganese Bronze	Al 0.10-1.0 Cu 56.0-59.0 Mn 0.50-1.5 Ni 0.50-1.5 Pb 0.50-1.5 Sn 0.50-1.5 Zn rem Other Ag included in Cu; Co included in Ni	
C67300	Manganese Bronze	Al 0.25 max Cu 58.0-63.0 Fe 0.50 max Mn 2.0-3.5 Ni 0.25 max Pb 0.40-3.0 Si 0.50-1.5 Sn 0.30 max Zn rem	SAE J461 (CA673); J463 (CA673)
C67400	Manganese Bronze	Al 0.50-2.0 Cu 57.0-60.0 Fe 0.35 max Mn 2.0-3.5 Ni 0.25 max Pb 0.50 max Si 0.50-1.5 Sn 0.30 max Zn rem	SAE J461 (CA674); J463 (CA674)
C67410	Manganese Bronze	Al 1.3-2.3 Cu 55.5-59.0 Fe 1.0 max Mn 1.0-2.4 Ni 2.0 max Pb 0.8 max Si 0.7-1.3 Sn 0.50 max Zn rem	
C67500	Manganese Bronze A	Al 0.25 max Cu 57.0-60.0 Fe 0.8-2.0 Mn 0.05-0.50 Pb 0.20 max Sn 0.50-1.5 Zn rem	ASTM B124 (675); B138 (675); B283 (675) FED QQ-B-728 SAE J461 (CA675); J463 (CA675)
C67600	Leaded Manganese Bronze	Cu 57.0-60.0 Fe 0.40-1.3 Mn 0.05-0.50 Pb 0.50-1.0 Sn 0.50-1.5 Zn rem	
C67620	Leaded Manganese Bronze	Cu 55.0-57.0 Fe 0.50-1.2 Mn 1.0-2.0 Pb 0.50-1.5 Zn rem Other Ag included in Cu	
C67700	Copper Alloy	As 0.40-0.8 Cu 55.5-58.0 Fe 0.7-1.5 Mn 0.05-0.30 Ni 1.5-2.3 Pb 0.50-1.0 Zn rem	
C67800	Copper Alloy	Al 0.50-1.5 Cu 56.0-59.0 Fe 0.7-1.5 Mn 0.20-0.6 Pb 0.30 max Sn 0.20 max Zn rem	

The chemical compositions listed are for identification purposes and should not be used in lieu of the cross referenced specifications.

UNIFIED NUMBER	DESCRIPTION	CHEMICAL COMPOSITION	CROSS REFERENCE SPECIFICATIONS
C67810	Manganese Bronze	Al 0.40-1.6 Cu 56.5-59.5 Fe 1.0 max Mn 0.40-1.8 Ni 1.5 max Pb 1.0 max Si 0.6 max Sn 0.50 max Zn rem	
C67820	Manganese Bronze	Al 0.30-1.3 Cu 56.0-60.0 Fe 0.50-1.2 N 0.30-2.0 Pb 0.10 max Sn 0.30-1.2 Zn rem Other Ag included in Cu	
C68000	Bronze, Low Fuming (Nickel)	Al 0.01 max Cu 56.0-60.0 Fe 0.25-1.2 Mn 0.01-0.50 Ni 0.20-0.8 Pb 0.05 max Si 0.04-0.15 Sn 0.8-1.1 Zn rem	ASME SFA5.27 (RCuZn-B) AWS A5.27 (RCuZn-B) FED QQ-R-571 (RCuZn-B)
C68100	Bronze, Low Fuming	Al 0.01 max Cu 56.0-60.0 Fe 0.25-1.2 Mn 0.01-0.50 Pb 0.05 max Si 0.04-0.15 Sn 0.8-1.1 Zn rem	ASME SFA5.8 (RBCuZn-C); SFA5.27 (RCuZn-C) AWS A5.8 (RBCuZn-C); A5.27 (RCuZn-C) FED QQ-R-571 (RCuZn-C)
C68200	Copper Alloy	Cu 58.0-60.0 Fe 0.6-1.0 Mn 0.6-1.0 Si 0.07-0.15 Zn rem	
C68600	Leaded Manganese Bronze	Al 0.30-1.5 Cu 56.0-60.0 Fe 0.50-1.2 Mn 0.30-2.0 Pb 0.50-1.5 Sn 0.20-1.0 Zn rem Other Ag included in Cu	
C68700	Aluminum Brass, Arsenical	Al 1.8-2.5 As 0.02-0.06 Cu 76.0-79.0 Fe 0.06 max Pb 0.07 max Zn rem	ASME SB111; SB359; SB395; SB543 ASTM B111 (687); B359 (687); B395 (687); B543 (687)
C68800	Copper Alloy	Al 3.0-3.8 Co 0.25-0.55 Cu rem includes Ag Fe 0.20 max Pb 0.05 max Zn 21.3-24.1 Other Al + Zn 25.1-27.1	ASTM B592 (688)
C69000	Miscellaneous Copper-Zinc	Al 3.0-3.8 Cu 72.0-74.5 Fe 0.05 max Ni 0.50-0.80 Pb 0.025 max Zn rem Other including Ag, including Co	
C69100	Copper Alloy	Al 0.7-1.2 Cu 81.0-84.0 Fe 0.25 max Mn 0.10 max Ni 0.8-1.4 Pb 0.05 max Si 0.8-1.3 Sn 0.10 max Zn rem Other including Co; including Ag	ASTM B111
C69400	Silicon Red Brass	Cu 80.0-83.0 Fe 0.20 max Pb 0.30 max Si 3.5-4.5 Zn rem	ASTM B371 (694)
C69430	Silicon Red Brass	As 0.03-0.06 Cu 80.0-83.0 Fe 0.20 max Pb 0.30 max Si 3.5-4.5 Zn rem	
C69440	Silicon Red Brass	Cu 80.0-83.0 Fe 0.20 max Pb 0.30 max Sb 0.03-0.06 Si 3.5-4.5 Zn rem	
C69450	Silicon Red Brass	Cu 80.0-83.0 Fe 0.20 max Mn 0.40 max P 0.03-0.06 Pb 0.30 max Si 3.5-4.5 Zn rem	
C69700	Leaded Silicon Brass	Cu 75.0-80.0 Fe 0.20 max Mn 0.40 max Pb 0.50-1.5 Si 2.5-3.5 Zn rem	ASTM B371 (697)
C69710	Silicon Brass	As 0.03-0.06 Cu 75.0-80.0 Fe 0.20 max Mn 0.40 max Pb 0.50-1.5 Si 2.5-3.5 Zn Rem	
C69720	Silicon Brass	Cu 75.0-80.0 Fe 0.20 max Mn 0.40 max Pb 0.50-1.5 Sb 0.03-0.06 Si 2.5-3.5	
C69730	Silicon Brass	Cu 75.0-80.0 Fe 0.20 max Mn 0.40 max P 0.03-0.06 Pb 0.50-1.5 Si 2.5-3.5	
C69800	Copper Zinc	Cu 66.0-70.0 Fe 0.4 max Ni 0.50 max Pb 0.8 max Si 0.7-1.3 Zn rem	
C69900	Copper Manganese Alloy	Ag 0.05 max Al 1.4-2.3 As 0.01 max C 0.05 max Cd 0.05 max Co 0.20 max Fe 0.10 max Mn 40.0-48.0 Ni 0.10 max Pb 0.02 max Zn 0.14 max Other Cu + all named elements 99.5 min	

The chemical compositions listed are for identification purposes and should not be used in lieu of the cross referenced specifications.

UNIFIED NUMBER	DESCRIPTION	CHEMICAL COMPOSITION	CROSS REFERENCE SPECIFICATIONS
C69910	Brazing Alloy (Cu-Mn-Fe)	**Al** 0.25-0.8 **Cu** rem **Fe** 1.0-1.4 **Mn** 28.0-32.0 **Pb** 0.01 max **Zn** 3.0-5.0 **Other** Cu + all named elements 99.5 min	
C69950	Miscellaneous Copper-Zinc	**Cu** 51.0-54.0 **Mn** 36.0-40.0 **Ni** 8.5-10.5 **Other** Cu + all named elements = 99.5 min	**AMS** 4764
C70100	Copper-Nickel Alloy	**Cu** 99.7 min **Fe** 0.05 max **Mn** 0.50 max **Ni** 3.0-4.0 **Zn** 0.25 max **Other** All named elements included in Cu	
C70200	Copper-Nickel	**Cu** 99.7 min **Fe** 0.10 max **Mn** 0.40 max **Ni** 2.0-3.0 **Pb** 0.05 max **Other** All named elements included in Cu	**MIL SPEC** MIL-B-18907; MIL-B-20292
* C70300	Copper-Nickel Alloy	**Cu** 99.5 min **Fe** 0.05 max **Mn** 0.50 max **Ni** 4.7-5.7 **Other** Fe + Mn + Ni included in Cu	
C70320	Copper-Nickel	**Al** 0.20-1.2 **Cr** 0.18-0.50 **Cu** rem **Ni** 2.5-5.0 **Si** 0.20-1.2	
C70400	Copper-Nickel, 5%	**Cu** 99.5 min **Fe** 1.3-1.7 **Mn** 0.30-0.8 **Ni** 4.8-6.2 **Pb** 0.05 max **Zn** 1.0 max **Other** All named elements included in Cu	**ASME** SB111; SB359; SB395; SB466; SB543 **ASTM** B111 (704); B359 (704); B395 (704); B466 (704); B543 (704)
C70440	95/5 Copper-Nickel	**C** 0.05 max **Cu** rem **Fe** 1.0-1.8 **Mn** 0.50-1.5 **Ni** 4.5-6.0 **Pb** 0.05 max **S** 0.05 max **Si** 0.35-0.45 **Sn** 0.10 max **Zn** 1.0 max **Other** Ag included in Cu; Co included in Ni	
C70500	Copper-Nickel, 7%	**Cu** 99.5 min **Fe** 0.10 max **Mn** 0.15 max **Ni** 5.8-7.8 **Pb** 0.05 max **Zn** 0.20 max **Other** All named elements included in Cu	
C70600	Copper-Nickel, 10%	**Cu** 86.5 min **Fe** 1.0-1.8 **Mn** 1.0 max **Ni** 9.0-11.0 **Pb** 0.05 max **Zn** 1.0 max **Other** Ag included in Cu; Cu 99.5 min including all named elements. For welded applications: C 0.05 max, P 0.02 max, Pb 0.02 max, S 0.02 max, Zn 0.50 max	**ASME** SB111; SB171; SB359; SB395; SB402; SB466; SB467; SB543 **ASTM** B111 (706); B122 (706); B151 (706); B171 (706); B359 (706); B395 (706); B402 (706); B432 (706); B466 (706); B467 (706); B543 (706); B552 (706) **MIL SPEC** MIL-T-15005; MIL-C-15726; MIL-T-16420; MIL-T-22214 **SAE** J461 (CA706); J463 (CA706)
C70610	90/10 Copper-Nickel	**C** 0.05 max **Cu** rem **Fe** 1.0-2.0 **Mn** 0.50-1.0 **Ni** 10.0-11.0 **Pb** 0.01 max **S** 0.05 max **Other** Ag included in Cu; Co included in Ni	**CDA** C70610
C70690	Copper-Nickel, 90Cu-10Ni	**Al** 0.002 max **As** 0.001 max **Bi** 0.001 max **C** 0.03 max **Co** 0.02 max **Cu** 89.0 min **Fe** 0.005 max **Hg** 0.0005 max **Mn** 0.001 max **Ni** 9.0-11.0 **P** 0.001 max **Pb** 0.001 max **S** 0.003 max **Sb** 0.001 max **Si** 0.02 max **Sn** 0.001 max **Ti** 0.001 max **Zn** 0.001 max **Other** Ag included in Cu; Co included in Ni, Cu + all named elements 99.5 min	**ASTM** F96
C70700	Copper-Nickel	**Cu** 99.5 min **Fe** 0.05 max **Mn** 0.50 max **Ni** 9.5-10.5 **Other** Fe + Mn + Ni included in Cu	
C70800	Copper-Nickel, 11%	**Cu** 99.5 min **Fe** 0.10 max **Mn** 0.15 max **Ni** 10.5-12.5 **Pb** 0.05 max **Zn** 0.20 max **Other** all named elements included in Cu	

*Boxed entries are no longer active and are retained for reference purposes only.

The chemical compositions listed are for identification purposes and should not be used in lieu of the cross referenced specifications.

UNIFIED NUMBER	DESCRIPTION	CHEMICAL COMPOSITION	CROSS REFERENCE SPECIFICATIONS
C70900	Copper-Nickel	**Cu** 99.5 min **Fe** 0.6 max **Mn** 0.6 max **Ni** 13.5-16.5 **Pb** 0.05 max **Zn** 1.0 max **Other** All named elements included in Cu	
C71000	Copper-Nickel, 20%	**Cu** 99.5 min **Fe** 1.0 max **Mn** 1.0 max **Ni** 19.0-23.0 **Pb** 0.05 max **Zn** 1.0 max **Other** All named elements included in Cu	**ASME** SB111; SB359; SB395; SB466; SB467 **ASTM** B111 (710); B122 (710); B206 (710); B359 (710); B395 (710); B466 (710); B467 (710) **SAE** J461 (CA710); J463 (CA710)
C71100	Copper-Nickel	**Cu** 99.5 min **Fe** 0.10 max **Mn** 0.15 max **Ni** 22.0-24.0 **Pb** 0.05 max **Zn** 0.20 max **Other** All named elements included in Cu	
C71110	Copper-Nickel Alloy (77.5 Cu - 22.5 Ni)	**Cu** rem **Mn** 0.35 max **Ni** 21.5-23.5 **S** 0.008 max **Ti** 0.05 max **NOTE:** When the product is for subsequent welding applications and so specified, Zn shall be 0.50 max, Pb 0.02 max, P 0.02 max, and C 0.05 max.	**AMS** 4732
C71300	Copper-Nickel	**Cu** 99.5 min **Fe** 0.20 max **Mn** 1.0 max **Ni** 23.5-26.5 **Pb** 0.05 max **Zn** 1.0 max **Other** All named elements included in Cu	
C71500	Copper-Nickel, 30%	**Cu** 99.5 min **Fe** 0.40-1.0 **Mn** 1.0 max **Ni** 29.0-33.0 **Pb** 0.05 max **Zn** 1.0 max **Other** Ag included in Cu; Cu 99.5 min including all named elements. For welded applications: C 0.05 max, P 0.02 max, Pb 0.02 max, S 0.02 max, Zn 0.50 max	**ASME** SB111; SB171; SB359; SB395; SB402; SB466; SB467; SB543 **ASTM** B111 (715); B122 (715); B151 (715); B171 (715); B359 (715); B395 (715); B402 (715); B432 (715); B466 (715); B467 (715); B543 (715); B552 (715) **MIL SPEC** MIL-T-15005; MIL-C-15726; MIL-T-16420; MIL-T-22214 **SAE** J461 (CA715); J463 (CA715)
C71580	Copper-Nickel	**Al** 0.05 max **C** 0.07 max **Cu** rem **Fe** 0.50 max **Mn** 0.30 max **Ni** 29.0-33.0 **P** 0.03 max **Pb** 0.05 max **S** 0.024 max **Si** 0.15 max **Zn** 0.05 max **Other** includes Ag, includes Co	
C71581	Copper-Nickel Welding Filler Metal	**Cu** rem **Fe** 0.40-0.7 **Mn** 1.0 max **Ni** 29.0-32.0 **P** 0.02 max **Pb** 0.02 max **S** 0.01 max **Si** 0.25 max **Ti** 0.20-0.50	**ASME** SFA5.7 (ERCuNi); SFA5.27 (ERCuNi); SFA5.30 (IN67) **AWS** A5.7 (ERCuNi); A5.27 (ERCuNi); A5.30 (IN67) **FED** QQ-R-571 (RCuNi) **MIL SPEC** MIL-R-19631 (MIL-RCuNi); MIL-E-21562 (EN67, RN67) MIL-I-23413 (MIL-67)
C71590	Copper-Nickel, 70Cu-30Ni	**Al** 0.002 max **As** 0.001 max **Bi** 0.001 max **C** 0.03 max **Co** 0.05 max **Cu** 67.0 min **Fe** 0.005 max **Hg** 0.0005 max **Mn** 0.001 max **Ni** 29.0-33.0 **P** 0.001 max **Pb** 0.001 max **S** 0.003 max **Sb** 0.001 max **Si** 0.02 max **Sn** 0.001 max **Ti** 0.001 max **Zn** 0.001 max **Other** Ag included in Cu; Co included in Ni; Cu + all named elements 99.5 min	**ASTM** F96
C71630	Copper-Nickel	**C** 0.06 max **Cu** rem **Fe** 0.40-1.0 **Mn** 0.50-1.5 **Ni** 30.0-32.0 **Pb** 0.01 max **S** 0.08 max **Other** Co included in Ni	**CDA** C71630
C71640	Copper-Nickel	**C** 0.06 max **Cu** rem **Fe** 1.7-2.3 **Mn** 1.5-2.5 **Ni** 29.0-32.0 **Pb** 0.01 max **S** 0.03 max **Other** Ag included in Cu; Co included in Ni	**ASTM** B111, B395
C71700	Copper-Nickel	**Be** 0.30-0.7 **Cu** 99.5 min **Fe** 0.40-1.0 **Ni** 29.0-33.0 **Other** Be + Fe + Ni included in Cu	
C71900	Copper-Nickel	**C** 0.04 max **Cr** 2.2-3.0 **Cu** rem **Fe** 0.50 max **Mn** 0.20-1.0 **Ni** 28.0-33.0 **P** 0.02 max **Pb** 0.015 max **Si** 0.25 max **Sn** 0.015 max **Ti** 0.01-0.20 **Zn** 0.05 max **Zr** 0.02-0.35 **Other** Ag included in Cu	

The chemical compositions listed are for identification purposes and should not be used in lieu of the cross referenced specifications.

UNIFIED NUMBER	DESCRIPTION	CHEMICAL COMPOSITION	CROSS REFERENCE SPECIFICATIONS
C72150	Copper-Nickel	**C** 0.10 max **Cu** rem **Fe** 0.10 max **Mn** 0.05 max **Ni** 43.0-46.0 **Pb** 0.05 max **Si** 0.50 max **Zn** 0.20 max **Other** Cu + all named elements = 99.5 min	
C72200	Copper-Nickel	**C** 0.03 max **Cr** 0.30-0.7 **Cu** rem **Fe** 0.5-1.0 **Mn** 1.0 max **Ni** 15.0-18.0 **Si** 0.03 max **Ti** 0.03 max **Other** Ag included in Cu; Co included in Ni	
C72400	Copper-Nickel Alloy	**Al** 1.5-2.5 **Cu** rem **Fe** 0.10 max **Mg** 0.05-0.40 **Mn** 1.0 max **Ni** 11.0-15.0 **Pb** 0.05 max **Sn** 0.05 max **Zn** 0.50 max **NOTE:** Cu includes Ag and Ni includes Co; Cu + all named elements = 99.5 min.	
C72500	Copper-Nickel	**Cu** 99.8 min **Fe** 0.6 max **Mn** 0.20 max **Ni** 8.5-10.5 **Pb** 0.05 max **Sn** 1.8-2.8 **Zn** 0.50 max **Other** All named elements included in Cu	
C72600	Copper-Nickel Tin Spinodal Alloy	**Cu** 91.0-93.0 **Fe** 0.20 max **Mn** 0.20 max **Ni** 3.5-4.5 **P** 0.05 max **Pb** 0.05 max **Sn** 3.5-4.5 **Zn** 0.50 max **Other** Ag included in Cu; Co included in Ni	**CDA** C72600
C72700	Copper-Nickel Tin Spinodal Alloy	**Cb** 0.10 max **Cu** rem **Fe** 0.50 max **Mg** 0.15 max **Mn** 0.05-0.30 **Ni** 8.5-9.5 **Pb** 0.02 max **Sn** 5.5-6.5 **Zn** 0.50 max **Other** Ag included in Cu; Co included in Ni; Fe 0.005 max for hot rolling	**CDA** C72700
C72800	Copper-Nickel Tin Spinodal Alloy	**Al** 0.10 max **B** 0.001 max **Bi** 0.001 max **Cb** 0.10-0.30 **Cu** rem **Fe** 0.50 max **Mg** 0.005-0.15 **Mn** 0.05-0.30 **Ni** 9.5-10.5 **P** 0.005 max **Pb** 0.005 max **S** 0.0025 max **Sb** 0.02 max **Si** 0.05 max **Sn** 7.5-8.5 **Ti** 0.01 max **Zn** 1.0 max **Other** Ag included in Cu; Co included in Ni	**CDA** C72800
C72900	Copper-Nickel Tin Spinodal Alloy	**Cb** 0.10 max **Cu** rem **Fe** 0.50 max **Mg** 0.15 max **Mn** 0.05-0.30 **Ni** 14.5-15.5 **Pb** 0.02 **Sn** 7.5-8.5 **Zn** 0.50 max **Other** Ag included in Cu; Co included in Ni; Fe 0.005 max for hot rolling	**CDA** C72900
C73200	Nickel Silver	**Cu** 70.0 min **Fe** 0.6 max **Mn** 1.0 max **Ni** 19.0-23.0 **Pb** 0.05 max **Zn** 3.0-6.0	**ASTM** B122 (732)
C73500	Nickel Silver	**Cu** 70.5-73.5 **Fe** 0.25 max **Mn** 0.50 max **Ni** 16.5-19.5 **Pb** 0.10 max **Zn** rem	**ASTM** B122 (735) **FED** QQ-C-585
C73800	Nickel Silver, 70-12	**Cu** 68.5-71.5 **Fe** 0.25 max **Mn** 0.50 max **Ni** 11.0-13.0 **Pb** 0.05 max **Zn** rem	
C74000	Nickel Silver	**Cu** 69.0-73.5 **Fe** 0.25 max **Mn** 0.50 max **Ni** 9.0-11.0 **Pb** 0.10 max **Zn** rem	**ASTM** B122 (740)
C74300	Nickel Silver	**Cu** 63.0-66.0 **Fe** 0.25 max **Mn** 0.50 max **Ni** 7.0-9.0 **Pb** 0.10 max **Zn** rem	
C74500	Nickel Silver, 65-10	**Cu** 63.5-66.5 **Fe** 0.25 max **Mn** 0.50 max **Ni** 9.0-11.0 **Pb** 0.10 max **Zn** rem **Other** For rod and wire, Pb 0.05 max	**ASTM** B122 (745); B151 (745); B206 (745) **FED** QQ-C-585; QQ-C-586; QQ-W-321
C75200	Nickel Silver, 65-18	**Cu** 63.5-66.5 **Fe** 0.25 max **Mn** 0.50 max **Ni** 16.5-19.5 **Pb** 0.05 max **Zn** rem **Other** Co included in Ni	**ASTM** B122 (752); B151 (752); B206 (752) **CDA** 752 **FED** QQ-C-585; QQ-C-586; QQ-W-321 **SAE** J461 (CA752); J463 (CA752)
C75400	Nickel Silver, 65-15	**Cu** 63.5-66.5 **Fe** 0.25 max **Mn** 0.50 max **Ni** 14.0-16.0 **Pb** 0.10 max **Zn** rem	

The chemical compositions listed are for identification purposes and should not be used in lieu of the cross referenced specifications.

COPPER AND COPPER ALLOYS

UNIFIED NUMBER	DESCRIPTION	CHEMICAL COMPOSITION	CROSS REFERENCE SPECIFICATIONS
C75700	Nickel Silver, 65-12	**Cu** 63.5-66.5 **Fe** 0.25 max **Mn** 0.50 max **Ni** 11.0-13.0 **Pb** 0.05 max **Zn** rem	**ASTM** B151 (757); B206 (757) **FED** QQ-W-321
C75720	12% Nickel Silver	**Cu** 60.0-65.0 **Fe** 0.25 max **Mn** 0.05-0.30 **Ni** 11.0-13.0 **Pb** 0.04 max **Zn** rem **Other** Co included in Ni	**CDA** C75720
C75900	Nickel Silver	**Cu** 60.0-63.0 **Fe** 0.25 max **Mn** 0.50 max **Ni** 17.0-19.0 **Pb** 0.10 max **Zn** rem **Other** Co included in Ni	
C76000	Nickel Silver	**Cu** 60.0-63.0 **Fe** 0.25 max **Mn** 0.50 max **Ni** 7.0-9.0 **Pb** 0.10 max **Zn** rem	
C76100	Nickel Silver	**Cu** 59.0-63.0 **Fe** 0.25 max **Mn** 0.50 max **Ni** 9.0-11.0 **Pb** 0.10 max **Zn** rem	
C76200	Nickel Silver	**Cu** 57.0-61.0 **Fe** 0.25 max **Mn** 0.50 max **Ni** 11.0-13.5 **Pb** 0.10 max **Zn** rem	**ASTM** B122 (762) **FED** QQ-C-585
C76300	Nickel Silver	**Cu** 60.0-64.0 **Fe** 0.50 max **Mn** 0.50 max **Ni** 17.0-19.0 **Pb** 0.50-2.0 **Zn** rem **Other** Co included with Ni	
C76390	Leaded Nickel Silver	**Cu** 59.0-63.0 **Fe** 0.25 max **Mn** 0.50 max **Ni** 23.0-26.0 **Pb** 0.8-1.1 **Sn** 0.40-0.60 **Zn** rem **Other** Co included in Ni	**CDA** C76390
C76400	Nickel Silver	**Cu** 58.5-61.5 **Fe** 0.25 max **Mn** 0.50 max **Ni** 16.5-19.5 **Pb** 0.05 max **Zn** rem	**ASTM** B151 (764); B206 (764) **FED** QQ-C-586; QQ-W-321
C76600	Nickel Silver	**Cu** 55.0-58.0 **Fe** 0.25 max **Mn** 0.50 max **Ni** 11.0-13.5 **Pb** 0.10 max **Zn** rem	**FED** QQ-C-585
C76700	Nickel Silver, 56.5-15	**Cu** 55.0-58.0 **Mn** 0.50 max **Ni** 14.0-16.0 **Zn** rem	
C77000	Nickel Silver, 55-18	**Cu** 53.5-56.5 **Fe** 0.25 max **Mn** 0.50 max **Ni** 16.5-19.5 **Pb** 0.05 max **Zn** rem **Other** Co included in Ni	**ASTM** B122 (770); B151 (770); B206 (770) **FED** QQ-C-585; QQ-C-586; QQ-W-321 **SAE** J461 (CA770); J463 (CA770)
C77010	Nickel Silver	**Cu** 54.0-56.0 **Fe** 0.30 max **Mn** 0.05-0.35 **Ni** 17.0-19.0 **Pb** 0.03 max **Zn** rem **Other** Co included in Ni	**CDA** C77010
C77300	Nickel Brass Welding and Brazing Filler Metal	**Al** 0.01 max **Cu** 46.0-50.0 **Ni** 9.0-11.0 **P** 0.25 max **Pb** 0.05 max **Si** 0.04-0.25 **Zn** rem	**ASME** SFA5.8 (RBCuZn-D); SFA5.27 (RBCuZn-D) **AWS** A5.8 (RBCuZn-D); A5.27 (RBCuZn-D) **FED** QQ-R-571 (RBCuZn-D)
C77310	Nickel Silver	**Al** 0.01 max **Cu** 46.0-56.0 **Mn** 0.50 max **Ni** 9.0-11.0 **P** 0.25 max **Pb** 0.05 **Si** 0.04-0.25 **Zn** rem **Other** Co included in Ni	**CDA** C77310
C77400	Nickel Silver	**Cu** 43.0-47.0 **Ni** 9.0-11.0 **Pb** 0.20 max **Zn** rem	**ASTM** B124 (774); B283 (774)
C77600	Nickel Silver, 43.5-13	**Cu** 42.0-45.0 **Fe** 0.20 max **Mn** 0.25 max **Ni** 12.0-14.0 **Pb** 0.25 max **Sn** 0.15 max **Zn** rem	
C78200	Leaded Nickel Silver	**Cu** 63.0-67.0 **Fe** 0.35 max **Mn** 0.50 max **Ni** 7.0-9.0 **Pb** 1.5-2.5 **Zn** rem	
C78800	Leaded Nickel Silver	**Cu** 63.0-67.0 **Fe** 0.25 max **Mn** 0.50 max **Ni** 9.0-11.0 **Pb** 1.5-2.0 **Zn** rem	
C79000	Leaded Nickel Silver	**Cu** 63.0-67.0 **Fe** 0.35 max **Mn** 0.50 max **Ni** 11.0-13.0 **Pb** 1.5-2.2 **Zn** rem	
C79200	Leaded Nickel Silver	**Cu** 59.0-66.5 **Fe** 0.25 max **Mn** 0.50 max **Ni** 11.0-13.0 **Pb** 0.8-1.4 **Zn** rem	**ASTM** B151 (792) **FED** QQ-C-586

The chemical compositions listed are for identification purposes and should not be used in lieu of the cross referenced specifications.

UNIFIED NUMBER	DESCRIPTION	CHEMICAL COMPOSITION	CROSS REFERENCE SPECIFICATIONS
C79300	Leaded Nickel Silver	Cu 55.0-59.00 Fe 0.50 max Mn 0.50 max Ni 11.0-13.0 Pb 0.50-2.0 Zn rem Other Co included with Ni	
C79600	Leaded Nickel Silver, 10%	Cu 43.5-46.5 Mn 1.5-2.5 Ni 9.0-11.0 Pb 0.8-1.2 Zn rem	
C79620	Leaded Nickel Silver	Cu 46.0-48.0 Mn 0.50 max Ni 8.0-11.0 Pb 0.50-2.0 Zn rem Other Co included in Ni	CDA C79620
C79800	Leaded Nickel Silver	Cu 45.5-48.5 Fe 0.25 max Mn 1.5-2.5 Ni 9.0-11.0 Pb 1.5-2.5 Zn rem	
C79820	Leaded Nickel Silver	Cu 46.0-48.0 Mn 0.50 max Ni 8.0-11.0 Pb 2.0-3.5 Zn rem Other Co included in Ni	CDA C79820
C79900	Leaded Nickel Silver	Cu 47.5-50.5 Fe 0.25 max Mn 0.50 max Ni 6.5-8.5 Pb 1.0-1.5 Zn rem	
C80100	Cast Copper	Cu 99.95 min Other total 0.05 max, Ag included in Cu	
C80300	Cast Copper, Silver Bearing	Ag 0.034 min Cu 99.95 min Other total 0.05 max, Ag included in Cu	
C80500	Cast Copper, Silver Bearing	Ag 0.034 min B 0.02 max Cu 99.75 min Other total 0.23 max, Ag included in Cu	
C80700	Cast Copper	B 0.02 max Cu 99.75 min Other total 0.23 max, Ag included in Cu	
C80900	Cast Copper	Ag 0.034 min Cu 99.7 min Other total 0.30 max, Ag included in Cu	
C81100	Cast Copper	Cu 99.7 min Other total 0.30 max, Ag included in Cu	
C81200	Phosphorus Deoxidized Copper	Cu 99.9 min P 0.040-0.065 Other Ag included in Cu	
C81300	Cast Beryllium Copper	Be 0.02-0.10 Co 0.6-1.0 Cu 98.5 min Other total named elements 99.5 min	
C81400	Cast Chromium Copper, 70C	Be 0.02-0.10 Cr 0.6-1.0 Cu 98.5 min Other total named elements 99.5 min	
C81500	Cast Chromium Copper	Al 0.10 max Cr 0.40-1.5 Cu 98.0 min Fe 0.10 max Pb 0.02 max Si 0.15 max Sn 0.10 max Zn 0.10 max Other total named elements 99.5 min	MIL SPEC MIL-C-19310
C81540	Cast Chromium Nickel Copper	Al 0.10 max Cr 0.10-0.6 Cu 95.1 min Fe 0.15 max Ni 2.0-3.0 Pb 0.02 max Si 0.40-0.8 Sn 0.10 max Zn 0.10 max Other Ag included in Cu; Co included in Ni	CDA C81540
C81700	Cast Beryllium Copper	Ag 0.8-1.2 Be 0.30-0.55 Co 0.25-1.5 Cu 94.2 min Ni 0.25-1.5 Other total named elements 99.5 min	
C81800	Cast Chromium Copper, 50C	Ag 0.8-1.2 Be 0.30-0.55 Co 1.4-1.7 Cu 95.6 min Other total named elements 99.5 min	
C82000	Cast Beryllium Copper	Al 0.10 max Be 0.45-0.8 Co 2.4-2.7 Cr 0.10 max Cu 95.0 min Fe 0.10 max Ni 0.20 max Pb 0.02 max Si 0.15 max Sn 0.10 max Zn 0.10 max Other total named elements 99.5 min, Co includes Ni	FED QQ-C-390

The chemical compositions listed are for identification purposes and should not be used in lieu of the cross referenced specifications.

UNIFIED NUMBER	DESCRIPTION	CHEMICAL COMPOSITION	CROSS REFERENCE SPECIFICATIONS
C82100	Cast Beryllium Copper	Be 0.35-0.8 Co 0.25-1.5 Cu 95.5 min Ni 0.25-1.5 Other total named elements 99.5 min	
C82200	Cast Beryllium Copper, 35B, 35C	Be 0.35-0.8 Cu 96.5 min Ni 1.0-2.0 Other total named elements 99.5 min	
C82400	Cast Beryllium Copper, 165C	Al 0.15 max Be 1.65-1.75 Co 0.20-0.40 Cr 0.10 max Cu 96.4 min Fe 0.20 max Ni 0.10 max Pb 0.02 max Sn 0.10 max Zn 0.10 Other total named elements 99.5 min	FED QQ-C-390
C82500	Cast Beryllium Copper, 20C	Al 0.15 max Be 1.90-2.15 Co 0.35-0.7 Cr 0.10 max Cu 95.5 min Fe 0.25 max Ni 0.20 max Pb 0.02 max Si 0.20-0.35 Sn 0.10 max Zn 0.10 max Other total named elements 99.5 min, Co includes Ni	AMS 4890 FED QQ-C-390 MIL SPEC MIL-C-11866 (17); MIL-C-22087 (10)
C82510	Copper-Beryllium	Al 0.15 max Be 1.90-2.15 Co 1.0-2.0 Cr 0.10 max Cu 95.5 min Fe 0.25 max Ni 0.20 max Pb 0.02 max Si 0.20-0.35 Sn 0.10 max Zn 0.10 max	
C82600	Cast Beryllium Copper, 245C	Al 0.15 max Be 2.25-2.45 Co 0.35-0.7 Cr 0.10 max Cu 95.2 min Fe 0.25 max Ni 0.20 max Pb 0.02 max Si 0.20-0.35 Sn 0.10 max Zn 0.10 max Other total named elements 99.5 min	FED QQ-C-390
C82700	Cast Beryllium Copper	Al 0.15 max Be 2.35-2.55 Cr 0.10 max Cu 94.6 min Fe 0.25 max Ni 1.0-1.5 Pb 0.02 max Si 0.15 max Sn 0.10 max Zn 0.10 max Other total named elements 99.5 min	FED QQ-C-390
C82800	Cast Beryllium Copper, 275C	Al 0.15 max Be 2.50-2.75 Co 0.35-0.7 Cr 0.10 max Cu 94.8 min Fe 0.25 max Ni 0.20 max Pb 0.02 max Si 0.20-0.35 Sn 0.10 max Zn 0.10 max Other total named elements 99.5 min, Co includes Ni	FED QQ-C-390
C83300	Cast Leaded Red Brass	Cu 92.0-94.0 Pb 1.0-2.0 Sn 1.0-2.0 Zn 2.0-6.0	
C83400	Cast Red Brass	Cu 88.0-92.0 Pb 0.50 max Sn 0.20 max Zn 8.0-12.0	MIL SPEC MIL-B-46066
C83410	Red Brass	Al 0.005 max Cu 88.0-91.0 Fe 0.05 max Ni 0.05 max Pb 0.10 max Si 0.005 max Sn 1.0-2.0 Zn rem Other Includes Co	
C83420	Cast Red Brass	Cu 88.0-92.0 Fe 0.10 max Pb 0.50 max Sn 0.25-0.7 Zn rem	
C83450	Leaded Red Brass	Al 0.005 max Cu 87.0-89.0 Fe 0.30 max Ni 0.8-2.0 P 0.03 max Pb 1.5-3.0 S 0.08 max Sb 0.25 max Si 0.005 max Sn 2.0-3.5 Zn 5.5-7.5 Other Includes Co	
C83500	Cast Leaded Red Brass	Cu 86.0-88.0 Fe 0.25 max Ni 0.50-1.0 P 0.03 max Pb 3.5-5.5 S 0.08 max Sb 0.25 max Si 0.005 max Sn 5.5-6.5 Zn 1.0-2.5 Other Cu may include Ni; for continuous castings, P 1.5 max	
C83520	Cast Leaded Red Brass	Cu rem Fe 0.30 max Ni 1.0 max Pb 3.5-4.5 Sb 0.25 max Sn 3.5-4.5 Zn 1.5-4.0	SAE J455

The chemical compositions listed are for identification purposes and should not be used in lieu of the cross referenced specifications.

UNIFIED NUMBER	DESCRIPTION	CHEMICAL COMPOSITION	CROSS REFERENCE SPECIFICATIONS
C83600	Cast Leaded Red Brass	Al 0.005 max Cu 84.0-86.0 Fe 0.30 max Ni 1.0 max P 0.05 max Pb 4.0-6.0 S 0.08 max Sb 0.25 max Si 0.005 max Sn 4.0-6.0 Zn 4.0-6.0 Other Cu may include Ni; for continuous castings, P 1.5 max	AMS 4855 ASME SB62; SB271 ASTM B30 (836); B62 (836); B271 (836); B505 (836); B584 (836) (formerly B145) FED QQ-C-390; QQ-C-525 MIL SPEC MIL-C-11866 (25); MIL-C-15345 (1); MIL-C-22087 (2); Mil-C-22229 (2) SAE J461 (CA836); J462 (CA836)
C83700	Red Brass	Al 0.005 max As 0.05-0.20 Cu 83.0-88.0 Fe 0.30 max Ni 0.30 max Pb 0.50 max Si 0.005 max Zn rem Other Includes Co, in determining copper, minimum copper may be calculated as Cu + Ni	
C83800	Cast Leaded Red Brass	Al 0.005 max Cu 82.0-83.8 Fe 0.30 max Ni 1.0 max P 0.03 max Pb 5.0-7.0 S 0.08 max Sb 0.25 max Si 0.005 max Sn 3.3-4.2 Zn 5.0-8.0 Other Cu may include Ni; for continuous castings, P 1.5 max	ASTM B30 (838); B271 (838); B505 (838); B584 (838) (formerly B145) FED QQ-C-390 SAE J461 (CA838); J462 (CA838)
C83810	Leaded Red Brass	Al 0.01 max Bi 0.10 max Cu rem Ni 2.0 max Pb 4.0-6.0 Si 0.02 max Sn 2.0-3.5 Zn 7.5-9.5 Other Fe+As+Sb=0.8 max, includes Co	
C84200	Cast Semi-Red Brass	Cu 78.0-82.0 Fe 0.40 max Ni 0.8 max Pb 2.0-3.0 Sn 4.0-6.0 Zn 10.0-16.0 Other for continuous castings, P 1.5 max	ASTM B30 (842); B505 (842) FED QQ-C-390; QQ-C-525
C84400	Cast Semi-Red Brass	Al 0.005 max Cu 78.0-82.0 Fe 0.40 max Ni 1.0 max P 0.02 max Pb 6.0-8.0 S 0.08 max Sb 0.25 max Si 0.005 max Sn 2.3-3.5 Zn 7.0-10.0 Other Cu may include Ni; for continuous castings, P 1.5 max	ASTM B30 (844); B271 (844); B505 (844); B584 (844) (formerly B145) FED QQ-C-390; QQ-C-525
C84500	Cast Semi-Red Brass	Cu 77.0-79.0 Fe 0.40 max Ni 1.0 max P 0.02 max Pb 6.0-7.5 Sn 2.0-4.0 Zn 10.0-14.0 Other Cu may include Ni; for continuous castings, P 1.5 max	
C84800	Cast Semi-Red Brass	Al 0.005 max Cu 75.0-77.0 Fe 0.40 max Ni 1.0 max P 0.02 max Pb 5.5-7.0 S 0.08 max Sb 0.25 max Si 0.005 max Sn 2.0-3.0 Zn 13.0-17.0 Other Cu may include Ni; for continuous castings, P 1.5 max	ASTM B30 (848); B271 (848); B505 (848); B584 (848) (formerly B145) FED QQ-C-390
C85200	Cast Leaded Yellow Brass	Al 0.005 max Cu 70.0-74.0 Fe 0.6 max Ni 1.0 max P 0.02 max Pb 1.5-3.8 S 0.05 max Sb 0.20 max Si 0.05 max Sn 0.7-2.0 Zn 20.0-27.0	ASTM B30 (852); B271 (852); B584 (852) (formerly B146) FED QQ-C-390 SAE J461 (CA852); J462 (CA852)
C85210	Yellow Brass	Al 0.005 max As 0.02-0.06 Cu 70.0-75.0 Fe 0.8 max Ni 1.0 max Pb 2.0-5.0 Si 0.005 max Sn 1.0-3.0 Zn rem Other Includes Co	
C85300	Cast Leaded Yellow Brass	Cu 68.0-72.0 Pb 0.50 max Sn 0.50 max Zn rem Other total (including Pb + Sn) 1.0 max	
C85310	Leaded Yellow Brass	Al 0.01 max As 0.02-0.06 Cu 68.0-73.0 Fe 0.8 max Ni 1.0 max Pb 2.0-5.0 Sn 1.5 max Zn rem Other Includes Co	
C85400	Cast Leaded Yellow Brass	Al 0.35 max Cu 65.0-70.0 Fe 0.7 max Ni 1.0 max Pb 1.5-3.8 Si 0.05 max Sn 0.50-1.5 Zn 24.0-32.0	ASTM B30 (854); B271 (854); B584 (854) (formerly B146) FED QQ-C-390 SAE J461 (CA854); J462 (CA854)

The chemical compositions listed are for identification purposes and should not be used in lieu of the cross referenced specifications.

UNIFIED NUMBER	DESCRIPTION	CHEMICAL COMPOSITION	CROSS REFERENCE SPECIFICATIONS
C85500	Cast Leaded Yellow Brass	Cu 59.0-63.0 Fe 0.20 max Mn 0.20 max Ni 0.20 max Pb 0.20 max Sn 0.20 max Zn rem Other total Fe + Mn + Ni + Pb + Sn 1.0 max	FED QQ-C-390
* C85600	Cast Leaded Yellow Brass	Cu 59.0-63.0 Fe 0.20 max Mn 0.20 max Ni 0.20 max Pb 0.20 max Sn 0.20 max Zn rem	
C85700	Cast Leaded Yellow Brass	Al 0.80 max Cu 58.0-64.0 Fe 0.70 max Ni 1.0 max Includes Co Pb 0.8-1.5 Si 0.05 max Sn 0.50-1.5 Zn 32.0-40.0	ASTM B30 (857); B271 (857); B584 (857) FED QQ-C-390 MIL SPEC MIL-C-15345 (3)
C85710	Leaded Yellow Brass	Al 0.20-0.8 Cu 58.0-63.0 Fe 0.8 max Mn 0.50 max Ni 1.0 max Pb 1.0-2.7 Si 0.05 max Sn 1.0 max Zn rem Other Includes Co	
C85800	Cast Leaded Yellow Brass	Al 0.50 max As 0.05 max Cu 57.0 min Fe 0.50 max Mn 0.25 max Ni 0.50 max P 0.01 max Pb 1.5 max S 0.05 max Sb 0.05 max Si 0.25 max Sn 1.5 max Zn 31.0-41.0 Other total named elements, 99.5 min	ASTM B30 (858); B176 (858) MIL SPEC MIL-B-15894 (1) SAE J461 (CA858); J462 (CA856)
C86100	Cast Manganese Bronze	Al 4.5-5.5 Cu 66.0-68.0 Fe 2.0-4.0 Mn 2.5-5.0 Pb 0.20 max Sn 0.20 max Zn rem	FED QQ-C-390; QQ-C-523 MIL SPEC MIL-C-15345 (5); MIL-C-22087 (7); MIL-C-22229 (10)
C86200	Cast Manganese Bronze	Al 3.0-4.9 Cu 60.0-66.0 Fe 2.0-4.0 Mn 2.5-5.0 Ni 1.0 max Pb 0.20 max Sn 0.20 max Zn 22.0-28.0	ASTM B30 (862); B271 (862); B505 (862); B584 (862) (formerly B147) FED QQ-C-390; QQ-C-523 MIL SPEC MIL-C-11866 (20); MIL-C-22087 (9); MIL-C-22229 (9) SAE J461 (CA862); J462 (CA862)
C86300	Cast Manganese Bronze	Al 5.0-7.5 Cu 60.0-66.0 Fe 2.0-4.0 Mn 2.5-5.0 Ni 1.0 max Pb 0.20 max Sn 0.20 max Zn 22.0-28.0	AMS 4862 ASTM B22 (863); B30 (863); B271 (863); B505 (863); B584 (863) (formerly B147) FED QQ-C-390; QQ-C-523 MIL SPEC MIL-C-11866 (21); MIL-C-15345 (6); MIL-C-22087 (9); MIL-C-22229 (8) SAE J461 (CA863); J462 (CA863)
C86400	Cast Leaded Manganese Bronze	Al 0.50-1.5 Cu 56.0-62.0 Fe 0.40-2.0 Mn 0.10-1.0 Ni 1.0 max Pb 0.50-1.5 Sn 0.50-1.5 Zn 34.0-42.0	ASTM B30 (864); B271 (864); B584 (864) (formerly B132, B147) FED QQ-C-390; QQ-C-523
C86500	Cast Manganese Bronze	Al 0.50-1.5 Cu 55.0-60.0 Fe 0.40-2.0 Mn 0.10-1.5 Ni 1.0 max Pb 0.40 max Sn 1.0 max Zn 36.0-42.0	AMS 4860 ASTM B30 (865); B271 (865); B505 (865); B584 (865) (formerly B147) FED QQ-C-390; QQ-C-523 MIL SPEC MIL-C-15345 (4); MIL-C-22087 (5); MIL-C-22229 (7) SAE J461 (CA865); J462 (CA865)
C86700	Cast Leaded Manganese Bronze	Al 1.0-3.0 Cu 55.0-60.0 Fe 1.0-3.0 Mn 1.0-3.5 Ni 1.0 max Pb 0.5-1.5 Sn 1.5 max Zn 30.0-38.0	ASTM B30 (867); B271 (867); B584 (867) (formerly B132)
C86800	Cast Manganese Bronze	Al 2.0 max Cu 53.5-57.0 Fe 1.0-2.5 Mn 2.5-4.0 Ni 2.5-4.0 Pb 0.20 max Sn 1.0 max Zn rem	FED QQ-C-390
C87200	Cast Copper Silicon Bronze	Al 1.5 max Cu 89.0 min Fe 2.5 max Mn 1.5 max Pb 0.50 max Si 1.0-5.0 Sn 1.0 max Zn 5.0 max Other total named elements 99.5 min	ASTM B271; B584 FED QQ-C-390; WW-V-51; WW-V-1967 MIL SPEC MIL-C-11866 (19); MIL-C-22087; MIL-C-22229 (4) SAE J461; J462
C87300	Copper Silicon Alloy	Cu 94.0 min Fe 0.20 max Mn 0.8-1.5 Pb 0.20 max Si 3.5-4.5 Zn 0.25 max	
C87400	Cast Silicon Brass	Al 0.8 max Cu 79.0 min Pb 1.0 max Si 2.5-4.0 Zn 12.0-16.0 Other total named elements 99.5 min	ASTM B30 (874); B271 (874); B584 (874) (formerly B198) FED QQ-C-390 SAE J461 (CA874); J462 (CA874)
C87410	Cast Silicon Brass	Al 0.8 max As 0.03-0.06 Cu 79.0 min Pb 1.0 max Si 2.5-4.0 Zn 12.0-16.0 Other total named elements 99.5 min	

*Boxed entries are no longer active and are retained for reference purposes only.

The chemical compositions listed are for identification purposes and should not be used in lieu of the cross referenced specifications.

UNIFIED NUMBER	DESCRIPTION	CHEMICAL COMPOSITION	CROSS REFERENCE SPECIFICATIONS
C87420	Cast Silicon Brass	**Al** 0.8 max **Cu** 79.0 min **Pb** 1.0 max **Sb** 0.03-0.06 **Si** 2.5-4.0 **Zn** 12.0-16.0 **Other** total named elements 99.5 min	
C87430	Cast Silicon Brass	**Al** 0.8 max **Cu** 79.0 min **P** 0.03-0.06 **Pb** 1.0 max **Si** 2.5-4.0 **Zn** 12.0-16.0 **Other** total named elements 99.5 min	
C87500	Cast Silicon Brass	**Al** 0.50 max **Cu** 79.0 min **Pb** 0.50 max **Si** 3.0-5.0 **Zn** 12.0-16.0 **Other** total named elements 99.5 min	**ASTM** B30 (875); B271 (875); B584 (875) (formerly B198) **FED** QQ-C-390 **MIL SPEC** MIL-C-22087 (4) **SAE** J461 (CA875); J462 (CA875)
C87510	Cast Silicon Brass	**Al** 0.50 max **As** 0.03-0.06 **Cu** 79.0 min **Pb** 0.50 **Si** 3.0-5.0 **Zn** 12.0-16.0 **Other** total named elements 99.5 min	
C87520	Cast Silicon Brass	**Al** 0.50 max **Cu** 79.0 min **Pb** 0.50 max **Sb** 0.03-0.06 **Si** 3.0-5.0 **Zn** 12.0-16.0 **Other** total named elements 99.5 min	
C87530	Cast Silicon Brass	**Al** 0.50 max **Cu** 79.0 min **P** 0.03-0.06 **Pb** 0.50 max **Si** 3.0-5.0 **Zn** 12.0-16.0 **Other** total named elements 99.5 min	
C87600	Cast Silicon Brass	**Cu** 88.0 min **Fe** 0.20 max **Mn** 0.25 max **Pb** 0.50 max **Si** 3.5-4.5 **Zn** 4.0-7.0	**ASTM** B30 (876); B584 (876)
C87800	Cast Silicon Bronze	**Al** 0.15 max **As** 0.05 max **Cu** 80.0 min **Fe** 0.15 max **Mg** 0.01 max **Mn** 0.15 max **Ni** 0.20 max **P** 0.01 max **Pb** 0.15 max **S** 0.05 max **Sb** 0.05 max **Si** 3.8-4.2 **Sn** 0.25 max **Zn** 12.0-16.0 **Other** total named elements 99.8 min	**ASTM** B30 (878); B176 (878) **MIL SPEC** MIL-B-15894 (3) **SAE** J461 (CA878); J462 (CA878)
C87900	Cast Silicon Bronze	**Al** 0.15 max **As** 0.05 max **Cu** 63.0 min **Fe** 0.40 max **Mn** 0.15 max **Ni** 0.50 max **P** 0.01 max **Pb** 0.25 max **S** 0.05 max **Sb** 0.05 max **Si** 0.8-1.2 **Sn** 0.25 max **Zn** 30.0-36.0 **Other** total named elements 99.5 min	**ASTM** B30 (879); B176 (879) **MIL SPEC** MIL-B-15894 (2) **SAE** J461 (CA879); J462 (CA879)
C90200	Cast Tin Bronze	**Al** 0.005 **Cu** 91.0-94.0 **Fe** 0.20 **Ni** 0.50 **P** 0.05 **S** 0.05 **Sb** 0.20 **Si** 0.005 **Sn** 6.0-8.0 **Zn** 0.50 max **Other** total named elements 99.0 min	
C90250	Cast Tin Bronze	**Al** 0.01 max **Cu** 89.0-91.0 **Fe** 0.15 max **Mn** 0.10 max **Ni** 0.8 max **P** 0.05 max **Pb** 0.8 max **S** 0.04 max **Sb** 0.20 max **Si** 0.01 max **Sn** 9.0-11.0 **Zn** 0.50 max	
C90300	Cast Tin Bronze	**Al** 0.005 max **Cu** 86.0-89.0 **Fe** 0.20 max **Ni** 1.0 max **P** 0.05 max **Pb** 0.30 max **S** 0.05 max **Sb** 0.20 max **Si** 0.005 **Sn** 7.5-9.0 **Zn** 3.0-5.0 **Other** Cu may include Ni	**ASTM** B30 (903); B271 (903); B505 (903); B584 (903) (formerly B143) **FED** QQ-C-390; QQ-C-525 **MIL SPEC** MIL-C-11866 (26); MIL-C-15345 (8); MIL-C-22087 (3); MIL-C-22229 (1) **SAE** J461 (CA903); J462 (CA903)
C90500	Cast Tin Bronze	**Al** 0.005 max **Cu** 86.0-89.0 **Fe** 0.20 max **Ni** 1.0 max **P** 0.05 max **Pb** 0.30 max **S** 0.05 max **Sb** 0.20 max **Si** 0.005 max **Sn** 9.0-11.0 **Zn** 1.0-3.0 **Other** Cu may include Ni	**AMS** 4845 **ASTM** B22 (905); B30 (905); B271 (905); B505 (905); B584 (905) (formerly B143) **FED** QQ-C-390 **SAE** J461 (CA905); J462 (CA905)
C90700	Cast Tin Bronze	**Al** 0.005 max **Cu** 88.0-90.0 **Fe** 0.15 max **Ni** 0.50 **P** 0.30 max **Pb** 0.50 max **S** 0.05 max **Sb** 0.20 max **Si** 0.005 max **Sn** 10.0-12.0 **Zn** 0.50 max **Other** total Pb + Ni + Zn 1.0 max	**ASTM** B30 (907); B505 (907) **FED** QQ-C-390 **SAE** J461 (CA907); J462 (CA907)
C90710	Tin Bronze	**Al** 0.005 max **Cu** rem **Fe** 0.10 max **Ni** 0.10 max **P** 0.50-1.2 **Pb** 0.25 max **Si** 0.005 max **Sn** 10.0-12.0 **Zn** 0.05 max **Other** Incl Co	

The chemical compositions listed are for identification purposes and should not be used in lieu of the cross referenced specifications.

UNIFIED NUMBER	DESCRIPTION	CHEMICAL COMPOSITION	CROSS REFERENCE SPECIFICATIONS
C90800	Cast Tin Bronze	**Al** 0.005 max **Cu** 85.0-89.0 **Fe** 0.15 max **Ni** 0.50 max **P** 0.30 max **Pb** 0.25 max **S** 0.05 max **Sb** 0.20 max **Si** 0.005 max **Sn** 11.0-13.0 **Zn** 0.25 max **Other** total named elements 99.5 min	**ASTM** B30 (908); B427 (908)
C90810	Tin Bronze	**Al** 0.005 max **Cu** 85.0-89.0 **Fe** 0.15 max **Ni** 0.50 **P** 0.15-0.8 **Pb** 0.25 max **S** 0.05 max **Sb** 0.20 max **Si** 0.005 max **Sn** 11.0-13.0 **Zn** 0.25 max **Other** Includes Co	
C90900	Cast Tin Bronze	**Al** 0.005 max **Cu** 86.0-89.0 **Fe** 0.15 max **Ni** 0.50 max **P** 0.05 max **Pb** 0.25 max **S** 0.05 max **Sb** 0.20 max **Si** 0.005 max **Sn** 12.0-14.0 **Zn** 0.25 max **Other** total named elements 99.5 min	
C91000	Cast Tin Bronze	**Al** 0.005 max **Cu** 84.0-86.0 **Fe** 0.10 max **Ni** 0.8 **P** 0.05 max **Pb** 0.20 max **S** 0.5 max **Sb** 0.20 max **Si** 0.005 max **Sn** 14.0-16.0 **Zn** 1.5 max	**ASTM** B30 (910); B505 (910) **FED** QQ-C-390; QQ-C-525
C91100	Cast Tin Bronze	**Al** 0.005 max **Cu** 82.0-85.0 **Fe** 0.25 max **Ni** 0.50 max **P** 1.0 max **Pb** 0.25 max **S** 0.05 max **Sb** 0.20 max **Si** 0.005 max **Sn** 15.0-17.0 **Zn** 0.25 max	**ASTM** B22 (911); B30 (911)
C91300	Cast Tin Bronze	**Al** 0.005 max **Cu** 79.0-82.0 **Fe** 0.25 max **Ni** 0.50 max **P** 1.0 max **Pb** 0.25 max **S** 0.05 max **Sb** 0.20 max **Si** 0.005 max **Sn** 18.0-20.0 **Zn** 0.25 max	**AMS** 7322 **ASTM** B22 (913); B30 (913); B505 (913) **FED** QQ-C-390; QQ-C-525
C91600	Cast Tin Bronze	**Al** 0.005 max **Cu** 86.0-89.0 **Fe** 0.20 max **Ni** 1.2-2.0 **P** 0.25 max **Pb** 0.25 max **S** 0.05 max **Sb** 0.20 max **Si** 0.005 max **Sn** 9.7-10.8 **Zn** 0.25 max **Other** total named elements 99.5 min	**ASTM** B30 (916); B427 (916) **FED** QQ-C-390 **MIL SPEC** MIL-C-15345 (23)
C91700	Cast Tin Bronze	**Al** 0.005 max **Cu** 84.0-87.0 **Fe** 0.20 max **Ni** 1.2-2.0 **P** 0.30 max **Pb** 0.25 max **S** 0.05 max **Sb** 0.20 max **Si** 0.005 max **Sn** 11.3-12.5 **Zn** 0.25 max **Other** total named elements 99.5 min	**ASTM** B30 (917); B427 (917)
C92200	Cast Leaded Tin Bronze	**Al** 0.005 max **Cu** 86.0-90.0 **Fe** 0.25 max **Ni** 1.0 max **P** 0.05 max **Pb** 1.0-2.0 **S** 0.05 max **Sb** 0.25 max **Si** 0.005 max **Sn** 5.5-6.5 **Zn** 3.0-5.0 **Other** Cu may include Ni	**ASME** SB61; SB271; SB584 **ASTM** B30 (922); B61 (922); B271 (922); B505 (922); B584 (922) (formerly B143) **FED** QQ-C-390; QQ-C-525 **MIL SPEC** MIL-C-15345 (9); MIL-B-16541 **SAE** J461 (CA922); J462 (CA922)
C92300	Cast Leaded Tin Bronze	**Al** 0.005 max **Cu** 85.0-89.0 **Fe** 0.25 max **Ni** 1.0 max **P** 0.05 max **Pb** 0.30-1.0 **S** 0.05 max **Sb** 0.20 max **Si** 0.005 max **Sn** 7.5-9.0 **Zn** 2.5-5.0 **Other** Cu may include Ni	**ASTM** B30 (923); B271 (923); B505 (923); B584 (923) (formerly B143) **FED** QQ-C-390 (D3) **MIL SPEC** MIL-C-15345 (10) **SAE** J461 (CA923); J462 (CA923)
C92310	Leaded Tin Bronze	**Al** 0.005 max **Bi** 0.03 max **Cu** 86.0-89.0 **No** 1.0 max **Pb** 0.30-1.5 **Si** 0.005 max **Sn** 7.5-8.5 **Zn** 3.5-4.5	
C92400	Cast Leaded Tin Bronze	**Al** 0.005 max **Cu** 86.0-89.0 **Fe** 0.25 max **Ni** 1.0 max **P** 0.05 max **Pb** 1.0-2.5 **S** 0.05 max **Sb** 0.25 max **Si** 0.005 max **Sn** 9.0-11.0 **Zn** 1.0-3.0 **CDA** 924	
C92410	Leaded Tin Bronze	**Al** 0.01 max **Bi** 0.03 max **Cu** rem **Fe** 0.20 max **Ni** 2.0 max **Pb** 2.5-3.5 **Sb** 0.25 max **Si** 0.01 max **Sn** 6.0-8.0 **Zn** 1.5-3.0 **Other** Incl Co	
C92500	Cast Leaded Tin Bronze	**Al** 0.005 max **Cu** 85.0-88.0 **Fe** 0.30 max **Ni** 0.8-1.5 **P** 0.20-0.30 **Pb** 1.0-1.5 **Sb** 0.25 max **Si** 0.005 max **Sn** 10.0-12.0 **Zn** 0.50 max	**ASTM** B30 (925); B505 (925) **FED** QQ-C-390 **SAE** J461 (CA925); J462 (CA925)

The chemical compositions listed are for identification purposes and should not be used in lieu of the cross referenced specifications.

UNIFIED NUMBER	DESCRIPTION	CHEMICAL COMPOSITION	CROSS REFERENCE SPECIFICATIONS
C92600	Cast Leaded Tin Bronze	**Al** 0.005 max **Cu** 86.0-88.5 **Fe** 0.20 max **Ni** 0.7 max **P** 0.03 max **Pb** 0.8-1.2 **S** 0.05 max **Sb** 0.25 max **Si** 0.005 max **Sn** 9.3-10.5 **Zn** 1.3-2.5	
C92610	Leaded Tin Bronze	**Al** 0.005 max **Bi** 0.03 max **Cu** rem **Fe** 0.15 max **Ni** 1.0 max **Pb** 0.30-1.5 **Si** 0.005 max **Sn** 9.5-10.5 **Zn** 1.7-2.8 **Other** Ni includes Co	
C92700	Cast Leaded Tin Bronze	**Al** 0.005 max **Cu** 86.0-89.0 **Fe** 0.20 max **Ni** 1.0 max **P** 0.25 max **Pb** 1.0-2.5 **S** 0.05 max **Sb** 0.25 max **Si** 0.005 max **Sn** 9.0-11.0 **Zn** 0.7 max	**ASTM** B30 (927); B505 (927) **FED** QQ-C-390 **SAE** J461 (CA927); J462 (CA927)
C92710	Leaded Tin Bronze	**Al** 0.005 max **Cu** rem **Fe** 0.20 max **Ni** 2.0 max **P** 0.10 max **Pb** 4.0-6.0 **S** 0.05 max **Sb** 0.25 max **Si** 0.005 max **Sn** 9.0-11.0 **Zn** 1.0 max **Other** Includes Co	
C92800	Cast Leaded Tin Bronze	**Al** 0.005 max **Cu** 78.0-82.0 **Fe** 0.20 max **Ni** 0.50 max **P** 0.25 max **Pb** 4.0-6.0 **S** 0.05 max **Sb** 0.25 max **Si** 0.005 max **Sn** 15.0-17.0 **Zn** 0.50 max	**AMS** 7320 **ASTM** B30 (928); B505 (928)
C92900	Cast Leaded Tin Bronze	**Al** 0.005 max **Cu** 82.0-86.0 **Fe** 0.20 max **Ni** 2.8-4.0 **P** 0.50 max **Pb** 2.8-4.0 **S** 0.05 max **Sb** 0.25 max **Si** 0.005 max **Sn** 9.0-11.0 **Zn** 0.25 max **Other** total named elements 99.5 min	**ASTM** B30 (928); B427 (928); B505 (928) **FED** QQ-C-390 **SAE** J461 (CA929); J462 (CA929)
C93100	High-Leaded Tin Bronze	**Al** 0.005 max **Cu** rem **Fe** 0.20 max **Ni** 1.0 max **P** 0.20 max **Pb** 2.0-5.0 **S** 0.05 max **Sb** 0.25 max **Si** 0.005 max **Sn** 6.5-8.5 **Zn** 2.0 max **Other** Includes Co	
C93200	Cast High-Leaded Tin Bronze	**Al** 0.005 max **Cu** 81.0-85.0 **Fe** 0.20 max **Ni** 1.0 max **P** 0.15 max **Pb** 6.0-8.0 **S** 0.08 max **Sb** 0.35 max **Si** 0.005 max **Sn** 6.3-7.5 **Zn** 2.0-4.0 **Other** Cu may include Ni; for continuous castings, P 1.5 max	**ASTM** B30 (932); B271 (932); B505 (932); B584 (932) (formerly B144) **FED** QQ-C-390; QQ-C-525 **MIL SPEC** MIL-C-15345 (12) **SAE** J461 (CA932); J462 (CA932)
C93400	Cast High-Leaded Tin Bronze	**Cu** 82.0-85.0 **Fe** 0.15 max **Ni** 1.0 max **P** 0.50 max **Pb** 7.0-9.0 **Sb** 0.50 max **Sn** 7.0-9.0 **Zn** 0.7 max	**ASTM** B30 (934); B505 (934) **FED** QQ-C-390; QQ-C-525 **MIL SPEC** MIL-C-15345 (11); MIL-C-22087 (1); MIL-C-22229 (1)
C93500	Cast High-Leaded Tin Bronze	**Al** 0.005 max **Cu** 83.0-86.0 **Fe** 0.20 max **Ni** 1.0 max **P** 0.05 max **Pb** 8.0-10.0 **S** 0.08 max **Sb** 0.30 max **Si** 0.005 max **Sn** 4.3-6.0 **Zn** 2.0 max **Other** Cu may include Ni; for continuous castings, P 1.5 max	**ASTM** B30 (935); B271 (935); B505 (935); B584 (935) (formerly B144) **FED** QQ-C-390 **SAE** J461 (CA935); J462 (CA935)
* C93600	Cast High-Leaded Tin Bronze	**Cu** 83.0 min **Fe** 0.35 max **Ni** 1.5 max **Pb** 7.0-9.0 **Sb** 0.30 max **Sn** 3.5-4.5 **Zn** 4.0 max **Other** total 0.30 max	
C93700	Cast High-Leaded Tin Bronze	**Al** 0.005 max **Cu** 78.0-82.0 **Fe** 0.15 max **Ni** 1.0 max **P** 0.15 max **Pb** 8.0-11.0 **S** 0.08 max **Sb** 0.55 max **Si** 0.005 max **Sn** 9.0-11.0 **Zn** 0.8 max **Other** Cu may include Ni; Fe 0.35 max for steel-backed bearings; P 1.5 max for continuous castings	**AMS** 4842 **ASME** SB271; SB584 (formerly SB144) **ASTM** B22 (937); B30 (937); B271 (937); B505 (937); B584 (937) (formerly B144) **FED** QQ-C-390 **SAE** J461 (CA937); J462 (CA937)
C93720	Cast High-Leaded Tin Bronze	**Cu** 83.0 min **Fe** 0.35 max **Ni** 0.50 max **Pb** 7.0-9.0 **Sb** 0.50 max **Sn** 3.5-4.5 **Zn** 4.0 max **Other** Cu includes Ni, Ni includes Co	**SAE** J460

*Boxed entries are no longer active and are retained for reference purposes only.

The chemical compositions listed are for identification purposes and should not be used in lieu of the cross referenced specifications.

UNIFIED NUMBER	DESCRIPTION	CHEMICAL COMPOSITION	CROSS REFERENCE SPECIFICATIONS
C93800	Cast High-Leaded Tin Bronze	**Al** 0.005 max **Cu** 75.0-79.0 **Fe** 0.15 max **Ni** 1.0 max **P** 0.05 max **Pb** 13.0-16.0 **S** 0.08 max **Sb** 0.08 max **Si** 0.005 max **Sn** 6.3-7.5 **Zn** 0.8 max **Other** Cu may include Ni; for continuous castings, P 1.5 max	**ASTM** B30 (938); B66 (938); B271 (938); B505 (938); B584 (938) **FED** QQ-C-390 **SAE** J461 (CA938); J462 (CA938)
C93900	Cast High-Leaded Tin Bronze	**Al** 0.005 max **Cu** 76.5-79.5 **Fe** 0.40 max **Ni** 0.8 max **P** 1.5 max **Pb** 14.0-18.0 **Sb** 0.50 max **Si** 0.005 max **Sn** 5.0-7.0 **Zn** 1.5 max **Other** Cu may include Ni	**ASTM** B30 (939); B505 (939) **FED** QQ-C-390; QQ-C-525 **SAE** J462 (CA939)
C94000	Cast High-Leaded Tin Bronze	**Al** 0.005 max **Cu** 69.0-72.0 **Fe** 0.25 max **Ni** 0.50-1.0 **P** 0.05 max **Pb** 14.0-16.0 **S** 0.08 max **Sb** 0.35 max **Si** 0.005 max **Sn** 12.0-14.0 **Zn** 0.50 max **Other** Cu may include Ni; for continuous castings, P 1.5 max; S 0.25 max; other total 0.35 max	**ASTM** B30 (940); B505 (940) **FED** QQ-C-390; QQ-C-525
C94100	Cast High-Leaded Tin Bronze	**Al** 0.005 max **Cu** 65.0-75.0 **Fe** 0.25 max **Ni** 0.8 max **P** 0.05 max **Pb** 15.0-22.0 **S** 0.08 max **Sb** 0.8 max **Si** 0.005 max **Sn** 4.5-6.5 **Zn** 3.0 max **Other** total 1.0 max	**ASTM** B30 (941); B67; B505 **FED** QQ-C-390
* C94200	Cast High-Leaded Tin Bronze	**Cu** 68.5-75.5 **Fe** 0.35 max **Ni** 0.50 max **Pb** 3.0-4.0 **Sb** 0.50 max **Sn** 3.0-4.0 **Zn** 3.0 max **Other** total 0.40 max	
C94300	Cast High-Leaded Tin Bronze	**Al** 0.005 max **Cu** 68.5-73.5 **Fe** 0.15 max **Ni** 1.0 max **P** 0.05 max **Pb** 22.0-25.0 **S** 0.08 max **Sb** 0.8 max **Si** 0.005 max **Sn** 4.5-6.0 **Zn** 0.8 max **Other** Cu may include Ni; for continuous castings, P 1.5 max	**ASTM** B30 (943); B66 (943); B271 (943); B505 (943); B584 (943) (formerly B144) **FED** QQ-C-390 **SAE** J461 (CA943); J462 (CA943)
C94310	High-Leaded Tin Bronze	**Cu** rem **Fe** 0.05 max **Ni** 0.25-1.0 **P** 0.05 max **Pb** 27.0-34.0 **Sb** 0.50 max **Sn** 1.5-3.0 **Zn** 0.50 max	
C94320	Cast High-Leaded Tin Bronze	**Cu** rem **Fe** 0.35 max **Pb** 24.0-32.0 **Sn** 4.0-7.0	**SAE** J460
C94330	Cast High-Leaded Tin Bronze	**Cu** 68.5-75.5 **Fe** 0.35 max **Ni** 0.50 max **Pb** 21.0-25.0 **Sb** 0.50 max **Sn** 3.0-4.0 **Zn** 3.0 max **Other** Cu includes Ni, Ni includes Co	**SAE** J460
C94400	Cast High-Leaded Tin Bronze	**Al** 0.005 max **Cu** rem **Fe** 0.15 max **Ni** 1.0 max **P** 0.50 max **Pb** 9.0-12.0 **S** 0.08 max **Sb** 0.8 max **Si** 0.005 max **Sn** 7.0-9.0 **Zn** 0.7 max **Other** total named elements 99.5 min	**ASTM** B30 (944); B66 (944)
C94500	Cast High-Leaded Tin Bronze	**Al** 0.005 max **Cu** rem **Fe** 0.15 max **Ni** 1.0 max **P** 0.05 max **Pb** 16.0-22.0 **S** 0.08 max **Sb** 0.8 max **Si** 0.005 max **Sn** 6.0-8.0 **Zn** 1.2 max **Other** total 0.7 max	**ASTM** B30 (945); B66 (945)
C94700	Cast Nickel-Tin Bronze	**Al** 0.005 max **Cu** 85.0-89.0 **Fe** 0.25 max **Mn** 0.20 max **Ni** 4.5-6.0 **P** 0.05 max **Pb** 0.10 max **S** 0.05 max **Sb** 0.15 max **Si** 0.005 max **Sn** 4.5-6.0 **Zn** 1.0-2.5 **Other** Pb 0.01 max for heat-treated products	**ASTM** B30 (947); B505 (947); B584 (947) (formerly B292) **FED** QQ-C-390 **SAE** J461 (CA947); J462 (CA947)

*Boxed entries are no longer active and are retained for reference purposes only.

The chemical compositions listed are for identification purposes and should not be used in lieu of the cross referenced specifications.

UNIFIED NUMBER	DESCRIPTION	CHEMICAL COMPOSITION	CROSS REFERENCE SPECIFICATIONS
C94800	Cast Nickel-Tin Bronze	**Al** 0.005 max **Cu** 84.0-89.0 **Fe** 0.25 max **Mn** 0.20 max **Ni** 4.5-6.0 **P** 0.05 max **Pb** 0.30-1.0 **S** 0.05 max **Sb** 0.15 max **Si** 0.005 max **Sn** 4.5-6.0 **Zn** 1.0-2.5	**ASTM** B30 (948); B505 (948); B584 (948) (formerly B292) **FED** QQ-C-390 **SAE** J461 (CA948); J462 (CA948)
C94900	Cast Nickel-Tin Bronze	**Al** 0.005 max **Cu** 79.0-81.0 **Fe** 0.30 max **Mn** 0.10 max **Ni** 4.0-6.0 **P** 0.05 max **Pb** 4.0-6.0 **S** 0.08 max **Sb** 0.25 max **Si** 0.005 max **Sn** 4.0-6.0 **Zn** 4.0-6.0	**ASTM** B30 (949); B584 (949) (formerly B292)
C95200	Cast Aluminum Bronze	**Al** 8.5-9.5 **Cu** 86.0 min **Fe** 2.5-4.0 **Other** total named elements 99.0 min	**ASME** SB148; SB271 **ASTM** B30 (952); B148 (952); B271 (952); B505 **FED** QQ-B-675; QQ-C-390 **MIL SPEC** MIL-C-22229 (5) **SAE** J461 (CA952); J462 (CA952)
C95210	Aluminum Bronze	**Al** 8.5-9.5 **Cu** 86.0 min **Fe** 2.5-4.0 **Mg** 0.05 max **Mn** 1.0 max **Ni** 1.0 **Pb** 0.05 max **Si** 0.25 max **Sn** 0.10 max **Zn** 0.50 max **Other** Includes Co	
C95300	Cast Aluminum Bronze	**Al** 9.0-11.0 **Cu** 86.0 min **Fe** 0.8-1.5 **Other** total named elements 99.0 min	**ASTM** B30 (953); B148 (953); B271 (953); B505 (953) **FED** QQ-B-675; QQ-C-390 **MIL SPEC** MIL-C-11866 (22) **SAE** J461 (CA953); J462 (CA953)
C95400	Cast Aluminum Bronze	**Al** 10.0-11.5 **Cu** 83.0 min **Fe** 3.0-5.0 **Mn** 0.50 max **Ni** 1.5 max Includes Co	**ASME** SB148; SB271 **ASTM** B30 (954); B148 (954); B271 (954); B505 (954) **FED** QQ-B-675; QQ-C-390 **MIL SPEC** MIL-C-I5345 (Alloy 13) **SAE** J461 (CA954); J462 (CA954)
C95410	Cast Aluminum Bronze	**Al** 10.0-11.5 **Cu** 83.0 min **Fe** 3.0-5.0 **Mn** 0.50 max **Ni** 1.5-2.5 **Other** total named elements 99.5 min	
C95420	Cast Aluminum Bronze	**Al** 10.5-12.0 **Cu** 83.5 min **Fe** 3.0-4.3 **Mn** 0.50 max **Ni** 0.50 max **Other** Ni includes Co	**AMS** 4870; 4871; 4872; 4873
C95500	Cast Aluminum Bronze	**Al** 10.0-11.5 **Cu** 78.0 min **Fe** 3.0-5.0 **Mn** 3.5 max **Ni** 3.0-5.5 **Other** total named elements 99.5 min	**ASTM** B30 (955); B148 (955); B271 (955); B505 (955) **FED** QQ-B-675; QQ-C-390 **MIL SPEC** MIL-C-15345 (14); MIL-C-22087 (8); MIL-C-22229 (6) **SAE** J461 (CA955); J462 (CA955)
C95510	Cast Aluminum Bronze	**Al** 9.7-10.9 **Cu** 78.0 min **Fe** 2.0-3.5 **Mn** 1.5 max **Ni** 4.5-5.5 **Sn** 0.20 max **Zn** 0.30 max **NOTE:** Cu includes Ag and Ni includes Co; Cu + all named elements = 99.8 min.	
C95520	Cast Aluminum Bronze	**Al** 10.5-11.5 **Co** 0.20 max **Cr** 0.05 max **Cu** 74.5 min **Fe** 4.0-5.5 **Mn** 1.5 max **Ni** 4.2-6.0 **Pb** 0.03 max **Si** 0.15 max **Sn** 0.25 max **Zn** 0.30 max	**AMS** 4881
C95600	Cast Aluminum Bronze	**Al** 6.0-8.0 **Cu** 88.0 min **Ni** 0.25 max **Si** 1.8-3.3 **Other** total named elements 99.0 min	**ASTM** B30 (956); B148 (956) **FED** QQ-B-675
C95700	Cast Aluminum Bronze	**Al** 7.0-8.5 **Cu** 71.0 min **Fe** 2.0-4.0 **Mn** 11.0-14.0 **Ni** 1.5-3.0 **Pb** 0.03 max **Si** 0.10 max **Other** total named elements 99.5 min	**ASTM** B30 (957); B148 (957) **FED** QQ-B-675; QQ-C-390 **MIL SPEC** MIL-B-24480
C95710	Aluminum Bronze	**Al** 7.0-8.5 **Cu** 71.0 min **Fe** 2.0-4.0 **Mn** 11.0-14.0 **Ni** 1.5-3.0 **P** 0.05 max **Pb** 0.05 max **Si** 0.15 max **Zn** 0.50 max **Other** Includes Co	
C95800	Cast Aluminum Bronze	**Al** 8.5-9.5 **Cu** 79.0 min **Fe** 3.5-4.5 **Mn** 0.8-1.5 **Ni** 4.0-5.0 **Pb** 0.03 max **Si** 0.10 max **Other** Fe less than Ni, total named elements 99.5 min	**ASTM** B30 (958); B148 (958); B271 (958); B505 (958) **FED** QQ-B-675; QQ-C-390 **MIL SPEC** MIL-C-15345 (28); MIL-B-24480 **SAE** J461 (CA958); J462 (CA958)

The chemical compositions listed are for identification purposes and should not be used in lieu of the cross referenced specifications.

UNIFIED NUMBER	DESCRIPTION	CHEMICAL COMPOSITION	CROSS REFERENCE SPECIFICATIONS
C95810	Aluminum Bronze	**Al** 8.5-9.5 **Cu** 79.0 min **Fe** 3.5-4.5 **Mg** 0.05 max **Mn** 1.5 max **Ni** 4.0-5.0 **Pb** 0.05 max **Si** 0.15 max **Zn** 0.50 max **Other** Fe content shall not exceed Ni content	
C96200	Cast Copper-Nickel	**C** 0.10 max **Cu** rem **Fe** 1.0-1.8 **Mn** 1.5 max **Nb** 0.50-1.0 **Ni** 9.0-11.0 **P** 0.02 max **Pb** 0.01 max **S** 0.02 max **Si** 0.30 max	**ASTM** B30 (962); B369 (962); B433 (962) **FED** QQ-C-390 **MIL SPEC** MIL-C-15345 (25); MIL-C-20159 (II) **SAE** J461 (CA962); J462 (CA962)
C96300	Cast Copper-Nickel	**Al** 0.05 max **Cu** rem **Fe** 0.50-1.5 **Mn** 0.25-1.5 **Nb** 1.0 max **Ni** 18.0-22.0 **P** 0.02 max **Pb** 0.01 max **Si** 0.03-0.8 **Zn** 0.10 max	**ASTM** B492
C96400	Cast Copper-Nickel	**C** 0.15 max **Cu** rem **Fe** 0.25-1.5 **Mn** 1.5 max **Nb** 0.50-1.5 **Ni** 28.0-32.0 **P** 0.02 max **Pb** 0.01 max **S** 0.50 max **Si** 0.50 max	**ASTM** B30 (964); B369 (964); B433 (964); B505 (964) **FED** QQ-C-390 **MIL SPEC** MIL-C-15345 (24); MIL-C-20159 (I)
C96600	Cast Copper-Nickel	**Be** 0.40-0.7 **Cu** rem **Fe** 0.8-1.1 **Mn** 1.0 max **Ni** 29.0-33.0 **Pb** 0.01 max **Si** 0.15 max	**MIL SPEC** MIL-C-81519
C96800	Cast Copper-Nickel Alloy	**Al** 0.10 max **B** 0.001 max **Bi** 0.001 max **Cb** 0.10-0.30 **Cu** rem **Fe** 0.50 max **Mg** 0.005-0.15 **Mn** 0.05-0.30 **Ni** 9.5-10.5 **P** 0.005 max **Pb** 0.005 max **S** 0.0025 max **Sb** 0.02 max **Si** 0.05 max **Sn** 7.5-8.5 **Ti** 0.01 max **Zn** 1.0 max	
C96900	Cast Copper-Nickel Alloy	**Cu** rem **Fe** 0.50 max **Mg** 0.15 max **Mn** 0.50-0.30 **Nb** 0.10 max **Ni** 14.5-15.5 **Pb** 0.02 max **Sn** 7.5-8.5 **Zn** 0.50 max **Other** Cu includes Ag, Ni includes Co	
C97300	Cast Nickel-Silver	**Al** 0.005 max **Cu** 53.0-58.0 **Fe** 1.5 max **Mn** 0.50 max **Ni** 11.0-14.0 **P** 0.05 max **Pb** 8.0-11.0 **S** 0.08 max **Sb** 0.35 max **Si** 0.15 max **Sn** 1.5-3.0 **Zn** 17.0-25.0	**ASTM** B30 (973); B271 (973); B505 (973); B584 (973) (formerly B149)
C97400	Cast Nickel-Silver	**Cu** 58.0-61.0 **Fe** 1.5 max **Mn** 0.50 max **Ni** 15.5-17.0 **Pb** 4.5-5.5 **Sn** 2.5-3.5 **Zn** rem	
C97600	Cast Nickel-Silver	**Al** 0.005 max **Cu** 63.0-67.0 **Fe** 1.5 max **Mn** 1.0 max **Ni** 19.0-21.5 **P** 0.05 max **Pb** 3.0-5.0 **S** 0.08 max **Sb** 0.25 max **Si** 0.15 max **Sn** 3.5-4.5 **Zn** 3.0-9.0	**ASME** SB271; SB584 **ASTM** B30 (976); B271 (976); B505 (976); B584 (976) (formerly B149) **MIL SPEC** MIL-C-17112
C97800	Cast Nickel-Silver	**Al** 0.005 max **Cu** 64.0-67.0 **Fe** 1.5 max **Mn** 1.0 max **Ni** 24.0-27.0 **P** 0.05 max **Pb** 1.0-2.5 **S** 0.08 max **Sb** 0.20 max **Si** 0.15 max **Sn** 4.0-5.5 **Zn** 1.0-4.0	**ASTM** B30 (978); B271 (978); B505 (978); B584 (978) (formerly B149)
C98200	Cast Leaded Copper	**Cu** 73.0-79.0 **Fe** 0.35 max **Pb** 21.0-27.0 **Sn** 0.50 max	**MIL SPEC** MIL-B-13506
C98400	Cast Leaded Copper	**Ag** 1.5 max **Cu** 67.0-74.0 **Fe** 0.35 max **P** 0.02 max **Pb** 25.0-32.0 **Sn** 0.25 max **Zn** 0.10 max	**MIL SPEC** MIL-B-13506
C98600	Cast Leaded Copper	**Ag** 1.5 max **Cu** 60.0-70.0 **Fe** 0.35 max **Pb** 30.0-40.0 **Sn** 0.50 max	**MIL SPEC** MIL-B-13506
C98800	Cast Leaded Copper	**Ag** 5.5 max **Cu** 56.5-62.5 **Fe** 0.35 max **P** 0.02 max **Pb** 37.5-42.5 **Sn** 0.25 max **Zn** 0.10 max	
C98820	Cast Leaded Copper	**Cu** rem **Fe** 0.35 max **Pb** 40.0-44.0 **Sn** 1.0-5.0	**SAE** J460
C98840	Cast Leaded Copper	**Cu** rem **Fe** 0.35 max **Pb** 44.0-58.0 **Sn** 1.0-5.0	**SAE** J460

The chemical compositions listed are for identification purposes and should not be used in lieu of the cross referenced specifications.

COPPER AND COPPER ALLOYS

UNIFIED NUMBER	DESCRIPTION	CHEMICAL COMPOSITION	CROSS REFERENCE SPECIFICATIONS
C99300	Incramet 800	**Al** 10.7-11.5 **Co** 1.0-2.0 **Cu** rem **Fe** 0.40-1.0 **Ni** 13.5-16.5 **Pb** 13.5-16.5 **Si** 0.02 max **Sn** 0.05 max **Other** total 0.25 max	
C99400	Cast Copper Alloy	**Al** 0.50-2.0 **Cu** rem **Fe** 1.0-3.0 **Mn** 0.50 max **Ni** 1.0-3.5 **Pb** 0.25 max **Si** 0.50-2.0 **Zn** 0.50-5.0	
C99500	Cast Copper Alloy	**Al** 0.50-2.0 **Cu** rem **Fe** 3.0-5.0 **Mn** 0.50 max **Ni** 3.5-5.5 **Pb** 0.25 max **Si** 0.50-2.0 **Zn** 0.50-2.0	
C99600	Cast Copper (Incramute I)	**Al** 1.0-2.8 **C** 0.05 max **Co** 0.20 max **Cu** rem **Fe** 0.20 max **Mn** 39.0-45.0 **Ni** 0.20 max **Pb** 0.02 max **Si** 0.10 max **Sn** 0.10 max **Zn** 0.20 max	
C99700	Cast Copper Alloy	**Al** 0.50-3.0 **Cb** 4.0-6.0 **Cu** 54.0 min **Fe** 1.0 max **Mn** 11.0-15.0 **Ni** 4.0-6.0 **Pb** 2.0 max **Sn** 1.0 max **Zn** 19.0-25.0 **Other** total named elements, 99.5 min	
C99750	Cast Copper Alloy	**Al** 0.25-3.0 **Cu** 55.0-61.0 **Fe** 1.0 max **Mn** 17.0-23.0 **Ni** 5.0 max **Pb** 0.50-2.5 **Zn** 17.0-23.0 **Other** Fe less than Ni	

The chemical compositions listed are for identification purposes and should not be used in lieu of the cross referenced specifications.

UNS NUMBERS ASSIGNED TO DATE

With Description of Each Material Covered and References To Documents In Which The Same or Similar Materials are Described

Exxxxx Number Series
Rare Earth and Similar Metals and Alloys

RARE EARTH AND SIMILAR METALS AND ALLOYS

UNIFIED NUMBER	DESCRIPTION	CHEMICAL COMPOSITION	CROSS REFERENCE SPECIFICATIONS
E00000	Commercial Grade Actinium	**Ac** 99.0 min	
E01000	Commercial Grade Cerium	**Ce** 99.0 min	
E01401	Cerium - Cobalt Rare Earth Alloy	**Ce** 70.0 nom **Co** 30.0 nom	
E21000	Mischmetal (Rare Earths) (Bastnasite Derived)	**Other** total mixed rare earths 99.0 min, consisting of Ce 50.0, La 38.0, Nd 12.0, Pr 4.0, other rare earth metals 1.0	
E21100	Mischmetal (Rare Earths) (Bastnasite Derived)-2.5 Magnesium	**Mg** 2.5 nom **Other** total mixed rare earths 97.5 nom, consisting of Ce 48.8, La 32.1, Nd 11.7, Pr 3.9, other rare earth metals 1.0	
E21101	Mischmetal (Rare Earths) (Bastnasite Derived)-5.0 Magnesium	**Mg** 5.0 nom **Other** total mixed rare earths 95.0 nom, consisting of Ce 47.5, La 31.3, Nd 11.4, Pr 3.9, other rare earth metals 1.0	
E21102	Mischmetal (Rare Earths) (Bastnasite Derived)-10.0 Magnesium	**Mg** 10.0 nom **Other** total mixed rare earths 90.0 nom, consisting of Ce 45.0, La 29.7, Nd 10.8, Pr 3.6, other rare earth metals 0.9	
E21103	Mischmetal (Rare Earths) (Bastnasite Derived)-20.0 Magnesium	**Mg** 20.0 nom **Other** total mixed rare earths 80.0 nom, consisting of Ce 40.0, La 26.4, Nd 9.6, Pr 3.6, other rare earth metals 0.8	
E21200	Mischmetal (Rare Earths) (Bastnasite Derived)-1.0 Al	**Al** 1.0 nom **Other** total mixed rare earths 99.0 nom	
E21201	Mischmetal (Rare Earths) (Bastnasite Derived)-20.0 Al	**Al** 20.0 nom **Other** total mixed rare earths 80.0 nom	
E21300	Mischmetal (Rare Earths) (Bastnasite Derived)-1.0 Fe	**Fe** 1.0 nom **Other** total mixed rare earths 99.0 nom, consisting of Ce 49.4, La 32.7, Nd 11.9, Pr 4.0, other rare earth metals 1.0	
E21301	Mischmetal (Rare Earths) (Bastnasite Derived)-5.0 Fe	**Fe** 5.0 nom **Other** total mixed rare earths 95.0 nom	
E21302	Mischmetal (Rare Earths) (Bastnasite Derived)-25.0 Fe	**Fe** 25.0 nom **Other** total mixed rare earths 75.0 nom	
E21500	Mischmetal (Rare Earths) (Bastnasite Derived)-1.0 Au	**Au** 1.0 nom **Other** total mixed rare earths 99.0 nom	
E23130	Mischmetal (Rare Earths) (Bastnasite Derived)-2.5 Mg, 1.0 Fe	**Fe** 1.0 nom **Mg** 2.5 nom **Other** total mixed rare earths 96.5 nom, consisting of Ce 48.2, La 31.8, Nd 11.6, Pr 3.9, other rare earth metals 1.0	
E23310	Mischmetal (Rare Earths) (Bastnasite Derived)-10.0 Fe-2.5 Mg	**Fe** 10.0 nom **Mg** 2.5 nom **Other** total mixed rare earths 87.5 nom	
E23311	Mischmetal (Rare Earths) (Bastnasite Derived)-22.0 Fe-2.0 Mg	**Fe** 22.0 nom **Mg** 2.0 nom **Other** total mixed rare earths 76.0 nom	
E23312	Mischmetal (Rare Earths) (Bastnasite Derived)-23.0 Fe-6.0 Mg	**Fe** 23.0 nom **Mg** 6.0 nom **Other** total mixed rare earths 71.0 nom	
E28531	Mischmetal (Rare Earths) (Bastnasite Derived)-36.0 Si-14.0 Fe-5.0 Ca	**Ca** 5.0 nom **Fe** 14.0 nom **Si** 36.0 nom **Other** total mixed rare earths 45.0 nom, consisting of Ce 22.5, La 14.8, Nd 5.4, Pr 1.8, other rare earth metals 0.5	

The chemical compositions listed are for identification purposes and should not be used in lieu of the cross referenced specifications.

UNIFIED NUMBER	DESCRIPTION	CHEMICAL COMPOSITION	CROSS REFERENCE SPECIFICATIONS
E31000	Mischmetal (Rare Earths) (Monazite Derived)	**Other** total mixed rare earths 99.0 min, consisting of Ce 45.0, La 20.0, Nd 19.0, Pr 6.0, Sm 4.0, Gd 2.0, Y 2.0, other rare earth metals 2.0	
E46000	Commercial Grade Dysprosium	**Dy** 99.0 min	
E48000	Commercial Grade Erbium	**Er** 99.0 min	
E50000	Commercial Grade Europium	**Eu** 99.0 min	
E52000	Commercial Grade Gadolinium	**Gd** 99.0 min	
E56000	Commercial Grade Holmium	**Ho** 99.0 min	
E58000	Commercial Grade Lanthanum	**La** 99.0 min	
E58401	Lanthanum-Cobalt Rare Earth Alloy	**Co** 25.0 nom **La** 75.0 nom	
E68000	Commercial Grade Lutetium	**Lu** 99.0 min	
E69000	Commercial Grade Neodymium	**Nd** 99.0 min	
E74000	Commercial Grade Praseodymium	**Pr** 99.0 min	
E74401	Praseodymium-Cobalt Rare Earth Alloy	**Co** 28.0 nom **Pr** 72.0 nom	
E78000	Commercial Grade Promethium	**Pm** 99.0 min	
E79000	Commercial Grade Samarium	**Sm** 99.0 min	
E79401	Samarium-Cobalt Rare Earth Alloy	**Co** 40.0 nom **Sm** 60.0 nom	
E83000	Commercial Grade Scandium	**Sc** 99.0 min	
E85000	Commercial Grade Terbium	**Tb** 99.0 min	
E87000	Commercial Grade Thulium	**Tm** 99.0 min	
E88000	Commercial Grade Ytterbium	**Yb** 99.0 min	
E90000	Commercial Grade Yttrium	**Y** 99.0 min	
E90401	Yttrium-Cobalt Rare Earth Alloy	**Co** 36.0 nom **Y** 64.0 nom	
E90501	Yttrium-Chromium Rare Earth Alloy	**Cr** 5.0 nom **Y** 95.0 nom	

The chemical compositions listed are for identification purposes and should not be used in lieu of the cross referenced specifications.

UNS NUMBERS ASSIGNED TO DATE

With Description of Each Material Covered and References To Documents In Which The Same or Similar Materials are Described

Fxxxxx Number Series
Cast Irons

CAST IRONS

UNIFIED NUMBER	DESCRIPTION	CHEMICAL COMPOSITION	CROSS REFERENCE SPECIFICATIONS
F10001	Cast Iron, Gray	**C** 3.50 min	**ASTM** A319 (I)
F10002	Cast Iron, Gray	**C** 3.20 min	**ASTM** A319 (II)
F10003	Cast Iron, Gray	**C** 2.80 min	**ASTM** A319 (III)
F10004	Cast Iron, Gray, HB: 187 max	**C** 3.40-3.70 **Mn** 0.50-0.80 **P** 0.25 max **S** 0.15 max **Si** 2.80-2.30	**ASTM** A159 (G1800) **SAE** J431 (G1800)
F10005	Cast Iron, Gray, HB: 170-229	**C** 3.20-3.50 **Mn** 0.60-0.90 **P** 0.20 max **S** 0.15 max **Si** 2.40-2.00	**ASTM** A159 (G2500) **SAE** J431 (G2500)
F10006	Cast Iron, Gray, HB: 187-241	**C** 3.10-3.40 **Mn** 0.60-0.90 **P** 0.15 max **S** 0.15 max **Si** 2.30-1.90	**ASTM** A159 (G3000) **SAE** J431 (G3000)
F10007	Cast Iron, Gray, HB: 207-255	**C** 3.00-3.30 **Mn** 0.60-0.90 **P** 0.12 max **S** 0.15 max **Si** 2.20-1.80	**ASTM** A159 (G3500) **SAE** J431 (G3500)
F10008	Cast Iron, Gray, HB: 217-269	**C** 3.00-3.30 **Mn** 0.70-1.00 **P** 0.10 max **S** 0.15 max **Si** 2.10-1.80	**ASTM** A159 (G4000) **SAE** J125 (G4000); J431 (G4000)
F10009	Cast Iron, Gray, HB: 170-229	**C** 3.40 min **Mn** 0.60-0.90 **P** 0.15 max **S** 0.12 max **Si** 1.60-2.10	**ASTM** A159 (G2500a) **SAE** J431 (G2500a)
F10010	Cast Iron, Gray, HB: 207-255	**C** 3.40 min **Mn** 0.60-0.90 **P** 0.15 max **S** 0.12 max **Si** 1.30-1.80	**ASTM** A159 (G3500b) **SAE** J431 (G3500b)
F10011	Cast Iron, Gray, HB: 207-255	**C** 3.50 min **Mn** 0.60-0.90 **P** 0.15 max **S** 0.12 max **Si** 1.30-1.80	**ASTM** A159 (G3500c) **SAE** J431 (G3500c)
F10012	Cast Iron, Gray, HB: 241-321	**C** 3.10-3.60 **Cr** 0.85-1.25 **Mn** 0.60-0.90 **Mo** 0.40-0.60 **P** 0.10 max **S** 0.15 max **Si** 1.95-2.40 **Other** Ni 0.20-0.45 (optional)	**ASTM** A159 (G4000d) **SAE** J431 (G4000d)
F10090	Cast Iron Welding Filler Metal (RCI)	**C** 3.25-3.50 **Mn** 0.60-0.75 **P** 0.50-0.75 **S** 0.10 max **Si** 2.75-3.00	**AWS** A5.15 (RCI)
F10091	Cast Iron Welding Filler Metal (RCI-A)	**C** 3.25-3.50 **Mn** 0.50-0.70 **Mo** 0.25-0.45 **Ni** 1.20-1.60 **P** 0.20-0.40 **S** 0.10 max **Si** 2.00-2.50	**AWS** A5.15 (RCI-A)
F10092	Cast Iron Welding Filler Metal (RCI-B)	**C** 3.25-4.00 **Mg** 0.04-0.1 **Mn** 0.10-0.40 **Ni** 0.50 max **P** 0.05 max **S** 0.015 max **Si** 3.25-3.75 **Other** Ce 0.20 max	**AWS** A5.15 (RCI-B)
F11401	Cast Iron, Gray, T.S.: 138 MPa (20 ksi) Min	**Other** None specified	**ASME** SA278 (20) **ASTM** A48 (20); A278 (20)
F11501	Cast Iron, Gray, T.S.: 145 MPa (21 ksi) Min	**Other** None specified	**ASTM** A126 (A)
F11701	Cast Iron, Gray, T.S.: 172 MPa (25 ksi) Min	**Other** None specified	**ASME** SA278 (25) **ASTM** A48 (25); A278 (25)
F12101	Cast Iron, Gray, T.S.: 207 MPa (30 ksi) Min	**Other** None specified	**ASME** SA278 (30) **ASTM** A48 (30); A278 (30)
F12102	Cast Iron, Gray, T.S.: 214 MPa (31 ksi) Min	**Other** None specified	**ASTM** A126 (B)
F12401	Cast Iron, Gray, T.S.: 241 MPa (35 ksi) Min	**Other** None specified	**ASME** SA278 (35) **ASTM** A48 (35); A278 (35)
F12801	Cast Iron, Gray, T.S.: 276 MPa (40 ksi) Min	**Other** None specified	**ASTM** A48 (40)
F12802	Cast Iron, Gray, T.S.: 283 MPa (41 ksi) Min	**Other** None specified	**ASTM** A126 (C)
F12803	Cast Iron, Gray, T.S.: 276 MPa (40 ksi) Min	**Other** None specified	**ASME** SA278 (40) **ASTM** A278 (40)
F13101	Cast Iron, Gray, T.S.: 310 MPa (45 ksi) Min	**Other** None specified	**ASTM** A48 (45)

The chemical compositions listed are for identification purposes and should not be used in lieu of the cross referenced specifications.

CAST IRONS

UNIFIED NUMBER	DESCRIPTION	CHEMICAL COMPOSITION	CROSS REFERENCE SPECIFICATIONS
F13102	Cast Iron, Gray, T.S.: 310 MPa (45 ksi) Min	**Other** None specified	**ASME** SA278 (45) **ASTM** A278 (45)
F13501	Cast Iron, Gray, T.S.: 345 MPa (50 ksi) Min	**Other** None specified	**ASTM** A48 (50)
F13502	Cast Iron, Gray, T.S.: 345 MPa (50 ksi) Min	**Other** None specified	**ASME** SA278 (50) **ASTM** A278 (50)
F13801	Cast Iron, Gray, T.S.: 379 MPa (55 ksi) Min	**Other** None specified	**ASTM** A48 (55)
F13802	Cast Iron, Gray, T.S.: 379 MPa (55 ksi) Min	**Other** None specified	**ASME** SA278 (55) **ASTM** A278 (55)
F14101	Cast Iron, Gray, T.S.: 414 MPa (60 ksi) Min	**Other** None specified	**ASTM** A48 (60)
F14102	Cast Iron, Gray, T.S.: 414 MPa (60 ksi) Min	**Other** None specified	**ASME** SA278 (60) **ASTM** A278 (60)
F14801	Cast Iron, Gray, T.S.: 483 MPa (70 ksi) Min	**Other** None specified	**ASME** SA278 (70) **ASTM** A278 (70)
F15501	Cast Iron, Gray, T.S.: 552 MPa (80 ksi) Min	**Other** None specified	**ASME** SA278 (80) **ASTM** A278 (80)
F20000	Cast Iron, Malleable, T.S.: 345 MPa (50 ksi) Min, Y.S.: 220.5 MPa (32 ksi) Min	**C** 2.20-2.90 **Mn** 0.15-1.25 **P** 0.02-0.15 **S** 0.02-0.20 **Si** 0.90-1.90	**ASTM** A602 (M3210) **SAE** J158 (M3210)
F20001	Cast Iron, Malleable, T.S.: 447.9 MPa (65 ksi) Min, Y.S.: 309.7 MPa (45 ksi) Min	**C** 2.20-2.90 **Mn** 0.15-1.25 **P** 0.02-0.15 **S** 0.02-0.20 **Si** 0.90-1.90	**SAE** J158 (M4504)
F20002	Cast Iron, Malleable, T.S.: 516.5 MPa (75 ksi) Min, Y.S.: 345 MPa (50 ksi) Min	**C** 2.20-2.90 **Mn** 0.15-1.25 **P** 0.02-0.15 **S** 0.02-0.20 **Si** 0.90-1.90	**SAE** J158 (M5003)
F20003	Cast Iron, Malleable, T.S.: 516.5 MPa (75 ksi) Min, Y.S.: 379.3 MPa (55 ksi) Min	**C** 2.20-2.90 **Mn** 0.15-1.25 **P** 0.02-0.15 **S** 0.02-0.20 **Si** 0.90-1.90	**SAE** J158 (M5503)
F20004	Cast Iron, Malleable, T.S.: 620.3 MPa (90 ksi) Min, Y.S.: 482.2 MPa (70 ksi) Min	**C** 2.20-2.90 **Mn** 0.15-1.25 **P** 0.02-0.15 **S** 0.02-0.20 **Si** 0.90-1.90	**SAE** J158 (M7002)
F20005	Cast Iron, Malleable, T.S.: 723.2 MPa (105 ksi) Min, Y.S.: 586.0 MPa (85 ksi) Min	**C** 2.20-2.90 **Mn** 0.15-1.25 **P** 0.02-0.15 **S** 0.02-0.20 **Si** 0.90-1.90	**SAE** J158 (M8501)
F22000	Cast Iron, Malleable, Cupola, T.S.: 276 MPa (40 ksi) Min, Y.S.: 207 MPa (30 ksi) Min	**Other** None specified	**ASTM** A197
F22200	Cast Iron, Malleable, T.S.: 345 MPa (50 ksi) Min, Y.S.: 224 MPa (32 ksi) Min	**Other** None specified	**ASTM** A47 (32510)
F22400	Cast Iron, Malleable, T.S.: 365 MPa (53 ksi) Min, Y.S.: 241 MPa (35 ksi) Min	**Other** None specified	**ASTM** A47 (35018)
F22830	Cast Iron, Pearlitic Malleable, T.S.: 414 MPa (60 ksi) Min, Y.S.: 276 MPa (40 ksi) Min	**Other** None specified	**ASTM** A220 (40010)
F23130	Cast Iron, Pearlitic Malleable, T.S.: 448 MPa (65 ksi) Min, Y.S.: 310 MPa (45 ksi) Min, Elongation: 8% Min	**Other** None specified	**ASTM** A220 (45008)

The chemical compositions listed are for identification purposes and should not be used in lieu of the cross referenced specifications.

UNIFIED NUMBER	DESCRIPTION	CHEMICAL COMPOSITION	CROSS REFERENCE SPECIFICATIONS
F23131	Cast Iron, Pearlitic Malleable, T.S.: 448 MPa (65 ksi) Min, Y.S.: 310 MPa (45 ksi) Min, Elongation: 6% Min	**Other** None specified	**ASTM** A220 (45006)
F23330	Cast Iron, Pearlitic Malleable, T.S.: 483 MPa (70 ksi) Min, Y.S.: 331 MPa (48 ksi) Min	**Other** None specified	**AMS** 5310
F23530	Cast Iron, Pearlitic Malleable, T.S.: 483 MPa (70 ksi) Min, Y.S.: 345 MPa (50 ksi) Min	**Other** None specified	**ASTM** A220 (50005)
F24130	Cast Iron, Pearlitic, T.S.: 552 MPa (80 ksi) Min, Y.S.: 414 MPa (60 ksi) Min	**Other** None specified	**ASTM** A220 (60004)
F24830	Cast Iron, Pearlitic Malleable, T.S.: 586 MPa (85 ksi) Min, Y.S.: 483 MPa (70 ksi) Min	**Other** None specified	**ASTM** A220 (70003)
F25530	Cast Iron, Pearlitic Malleable, T.S.: 655 MPa (95 ksi) Min, Y.S.: 552 MPa (80 ksi) Min	**Other** None specified	**ASTM** A220 (80002)
F26230	Cast Iron, Pearlitic Malleable, T.S.: 724 MPa (105 ksi) Min, Y.S.: 621 MPa (90 ksi) Min	**Other** None specified	**ASTM** A220 (90001)
F30000	Cast Iron, Ductile, Martensitic, HB: As specified (Q and T)	**Other** None specified	**SAE** J434 (DQ and T)
F32800	Cast Iron, Ductile, T.S.: 414 MPa (60 ksi) Min, Y.S.: 276 MPa (40 ksi) Min, HB: 170	**Other** None specified	**ASME** SA395 (60-40-18) **ASTM** A395 (60-40-18); A536 (60-40-18) **SAE** J434 (D4018)
F32900	Cast Iron, Ductile, T.S.: 414 MPa (60 ksi) Min, Y.S.: 290 MPa (42 ksi) Min		**ASTM** A716
F33100	Cast Iron, Ductile, T.S.: 448 MPa (65 ksi) Min, Y.S.: 310 MPa (45 ksi) Min, HB: 156-217	**Other** None specified	**ASTM** A536 (65-45-12) **SAE** J434 (D4512)
F33101	Cast Iron, Ductile, T.S.: 414 MPa (60 ksi) Min, Y.S.: 310 MPa (45 ksi) Min, HB: 190	**C** 3.0 min **Si** 2.50 max	**AMS** 5315 **MIL SPEC** MIL-I-24137 (A)
F33800	Cast Iron, Ductile, T.S.: 552 MPa (80 ksi) Min, Y.S.: 379 MPa (55 ksi) Min, HB: 187-255	**Other** None specified	**ASTM** A536 (80-55-06) **SAE** J434 (D5506)
F34100	Cast Iron, Ductile, T.S.: 552 MPa (80 ksi) Min, Y.S.: 414 MPa (60 ksi) Min	**C** 3.0 min **P** 0.08 max **S** 0.05 max **Si** 3.0 max	**AMS** 5316 **ASTM** A476 (80-60-03)
F34800	Cast Iron, Ductile, T.S.: 689 MPa (100 ksi) Min, Y.S.: 483 MPa (70 ksi) Min, HB: 241-302	**Other** None specified	**ASTM** A536 (100-70-03) **SAE** J434 (D7003)
F36200	Cast Iron, Ductile, T.S.: 827 MPa (120 ksi) Min, Y.S.: 621 MPa (90 ksi) Min	**Other** None specified	**ASTM** A536 (120-90-02)
F41000	Cast Iron, Gray, Austenitic, T.S.: 172 MPa (25 ksi) Min	**C** 3.00 max **Cr** 1.5-2.5 **Cu** 5.50-7.50 **Mn** 0.5-1.5 **Ni** 13.50-17.50 **S** 0.12 max **Si** 1.00-2.80	**ASTM** A436 (1)
F41001	Cast Iron, Gray, Austenitic, T.S.: 207 MPa (30 ksi) Min	**C** 3.00 max **Cr** 2.50-3.50 **Cu** 5.50-7.50 **Mn** 0.5-1.5 **Ni** 13.50-17.50 **S** 0.12 max **Si** 1.00-2.80	**ASTM** A436 (1b)

The chemical compositions listed are for identification purposes and should not be used in lieu of the cross referenced specifications.

UNIFIED NUMBER	DESCRIPTION	CHEMICAL COMPOSITION	CROSS REFERENCE SPECIFICATIONS
F41002	Cast Iron, Gray, Austenitic, Y.S.: 172 MPa (25 ksi) Min	C 3.00 max Cr 1.5-2.5 Cu 0.50 max Mn 0.50-1.50 Ni 18.00-22.00 S 0.12 max Si 1.00-2.80	ASTM A436 (2)
F41003	Cast Iron, Gray, Austenitic, T.S.: 207 MPa (30 ksi) Min	C 3.00 max Cr 3.00-6.00 Cu 0.50 max Mn 0.5-1.5 Ni 18.00-22.00 S 0.12 max Si 1.00-2.80	ASTM A436 (2b)
F41004	Cast Iron, Gray, Austenitic, T.S.: 172 MPa (25 ksi) Min	C 2.60 max Cr 2.50-3.50 Cu 0.50 max Mn 0.5-1.5 Ni 28.00-32.00 S 0.12 max Si 1.00-2.00	ASTM A436 (3)
F41005	Cast Iron, Gray, Austenitic, T.S.: 172 MPa (25 ksi) Min	C 2.60 max Cr 4.50-5.50 Cu 0.50 max Mn 0.5-1.5 Ni 29.00-32.00 S 0.12 max Si 5.00-6.00	ASTM A436 (4)
F41006	Cast Iron, Gray, Austenitic, T.S.: 138 MPa (20 ksi) Min	C 2.40 max Cr 0.10 max Cu 0.50 max Mn 0.5-1.5 Ni 34.00-36.00 S 0.12 max Si 1.00-2.00	ASTM A436 (5)
F41007	Cast Iron, Gray, Austenitic, T.S.: 172 MPa (25 ksi) Min	C 3.00 max Cr 1.00-2.00 Cu 3.50-5.50 Mn 0.5-1.5 Mo 1.00 max Ni 18.00-22.00 S 0.12 Si 1.50-2.50	ASTM A436 (6)
F43000	Cast Iron, Ductile, Austenitic, T.S.: 400 MPa (58 ksi) Min, Y.S.: 207 MPa (30 ksi) Min	C 3.00 max Cr 1.75-2.75 Mn 0.70-1.25 Ni 18.00-22.00 P 0.08 max Si 1.50-3.00	ASTM A439 (D-2)
F43001	Cast Iron, Ductile, Austenitic, T.S.: 400 MPa (58 ksi) Min, Y.S.: 207 MPa (30 ksi) Min	C 3.00 max Cr 2.75-4.00 Mn 0.70-1.25 Ni 18.00-22.00 P 0.08 max Si 1.50-3.00	ASTM A439 (D-2B)
F43002	Cast Iron, Ductile, Austenitic, T.S.: 400 MPa (58 ksi) Min, Y.S.: 193 MPa (28 ksi) Min	C 2.90 max Cr 0.50 max Mn 1.80-2.40 Ni 21.00-24.00 P 0.08 max Si 1.00-3.00	ASTM A439 (D-2C)
F43003	Cast Iron, Ductile, Austenitic, T.S.: 379 MPa (55 ksi) Min, Y.S.: 207 MPa (30 ksi) Min	C 2.60 max Cr 2.50-3.50 Mn 1.00 max Ni 28.00-32.00 P 0.80 max Si 1.00-2.80	ASTM A439 (D-3)
F43004	Cast Iron, Ductile, Austenitic, T.S.: 379 MPa (55 ksi) Min, Y.S.: 207 MPa (30 ksi) Min	C 2.60 max Cr 1.00-1.50 Mn 1.00 max Ni 28.00-32.00 P 0.08 max Si 1.00-2.80	ASTM A439 (D-3A)
F43005	Cast Iron, Ductile, Austenitic, T.S.: 414 MPa (60 ksi) Min, HB: 202-273	C 2.60 max Cr 4.50-5.50 Mn 1.00 max Ni 28.00-32.00 P 0.08 max Si 5.00-6.00	ASTM A439 (D-4)
F43006	Cast Iron, Ductile, Austenitic, T.S.: 379 MPa (55 ksi) Min, Y.S.: 207 MPa (30 ksi) Min	C 2.40 max Cr 0.10 max Mn 1.00 max Ni 34.00-36.00 P 0.08 max Si 1.00-2.80	ASTM A439 (D-5)
F43007	Cast Iron, Ductile, Austenitic, T.S.: 379 MPa (55 ksi) Min, Y.S.: 207 MPa (30 ksi) Min	C 2.40 max Cr 2.00-3.00 Mn 1.00 max Ni 34.00-36.00 P 0.08 max Si 1.00-2.80	ASTM A439 (D-5B)
F43010	Cast Iron, Ductile, Austenitic, T.S.: 448 MPa (65 ksi) Min, Y.S.: 207 MPa (30 ksi) Min	C 2.20-2.70 Cr 0.20 max Mn 3.75-4.50 Ni 21.0-24.0 P 0.08 max Si 1.50-2.50	ASTM A571 (D-2M)
F43020	Cast Iron, Ductile, Austenitic, T.S.: 379 MPa (50 ksi) Min, Y.S.: 207 MPa (30 ksi) Min	C 2.40-3.00 Cr 1.70-2.40 Mn 0.80-1.50 Ni 18.00-22.00 P 0.20 max Si 1.80-3.20	MIL SPEC MIL-I-24137 (B)
F43021	Cast Iron, Ductile, Austenitic, T.S.: 345 MPa (50 ksi) Min, Y.S.: 172 MPa (25 ksi) Min	C 2.70-3.10 Cr 0.50 max Mn 1.90-2.50 Ni 20.00-23.00 P 0.15 Si 2.00-3.00	MIL SPEC MIL-I-24137 (C)
F43030	Cast Iron, Ductile, T.S.: 345 MPa (50 ksi) Min, Y.S.: 173 MPa (25 ksi) Min	C 2.5-3.0 Cr 0.50 max Mn 1.9-2.5 Mo 0.30 max Ni 20.0-24.0 P 0.15 max S 0.05 max Si 2.0-3.0	AMS 5395
F45000	Cast Iron, White, HB (Sand Cast): 550 Min, HB (Chill Cast): 600 Min	C 3.0-3.6 Cr 1.4-4.0 Mn 1.3 max Mo 1.0 max Ni 3.3-5.0 P 0.30 max S 0.15 max Si 0.8 max	ASTM A532 (IA)

The chemical compositions listed are for identification purposes and should not be used in lieu of the cross referenced specifications.

UNIFIED NUMBER	DESCRIPTION	CHEMICAL COMPOSITION	CROSS REFERENCE SPECIFICATIONS
F45001	Cast Iron, White, HB (Sand Cast): 550 Min, HB (Chill Cast): 600 Min	**C** 2.5-3.0 **Cr** 1.4-4.0 **Mn** 1.3 max **Mo** 1.0 max **Ni** 3.3-5.0 **P** 0.30 max **S** 0.15 max **Si** 0.8 max	**ASTM** A532 (IB)
F45002	Cast Iron, White, HB (Sand Cast): 550 Min, HB (Chill Cast): 600 Min	**C** 2.9-3.7 **Cr** 1.1-1.5 **Mn** 1.3 max **Mo** 1.0 max **Ni** 2.7-4.0 **P** 0.30 max **S** 0.15 max **Si** 0.8 max	**ASTM** A532 (IC)
F45003	Cast Iron, White, HB: 400-600	**C** 2.5-3.6 **Cr** 7.0-11.0 **Mn** 1.3 max **Mo** 1.0 max **Ni** 4.5-7.0 **P** 0.10 max **S** 0.15 max **Si** 1.0-2.2	**ASTM** A532 (ID)
F45004	Cast Iron, White, HB: 400-600	**C** 2.4-2.8 **Cr** 11.0-14.0 **Cu** 1.2 max **Mn** 0.5-1.5 **Mo** 0.5-1.0 **Ni** 0.5 max **P** 0.10 max **S** 0.06 max **Si** 1.0 max	**ASTM** A532 (IIA)
F45005	Cast Iron, White, HB: 400-600	**C** 2.4-2.8 **Cr** 14.0-18.0 **Cu** 1.2 max **Mn** 0.5-1.5 **Mo** 1.0-3.0 **Ni** 0.5 max **P** 0.10 max **S** 0.06 max **Si** 1.0 max	**ASTM** A532 (IIB)
F45006	Cast Iron, White, HB: 400-600	**C** 2.8-3.6 **Cr** 14.0-18.0 **Cu** 1.2 max **Mn** 0.5-1.5 **Mo** 2.3-3.5 **Ni** 0.5 max **P** 0.10 max **S** 0.06 max **Si** 1.0 max	**ASTM** A532 (IIC)
F45007	Cast Iron, White, HB: 400-600	**C** 2.0-2.6 **Cr** 18.0-23.00 **Cu** 1.2 max **Mn** 0.5-1.5 **Mo** 1.5 max **Ni** 1.5 max **P** 0.10 max **S** 0.06 max **Si** 1.0 max	**ASTM** A532 (IID)
F45008	Cast Iron, White, HB: 400-600	**C** 2.6-3.2 **Cr** 18.0-23.0 **Cu** 1.2 max **Mn** 0.5-1.5 **Mo** 1.0-2.0 **Ni** 1.5 max **P** 0.10 max **S** 0.06 max **Si** 1.0 max	**ASTM** A532 (IIE)
F45009	Cast Iron, White, HB: 400-600	**C** 2.3-3.0 **Cr** 23.0-28.0 **Cu** 1.2 max **Mn** 0.5-1.5 **Mo** 1.5 max **Ni** 1.5 max **P** 0.10 max **S** 0.06 max **Si** 1.0 max	**ASTM** A532 (IIIA)
F45100	Cast Iron, White, Welding Rod for Surfacing	**C** 3.7-5.0 **Cr** 27.0-35.0 **Mn** 2.0-6.0 **Si** 1.10-2.5	**ASME** SFA5.13 (RFeCr-A1) **AWS** A5.13 (RFeCr-A1)
F47001	Cast Iron, Corrosion Resistant, T.S.: 172 MPa (25 ksi) Min	**C** 2.60-3.00 **Cr** 1.80-3.50 **Cu** 5.50-7.50 **Mn** 1.00-1.50 **Ni** 13.5-17.5 **P** 0.20 max **S** 0.10 max **Si** 1.25-2.20	
F47002	Cast Iron, Corrosion Resistant, T.S.: 172 MPa (25 ksi) Min	**C** 2.60-3.00 **Cr** 1.75-3.50 **Cu** 0.50 max **Mn** 0.80-1.30 **Ni** 18.0-22.0 **P** 0.20 max **S** 0.10 max **Si** 1.25-2.20	
F47003	Cast Iron, Corrosion Resistant	**C** 0.70-1.10 **Cr** 0.50 max **Cu** 0.50 max **Mn** 1.50 max **Mo** 0.50 max **Si** 14.20-14.75	**ASTM** A518 G Rate 2
F47004	Cast Iron, Corrosion Resistant, T.S.: 207 MPa (30 ksi) Min	**C** 2.4-2.8 **Cr** 1.8-2.4 **Cu** 6.0-7.0 **Mn** 1.0-1.5 **Ni** 14.0-16.0 **P** 0.30 max **Pb** 0.003 max **S** 0.12 max **Si** 1.5-2.5	**AMS** 5392
F47005	Cast Iron, Corrosion and Heat Resistant, T.S.: 207 MPa (30 ksi) Min	**C** 2.4-2.8 **Cr** 1.7-2.4 **Cu** 0.5 **Mn** 0.8-1.6 **Ni** 18.0-22.0 **P** 0.30 **Pb** 0.003 **S** 0.12 **Si** 1.5-2.8	**AMS** 5393
F47006	Cast Iron, Ductile, Corrosion and Heat Resistant, T.S.: 379 MPa (55 ksi) Min, Y.S.: 227 MPa (32 ksi) Min	**C** 2.4-3.0 **Cr** 1.7-2.4 **Cu** 0.50 max **Mn** 0.8-1.6 **Ni** 18.0-22.0 **P** 0.25 max **Pb** 0.003 max **Si** 2.0-3.2	**AMS** 5394

The chemical compositions listed are for identification purposes and should not be used in lieu of the cross referenced specifications.

UNS NUMBERS ASSIGNED TO DATE

With Description of Each Material Covered and References To Documents In Which The Same or Similar Materials are Described

Gxxxxx Number Series
AISI and SAE Carbon and Alloy Steels

AISI AND SAE CARBON AND ALLOY STEELS

UNIFIED NUMBER	DESCRIPTION	CHEMICAL COMPOSITION	CROSS REFERENCE SPECIFICATIONS
G10050	Carbon Steel	**C** 0.06 max **Mn** 0.35 max **P** 0.040 max **S** 0.50 max	**AISI** 1005 **ASTM** A29 (1005); A510 (1005) **MIL SPEC** MIL-S-11310 (CS1005) **SAE** J403 (1005); J412 (1005)
G10060	Carbon Steel	**C** 0.08 max **Mn** 0.25-0.40 **P** 0.040 max **S** 0.050 max **Other** Sheets and Plates, Mn 0.25-0.45	**AISI** 1006 **AMS** 5041 **ASTM** A29 (1006); A510 (1006); A545 (1006) **FED** QQ-W-461 (1006) **MIL SPEC** MIL-S-11310 (CS1006) **SAE** J403 (1006); J412 (1005); J414 (1006)
G10080	Carbon Steel	**C** 0.10 max **Mn** 0.30-0.50 **P** 0.040 max **S** 0.050 max **Other** Sheets, Mn 0.25-0.50; ERW Tubing, Mn 0.25-0.50	**AISI** 1008 **ASTM** A29 (1008); A108 (1008); A510 (1008); A519 (1008); A545 (1008); A549 (1008); A575 (M1008); A576 (1008) **FED** QQ-W-461 (1008) **MIL SPEC** MIL-S-11310 (CS1008) **SAE** J403 (1008); J412 (1008); J414 (1008)
G10090	Carbon Steel	**C** 0.15 max **Mn** 0.60 max **P** 0.040 max **S** 0.050 max	**AISI** 1009 **FED** QQ-S-635 (C1009) **SAE** J118 (1009)
* G10100	Carbon Steel	**C** 0.08-0.13 **Mn** 0.30-0.60 **P** 0.040 max **S** 0.050 max	**AISI** 1010 **AMS** 5040; 5042; 5044; 5047; 5050; 5053; 7225 **ASTM** A29 (1010); A108 (1010); A510 (1010); A519 (1010); A545 (1010); A549 (1010); A575 (1010); A576 (1010) **FED** QQ-W-461 **MIL SPEC** MIL-S-11310 (CS1010) **SAE** J403 (1010); J412 (1010); J414 (1010)
G10110	Carbon Steel	**C** 0.08-0.13 **Mn** 0.60-0.90 **P** 0.040 max **S** 0.050 max	**AISI** 1011 **ASTM** A29 (1011) **SAE** J118 (1011); J412 (1011)
G10120	Carbon Steel	**C** 0.10-0.15 **Mn** 0.30-0.60 **P** 0.040 max **S** 0.050 max	**AISI** 1012 **ASTM** A29 (1012); A510 (1012); A519 (1012); A545 (1012); A549 (1012); A575 (M1012); A576 (1012) **MIL SPEC** MIL-S-11310 (CS1012) **SAE** J403 (1012); J412 (1012); J414 (1012)
G10130	Carbon Steel	**C** 0.11-0.16 **Mn** 0.50-0.80 **P** 0.040 max **S** 0.050 max	**ASTM** A29 (1013) **SAE** J403 (1013); J412 (1013)
G10150	Carbon Steel	**C** 0.13-0.18 **Mn** 0.30-0.60 **P** 0.040 max **S** 0.050 max **Other** Sheets, C 0.12-0.18	**AISI** 1015 **AMS** 5060 **ASTM** A29 (1015); A510 (1015); A519 (1015); A545 (1015); A549 (1015); A575 (M1015); A576 (1015); A659 (1015) **FED** QQ-S-498 (C1015); QQ-W-461 **MIL SPEC** MIL-S-16974 (1015) **SAE** J403 (1015); J412 (1015); J414 (1015)
G10160	Carbon Steel	**C** 0.13-0.18 **Mn** 0.60-0.90 **P** 0.040 max **S** 0.050 max **Other** Sheets, C 0.12-0.18; ERW Tubing, C 0.12-0.19	**AISI** 1016 **ASTM** A29 (1016); A108 (1016); A510 (1016); A513 (1016); A545 (1016); A548 (1016); A549 (1016); A576 (1016); A659 (1016) **MIL SPEC** MIL-S-866 (1016) **SAE** J403 (1016); J412 (1016); J414 (1016)
G10170	Carbon Steel	**C** 0.15-0.20 **Mn** 0.30-0.60 **P** 0.040 max **S** 0.050 max **Other** Sheets and Plates, C 0.14-0.20; ERW Tubing, C 0.14-0.21	**AISI** 1017 **ASTM** A29 (1017); A108 (1017); A510 (1017); A513 (1017); A519 (1017); A544 (1017); A549 (1017); A575 (M1017); A576 (1017); A659 (1017) (1017) **MIL SPEC** MIL-S-11310 (CS1017) **SAE** J403 (1017); J412 (1017); J414 (1017)
G10180	Carbon Steel	**C** 0.15-0.20 **Mn** 0.60-0.90 **P** 0.040 max **S** 0.050 max **Other** Sheets and Plates, C 0.14-0.20; ERW Tubing, C 0.14-0.21	**AISI** 1018 **AMS** 5069 **ASTM** A29 (1018); A108 (1018); A510 (1018); A513 (1018); A519 (1018); A544 (1018); A545 (1018); A548 (1018); A549 (1018); A576 (1018); A659 (1018) **FED** QQ-W-461 (1018) **MIL SPEC** MIL-S-11310 (CS1018) **SAE** J403 (1018); J412 (1018); J414 (1018)
G10190	Carbon Steel	**C** 0.15-0.20 **Mn** 0.70-1.00 **P** 0.040 max **S** 0.050 max **Other** Sheets and Plates, C 0.14-0.20; ERW Tubing, C 0.14-0.21	**AISI** 1019 **ASTM** A29 (1019); A510 (1019); A513 (1019); A519 (1019); A545 (1019); A548 (1019); A576 (1019) **SAE** J403 (1019); J412 (1019); J414 (1019)
G10200	Carbon Steel	**C** 0.18-0.23 **Mn** 0.30-0.60 **P** 0.040 max **S** 0.050 max **Other** Sheets and Plates, C 0.17-0.23	**AISI** 1020 **AMS** 5032; 5045 **ASTM** A29 (1020); A510 (1020); A519 (1020); A544 (1020); A575 (M1020); A576 (1020); A659 (1020) **FED** QQ-W-461 (1020) **MIL SPEC** MIL-T-3520; MIL-S-7952; MIL-S-11310 (CS1020); MIL-S-16788; MIL-S-16974; MIL-S-46059 **SAE** J403 (1020); J412 (1020); J414 (1020)
G10210	Carbon Steel	**C** 0.18-0.23 **Mn** 0.60-0.90 **P** 0.040 max **S** 0.050 max **Other** Sheets, C 0.17-0.23; ERW Tubing, C 0.17-0.24	**AISI** 1021 **ASTM** A29 (1021); A510 (1021); A519 (1021); A545 (1021); A548 (1021); A576 (1021); A659 (1021) **SAE** J403 (1021); J412 (1021); J414 (1021)
G10220	Carbon Steel	**C** 0.18-0.23 **Mn** 0.70-1.00 **P** 0.040 max **S** 0.050 max **Other** Sheets, C 0.17-0.23; ERW Tubing, C 0.17-0.24	**AISI** 1022 **AMS** 5070 **ASTM** A29 (1022); A510 (1022); A519 (1022); A544 (1022); A545 (1022); A548 (1022); A576 (1022) **MIL SPEC** MIL-S-11310 (CS1022); MIL-S-16974 **SAE** J403 (1022); J412 (1022); J414 (1022)

*Boxed entries are no longer active and are retained for reference purposes only.

The chemical compositions listed are for identification purposes and should not be used in lieu of the cross referenced specifications.

UNIFIED NUMBER	DESCRIPTION	CHEMICAL COMPOSITION	CROSS REFERENCE SPECIFICATIONS
G10230	Carbon Steel	**C** 0.20-0.25 **Mn** 0.30-0.60 **P** 0.040 max **S** 0.050 max **Other** Sheets, C 0.19-0.25; ERW Tubing, C 0.19-0.26	**AISI** 1023 **ASTM** A29 (1023); A510 (1023); A575 (1023); A576 (1023); A659 (1023) **SAE** J403 (1023); J412 (1023); J414 (1023)
G10250	Carbon Steel	**C** 0.22-0.28 **Mn** 0.30-0.60 **P** 0.040 max **S** 0.050 max **Other** ERW Tubing, C 0.21-0.28	**AISI** 1025 **AMS** 5075; 5077 **ASTM** A29 (1025); A510 (1025); A512 (1025); A519 (1025); A575 (M1025); A576 (1025) **FED** QQ-S-700 (C1025) **MIL SPEC** MIL-T-3520; MIL-T-5066; MIL-S-7952; MIL-S-11310 (CS1025) **SAE** J403 (1025); J412 (1025); J414 (1025)
G10260	Carbon Steel	**C** 0.22-0.28 **Mn** 0.60-0.90 **P** 0.040 max **S** 0.050 max **Other** ERW Tubing, C 0.21-0.28	**AISI** 1026 **ASTM** A29 (1026); A273 (1026); A510 (1026); A519 (1026); A545 (1026); A576 (1026) **MIL SPEC** MIL-T-20157; MIL-F-20670; MIL-S-22698; MIL-S-24093 **SAE** J403 (1026); J412 (1026); J414 (1026)
G10290	Carbon Steel	**C** 0.25-0.31 **Mn** 0.60-0.90 **P** 0.040 max **S** 0.050 max	**AISI** 1029 **ASTM** A29 (1029); A273 (1029); A510 (1029); A576 (1029) **SAE** J403 (1029); J412 (1029)
G10300	Carbon Steel	**C** 0.28-0.34 **Mn** 0.60-0.90 **P** 0.040 max **S** 0.050 max **Other** Sheets, C 0.27-0.34, ERW Tubing, C 0.27-0.35	**AISI** 1030 **ASTM** A29 (1030); A510 (1030); A512 (1030); A519 (1030); A544 (1030); A545 (1035); A546 (1030); A576 (1030); A682 (1030) **FED** QQ-S-700 (C1030) **MIL SPEC** MIL-S-11310 (CS1030); MIL-S-46070 **SAE** J403 (1030); J412 (1030); J414 (1030)
G10330	Carbon Steel	**C** 0.29-0.36 **Mn** 0.70-1.00 **P** 0.040 max **S** 0.050 max	**AISI** 1033 **ASTM** A513 (1033) **SAE** J118 (1033)
G10340	Carbon Steel	**C** 0.32-0.38 **Mn** 0.50-0.80 **P** 0.040 max **S** 0.050 max	**AISI** 1034 **ASTM** A29 (1034) **SAE** J118 (1034); J412 (1034)
G10350	Carbon Steel	**C** 0.32-0.38 **Mn** 0.60-0.90 **P** 0.040 max **S** 0.050 max **Other** Sheets, C 0.31-0.38; ERW Tubing, C 0.31-0.39	**AISI** 1035 **AMS** 5080; 5082 **ASTM** A29 (1035); A510 (1035); A519 (1035); A544 (1035); A545 (1035); A546 (1035); A576 (1035); A682 (1035) **FED** QQ-S-635 (C1035); QQ-S-700 (C1035); QQ-W-461 (1035) **MIL SPEC** MIL-S-3289; MIL-S-19434; MIL-S-46070 **SAE** J403 (1035); J412 (1035); J414 (1035)
G10370	Carbon Steel	**C** 0.32-0.38 **Mn** 0.70-1.00 **P** 0.040 max **S** 0.050 max **Other** Sheets, C 0.31-0.38	**AISI** 1037 **ASTM** A29 (1037); A510 (1037); A576 (1037) **SAE** J403 (1037); J412 (1037); J414 (1037)
G10380	Carbon Steel	**C** 0.35-0.42 **Mn** 0.60-0.90 **P** 0.040 max **S** 0.050 max **Other** Sheets, C 0.34-0.42	**AISI** 1038 **ASTM** A29 (1038); A510 (1038); A544 (1038); A545 (1038); A546 (1038); A576 (1038) **SAE** J403 (1038); J412 (1038); J414 (1038)
G10390	Carbon Steel	**C** 0.37-0.44 **Mn** 0.70-1.00 **P** 0.040 max **S** 0.050 max **Other** Sheets, C 0.36-0.44	**AISI** 1039 **ASTM** A29 (1039); A510 (1039); A546 (1039); A576 (1039) **SAE** J403 (1039); J412 (1039); J414 (1039)
G10400	Carbon Steel	**C** 0.37-0.44 **Mn** 0.60-0.90 **P** 0.040 max **S** 0.050 max **Other** Sheets, C 0.36-0.44	**AISI** 1040 **ASTM** A29 (1040); A510 (1040); A519 (1040); A546 (1040); A576 (1040); A682 (1040) **FED** QQ-S-635 **MIL SPEC** MIL-S-11310 (CS1040); MIL-S-16788; MIL-S-16974; MIL-S-46070 **SAE** J403 (1040); J412 (1040); J414 (1040)
G10420	Carbon Steel	**C** 0.40-0.47 **Mn** 0.60-0.90 **P** 0.040 max **S** 0.050 max **Other** Sheets, C 0.39-0.47	**AISI** 1042 **ASTM** A29 (1042); A273 (1042); A510 (1042); A576 (1042) **SAE** J403 (1042); J412 (1042); J414 (1042)
G10430	Carbon Steel	**C** 0.40-0.47 **Mn** 0.70-1.00 **P** 0.040 max **S** 0.050 max **Other** Sheets, C 0.39-0.47	**AISI** 1043 **ASTM** A29 (1043); A510 (1043); A576 (1043) **SAE** J403 (1043); J412 (1043); J414 (1043)
G10440	Carbon Steel	**C** 0.43-0.50 **Mn** 0.30-0.60 **P** 0.040 max **S** 0.050 max	**AISI** 1044 **ASTM** A510 (1044); A575 (M1044); A576 (1044) **SAE** J403 (1044); J412 (1044); J414 (1044)
G10450	Carbon Steel	**C** 0.43-0.50 **Mn** 0.60-0.90 **P** 0.040 max **S** 0.050 max **Other** Sheets, C 0.42-0.50	**AISI** 1045 **ASTM** A29 (1045); A67 (1046); A236 (1046); A266 (1046); A510 (1045); A519 (1045); A576 (1045); A682 (1045) **FED** QQ-S-635 (C1045); QQ-S-700 (C1045); QQ-W-461 (1045) **SAE** J403 (1045); J412 (1045); J414 (1045)
G10460	Carbon Steel	**C** 0.43-0.50 **Mn** 0.70-1.00 **P** 0.040 max **S** 0.050 max	**AISI** 1046 **ASTM** A29 (1046); A510 (1046); A576 (1046) **SAE** J403 (1046); J412 (1046); J414 (1046)
G10490	Carbon Steel	**C** 0.46-0.53 **Mn** 0.60-0.90 **P** 0.040 max **S** 0.050 max	**AISI** 1049 **ASTM** A29 (1049); A510 (1049); A576 (1049) **SAE** J403 (1049); J412 (1049); J414 (1049)

The chemical compositions listed are for identification purposes and should not be used in lieu of the cross referenced specifications.

UNIFIED NUMBER	DESCRIPTION	CHEMICAL COMPOSITION	CROSS REFERENCE SPECIFICATIONS
G10500	Carbon Steel	C 0.48-0.55 Mn 0.60-0.90 P 0.040 max S 0.050 max	AISI 1050 AMS 5085 ASTM A29 (1050); A510 (1050); A519 (1050); A576 (1050); A682 (1050) FED QQ-S-635 (C1050); QQ-S-700 (C1050) MIL SPEC MIL-S-16974 (1050); MIL-S-46059 SAE J403 (1050); J412 (1050); J414 (1050)
G10530	Carbon Steel	C 0.48-0.55 Mn 0.70-1.00 P 0.040 max S 0.050 max	AISI 1053 ASTM A510 (1053); A576 (1053) SAE J403 (1053); J412 (1053)
G10550	Carbon Steel	C 0.50-0.60 Mn 0.60-0.90 P 0.040 max S 0.050 max	AISI 1055 ASTM A510 (1055); A576 (1055); A682 (1055) FED QQ-S-700 (C1055) MIL SPEC MIL-S-10520 SAE J403 (1055); J412 (1055); J414 (1055)
G10590	Carbon Steel	C 0.55-0.65 Mn 0.50-0.80 P 0.040 max S 0.050 max	AISI 1059 ASTM A29 (1059) SAE J118 (1059); J412 (1059)
G10600	Carbon Steel	C 0.55-0.65 Mn 0.60-0.90 P 0.040 max S 0.050 max	AISI 1060 AMS 7240 ASTM A29 (1060); A510 (1060); A576 (1060); A682 (1060) MIL SPEC MIL-S-16974 (Gr.1060) SAE J403 (1060); J412 (1060); J414 (1060)
G10640	Carbon Steel	C 0.60-0.70 Mn 0.50-0.80 P 0.040 max S 0.050 max Other Sheets and Plates, C 0.59-0.70	AISI 1064 ASTM A26 (1064); A29 (1064); A57 (1064); A230 (1064); A682 (1064) SAE J403 (1064); J412 (1064); J414 (1064)
G10650	Carbon Steel	C 0.60-0.70 Mn 0.60-0.90 P 0.040 max S 0.050 max Other Sheets and Plates, C 0.54-0.70	AISI 1065 ASTM A29 (1065); A229 (1065); A682 (1065) FED QQ-S-700 (C1065) MIL SPEC MIL-S-46049 (1065); MIL-S-46409 (1065) SAE J403 (1065); J412 (1065); J414 (1065)
G10690	Carbon Steel	C 0.65-0.75 Mn 0.40-0.70 P 0.040 max S 0.050 max	AISI 1069 ASTM A29 (1069) MIL SPEC MIL-S-11713 SAE J403 (1069); J412 (1069)
G10700	Carbon Steel	C 0.65-0.75 Mn 0.60-0.90 P 0.040 max S 0.050 max Other Plates, C 0.65-0.76	AISI 1070 AMS 5115 ASTM A29 (1070); A510 (1070); A576 (1070); A682 (1070) MIL SPEC MIL-S-11713 (2); MIL-S-12504 SAE J403 (1070); J412 (1070); J414 (1070)
G10740	Carbon Steel	C 0.70-0.80 Mn 0.50-0.80 P 0.040 max S 0.050 max Other Sheets and Plates, C 0.69-0.80	AISI 1074 AMS 5120 ASTM A29 (1074); A682 (1074) FED QQ-S-700 (C1074) MIL SPEC MIL-S-46049 (1074) SAE J403 (1074); J412 (1074); J414 (1074)
G10750	Carbon Steel	C 0.70-0.80 Mn 0.40-0.70 P 0.040 max S 0.050 max	AISI 1075 ASTM A29 (1075) SAE J403 (1075); J412 (1075)
G10780	Carbon Steel	C 0.72-0.85 Mn 0.30-0.60 P 0.040 max S 0.050 max Other Plates, C 0.72-0.86	AISI 1078 ASTM A29 (1078); A510 (1078); A576 (1078) SAE J403 (1078); J412 (1078); J414 (1078)
G10800	Carbon Steel	C 0.75-0.88 Mn 0.60-0.90 P 0.040 max S 0.050 max Other Plates, C 0.74-0.88	AISI 1080 AMS 5110 ASTM A29 (1080); A510 (1080); A576 (1080); A682 (1080) FED QQ-S-700 (C1080); QQ-W-470 MIL SPEC MIL-S-16974 (1080) SAE J403 (1080); J412 (1080); J414 (1080)
G10840	Carbon Steel	C 0.80-0.93 Mn 0.60-0.90 P 0.040 max S 0.050 max Other Plates, C 0.80-0.94	AISI 1084 ASTM A29 (1084); A510 (1084); A576 (1084) FED QQ-S-700 (C1084) SAE J403 (1084); J412 (1084); J414 (1084)
G10850	Carbon Steel	C 0.80-0.94 Mn 0.70-1.00 P 0.040 max S 0.050 max	AISI 1085 ASTM A682 (1085) FED QQ-S-700 (C1085) SAE J403 (1085); J412 (1085); J414 (1085)
G10860	Carbon Steel	C 0.80-0.93 Mn 0.30-0.50 P 0.040 max S 0.050 max Other Sheets and Plates, C 0.80-0.94	AISI 1086 AMS 5112 ASTM A29 (1086); A228 (1086); A682 (1086) FED QQ-S-700 (C1086) SAE J403 (1086); J412 (1086); J414 (1086)
G10900	Carbon Steel	C 0.85-0.98 Mn 0.60-0.90 P 0.040 max S 0.050 max Other Plates, C 0.84-0.98	AISI 1090 AMS 5112 ASTM A29 (1090); A510 (1090); A576 (1090) SAE J403 (1090); J412 (1090); J414 (1090)
G10950	Carbon Steel	C 0.90-1.03 Mn 0.30-0.50 P 0.040 max S 0.050 max Other Plates, C 0.90-1.04	AISI 1095 AMS 5121; 5122; 5132; 7304 ASTM A29 (1095); A510 (1095); A576 (1095); A682 (1095) FED QQ-S-700 (C1095) MIL SPEC MIL-S-7947; MIL-S-8559; MIL-S-16788 (C10) SAE J403 (1095); J412 (1095); J414 (1095)
G11080	Resulfurized Carbon Steel	C 0.08-0.13 Mn 0.50-0.80 P 0.040 max S 0.08-0.13	AISI 1108 ASTM A29 (1108) SAE J403 (1108); J412 (1108)
G11090	Resulfurized Carbon Steel	C 0.08-0.13 Mn 0.60-0.90 P 0.040 max S 0.08-0.13	AISI 1109 ASTM A29 (1109); A576 (1109) SAE J118 (1109); J412 (1109); J414 (1109)
G11100	Resulfurized Carbon Steel	C 0.08-0.13 Mn 0.30-0.60 P 0.040 max S 0.08-0.13	AISI 1110 ASTM A29 (1110); A576 (1110) SAE J403 (1110); J412 (1110)

The chemical compositions listed are for identification purposes and should not be used in lieu of the cross referenced specifications.

UNIFIED NUMBER	DESCRIPTION	CHEMICAL COMPOSITION	CROSS REFERENCE SPECIFICATIONS
G11160	Resulfurized Carbon Steel	C 0.14-0.20 Mn 1.10-1.40 P 0.040 max S 0.16-0.23	**AISI** 1116 **ASTM** A29 (1116); A516 (1116) **SAE** J412 (1116)
G11170	Resulfurized Carbon Steel	C 0.14-0.20 Mn 1.00-1.30 P 0.040 max S 0.08-0.13	**AISI** 1117 **AMS** 5022 **ASTM** A29 (1117); A108 (1117); A576 (1117) **SAE** J403 (1117); J412 (1117); J414 (1117)
G11180	Resulfurized Carbon Steel	C 0.14-0.20 Mn 1.30-1.60 P 0.040 max S 0.08-0.13	**AISI** 1118 **ASTM** A29 (1118); A107 (1118); A108 (1118); A576 (1118) **SAE** J403 (1118); J412 (1118); J414 (1118)
G11190	Resulfurized Carbon Steel	C 0.14-0.20 Mn 1.00-1.30 P 0.040 max S 0.24-0.33	**AISI** 1119 **ASTM** A29 (1119); A576 (1119) **SAE** J118 (1119); J412 (1119); J414 (1119)
G11230	Resulfurized Carbon Steel	C 0.20-0.27 Mn 1.20-1.50 P 0.040 max S 0.06-0.09	**AISI** 1123 **SAE** J403 (1123)
G11320	Resulfurized Carbon Steel	C 0.27-0.34 Mn 1.35-1.65 P 0.040 max S 0.09-0.13	**AISI** 1132 **ASTM** A29 (1132); A576 (1132) **SAE** J118 (1132); J412 (1132); J414 (1132)
G11370	Resulfurized Carbon Steel	C 0.32-0.39 Mn 1.35-1.65 P 0.040 max S 0.08-0.13	**AISI** 1137 **AMS** 5024 **ASTM** A29 (1137); A109 (1137); A311 (1137); A576 (1137) **SAE** J403 (1137); J412 (1137); J414 (1137)
G11374	Steel, Free Machining	C 0.32-0.39 Mn 1.35-1.65 P 0.040 max Pb 0.15-0.35 S 0.08-0.15	**AMS** 5020
G11390	Resulfurized Carbon Steel	C 0.35-0.43 Mn 1.35-1.65 P 0.040 max S 0.13-0.20	**AISI** 1139 **ASTM** A29 (1139); A576 (1139) **SAE** J403 (1139)
G11400	Resulfurized Carbon Steel	C 0.37-0.44 Mn 0.70-1.00 P 0.040 max S 0.08-0.13	**AISI** 1140 **ASTM** A29 (1140); A576 (1140) **SAE** J403 (1140); J412 (1140); J414 (1140)
G11410	Resulfurized Carbon Steel	C 0.37-0.45 Mn 1.35-1.65 P 0.040 max S 0.08-0.13	**AISI** 1141 **ASTM** A29 (1141); A108 (1141); A311 (1141); A576 (1141) **SAE** J403 (1141); J412 (1141); J414 (1141)
G11440	Resulfurized Carbon Steel	C 0.40-0.48 Mn 1.35-1.65 P 0.040 max S 0.24-0.33	**AISI** 1144 **ASTM** A29 (1144); A108 (1144); A311 (1144) **SAE** J403 (1144); J412 (1144); J414 (1144)
G11450	Resulfurized Carbon Steel	C 0.41-0.49 Mn 0.70-1.00 P 0.040 max S 0.08-0.13	**AISI** 1145 **ASTM** A29 (1145) **SAE** J118 (1145); J412 (1145); J414 (1145)
G11460	Resulfurized Carbon Steel	C 0.42-0.49 Mn 0.70-1.00 P 0.040 max S 0.08-0.13	**AISI** 1146 **ASTM** A29 (1146); A576 (1146) **SAE** J403 (1146); J412 (1146); J414 (1146)
G11510	Resulfurized Carbon Steel	C 0.48-0.55 Mn 0.70-1.00 P 0.040 max S 0.08-0.13	**AISI** 1151 **ASTM** A29 (1151); A108 (1151); A311 (1151); A576 (1151) **SAE** J403 (1151); J412 (1151); J414 (1151)
G12110	Rephosphorized and Resulfurized Carbon Steel	C 0.13 max Mn 0.60-0.90 P 0.07-0.12 S 0.10-0.15	**AISI** 1211 **ASTM** A29 (1211); A108 (1211); A576 (1211) **SAE** J403 (1211)
G12120	Rephosphorized and Resulfurized Carbon Steel	C 0.13 max Mn 0.70-1.00 P 0.07-0.12 S 0.16-0.23	**AISI** 1212 **AMS** 5010 **ASTM** A29 (1212); A108 (1212); A576 (1212) **SAE** J403 (1212)
G12130	Rephosphorized and Resulfurized Carbon Steel	C 0.13 max Mn 0.70-1.00 P 0.07-0.12 S 0.24-0.33	**AISI** 1213 **ASTM** A29 (1213); A108 (1213); A576 (1213) **SAE** J403 (1213)
G12134	Rephosphorized, Resulphurized and Leaded Carbon Steel	C 0.13 max Mn 0.70-1.00 P 0.07-0.12 Pb 0.15-0.35 S 0.24-0.33	**AISI** 12L13 **ASTM** A29 (12L13); A576 (12L13) **SAE** J403 12L13
G12144	Rephosphorized and Resulfurized Carbon Steel	C 0.15 max Mn 0.85-1.15 P 0.04-0.09 Pb 0.15-0.35 S 0.26-0.35	**AISI** 12L14 **ASTM** A108 (12L14); A576 (12L14) **SAE** J403 (12L14); J412 (12L14); J414 (12L14)
G12150	Rephosphorized and Resulfurized Carbon Steel	C 0.09 max Mn 0.75-1.05 P 0.04-0.09 S 0.26-0.35	**AISI** 1215 **ASTM** A29 (1215); A108 (1215); A576 (1215) **SAE** J403 (1215); J412 (1215)
G13300	Mn Alloy Steel	C 0.28-0.33 Mn 1.60-1.90 P 0.035 max S 0.040 max Si 0.15-0.35	**AISI** 1330 **ASTM** A322 (1330); A331 (1330); A519 (1330); A711 **MIL SPEC** MIL-S-16974 (1330) **SAE** J404 (1330); J412 (1330); J770 (1330)
G13350	Mn Alloy Steel	C 0.33-0.38 Mn 1.60-1.90 P 0.035 max S 0.040 max Si 0.15-0.35	**AISI** 1335 **ASTM** A331 (1335); A519 (1335); A547 (1335); A711 (1335) **MIL SPEC** MIL-S-16974 (1335) **SAE** J404 (1335); J412 (1335); J770 (1335)

The chemical compositions listed are for identification purposes and should not be used in lieu of the cross referenced specifications.

UNIFIED NUMBER	DESCRIPTION	CHEMICAL COMPOSITION	CROSS REFERENCE SPECIFICATIONS
G13400	Mn Alloy Steel	**C** 0.38-0.43 **Mn** 1.60-1.90 **P** 0.035 max **S** 0.040 max **Si** 0.15-0.35	**AISI** 1340 **ASTM** A322 (1340); A331 (1340); A519 (1340); A547 (1340) **MIL SPEC** MIL-S-16974 (1340) **SAE** J404 (1340); J412 (1340); J770 (1340)
G13450	Mn Alloy Steel	**C** 0.43-0.48 **Mn** 1.60-1.90 **P** 0.035 max **S** 0.040 max **Si** 0.15-0.35	**AISI** 1345 **ASTM** A322 (1345); A331 (1345); A372; A519 (1345) **SAE** J404 (1345); J412 (1345); J770 (1345)
G15116	Alloy Steel replaced by G51986		
G15130	Carbon Steel	**C** 0.10-0.16 **Mn** 1.10-1.40 **P** 0.040 max **S** 0.050 max	**AISI** 1513 **ASTM** A29 (1513); A108 **SAE** J403 (1513); J412 (1513)
G15180	Carbon Steel	**C** 0.15-0.21 **Mn** 1.10-1.40 **P** 0.040 max **S** 0.050 max	**AISI** 1518 **ASTM** A29 (1518); A108; A510 (1518); A519 (1518) **SAE** J118 (1518); J412 (1518)
G15216	Alloy Steel replaced by G52986		
G15220	Carbon Steel	**C** 0.18-0.24 **Mn** 1.10-1.40 **P** 0.040 max **S** 0.050 max	**AISI** 1522 **ASTM** A29 (1522); A108; A510 (1522) **SAE** J403 (1522); J412 (1522)
G15240	Carbon Steel	**C** 0.19-0.25 **Mn** 1.35-1.65 **P** 0.040 max **S** 0.050 max **Other** Sheets and Plates, C 0.18-0.25; ERW Tubing, C 0.18-0.25 Mn 1.30-1.65	**AISI** 1024; 1524 **ASTM** A29 (1524); A108; A510 (1524); A513 (1024); A519 (1524); A545 (1524) **SAE** J403 (1524); J412 (1524); J414 (1524)
G15250	Carbon Steel	**C** 0.23-0.29 **Mn** 0.80-1.10 **P** 0.040 max **S** 0.050 max	**AISI** 1525 **ASTM** A29 (1525); A108; A510 (1525) **SAE** J118 (1525); J412 (1525)
G15260	Carbon Steel	**C** 0.22-0.29 **Mn** 1.10-1.40 **P** 0.040 max **S** 0.050 max	**AISI** 1526 **ASTM** A29 (1526); A108; A510 (1526) **SAE** J403 (1526); J412 (1526)
G15270	Carbon Steel	**C** 0.22-0.29 **Mn** 1.20-1.50 **P** 0.040 max **S** 0.050 max **Other** Sheets and Plates, Mn 1.20-1.55; ERW Tubing, C 0.21-0.29 Mn 1.20-1.55	**AISI** 1027; 1527 **ASTM** A29 (1527); A108; A510 (1527); A513 (1027) **SAE** J403 (1527); J412 (1527)
G15330	Carbon Steel	**C** 0.30-0.37 **Mn** 1.20-1.50 **P** 0.040 max **S** 0.050 max	**AISI** 1533 **SAE** J403 (1533)
G15340	Carbon Steel	**C** 0.30-0.37 **Mn** 1.20-1.50 **P** 0.040 max **S** 0.050 max	**AISI** 1534 **SAE** J403 (1534)
G15360	Carbon Steel	**C** 0.30-0.37 **Mn** 1.20-1.50 **P** 0.040 max **S** 0.050 max **Other** Sheets and Plates, C 0.30-0.38 Mn 1.20-1.55	**AISI** 1036; 1536 **ASTM** A29 (1536); A108; A510 (1536) **SAE** J403 (1536); J412 (1536); J414 (1536)
G15410	Carbon Steel	**C** 0.36-0.44 **Mn** 1.35-1.65 **P** 0.040 max **S** 0.050 max **Other** Sheets and Plates, C 0.36-0.45 Mn 1.30-1.65	**AISI** 1041; 1541 **ASTM** A29 (1541); A108; A510 (1541); A519 (1541); A545 (1541); A546 (1541) **SAE** J403 (1541); J412 (1541); J414 (1541)
G15470	Carbon Steel	**C** 0.43-0.51 **Mn** 1.35-1.65 **P** 0.040 max **S** 0.050 max	**AISI** 1547 **ASTM** A29 (1547); A108; A510 (1547) **SAE** J118 (1547); J403 (1547); J412 (1547); J414 (1547)
G15480	Carbon Steel	**C** 0.44-0.52 **Mn** 1.10-1.40 **P** 0.040 max **S** 0.050 max **Other** Sheets and Plates, C 0.43-0.52 Mn 1.05-1.40	**AISI** 1048; 1548 **ASTM** A29 (1548); A108; A510 (1548) **SAE** J403 (1548); J412 (1548); J414 (1548)
G15510	Carbon Steel	**C** 0.45-0.56 **Mn** 0.85-1.15 **P** 0.040 max **S** 0.050 max	**AISI** 1051; 1551 **ASTM** A29 (1551); A108; A510 (1551) **SAE** J403 (1551); J412 (1551)
G15520	Carbon Steel	**C** 0.47-0.55 **Mn** 1.20-1.50 **P** 0.040 max **S** 0.050 max **Other** Sheets and Plates, C 0.46-0.55 Mn 1.20-1.55	**AISI** 1052; 1552 **ASTM** A29 (1552); A108; A510 (1552) **SAE** J403 (1552); J412 (1552); J414 (1552)
G15530	Carbon Steel	**C** 0.48-0.56 **Mn** 0.80-1.10 **P** 0.040 max **S** 0.050 max	**AISI** 1553 **SAE** J403 (1553)
G15610	Carbon Steel	**C** 0.55-0.65 **Mn** 0.75-1.05 **P** 0.040 max **S** 0.050 max	**AISI** 1061; 1561 **ASTM** A29 (1561); A108; A510 (1561) **SAE** J403 (1561); J412 (1561)
G15660	Carbon Steel	**C** 0.60-0.71 **Mn** 0.85-1.15 **P** 0.040 max **S** 0.050 max	**AISI** 1066; 1566 **ASTM** A29 (1566); A108; A510 (1566) **SAE** J403 (1566); J412 (1566)

The chemical compositions listed are for identification purposes and should not be used in lieu of the cross referenced specifications.

UNIFIED NUMBER	DESCRIPTION	CHEMICAL COMPOSITION	CROSS REFERENCE SPECIFICATIONS
G15700	Carbon Steel	C 0.65-0.75 Mn 0.80-1.10 P 0.040 max S 0.050 max	AISI 1570 SAE J403 (1570)
G15720	Carbon Steel	C 0.65-0.76 Mn 1.00-1.30 P 0.040 max S 0.050 max	AISI 1072; 1572 ASTM A29 (1572); A108; A510 (1572) SAE J403 (1572); J412 (1572)
G15800	Carbon Steel	C 0.75-0.85 Mn 0.80-1.10 P 0.040 max S 0.050 max	AISI 1580 SAE J403 (1580)
G15900	Carbon Steel	C 0.85-0.98 Mn 0.80-1.10 P 0.040 max S 0.050 max	AISI 1590 SAE J403 (1590)
G31400	Alloy Steel	C 0.38-0.43 Cr 0.55-0.75 Mn 0.70-0.90 Ni 1.10-1.40 P 0.035 max S 0.040 max Si 0.15-0.35	AISI 3140 ASTM A274 (3140); A331 (3140); A519 (3140) MIL SPEC MIL-S-16974 (3140) SAE J778 (3140)
G33106	Alloy Steel	C 0.08-0.13 Cr 1.40-1.75 Mn 0.45-0.60 Ni 3.25-3.75 P 0.025 max S 0.025 max Si 0.15-0.30	AISI E3310 ASTM A331 (E3310); A519 (E3310) MIL SPEC MIL-S-7393; MIL-S-8503 SAE J778 (3310)
G40120	Mo Alloy Steel	C 0.09-0.14 Mn 0.75-1.00 Mo 0.15-0.25 P 0.035 max S 0.040 max Si 0.15-0.35	AISI 4012 ASTM A331 (4012); A505 (4012); A519 (4012) SAE J412 (4012); J770 (4012); J778 (4012)
G40230	Mo Alloy Steel	C 0.20-0.25 Mn 0.70-0.90 Mo 0.20-0.30 P 0.035 max S 0.040 max Si 0.15-0.35	AISI 4023 ASTM A29 (4023); A322 (4023); A331 (4023); A519 (4023); A534 (4023) SAE J404 (4023); J412 (4023); J770 (4023)
G40240	Mo Alloy Steel	C 0.20-0.25 Mn 0.70-0.90 Mo 0.20-0.30 P 0.035 max S 0.035-0.050 Si 0.15-0.35	AISI 4024 ASTM A29 (4024); A322 (4024); A331 (4024); A519 (4024) SAE J404 (4024); J412 (4024); J770 (4024)
G40270	Mo Alloy Steel	C 0.25-0.30 Mn 0.70-0.90 Mo 0.20-0.30 P 0.035 max S 0.040 max Si 0.15-0.35	AISI 4027 ASTM A29 (4027); A322 (4027); A331 (4027); A519 (4027) SAE J404 (4027); J412 (4027); J770 (4027)
G40280	Mo Alloy Steel	C 0.25-0.30 Mn 0.70-0.90 Mo 0.20-0.30 P 0.035 max S 0.035-0.050 Si 0.15-0.35	AISI 4028 ASTM A29 (4028); A322 (4028); A331 (4028); A519 (4028) SAE J404 (4028); J412 (4028); J770 (4028)
G40320	Mo Alloy Steel	C 0.30-0.35 Mn 0.70-0.90 Mo 0.20-0.30 P 0.035 max S 0.040 max Si 0.15-0.35	AISI 4032 SAE J404 (4032); J412 (4032); J770 (4032)
G40370	Mo Alloy Steel	C 0.35-0.40 Mn 0.70-0.90 Mo 0.20-0.30 P 0.035 max S 0.040 max Si 0.15-0.35	AISI 4037 AMS 6300 ASTM A29 (4037); A322 (4037); A331 (4037); A519 (4037); A547 (4037) SAE J404 (4037); J412 (4037); J770 (4037)
G40420	Mo Alloy Steel	C 0.40-0.45 Mn 0.70-0.90 Mo 0.20-0.30 P 0.035 max S 0.040 max Si 0.15-0.30	AISI 4042 ASTM A29 (4042); A331 (4042); A519 (4042) SAE J404 (4042); J412 (4042); J770 (4042)
G40470	Mo Alloy Steel	C 0.45-0.50 Mn 0.70-0.90 Mo 0.20-0.30 P 0.035 max S 0.040 max Si 0.15-0.35	AISI 4047 ASTM A29 (4047); A322 (4047); A331 (4047); A519 (4047) SAE J404 (4047); J412 (4047); J770 (4047)
G40630	Mo Alloy Steel	C 0.60-0.67 Mn 0.75-1.00 Mo 0.20-0.30 P 0.035 max S 0.040 max Si 0.15-0.35	AISI 4063 ASTM A331 (4063); A519 (4063) SAE J778 (4063)
G41180	Cr-Mo Alloy Steel	C 0.18-0.23 Cr 0.40-0.60 Mn 0.70-0.90 Mo 0.08-0.15 P 0.035 max S 0.040 max Si 0.15-0.35	AISI 4118 ASTM A29 (4118); A322 (4118); A331 (4118); A505 (4118); A519 (4118) SAE J404 (4118); J412 (4118); J770 (4118)
G41200	Cr-Mo Alloy Steel	C 0.18-0.23 Cr 0.40-0.60 Mn 0.80-1.20 Mo 0.15-0.25 P 0.035 max S 0.040 max Si 0.15-0.35	AISI 4120 SAE J403 (EX15)
G41210	Cr-Mo Alloy Steel	C 0.18-0.23 Cr 0.45-0.65 Mn 0.75-1.00 Mo 0.15-0.25 P 0.035 max S 0.040 max Si 0.15-0.35	AISI 4121 SAE J403 (EX24)

The chemical compositions listed are for identification purposes and should not be used in lieu of the cross referenced specifications.

UNIFIED NUMBER	DESCRIPTION	CHEMICAL COMPOSITION	CROSS REFERENCE SPECIFICATIONS
G41300	Cr-Mo Alloy Steel	**C** 0.28-0.33 **Cr** 0.80-1.10 **Mn** 0.40-0.60 **Mo** 0.15-0.25 **P** 0.035 max **S** 0.040 max **Si** 0.15-0.35	**AISI** 4130 **AMS** 6348; 6350; 6351; 6360; 6361; 6362; 6370; 6371; 6373; 6374 **ASTM** A29 (4130); A322 (4130); A331 (4130); A505 (4130); A513 (4130); A519 (4130); A646 (4130) **MIL SPEC** MIL-S-6758; MIL-S-16974 (4130); MIL-S-18729; MIL-S-46059 **SAE** J404 (4130); J412 (4130); J770 (4130)
G41350	Cr-Mo Alloy Steel	**C** 0.33-0.38 **Cr** 0.80-1.10 **Mn** 0.70-0.90 **Mo** 0.15-0.25 **P** 0.035 max **S** 0.040 max **Si** 0.15-0.35	**AISI** 4135 **AMS** 6352; 6365; 6372 **ASTM** A29 (4135); A331 (4135); A372; A519 (4135) **MIL SPEC** MIL-S-16974 (4135) **SAE** J404 (4135); J412 (4135); J770 (4135)
G41370	Cr-Mo Alloy Steel	**C** 0.35-0.40 **Cr** 0.80-1.10 **Mn** 0.70-0.90 **Mo** 0.15-0.25 **P** 0.035 max **S** 0.040 max **Si** 0.15-0.35	**AISI** 4137 **ASTM** A29 (4137); A322 (4137); A331 (4137); A372; A505 (4137); A519 (4137); A547 (4137) **SAE** J404 (4137); J412 (4137); J770 (4137)
G41400	Cr-Mo Alloy Steel	**C** 0.38-0.43 **Cr** 0.80-1.10 **Mn** 0.75-1.00 **Mo** 0.15-0.25 **P** 0.035 max **S** 0.040 max **Si** 0.15-0.35	**AISI** 4140 **AMS** 6349; 6381; 6382; 6390; 6395 **ASTM** A29 (4140); A322 (4140); A331 (4140); A505 (4140); A519 (4140); A547 (4140); A646 (4140); A711 **MIL SPEC** MIL-S-5626; MIL-S-16974 (4140); MIL-S-46059 **SAE** J404 (4140); J412 (4140); J770 (4140)
G41420	Cr-Mo Alloy Steel	**C** 0.40-0.45 **Cr** 0.80-1.10 **Mn** 0.75-1.00 **Mo** 0.15-0.25 **P** 0.035 max **S** 0.040 max **Si** 0.15-0.35	**AISI** 4142 **ASTM** A29 (4142); A322 (4142); A331 (4142); A372; A505 (4142); A519 (4142); A547 (4142); A711 **SAE** J404 (4142); J412 (4142); J770 (4142)
G41450	Cr-Mo Alloy Steel	**C** 0.43-0.48 **Cr** 0.80-1.10 **Mn** 0.75-1.00 **Mo** 0.15-0.25 **P** 0.035 max **S** 0.040 max **Si** 0.15-0.35	**AISI** 4145 **ASTM** A29 (4145); A322 (4145); A331 (4145); A505 (4145); A519 (4145); A711 **MIL SPEC** MIL-S-16974 (4145) **SAE** J404 (4145); J412 (4145); J770 (4145)
G41470	Cr-Mo Alloy Steel	**C** 0.45-0.50 **Cr** 0.80-1.10 **Mn** 0.75-1.00 **Mo** 0.15-0.25 **P** 0.035 max **S** 0.040 max **Si** 0.15-0.35	**AISI** 4147 **ASTM** A29 (4147); A322 (4147); A331 (4147); A372; A505 (4147); A519 (4147) **SAE** J404 (4147); J412 (4147); J770 (4147)
G41500	Cr-Mo Alloy Steel	**C** 0.48-0.53 **Cr** 0.80-1.10 **Mn** 0.75-1.00 **Mo** 0.15-0.25 **P** 0.035 max **S** 0.040 max **Si** 0.15-0.35	**AISI** 4150 **ASTM** A29 (4150); A322 (4150); A331 (4150); A505 (4150); A519 (4150); A711 **MIL SPEC** MIL-S-11595 (ORD4150) **SAE** J404 (4150); J412 (4150); J770 (4150)
G41610	Cr-Mo Alloy Steel	**C** 0.56-0.64 **Cr** 0.70-0.90 **Mn** 0.75-1.00 **Mo** 0.25-0.35 **P** 0.035 max **S** 0.40 max **Si** 0.15-0.35	**AISI** 4161 **ASTM** A29 (4161); A322 (4161); A331 (4161) **SAE** J404 (4161); J412 (4161); J770 (4161)
G43200	Ni-Cr-Mo Alloy Steel	**C** 0.17-0.22 **Cr** 0.40-0.60 **Mn** 0.45-0.65 **Mo** 0.20-0.30 **Ni** 1.65-2.00 **P** 0.035 max **S** 0.040 max **Si** 0.15-0.35	**AISI** 4320 **ASTM** A29 (4320); A322 (4320); A331 (4320); A505 (4320); A519 (4320); A535 (4320) **SAE** J404 (4320); J412 (4320); J770 (432J)
G43370	Ni-Cr-Mo Alloy Steel	**C** 0.35-0.40 **Cr** 0.70-0.90 **Mn** 0.60-0.80 **Mo** 0.20-0.30 **Ni** 1.65-2.00 **P** 0.035 max **S** 0.040 max **Si** 0.15-0.35	**AISI** 4337 **AMS** 6412; 6413 **ASTM** A519 (4337) **SAE** J778 (4337)
G43376	Ni-Cr-Mo Alloy Steel	**C** 0.35-0.40 **Cr** 0.70-0.90 **Mn** 0.65-0.85 **Mo** 0.20-0.30 **Ni** 1.65-2.00 **P** 0.025 max **S** 0.025 max **Si** 0.15-0.35	**AISI** E4337 **ASTM** A519 (E4337)
G43400	Ni-Cr-Mo Alloy Steel	**C** 0.38-0.43 **Cr** 0.70-0.90 **Mn** 0.60-0.80 **Mo** 0.20-0.30 **Ni** 1.65-2.00 **P** 0.035 max **S** 0.040 max **Si** 0.15-0.30	**AISI** 4340 **AMS** 6359; 6414; 6415; 6454 **ASTM** A29 (4340); A322 (4340); A331 (4340); A505 (4340); A519 (4340); A547 (4340); A646 (4340); A711 **FED** QQ-S-681 **MIL SPEC** MIL-S-5000; MIL-S-16974 (4340); MIL-S-46059 **SAE** J404 (4340); J412 (4340); J770 (4340)
G43406	Ni-Cr-Mo Alloy Steel	**C** 0.38-0.43 **Cr** 0.70-0.90 **Mn** 0.65-0.85 **Mo** 0.20-0.30 **Ni** 1.65-2.00 **P** 0.025 max **S** 0.025 max **Si** 0.15-0.35	**AISI** E4340 **AMS** 6415 **ASTM** A331 (E4340); A505 (E4340); A519 (E4340) **MIL SPEC** MIL-S-5000; MIL-S-83135 **SAE** J404 (E4340); J770 (E4340)
G44190	Mo Alloy Steel	**C** 0.18-0.23 **Mn** 0.45-0.65 **Mo** 0.45-0.60 **P** 0.035 max **S** 0.040 max **Si** 0.15-0.35	**AISI** 4419 **ASTM** A29 (4419); A331 (4419) **SAE** J404 (4419); J412 (4419); J770 (4419)
G44220	Mo Alloy Steel	**C** 0.20-0.25 **Mn** 0.70-0.90 **Mo** 0.35-0.45 **P** 0.035 max **S** 0.040 max **Si** 0.15-0.35	**AISI** 4422 **ASTM** A29 (4422); A331 (4422); A519 (4422) **SAE** J404 (4422); J412 (4422); J770 (4422)
G44270	Mo Alloy Steel	**C** 0.24-0.29 **Mn** 0.70-0.90 **Mo** 0.35-0.45 **P** 0.035 max **S** 0.040 max **Si** 0.15-0.35	**AISI** 4427 **ASTM** A29 (4427); A331 (4427); A519 (4427) **SAE** J404 (4427); J412 (4427); J470 (4427)

The chemical compositions listed are for identification purposes and should not be used in lieu of the cross referenced specifications.

AISI AND SAE CARBON AND ALLOY STEELS

UNIFIED NUMBER	DESCRIPTION	CHEMICAL COMPOSITION	CROSS REFERENCE SPECIFICATIONS
G45200	Mo Alloy Steel	C 0.18-0.23 Mn 0.45-0.65 Mo 0.45-0.60 P 0.035 max S 0.040 max Si 0.15-0.35	AISI 4520 ASTM A519 (4520)
G46150	Ni-Mo Alloy Steel	C 0.13-0.18 Mn 0.45-0.65 Mo 0.20-0.30 Ni 1.65-2.00 P 0.035 max S 0.040 max Si 0.15-0.35	AISI 4615 AMS 6290 ASTM A29 (4615); A322 (4615); A331 (4615); A505 (4615) SAE J404 (4615); J412 (4615); J770 (4615)
G46170	Ni-Mo Alloy Steel	C 0.15-0.20 Mn 0.45-0.65 Mo 0.20-0.30 Ni 1.65-2.00 P 0.035 max S 0.040 max Si 0.15-0.35	AISI 4617 AMS 6292 ASTM A519 (4617) SAE J404 (4617); J412 (4617); J770 (4617)
G46200	Ni-Mo Alloy Steel	C 0.17-0.22 Mn 0.45-0.65 Mo 0.20-0.30 Ni 1.65-2.00 P 0.035 max S 0.040 max Si 0.15-0.35	AISI 4620 AMS 6294 ASTM A29 (4620); A322 (4620); A331 (4620); A505 (4620); A535 (4620) SAE J404 (4620); J412 (4620); J770 (4620)
G46210	Ni-Mo Alloy Steel	C 0.18-0.23 Mn 0.70-0.90 Mo 0.20-0.30 Ni 1.65-2.00 P 0.035 max S 0.040 max Si 0.15-0.35	AISI 4621 ASTM A322 (4621); A331 (4621) SAE J404 (4621); J412 (4621); J770 (4621)
G46260	Ni-Mo Alloy Steel	C 0.24-0.29 Mn 0.45-0.65 Mo 0.15-0.25 Ni 0.70-1.00 P 0.035 max S 0.040 max Si 0.15-0.35	AISI 4626 ASTM A29 (4626); A322 (4626); A331 (4626) SAE J404 (4626); J412 (4626); J770 (4626)
G47150	Ni-Cr-Mo Alloy Steel	C 0.13-0.18 Cr 0.45-0.65 Mn 0.70-0.90 Mo 0.45-0.65 Ni 0.70-1.00 P 0.035 max S 0.040 max Si 0.15-0.35	AISI 4715 SAE J403 (EX30)
G47180	Ni-Cr-Mo Alloy Steel	C 0.16-0.21 Cr 0.35-0.55 Mn 0.70-0.90 Mo 0.30-0.40 Ni 0.90-1.20 P 0.035 max S 0.040 max Si 0.15-0.35	AISI 4718 ASTM A29 (4718); A322 (4718); A331 (4718); A505 (4718) SAE J404 (4718); J412 (4718); J770 (4718)
G47200	Ni-Cr-Mo Alloy Steel	C 0.17-0.22 Cr 0.35-0.55 Mn 0.50-0.70 Mo 0.15-0.25 Ni 0.90-1.20 P 0.035 max S 0.040 max Si 0.15-0.35	AISI 4720 ASTM A29 (4720); A322 (4720); A331 (4720); A519 (4720); A535 (4720); A711 SAE J404 (4720); J412 (4720); J770 (4720)
G48150	Ni-Mo Alloy Steel	C 0.13-0.18 Mn 0.40-0.60 Mo 0.20-0.30 Ni 3.25-3.75 P 0.035 max S 0.040 max Si 0.15-0.35	AISI 4815 ASTM A29 (4815); A322 (4815); A331 (4815); A505 (4815) SAE J404 (4815); J412 (4815); J770 (4815)
G48170	Ni-Mo Alloy Steel	C 0.15-0.20 Mn 0.40-0.60 Mo 0.20-0.30 Ni 3.25-3.75 P 0.035 max S 0.040 max Si 0.15-0.35	AISI 4817 ASTM A29 (4817); A322 (4817); A331 (4817); A519 (4817) SAE J404 (4817); J412 (4817); J770 (4817)
G48200	Ni-Mo Alloy Steel	C 0.18-0.23 Mn 0.50-0.70 Mo 0.20-0.30 Ni 3.25-3.75 P 0.035 max S 0.040 max Si 0.15-0.35	AISI 4820 ASTM A29 (4820); A322 (4820); A331 (4820); A505 (4820); A519 (4820); A535 (4820) SAE J404 (4820); J412 (4820); J770 (4820)
G50150	Cr Alloy Steel	C 0.12-0.17 Cr 0.30-0.50 Mn 0.30-0.50 P 0.035 max S 0.40 max Si 0.15-0.35	AISI 5015 ASTM A29 (5015); A322 (5015); A331 (5015); A505 (5015); A519 (5015) SAE J404 (5015); J412 (5015); J770 (5015)
G50401	Cr-B Alloy Steel	B 0.0005-0.003 C 0.38-0.42 Cr 0.40-0.60 Mn 0.75-1.00 P 0.035 max S 0.040 max Si 0.15-0.35	AISI 50B40 ASTM A519 (50B40) SAE J404 (50B40); J412 (50B40); J770 (50B40)
G50441	Cr-B Alloy Steel	B 0.0005-0.003 C 0.43-0.48 Cr 0.40-0.60 Mn 0.75-1.00 P 0.035 max S 0.040 max Si 0.15-0.35	AISI 50B44 ASTM A322 (50B44); A331 (50B44); A519 (50B44) SAE J404 (50B44); J412 (50B44); J770 (50B44)
G50460	Cr Alloy Steel	C 0.43-0.50 Cr 0.20-0.35 Mn 0.75-1.00 P 0.035 max S 0.040 max Si 0.15-0.35	AISI 5046 ASTM A29 (5046); A519 (5046) SAE J404 (5046); J412 (5046); J770 (5046)
G50461	Cr-B Alloy Steel	B 0.0005-0.003 C 0.44-0.49 Cr 0.20-0.35 Mn 0.75-1.00 P 0.035 max S 0.040 max Si 0.15-0.35	AISI 50B46 ASTM A322 (50B46); A331 (50B46); A519 (50B46) SAE J404 (50B46); J412 (50B46); J770 (50B46)
G50501	Cr-B Alloy Steel	B 0.0005-0.003 C 0.48-0.53 Cr 0.40-0.60 Mn 0.75-1.00 P 0.035 max S 0.040 max Si 0.15-0.35	AISI 50B50 ASTM A322 (50B50); A331 (50B50); A519 (50B50) SAE J404 (50B50); J412 (50B50); J770 (50B50)
G50600	Cr Alloy Steel	C 0.56-0.64 Cr 0.40-0.60 Mn 0.75-1.00 P 0.035 max S 0.040 max Si 0.15-0.35	AISI 5060 SAE J404 (5060); J770 (5060)

The chemical compositions listed are for identification purposes and should not be used in lieu of the cross referenced specifications.

UNIFIED NUMBER	DESCRIPTION	CHEMICAL COMPOSITION	CROSS REFERENCE SPECIFICATIONS
G50601	Cr-B Alloy Steel	**B** 0.0005-0.003 **C** 0.56-0.64 **Cr** 0.40-0.60 **Mn** 0.75-1.00 **P** 0.035 max **S** 0.040 max **Si** 0.15-0.35	**AISI** 50B60 **ASTM** A322 (50B60); A331 (50B60); A519 (50B60) **SAE** J404 (50B60); J412 (50B60); J770 (50B60)
G50986	Cr Alloy Steel	**C** 0.95-1.10 **Cr** 0.40-0.60 **Mn** 0.25-0.45 **P** 0.025 max **S** 0.025 max **Si** 0.15-0.35	**AISI** E50100 **AMS** 6442 **ASTM** A29 (E50100); A295 (50100); A519 (E50100) **SAE** J404 (50100); J412 (50100); J770 (50100)
G51150	Cr Alloy Steel	**C** 0.13-0.18 **Cr** 0.70-0.90 **Mn** 0.70-0.90 **P** 0.035 max **S** 0.040 max **Si** 0.15-0.35	**AISI** 5115 **ASTM** A29 (5515); A519 (5515) **SAE** J404 (5115); J770 (5115)
G51170	Cr Alloy Steel	**C** 0.15-0.20 **Cr** 0.70-0.90 **Mn** 0.70-0.90 **P** 0.035 max **S** 0.040 max **Si** 0.15-0.35	**AISI** 5117 **ASTM** A322 (5117); A331 (5117) **SAE** J404 (5117)
G51200	Cr Alloy Steel	**C** 0.17-0.22 **Cr** 0.70-0.90 **Mn** 0.70-0.90 **P** 0.035 max **S** 0.040 max **Si** 0.15-0.35	**AISI** 5120 **ASTM** A29 (5120); A322 (5120); A331 (5120); A519 (5120) **SAE** J404 (5120); J770 (5120)
G51300	Cr Alloy Steel	**C** 0.28-0.33 **Cr** 0.80-1.10 **Mn** 0.70-0.90 **P** 0.035 max **S** 0.040 max **Si** 0.15-0.35	**AISI** 5130 **ASTM** A29 (5130); A322 (5130); A331 (5130) **SAE** J404 (5130); J412 (5130); J770 (5130)
G51320	Cr Alloy Steel	**C** 0.30-0.35 **Cr** 0.75-1.00 **Mn** 0.60-0.80 **P** 0.035 max **S** 0.040 max **Si** 0.15-0.35	**AISI** 5132 **ASTM** A29 (5132); A322 (5132); A331 (5132); A505 (5132); A519 (5132) **SAE** J404 (5132); J412 (5132); J770 (5132)
G51350	Cr Alloy Steel	**C** 0.33-0.38 **Cr** 0.80-1.05 **Mn** 0.60-0.80 **P** 0.035 max **S** 0.040 max **Si** 0.15-0.35	**AISI** 5135 **ASTM** A29 (5135); A322 (5135); A331 (5135); A519 (5135) **SAE** J404 (5135); J412 (5135); J770 (5135)
G51400	Cr Alloy Steel	**C** 0.38-0.43 **Cr** 0.70-0.90 **Mn** 0.70-0.90 **P** 0.035 max **S** 0.040 max **Si** 0.15-0.35	**AISI** 5140 **ASTM** A29 (5140); A322 (5140); A331 (5140); A505 (5140); A519 (5140) **SAE** J404 (5140); J412 (5140); J770 (5140)
G51450	Cr Alloy Steel	**C** 0.43-0.48 **Cr** 0.70-0.90 **Mn** 0.70-0.90 **P** 0.035 max **S** 0.040 max **Si** 0.15-0.35	**AISI** 5145 **ASTM** A29 (5145); A322 (5145); A331 (5145); A519 (5145) **SAE** J404 (5145); J412 (5145); J770 (5145)
G51470	Cr Alloy Steel	**C** 0.46-0.51 **Cr** 0.85-1.15 **Mn** 0.70-0.95 **P** 0.035 max **S** 0.040 max **Si** 0.15-0.35	**AISI** 5147 **ASTM** A29 (5147); A322 (5147); A331 (5147); A519 (5147) **SAE** J404 (5147); J412 (5147); J770 (5147)
G51500	Cr Alloy Steel	**C** 0.48-0.53 **Cr** 0.70-0.90 **Mn** 0.70-0.90 **P** 0.035 max **S** 0.040 max **Si** 0.15-0.35	**AISI** 5150 **ASTM** A29 (5150); A322 (5150); A331 (5150); A505 (5150); A519 (5150) **SAE** J404 (5150); J412 (5150); J770 (5150)
G51550	Cr Alloy Steel	**C** 0.51-0.59 **Cr** 0.70-0.90 **Mn** 0.70-0.90 **P** 0.035 max **S** 0.040 max **Si** 0.15-0.35	**AISI** 5155 **ASTM** A29 (5155); A322 (5155); A331 (5155); A519 (5155) **SAE** J404 (5155); J412 (5155); J770 (5155)
G51600	Cr Alloy Steel	**C** 0.56-0.64 **Cr** 0.70-0.90 **Mn** 0.75-1.00 **P** 0.035 max **S** 0.040 max **Si** 0.15-0.35	**AISI** 5160 **ASTM** A29 (5160); A322 (5160); A331 (5160); A505 (5160); A519 (5160) **SAE** J404 (5160); J412 (5160); J770 (5160)
G51601	Cr-B Alloy Steel	**B** 0.0005 min **C** 0.56-0.64 **Cr** 0.70-0.90 **Mn** 0.75-1.00 **P** 0.035 max **S** 0.040 max **Si** 0.15-0.35	**AISI** 51B60 **ASTM** A322 (51B60); A331 (51B60); A519 (51B60) **SAE** J404 (51B60); J412 (51B60); J770 (51B60)
G51986	Cr Alloy Steel	**C** 0.98-1.10 **Cr** 0.90-1.15 **Mn** 0.25-0.45 **P** 0.025 max **S** 0.025 max **Si** 0.15-0.35	**AISI** E51100 **AMS** 6443; 6446; 6447; 6449 **ASTM** A29 (E51100); A295 (E51100); A322 (E51100); A505 (E51100); A519 (E51100); A711 **SAE** J404 (51100); J412 (51100); J770 (51100)
G52986	Cr Alloy Steel	**C** 0.98-1.10 **Cr** 1.30-1.60 **Mn** 0.25-0.45 **P** 0.025 max **S** 0.025 max **Si** 0.15-0.35	**AISI** E52100 **AMS** 6440; 6444; 6447 **ASTM** A29 (E52100); A322 (E52100); A331 (E52100); A505 (E52100); A519 (E52100); A535 (E52100); A646 (E52100) **MIL SPEC** MIL-S-980 (52100); MIL-S-7420; MIL-S-22141 (52100) **SAE** J404 (52100); J412 (52100); J770 (52100)
G61180	Cr-V Alloy Steel	**C** 0.16-0.21 **Cr** 0.50-0.70 **Mn** 0.50-0.70 **P** 0.035 max **S** 0.040 max **Si** 0.15-0.35 **V** 0.10-0.15	**AISI** 6118 **ASTM** A29 (6118); A322 (6118); A331 (6118) **SAE** J404 (6118); J770 (6118)
G61200	Cr-V Alloy Steel	**C** 0.17-0.22 **Cr** 0.70-0.90 **Mn** 0.70-0.90 **P** 0.035 max **S** 0.040 max **Si** 0.15-0.35 **V** 0.10 min	**AISI** 6120 **ASTM** A331 (6120); A519 (6120); A711 **SAE** J778 (6120)
G61500	Cr-V Alloy Steel	**C** 0.48-0.53 **Cr** 0.80-1.10 **Mn** 0.70-0.90 **P** 0.035 max **S** 0.040 max **Si** 0.15-0.35 **V** 0.15 min	**AISI** 6150 **AMS** 6448; 6450; 6455; 7301 **ASTM** A29 (6150); A322 (6150); A331 (6150); A519 (6150) **MIL SPEC** MIL-S-8503 **SAE** J404 (6150); J412 (6150); J770 (6150)

The chemical compositions listed are for identification purposes and should not be used in lieu of the cross referenced specifications.

UNIFIED NUMBER	DESCRIPTION	CHEMICAL COMPOSITION	CROSS REFERENCE SPECIFICATIONS
G71406	Cr-V-Al Alloy Steel	**Al** 0.95-1.30 **C** 0.38-0.43 **Cr** 1.40-1.80 **Mn** 0.50-0.70 **P** 0.025 max **S** 0.025 max **Si** 0.15-0.35 **V** 0.30-0.40	**AISI** E71400 **ASTM** A519 (E7140)
G81150	Ni-Cr-Mo Alloy Steel	**C** 0.13-0.18 **Cr** 0.30-0.50 **Mn** 0.70-0.90 **Mo** 0.08-0.15 **Ni** 0.20-0.40 **P** 0.035 max **S** 0.040 max **Si** 0.15-0.35	**AISI** 8115 **ASTM** A29 (8115); A519 (8115) **SAE** J404 (8115); J770 (8115)
G81451	Ni-Cr-Mo-B Alloy Steel	**B** 0.0005 min **C** 0.43-0.48 **Cr** 0.35-0.55 **Mn** 0.75-1.00 **Mo** 0.08-0.15 **Ni** 0.20-0.40 **P** 0.035 max **S** 0.040 max **Si** 0.15-0.35	**AISI** 81B45 **ASTM** A322; A331 (81B45); A519 (81B45) **SAE** J404 (81B45); J412 (81B45); J770 (81B45)
G86150	Ni-Cr-Mo Alloy Steel	**C** 0.13-0.18 **Cr** 0.40-0.60 **Mn** 0.70-0.90 **Mo** 0.15-0.25 **Ni** 0.40-0.70 **P** 0.035 max **S** 0.040 max **Si** 0.15-0.35	**AISI** 8615 **AMS** 6270 **ASTM** A29 (8615); A322 (8615); A331 (8615); A519 (8615) **MIL SPEC** MIL-S-866 (8615) **SAE** J404 (8615); J770 (8615)
G86170	Ni-Cr-Mo Alloy Steel	**C** 0.15-0.20 **Cr** 0.40-0.60 **Mn** 0.70-0.90 **Mo** 0.15-0.25 **Ni** 0.40-0.70 **P** 0.035 max **S** 0.040 max **Si** 0.15-0.35	**AISI** 8617 **AMS** 6272 **ASTM** A29 (8617); A322 (8617); A331 (8617); A519 (8617) **SAE** J404 (8617); J770 (8617)
G86200	Ni-Cr-Mo Alloy Steel	**C** 0.18-0.23 **Cr** 0.40-0.60 **Mn** 0.70-0.90 **Mo** 0.15-0.25 **Ni** 0.40-0.70 **P** 0.035 max **S** 0.040 max **Si** 0.15-0.35	**AISI** 8620 **AMS** 6274; 6276; 6277 **ASTM** A29 (8620); A322 (8620); A331 (8620); A513 (8620); A519 (8620) **FED** QQ-S-626 **MIL SPEC** MIL-S-8690; MIL-S-16974 (8620) **SAE** J404 (8620); J770 (8620)
G86220	Ni-Cr-Mo Alloy Steel	**C** 0.20-0.25 **Cr** 0.40-0.60 **Mn** 0.70-0.90 **Mo** 0.15-0.25 **Ni** 0.40-0.70 **P** 0.035 max **S** 0.040 max **Si** 0.15-0.35	**AISI** 8622 **ASTM** A29 (8622); A322 (8622); A331 (8622); A519 (8622) **MIL SPEC** MIL-S-16974 **SAE** J404 (8622); J770 (8622)
G86250	Ni-Cr-Mo Alloy Steel	**C** 0.23-0.28 **Cr** 0.40-0.60 **Mn** 0.70-0.90 **Mo** 0.15-0.25 **Ni** 0.40-0.70 **P** 0.035 max **S** 0.040 max **Si** 0.15-0.35	**AISI** 8625 **ASTM** A29 (8625); A322 (8625); A331 (8625); A519 (8625) **MIL SPEC** MIL-S-16974 (8625) **SAE** J404 (8625); J770 (8625)
G86270	Ni-Cr-Mo Alloy Steel	**C** 0.25-0.30 **Cr** 0.40-0.60 **Mn** 0.70-0.90 **Mo** 0.15-0.25 **Ni** 0.40-0.70 **P** 0.035 max **S** 0.040 max **Si** 0.15-0.35	**AISI** 8627 **ASTM** A29 (8627); A322 (8627); A331 (8627); A519 (8627) **MIL SPEC** MIL-S-16974 **SAE** J404 (8627); J770 (8627)
G86300	Ni-Cr-Mo Alloy Steel	**C** 0.28-0.33 **Cr** 0.40-0.60 **Mn** 0.70-0.90 **Mo** 0.15-0.25 **Ni** 0.40-0.70 **P** 0.035 max **S** 0.040 max **Si** 0.15-0.35	**AISI** 8630 **AMS** 6280; 6281; 6355; 6530 **ASTM** A29 (8630); A322 (8630); A331 (8630); A519 (8630) **FED** QQ-S-681 **MIL SPEC** MIL-S-6050; MIL-S-16974 (8630); MIL-S-18728; MIL-S-46059 **SAE** J404 (8630); J412 (8630); J770 (8630)
G86370	Ni-Cr-Mo Alloy Steel	**C** 0.35-0.40 **Cr** 0.40-0.60 **Mn** 0.75-1.00 **Mo** 0.15-0.25 **Ni** 0.40-0.70 **P** 0.035 max **S** 0.040 max **Si** 0.15-0.35	**AISI** 8637 **ASTM** A29 (8637); A322 (8637); A331 (8637); A689 (8637) **SAE** J404 (8637); J412 (8637); J770 (8637)
G86400	Ni-Cr-Mo Alloy Steel	**C** 0.38-0.43 **Cr** 0.40-0.60 **Mn** 0.75-1.00 **Mo** 0.15-0.25 **Ni** 0.40-0.70 **P** 0.035 max **S** 0.040 max **Si** 0.15-0.35	**AISI** 8640 **ASTM** A29 (8640); A322 (8640); A331 (8640); A519 (8640) **MIL SPEC** MIL-S-16974 (8640) **SAE** J404 (8640); J412 (8640); J770 (8640)
G86420	Ni-Cr-Mo Alloy Steel	**C** 0.40-0.45 **Cr** 0.40-0.60 **Mn** 0.75-1.00 **Mo** 0.15-0.25 **Ni** 0.40-0.70 **P** 0.035 max **S** 0.40 max **Si** 0.15-0.35	**AISI** 8642 **ASTM** A29 (8642); A322 (8642); A331 (8642); A519 (8642) **SAE** J404 (8642); J412 (8642); J770 (8642)
G86450	Ni-Cr-Mo Alloy Steel	**C** 0.43-0.48 **Cr** 0.40-0.60 **Mn** 0.75-1.00 **Mo** 0.15-0.25 **Ni** 0.40-0.70 **P** 0.035 max **S** 0.040 max **Si** 0.15-0.35	**AISI** 8645 **ASTM** A29 (8645); A322 (8645); A331 (8645); A519 (8645) **MIL SPEC** MIL-S-16974 (8645) **SAE** J404 (8645); J412 (8645); J770 (8645)
G86451	Ni-Cr-Mo-B Alloy Steel	**B** 0.0005-0.003 **C** 0.43-0.48 **Cr** 0.40-.060 **Mn** 0.75-1.00 **Mo** 0.15-0.25 **Ni** 0.40-0.70 **P** 0.035 max **S** 0.040 max **Si** 0.15-0.35	**AISI** 86B45 **ASTM** A519 (86B45) **SAE** J404 (86B45); J412 (86B45); J770 (86B45)
G86500	Ni-Cr-Mo Alloy Steel	**C** 0.48-0.53 **Cr** 0.40-0.60 **Mn** 0.75-1.00 **Mo** 0.15-0.25 **Ni** 0.40-0.70 **P** 0.035 max **S** 0.040 max **Si** 0.15-0.35	**AISI** 8650 **ASTM** A29 (8650); A322 (8650); A331 (8650); A519 (8650); A689 **SAE** J404 (8650); J412 (8650); J770 (8650)
G86550	Ni-Cr-Mo Alloy Steel	**C** 0.51-0.59 **Cr** 0.40-0.60 **Mn** 0.75-1.00 **Mo** 0.15-0.25 **Ni** 0.40-0.70 **P** 0.035 max **S** 0.040 max **Si** 0.15-0.35	**AISI** 8655 **ASTM** A29 (8655); A322 (8655); A331 (8655); A519 (8655); A689 **SAE** J404 (8655); J412 (8655); J770 (8655)
G86600	Ni-Cr-Mo Alloy Steel	**C** 0.55-0.65 **Cr** 0.40-0.60 **Mn** 0.75-1.00 **Mo** 0.15-0.25 **Ni** 0.40-0.70 **P** 0.035 max **S** 0.040 max **Si** 0.15-0.35	**AISI** 8660 **ASTM** A322 (8660); A519 (8660); A689; A711 **SAE** J404 (8660); J412 (8660); J770 (8660)

The chemical compositions listed are for identification purposes and should not be used in lieu of the cross referenced specifications.

UNIFIED NUMBER	DESCRIPTION	CHEMICAL COMPOSITION	CROSS REFERENCE SPECIFICATIONS
G87200	Ni-Cr-Mo Alloy Steel	**C** 0.18-0.23 **Cr** 0.40-0.60 **Mn** 0.70-0.90 **Mo** 0.20-0.30 **Ni** 0.40-0.70 **P** 0.035 max **S** 0.040 max **Si** 0.15-0.35	**AISI** 8720 **ASTM** A29 (8720); A322 (8720); A331 (8720); A519 (8720) **SAE** J404 (8720); J770 (8720)
G87350	Ni-Cr-Mo Alloy Steel	**C** 0.33-0.38 **Cr** 0.40-0.60 **Mn** 0.75-1.00 **Mo** 0.20-0.30 **Ni** 0.40-0.70 **P** 0.035 max **S** 0.040 max **Si** 0.15-0.35	**AISI** 8735 **AMS** 6282; 6320; 6357; 6535 **ASTM** A519 (8735) **MIL SPEC** MIL-S-6098; MIL-T-6733 **SAE** J778 (8735)
G87400	Ni-Cr-Mo Alloy Steel	**C** 0.38-0.43 **Cr** 0.40-0.60 **Mn** 0.75-1.00 **Mo** 0.20-0.30 **Ni** 0.40-0.70 **P** 0.035 max **S** 0.040 max **Si** 0.15-0.35	**AISI** 8740 **AMS** 6322; 6323; 6325; 6327; 6358 **ASTM** A29 (8740); A322 (8740); A331 (8740); A519 (8740) **MIL SPEC** MIL-S-6049 **SAE** J404 (8740); J412 (8740); J770 (8740)
G87420	Ni-Cr-Mo Alloy Steel	**C** 0.40-0.45 **Cr** 0.40-0.60 **Mn** 0.75-1.00 **Mo** 0.20-0.30 **Ni** 0.40-0.70 **P** 0.035 max **S** 0.040 max **Si** 0.15-0.35	**AISI** 8742 **ASTM** A331 (8742); A519 (8742) **SAE** J778 (8742)
G88220	Ni-Cr-Mo Alloy Steel	**C** 0.20-0.25 **Cr** 0.40-0.60 **Mn** 0.75-1.00 **Mo** 0.30-0.40 **Ni** 0.40-0.70 **P** 0.035 max **S** 0.040 max **Si** 0.15-0.35	**AISI** 8822 **ASTM** A29 (8622); A322 (8622); A331 (8622); A519 (8622) **SAE** J404 (8822); J770 (8822)
G92540	Cr-Si Alloy Steel	**C** 0.51-0.59 **Cr** 0.60-0.80 **Mn** 0.60-0.80 **P** 0.035 max **S** 0.040 max **Si** 1.20-1.60	**AISI** 9254 **AMS** 6451 **ASTM** A29 (9254); A401 **SAE** J404 (9254); J412 (9254); J770 (9254)
G92550	Si-Mn Alloy Steel	**C** 0.51-0.59 **Mn** 0.70-0.95 **P** 0.035 max **S** 0.040 max **Si** 1.80-2.20	**AISI** 9255 **ASTM** A29 (9255) **SAE** J404 (9255); J412 (9255); J770 (9255)
G92600	Si-Mn Alloy Steel	**C** 0.56-0.64 **Mn** 0.75-1.00 **P** 0.035 max **S** 0.040 max **Si** 1.80-2.20	**AISI** 9260 **ASTM** A29 (9260); A59 (9260); A322 (9260); A331 (9260); A689 **SAE** J404 (9260); J412 (9260); J770 (9260)
G92620	Si-Cr Alloy Steel	**C** 0.55-0.65 **Cr** 0.25-0.40 **Mn** 0.75-1.00 **P** 0.035 max **S** 0.040 max **Si** 1.80-2.20	**AISI** 9262 **ASTM** A304 (9262); A519 (9262) **SAE** J778 (9262)
G93106	Ni-Cr-Mo Alloy Steel	**C** 0.08-0.13 **Cr** 1.00-1.40 **Mn** 0.45-0.65 **Mo** 0.08-0.15 **Ni** 3.00-3.50 **P** 0.025 max **S** 0.025 max **Si** 0.15-0.35	**AISI** E9310 **AMS** 6260; 6265; 6267 **ASTM** A29 (E9310); A274 (E9310); A322 (E9310); A331 (E9310); A519 (E9310) **SAE** J404 (9310); J770 (9310)
G94151	Ni-Cr-Mo-B Alloy Steel	**B** 0.0005 min **C** 0.13-0.18 **Cr** 0.30-0.50 **Mn** 0.75-1.00 **Mo** 0.08-0.15 **Ni** 0.30-0.60 **P** 0.035 max **S** 0.040 max **Si** 0.15-0.35	**AISI** 94B15 **ASTM** A519 (94B15) **SAE** J404 (94B15); J770 (94B15)
G94171	Ni-Cr-Mo-B Alloy Steel	**B** 0.0005-0.003 **C** 0.15-0.20 **Cr** 0.30-0.50 **Mn** 0.75-1.00 **Mo** 0.08-0.15 **Ni** 0.30-0.60 **P** 0.035 max **S** 0.040 max **Si** 0.15-0.35	**AISI** 94B17 **AMS** 6275 **ASTM** A519 (94B17) **SAE** J404 (94B17); J770 (94B17)
G94301	Ni-Cr-Mo-B Alloy Steel	**B** 0.0005-0.003 **C** 0.28-0.33 **Cr** 0.30-0.50 **Mn** 0.75-1.00 **Mo** 0.08-0.15 **Ni** 0.30-0.60 **P** 0.035 max **S** 0.040 max **Si** 0.15-0.35	**AISI** 94B30 **ASTM** A322 (94B30); A331 (94B30); A519 (94B30) **SAE** J404 (94B30); J412 (94B30); J770 (94B30)
G94401	Ni-Cr-Mo-B Alloy Steel	**B** 0.0005-0.003 **C** 0.38-0.43 **Cr** 0.30-0.50 **Mn** 0.75-1.00 **Mo** 0.08-0.15 **Ni** 0.30-0.60 **P** 0.035 max **S** 0.040 max **Si** 0.15-0.35	**AISI** 94B40 **ASTM** A519 (94B40) **SAE** J778 (94B40)
G98400	Ni-Cr-Mo Alloy Steel	**C** 0.38-0.43 **Cr** 0.70-0.90 **Mn** 0.70-0.90 **Mo** 0.20-0.30 **Ni** 0.85-1.15 **P** 0.035 max **S** 0.040 max **Si** 0.15-0.35	**AISI** 9840 **AMS** 6342 **ASTM** A519 (9840); A711 **SAE** J778 (9840)
G98500	Ni-Cr-Mo Alloy Steel	**C** 0.48-0.53 **Cr** 0.70-0.90 **Mn** 0.70-0.90 **Mo** 0.20-0.30 **Ni** 0.85-1.15 **P** 0.035 max **S** 0.040 max **Si** 0.15-0.35	**AISI** 9850 **ASTM** A519 (9850) **MIL SPEC** MIL-S-19434 **SAE** J778 (9850)

The chemical compositions listed are for identification purposes and should not be used in lieu of the cross referenced specifications.

UNS NUMBERS ASSIGNED TO DATE

With Description of Each Material Covered and References To Documents In Which The Same or Similar Materials are Described

Hxxxxx Number Series
AISI and SAE H-Steels

AISI AND SAE H-STEELS

UNIFIED NUMBER	DESCRIPTION	CHEMICAL COMPOSITION	CROSS REFERENCE SPECIFICATIONS
H10380	H-Carbon Steel	C 0.34-0.43 Mn 0.50-1.00 P 0.040 max S 0.050 max Si 0.15-0.35	AISI 1038 H ASTM A304 (1038H) SAE J1268 (1038 H)
H10450	H-Carbon Steel	C 0.42-0.51 Mn 0.50-1.00 P 0.040 max S 0.050 max Si 0.15-0.35	AISI 1045 H ASTM A304 (1045H) SAE J1268 (1045 H)
H13300	C-Mn H-Alloy Steel	C 0.27-0.33 Mn l.45-2.05 P 0.035 max S 0.040 max Si 0.15-0.35	AISI 1330 H ASTM A304 (1330 H) SAE J1268 (1330 H)
H13350	C-Mn H-Alloy Steel	C 0.32-0.38 Mn 1.45-2.05 P 0.035 max S 0.040 max Si 0.15-0.35	AISI 1335 H ASTM A304 (1335 H) SAE J1268 (1335 H)
H13400	C-Mn H-Alloy Steel	C 0.37-0.44 Mn 1.45-2.05 P 0.035 max S 0.040 max Si 0.15-0.35	AISI 1340 H ASTM A304 (1340 H) SAE J1268 (1340 H)
H13450	C-Mn H-Alloy Steel	C 0.42-0.49 Mn 1.45-2.05 P 0.035 max S 0.040 max Si 0.15-0.35	AISI 1345 H ASTM A304 (1345 H) SAE J1268 (1345 H)
H15211	C-B H-Carbon Steel	B 0.0005-0.003 C 0.17-0.24 Mn 0.70-1.20 P 0.040 max S 0.050 max Si 0.15-0.35	AISI 15B21 H SAE J1268 (15B21 H)
H15220	Mn H-Carbon Steel	C 0.17-0.25 Mn 1.00-1.50 P 0.040 max S 0.050 max Si 0.15-0.35	AISI 1522 H ASTM A304 (1522H) SAE J1268 (1522 H)
H15240	Mn H-Carbon Steel	C 0.18-0.26 Mn 1.25-1.75 P 0.040 max S 0.050 max Si 0.15-0.35	AISI 1524 H ASTM A304 (1526H) SAE J1268 (1524 H)
H15260	Mn H-Carbon Steel	C 0.21-0.30 Mn 1.00-1.50 P 0.040 max S 0.050 max Si 0.15-0.35	AISI 1526 H ASTM A304 (1526H) SAE J1268 (1526 H)
H15351	B-Mn H-Carbon Steel	B 0.0005-0.003 C 0.31-0.39 Mn 0.70-1.20 P 0.040 max S 0.050 max Si 0.15-0.35	AISI 15B35 H ASTM A304 (15B35H) SAE J1268 (15B35 H)
H15371	B-Mn H-Carbon Steel	B 0.0005-0.003 C 0.30-0.39 Mn 1.00-1.50 P 0.040 max S 0.050 max Si 0.15-0.35	AISI 15B37 H ASTM A304 (15B37H) SAE J1268 (15B37 H)
H15410	Mn H-Carbon Steel	C 0.35-0.45 Mn 1.25-1.75 P 0.040 max S 0.050 max Si 0.15-0.35	AISI 1541 H ASTM A304 (1541H) SAE J1268 (1541 H)
H15411	C-B H-Carbon Steel	B 0.0005-0.003 C 0.35-0.45 Mn 1.25-1.75 P 0.040 max S 0.050 max Si 0.15-0.35	AISI 15B41 H ASTM A304 (15B41H) SAE J1268 (15B41 H)
H15481	C-B H-Carbon Steel	B 0.0005-0.003 C 0.43-0.55 Mn 1.00-1.50 P 0.040 max S 0.050 max Si 0.15-0.35	AISI 15B48 H ASTM A304 (15B48H) SAE J1268 (15B48 H)
H15621	C-B H-Carbon Steel	B 0.0005-0.003 C 0.54-0.67 Mn 1.00-1.50 P 0.040 max S 0.050 max Si 0.40-0.60	AISI 15B62 H ASTM A304 (15B62H) SAE J1268 (15B62 H)
H40270	C-Mo H-Alloy Steel	C 0.24-0.30 Mn 0.60-1.00 Mo 0.20-0.30 P 0.035 max S 0.040 max Si 0.15-0.35	AISI 4027 H ASTM A304 (4027 H) SAE J1268 (4027 H)
H40280	C-Mo H-Alloy Steel	C 0.24-0.30 Mn 0.60-1.00 Mo 0.20-0.30 P 0.035 max S 0.035-0.050 Si 0.15-0.35	AISI 4028 H ASTM A304 (4028 H) SAE J1268 (4028 H)
H40320	C-Mo H-Alloy Steel	C 0.29-0.35 Mn 0.60-1.00 Mo 0.20-0.30 Si 0.15-0.35	AISI 4032H ASTM A304 (4032 H) SAE J1268 (4032 H)
H40370	C-Mo H-Alloy Steel	C 0.34-0.41 Mn 0.60-1.00 Mo 0.20-0.30 P 0.035 max S 0.040 max Si 0.15-0.35	AISI 4037 H ASTM A304 (4037 H) SAE J1268 (4037 H)
H40420	C-Mo H-Alloy Steel	C 0.39-0.46 Mn 0.60-1.00 Mo 0.20-0.30 Si 0.15-0.35	AISI 4042 H ASTM A304 (4042 H) SAE J1268 (4042 H)
H40470	C-Mo H-Alloy Steel	C 0.44-0.51 Mn 0.60-1.00 Mo 0.20-0.30 P 0.035 max S 0.040 max Si 0.15-0.35	AISI 4047 H ASTM A304 (4047 H) SAE J1268 (4047 H)

The chemical compositions listed are for identification purposes and should not be used in lieu of the cross referenced specifications.

AISI AND SAE H-STEELS

UNIFIED NUMBER	DESCRIPTION	CHEMICAL COMPOSITION	CROSS REFERENCE SPECIFICATIONS
H41180	Cr-Mo H-Alloy Steel	C 0.17-0.23 Cr 0.30-0.70 Mn 0.60-1.00 Mo 0.08-0.15 P 0.035 max S 0.040 max Si 0.15-0.35	AISI 4118 H ASTM A304 (4118 H) SAE J1268 (4118 H)
H41300	Cr-Mo H-Alloy Steel	C 0.27-0.33 Cr 0.75-1.20 Mn 0.30-0.70 Mo 0.15-0.25 P 0.035 max S 0.040 max Si 0.15-0.35	AISI 4130 H ASTM A304 (4130 H) SAE J1268 (4130 H)
H41350	Cr-Mo H-Alloy Steel	C 0.32-0.38 Cr 0.75-1.20 Mn 0.60-1.00 Mo 0.15-0.25 Si 0.15-0.35	AISI 4135 H ASTM A304 (4135 H) SAE J1268 (4135 H)
H41370	Cr-Mo H-Alloy Steel	C 0.34-0.41 Cr 0.75-1.20 Mn 0.60-1.00 Mo 0.15-0.25 P 0.035 max S 0.040 max Si 0.15-0.35	AISI 4137 H ASTM A304 (4137 H) SAE J1268 (4137 H)
H41400	Cr-Mo H-Alloy Steel	C 0.37-0.44 Cr 0.75-1.20 Mn 0.65-1.10 Mo 0.15-0.25 P 0.035 max S 0.040 max Si 0.15-0.35	AISI 4140 H ASTM A304 (4140 H) SAE J1268 (4140 H)
H41420	Cr-Mo H-Alloy Steel	C 0.39-0.46 Cr 0.75-1.20 Mn 0.65-1.10 Mo 0.15-0.25 P 0.035 max S 0.040 max Si 0.15-0.35	AISI 4142 H ASTM A304 (4142 H) SAE J1268 (4142 H)
H41450	Cr-Mo H-Alloy Steel	C 0.42-0.49 Cr 0.75-1.20 Mn 0.65-1.10 Mo 0.15-0.25 P 0.035 max S 0.040 max Si 0.15-0.35	AISI 4145 H ASTM A304 (4145 H) SAE J1268 (4145 H)
H41470	Cr-Mo H-Alloy Steel	C 0.44-0.51 Cr 0.75-1.20 Mn 0.65-1.10 Mo 0.15-0.25 P 0.035 max S 0.040 max Si 0.15-0.35	AISI 4147 H ASTM A304 (4147 H) SAE J1268 (4147 H)
H41500	Cr-Mo H-Alloy Steel	C 0.47-0.54 Cr 0.75-1.20 Mn 0.65-1.10 Mo 0.15-0.25 P 0.035 max S 0.040 max Si 0.15-0.35	AISI 4150 H ASTM A304 (4150 H) SAE J1268 (4150 H)
H41610	Cr-Mo H-Alloy Steel	C 0.55-0.65 Cr 0.65-0.95 Mn 0.65-1.10 Mo 0.25-0.35 P 0.035 max S 0.040 max Si 0.15-0.35	AISI 4161 H ASTM A304 (4161 H) SAE J1268 (4161 H)
H43200	Ni-Cr-Mo H-Alloy Steel	C 0.17-0.23 Cr 0.35-0.65 Mn 0.40-0.70 Mo 0.20-0.30 Ni 1.55-2.00 P 0.035 max S 0.040 max Si 0.15-0.35	AISI 4320 H AMS 6299 ASTM A304 (4320 H) SAE J1268 (4320 H)
H43400	Ni-Cr-Mo H-Alloy Steel	C 0.37-0.44 Cr 0.65-0.95 Mn 0.55-0.90 Mo 0.20-0.30 Ni 1.55-2.00 P 0.035 max S 0.040 max Si 0.15-0.35	AA 4340 H ASTM A304 (4340 H) SAE J1268 (4340 H)
H43406	Ni-Cr-Mo H-Alloy Steel	C 0.37-0.44 Cr 0.65-0.95 Mn 0.60-0.95 Mo 0.20-0.30 Ni 1.55-2.00 P 0.025 max S 0.025 max Si 0.15-0.35	AISI E4340 H ASTM A304 (E4340 H) SAE J1268 (E4340 H)
H44190	C-Mo H-Alloy Steel	C 0.17-0.23 Mn 0.35-0.75 Mo 0.45-0.60 P 0.035 max S 0.040 max Si 0.15-0.35	AISI 4419 H ASTM A304 (4419 H) SAE J1268 (4419 H)
H46200	Ni-Mo H-Alloy Steel	C 0.17-0.23 Mn 0.35-0.75 Mo 0.20-0.30 Ni 1.55-2.00 P 0.035 max S 0.040 max Si 0.15-0.35	AISI 4620 H ASTM A304 (4620 H) SAE J1268 (4620 H)
H46210	Ni-Mo H-Alloy Steel	C 0.17-0.23 Mn 0.60-1.00 Mo 0.20-0.30 Ni 1.55-2.00 P 0.035 max S 0.040 max Si 0.15-0.35	AISI 4621 H ASTM A304 (4621 H) SAE J1268 (4621 H)
H46260	Ni-Mo H-Alloy Steel	C 0.23-0.29 Mn 0.40-0.70 Mo 0.15-0.25 Ni 0.65-1.05 P 0.035 max S 0.040 max Si 0.15-0.35	AISI 4626 H ASTM A304 (4626H)
H47180	Ni-Cr-Mo H-Alloy Steel	C 0.15-0.21 Cr 0.30-0.60 Mn 0.60-0.95 Mo 0.30-0.40 Ni 0.85-1.25 P 0.035 max S 0.040 max Si 0.15-0.35	AISI 4718 H ASTM A304 (4718 H) SAE J1268 (4718 H)
H47200	Ni-Cr-Mo H-Alloy Steel	C 0.17-0.23 Cr 0.30-0.60 Mn 0.45-0.75 Mo 0.15-0.25 Ni 0.85-1.25 P 0.035 max S 0.040 max Si 0.15-0.35	AISI 4720 H ASTM A304 (4720 H) SAE J1268 (4720 H)

The chemical compositions listed are for identification purposes and should not be used in lieu of the cross referenced specifications.

UNIFIED NUMBER	DESCRIPTION	CHEMICAL COMPOSITION	CROSS REFERENCE SPECIFICATIONS
H48150	Ni-Mo H-Alloy Steel	**C** 0.12-0.18 **Mn** 0.30-0.70 **Mo** 0.20-0.30 **Ni** 3.20-3.80 **P** 0.035 max **S** 0.040 max **Si** 0.15-0.35	**AISI** 4815 H **ASTM** A304 (4815 H) **SAE** J1268 (4815 H)
H48170	Ni-Mo H-Alloy Steel	**C** 0.14-0.20 **Mn** 0.30-0.70 **Mo** 0.20-0.30 **Ni** 3.20-3.80 **P** 0.035 max **S** 0.040 max **Si** 0.15-0.35	**AISI** 4817 H **ASTM** A304 (4817 H) **SAE** J1268 (4817 H)
H48200	Ni-Mo H-Alloy Steel	**C** 0.17-0.23 **Mn** 0.40-0.80 **Mo** 0.20-0.30 **Ni** 3.20-3.80 **P** 0.035 max **S** 0.040 max **Si** 0.15-0.35	**AISI** 4820 H **ASTM** A304 (4820 H) **SAE** J1268 (4820 H)
H50401	C-Cr-B H-Alloy Steel	**B** 0.0005-0.003 **C** 0.37-0.44 **Cr** 0.30-0.70 **Mn** 0.65-1.10 **Si** 0.15-0.35	**AISI** 50B40 H **ASTM** A304 (50B40 H) **SAE** J1268 (50B40 H)
H50441	C-Cr-B H-Alloy Steel	**B** 0.0005-0.003 **C** 0.42-0.49 **Cr** 0.30-0.70 **Mn** 0.65-1.10 **P** 0.035 max **S** 0.040 max **Si** 0.15-0.35	**AISI** 50B44 H **ASTM** A304 (50B44 H) **SAE** J1268 (50B44 H)
H50460	C-Cr H-Alloy Steel	**C** 0.43-0.50 **Cr** 0.13-0.43 **Mn** 0.65-1.10 **Si** 0.15-0.35	**AISI** 5046 H **ASTM** A304 (5046 H) **SAE** J1268 (5046 H)
H50461	C-Cr-B H-Alloy Steel	**B** 0.0005-0.003 **C** 0.43-0.50 **Cr** 0.13-0.43 **Mn** 0.65-1.10 **P** 0.035 max **S** 0.040 max **Si** 0.15-0.35	**AISI** 50B46 H **ASTM** A304 (50B46 H) **SAE** J1268 (50B46 H)
H50501	C-Cr-B H-Alloy Steel	**B** 0.0005-0.003 **C** 0.47-0.54 **Cr** 0.30-0.70 **Mn** 0.65-1.10 **P** 0.035 max **S** 0.040 max **Si** 0.15-0.35	**AISI** 50B50 H **ASTM** A304 (50B50 H) **SAE** J1268 (50B50 H)
H50601	C-Cr-B H-Alloy Steel	**B** 0.0005-0.003 **C** 0.55-0.65 **Cr** 0.30-0.70 **Mn** 0.65-1.10 **P** 0.035 max **S** 0.040 max **Si** 0.15-0.35	**AISI** 50B60 H **ASTM** A304 (50B60 H) **SAE** J1268 (50B60 H)
H51200	C-Cr H-Alloy Steel	**C** 0.17-0.23 **Cr** 0.60-1.00 **Mn** 0.60-1.00 **P** 0.035 max **S** 0.040 max **Si** 0.15-0.35	**AISI** 5120 H **ASTM** A304 (5120 H) **SAE** J1268 (5120 H)
H51300	C-Cr H-Alloy Steel	**C** 0.27-0.33 **Cr** 0.75-1.20 **Mn** 0.60-1.00 **P** 0.035 max **S** 0.040 max **Si** 0.15-0.35	**AISI** 5130 H **ASTM** A304 (5130 H) **SAE** J1268 (5130 H)
H51320	C-Cr H-Alloy Steel	**C** 0.29-0.35 **Cr** 0.65-1.10 **Mn** 0.50-0.90 **P** 0.035 max **S** 0.040 max **Si** 0.15-0.35	**AISI** 5132 H **ASTM** A304 (5132 H) **SAE** J1268 (5132 H)
H51350	C-Cr H-Alloy Steel	**C** 0.32-0.38 **Cr** 0.70-1.15 **Mn** 0.50-0.90 **P** 0.035 max **S** 0.040 max **Si** 0.15-0.35	**AISI** 5135 H **ASTM** A304 (5135 H) **SAE** J1268 (5135 H)
H51400	C-Cr H-Alloy Steel	**C** 0.37-0.44 **Cr** 0.60-1.00 **Mn** 0.60-1.00 **P** 0.035 max **S** 0.040 max **Si** 0.15-0.35	**AISI** 5140 H **ASTM** A304 (5140 H) **SAE** J1268 (5140 H)
H51450	C-Cr H-Alloy Steel	**C** 0.42-0.49 **Cr** 0.60-1.00 **Mn** 0.60-1.00 **P** 0.035 max **S** 0.040 max **Si** 0.15-0.35	**AISI** 5145 H **ASTM** A304 (5145 H)
H51470	C-Cr H-Alloy Steel	**C** 0.45-0.52 **Cr** 0.80-1.25 **Mn** 0.60-1.05 **P** 0.035 max **S** 0.040 max **Si** 0.15-0.35	**AISI** 5147 H **ASTM** A304 (5147 H) **SAE** J1268 (5147 H)
H51500	C-Cr H-Alloy Steel	**C** 0.47-0.54 **Cr** 0.60-1.00 **Mn** 0.60-1.00 **P** 0.035 max **S** 0.040 max **Si** 0.15-0.35	**AISI** 5150 H **ASTM** A304 (5150 H) **SAE** J1268 (5150 H)
H51550	C-Cr H-Alloy Steel	**C** 0.50-0.60 **Cr** 0.60-1.00 **Mn** 0.60-1.00 **P** 0.035 max **S** 0.040 max **Si** 0.15-0.35	**AISI** 5155 H **ASTM** A304 (5155 H) **SAE** J1268 (5155 H)
H51600	C-Cr H-Alloy Steel	**C** 0.55-0.65 **Cr** 0.60-1.00 **Mn** 0.65-1.10 **P** 0.035 max **S** 0.040 max **Si** 0.15-0.35	**AISI** 5160 H **ASTM** A304 (5160 H) **SAE** J1268 (5160 H)
H51601	C-Cr-B H-Alloy Steel	**B** 0.0005-0.003 **C** 0.55-0.65 **Cr** 0.60-1.00 **Mn** 0.65-1.10 **Si** 0.15-0.35	**AISI** 51B60 H **ASTM** A304 (51B60 H) **SAE** J1268 (51B60 H)
H61180	Cr-V H-Alloy Steel	**C** 0.15-0.21 **Cr** 0.40-0.80 **Mn** 0.40-0.80 **P** 0.035 max **S** 0.040 max **Si** 0.15-0.35 **V** 0.10-0.15	**AISI** 6118 H **ASTM** A304 (6118 H) **SAE** J1268 (6118 H)
H61500	Cr-V H-Alloy Steel	**C** 0.47-0.54 **Cr** 0.75-1.20 **Mn** 0.60-1.00 **P** 0.035 max **S** 0.040 max **Si** 0.15-0.35 **V** 0.15 min	**AISI** 6150 H **ASTM** A304 (6150 H) **SAE** J1268 (6150 H)

The chemical compositions listed are for identification purposes and should not be used in lieu of the cross referenced specifications.

UNIFIED NUMBER	DESCRIPTION	CHEMICAL COMPOSITION	CROSS REFERENCE SPECIFICATIONS
H81451	Ni-Cr-Mo-B H-Alloy Steel	**B** 0.0005-0.003 **C** 0.42-0.49 **Cr** 0.30-0.60 **Mn** 0.70-1.05 **Mo** 0.08-0.15 **Ni** 0.15-0.45 **P** 0.035 max **S** 0.040 max **Si** 0.15-0.35	**AISI** 81B45 H **ASTM** A304 (81B45 H) **SAE** J1268 (81B45 H)
H86170	Ni-Cr-Mo H-Alloy Steel	**C** 0.14-0.20 **Cr** 0.35-0.65 **Mn** 0.60-0.95 **Mo** 0.15-0.25 **Ni** 0.35-0.75 **P** 0.035 max **S** 0.040 max **Si** 0.15-0.30	**AISI** 8617 H **ASTM** A304 (8617 H) **SAE** J1268 (8617 H)
H86200	Ni-Cr-Mo H-Alloy Steel	**C** 0.17-0.23 **Cr** 0.35-0.65 **Mn** 0.60-0.95 **Mo** 0.15-0.25 **Ni** 0.35-0.75 **P** 0.035 max **S** 0.040 max **Si** 0.15-0.35	**AISI** 8620 H **ASTM** A304 (8620 H) **SAE** J1268 (8620 H)
H86220	Ni-Cr-Mo H-Alloy Steel	**C** 0.19-0.25 **Cr** 0.35-0.65 **Mn** 0.60-0.95 **Mo** 0.15-0.25 **Ni** 0.35-0.75 **P** 0.035 max **S** 0.040 max **Si** 0.15-0.35	**AISI** 8622 H **ASTM** A304 (8622 H) **SAE** J1268 (8622 H)
H86250	Ni-Cr-Mo H-Alloy Steel	**C** 0.22-0.28 **Cr** 0.35-0.65 **Mn** 0.60-0.95 **Mo** 0.15-0.25 **Ni** 0.35-0.75 **P** 0.035 max **S** 0.040 max **Si** 0.15-0.35	**AISI** 8625 H **ASTM** A304 (8625 H) **SAE** J1268 (8625 H)
H86270	Ni-Cr-Mo H-Alloy Steel	**C** 0.24-0.30 **Cr** 0.35-0.65 **Mn** 0.60-0.95 **Mo** 0.15-0.25 **Ni** 0.35-0.75 **P** 0.035 max **S** 0.040 max **Si** 0.15-0.35	**AISI** 8627 H **ASTM** A304 (8627 H) **SAE** J1268 (8627 H)
H86300	Ni-Cr-Mo H-Alloy Steel	**C** 0.27-0.33 **Cr** 0.35-0.65 **Mn** 0.60-0.95 **Mo** 0.15-0.25 **Ni** 0.35-0.75 **P** 0.035 max **S** 0.040 max **Si** 0.15-0.35	**AISI** 8630 H **ASTM** A304 (8630 H) **SAE** J1268 (8630 H)
H86301	Ni-Cr-Mo-B H-Alloy Steel	**B** 0.0005-0.003 **C** 0.27-0.33 **Cr** 0.35-0.65 **Mn** 0.60-0.95 **Mo** 0.15-0.25 **Ni** 0.35-0.75 **Si** 0.15-0.35	**AISI** 86B30 H **ASTM** A304 (86B30 H) **SAE** J1268 (86B30 H)
H86370	Ni-Cr-Mo H-Alloy Steel	**C** 0.34-0.41 **Cr** 0.35-0.65 **Mn** 0.70-1.05 **Mo** 0.15-0.25 **Ni** 0.35-0.75 **P** 0.035 max **S** 0.040 max **Si** 0.15-0.35	**AISI** 8637 H **ASTM** A304 (8637 H) **SAE** J1268 (8637 H)
H86400	Ni-Cr-Mo H-Alloy Steel	**C** 0.37-0.44 **Cr** 0.35-0.65 **Mn** 0.70-1.05 **Mo** 0.15-0.25 **Ni** 0.35-0.75 **P** 0.035 max **S** 0.040 max **Si** 0.15-0.35	**AISI** 8640 H **ASTM** A304 (8640 H) **SAE** J1268 (8640 H)
H86420	Ni-Cr-Mo H-Alloy Steel	**C** 0.39-0.46 **Cr** 0.35-0.65 **Mn** 0.70-1.05 **Mo** 0.15-0.25 **Ni** 0.35-0.75 **P** 0.035 max **S** 0.040 max **Si** 0.15-0.35	**AISI** 8642 H **ASTM** A304 (8642 H) **SAE** J1268 (8642 H)
H86450	Ni-Cr-Mo H-Alloy Steel	**C** 0.42-0.49 **Cr** 0.35-0.65 **Mn** 0.70-1.05 **Mo** 0.15-0.25 **Ni** 0.35-0.75 **P** 0.035 max **S** 0.040 max **Si** 0.15-0.35	**AISI** 8645 H **ASTM** A304 (8645 H) **SAE** J1268 (8645 H)
H86451	Ni-Cr-Mo-B H-Alloy Steel	**B** 0.0005-0.003 **C** 0.42-0.49 **Cr** 0.35-0.65 **Mn** 0.70-1.05 **Mo** 0.15-0.25 **Ni** 0.35-0.75 **Si** 0.15-0.35	**AISI** 86B45 H **ASTM** A304 (86B45 H) **SAE** J1268 (86B45 H)
H86500	Ni-Cr-Mo H-Alloy Steel	**C** 0.47-0.54 **Cr** 0.35-0.65 **Mn** 0.70-1.05 **Mo** 0.15-0.25 **Ni** 0.35-0.75 **Si** 0.15-0.35	**AISI** 8650 H **ASTM** A304 (8650 H) **SAE** J1268 (8650 H)
H86550	Ni-Cr-Mo H-Alloy Steel	**C** 0.50-0.60 **Cr** 0.35-0.65 **Mn** 0.70-1.05 **Mo** 0.15-0.25 **Ni** 0.35-0.75 **P** 0.035 max **S** 0.040 max **Si** 0.15-0.35	**AISI** 8655 H **ASTM** A304 (8655 H) **SAE** J1268 (8655 H)
H86600	Ni-Cr-Mo H-Alloy Steel	**C** 0.55-0.65 **Cr** 0.35-0.65 **Mn** 0.70-1.05 **Mo** 0.15-0.25 **Ni** 0.35-0.75 **Si** 0.15-0.35	**AISI** 8660 H **ASTM** A304 (8660 H) **SAE** J1268 (8660 H)
H87200	Ni-Cr-Mo H-Alloy Steel	**C** 0.17-0.23 **Cr** 0.35-0.65 **Mn** 0.60-0.95 **Mo** 0.20-0.30 **Ni** 0.35-0.75 **P** 0.035 max **S** 0.040 max **Si** 0.15-0.35	**AISI** 8720 H **ASTM** A304 (8720 H) **SAE** J1268 (8720 H)
H87400	Ni-Cr-Mo H-Alloy Steel	**C** 0.37-0.44 **Cr** 0.35-0.65 **Mn** 0.70-1.05 **Mo** 0.20-0.30 **Ni** 0.35-0.75 **P** 0.035 max **S** 0.040 max **Si** 0.15-0.35	**AISI** 8740 H **ASTM** A304 (8740 H) **SAE** J1268 (8740 H)
H88220	Ni-Cr-Mo H-Alloy Steel	**C** 0.19-0.25 **Cr** 0.35-0.65 **Mn** 0.70-1.05 **Mo** 0.30-0.40 **Ni** 0.35-0.75 **Si** 0.15-0.35	**AISI** 8822 H **ASTM** A304 (8822 H) **SAE** J1268 (8822 H)
H92600	C-Si H-Alloy Steel	**C** 0.55-0.65 **Mn** 0.65-1.10 **Si** 1.70-2.20	**AISI** 9260 H **ASTM** A304 (9260 H) **SAE** J1268 (9260 H)

The chemical compositions listed are for identification purposes and should not be used in lieu of the cross referenced specifications.

UNIFIED NUMBER	DESCRIPTION	CHEMICAL COMPOSITION	CROSS REFERENCE SPECIFICATIONS
H93100	Ni-Cr-Mo H-Alloy Steel	**C** 0.07-0.13 **Cr** 1.00-1.45 **Mn** 0.40-0.70 **Mo** 0.08-0.15 **Ni** 2.95-3.55 **Si** 0.15-0.35	**AISI** 9310 H **ASTM** A304 (9310 H) **SAE** J1268 (9310 H)
H94151	Ni-Cr-Mo-B H-Alloy Steel	**B** 0.0005-0.003 **C** 0.12-0.18 **Cr** 0.25-0.55 **Mn** 0.70-1.05 **Mo** 0.08-0.15 **Ni** 0.25-0.65 **Si** 0.15-0.35	**AISI** 94B15 H **ASTM** A304 (94B15 H) **SAE** J1268 (94B15 H)
H94171	Ni-Cr-Mo-B H-Alloy Steel	**B** 0.0005-0.003 **C** 0.14-0.20 **Cr** 0.25-0.55 **Mn** 0.70-1.05 **Mo** 0.08-0.15 **Ni** 0.25-0.65 **P** 0.035 max **S** 0.040 max **Si** 0.15-0.35	**AISI** 94B17 H **ASTM** A304 (94B17 H) **SAE** J1268 (94B17 H)
H94301	Ni-Cr-Mo-B H-Alloy Steel	**B** 0.0005-0.003 **C** 0.27-0.33 **Cr** 0.25-0.55 **Mn** 0.70-1.05 **Mo** 0.08-0.15 **Ni** 0.25-0.65 **P** 0.035 max **S** 0.040 max **Si** 0.15-0.35	**AISI** 94B30 H **ASTM** A304 (94B30 H) **SAE** J1268 (94B30 H)

The chemical compositions listed are for identification purposes and should not be used in lieu of the cross referenced specifications.

UNS NUMBERS ASSIGNED TO DATE

With Description of Each Material Covered and References To Documents In Which The Same or Similar Materials are Described

Jxxxxx Number Series
Cast Steels (Except Tool Steels)

CAST STEELS (EXCEPT TOOL STEELS)

UNIFIED NUMBER	DESCRIPTION	CHEMICAL COMPOSITION	CROSS REFERENCE SPECIFICATIONS
J01700	Carbon Steel Casting	C 0.12-0.22 Mn 0.50-0.90 P 0.040 max S 0.045 max Si 0.60 max	SAE J435 (0022)
J02000	Carbon Steel Casting	C 0.15-0.25 Mn 0.30-0.60 P 0.04 S 0.04 Si 0.20-1.00	MIL SPEC MIL-S-81591 (1020)
J02001	Carbon Steel Casting	C 0.15-0.25 Mn 0.30-0.60 P 0.025 max S 0.025 max Si 0.20-1.00	MIL SPEC MIL-S-22141 (1020)
J02002	Carbon Steel Casting	C 0.15-0.25 Cu 0.50 max Mn 0.20-0.60 P 0.04 max S 0.045 max Si 0.20-1.00 Other Mo+W 0.25 max	ASTM A732 (1A)
J02500	Carbon Steel Casting	C 0.25 max Mn 0.75 max P 0.05 max S 0.06 max Si 0.80 max	ASTM A27 (N1, U-60-30)
J02501	Carbon Steel Casting	C 0.25 max Mn 1.20 max P 0.05 max S 0.06 max Si 0.80 max	ASTM A27 (70-40)
J02502	Carbon Steel Casting	C 0.25 max Cr 0.40 max Cu 0.50 max Mn 0.70 max Mo 0.20 max Ni 0.50 max P 0.04 max S 0.045 max Si 0.60 max V 0.03 max Other Cr+Cu+Mo+Ni+V 1.00 max	ASTM A216 (WCA); A487 (A, AN, AQ)
J02503	Carbon Steel Casting	C 0.25 max Cr 0.50 max Cu 0.30 max Mn 1.20 max Mo 0.20 max Ni 0.50 max P 0.04 max S 0.045 max Si 0.60 max V 0.03 max Other Cr+Cu+Mo+Ni+V 1.00 max	ASTM A216 (WCC); A487 (C, CN, CQ); A643 (A); A757 (A2Q)
J02504	Carbon Steel Casting	C 0.25 max Mn 0.70 max P 0.04 max S 0.045 max Si 0.60 max	ASTM A352 (LCA); A660 (WCA)
J02505	Carbon Steel Casting	C 0.25 max Mn 1.20 max P 0.04 max S 0.045 max Si 0.60 max	ASTM A352 (LCC); A660 (WCC)
J02506	Carbon Steel Casting replaced by J22500		
J02507	Carbon Steel Casting	C 0.25 max Mn 0.75 max P 0.040 max S 0.045 max Si 0.80 max	SAE J435 (0025)
J03000	Carbon Steel Casting	C 0.30 max Mn 0.60 max P 0.05 max S 0.06 max Si 0.80 max	ASTM A27 (60-30)
J03001	Carbon Steel Casting	C 0.30 max Mn 0.70 max P 0.05 max S 0.06 max Si 0.80 max	ASTM A27 (65-35)
J03002	Carbon Steel Casting	C 0.30 max Cr 0.50 max Cu 0.30 max Mn 1.00 max Mo 0.20 max Ni 0.50 max P 0.04 max S 0.045 max Si 0.60 max V 0.03 max Other total Cr+Cu+Mo+Ni+V 1.00 max	ASTM A216 (WCB); A487 (B, BN, BQ); A774 (A7Q); A757 (A7Q)
J03003	Carbon Steel Casting	C 0.30 max Mn 1.00 max P 0.04 max S 0.045 max Si 0.60 max	ASTM A352 (LCB); A660 (WCB)
J03004	Carbon Steel Casting replaced by J13002		
J03005	Carbon Steel Casting	C 0.25-0.35 Mn 0.70-1.00 P 0.025 max S 0.025 max Si 0.20-1.00	MIL SPEC MIL-S-22141 (IC-1030)
J03006	Carbon Steel Casting	C 0.25-0.35 Mn 0.70-1.00 P 0.04 max S 0.04 max Si 0.20-1.00	MIL SPEC MIL-S-81591 (IC-1030)
J03007	Carbon Steel Casting	C 0.30 max Mn 0.70 max P 0.07 max S 0.06 max	MIL SPEC MIL-S-15083 (CW)
J03008	Carbon Steel Casting	C 0.30 max Cr 0.20 max Cu 0.30 max Mn 0.60 max Mo 0.20 max Ni 0.50 max P 0.05 max S 0.05 max Si 0.20-0.60	MIL SPEC MIL-S-15083 (B)

The chemical compositions listed are for identification purposes and should not be used in lieu of the cross referenced specifications.

CAST STEELS (EXCEPT TOOL STEELS)

UNIFIED NUMBER	DESCRIPTION	CHEMICAL COMPOSITION	CROSS REFERENCE SPECIFICATIONS
J03009	Carbon Steel Casting	C 0.30 max Cr 0.35 max Cu 0.30 max Mn 0.70 max Mo 0.20 max Ni 0.50 max P 0.05 max S 0.06 max Si 0.80 max Other total Cu+Mo+Ni 1.00 max	MIL SPEC MIL-S-15083 (65-35)
J03010	Carbon Steel Casting	C 0.30 max Mn 0.70 max P 0.040 max S 0.045 max Si 0.80 max	SAE J435 (0030)
J03011	Carbon Steel Casting	C 0.25-0.35 Cr 0.35 max Cu 0.50 max Mn 0.70-1.00 Ni 0.50 max P 0.04 max S 0.045 max Si 0.20-1.00 W 0.10 max	ASTM A732 (2A, 2Q)
J03500	Carbon Steel Casting	C 0.35 max Mn 0.60 max P 0.05 max S 0.06 max Si 0.80 max	ASTM A27 (N2)
J03501	Carbon Steel Casting	C 0.35 max Mn 0.70 max P 0.05 max S 0.06 max Si 0.80 max	ASTM A27 (70-36)
J03502	Carbon Steel Casting	C 0.35 max Mn 0.70 max P 0.035 max S 0.030 max Si 0.60 max	ASTM A356 (1)
J03503	Carbon Steel Casting	C 0.35 max Mn 0.90 max P 0.05 max S 0.06 max Si 0.80 max	ASTM A486 (70)
J03504	Carbon Steel Casting	C 0.35 max P 0.05 max S 0.05 max	MIL SPEC MIL-S-15083 (70-36)
J04000	Carbon Steel Casting	C 0.35-0.45 Mn 0.70-1.00 P 0.04 max S 0.04 max Si 0.20-1.00	MIL SPEC MIL-S-81591 (1040)
J04001	Carbon Steel Casting	C 0.35-0.45 Mn 0.70-1.00 P 0.025 max S 0.025 max Si 0.20-1.00	MIL SPEC MIL-S-22141 (1040)
J04002	Carbon Steel Casting	C 0.35-0.45 Cr 0.30 max Cu 0.50 max Mn 0.70-1.00 Ni 0.50 max P 0.04 max S 0.045 max Si 0.20-1.00 W 0.10 max	ASTM A732 (3A, 3Q)
J04500	Carbon Steel Casting	C 0.40-0.50 Cr 0.35 max Cu 0.35 max Mn 0.50-0.90 Ni 0.50 max P 0.04 max S 0.045 max Si 0.80 max V 0.03 max W 0.10 max Other Mo+W 0.10 max	ASTM A487 (DN)
J04501	Carbon Steel Casting	C 0.40-0.50 Mn 0.50-0.90 P 0.040 max S 0.045 max Si 0.80 max	SAE J435 (0050A, 0050B)
J05000	Carbon Steel Casting	C 0.45-0.55 Mn 0.70-1.00 P 0.04 max S 0.04 max Si 0.20-1.00	MIL SPEC MIL-S-81591 (1050)
J05001	Carbon Steel Casting	C 0.45-0.55 Mn 0.70-1.00 P 0.025 max S 0.025 max Si 0.20-1.00	MIL SPEC MIL-S-22141 (1050)
J05002	Carbon Steel Casting	C 0.50 max Mn 0.90 max P 0.05 max S 0.06 max Si 0.80 max	MIL SPEC MIL-S-15083 (80-40)
J05003	Carbon Steel Casting	C 0.30 max Cu 0.50 max Mn 0.70-1.00 P 0.04 max S 0.045 max Si 0.80 max W 0.10 max	ASTM A487 (4A, 4N, 4Q, 4QA)
J11442	Alloy Steel Casting (8615)	C 0.11-0.17 Cr 0.35-0.65 Cu 0.35 max Mn 0.65-1.00 Mo 0.15-0.35 Ni 0.35-0.75 P 0.04 max S 0.04 max Si 0.50-1.00	AMS 5333
J11522	Alloy Steel Casting	C 0.15 max Cr 0.35 max Cu 0.50 max Mn 0.30-0.60 Mo 0.44-0.65 Ni 0.50 max P 0.040 max S 0.045 max Si 0.15-1.65 W 0.10 max	ASTM A426 (CP15)
J11547	Alloy Steel Casting	C 0.10-0.20 Cr 0.50-0.81 Cu 0.50 max Mn 0.30-0.61 Mo 0.44-0.65 Ni 0.50 max P 0.040 max S 0.045 max Si 0.10-0.50 W 0.10 max	ASTM A426 (CP2)

The chemical compositions listed are for identification purposes and should not be used in lieu of the cross referenced specifications.

UNIFIED NUMBER	DESCRIPTION	CHEMICAL COMPOSITION	CROSS REFERENCE SPECIFICATIONS
J11562	Alloy Steel Casting	C 0.15 max Cr 0.80-1.25 Cu 0.50 max Mn 0.30-0.61 Mo 0.44-0.65 Ni 0.50 max P 0.040 max S 0.045 max Si 0.50 max W 0.10 max	ASTM A426 (CP12)
J11697	Alloy Steel Casting	C 0.13-0.20 Cr 1.00-1.50 Mn 0.50-0.90 Mo 0.90-1.20 P 0.035 max S 0.030 max Si 0.20-0.60 V 0.05-0.15	ASTM A356 (8)
J11872	Alloy Steel Casting	Al 0.01 max C 0.15-0.21 Cr 1.00-1.50 Cu 0.35 max Mn 0.50-0.80 Mo 0.45-0.65 Ni 0.50 max P 0.020 max S 0.015 max Si 0.30-0.60 V 0.03 max	ASTM A217 (WC11)
J11875	Alloy Steel Casting	C 0.18 max Cr 1.00-1.60 max Cu 0.50 max Mn 0.40-0.70 Mo 0.40-0.60 Ni 0.60 max P 0.05 max S 0.05 max Si 0.60 max V 0.10-0.20	MIL SPEC MIL-S-15464 (3)
J12047	Alloy Steel Casting	C 0.15-0.25 Cr 0.40-0.60 Mn 0.65-0.95 Mo 0.15-0.25 Ni 0.40-0.70 P 0.025 max S 0.025 max Si 0.20-0.80	MIL SPEC MIL-S-22141 (8620)
J12048	Alloy Steel Casting	C 0.15-0.25 Cr 0.40-0.70 Cu 0.50 max Mn 0.65-0.95 Mo 0.15-0.25 Ni 0.40-0.70 P 0.04 max S 0.045 max Si 0.20-0.80 W 0.10 max	ASTM A732 (13Q)
J12070	Alloy Steel Casting	C 0.20 max Cr 1.00-1.60 max Cu 0.50 max Mn 0.50-0.80 Mo 0.40-0.60 Ni 0.60 max P 0.05 max S 0.05 max Si 0.20-0.60	MIL SPEC MIL-S-15464 (1)
J12072	Alloy Steel Casting	C 0.20 max Cr 1.00-1.50 Cu 0.50 max Mn 0.50-0.80 Mo 0.45-0.65 Ni 0.50 max P 0.04 max S 0.045 max Si 0.60 max W 0.10 max	ASTM A217 (WC 6); A426 (CP11)
J12073	Alloy Steel Casting	C 0.20 max Cr 1.00-1.50 Cu 0.50 max Mn 0.50-0.80 Mo 0.45-0.65 Ni 0.50 max P 0.035 max S 0.030 max Si 0.60 max W 0.10 max	ASTM A356 (6)
J12080	Alloy Steel Casting	C 0.20 max Cr 1.00-1.50 Mn 0.30-0.80 Mo 0.45-0.65 P 0.04 max S 0.045 max Si 0.60 max V 0.15-0.25	ASTM A389 (C23)
J12082	Alloy Steel Casting	C 0.20 max Cr 0.50-0.80 Cu 0.50 max Mn 0.50-0.80 Mo 0.45-0.65 Ni 0.70-1.10 P 0.04 max S 0.045 max Si 0.60 max V 0.03 max W 0.10 max	ASTM A217 (WC 4); A487 (11N, 11Q)
J12084	Alloy Steel Casting	B 0.002-0.006 C 0.20 max Cr 0.40-0.80 Cu 0.15-0.50 Mn 0.60-1.00 Mo 0.40-0.60 Ni 0.70-1.00 P 0.04 max S 0.045 max Si 0.80 max V 0.03-0.10 W 0.10 max	ASTM A487 (7Q)
J12092	Alloy Steel Casting	C 0.20 max Cr 0.80-1.25 Mn 0.30-0.80 Mo 0.90-1.20 P 0.04 max S 0.045 max Si 0.60 max V 0.15-0.25	ASTM A389 (C24)
J12093	Alloy Steel Casting (4620)	C 0.15-0.25 Mn 0.40-0.70 Mo 0.20-0.30 Ni 1.65-2.00 P 0.025 max S 0.025 max Si 0.20-0.80	MIL SPEC MIL-S-22141 (4620)
J12094	Alloy Steel Casting	C 0.15-0.25 Cr 0.35 max Cu 0.50 max Mn 0.40-0.70 Mo 0.20-0.30 Ni 1.65-2.00 P 0.04 max S 0.045 max Si 0.20-0.80 W 0.10 max	ASTM A732 (11Q)
J12520	Alloy Steel Casting	C 0.25 max Cr 0.35 max Cu 0.50 max Mn 0.50-0.80 Mo 0.45-0.65 Ni 0.50 max P 0.04 max S 0.45 max Si 0.60	ASTM A217 (WCI) MIL SPEC MIL-S-870

The chemical compositions listed are for identification purposes and should not be used in lieu of the cross referenced specifications.

CAST STEELS (EXCEPT TOOL STEELS)

UNIFIED NUMBER	DESCRIPTION	CHEMICAL COMPOSITION	CROSS REFERENCE SPECIFICATIONS
J12521	Alloy Steel Casting	**C** 0.25 max **Mn** 0.30-0.80 **Mo** 0.44-0.65 **P** 0.040 max **S** 0.045 max **Si** 0.10-0.50	**ASTM** A426 (CP1)
J12522	Alloy Steel Casting	**C** 0.25 max **Mn** 0.50-0.80 **Mo** 0.45-0.65 **P** 0.04 max **S** 0.045 max **Si** 0.60 max	**ASTM** A352 (LC1)
J12523	Alloy Steel Casting	**C** 0.25 max **Mn** 0.70 max **Mo** 0.45-0.65 **P** 0.035 max **S** 0.030 max **Si** 0.60 max	**ASTM** A356 (2)
J12524	Alloy Steel Casting	**C** 0.25 max **Cu** 0.50 max **Mn** 0.50-0.80 **Mo** 0.45-0.65 **N** 0.50 max **P** 0.04 max **S** 0.045 max **Si** 0.60 max **W** 0.10 max	**ASTM** A217 (WC1)
J12540	Alloy Steel Casting	**C** 0.25 max **Cr** 0.40-0.70 **Mn** 0.70 max **Mo** 0.40-0.60 **P** 0.035 max **S** 0.030 max **Si** 0.60 max	**ASTM** A356 (5)
J12545	Alloy Steel Casting	**C** 0.25 max **Cr** 0.40 max **Cu** 0.50 max **Mn** 1.15-1.50 **Mo** 0.45-0.60 **Ni** 0.45-1.00 **P** 0.035 max **S** 0.035 max **Si** 0.60 max **V** 0.03 max	
J12582	Alloy Steel Casting	**C** 0.25 max **Cr** 0.40 max **Cu** 0.50 max **Mn** 1.20 max **Ni** 1.5-2.0 **P** 0.025 max **S** 0.025 max **Si** 0.60 max **V** 0.03 max	**ASTM** A757 (C1Q)
J13002	Alloy Steel Casting	**C** 0.30 max **Cr** 0.35 max **Cu** 0.50 max **Mn** 1.00 max **Ni** 0.50 max **P** 0.04 max **S** 0.045 max **Si** 0.80 max **V** 0.04-0.12 **Other** Mo+W 0.25 max	**ASTM** A487 (1N, 1Q)
J13005	Alloy Steel Casting	**C** 0.30 max **Cr** 0.35 max **Cu** 0.50 max **Mn** 1.00-1.40 **Mo** 0.10-0.30 **Ni** 0.50 max **P** 0.04 max **S** 0.045 max **Si** 0.80 max **V** 0.03 max **W** 0.10 max	**ASTM** A487 (2N, 2Q)
J13025	Alloy Steel Casting	**C** 0.30 max **Cr** 0.50-0.70 **Mn** 0.30-0.50 **Mo** 0.20 max **Ni** 0.70 max **Si** 0.50 max **V** 0.10 max **Other** P+S 0.07 max	**MIL SPEC** MIL-A-11356
J13042	Alloy Steel Casting	**C** 0.25-0.35 **Cr** 0.35-0.65 **Cu** 0.35 max **Mn** 0.60-0.95 **Mo** 0.15-0.30 **Ni** 0.35-0.75 **P** 0.04 max **S** 0.04 max **Si** 1.00 max	**AMS** 5334
J13045	Alloy Steel Casting	**C** 0.25-0.35 **Cr** 0.80-1.10 **Cu** 0.50 max **Mn** 0.40-0.70 **Mo** 0.15-0.25 **P** 0.04 max **S** 0.045 max **Si** 0.20-0.80 **W** 0.10 max	**ASTM** A732 (7Q)
J13046	Alloy Steel Casting	**C** 0.25-0.35 **Cr** 0.80-1.10 **Cu** 0.35 max **Mn** 0.40-0.80 **Mo** 0.15-0.25 **Ni** 0.25 max **P** 0.04 max **S** 0.04 max **Si** 1.00 max	**AMS** 5336
J13047	Alloy Steel Casting	**C** 0.30 max **Cr** 0.40-0.80 **Cu** 0.50 max **Mn** 1.00 max **Mo** 0.15-0.30 **Ni** 0.40-0.80 **P** 0.04 max **S** 0.045 max **Si** 0.80 max **V** 0.03 max **W** 0.10 max	**ASTM** A487 (4N, 4Q, 4QA)
J13048	Alloy Steel Casting (4130)	**C** 0.25-0.35 **Cr** 0.80-1.10 **Mn** 0.40-0.70 **Mo** 0.15-0.25 **P** 0.025 max **S** 0.025 max **Si** 0.20-0.80	**MIL SPEC** MIL-S-22141 (4130)
J13049	Alloy Steel Casting (8630)	**C** 0.25-0.35 **Cr** 0.40-0.60 **Mn** 0.65-0.95 **Mo** 0.15-0.25 **Ni** 0.40-0.70 **P** 0.025 max **S** 0.025 max **Si** 0.20-0.80	**MIL SPEC** MIL-S-22141 (8630)
J13050	Alloy Steel Casting (8630 Modified)	**C** 0.25-0.33 **Cr** 0.40-0.90 **Cu** 0.35 max **Mn** 0.60-0.95 **Mo** 0.15-0.25 **Ni** 0.40-1.10 **P** 0.025 max **S** 0.025 max **Si** 0.50-0.90	**AMS** 5335

The chemical compositions listed are for identification purposes and should not be used in lieu of the cross referenced specifications.

UNIFIED NUMBER	DESCRIPTION	CHEMICAL COMPOSITION	CROSS REFERENCE SPECIFICATIONS
J13051	Alloy Steel Casting	C 0.25-0.35 Cr 0.40-0.70 Cu 0.50 max Mn 0.65-0.95 Mo 0.15-0.25 Ni 0.40-0.70 P 0.04 max S 0.045 max Si 0.20-0.80 W 0.10 max	ASTM A732 (14Q)
J13052	Alloy Steel Casting	C 0.30 max Cr 0.35 max Cu 0.50 max Mn 0.70-1.00 Mo 0.15-0.25 P 0.04 max S 0.045 max Si 0.20-0.80 W 0.10 max	ASTM A732 (5N)
J13080	Alloy Steel Casting	C 0.30 max Cr 0.40 max Cu 0.50 max Mn 0.80-1.10 Mo 0.20-0.30 Ni 1.40-1.75 P 0.04 max S 0.045 max Si 0.60 max V 0.03 max W 0.10 max	ASTM A487 (13N, 13Q)
J13345	Alloy Steel Casting	C 0.33 max Cr 0.75-1.10 Cu 0.50 max Mn 0.60-0.90 Mo 0.15-0.30 Ni 0.50 max P 0.04 max S 0.045 max Si 0.80 max V 0.03 max W 0.10 max	ASTM A487 (9N, 9Q)
J13432	Alloy Steel Casting (4335M)	Al 0.05 max C 0.30-0.38 Mn 0.60-1.00 Mo 0.65-1.00 P 0.04 max S 0.04 max Si 0.50-1.00 V 0.14 max	MIL SPEC MIL-S-22141 (4335M)
J13442	Alloy Steel Casting (8735)	C 0.30-0.38 Cr 0.35-0.90 Mn 0.30-0.70 Mo 0.15-0.40 Ni 0.35-0.75 P 0.04 max S 0.04 max Si 0.20-1.00	MIL SPEC MIL-S-22141 (8735)
J13512	Alloy Steel Casting	C 0.35 max Mn 1.35-1.75 Mo 0.25-0.55 P 0.04 max S 0.045 max Si 0.20-0.80	ASTM A732 (6N)
J13855	Alloy Steel Casting	C 0.38 max Cr 0.40-0.80 Cu 0.50 max Mn 1.30-1.70 Mo 0.30-0.40 Ni 0.40-0.80 P 0.04 max S 0.045 max Si 0.80 max V 0.03 max W 0.10 max	ASTM A487 (6N, 6Q)
J14046	Alloy Steel Casting	C 0.35-0.45 Cr 0.80-1.10 Cu 0.35 max Mn 0.75-1.00 Mo 0.15-0.25 Ni 0.25 max P 0.04 max S 0.04 max Si 1.00 max	AMS 5338
J14047	Alloy Steel Casting (4140)	C 0.35-0.45 Cr 0.80-1.10 Mn 0.70-1.35 Mo 0.15-0.25 P 0.025 max S 0.025 max Si 0.20-0.80	MIL SPEC MIL-S-22141 (4140)
J14048	Alloy Steel Casting (8640)	C 0.35-0.45 Cr 0.40-0.60 Mn 0.70-1.05 Mo 0.15-0.25 Ni 0.40-0.70 P 0.025 max S 0.025 max Si 0.20-0.80	MIL SPEC MIL-S-22141 (8640)
J14049	Alloy Steel Casting	C 0.35-0.45 Cr 0.80-1.10 Cu 0.50 max Mn 0.70-1.00 Mo 0.15-0.25 Ni 0.50 max P 0.04 max S 0.045 max Si 0.20-0.80 W 0.10 max	ASTM A732 (8Q)
J15047	Alloy Steel Casting (6150)	C 0.45-0.55 Cr 0.80-1.10 Mn 0.65-0.95 P 0.025 max S 0.025 max Si 0.20-0.80 V 0.15 min	MIL SPEC MIL-S-22141 (6150)
J15048	Alloy Steel Casting	C 0.45-0.55 Cr 0.80-1.10 Cu 0.50 max Mn 0.65-0.95 Ni 0.50 max P 0.04 max S 0.045 max Si 0.20-0.80 V 0.15 min W 0.10 max Other Mo+W 0.10 max	ASTM A732 (12Q)
J15580	Alloy Steel Casting	C 0.55 max Cr 0.40 max Cu 0.50 max Mn 0.80-1.10 Mo 0.20-0.30 Ni 1.40-1.75 P 0.04 max S 0.045 max Si 0.60 max V 0.03 max W 0.10 max	ASTM A487 (14Q)
J19965	Alloy Steel Casting (52100)	C 0.95-1.10 Cr 1.30-1.60 Mn 0.25-0.55 P 0.025 max S 0.025 max Si 0.20-0.80	MIL SPEC MIL-S-22141 (52100)
J19966	Alloy Steel Casting	C 0.95-1.10 Cr 1.30-1.60 Cu 0.50 max Mn 0.25-0.55 Ni 0.50 max P 0.04 max S 0.045 max Si 0.20-0.80 W 0.10 max	ASTM A732 (15A)

The chemical compositions listed are for identification purposes and should not be used in lieu of the cross referenced specifications.

CAST STEELS (EXCEPT TOOL STEELS)

UNIFIED NUMBER	DESCRIPTION	CHEMICAL COMPOSITION	CROSS REFERENCE SPECIFICATIONS
J21610	Alloy Steel Casting	**C** 0.13-0.20 **Cr** 1.00-1.50 **Mn** 0.50-0.90 **Mo** 0.90-1.20 **P** 0.035 max **S** 0.030 max **Si** 0.20-0.60 **V** 0.20-0.35	**ASTM** A356 (9)
J21880	Alloy Steel Casting	**C** 0.18 max **Cr** 2.00-2.75 **Cu** 0.50 max **Mn** 0.40-0.70 **Mo** 0.80-1.10 **Ni** 0.60 max **P** 0.05 max **S** 0.05 max **Si** 0.20-0.60	**MIL SPEC** MIL-S-15464 (2)
J21890	Alloy Steel Casting	**C** 0.18 max **Cr** 2.00-2.75 **Cu** 0.50 max **Mn** 0.40-0.70 **Mo** 0.90-1.20 **Ni** 0.50 max **P** 0.04 max **S** 0.045 max **Si** 0.60 max **W** 0.10 max	**ASTM** A217 (WC9); A426 (CP22)
J22000	Alloy Steel Casting	**C** 0.20 max **Cr** 0.50-0.90 **Cu** 0.50 max **Mn** 0.40-0.70 **Mo** 0.90-1.20 **Ni** 0.60-1.00 **P** 0.04 max **S** 0.045 max **Si** 0.60 max **V** 0.03 max **W** 0.10 max	**ASTM** A217 (WC5); A487 (12N, 12Q)
J22090	Alloy Steel Casting	**C** 0.20 max **Cr** 2.00-2.75 **Mn** 0.50-0.80 **Mo** 0.90-1.20 **P** 0.035 max **S** 0.030 max **Si** 0.60 max	**ASTM** A356 (10)
J22091	Alloy Steel Casting	**C** 0.20 max **Cr** 2.00-2.75 **Cu** 0.50 max **Mn** 0.50-0.90 **Mo** 0.90-1.10 **P** 0.04 max **S** 0.045 max **Si** 0.80 max **V** 0.03 max **W** 0.10 max	**ASTM** A487 (8N, 8Q)
J22092	Alloy Steel Casting	**C** 0.20 max **Cr** 2.00-2.75 **Cu** 0.50 max **Mn** 0.40-0.80 **Mo** 0.90-1.20 **Ni** 0.50 max **P** 0.035 max **S** 0.035 max **Si** 0.60 max **V** 0.03 max **W** 0.10 max	**ASTM** A643 (C); A757 (D1N, D1Q)
J22500	Alloy Steel Casting	**C** 0.25 max **Mn** 0.50-0.80 **Ni** 2.0-3.0 **P** 0.04 max **S** 0.045 max **Si** 0.60 max	**ASTM** A352 (LC2)
J22501	Alloy Steel Casting	**C** 0.25 max **Cr** 0.40 max **Cu** 0.50 max **Mn** 0.50-0.80 **Mo** 0.25 max **Ni** 2.0-3.0 **P** 0.025 max **S** 0.025 max **Si** 0.60 max **V** 0.03 max	**ASTM** A757 (B2N, B2Q)
J23015	Alloy Steel Casting	**C** 0.30 max **Cr** 0.55-0.90 **Cu** 0.50 max **Mo** 0.20-0.40 **Ni** 1.40-2.00 **P** 0.04 max **S** 0.045 max **Si** 0.80 max **V** 0.03 max **W** 0.10 max	**ASTM** A487 (10N, 10Q)
J23055	Alloy Steel Casting	**C** 0.25-0.35 **Cr** 0.70-0.90 **Cu** 0.50 max **Mn** 0.40-0.70 **Mo** 0.20-0.30 **Ni** 1.65-2.00 **P** 0.04 max **S** 0.045 max **Si** 0.20-0.80 **W** 0.10 max	**ASTM** A732 (9Q)
J23260	Alloy Steel Casting (4330 Modified)	**C** 0.28-0.36 **Cr** 0.65-1.00 **Mn** 0.60-0.90 **Mo** 0.30-0.45 **Ni** 1.65-2.00 **P** 0.025 max **S** 0.025 max **Si** 0.50-1.00	**AMS** 5328; 5329
J24054	Alloy Steel Casting	**C** 0.35-0.45 **Cr** 0.70-0.90 **Cu** 0.50 max **Mn** 0.70-1.00 **Mo** 0.20-0.30 **Ni** 1.65-2.00 **P** 0.04 max **S** 0.045 max **Si** 0.20-0.80 **W** 0.10 max	**ASTM** A732 (10Q)
J24055	Alloy Steel Casting (4340)	**C** 0.36-0.44 **Cr** 0.70-0.90 **Mn** 0.60-0.90 **Mo** 0.20-0.30 **Ni** 1.65-2.00 **P** 0.04 max **S** 0.04 max **Si** 0.20-0.80	**MIL SPEC** MIL-S-22141 (4340)
J24056	Alloy Steel Casting (Nitralloy)	**Al** 0.85-1.20 **C** 0.35-0.45 **Cr** 1.40-1.80 **Mn** 0.40-0.70 **Mo** 0.30-0.45 **P** 0.025 max **S** 0.025 max **Si** 0.20-0.80	**MIL SPEC** MIL-S-22141 (Nitralloy)
J24060	Alloy Steel Casting (4340 Modified)	**C** 0.38-0.46 **Cr** 0.65-1.00 **Mn** 0.60-1.00 **Mo** 0.30-0.45 **Ni** 1.65-2.00 **P** 0.025 max **S** 0.025 max **Si** 0.50-1.50	**AMS** 5330; 5331

The chemical compositions listed are for identification purposes and should not be used in lieu of the cross referenced specifications.

UNIFIED NUMBER	DESCRIPTION	CHEMICAL COMPOSITION	CROSS REFERENCE SPECIFICATIONS
J31500	Alloy Steel Casting	**C** 0.15 max **Cr** 0.40 max **Cu** 0.50 max **Mn** 0.50-0.80 **Mo** 0.25 max **Ni** 3.0-4.0 **P** 0.025 max **S** 0.025 max **Si** 0.60 max **V** 0.03 max	**ASTM** A757 (B3N, B3Q)
J31545	Alloy Steel Casting	**C** 0.15 max **Cr** 2.65-3.35 **Cu** 0.50 max **Mn** 0.30-0.60 **Mo** 0.80-1.06 **Ni** 0.50 max **P** 0.040 max **S** 0.045 max **Si** 0.50 max **W** 0.10 max	**ASTM** A426 (CP21)
J31550	Alloy Steel Casting	**C** 0.15 max **Mn** 0.50-0.80 **Ni** 3.00-4.00 **P** 0.04 max **S** 0.045 max **Si** 0.60 max	**ASTM** A352 (LC3)
J32075	Alloy Steel Casting	**C** 0.20 max **Cr** 1.15-1.65 **Cu** 0.20 max **Mn** 0.55-0.75 **Mo** 0.10-0.60 **Ni** 2.50-3.25 **P** 0.02 max **S** 0.015 max **Si** 0.50 max	**MIL SPEC** MIL-S-15083 (100-70)
J41500	Alloy Steel Casting	**C** 0.15 max **Mn** 0.50-0.80 **Ni** 4.00-5.00 **P** 0.04 max **S** 0.045 max **Si** 0.60 max	**ASTM** A352 (LC4)
J41501	Alloy Steel Casting	**C** 0.15 max **Cr** 0.40 max **Cu** 0.50 max **Mn** 0.50-0.80 **Mo** 0.25 max **Ni** 4.0-5.0 **P** 0.025 max **S** 0.025 max **Si** 0.60 max **V** 0.03 max	**ASTM** A757 (B4N, B4Q)
J42015	Alloy Steel Casting (HY80)	**C** 0.20 max **Cr** 1.35-1.65 **Mn** 0.55-0.75 **Mo** 0.30-0.60 **Ni** 2.50-3.25 **P** 0.02 max **S** 0.015 max **Si** 0.50 max	**MIL SPEC** MIL-S-16216 (HY80); MIL-S-23008 (HY80); MIL-S-23009 (HY80)
J42045	Alloy Steel Casting	**C** 0.20 max **Cr** 4.00-6.50 **Cu** 0.50 max **Mn** 0.40-0.70 **Mo** 0.45-0.65 **Ni** 0.50 max **P** 0.04 max **S** 0.045 max **Si** 0.75 max **W** 0.10 max	**ASTM** A217 (C5); A426 (CP5)
J42065	Alloy Steel Casting	**C** 0.20 max **Cr** 1.50-2.00 **Cu** 0.50 max **Mn** 0.40-0.70 **Mo** 0.40-0.60 **Ni** 2.75-3.90 **P** 0.020 max **S** 0.020 max **Si** 0.60 max **V** 0.03 max **W** 0.10 max	**ASTM** A643 (D); A757 (E2N, E2Q)
J42215	Alloy Steel Casting	**C** 0.22 max **Cr** 1.35-1.85 **Mn** 0.55-0.75 **Mo** 0.30-0.60 **Ni** 2.50-3.50 **P** 0.04 max **S** 0.045 max **Si** 0.50 max	**ASTM** A352 (LC2-1)
J42220	Alloy Steel Casting	**C** 0.22 max **Cr** 1.35-1.85 **Cu** 0.50 max **Mn** 0.50-0.80 **Mo** 0.35-0.60 **Ni** 2.5-3.5 **P** 0.025 max **S** 0.025 max **Si** 0.60 max **V** 0.03 max	**ASTM** A757 (E1Q)
J42240	Alloy Steel Casting (HY100)	**C** 0.22 max **Cr** 1.35-1.85 **Mn** 0.55-0.75 **Mo** 0.30-0.60 **Ni** 2.75-3.50 **P** 0.02 max **S** 0.015 max **Si** 0.50 max	**MIL SPEC** MIL-S-16216 (HY100); MIL-S-23008 (HY100); MIL-S-23009 (HY100)
J51545	Alloy Steel Casting	**C** 0.15 max **Cr** 4.00-6.00 **Cu** 0.50 max **Mn** 0.30-0.60 **Mo** 0.45-0.65 **Ni** 0.50 max **P** 0.040 max **S** 0.045 max **Si** 1.00-2.00 **W** 0.10 max	**ASTM** A426 (CP5b)
J61594	Alloy Steel Casting	**C** 0.15 max **Cr** 6.00-8.00 **Cu** 0.50 max **Mn** 0.30-0.60 **Mo** 0.44-0.65 **Ni** 0.50 max **P** 0.040 max **S** 0.045 max **Si** 0.50-1.00 **W** 0.10 max	**ASTM** A426 (CP7)
J82090	Alloy Steel Casting	**C** 0.20 max **Cr** 8.00-10.00 **Cu** 0.50 max **Mn** 0.35-0.65 **Mo** 0.90-1.20 **Ni** 0.50 max **P** 0.04 max **S** 0.045 max **Si** 1.00 max **W** 0.10 max	**ASTM** A217 (C12); A426 (CP9)
J91109	Austenitic Manganese Steel Casting	**C** 1.05-1.35 **Mn** 11.0 min **P** 0.07 max **Si** 1.00 max	**ASTM** A128 (A)
J91119	Austenitic Manganese Steel Casting	**C** 0.9-1.05 **Mn** 11.5-14.0 **P** 0.07 max **Si** 1.00 max	**ASTM** A128 (B-1)

The chemical compositions listed are for identification purposes and should not be used in lieu of the cross referenced specifications.

CAST STEELS (EXCEPT TOOL STEELS)

UNIFIED NUMBER	DESCRIPTION	CHEMICAL COMPOSITION	CROSS REFERENCE SPECIFICATIONS
J91129	Austenitic Manganese Steel Casting	C 1.05-1.2 Mn 11.5-14.0 P 0.07 max Si 1.00 max	ASTM A128 (B-2)
J91139	Austenitic Manganese Steel Casting	C 1.12-1.28 Mn 11.5-14.0 P 0.07 max Si 1.00 max	ASTM A128 (B-3)
J91149	Austenitic Manganese Steel Casting	C 1.2-1.35 Mn 11.5-14.0 P 0.07 max Si 1.00 max	ASTM A128 (B-4)
J91150	Alloy Steel Casting (CA-15)	C 0.15 max Cr 11.5-14.0 Mn 1.00 max Mo 0.50 max Ni 1.00 max P 0.040 max S 0.040 max Si 1.50 max	ACI CA-15 AMS 5351 ASTM A217 (CA-15); A426 (CPCA-15); A487 (CA-15a); A743 (CA-15) MIL SPEC MIL-S-16993 (Class 1)
J91151	Alloy Steel Casting (CA-15M)	C 0.15 max Cr 11.5-14.0 Mn 1.00 max Mo 0.15-1.0 Ni 1.0 max P 0.040 max S 0.040 max Si 0.65 max	ACI CA-15M ASTM A487 (CA-15M); A743 (CA-15M)
J91152	Alloy Steel Casting (410)	C 0.05-0.15 Cr 11.5-13.5 Cu 0.50 max Mn 1.00 max Mo 0.50 max Ni 0.50 max P 0.04 max S 0.03 max Si 1.00 max	AMS 5350 MIL SPEC MIL-S-81591 (410)
J91153	Alloy Steel Casting (CA-40)	C 0.20-0.40 Cr 11.5-14.0 Mn 1.00 max Mo 0.5 max Ni 1.0 max P 0.04 max S 0.04 max Si 1.50 max	ACI CA-40 ASTM A743 (CA-40)
J91161	Alloy Steel Casting (60416)	C 0.15 max Cr 11.5-14.0 Cu 0.5 max Mn 1.25 max Mo 0.5 max Ni 0.5 max P 0.05 max S 0.15-0.35 Si 1.5 max Zr 0.5 max	AMS 5349
J91171	Alloy Steel Casting	C 0.15 max Cr 11.5-14.0 Cu 0.50 max Mn 1.00 max Mo 0.50 max N 1.00 max P 0.040 max S 0.040 max Si 1.50 max V 0.03 max W 0.10 max	ASTM A426 (CPCA15); A487 (CA15)
J91201	Alloy Steel Casting (420)	C 0.15 min Cr 12.0-14.0 Mn 1.00 max P 0.04 max S 0.03 max Si 1.00 max	MIL SPEC MIL-S-81591 (420)
J91209	Austenitic Manganese Steel Casting	C 1.00-1.35 Cr 0.75 max Mn 12.00-14.00 Mo 0.50 max Ni 1.00 max P 0.060 max Si 0.40-1.00	MIL SPEC MIL-S-17249
J91249	Austenitic Manganese Steel Casting	C 0.7-1.3 Mn 11.5-14.0 Mo 0.9-1.2 P 0.07 max Si 1.00 max	ASTM A128 (E-1)
J91261	Alloy Steel Casting	C 0.15 max Cr 11.50-14.00 Mn 1.00 max Mo 0.50-0.70 Ni 0.65-1.00 P 0.05 max S 0.05 max Si 0.50 max	MIL SPEC MIL-S-16993 (2)
J91309	Austenitic Manganese Steel Casting	C 1.05-1.35 Cr 1.5-2.5 Mn 11.5-14.0 P 0.07 max Si 1.00 max	ASTM A128 (C)
J91339	Austenitic Manganese Steel Casting	C 1.05-1.45 Mn 11.5-14.0 Mo 1.8-2.1 P 0.07 max Si 1.00 max	ASTM A128 (E-2)
J91459	Austenitic Manganese Steel Casting	C 0.7-1.3 Mn 11.5-14.0 Ni 3.0-4.0 P 0.07 max Si 1.00 max	ASTM A128 (D)
J91540	Alloy Steel Casting (CA-6NM)	C 0.06 max Cr 11.5-14.0 Mn 1.00 max Mo 0.40-1.0 Ni 3.5-4.5 P 0.04 max S 0.03 max	ACI CA-6NM ASTM A487 (CA-6NM)
J91550	Alloy Steel Casting	C 0.06 max Cr 11.5-14.0 Cu 0.50 max Mn 1.00 max Mo 0.40-1.0 Ni 3.5-4.5 P 0.030 max S 0.030 max Si 1.00 max W 0.10 max	ASTM A757 (E3N)
J91601	Alloy Steel Casting	C 0.12-0.20 Cr 14.5-17.0 Cu 0.50 max Mn 1.0 max Mo 0.50 max Ni 1.5-2.25 P 0.04 max S 0.04 max Si 1.0 max	AMS 5372
J91606	Alloy Steel Casting (440A)	C 0.60-0.75 Cr 16.0-18.0 Mn 1.00 max Mo 0.75 max P 0.04 max S 0.03 max Si 1.00 max	MIL SPEC MIL-S-81591 (440A)

The chemical compositions listed are for identification purposes and should not be used in lieu of the cross referenced specifications.

UNIFIED NUMBER	DESCRIPTION	CHEMICAL COMPOSITION	CROSS REFERENCE SPECIFICATIONS
J91631	Alloy Steel Casting	C 0.12-0.20 Cr 12.0-14.0 Cu 0.50 max Mn 1.0 max Mo 0.50 max Ni 1.8-2.2 P 0.04 max S 0.03 max Si 1.0 max W 2.5-3.5	AMS 5354
J91639	Alloy Steel Casting (440C)	C 0.95-1.20 Cr 16.0-18.0 Mn 1.00 max Mo 0.35-0.75 Ni 0.75 max P 0.04 max S 0.03 max Si 1.00 max	AMS 5352 MIL SPEC MIL-S-81591 (440C)
J91650	Alloy Steel Casting	C 0.06 max Cr 10.5-12.5 Mn 0.50 max Ni 6.0-8.0 P 0.02 max S 0.02 max Si 1.00 max	ASTM A743 (CA-6N)
J91651	Alloy Steel Casting (431)	C 0.08-0.15 Cr 15.0-17.0 Mn 1.0 max N 0.03-0.12 Ni 1.5-2.2 P 0.04 max S 0.04 max Si 1.0 max	AMS 5353 MIL SPEC MIL-S-81591 (431)
J91803	Alloy Steel Casting (CB-30)	C 0.30 max Cr 18.0-21.0 Mn 1.00 max Ni 2.00 max P 0.04 max S 0.04 max Si 1.50 max	ACI CB-30 ASTM A743 (CB-30)
J92001	Alloy Steel Casting	C 0.08-0.15 Cr 14.5-15.5 Mn 0.40-1.10 Mo 2.0-2.6 N 0.05-0.13 Ni 3.5-4.5 P 0.04 max S 0.03 max Si 0.75 max	AMS 5359; 5368
J92110	Alloy Steel Casting, Precipitation Hardening (15-5 PH)	C 0.07 max Cb 0.15-0.35 Cr 14.0-15.50 Cu 2.50-3.20 Mn 0.70 max N 0.05 max Ni 4.20-5.50 P 0.035 max S 0.03 max Si 1.00 max	AMS 5346; 5347; 5348; 5356; 5357; 5400 ASTM A747 (CB7Cu-2)
J92130	Alloy Steel Casting (15-5 PH) combined with J92110		
J92150	Alloy Steel Casting (17-4)	C 0.08 max Cr 15.5-17.5 Cu 3.0-5.0 Mn 1.00 max Ni 3.0-5.0 P 0.04 max S 0.04 max Si 1.00 max	MIL SPEC MIL-S-81591 (17-4)
J92170	Alloy Steel Casting, Precipitation Hardening	C 0.06 max Cr 15.5-16.7 Cu 2.5-3.2 Mn 0.70 max Ni 3.6-4.6 P 0.04 max S 0.03 max Si 0.50-1.0	
J92180	Alloy Steel Casting, Precipitation Hardening	C 0.07 max Cb 0.15-0.35 Cr 15.50-17.70 Cu 2.50-3.20 Mn 0.70 max N 0.05 max Ni 3.60-4.60 P 0.035 max S 0.03 max Si 1.00 max	ASTM A747 (CB7Cu-1)
J92200	Alloy Steel Casting (17-4)	C 0.06 max Cr 15.5-16.7 Cu 2.8-3.5 Mn 0.70 max N 0.05 max Ni 3.6-4.6 P 0.04 max S 0.03 max Si 0.5-1.0 Other Cb+Ta 0.15-0.40	AMS 5342; 5343; 5344; 5355; 5398
J92240	Alloy Steel Casting (14-4PH)	C 0.06 max Cr 13.5-14.7 Cu 3.0-3.5 Mn 0.7 max Mo 2.0-2.75 N 0.05 max Ni 3.75-4.75 P 0.02 max S 0.025 max Si 0.5-1.0 Other Cb+Ta 0.15-0.35	AMS 5340
J92500	Alloy Steel Casting (CF-3)	C 0.03 max Cr 17.0-21.0 Mn 1.50 max Ni 8.0-12.0 P 0.040 max S 0.040 max Si 2.00 max	ACI CF-3 ASTM A351 (CF-3); A743 (CF-3); A744 (CF-3)
J92501	Alloy Steel Casting (302)	C 0.15 max Cr 17.0-19.0 Mn 2.00 max Ni 8.0-10.0 P 0.04 max S 0.03 max Si 1.00 max	MIL SPEC MIL-S-81591 (302)
J92502	Alloy Steel Casting	C 0.15-0.30 Cr 17.00-20.00 Mn 1.50 max Mo 0.50 max Ni 8.00-11.00 P 0.05 max S 0.05 max Si 1.50 max	MIL SPEC MIL-S-17509 (I)
J92511	Alloy Steel Casting (303)	C 0.12 max Cr 17.0-20.0 Cu 0.50 max Mn 2.00 max Mo 0.60 max Ni 8.0-10.0 P 0.17 max S 0.15-0.35 Si 1.00 max	MIL SPEC MIL-S-81591 (IC-303)
J92512	Alloy Steel Casting	C 0.25 max Cr 17.0-19.0 Cu 0.5 max Mn 2.0 max Mo 0.5 max Ni 8.0-10.0 P 0.04 max S 0.03 max Si 1.0 max	AMS 5358

The chemical compositions listed are for identification purposes and should not be used in lieu of the cross referenced specifications.

CAST STEELS (EXCEPT TOOL STEELS)

UNIFIED NUMBER	DESCRIPTION	CHEMICAL COMPOSITION	CROSS REFERENCE SPECIFICATIONS
J92590	Alloy Steel Casting (304H)	C 0.04-0.10 Cr 18.0-20.0 Mn 2.00 max Ni 8.00-11.0 P 0.040 max S 0.030 max Si 0.75 max	ASTM A452 (TP304H)
J92600	Alloy Steel Casting (CF-8)	C 0.08 max Cr 18.0-21.0 Mn 1.50 max Ni 8.0-11.0 P 0.04 max S 0.04 max Si 2.00 max	ACI CF-8 ASTM A351 (CF-8); A451 (CPF8); A743 (CF-8); A744 (CF-8)
J92602	Alloy Steel Casting (CF-20)	C 0.20 max Cr 18.0-21.0 Mn 1.50 max Ni 8.0-11.0 P 0.04 max S 0.04 max Si 2.00 max	ACI CF-20 AMS 5358 ASTM A743 (CF-20)
J92603	Alloy Steel Casting (HF)	C 0.20-0.40 Cr 18.0-23.0 Mn 2.00 max Mo 0.50 max Ni 8.0-12.0 P 0.04 max S 0.04 max Si 2.00 max	ACI HF ASTM A297 (HF)
J92605	Alloy Steel Casting (HC)	C 0.50 max Cr 26.0-30.0 Mn 1.00 max Mo 0.50 max Ni 4.00 max P 0.04 max S 0.04 max Si 2.00 max	ACI HC ASTM A297 (HC)
J92610	Alloy Steel Casting (304)	C 0.08 max Cr 18.0-20.0 Mn 2.00 max Ni 8.0-12.0 P 0.04 max S 0.03 max Si 1.00 max	MIL SPEC MIL-S-81591 (304)
J92613	Alloy Steel Casting (HC-30)	C 0.25-0.35 Cr 26.0-30.0 Mn 0.5-1.0 Mo 0.50 max Ni 4.0 max P 0.04 max S 0.04 max Si 0.50-2.00	ACI HC-30 ASTM A608 (HC30)
J92615	Alloy Steel Casting (CC-50)	C 0.50 max Cr 26.0-30.0 Mn 1.00 max Ni 4.00 max P 0.04 max S 0.04 max Si 1.50 max	ACI CC-50 ASTM A743 (CC-50)
J92620	Alloy Steel Casting (304L)	C 0.05 max Cr 18.0-21.0 Cu 0.50 max Mn 1.00-2.00 Mo 0.50 max Ni 8.0-11.0 P 0.04 max S 0.03 max Si 1.00 max	AMS 5370; 5371 MIL SPEC MIL-S-81591 (304L)
J92630	Alloy Steel Casting (321)	C 0.08 max Cr 17.0-19.0 Mn 2.00 max Ni 9.0-12.0 P 0.04 max S 0.03 max Si 1.00 max Other Ti 5×C min	MIL SPEC MIL-S-81591 (321)
J92640	Alloy Steel Casting (347)	C 0.08 max Cr 17.0-19.5 Mn 2.00 max Ni 9.0-13.0 P 0.04 max S 0.03 max Si 1.00 max Other Cb+Ta 10xC min, 1.5 max	MIL SPEC MIL-S-81591 (347)
J92641	Alloy Steel Casting	C 0.10 max Cr 17.00-20.00 Cu 0.50 max Mn 2.00 max Mo 0.50 max Ni 9.00-12.00 P 0.04 max S 0.04 max Si 1.50 max Other Cb+Ta 10xC min-1.35 max	AMS 5363; 5364
J92650	Alloy Steel Casting	C 0.08 max Cr 18.00-21.00 Mn 1.50 max Ni 8.00-11.00 P 0.05 max S 0.05 max Si 2.00 max	MIL SPEC MIL-S-867 (I)
J92660	Alloy Steel Casting (347H)	C 0.04-0.10 Cr 17.0-20.0 Mn 2.00 max Ni 9.00-13.0 P 0.040 max S 0.030 max Si 0.75 max Other (Cb÷Ta) 8xC min-1.0 max	ASTM A452 (TP347H)
J92700	Alloy Steel Casting (CF-3)	C 0.03 max Cr 17.0-21.0 Mn 1.50 max Mo 2.0-3.0 Ni 8.0-12.0 P 0.04 max S 0.04 max Si 1.50 max	ACI CF-3 ASTM A296 (CF-3)
J92701	Alloy Steel Casting (CF-16F)	C 0.16 max Cr 18.0-21.0 Mn 1.50 max Ni 9.0-12.0 P 0.04 max S 0.04 max Si 2.00 max	ACI CF-16F ASTM A743 (CF-16F)
J92710	Alloy Steel Casting (CF-8C)	C 0.08 max Cr 18.0-21.0 Mn 1.50 max Ni 9.0-12.0 P 0.04 max S 0.04 max Si 2.00 max	ACI CF-8C ASTM A351 (CF-8C); A451 (CPF8C); A743 (CF-8C); A744 (CF-8C)

The chemical compositions listed are for identification purposes and should not be used in lieu of the cross referenced specifications.

UNIFIED NUMBER	DESCRIPTION	CHEMICAL COMPOSITION	CROSS REFERENCE SPECIFICATIONS
J92711	Alloy Steel Casting (60303)	**C** 0.16 max **Cr** 18.00-21.00 **Cu** 0.50 max **Mn** 2.00 max **Mo** 0.80 max **Ni** 9.00-12.00 **P** 0.04 max **S** 0.15-0.35 **Si** 2.00 max	**AMS** 5341
J92720	Alloy Steel Casting	**C** 0.08 max **Cr** 17.00-20.00 **Mn** 1.50 max **Mo** 0.50 max **Ni** 10.00 min **P** 0.05 max **S** 0.05 max **Si** 1.50 max	**MIL SPEC** MIL-S-17509 (III)
J92730	Alloy Steel Casting	**C** 0.08 max **Cr** 18.00-21.00 **Mn** 1.50 max **Ni** 9.00-12.00 **P** 0.05 max **S** 0.05 max **Si** 2.00 max	**MIL SPEC** MIL-S-867 (II)
J92740	Alloy Steel Casting (Durcomet 101)	**C** 0.07 max **Cr** 22.00-24.00 **Cu** 0.50 max **Fe** balance **Mg** 0.50 max **Mo** 0.85-1.15 **Ni** 4.50-5.50 **P** 0.04 max **S** 0.04 max **Si** 1.70 max	
J92800	Alloy Steel Casting (CF-3M)	**C** 0.03 max **Cr** 17.0-21.0 **Mn** 1.50 max **Mo** 2.0-3.0 **Ni** 9.0-13.0 **P** 0.04 max **S** 0.04 max **Si** 1.50 max	**ACI** CF-3M **ASTM** A351 (CF3M, CF3MA); A743 (CF-3F); A744 (CF-3M)
J92801	Alloy Steel Casting	**C** 0.10 max **Cr** 17.00-20.00 **Mn** 1.50 max **Mo** 0.50 max **Ni** 11.00 min **P** 0.05 max **S** 0.05 max **Si** 1.50 max	**MIL SPEC** MIL-S-17509 (II)
J92803	Alloy Steel Casting (HF-30)	**C** 0.25-0.35 **Cr** 19.0-23.0 **Mn** 1.50 max **Mo** 0.50 max **Ni** 9.0-12.0 **P** 0.04 max **S** 0.04 max **Si** 0.50-2.00	**ASTM** A608 (HF30)
J92810	Alloy Steel Casting (316)	**C** 0.08 max **Cr** 16.0-18.0 **Mn** 2.00 max **Mo** 2.0-3.0 **Ni** 10.0-14.0 **P** 0.04 max **S** 0.03 max **Si** 1.00 max	**MIL SPEC** MIL-S-81591 (316)
J92811	Alloy Steel Casting	**C** 0.12 max **Cr** 18.0-19.5 **Cu** 0.5 max **Mn** 2.0 max **Mo** 0.5 max **Ni** 10.0-14.0 **P** 0.04 max **S** 0.03 max **Si** 1.5 max **Other** Cb+Ta 10xC min-1.5 max	**AMS** 5362
J92843	Alloy Steel Casting	**C** 0.28-0.35 **Cr** 18.00-21.00 **Cu** 0.50 max **Mn** 0.75-1.50 **Mo** 1.00-1.75 **Ni** 8.00-11.00 **P** 0.04 max **S** 0.04 max **Si** 1.00 max **Ti** 0.15-0.50 **W** 1.00-1.75 **Other** Cb+Ta 0.30-0.70	**AMS** 5369
J92900	Alloy Steel Casting (CF-8M)	**C** 0.08 max **Cr** 18.0-21.0 **Mn** 1.50 max **Mo** 2.0-3.0 **Ni** 9.0-12.0 **P** 0.04 max **S** 0.04 max **Si** 2.00 max	**ACI** CF-8M **ASTM** A351 (CF-8M); A451 (CPF8M); A743 (CF-8M); A744 (CF-8M)
J92910	Alloy Steel Casting	**C** 0.08 max **Cr** 18.00-21.00 **Mn** 1.50 max **Mo** 2.00-5.00 **Ni** 9.00-12.00 **P** 0.05 max **S** 0.05 max **Si** 2.00 max	**MIL SPEC** MIL-S-867 (III)
J92920	Alloy Steel Casting (316H)	**C** 0.04-0.10 **Cr** 16.0-18.0 **Mn** 2.00 max **Mo** 2.00-3.00 **Ni** 11.0-14.0 **P** 0.040 max **S** 0.030 max **Si** 0.75 max	**ASTM** A452 (TP316H)
J92951	Alloy Steel Casting	**C** 0.15 max **Cr** 16.0-18.0 **Cu** 0.50 max **Mn** 2.0 max **Mo** 1.5-2.25 **Ni** 12.0-14.0 **P** 0.04 max **S** 0.03 max **Si** 0.75 max	**AMS** 5360
J92971	Alloy Steel Casting (CF-10MC)	**C** 0.10 max **Cr** 15.0-18.0 **Mn** 1.50 max **Mo** 1.75-2.25 **Ni** 13.0-16.0 **P** 0.040 max **S** 0.040 max **Si** 1.50 max **Other** Cb 10×C min, 1.20 max	**ACI** CF-10MC **ASTM** A351 (CF-10MC)
J93000	Alloy Steel Casting (CG-8M)	**C** 0.08 max **Cr** 18.0-21.0 **Mn** 1.50 max **Mo** 3.0-4.0 **Ni** 9.0-13.0 **P** 0.04 max **S** 0.04 max **Si** 1.50 max	**ASTM** A743 (CG-8M); A744 (CG-8M)
J93001	Alloy Steel Casting (CG-12)	**C** 0.12 max **Cr** 20.0-23.0 **Mn** 1.50 max **Ni** 10.0-13.0 **P** 0.04 max **S** 0.04 max **Si** 2.00 max	**ACI** CG-12 **ASTM** A743 (CG-12)

The chemical compositions listed are for identification purposes and should not be used in lieu of the cross referenced specifications.

CAST STEELS (EXCEPT TOOL STEELS)

UNIFIED NUMBER	DESCRIPTION	CHEMICAL COMPOSITION	CROSS REFERENCE SPECIFICATIONS
J93005	Alloy Steel Casting (HD)	C 0.50 max Cr 26.0-30.0 Mn 1.50 max Mo 0.50 max Ni 4.0-7.0 P 0.04 max S 0.04 max Si 2.00 max	ACI HD ASTM A297 (HD)
J93010	Alloy Steel Casting, Maraging	Al 0.02-0.10 C 0.03 max Co 9.50-11.00 Fe rem Mn 0.10 max Mo 4.40-4.80 Ni 16.00-17.50 P 0.01 maX S 0.01 max Si 0.10 max Ti 0.15-0.45	AMS 5339
J93015	Alloy Steel Casting (HD-50)	C 0.45-0.55 Cr 26.0-30.0 Mn 1.50 max Mo 0.50 max Ni 4.0-7.0 P 0.04 max S 0.04 max Si 0.50-2.00	ASTM A608 (HD-50)
J93072	Alloy Steel Casting	C 0.15-0.25 Cr 17.0-20.0 Mn 2.0 max Mo 1.75-2.5 Ni 12.0-15.0 P 0.04 max S 0.04 max Si 1.0 max	AMS 5361
J93150	Alloy Steel Casting, Maraging	Al 0.05-0.20 C 0.03 max Co 8.50-9.50 Mn 0.10 max Mo 4.50-5.50 Ni 18.0-19.0 P 0.010 max S 0.010 max Si 0.10 max Ti 0.55-0.85	AMS 5337
J93183	Cast Ferritic-Austenitic Stainless Steel (KCR-D183)	C 0.03 max Co 0.5-1.5 Cr 20.0-23.0 Cu 1.0 max Mn 2.0 max Mo 2.0-4.0 N 0.08-0.25 Ni 4.0-6.0 P 0.040 max S 0.03 max Si 2.0 max	
J93303	Alloy Steel Casting	C 0.20-0.45 Cr 23.00-28.00 Mn 2.50 max N 0.20 max Ni 10.0-14.0 P 0.05 max S 0.05 max Si 1.75 max	ASTM A447
J93370	Alloy Steel Casting	C 0.04 max Cr 24.5-26.5 Cu 2.75-3.25 Mn 1.00 max Mo 1.75-2.25 Ni 4.75-6.00 P 0.04 max S 0.04 max Si 1.00 max	ASTM A744 (CD-4MCu)
J93400	Alloy Steel Casting (CH-8)	C 0.08 max Cr 22.0-26.0 Mn 1.50 max Ni 12.0-15.0 P 0.040 max S 0.040 max Si 1.50 max	ACI CH-8 ASTM A351 (CH-8); A451 (CPH-8)
J93401	Alloy Steel Casting (CH-10)	C 0.10 max Cr 22.0-26.0 Mn 1.50 max Ni 12.0-15.0 P 0.040 max S 0.040 max Si 2.00 max	ACI CH-10 ASTM A351 (CH-10)
J93402	Alloy Steel Casting (CH-20)	C 0.20 max Cr 22.0-26.0 Mn 1.50 max Ni 12.0-15.0 P 0.04 max S 0.04 max Si 2.00 max	ACI CH-20 ASTM A351 (CH-20); A451 (CH-2); A743 (CH-20)
J93403	Alloy Steel Casting (HE)	C 0.20-0.50 Cr 26.0-30.0 Mn 2.00 max Mo 0.50 max Ni 8.0-11.0 P 0.04 max S 0.04 max Si 2.00 max	ACI HE ASTM A297 (HE)
J93413	Alloy Steel Casting (HE-35)	C 0.30-0.40 Cr 26.0-30.0 Mn 1.50 max Mo 0.50 max Ni 8.0-11.0 P 0.04 max S 0.04 max Si 0.50-2.00	ACI HE-35 ASTM A608 (HE-35)
J93423	Alloy Steel Casting (CE-30)	C 0.30 max Cr 26.0-30.0 Mn 1.50 max Ni 8.0-11.0 P 0.04 max S 0.04 max Si 2.00 max	ACI CE-30 ASTM A743 (CE-30)
J93503	Alloy Steel Casting (HH)	C 0.20-0.50 Cr 24.0-28.0 Mn 2.00 max Mo 0.50 max Ni 11.0-14.0 P 0.04 max S 0.04 max Si 2.00 max	ACI HH ASTM A297 (HH)
J93513	Alloy Steel Casting (HH-30)	C 0.25-0.35 Cr 24.0-28.0 Mn 1.50 max Mo 0.50 max Ni 11.0-14.0 P 0.04 max S 0.04 max Si 0.50-2.00	ACI HH-30 ASTM A608 (HH30)
J93550	Cast Ferritic-Austenitic Stainless Steel (KCR-D283)	C 0.03 max Co 0.5-1.5 Cr 23.0-26.0 Cu 1.0 max Mn 2.0 max Mo 5.0-8.0 N 0.08-0.25 P 0.040 max S 0.03 max Si 2.0 max	

The chemical compositions listed are for identification purposes and should not be used in lieu of the cross referenced specifications.

UNIFIED NUMBER	DESCRIPTION	CHEMICAL COMPOSITION	CROSS REFERENCE SPECIFICATIONS
J93633	Alloy Steel Casting (HH-33)	**C** 0.28-0.38 **Cr** 24.0-26.0 **Mn** 1.50 max **Mo** 0.50 max **Ni** 12.0-14.0 **P** 0.04 max **S** 0.04 max **Si** 0.50-2.00	**ASTM** A608 (HH33)
J93790	Alloy Steel Casting (CG6MMN)	**C** 0.06 max **Cb** 0.10-0.30 **Cr** 20.50-23.50 **Mn** 4.00-6.00 **Mo** 1.50-3.00 **N** 0.20-0.40 **Ni** 11.50-13.50 **P** 0.040 max **S** 0.030 max **Si** 1.00 max **V** 0.10-0.30	**ASTM** A351 (CG6MMN)
J93900	Stainless Steel Casting (Durcomet 5)	**C** 0.025 max **Cr** 20.0-22.0 **Mn** 1.50 max **N** 0.08-0.20 **Ni** 15.0-17.0 **P** 0.040 max **S** 0.040 max **Si** 4.00-6.00	
J94003	Alloy Steel Casting (HI)	**C** 0.20-0.50 **Cr** 26.0-30.0 **Mn** 2.00 max **Mo** 0.50 max **Ni** 14.0-18.0 **P** 0.04 max **S** 0.04 max **Si** 2.00 max	**ACI** HI **ASTM** A297 (HI)
J94013	Alloy Steel Casting (HI-35)	**C** 0.30-0.40 **Cr** 26.0-30.0 **Mn** 1.50 max **Mo** 0.50 max **Ni** 14.0-18.0 **P** 0.04 max **S** 0.04 max **Si** 0.50-2.00	**ASTM** A608 (HI35)
J94202	Alloy Steel Casting (CK-20)	**C** 0.20 max **Cr** 23.0-27.0 **Mn** 2.00 max **Ni** 19.0-22.0 **P** 0.04 max **S** 0.04 max **Si** 2.00 max	**ACI** CK-20 **ASTM** A351 (CK-20); A451 (CPK20); A743 (CK-20)
J94203	Alloy Steel Casting (HK-30)	**C** 0.25-0.35 **Cr** 23.0-27.0 **Mn** 1.50 max **Ni** 19.0-22.0 **P** 0.040 max **S** 0.040 max **Si** 1.75 max	**ACI** HK-30 **ASTM** A351 (HK-30); A608 (HK-30)
J94204	Alloy Steel Casting (HK-40)	**C** 0.35-0.45 **Cr** 23.0-27.0 **Mn** 1.50 max **Ni** 19.0-22.0 **P** 0.040 max **S** 0.040 max **Si** 1.75 max	**ACI** HK-40 **ASTM** A351 (HK-40); A608 (HK40)
J94211	Alloy Steel Casting	**C** 0.10-0.18 **Cr** 23.0-26.0 **Cu** 0.50 max **Mn** 2.0 max **Mo** 0.50 max **Ni** 19.0-22.0 **P** 0.04 max **S** 0.04 max **Si** 0.50-1.50	**AMS** 5365; 5366
J94213	Alloy Steel Casting (HN)	**C** 0.20-0.50 **Cr** 19.0-23.0 **Mn** 2.00 max **Mo** 0.50 max **Ni** 23.0-27.0 **P** 0.04 max **S** 0.04 max **Si** 2.00 max	**ACI** HN **ASTM** A297 (HN)
J94214	Alloy Steel Casting (HN-40)	**C** 0.35-0.45 **Cr** 19.0-23.0 **Mn** 1.50 max **Mo** 0.50 max **Ni** 23.0-27.0 **P** 0.04 max **S** 0.04 max **Si** 0.50-2.00	**ASTM** A608 (HN40)
J94224	Alloy Steel Casting (HK)	**C** 0.20-0.60 **Cr** 24.0-28.0 **Mn** 2.00 max **Mo** 0.50 max **Ni** 18.0-22.0 **P** 0.04 max **S** 0.04 max **Si** 2.00 max	**ACI** HK **ASTM** A297 (HK)
J94302	Alloy Steel Casting (310)	**C** 0.25 max **Cr** 24.0-26.0 **Mn** 2.00 max **Ni** 19.0-22.0 **P** 0.04 max **S** 0.03 max **Si** 1.00 max	**MIL SPEC** MIL-S-81591 (310)
J94603	Alloy Steel Casting (HT-30)	**C** 0.25-0.35 **Cr** 13.0-17.0 **Mn** 2.00 max **Mo** 0.50 max **Ni** 33.0-37.0 **P** 0.040 max **S** 0.040 max **Si** 2.50 max	**ACI** HT-30 **ASTM** A351 (HT-30)
J94604	Alloy Steel Casting (HL)	**C** 0.20-0.60 **Cr** 28.0-32.0 **Mn** 2.00 max **Mo** 0.50 max **Ni** 18.0-22.0 **P** 0.04 max **S** 0.04 max **Si** 2.00 max	**ACI** HL **ASTM** A297 (HL)
J94605	Alloy Steel Casting (HT)	**C** 0.35-0.75 **Cr** 15.0-19.0 **Mn** 2.00 max **N** 33.0-37.0 **P** 0.04 max **S** 0.04 max **Si** 2.50 max	**ACI** HT **ASTM** A297 (HT); A448
J94613	Alloy Steel Casting (HL-30)	**C** 0.25-0.35 **Cr** 28.0-32.0 **Mn** 1.50 max **Mo** 0.50 max **Ni** 18.0-22.0 **P** 0.04 max **S** 0.04 max **Si** 0.50-2.00	**ACI** HL-30 **ASTM** A608 (HL-30)
J94614	Alloy Steel Casting (HL-40)	**C** 0.35-0.45 **Cr** 28.0-32.0 **Mn** 1.50 max **Mo** 0.50 max **Ni** 18.0-22.0 **P** 0.04 max **S** 0.04 max **Si** 0.50-2.00	**ASTM** A608 (HL40)

The chemical compositions listed are for identification purposes and should not be used in lieu of the cross referenced specifications.

CAST STEELS (EXCEPT TOOL STEELS)

UNIFIED NUMBER	DESCRIPTION	CHEMICAL COMPOSITION	CROSS REFERENCE SPECIFICATIONS
J94650	Alloy Steel Casting (CN-7MS)	C 0.07 max Cr 18.0-20.0 Cu 1.5-2.0 Mn 1.00 max Mo 2.5-3.0 Ni 22.0-25.0 P 0.04 max S 0.03 max Si 2.50-3.50	ASTM A743 (CN-7MS); A744 (CN-7MS)
J94805	Alloy Steel Casting (HT50)	C 0.40-0.60 Cr 15.0-19.0 Mn 1.50 max Mo 0.50 max Ni 33.0-37.0 P 0.04 max S 0.04 max Si 0.50-2.00	ACI HT50 ASTM A608 (HT50)
J95150	Alloy Steel Casting (CN-7M)	C 0.07 max Cr 19.0-22.0 Cu 3.0-4.0 Mn 1.50 max Mo 2.0-3.0 Ni 27.5-30.5 P 0.04 max S 0.04 max Si 1.50 max	ACI CN-7M ASTM A351 (CN-7M); A743 (CN-7M); A744 (CN-7M)
J95151	Alloy Steel Casting, 46 Fe-32 Ni	C 0.05-0.15 Cb 0.50-1.50 Cr 19.0-21.0 Mn 0.50-1.50 Ni 31.0-34.0 P 0.03 max S 0.03 max Si 0.50-1.50	ASME SA351 ASTM A351
J95404	Alloy Steel Casting (HU50)	C 0.40-0.60 Cr 17.0-21.0 Mn 1.50 max Mo 0.50 max Ni 37.0-41.0 P 0.04 max S 0.04 max Si 0.50-2.00	ACI HU50 ASTM A608 (HU50)
J95405	Alloy Steel Casting (HU)	C 0.35-0.75 Cr 17.0-21.0 Mn 2.00 max Mo 0.50 max Ni 37.0-41.0 P 0.04 max S 0.04 max Si 2.50 max	ASTM A297 (HU)
J95705	Alloy Steel Casting (HP)	C 0.35-0.75 Cr 24.0-28.0 Mn 2.00 max Mo 0.50 max Ni 33.0-37.0 P 0.04 max S 0.04 max Si 2.50 max	ASTM A297 (HP)

The chemical compositions listed are for identification purposes and should not be used in lieu of the cross referenced specifications.

UNS NUMBERS ASSIGNED TO DATE

With Description of Each Material Covered and References To Documents In Which The Same or Similar Materials are Described

Kxxxxx Number Series
Miscellaneous Steels and Ferrous Alloys

MISCELLANEOUS STEELS AND FERROUS ALLOYS

UNIFIED NUMBER	DESCRIPTION	CHEMICAL COMPOSITION	CROSS REFERENCE SPECIFICATIONS
K00045	Steel Welding Rod (RG45)	**Al** 0.02 max **C** 0.08 max **Cu** 0.25 max **Mn** 0.50 max **P** 0.040 max **S** 0.040 max **Si** 0.10 max	**ASME** SFA5.2 (RG45) **AWS** A5.2 (RG45)
K00060	Steel Welding Rod (RG60)	**Al** 0.02 max **C** 0.15 max **Cu** 0.25 max **Mn** 0.90-1.40 **P** 0.040 max **S** 0.040 max **Si** 0.10-0.35	**ASME** SFA5.2 (RG60) **AWS** A5.2 (RG60)
K00065	Steel Welding Rod (RG65)	**Al** 0.02 max **P** 0.040 max **S** 0.040 max	**ASME** SFA5.2 (RG65) **AWS** A5.2 (RG65)
K00095	Commercially Pure Iron	**Cu** 0.15 max **P** 0.010 max **S** 0.030 max **Other** C + Mn + Si + P + S 0.10 max	**AMS** 7707
K00100	Enameling Steel	**C** 0.008 max **Mn** 0.60 max **S** 0.040 max	**ASTM** A424 (I)
K00400	Enameling Steel	**C** 0.04 max **Mn** 0.12 max **P** 0.015 max **S** 0.040 max	**ASTM** A424 (IIA)
K00600	Carbon Steel, Special Magnetic Properties	**Al** 0.015 max **C** 0.06 max **Mn** 0.40 max **P** 0.015 max **S** 0.015 max **Si** 0.20 max	**ASTM** A594 (4)
K00606	Carbon Steel Welding Wire	**C** 0.06 max **Cu** 0.15 max **Mn** 0.25 max **P** 0.040 max **S** 0.040 max **Si** 0.08 max	**AMS** 5030
K00800	Carbon Steel, Special Magnetic Properties	**Al** 0.02 max **C** 0.08 max **Mn** 0.40 max **P** 0.025 max **S** 0.025 max **Si** 0.20 max	**ASTM** A594 (3)
K00801	Enameling Steel	**C** 0.08 max **Mn** 0.20 max **P** 0.015 max **S** 0.040 max	**ASTM** A424 (IIB)
K00802	Low Carbon Steel	**C** 0.08-0.20 **Mn** 0.40-0.80 **P** 0.040 max **S** 0.050 max	**AMS** 5061
K00912	Steel Electrode (EL12N)	**C** 0.04-0.14 **Cu** 0.08 max **Mn** 0.25-0.60 **P** 0.012 max **S** 0.030 max **Si** 0.10 max **V** 0.05 max **Other** total other elements (except iron) 0.50 max	**ASME** SFA5.23 (EL12N) **AWS** A5.23 (EL12N)
K01000	Carbon Steel, Special Magnetic Properties	**Al** 0.02 max **C** 0.10 max **Mn** 0.60 max **P** 0.04 max **S** 0.04 max **Si** 0.20 max	**ASTM** A594 (2)
K01001	Carbon Steel	**C** 0.05-0.15 **Mn** 0.27-0.63 **P** 0.050 max **S** 0.060 max	**ASTM** A254
K01008	Steel Electrode (EL8)	**C** 0.10 max **Cu** 0.35 max **Mn** 0.25-0.60 **P** 0.030 max **S** 0.030 max **Si** 0.07 max **Other** total other elements (except iron) 0.50 max	**ASME** SFA5.17 (EL8) **AWS** A5.17 (EL8)
K01009	Steel Electrode (EL8K)	**C** 0.10 max **Cu** 0.35 max **Mn** 0.25-0.60 **P** 0.030 max **S** 0.030 max **Si** 0.10-0.25 **Other** total other elements (except iron) 0.50 max	**ASME** SFA5.17 (EL8K) **AWS** A5.17 (EL8K)
K01010	Steel Electrode (EH10K-EW)	**C** 0.06-0.14 **Mn** 1.40-2.0 **P** 0.025 max **S** 0.030 max **Si** 0.15-0.30 **Other** total other elements 0.50 max	**ASME** SFA5.25 (EH10K-EW) **AWS** A5.25 (EH10K-EW)
K01012	Steel Electrode (EL12)	**C** 0.04-0.14 **Cu** 0.35 max **Mn** 0.25-0.60 **P** 0.030 max **S** 0.030 max **Si** 0.10 max **Other** total other elements (except iron) 0.50 max	**ASME** SFA5.17 (EL12); SFA5.23 (EL12) **AWS** A5.17 (EL12); A5.23 (EL12)
K01013	Steel Electrode (EM12KN)	**C** 0.05-0.15 **Cu** 0.08 max **Mn** 0.80-1.25 **P** 0.012 max **S** 0.030 max **Si** 0.10-0.35 **V** 0.05 max **Other** total other elements (except iron) 0.50 max	**ASME** SFA5.23 (EM12KN) **AWS** A5.23 (EM12KN)
K01112	Steel Electrode (EM12)	**C** 0.06-0.15 **Cu** 0.35 max **Mn** 0.80-1.25 **P** 0.030 max **S** 0.030 max **Si** 0.10 max **Other** total other elements (except iron) 0.50 max	**ASME** SFA5.17 (EM12); SFA5.25 (EM12-EW) **AWS** A5.17 (EM12); A5.25 (EM12-EW)

The chemical compositions listed are for identification purposes and should not be used in lieu of the cross referenced specifications.

MISCELLANEOUS STEELS AND FERROUS ALLOYS

UNIFIED NUMBER	DESCRIPTION	CHEMICAL COMPOSITION	CROSS REFERENCE SPECIFICATIONS
K01113	Steel Electrode (EM12K) (EM12K-EW)	**C** 0.05-0.15 **Cu** 0.35 max **Mn** 0.80-1.25 **P** 0.030 max **S** 0.030 max **Si** 0.10-0.35 **Other** total other elements (except iron) 0.50 max	**ASME** SFA5.17 (EM12K); SFA5.23 (EM12K); SFA5.25 (EM12K-EW) **AWS** A5.17 (EM12K); A5.23 (EM12K); A5.25 (EM12K-EW)
K01200	Carbon Steel	**C** 0.06-0.18 **Mn** 0.27-0.63 **P** 0.050 max **S** 0.060 max	**ASTM** A178 (A); A179
K01201	Carbon Steel	**C** 0.06-0.18 **Mn** 0.27-0.63 **P** 0.048 max **S** 0.058 max **Si** 0.25 max	**ASTM** A192; A226
K01313	Steel Electrode (EM13K)	**C** 0.06-0.16 **Cu** 0.35 max **Mn** 0.90-1.40 **P** 0.030 max **S** 0.030 max **Si** 0.35-0.75 **Other** total other elements (except iron) 0.50 max	**ASME** SFA5.17 (EM13K); SFA5.25 (EM13K-EW); SFA5.26 (EGxxS-1); SFA5.30 (INMs2) **AWS** A5.17 (EM13K); A5.25 (EM13K-EW); A5.26 (EGxxS-1); A5.30 (INMs2) **MIL SPEC** MIL-I-23413 (MS2); MIL-E-23765/1C (70S-3)
K01314	Steel Electrode (EM14K)	**C** 0.06-0.10 **Cu** 0.35 max **Mn** 0.90-1.40 **P** 0.025 max **S** 0.025 max **Si** 0.35-0.75 **Ti** 0.03-0.17	**ASME** SFA5.17 (EM14K) **AWS** A5.17 (EM14K)
K01500	Carbon Steel, Special Magnetic Properties	**Al** 0.02 max **C** 0.15 max **Mn** 0.50 max **P** 0.04 max **S** 0.04 max **Si** 0.20 max	**ASTM** A594 (1)
K01501	Carbon Steel	**C** 0.15 max **Mn** 0.90 max **P** 0.035 max **S** 0.040 max **Other** Cu 0.20 min (when specified)	**ASTM** A414 (A)
K01502	Carbon Steel	**C** 0.15 max **Mn** 0.30-0.60 **P** 0.045 max **S** 0.050 max	**ASTM** A730 (A)
K01503	Carbon Steel	**C** 0.15 min	**ASTM** A194 (1)
K01504	Carbon Steel	**C** 0.10-0.20 **Mn** 0.30-0.80 **P** 0.048 max **S** 0.058 max **Si** 0.25 max	**ASTM** A161 (LCST)
K01505	Carbon Steel	**C** 0.15 min **P** 0.05 max **Other** Cu 0.20 min (when specified)	**ASTM** A67
K01506	Carbon Steel	**C** 0.15 max **Mn** 0.63 max **P** 0.050 max **S** 0.060 max	**ASTM** A539, A587
K01507	Carbon Steel	**C** 0.15 max **Mn** 0.60 max **P** 0.035 max **S** 0.04 max **Other** CU0.20 min (when specified)	**ASTM** A109 (4,5)
K01515	Steel Electrode (EM15K) (EM15K-EW)	**C** 0.10-0.20 **Cu** 0.35 max **Mn** 0.80-1.25 **P** 0.035 max **S** 0.035 max **Si** 0.10-0.35 **Other** total other elements (except iron) 0.50 max	**ASME** SFA5.17 (EM15K); SFA5.25 (EM15K-EW) **AWS** A5.17 (EM15K); A5.25 (EM15K-EW) **MIL SPEC** MIL-E-23765/1C (MIL-70S-1)
K01520	Steel Electrode	**C** 0.15 max **Fe** rem **Mn** 0.30-0.60 **P** 0.04 max **S** 0.04 max **Si** 0.03 max	**AWS** A5.15 (ESt)
K01600	Carbon Steel	**C** 0.16 max **Mn** 0.90-1.50 **P** 0.04 max **S** 0.05 max **Si** 0.15-0.50 **Other** Cu 0.20 min (when specified)	**ASTM** A678 (A)
K01601	Carbon Steel	**C** 0.16 max **Mn** 1.00-1.35 **P** 0.04 max **S** 0.04 max **Si** 0.10-0.35	**ASTM** A131 (CS) (DS)
K01602	Carbon Steel	**C** 0.13-0.19 **Mn** 0.30-0.60 **P** 0.04 max **S** 0.05 max **Si** 0.10 max	**MIL SPEC** MIL-S-645
K01700	Carbon Steel	**C** 0.17 max **Mn** 0.90 max **P** 0.035 max **S** 0.045 max	**ASTM** A285/A285M (A)
K01701	Carbon Steel	**C** 0.17 max **Mn** 0.84-1.46 **P** 0.035 max **S** 0.040 max **Si** 0.13-0.45	**ASTM** A662/A662M (A)
K01800	Carbon Steel	**C** 0.18 max **Mn** 0.55-0.98 **P** 0.035 max **S** 0.04 max **Si** 0.13-0.45	**ASTM** A516/A516M (55)
K01801	Carbon Steel	**C** 0.18 max **Mn** 0.70-1.35 **P** 0.04 max **S** 0.04 max **Si** 0.10-0.35	**ASTM** A131 (E)

The chemical compositions listed are for identification purposes and should not be used in lieu of the cross referenced specifications.

UNIFIED NUMBER	DESCRIPTION	CHEMICAL COMPOSITION	CROSS REFERENCE SPECIFICATIONS
K01802	Carbon Steel	C 0.18 max Cb 0.05 max Mn 1.00-1.35 P 0.04 max S 0.05 max Si 0.15-0.50	ASTM A633 (A)
K01803	Carbon Steel	C 0.18 max Mn 1.00-1.35 P 0.04 max S 0.05 max Si 0.15-0.50 V 0.10 max	ASTM A633
K01804	Carbon Steel	C 0.18 max Mn 0.90 max P 0.04 max S 0.05 max Si 0.10-0.30	ASTM A284
K01807	Carbon Steel	C 0.18 max Mn 0.27-0.63 P 0.048 max S 0.058 max	ASTM A214; A556 (A2); A557 (A2)
K01900	Carbon Steel	C 0.11-0.27 Mn 0.27-0.93 P 0.048 max S 0.058 max Other Cu 0.18 min (when specified)	ASTM A502 (1)
K02000	Carbon Steel	C 0.15-0.25 Mn 0.30-0.60 P 0.045 max S 0.050 max	ASTM A730 (B)
K02001	Carbon Steel	C 0.20 max Mn 0.90 max P 0.04 max S 0.05 max Si 0.15-0.30	ASTM A284; A515 (55)
K02002	Carbon Steel	C 0.20 max Mn 0.70-1.35 P 0.04 max S 0.05 max Si 0.15-0.50 Other Cu 0.20 min (when specified)	ASTM A678 (B)
K02003	Steel, High Strength Low Alloy, replaced by K12037		
K02004	Carbon Steel	C 0.12-0.29 Mn 0.26-0.94 P 0.05 max S 0.06 max	ASTM A595 (A)
K02005	Carbon Steel	C 0.12-0.29 Mn 0.35-1.40 P 0.05 max S 0.06 max	ASTM A595 (B)
K02006	Carbon Steel	C 0.20 max Mn 0.90-1.45 P 0.035 max S 0.040 max Si 0.15-0.35	
K02007	Carbon Steel	C 0.20 max Mn 1.00-1.60 P 0.035 max S 0.040 max Si 0.15-0.50	ASTM A662/A662M (C)
K02100	Carbon Steel	C 0.21 max Mn 0.55-0.98 P 0.035 max S 0.04 max Si 0.13-0.45	ASTM A516/A516M (60)
K02101	Carbon Steel	C 0.21 max Mn 0.70-1.35 P 0.04 max S 0.04 max Si 0.10-0.35	ASTM A131 (D)
K02102	Carbon Steel	C 0.21 max Mn 0.80-1.10 P 0.04 max S 0.04 max Si 0.35 max	ASTM A131 (B) MIL SPEC MIL-S-22698 (B); MIL-S-23495 (B)
K02104	Carbon Steel	C 0.21 max Mn 0.90-1.35 P 0.048 max S 0.058 max Si 0.10-0.40	ASTM A524
K02200	Carbon Steel	C 0.22 max Mn 0.90 max P 0.035 max S 0.045 max	ASTM A285/A285M (B)
K02201	Carbon Steel	C 0.22 max Mn 0.90 max P 0.035 max S 0.040 max Other Cu 0.20 min (when specified)	ASTM A414 (B)
K02202	Carbon Steel	C 0.22 max Mn 0.80-1.10 P 0.035 max S 0.04 max Si 0.13-0.40	ASTM A442/A442M (55)
K02203	Carbon Steel	C 0.22 max Mn 0.80-1.55 P 0.035 max S 0.040 max Si 0.13-0.33	ASTM A662/A662M (B)
K02204	Carbon Steel	C 0.22 max Mn 1.00-1.60 P 0.04 max S 0.05 max Si 0.20-0.50 Other Cu 0.20 min (when specified)	ASTM A678 (C)
K02300	Carbon Steel	C 0.23 max P 0.05 max S 0.05 max	ASTM A131 (A)
K02301	Carbon Steel	C 0.23 max Mn 0.60-0.90 P 0.04 max S 0.05 max Si 0.10-0.35	ASTM A573 (58)

The chemical compositions listed are for identification purposes and should not be used in lieu of the cross referenced specifications.

MISCELLANEOUS STEELS AND FERROUS ALLOYS

UNIFIED NUMBER	DESCRIPTION	CHEMICAL COMPOSITION	CROSS REFERENCE SPECIFICATIONS
K02302	Carbon Steel	**C** 0.23 max **Mn** 1.60 max **P** 0.035 max **S** 0.040 max **Si** 0.37 max	**ASTM** A707 (L1)
K02400	Carbon Steel	**C** 0.24 max **Cr** 0.25 max **Cu** 0.35 max **Mn** 0.65-1.40 **Mo** 0.08 max **Ni** 0.25 max **P** 0.035 max **S** 0.040 max **Si** 0.13-0.55	**ASTM** A537
K02401	Carbon Steel	**C** 0.24 max **Mn** 0.90 max **P** 0.04 max **S** 0.05 max **Si** 0.15-0.30	**ASTM** A284 (C); A515 (60) **MIL SPEC** MIL-S-23495 (C)
K02402	Carbon Steel	**C** 0.24 max **Mn** 0.80-1.10 **P** 0.35 max **S** 0.04 max **Si** 0.15-0.40	**ASTM** A442/A442M (60)
K02403	Carbon Steel	**C** 0.24 max **Mn** 0.79-1.30 **P** 0.035 max **S** 0.04 max **Si** 0.13-0.45	**ASTM** A516/A516M (65)
K02404	Carbon Steel	**C** 0.24 max **Mn** 0.85-1.20 **P** 0.04 max **S** 0.05 max **Si** 0.15-0.40	**ASTM** A573 (65)
K02405	Carbon Steel	**C** 0.16-0.33 **Mn** 1.14-1.71 **P** 0.048 max **S** 0.058 max **Si** 0.08-0.32 **Other** Cu 0.18 min (when specified)	**ASTM** A502 (2)
K02500	Carbon Steel	**C** 0.25 max **Mn** 0.60 **P** 0.035 max **S** 0.04 max **Other** Cu 0.20 min (when specified)	**ASTM** A109 (1, 2, 3)
K02501	Carbon Steel	**C** 0.25 max **Mn** 0.27-0.93 **P** 0.048 max **S** 0.058 max **Si** 0.10 min	**ASTM** A106 (A); A369 (FPA)
K02502	Carbon Steel	**C** 0.25 max **Mn** 0.90 max **P** 0.04 max **S** 0.05 max **Other** Cu 0.20 min (when specified)	**ASTM** A570 (30, 33, 36, 40)
K02503	Carbon Steel	**C** 0.25 max **Mn** 0.90 max **P** 0.035 max **S** 0.040 max **Other** Cu 0.20 min (when specified)	**ASTM** A414 (C)
K02504	Carbon Steel	**C** 0.25 max **Mn** 0.95 max **P** 0.05 max **S** 0.06 max	**ASTM** A53 (E-A) (S-A); A523 (A)
K02505	Carbon Steel	**C** 0.25 max **Mn** 1.20 max **P** 0.035 max **S** 0.040 max **Other** Cu 0.20 min (when specified)	**ASTM** A414 (D)
K02506	Carbon Steel	**C** 0.25 max **Mn** 0.90-1.35 **P** 0.035 max **S** 0.025 max **Si** 0.15-0.30	**ASTM** A727
K02507	Carbon Steel	**C** 0.25 max **Mn** 1.35 max **P** 0.040 max **S** 0.05 max **Other** Cu 0.20 min (when specified)	
K02508	Carbon Steel	**C** 0.25 max **Mn** 1.00 max **P** 0.040 max **S** 0.050 max	**AMS** 5062
K02600	Carbon Steel	**C** 0.26 max **P** 0.04 max **S** 0.05 max **Other** Cu 0.20 min (when Copper Steel is specified)	**ASTM** A36 (SHAPES)
K02601	Carbon Steel	**C** 0.26 max **Mn** 1.30 max **S** 0.063 max	**ASTM** A618 (I)
K02603	Carbon Steel	**C** 0.26 max **Mn** 0.60-0.90 **P** 0.035 max **S** 0.040 max	
K02700	Carbon Steel	**C** 0.27 max **Mn** 0.79-1.30 **P** 0.035 max **S** 0.04 max **Si** 0.13-0.45	**ASTM** A516/A516M (70)
K02701	Carbon Steel	**C** 0.27 max **Mn** 0.85-1.20 **P** 0.04 max **S** 0.05 max **Si** 0.15-0.40	**ASTM** A573 (70)
K02702	Carbon Steel	**C** 0.27 max **Mn** 0.90 max **P** 0.04 max **S** 0.05 max **Si** 0.15-0.40	**ASTM** A284 (D)

The chemical compositions listed are for identification purposes and should not be used in lieu of the cross referenced specifications.

UNIFIED NUMBER	DESCRIPTION	CHEMICAL COMPOSITION	CROSS REFERENCE SPECIFICATIONS
K02703	Carbon Steel	**C** 0.27 max **Mn** 1.20 max **P** 0.04 max **S** 0.05 max **Other** Cu 0.20 min (when specified)	**ASTM** A529
K02704	Carbon Steel	**C** 0.27 max **Mn** 1.20 max **P** 0.035 max **S** 0.040 max **Other** Cu 0.20 min (when specified)	**ASTM** A414 (E)
K02705	Carbon Steel	**C** 0.27 max **Mn** 1.40 max **P** 0.05 max **S** 0.063 max **Other** Cu 0.18 min (when specified)	**ASTM** A500 (C)
K02706	Carbon Steel	**C** 0.27 min **Mn** 0.47 min **P** 0.048 max **S** 0.058 max	**ASTM** A325 (1)
K02707	Carbon Steel	**C** 0.27 max **Mn** 0.93 max **P** 0.048 max **S** 0.058 max **Si** 0.10 min	**ASTM** A210 (A-1); A556 (B2)
K02708	Carbon Steel	**C** 0.24-0.30 **Mn** 0.60-0.90 **P** 0.04 max **S** 0.05 max **Si** 0.15-0.30	**MIL SPEC** MIL-S-22698 (C)
K02741	Carbon Steel	**C** 0.27 max **Cr** 0.25 max **Cu** 0.35 max **Mn** 0.85-1.20 **Mo** 0.08 max **Ni** 0.25 max **P** 0.035 max **Pb** 0.05 max **S** 0.035 max **Si** 0.15-0.30 **V** 0.05 max	**ASTM** A758
K02800	Carbon Steel	**C** 0.28 max **Mn** 0.90 max **P** 0.035 max **S** 0.04 max **Si** 0.13-0.45	**ASTM** A515/A515M (65)
K02801	Carbon Steel	**C** 0.28 max **Mn** 0.90 max **P** 0.035 max **S** 0.040 max	**ASTM** A285/A285M (C)
K02802	Carbon Steel	**C** 0.28 max **Mn** 0.81-1.25 **P** 0.040 max **S** 0.050 max **Si** 0.13-0.33	**ASTM** A445/A445M
K02803	Carbon Steel	**C** 0.28 max **Mn** 0.84-1.52 **P** 0.035 max **S** 0.040 max **Si** 0.13-0.45	**ASTM** A299/A299M
K02900	Carbon Steel	**C** 0.29 max **Mn** 0.92-1.46 **P** 0.035 max **S** 0.040 max **Si** 0.13-0.45	**ASTM** A612/A612M
K02901	Carbon Steel	**C** 0.26-0.33 **Mn** 0.60-0.90 **P** 0.040 max **S** 0.040 max **Si** 0.10 max	**MIL SPEC** MIL-S-3289
K03000	Carbon Steel	**C** 0.30 max **P** 0.05 max **S** 0.063 max **Other** Cu 0.18 min (when Copper Steel is specified)	**ASTM** A500 (A, B); A501
K03002	Carbon Steel	**C** 0.30 max **Mn** 1.00 max **P** 0.04 max **S** 0.05 max **Si** 0.15-0.30	**ASTM** A372 (1)
K03003	Carbon Steel	**C** 0.30 max **Mn** 1.00 max **P** 0.040 max **S** 0.050 max	**ASTM** A139 (B)
K03004	Carbon Steel	**C** 0.30 max **Mn** 1.20 max **P** 0.040 max **S** 0.050 max	**ASTM** A139 (C)
K03005	Carbon Steel	**C** 0.30 max **Mn** 1.20 max **P** 0.05 max **S** 0.06 max	**ASTM** A53 (E-B) (S-B); A523 (B)
K03006	Carbon Steel	**C** 0.30 max **Mn** 0.29-1.06 **P** 0.048 max **S** 0.058 max **Si** 0.10 min	**ASTM** A106 (B); A234 (WPB); A333 (6); A334 (6); A369 (FPB); A556 (C2)
K03007	Carbon Steel	**C** 0.30 max **Mn** 0.27-0.93 **P** 0.050 max **S** 0.060 max	**ASTM** A557 (B2)
K03008	Carbon Steel	**C** 0.30 max **Mn** 0.40-1.06 **P** 0.05 max **S** 0.06 max	**ASTM** A333 (1); A334 (1)
K03009	Carbon Steel	**C** 0.30 max **Mn** 1.35 max **P** 0.035 max **S** 0.040 max **Si** 0.15-0.30	**ASTM** A350 (LF1)

The chemical compositions listed are for identification purposes and should not be used in lieu of the cross referenced specifications.

UNIFIED NUMBER	DESCRIPTION	CHEMICAL COMPOSITION	CROSS REFERENCE SPECIFICATIONS
K03010	Carbon Steel	**C** 0.30 max **Mn** 1.30 max **P** 0.040 max **S** 0.050 max	**ASTM** A139 (D)
K03011	Carbon Steel	**C** 0.30 max **Mn** 1.35 max **P** 0.035 max **S** 0.040 max **Si** 0.15-0.30	**ASTM** A350 (LF2)
K03012	Carbon Steel	**C** 0.30 max **Mn** 1.40 max **P** 0.040 max **S** 0.050 max	**ASTM** A139 (E)
K03013	Carbon Steel	**C** 0.30 max **Mn** 1.50 max **P** 0.050 max **S** 0.060 max	**ASTM** A381
K03014	Carbon Steel	**C** 0.30 max **Mn** 1.50 max **P** 0.050 max **S** 0.060 max **Si** 0.13-0.37	**ASTM** A694
K03015	Carbon Steel	**C** 0.30 min **P** 0.04 max **S** 0.06 max	**ASTM** A183
K03016	Carbon Steel	**C** 0.30 max **Mn** 0.60-0.90 **P** 0.04 max **S** 0.05 max **Si** 0.15-0.30	**MIL SPEC** MIL-S-23495
K03017	Carbon Steel	**C** 0.30 max **Mn** 0.80-1.35 **P** 0.035 max **S** 0.040 max **Si** 0.15-0.35	**ASTM** A266 (4)
K03046	Carbon Steel	**Al** 0.05 max **C** 0.30 max **Cr** 0.40 max **Cu** 0.35 max **Mn** 0.60-1.05 **Mo** 0.25 max **Ni** 0.50 max **P** 0.020 max **S** 0.020 max **Si** 0.15-0.35 **V** 0.05 max	**ASTM** A765 (I)
K03047	Carbon Steel	**Al** 0.05 max **C** 0.30 max **Cr** 0.40 max **Cu** 0.35 max **Mn** 0.60-1.35 **Mo** 0.25 max **Ni** 0.50 max **P** 0.020 max **S** 0.020 max **Si** 0.15-0.35 **V** 0.05 max	**ASTM** A765 (II)
K03100	Carbon Steel	**C** 0.31 max **Mn** 0.27-0.83 **P** 0.048 max **S** 0.058 max	**ASTM** A31 (B)
K03101	Carbon Steel	**C** 0.31 max **Mn** 0.90 max **P** 0.035 max **S** 0.04 max **Si** 0.13-0.33	**ASTM** A515/A515M (70)
K03102	Carbon Steel	**C** 0.31 max **Mn** 1.20 max **P** 0.035 max **S** 0.040 max **Other** Cu 0.20 min (when specified)	**ASTM** A414 (F)
K03103	Carbon Steel	**C** 0.31 max **Mn** 1.35 max **P** 0.035 max **S** 0.040 max **Other** Cu 0.20 min (when specified)	**ASTM** A414 (G)
K03104	Carbon Steel	**C** 0.31 min **P** 0.045 max **S** 0.040 max	**ASTM** A574
K03200	Carbon Steel	**C** 0.32 max **Mn** 1.04 max **P** 0.035 max **S** 0.045 max **Si** 0.15-0.30	**ASTM** A696 (B, C)
K03201	Carbon Steel	**C** 0.25-0.40 **Mn** 1.15-1.55 **P** 0.040 max **S** 0.08-0.13 **Si** 0.10-0.20	**MIL SPEC** MIL-S-43
K03300	Carbon Steel	**C** 0.33 max **Mn** 0.81-1.25 **P** 0.040 max **S** 0.050 max **Si** 0.13 max	**ASTM** A455 (I)
K03301	Carbon Steel	**C** 0.33 max **Mn** 1.45 max **P** 0.035 max **S** 0.040 max **Si** 0.37 max	**ASTM** A707 (L2)
K03500	Carbon Steel	**C** 0.35 max **P** 0.040 max **S** 0.045 max	**ASTM** A454 (3)
K03501	Carbon Steel	**C** 0.35 max **Mn** 0.29-1.06 **P** 0.048 max **S** 0.058 max **Si** 0.10 min	**ASTM** A106 (C); A210 (C); A234 (WPC)
K03502	Carbon Steel	**C** 0.35 max **Mn** 1.10 max **P** 0.05 max **S** 0.05 max	**ASTM** A181
K03503	Carbon Steel	**C** 0.35 max **Mn** 0.80 max **P** 0.050 max **S** 0.060 max	**ASTM** A178 (C)
K03504	Carbon Steel	**C** 0.35 max **Mn** 0.60-1.05 **P** 0.040 max **S** 0.050 max **Si** 0.35 max	**ASTM** A105; A695 (B)

The chemical compositions listed are for identification purposes and should not be used in lieu of the cross referenced specifications.

UNIFIED NUMBER	DESCRIPTION	CHEMICAL COMPOSITION	CROSS REFERENCE SPECIFICATIONS
K03505	Carbon Steel	**C** 0.35 max **Mn** 0.27-1.06 **P** 0.050 max **S** 0.060 max	**ASTM** A557 (C2)
K03506	Carbon Steel	**C** 0.35 max **Mn** 0.40-1.05 **P** 0.040 max **S** 0.040 max **Si** 0.15-0.35	**ASTM** A266 (1, 2); A541 (1)
K03700	Carbon Steel	**C** 0.37 max **P** 0.048 max **S** 0.058 max	**ASTM** A413; A454 (1, 2)
K03800	Carbon Steel	**C** 0.18-0.58 **Mn** 0.57 min **P** 0.048 max **S** 0.058 max	**ASTM** A563 (DH)
K03810	Carbon Steel	**B** 0.0005-0.005 **C** 0.38-0.43 **Cr** 0.30-0.55 **Cu** 0.35 max **Mn** 0.75-1.00 **Mo** 0.08-0.15 **Ni** 0.20-0.40 **P** 0.025 max **S** 0.025 max **Si** 0.15-0.35	**AMS** 6321
K03900	Carbon Steel	**C** 0.28-0.50 **P** 0.045 max **S** 0.045 max	**ASTM** A490
K04000	Carbon Steel	**C** 0.35-0.45 **Cr** 0.25 max **Cu** 0.35 **Mn** 0.60-0.90 **Mo** 0.10 max **Ni** 0.30 max **P** 0.040 max **S** 0.040 max **Si** 0.15-0.30 **V** 0.06 max	**ASTM** A290 (A, B)
K04001	Carbon Steel	**C** 0.35 max **Mn** 1.35 max **P** 0.04 max **S** 0.05 max **Si** 0.15-0.35	**ASTM** A372 (2)
K04002	Carbon Steel	**C** 0.40 min **P** 0.040 max **S** 0.050 max	**ASTM** A194 (2, 2HM, 2H)
K04100	Carbon Steel	**C** 0.28-0.55 **P** 0.040 max **S** 0.045 max	**ASTM** A354
K04200	Carbon Steel	**C** 0.25-0.58 **Mn** 0.57 min **P** 0.048 max **S** 0.058 max	**ASTM** A449
K04500	Carbon Steel	**C** 0.40-0.50 **Cr** 0.25 max **Cu** 0.35 **Mn** 0.60-0.90 **Mo** 0.10 max **Ni** 0.30 max **P** 0.040 max **S** 0.040 max **Si** 0.15-0.35 **V** 0.06 max	**ASTM** A290 (C, D)
K04600	Carbon Steel	**C** 0.43-0.50 **Mn** 0.45-0.75 **P** 0.030 max **S** 0.035 max **Si** 0.15-0.35	**MIL SPEC** MIL-S-3039
K04700	Carbon Steel	**C** 0.40-0.55 **Mn** 0.60-0.90 **P** 0.045 max **S** 0.050 max **Si** 0.15 min	**ASTM** A21 (U); A383; A730 (C, D, E)
K04701	Carbon Steel	**C** 0.35-0.60 **Mn** 1.00 max **P** 0.04 max	**ASTM** A49
K04800	Carbon Steel	**C** 0.48 max **Mn** 1.00 max **P** 0.040 max **S** 0.050 max **Si** 0.15-0.35	**ASTM** A489
K04801	Carbon Steel	**C** 0.48 max **Mn** 1.65 max **P** 0.04 max **S** 0.05 max **Si** 0.15-0.35	**ASTM** A372 (3)
K05000	Carbon Steel	**C** 0.50 max **Cu** 0.35 **Mn** 0.40-0.90 **P** 0.040 max **S** 0.040 max **Si** 0.15-0.35 **V** 0.10 max	**ASTM** A291 (2)
K05001	Carbon Steel	**C** 0.50 max **Mn** 0.50-0.90 **P** 0.040 max **S** 0.040 max **Si** 0.35 max	**ASTM** A266 (3); A649 (2)
K05002	Carbon Steel	**C** 0.50 max **Mn** 0.60-1.00 **P** 0.025 max **S** 0.025 max **Si** 0.15-0.30	**ASTM** A288 (1)
K05003	Carbon Steel	**C** 0.50 max **P** 0.05 max **S** 0.05 max	
K05200	Carbon Steel	**C** 0.45-0.59 **Mn** 0.60-0.90 **P** 0.045 max **S** 0.050 max **Si** 0.15 min	**ASTM** A21 (F); A730 (F)
K05500	Carbon Steel	**C** 0.55 max **Cr** 0.25 **Cu** 0.35 **Mn** 0.60-0.90 **Mo** 0.10 **Ni** 0.30 **P** 0.040 max **S** 0.040 max **Si** 0.15-0.35 **V** 0.06	**ASTM** A291 (1)
K05501	Carbon Steel	**C** 0.55 max **Mn** 0.60-0.90 **P** 0.040 max **S** 0.050 max **Si** 0.15-0.35 **Other** Pb 0.15-0.35 (when specified)	**ASTM** A321

The chemical compositions listed are for identification purposes and should not be used in lieu of the cross referenced specifications.

UNIFIED NUMBER	DESCRIPTION	CHEMICAL COMPOSITION	CROSS REFERENCE SPECIFICATIONS
K05700	Carbon Steel	**C** 0.57 max **Mn** 0.60-0.85 **P** 0.05 max **S** 0.05 max **Si** 0.15 min	**ASTM** A25 (A); A504 (A); A583 (A); A631 (A)
K05701	Carbon Steel	**C** 0.50-0.65 **Mn** 0.60-0.90 **P** 0.050 max **S** 0.050 max **Si** 0.15-0.35	**ASTM** A551 (A, AHT)
K05800	Carbon Steel	**C** 0.35-0.82 **P** 0.050 max **Other** Cu 0.20 min (when specified)	**ASTM** A241
K05801	Carbon Steel	**C** 0.58 max **Mn** 0.27 min **P** 0.048 max **S** 0.058 max	**ASTM** A563 (D)
K05802	Carbon Steel	**C** 0.58 max **P** 0.13 **S** 0.15 max	**ASTM** A563 (O, A, B, C)
K06000	Carbon Steel	**C** 0.45-0.75 **Mn** 0.60-1.10 **P** 0.040 max **S** 0.050 max **Si** 0.10-0.35	**ASTM** A648 (I) **MIL SPEC** MIL-W-21425 (II)
K06001	Carbon Steel	**C** 0.60 max **Mn** 1.30-1.70 **P** 0.045 max **S** 0.050 max **Si** 0.15 min	**ASTM** A729
K06002	Carbon Steel	**C** 0.35-0.85 **P** 0.05 max **Other** Cu 0.20 min (when specified)	**ASTM** A67 (2)
K06100	Carbon Steel	**C** 0.55-0.68 **Mn** 0.60-0.90 **P** 0.04 max **S** 0.05 **Si** 0.10-0.25	**ASTM** A1 (61-80)
K06200	Carbon Steel	**C** 0.57-0.67 **Mn** 0.60-0.85 **P** 0.05 max **S** 0.05 max **Si** 0.15 min	**ASTM** A25 (B); A504 (B); A583 (B); A631 (B)
K06201	Carbon Steel	**C** 0.50-0.75 **Mn** 0.60-1.20 **P** 0.040 max **S** 0.050 max	**ASTM** A417
K06500	Carbon Steel	**C** 0.55-0.75 **Mn** 0.60-1.20 **P** 0.04 max **S** 0.05 max **Si** 0.10-0.30	**MIL SPEC** MIL-W-21425 (I, III)
K06501	Carbon Steel	**C** 0.45-0.85 **Mn** 0.30-1.30 **P** 0.040 max **S** 0.050 max **Si** 0.10-0.35	**ASTM** A227 **SAE** J113
K06700	Carbon Steel	**C** 0.50-0.85 **Mn** 0.60-1.10 **P** 0.040 max **S** 0.050 max **Si** 0.10-0.35	**ASTM** A648 (II)
K06701	Carbon Steel	**C** 0.60-0.75 **Mn** 0.60-0.90 **P** 0.025 max **S** 0.030 max **Si** 0.15-0.35	**ASTM** A230 **SAE** J172; J316
K06702	Carbon Steel	**C** 0.60-0.75 **Mn** 0.60-0.90 **P** 0.050 max **S** 0.050 max **Si** 0.15-0.35	**ASTM** A551 (B, BHT)
K06703	Carbon Steel	**C** 0.60-0.75 **Mn** 0.60-0.90 **P** 0.04 max **Si** 0.10-0.40	**ASTM** A2 (A)
K07000	Carbon Steel	**C** 0.64-0.77 **Mn** 0.60-0.90 **P** 0.04 max **S** 0.05 **Si** 0.10-0.25	**ASTM** A1 (81-90)
K07001	Carbon Steel	**C** 0.55-0.85 **Mn** 0.60-1.20 **P** 0.040 max **S** 0.050 max **Si** 0.10-0.35	**ASTM** A229 **SAE** J316
K07100	Carbon Steel	**C** 0.55-0.88 **Mn** 0.60-1.10 **P** 0.040 max **S** 0.050 max **Si** 0.10-0.35	**ASTM** A648 (III)
K07200	Carbon Steel	**C** 0.65-0.80 **Mn** 0.60-0.85 **P** 0.05 max **S** 0.05 max **Si** 0.15 min	**ASTM** A25 (U); A504 (U); A583 (U); A631 (U)
K07201	Carbon Steel	**C** 0.67-0.77 **Mn** 0.60-0.85 **P** 0.05 max **S** 0.05 max **Si** 0.15 min	**ASTM** A25 (C); A504 (C); A583 (C); A631 (C)
K07301	Carbon Steel	**C** 0.67-0.80 **Mn** 0.70-1.00 **P** 0.04 max **S** 0.05 **Si** 0.10-0.25	**ASTM** A1 (91-120)
K07500	Carbon Steel	**C** 0.69-0.82 **Mn** 0.70-1.00 **P** 0.04 max **S** 0.05 max **Si** 0.10-0.25	**ASTM** A1 (121 and over); A759
K07700	Carbon Steel	**C** 0.70-0.85 **Mn** 0.60-0.90 **P** 0.04 max **Si** 0.10-0.40	**ASTM** A2 (B)

The chemical compositions listed are for identification purposes and should not be used in lieu of the cross-referenced specifications.

UNIFIED NUMBER	DESCRIPTION	CHEMICAL COMPOSITION	CROSS REFERENCE SPECIFICATIONS
K07701	Carbon Steel	C 0.70-0.85 Mn 0.60-0.90 P 0.050 max S 0.050 max Si 0.15-0.35	**ASTM** A551 (C, CHT, DHT)
K08200	Carbon Steel	C 0.65-1.00 Mn 0.20-1.30 P 0.040 max S 0.050 max Si 0.10-0.40	**ASTM** A679
K08201	Carbon Steel	C 0.75-0.90 Mn 0.60-0.90 P 0.04 max Si 0.10-0.40	**ASTM** A2 (C)
K08500	Carbon Steel	C 0.70-1.00 Mn 0.20-0.60 P 0.025 max S 0.03 max Si 0.12-0.30	**ASTM** A228 **FED** QQ-W-470 **SAE** J178
K08700	Carbon Steel	C 0.85-0.90 Cr 0.04 max Cu 0.08 max Mn 0.25-0.40 Mo 0.04 max Ni 0.03 max P 0.025 max S 0.020 max Si 0.15-0.25	**MIL SPEC** MIL-W-8957
K10614	Alloy Steel	C 0.06 max Cb 0.03-0.09 Mn 1.20-1.90 Mo 0.25-0.35 P 0.04 max S 0.025 max Si 0.40 max	**ASTM** A699
K10623	Alloy Steel	C 0.06 max Cb 0.03-0.09 Cu 0.20-0.35 Mn 1.20-1.90 Mo 0.23-0.47 P 0.04 max S 0.025 max Si 0.40 max	**ASTM** A735/A735M
K10726	Steel Electrode	Al 0.05-0.15 C 0.07 max Cu 0.50 max Mn 0.90-1.40 P 0.025 max S 0.035 max Si 0.40-0.70 Ti 0.05-0.15 Zr 0.02-0.12	**ASME** SFA5.18 (ER70S-2); SFA5.25 (EM5K-EW); SFA5.26 (EGxxS-2); SFA5.30 (INMs1) **AWS** A5.18 (ER70S-2); A5.25 (EM5K-EW); A5.26 (EGxxS-2); A5.30 (INMs1) **MIL SPEC** MIL-I-23413B (MIL-Ms-1); MIL-E-23765/1C (MIL-705-2)
K10882	Alloy Steel Electrode (ER100S-1)	Al 0.10 max C 0.08 max Cr 0.30 max Cu 0.25 max Mn 1.25-1.80 Mo 0.25-0.55 Ni 1.40-2.10 P 0.010 max S 0.010 max Si 0.20-0.50 Ti 0.10 max V 0.05 max Zr 0.10 max Other total other elements (except iron) 0.50 max	**ASME** SFA5.23 (EM2); SFA5.28 (ER100S-1) **AWS** A5.23 (EM2); A5.28 (ER100S-1) **MIL SPEC** MIL-E-23765/2A (MIL-100S-1)
K10940	Alloy Steel Electrode (ENi1N)	C 0.12 max Cr 0.15 max Cu 0.08 max Mn 0.75-1.25 Mo 0.30 max Ni 0.75-1.25 P 0.012 max S 0.010 max Si 0.05-0.30 V 0.05 max Other total other elements (except iron) 0.50 percent max	**ASME** SFA5.23 (ENi1N) **AWS** A5.23 (ENi1N)
K10943	Alloy Steel Electrode (EB1N)	C 0.10 max Cr 0.40-0.75 Cu 0.08 max Mn 0.40-0.80 Mo 0.45-0.65 P 0.012 max S 0.025 max Si 0.05-0.30 V 0.05 max	**ASME** SFA5.23 (EB1N) **AWS** A5.23 (EB1N)
K10945	Alloy Steel Electrode	C 0.07-0.12 Cu 0.15 max Mn 1.60-2.10 Mo 0.40-0.60 P 0.03 max S 0.035 max Si 0.50-0.80 Other total other elements 0.50 max	**ASME** SFA5.25 (EH10Mo-EW); SFA5.26 (EGxxS-1B); SFA5.28 (ER80S-D2) **AWS** A5.25 (EH10Mo-EW); A5.26 (EGxxS-1B); A5.28 (ER80S-D2)
K10958	Alloy Steel Electrode (ENi1KN)	C 0.12 max Cu 0.08 max Mn 0.80-1.40 Ni 0.75-1.25 P 0.012 max S 0.020 max Si 0.40-0.80 V 0.05 max	**ASME** SFA5.23 (ENi1KN) **AWS** A5.23 (ENi1KN)
K10982	Alloy Steel Electrode (EM2N)	Al 0.10 max C 0.10 max Cr 0.30 max Cu 0.08 max Mn 1.25-1.80 Mo 0.25-0.55 Ni 1.40-2.10 P 0.012 max S 0.010 max Si 0.20-0.60 Ti 0.10 max V 0.05 max Zr 0.10 max Other total other elements (except iron) 0.50 max	**ASME** SFA5.23 (EM2N) **AWS** A5.23 (EM2N)
K11022	Steel Electrode	C 0.06-0.15 Cu 0.50 max Mn 0.90-1.40 P 0.025 max S 0.035 max Si 0.45-0.70	**ASME** SFA5.18 (ER70S-3); SFA5.26 (EGxxS-3) **AWS** A5.18 (ER70S-3); A5.26 (EGxxS-3)
K11040	Alloy Steel Electrode (ENi1)	C 0.12 max Cr 0.15 max Cu 0.35 max Mn 0.75-1.25 Mo 0.30 max Ni 0.75-1.25 P 0.020 max S 0.020 max Si 0.05-0.30 Other total other elements (except iron) 0.50 max	**ASME** SFA5.23 (ENi1) **AWS** A5.23 (ENi1)

The chemical compositions listed are for identification purposes and should not be used in lieu of the cross referenced specifications.

MISCELLANEOUS STEELS AND FERROUS ALLOYS

UNIFIED NUMBER	DESCRIPTION	CHEMICAL COMPOSITION	CROSS REFERENCE SPECIFICATIONS
K11043	Alloy Steel Electrode (EB1)	C 0.10 max Cr 0.40-0.75 Cu 0.35 max Mn 0.40-0.80 Mo 0.45-0.65 P 0.025 max S 0.025 max Si 0.05-0.30	ASME SFA5.23 (EB1) AWS A5.23 (EB1)
K11058	Alloy Steel Electrode (ENi1K)	C 0.18 max Cu 0.35 max Mn 0.80-1.40 Ni 0.75-1.25 P 0.020 max S 0.020 max Si 0.40-0.80	ASME SFA5.23 (ENi1K) AWS A5.23 (ENi1K)
K11060	Alloy Steel Electrode (EF1N)	C 0.07-0.15 Cu 0.08 max Mn 0.90-1.70 Mo 0.25-0.55 Ni 0.95-1.60 P 0.012 max S 0.025 max Si 0.15-0.35 V 0.05 max Other total other elements (except iron) 0.50 max	ASME SFA5.23 (EF1N) AWS A5.23 (EF1N)
K11072	Alloy Steel Electrode (EB2N)	C 0.07-0.15 Cr 1.00-1.75 Cu 0.08 max Mn 0.45-1.00 Mo 0.45-0.65 P 0.012 max S 0.030 max Si 0.05-0.30 V 0.05 max Other total other elements (except iron) 0.50 max	ASME SFA5.23 (EB2N) AWS A5.23 (EB2N)
K11122	Alloy Steel Electrode (EA1N)	C 0.07-0.17 Cu 0.08 max Mn 0.65-1.00 Mo 0.45-0.65 P 0.012 max S 0.030 max Si 0.20 max V 0.05 max Other total other elements (except iron) 0.50 max	ASME SFA5.23 (EA1N) AWS A5.23 (EA1N)
K11123	Alloy Steel Electrode (EA2N)	C 0.07-0.17 Cu 0.08 max Mn 0.95-1.35 Mo 0.45-0.65 P 0.012 max S 0.030 max Si 0.20 max V 0.05 max Other total other elements (except iron) 0.50 max	ASME SFA5.23 (EA2N) AWS A5.23 (EA2N) MIL SPEC MIL-E-23765/1C MIL-70S-8
K11125	Steel Electrode (ER70S-7)	C 0.07-0.15 Cu 0.50 max Mn 1.50-2.00 P 0.025 max S 0.035 max Si 0.50-0.80	ASME SFA5.18 (ER70S-7) AWS A5.18 (ER70S-7)
K11132	Steel Electrode (ER70S-4)	C 0.07-0.15 Cu 0.50 max Mn 1.00-1.50 P 0.025 max S 0.035 max Si 0.65-0.85	ASME SFA5.18 (ER70S-4) AWS A5.18 (ER70S-4) MIL SPEC MIL-E-23765 (MIL-70S-4)
K11140	Steel Electrode (ER70S-6)	C 0.07-0.15 Cu 0.50 max Mn 1.40-1.85 P 0.025 max S 0.035 max Si 0.80-1.15	ASME SFA5.17 (EH11K) SFA5.18 (ER70S-6); SFA5.25 (EH11K-EW); SFA5.26 (EGxxS-6); SFA5.30 (IN-Ms3) AWS A5.17 (EH11K) A5.18 (ER70S-6); A5.25 (EH11K-EW); A5.26 (EGxxS-6); A5.30 (IN-Ms3) MIL SPEC MIL-E-23765 (MIL-70S-6)
K11160	Alloy Steel Electrode (EF1)	C 0.07-0.15 Cu 0.35 max Mn 0.90-1.70 Mo 0.25-0.55 Ni 0.95-1.60 P 0.025 max S 0.025 max Si 0.15-0.35 Other total other elements (except iron) 0.50 max	ASME SFA5.23 (EF1) AWS A5.23 (EF1)
K11172	Alloy Steel Electrode (EB2)	C 0.07-0.15 Cr 1.00-1.75 Cu 0.35 max Mn 0.45-1.00 Mo 0.45-0.65 P 0.025 max S 0.030 max Si 0.05-0.30 Other total other elements (except iron) 0.50 max	ASME SFA5.23 (EB2) AWS A5.23 (EB2)
K11201	Alloy Steel	C 0.12 max Mn 1.30 max P 0.035 max S 0.040 max Si 0.15-0.35 max V 0.020 min	MIL SPEC MIL-S-24412
K11210	Alloy Steel	C 0.12 min Cu 0.20 min	ASTM A65
K11222	Alloy Steel Electrode (EA1)	C 0.07-0.17 Cu 0.35 max Mn 0.65-1.00 Mo 0.45-0.65 P 0.025 max S 0.030 max Si 0.20 max Other total of all other elements (except iron) shall not exceed 0.50 percent	ASME SFA5.23 (EA1) AWS A5.23 (EA1)
K11223	Alloy Steel Electrode (EA2)	C 0.07-0.17 Cu 0.35 max Mn 0.95-1.35 Mo 0.45-0.65 P 0.025 max S 0.030 max Si 0.20 max Other total other elements (except iron) 0.50 max	ASME SFA5.23 (EA2) AWS A5.23 (EA2)
K11224	Alloy Steel	C 0.12 max Cu 0.15 max Mn 1.20 max P 0.035 max S 0.04 max Si 0.15-0.50 Other Ti 4×C min	ASTM A562/A562M

The chemical compositions listed are for identification purposes and should not be used in lieu of the cross referenced specifications.

UNIFIED NUMBER	DESCRIPTION	CHEMICAL COMPOSITION	CROSS REFERENCE SPECIFICATIONS
K11245	Alloy Steel Electrode (EW, EWS)	**C** 0.12 max **Cr** 0.50-0.80 **Cu** 0.30-0.80 **Mn** 0.35-0.65 **Ni** 0.40-0.80 **P** 0.030 max **S** 0.040 max **Si** 0.20-0.35 **Other** total other elements (except iron) 0.50 max	**ASME** SFA5.23 (EW); SFA5.25 (EWS-EW) **AWS** A5.23 (EW); A5.25 (EWS-EW)
K11250	Alloy Steel Electrode (ER100S-2)	**Al** 0.10 max **C** 0.12 max **Cr** 0.30 max **Cu** 0.35-0.65 **Mn** 1.25-1.80 **Mo** 0.20-0.55 **Ni** 0.80-1.25 **P** 0.010 max **S** 0.010 max **Si** 0.20-0.60 **Ti** 0.10 max **V** 0.05 max **Zr** 0.10 max **Other** total other elements (except iron) 0.50 max	**ASME** SFA5.28 (ER100S-2) **AWS** A5.28 (ER100S-2)
K11260	Alloy Steel Electrode (ER80S-Ni1)	**C** 0.12 max **Cr** 0.15 max **Cu** 0.35 max **Mn** 1.25 max **Mo** 0.35 max **Ni** 0.80-1.10 **P** 0.025 max **S** 0.025 max **Si** 0.40-0.80 **V** 0.05 max **Other** total other elements (except iron) 0.50 max	**ASME** SFA5.28 (ER80S-Ni1) **AWS** A5.28 (ER80S-Ni1)
K11267	High Strength, Low Alloy, Precipitation Hardening Steel	**Al** 0.04-0.30 **C** 0.12 max **Cr** 0.44-1.01 **Cu** 0.40-0.75 **Mn** 0.50-1.05 **Ni** 0.47-0.98 **P** 0.04 max **S** 0.04 max **Si** 0.08-0.37	**ASTM** A333 (4)
K11268	High Strength, Low Alloy, Precipitation Hardening Steel	**Al** 0.02-0.23 **C** 0.12 max **Cu** 0.85-1.30 **Mn** 0.50-1.00 **Mo** 0.25 max **Ni** 0.50-1.00 **P** 0.05 max **S** 0.05 max **Si** 0.15 max	**MIL SPEC** MIL-S-47038
K11323	Alloy Steel Electrode Nuclear Grade (EA3N)	**C** 0.07-0.17 **Cu** 0.08 max **Mn** 1.65-2.20 **Mo** 0.45-0.65 **P** 0.012 max **S** 0.030 max **Si** 0.20 max **V** 0.05 max **Other** total other elements (except iron) 0.50 max	**ASME** SFA5.23 (EA3N) **AWS** A5.23 (EA3N)
K11324	Alloy Steel Electrode Nuclear Grade (EA4N)	**C** 0.07-0.17 **Cu** 0.08 max **Mn** 1.20-1.70 **Mo** 0.45-0.65 **P** 0.012 max **S** 0.030 max **Si** 0.20 max **V** 0.05 max **Other** total other elements (except iron) 0.50 max	**ASME** SFA5.23 (EA4N) **AWS** A5.23 (EA4N)
K11325	Steel Electrode	**Al** 0.50-0.90 **C** 0.07-0.19 **Mn** 0.90-1.40 **P** 0.025 max **S** 0.035 max **Si** 0.30-0.60	**ASME** SFA5.26 (EGxxS-5) **AWS** A5.26 (EGxxS-5)
K11356	Alloy Steel	**C** 0.13 max **Cr** 0.74-1.26 **Cu** 0.22-0.48 **Mn** 0.65 max **Ni** 0.17-0.53 **S** 0.06 max	**ASTM** A714 (IV)
K11357	Steel Electrode (ER70S-5)	**Al** 0.50-0.90 **C** 0.07-0.19 **Cu** 0.50 max **Mn** 1.00-1.50 **P** 0.025 **S** 0.035 **Si** 0.65-0.85	**ASME** SFA5.18 (ER70S-5) **AWS** A5.18 (ER70S-5) **MIL SPEC** MIL-E-23765 (MIL-70S-5)
K11365	Alloy Steel Welding Wire	**C** 0.10-0.17 **Cr** 0.50-0.75 **Cu** 0.35 max **Mn** 0.50-0.80 **Mo** 0.15-0.25 **Ni** 0.25 max **P** 0.040 max **S** 0.040 max **Si** 0.60-0.90 **Zr** 0.05-0.15	**AMS** 6460
K11385	Alloy Steel Electrode (ENi4N)	**C** 0.12-0.19 **Cu** 0.08 max **Mn** 0.60-1.00 **Mo** 0.10-0.30 **Ni** 1.60-2.10 **P** 0.012 max **S** 0.020 max **Si** 0.10-0.30 **V** 0.05 max **Other** total other elements (except iron) 0.50 max	**ASME** SFA5.23 (ENi4N) **AWS** A5.23 (ENi4N)
K11422	Alloy Steel	**C** 0.14 max **Mn** 0.30-0.80 **Mo** 0.44-0.65 **P** 0.045 max **S** 0.045 max **Si** 0.10-0.50	**ASTM** A209 (T1b); A250 (T1b)
K11423	Alloy Steel Electrode (EA3)	**C** 0.07-0.17 **Cu** 0.35 max **Mn** 1.65-2.20 **Mo** 0.45-0.65 **P** 0.025 max **S** 0.030 max **Si** 0.20 max **Other** total other elements (except iron) 0.50 max	**ASME** SFA5.23 (EA3) **AWS** A5.23 (EA3)
K11424	Alloy Steel Electrode (EA4)	**C** 0.07-0.17 **Cu** 0.35 max **Mn** 1.20-1.70 **Mo** 0.45-0.65 **P** 0.025 max **S** 0.030 max **Si** 0.20 max **Other** total other elements (except iron) 0.50 max	**ASME** SFA5.23 (EA4) **AWS** A5.23 (EA4)

The chemical compositions listed are for identification purposes and should not be used in lieu of the cross referenced specifications.

MISCELLANEOUS STEELS AND FERROUS ALLOYS

UNIFIED NUMBER	DESCRIPTION	CHEMICAL COMPOSITION	CROSS REFERENCE SPECIFICATIONS
K11430	Alloy Steel	**C** 0.09-0.20 **Cr** 0.37-0.68 **Cu** 0.22-0.43 **Mn** 0.86-1.24 **P** 0.045 max **S** 0.055 max **Si** 0.13-0.32 **V** 0.01-0.11	**ASTM** A502 (3A); A588 (A)
K11485	Alloy Steel Electrode (ENi4)	**C** 0.12-0.19 **Cu** 0.35 max **Mn** 0.60-1.00 **Mo** 0.10-0.30 **Ni** 1.60-2.10 **P** 0.015 max **S** 0.020 max **Si** 0.10-0.30 **Other** total other elements (except iron) 0.50 max	**ASME** SFA5.23 (ENi4) **AWS** A5.23 (ENi4)
K11500	Alloy Steel	**Al** 0.02 min **C** 0.15 max **Mn** 0.27-0.63 **P** 0.048 max **S** 0.058 max	**ASTM** A587
K11501	High-Strength Low-Alloy Steel	**C** 0.15 max **Mn** 1.65 max **P** 0.025 max **S** 0.035 max **Si** 0.10 max **Ti** 0.05 min	**ASTM** A715 (1)
K11502	High-Strength Low-Alloy Steel	**C** 0.15 max **Mn** 1.65 max **N** 0.005 min **P** 0.025 max **S** 0.035 max **Si** 0.60 max **V** 0.02 min	**ASTM** A715 (2)
K11503	High-Strength Low-Alloy Steel	**C** 0.15 max **Cb** 0.005 min **Mn** 1.65 min **N** 0.020 max **P** 0.025 max **S** 0.035 max **Si** 0.60 max **V** 0.08 max	**ASTM** A715 (3)
K11503	Alloy Steel	**C** 0.18 max **Mn** 1.65 max **P** 0.025 max **S** 0.035 max **Si** 0.30 max **Ti** 0.05-0.40	**ASTM** A656 (2)
K11504	High-Strength Low-Alloy Steel	**B** 0.0025 max **C** 0.15 max **Cb** 0.005-0.06 **Cr** 0.80 max **Mn** 1.65 max **P** 0.025 max **S** 0.035 max **Si** 0.90 max **Ti** 0.10 max **Zr** 0.05 min	**ASTM** A715 (4)
K11505	High-Strength Low-Alloy Steel	**C** 0.15 max **Cb** 0.03 min **Mn** 1.65 max **Mo** 0.20 min **P** 0.025 max **S** 0.035 max **Si** 0.30 max	**ASTM** A715 (5)
K11506	High-Strength Low-Alloy Steel	**C** 0.15 max **Cb** 0.005-0.10 **Mn** 1.65 max **P** 0.025 max **S** 0.035 max **Si** 0.90 max	**ASTM** A715 (6)
K11507	High-Strength Low-Alloy Steel	**C** 0.15 max **Mn** 1.65 max **N** 0.020 max **P** 0.025 max **S** 0.035 max **Si** 0.60 max **Other** columbium or vanadium 0.005 min; or columbium 0.005 min and vanadium 0.005 min	**ASTM** A715 (7)
K11508	High-Strength Low-Alloy Steel	**C** 0.15 max **Cb** 0.005-0.15 **Mn** 1.65 max **P** 0.025 max **S** 0.035 max **Zr** 0.05 min	**ASTM** A715 (8)
K11510	Alloy Steel	**C** 0.15 max **Cu** 0.20 min **Mn** 1.00 max **P** 0.15 max **S** 0.05 max	**ASTM** A242 (1)
K11511	Alloy Steel	**B** 0.001-0.005 **C** 0.10-0.20 **Mn** 1.10-1.50 **Mo** 0.20-0.30 **P** 0.035 max **S** 0.04 max **Si** 0.15-0.30	**AMS** 6386 (3) **ASTM** A514 (C); A517 (C)
K11522	Alloy Steel	**C** 0.10-0.20 **Mn** 0.30-0.80 **Mo** 0.44-0.65 **P** 0.045 max **S** 0.045 max **Si** 0.10-0.50	**ASTM** A161 (T1); A209 (T1); A250 (T1); A335 (P1); A369 (FP1)
K11523	Alloy Steel	**B** 0.001-0.005 **C** 0.10-0.20 **Mn** 1.10-1.50 **Mo** 0.45-0.55 **P** 0.035 max **S** 0.04 max **Si** 0.15-0.30	**ASTM** A514 (K); A517 (K)
K11526	Alloy Steel	**C** 0.15 max **Cr** 0.24-1.31 **Cu** 0.22-0.58 **Mn** 0.17-0.53 **Ni** 0.68 max **P** 0.06-0.16 **S** 0.06 max **Si** 0.19-0.81	**ASTM** A595 (C)
K11535	Alloy Steel	**C** 0.15 max **Cr** 0.24-1.31 **Cu** 0.20-0.60 **Mn** 0.55 max **Ni** 0.20-0.70 **P** 0.06-0.16 **S** 0.060 max **Si** 0.10 min	**ASTM** A423 (1)

The chemical compositions listed are for identification purposes and should not be used in lieu of the cross referenced specifications.

UNIFIED NUMBER	DESCRIPTION	CHEMICAL COMPOSITION	CROSS REFERENCE SPECIFICATIONS
K11538	Alloy Steel	**C** 0.15 max **Cr** 0.30-0.50 **Cu** 0.20-0.50 **Mn** 0.80-1.35 **Ni** 0.25-0.50 **P** 0.04 max **S** 0.05 max **Si** 0.15-0.30 **V** 0.01-0.10	**ASTM** A588 (C)
K11540	Alloy Steel	**C** 0.15 max **Cu** 0.30-1.00 **Mn** 0.50-1.00 **Mo** 0.10 min **Ni** 0.40-1.10 **P** 0.04 **S** 0.05 max	**ASTM** A423 (2)
K11541	Alloy Steel	**C** 0.10-0.20 **Cr** 0.30 max **Cu** 0.30-1.00 **Mn** 0.50-1.00 **Mo** 0.10-0.20 **Ni** 0.40-1.10 **P** 0.04 max **S** 0.05 max **Si** 0.30 max **V** 0.01-0.10	**ASTM** A588 (F)
K11542	Alloy Steel	**C** 0.38-0.45 **Cr** 0.80-1.10 **Cu** 0.35 max **Mn** 0.75-1.00 **Mo** 0.15-0.25 **Ni** 0.25 max **P** 0.040 max **S** 0.040 max **Si** 0.20-0.35 **Te** 0.035-0.060	**AMS** 6378
K11546	Alloy Steel (4140 Modified)	**C** 0.40-0.53 **Cr** 0.80-1.10 **Cu** 0.35 max **Mn** 0.75-1.00 **Mo** 0.15-0.25 **Ni** 0.25 max **P** 0.040 max **S** 0.040 max **Si** 0.20-0.35 **Te** 0.035-0.060	**AMS** 6379
K11547	Alloy Steel	**C** 0.10-0.20 **Cr** 0.50-0.81 **Mn** 0.30-0.61 **Mo** 0.44-0.65 **P** 0.045 max **S** 0.045 max **Si** 0.10-0.30	**ASTM** A213 (T2); A335 (P2); A369 (FP2)
K11552	Alloy Steel	**C** 0.10-0.20 **Cb** 0.04 max **Cr** 0.50-0.90 **Cu** 0.30 max **Mn** 0.75-1.25 **P** 0.04 max **S** 0.05 max **Si** 0.50-0.90 **Zr** 0.05-0.15	**ASTM** A588 (D)
K11562	Alloy Steel	**C** 0.15 max **Cr** 0.80-1.25 **Mn** 0.30-0.61 **Mo** 0.44-0.65 **P** 0.045 max **S** 0.045 max **Si** 0.50 max	**ASTM** A213 (T12); A335 (P12); A369 (FP12)
K11564	Alloy Steel	**C** 0.10-0.20 **Cr** 0.80-1.10 **Mn** 0.30-0.80 **Mo** 0.45-0.65 **P** 0.040 max **S** 0.040 max **Si** 0.10-0.60	**ASTM** A182 (F12); A336 (F12)
K11567	Alloy Steel	**C** 0.15 max **Cu** 0.50-0.80 **Mn** 1.20 max **Mo** 0.08-0.25 **Ni** 0.75-1.25 **P** 0.04 max **S** 0.05 max **Si** 0.30 max **V** 0.05 max	**ASTM** A588 (E)
K11572	Alloy Steel	**C** 0.10-0.20 **Cr** 1.00-1.50 **Mn** 0.30-0.80 **Mo** 0.44-0.65 **P** 0.040 max **S** 0.040 max **Si** 0.50-1.00	**ASTM** A182 (F11); A336 (FN, F11A)
K11576	Alloy Steel	**B** 0.0005-0.006 **C** 0.10-0.20 **Cr** 0.40-0.65 **Cu** 0.15-0.50 **Mn** 0.60-1.00 **Mo** 0.40-0.60 **Ni** 0.70-1.00 **P** 0.035 max **S** 0.040 max **Si** 0.15-0.35 **V** 0.03-0.08	**ASTM** A514 (F); A517 (F); A592 (F)
K11578	Alloy Steel	**C** 0.15 max **Mn** 0.30-0.60 **Mo** 0.44-0.65 **P** 0.030 max **S** 0.030 max **Si** 1.15-1.65	**ASTM** A335 (P15)
K11585	Alloy Steel Electrode (EH-14)	**C** 0.10-0.20 **Cu** 0.35 max **Mn** 1.70-2.20 **P** 0.030 max **S** 0.030 max **Si** 0.10 max **Other** total other elements (except iron) 0.50 max	**ASME** SFA5.17 (EH14); SFA5.25 (EH14-EW) **AWS** A5.17 (EH14); A5.25 (EH14-EW) **MIL SPEC** MIL-E-23765(MIL-70S-7)
K11591	Alloy Steel	**C** 0.15 max **Cr** 0.80-1.25 **Mn** 0.30-0.60 **Mo** 0.87-1.13 **P** 0.030 max **S** 0.030 max **Si** 0.10-0.35 **V** 0.15-0.25	**ASTM** A405 (P24)
K11597	Alloy Steel	**C** 0.15 max **Cr** 1.00-1.50 **Mn** 0.30-0.60 **Mo** 0.44-0.65 **P** 0.030 max **S** 0.030 max **Si** 0.50-1.00	**ASTM** A199 (T11); A200 (T11); A213 (T11); A335 P11; A369 (FP11)
K11598	Alloy Steel	**C** 0.10-0.20 **Cr** 1.00-1.50 **Mn** 0.30-0.80 **Mo** 0.45-0.65 **P** 0.035 max **S** 0.040 max **Si** 0.50-1.00	**ASTM** A541 (5) **MIL SPEC** MIL-S-18410 (A)

The chemical compositions listed are for identification purposes and should not be used in lieu of the cross referenced specifications.

MISCELLANEOUS STEELS AND FERROUS ALLOYS

UNIFIED NUMBER	DESCRIPTION	CHEMICAL COMPOSITION	CROSS REFERENCE SPECIFICATIONS
K11625	Alloy Steel	B 0.001-0.005 C 0.12-0.21 Mn 0.45-0.70 Mo 0.50-0.65 P 0.035 max S 0.04 max Si 0.20-0.35	AMS 6386 (5) ASTM A514 (J); A517 (J)
K11630	Alloy Steel	B 0.0005-0.005 C 0.12-0.21 Cr 0.40-0.65 Mn 0.70-1.00 Mo 0.15-0.25 P 0.035 max S 0.04 max Si 0.20-0.35 Ti 0.01-0.03 V 0.03-0.08	AMS 6386 (2) ASTM A514 (B); A517 (B)
K11640	Alloy Steel	C 0.38-0.43 Cr 0.55-0.75 Cu 0.35 max Mn 0.75-1.00 Mo 0.20-0.30 Ni 0.55-0.85 P 0.025 max S 0.025 max Si 0.20-0.35	AMS 6324
K11646	Alloy Steel	B 0.0005-0.005 C 0.12-0.21 Cr 0.40-0.65 Mn 0.95-1.30 Mo 0.20-0.30 Ni 0.30-0.70 P 0.035 max S 0.04 max Si 0.20-0.35 V 0.03-0.08	ASTM A514 (H); A517 (H)
K11662	Alloy Steel	B 0.0015-0.005 C 0.13-0.20 Cr 0.85-1.20 Cu 0.20-0.40 Mn 0.40-0.70 Mo 0.15-0.25 P 0.035 max S 0.04 max Si 0.20-0.35 Ti 0.04-0.10	AMS 6386 (4) ASTM A514 (D); A517 (D)
K11682	Alloy Steel	B 0.0015-0.005 C 0.13-0.20 Cr 1.15-1.65 Cu 0.20-0.40 Mn 0.40-0.70 Mo 0.25-0.40 P 0.035 max S 0.04 max Si 0.20-0.35 Ti 0.04-0.10	ASTM A514 (L); A517 (L)
K11683	Alloy Steel	B 0.001-0.005 C 0.12-0.21 Mn 0.45-0.70 Mo 0.45-0.60 Ni 1.20-1.50 P 0.035 max S 0.04 max Si 0.20-0.35	ASTM A514 (M); A517 (M)
K11695	Alloy Steel	B 0.0015-0.005 C 0.10-0.22 Cr 1.34-2.06 Cu 0.17-0.43 Mn 0.37-0.74 Mo 0.36-0.64 P 0.035 max S 0.040 max Si 0.18-0.37 Ti 0.03-0.11	ASTM A592 (E)
K11711	Leaded-Resulfurized Steel	C 0.14-0.20 Mn 1.00-1.30 P 0.04 max Pb 0.15-0.35 S 0.08-0.13 Si 0.35 max	ASTM A766
K11720	High-Strength Low-Alloy Steel	Al 0.06 max C 0.17 max Cb 0.050 max Cu 0.35 max Mn 1.60 max P 0.035 max S 0.015 max Si 0.40 max V 0.11 max	ASTM A734/A734M (B)
K11742	Alloy Steel	C 0.17 max Cr 0.31-0.64 Mn 0.97-1.52 P 0.035 max S 0.040 max Si 0.54-0.96	ASTM A202 (A)
K11757	Alloy Steel	C 0.17 max Cr 0.74-1.21 Mn 0.35-0.73 Mo 0.40-0.65 P 0.035 max S 0.040 max Si 0.13-0.32	ASTM A387/A387M (12)
K11789	Alloy Steel	C 0.17 max Cr 0.94-1.56 Mn 0.35-0.73 Mo 0.40-0.70 P 0.035 max S 0.040 max Si 0.44-0.86	ASTM A387/A387M (11)
K11797	Alloy Steel	C 0.20 max Cr 1.00-1.50 Mn 0.40-0.65 Mo 0.45-0.65 P 0.035 max S 0.040 max Si 0.50-0.80	ASTM A739 (B11)
K11800	Alloy Steel	C 0.18 max Cr 0.15 max Mn 1.30 max Mo 0.05 max Ni 0.25 max P 0.035 max S 0.040 max Si 0.15-0.35 V 0.02-0.12	ASTM A541 (4)
K11801	Alloy Steel	C 0.18 max Cr 0.15 max Cu 0.35 max Mn 1.30 max Mo 0.06 max Ni 0.25 max P 0.040 max S 0.050 max Si 0.15-0.35 Ti 0.005 min V 0.02 min	MIL SPEC MIL-S-16113 (I)
K11802	Alloy Steel	C 0.18 max Cr 0.15 max Cu 0.35 max Mn 1.30 max Mo 0.06 max P 0.035 max S 0.040 max Si 0.15-0.35 Ti 0.005 max V 0.02-0.13	MIL SPEC MIL-S-16113 (II)

The chemical compositions listed are for identification purposes and should not be used in lieu of the cross referenced specifications.

UNIFIED NUMBER	DESCRIPTION	CHEMICAL COMPOSITION	CROSS REFERENCE SPECIFICATIONS
K11803	Alloy Steel	**C** 0.18 max **Mn** 1.45 max **P** 0.035 max **S** 0.040 max **Si** 0.13-0.32 **V** 0.07-0.16	**ASTM** A225/A225M
K11804	Alloy Steel	**C** 0.18 max **Mn** 1.60 max **N** 0.005-0.030 **P** 0.025 max **S** 0.035 max **Si** 0.60 max **V** 0.030-0.20	**ASTM** A656 (1)
K11805	Alloy Steel	**C** 0.18 max **Mn** 1.30 max **P** 0.035 max **S** 0.040 max **Si** 0.15-0.35 **V** 0.02 min	
K11820	Alloy Steel	**C** 0.18 max **Mn** 0.90 max **Mo** 0.41-0.64 **P** 0.035 max **S** 0.040 max **Si** 0.13-0.32	**ASTM** A204/A204M (A)
K11831	High-Strength Low-Alloy Steel	**C** 0.18 max **Cr** 0.25 max **Cu** 0.35 max **Mn** 1.00-1.60 **Mo** 0.08 max **Ni** 0.25 max **P** 0.035 max **S** 0.040 max **Si** 0.55 max **V** 0.08 max	**ASTM** A724/A724M (A)
K11835	Alloy Steel	**C** 0.18 max **Cr** 0.33 max **Cu** 0.27-1.03 **Mn** 0.45-1.05 **Mo** 0.09-0.21 **Ni** 0.35-1.15 **P** 0.045 max **S** 0.055 max	**ASTM** A714 (VI)
K11846	High-Strength Low-Alloy Steel	**Al** 0.065 max **C** 0.18 max **Cr** 0.25 max **Cu** 0.35 max **Mn** 0.90-10.60 **Mo** 0.08 max **Ni** 0.40 max **P** 0.04 max **S** 0.04 max **Si** 0.10-0.50	**ASTM** A131 (AH32, DH32, EH32)
K11847	Alloy Steel	**B** 0.0005-0.0025 **C** 0.15-0.21 **Cr** 0.50-0.80 **Mn** 0.80-1.10 **Mo** 0.25 max **P** 0.035 max **S** 0.04 max **Si** 0.40-0.90 **Zr** 0.05-0.15	**ASTM** A514 (N)
K11852	High-Strength Low-Alloy Steel	**C** 0.18 max **Cb** 0.05 max **Cr** 0.25 max **Cu** 0.35 max **Mn** 0.90-1.60 **Mo** 0.08 max **Ni** 0.40 max **P** 0.04 max **S** 0.04 max **Si** 0.10-0.50 **V** 0.10 max	**ASTM** A131 (AH36, DH36, EH36)
K11856	Alloy Steel	**B** 0.0025 max **C** 0.15-0.21 **Cr** 0.50-0.80 **Mn** 0.80-1.10 **Mo** 0.18-0.28 **P** 0.035 max **S** 0.04 max **Si** 0.40-0.80 **Zr** 0.05-0.15	**AMS** 6386 (1) **ASTM** A514 (A); A517 (A); A592 (A)
K11872	Alloy Steel	**B** 0.0025 max **C** 0.15-0.21 **Cr** 0.50-0.90 **Mn** 0.80-1.10 **Mo** 0.40-0.60 **P** 0.035 max **S** 0.04 max **Si** 0.50-0.90 **Zr** 0.05-0.15	**ASTM** A514 (G); A517 (G)
K11900	Alloy Steel	**B** 0.0005 min **C** 0.13-0.25 **Mn** 0.67 min **P** 0.048 max **S** 0.058 max	**ASTM** A325 (2)
K11914	Alloy Steel (NAX 9115-AC)	**C** 0.10-0.17 **Cr** 0.50-0.75 **Cu** 0.35 max **Mn** 0.50-0.80 **Mo** 0.15-0.25 **Ni** 0.25 max **P** 0.025 max **S** 0.035 max **Si** 0.60-0.90 **Zr** 0.05-0.10	**AMS** 6354
*K11918	Alloy Steel	**C** 0.10-0.28 **Cr** 0.30-0.40 **Mn** 0.30-0.40 **Mo** 0.07-0.15 **Ni** 0.50 max **Si** 0.20-0.40 **V** 0.10 max	**MIL SPEC** MIL-A-12560
K11940	Alloy Steel	**B** 0.0005-0.005 **C** 0.38-0.43 **Cr** 0.70-0.90 **Cu** 0.35 max **Mn** 0.65-0.85 **Mo** 0.15-0.25 **Ni** 0.70-1.00 **P** 0.025 max **S** 0.025 max **Si** 0.15-0.35 **V** 0.01-0.06	**AMS** 6422
K11948	Alloy Steel Electrode (EF4N)	**C** 0.16-0.23 **Cr** 0.40-0.60 **Cu** 0.08 max **Mn** 0.60-0.90 **Mo** 0.15-0.30 **Ni** 0.40-0.80 **P** 0.012 max **S** 0.035 max **Si** 0.15-0.35 **V** 0.05 max **Other** total other elements (except iron) 0.50 max	**ASME** SFA5.23 (EF4N) **AWS** A5.23 (EF4N)

*Boxed entries are no longer active and are retained for reference purposes only.

The chemical compositions listed are for identification purposes and should not be used in lieu of the cross referenced specifications.

UNIFIED NUMBER	DESCRIPTION	CHEMICAL COMPOSITION	CROSS REFERENCE SPECIFICATIONS
K12000	High-Strength Low-Alloy Steel	**C** 0.20 max **Cb** 0.01-0.05 **Mn** 1.15-1.50 **P** 0.04 max **S** 0.05 max **Si** 0.15-0.50	**ASTM** A633 (C)
K12001	High-Strength Low-Alloy Steel	**C** 0.20 max **Cb** 0.05 max **Mn** 1.15-1.50 **P** 0.035 max **S** 0.030 max **Si** 0.15-0.50	**ASTM** A737/A737M (B)
K12003	Alloy Steel	**C** 0.20 max **Mn** 1.45 max **P** 0.035 max **S** 0.040 max **Si** 0.13-0.32 **V** 0.07-0.16	**ASTM** A225/A225M
K12010	High-Strength Low-Alloy Steel	**C** 0.20 max **Cu** 0.20 min **Mn** 1.35 max **P** 0.04 max **S** 0.05 max	**ASTM** A242 (2)
K12020	Alloy Steel	**C** 0.20 max **Mn** 0.90 max **Mo** 0.41-0.64 **P** 0.035 max **S** 0.040 max **Si** 0.13-0.45	**ASTM** A204/A204M (B)
K12021	Alloy Steel	**C** 0.20 max **Mn** 0.87-1.41 **Mo** 0.41-0.64 **P** 0.035 max **S** 0.040 max **Si** 0.13-0.45	**ASTM** A302/A302M (A)
K12022	Alloy Steel	**C** 0.20 max **Mn** 1.07-1.62 **Mo** 0.41-0.64 **P** 0.035 max **S** 0.040 max **Si** 0.13-0.45	**ASTM** A302 (B)
K12023	Alloy Steel	**C** 0.15-0.25 **Mn** 0.30-0.80 **Mo** 0.44-0.65 **P** 0.045 max **S** 0.045 max **Si** 0.10-0.50	**ASTM** A209 (T1a)
K12031	High-Strength Low-Alloy Steel	**C** 0.20 max **Cr** 0.25 max **Cu** 0.35 max **Mn** 1.00-1.60 **Mo** 0.08 max **Ni** 0.25 max **P** 0.035 max **Si** 0.50 max **V** 0.08 max	**ASTM** A724/A724M (B)
K12032	Alloy Steel	**C** 0.20 max **Cr** 0.10-0.25 **Cu** 0.20-0.35 **Mn** 1.25 max **Mo** 0.15 max **Ni** 0.30-0.60 **P** 0.035 max **S** 0.040 max **Si** 0.25-0.75 **Ti** 0.005-0.030 **V** 0.02-0.10	**ASTM** A588 (H)
K12033	Alloy Steel	**C** 0.14-0.26 **Cr** 0.27-0.53 **Cu** 0.17-0.53 **Mn** 0.76-1.39 **Ni** 0.22-0.53 **P** 0.040 max **S** 0.045 max **Si** 0.13-0.32 **V** 0.010 min	**ASTM** A325 (C); A563 (C3C)
K12037	High-Strength Low-Alloy Steel	**C** 0.20 max **Cr** 0.25 max **Cu** 0.35 max **Mn** 0.70-1.35 **Mo** 0.08 max **Ni** 0.25 max **P** 0.04 max **S** 0.05 max **Si** 0.15-0.50	**ASTM** A633 (D)
K12039	Alloy Steel	**C** 0.20 max **Mn** 1.07-1.62 **Mo** 0.41-0.64 **Ni** 0.37-0.73 **P** 0.035 max **S** 0.040 max **Si** 0.13-0.45	**ASTM** A302/A302M (C)
K12040	Alloy Steel	**C** 0.20 max **Cr** 0.50-1.00 **Cu** 0.30-0.50 **Mn** 1.20 max **Mo** 0.10 max **Ni** 0.80 max **P** 0.04 max **S** 0.05 max **Si** 0.25-0.70 **Ti** 0.07 max	**ASTM** A588
K12042	Alloy Steel	**C** 0.25 max **Cr** 0.25 max **Mn** 1.20-1.50 **Mo** 0.45-0.60 **Ni** 0.40-1.00 **P** 0.025 max **S** 0.025 max **Si** 0.15-0.40 **V** 0.05 max	**ASTM** A508 (3 and 3a) **MIL SPEC** MIL-S-24238 (A)
K12043	Alloy Steel	**C** 0.20 max **Cr** 0.40-0.70 **Cu** 0.20-0.40 **Mn** 0.75-1.25 **Ni** 0.50 max **P** 0.04 max **S** 0.05 max **Si** 0.15-0.30 **V** 0.01-0.10	**ASTM** A588 (B)
K12044	Alloy Steel	**C** 0.20 max **Cu** 0.30 min **Mn** 0.60-1.00 **Ni** 0.50-0.70 **P** 0.04 max **S** 0.05 max **Si** 0.30-0.50 **Ti** 0.05	**ASTM** A588 (J)
K12045	Alloy Steel	**C** 0.25 max **Mn** 1.20-1.50 **Mo** 0.45-0.60 **Ni** 0.40-1.00 **P** 0.035 max **S** 0.040 max **Si** 0.15-0.35 **V** 0.05 max	**ASTM** A541 (3 and 3a)

The chemical compositions listed are for identification purposes and should not be used in lieu of the cross referenced specifications.

UNIFIED NUMBER	DESCRIPTION	CHEMICAL COMPOSITION	CROSS REFERENCE SPECIFICATIONS
K12047	Alloy Steel	C 0.15-0.25 Cr 0.80-1.25 Mn 0.30-0.61 P 0.045 max S 0.045 max Si 0.15-0.35 V 0.15 min	**ASTM** A213 (T17)
K12048	Alloy Steel Electrode (EF4)	C 0.16-0.23 Cr 0.40-0.60 Cu 0.35 max Mn 0.60-0.90 Mo 0.15-0.30 Ni 0.40-0.80 P 0.025 max S 0.035 max Si 0.15-0.35 Other total other elements (except iron) 0.50 max	**ASME** SFA5.2 (R100); SFA5.23 (EF4) **AWS** A5.2 (R100); A5.23 (EF4)
K12054	Alloy Steel	C 0.20 max Mn 1.07-1.62 Mo 0.41-0.64 Ni 0.67-1.03 P 0.035 max S 0.040 max Si 0.13-0.45	**ASTM** A302/A302M (D)
K12059	Alloy Steel	C 0.14-0.26 Cr 0.45-1.05 Cu 0.27-0.53 Mn 0.36-1.24 Mo 0.11 max Ni 0.47-0.83 P 0.045 max S 0.055 max Si 0.20-0.55 Ti 0.05 max	**ASTM** A325 (D); A563 (C3D)
K12062	Alloy Steel	C 0.20 max Cr 0.80-1.25 Mn 0.30-0.80 Mo 0.44-0.65 P 0.045 max S 0.045 max Si 0.60 max	**ASTM** A234 (WP12)
K12087	Alloy Steel Electrode (EB5N)	C 0.18-0.23 Cr 0.45-0.65 Cu 0.08 max Mn 0.40-0.70 Mo 0.90-1.10 P 0.012 max S 0.025 max Si 0.40-0.60 V 0.05 max Other total other elements (except iron) 0.50 max	**ASME** SFA5.23 (EB5N) **AWS** A5.23 (EB5N)
K12089	Alloy Steel	C 0.20 max Mn 0.40-0.70 Mo 0.19-0.33 Ni 1.60-2.05 P 0.030 max S 0.035 max Si 0.37 max	**ASTM** A707 (L4)
K12103	Alloy Steel	C 0.10-0.32 Cr 0.30-0.40 Mn 0.30-0.40 Mo 0.07-0.15 Ni 0.50 max P 0.04 max S 0.04 max Si 0.20-0.40 V 0.10 max	
K12121	Alloy Steel	C 0.17-0.26 Mn 0.46-0.94 Mo 0.42-0.68 P 0.045 max S 0.045 max Si 0.18-0.37	**ASTM** A692
K12122	Alloy Steel	C 0.21 max Cr 0.50-0.81 Mn 0.30-0.80 Mo 0.44-0.65 P 0.040 max S 0.040 max Si 0.10-0.60	**ASTM** A182 (F2)
K12125	Cr-Mo Alloy Steel	C 0.16-0.26 Cr 0.40-0.60 Mn 1.10-1.60 Mo 0.10-0.25 P 0.025 max S 0.025 max Si 0.10-0.40 NOTE: Steel is Boron treated	
K12143	Alloy Steel	C 0.21 max Cr 0.46-0.85 Mn 0.50-0.88 Mo 0.40-0.65 P 0.035 max S 0.040 max Si 0.13-0.45	**ASTM** A387/A387M (2)
K12187	Alloy Steel Electrode (EB5)	C 0.18-0.23 Cr 0.45-0.65 Cu 0.30 max Mn 0.40-0.70 Mo 0.90-1.10 P 0.025 max S 0.025 max Si 0.40-0.60 Other total other elements (except iron) 0.50 max	**ASME** SFA5.23 (EB5) **AWS** A5.23 (EB5)
K12202	Alloy Steel	C 0.22 max Mn 1.15-1.50 N 0.01-0.03 P 0.04 max S 0.05 max Si 0.15-0.50 V 0.04-0.11	**ASTM** A633 (E); A737 (C)
K12211	Alloy Steel	C 0.22 max Cu 0.20 min Mn 0.85-1.25 P 0.04 max S 0.05 max Si 0.40 max V 0.02 min	**ASTM** A441
K12220	Alloy Steel	C 0.22 max Cr 0.25 max Cu 0.25 max Mn 0.80 max Mo 0.40-0.60 Ni 0.25 max P 0.04 max S 0.04 max Si 0.15-0.40	**MIL SPEC** MIL-S-872 (A)

The chemical compositions listed are for identification purposes and should not be used in lieu of the cross referenced specifications.

UNIFIED NUMBER	DESCRIPTION	CHEMICAL COMPOSITION	CROSS REFERENCE SPECIFICATIONS
K12238	Alloy Steel	C 0.19-0.26 Cr 0.42-0.68 Cu 0.17-0.43 Mn 0.86-1.24 Ni 0.17-0.43 P 0.045 max S 0.045 max Si 0.13-0.32	ASTM A325 (F); A563 (C3F)
K12244	Alloy Steel	C 0.21 max Cr 0.37-0.73 Cu 0.17-0.43 Mn 0.71-1.29 Ni 0.22-0.53 P 0.045 max S 0.055 max Si 0.13-0.32 V 0.11 max	ASTM A502 (3B)
K12249	Alloy Steel	C 0.22 max Cu 0.50 min Mn 0.60-0.90 Ni 0.40-0.75 P 0.08-0.15 S 0.05 max Si 0.10 max	ASTM A690
K12254	Alloy Steel	C 0.18-0.27 Cr 0.55-0.95 Cu 0.27-0.63 Mn 0.56-1.04 Ni 0.27-0.63 P 0.045 max S 0.045 max Si 0.13-0.32	ASTM A325 (E); A563 (C3E)
K12320	Alloy Steel	C 0.23 max Mn 0.90 max Mo 0.41-0.64 P 0.035 max S 0.040 max Si 0.13-0.45	ASTM A204/A204M (C)
K12437	High-Strength Low-Alloy Steel	C 0.24 max Cr 0.25 max Cu 0.35 max Mn 0.70-1.35 Mo 0.08 max Ni 0.25 max P 0.035 max S 0.040 max Si 0.13-0.55	ASTM A537/A537M
K12447	Carbon-Manganese-Silicon Steel	C 0.24 max Cr 0.25 max Cu 0.35 max Mn 1.50 max Mo 0.08 max Ni 0.50 max P 0.035 max S 0.040 max Si 0.15-0.50	ASTM A738/A738M
K12510	Alloy Steel	C 0.25 max Cu 0.18 min Mn 1.05-1.60 N 0.005-0.035 P 0.030 max S 0.035 max Si 0.32 max V 0.03-0.13	ASTM A707 (L3)
K12520	Alloy Steel	C 0.20-0.30 Mn 0.60-0.80 Mo 0.40-0.60 P 0.040 max S 0.040 max Si 0.20-0.35	ASTM A336 (F1)
K12521	Alloy Steel	C 0.25 max Mn 1.07-1.62 Mo 0.41-0.64 P 0.035 max S 0.040 max Si 0.13-0.45	ASTM A533/A533M (A)
K12524	Alloy Steel	C 0.25 max Mn 1.60 max Ni 0.37-0.73 P 0.035 max S 0.040 max Si 0.13-0.45 V 0.11-0.20	ASTM A225/A225M (C)
K12529	Alloy Steel	C 0.25 max Mn 1.07-1.62 Mo 0.41-0.64 Ni 0.17-0.43 P 0.035 max S 0.040 max Si 0.13-0.45	ASTM A533 (D)
K12539	Alloy Steel	C 0.25 max Mn 1.07-1.62 Mo 0.41-0.64 Ni 0.37-0.73 P 0.035 max S 0.040 max Si 0.13-0.45	ASTM A533 (B)
K12542	Alloy Steel	C 0.25 max Cr 0.31-0.64 Mn 0.97-1.52 P 0.035 max S 0.040 max Si 0.54-0.96	ASTM A202/A202M (B)
K12554	Alloy Steel	C 0.25 max Mn 1.07-1.62 Mo 0.41-0.64 Ni 0.67-1.03 P 0.035 max S 0.040 max Si 0.13-0.45	ASTM A533 (C)
K12608	Alloy Steel	C 0.26 max Cu 0.18 min Mn 1.30 max S 0.063 max	ASTM A714 (I)
K12609	High-Strength Low-Alloy Steel	C 0.26 max Cu 0.18 min Mn 1.30 max P 0.05 max S 0.063 max Si 0.33 max V 0.01 min	ASTM A618 (II); A714 (II)
K12700	High-Strength Low-Alloy Steel	C 0.27 max Mn 1.40 max P 0.05 max S 0.06 max Si 0.35 max V 0.01 min	ASTM A618 (III)
K12709	High-Strength Low-Alloy Steel	C 0.27 max Cu 0.18 min Mn 1.40 max P 0.05 max S 0.06 max Si 0.35 max V 0.01 min	ASTM A714 (III)

The chemical compositions listed are for identification purposes and should not be used in lieu of the cross referenced specifications.

UNIFIED NUMBER	DESCRIPTION	CHEMICAL COMPOSITION	CROSS REFERENCE SPECIFICATIONS
K12765	Alloy Steel	**C** 0.27 max **Cr** 0.25-0.45 **Mn** 0.50-0.90 **Mo** 0.55-0.70 **Ni** 0.50-1.00 **P** 0.035 max **S** 0.040 max **Si** 0.15-0.35 **V** 0.05 max	**ASTM** A541 (2, 2A)
K12766	Alloy Steel	**C** 0.27 max **Cr** 0.25-0.45 **Mn** 0.50-1.00 **Mo** 0.55-0.70 **Ni** 0.50-1.00 **P** 0.025 max **S** 0.025 max **Si** 0.15-0.40 **V** 0.05 max	**ASTM** A508 (2, 2a)
K12810	Alloy Steel	**C** 0.28 max **Cu** 0.20 min **Mn** 1.10-1.60 **P** 0.04 max **S** 0.05 max **Si** 0.30 max	
K12821	Alloy Steel	**C** 0.28 max **Mn** 0.30-0.90 **Mo** 0.44-0.65 **P** 0.045 max **S** 0.045 max **Si** 0.10-0.50	**ASTM** A234 (WP1)
K12822	Alloy Steel	**C** 0.28 max **Mn** 0.60-0.90 **Mo** 0.44-0.65 **P** 0.045 max **S** 0.045 max **Si** 0.15-0.35	**ASTM** A182 (F1)
K13020	Alloy Steel	**C** 0.30 max **Cr** 0.25 max **Cu** 0.25 max **Mn** 0.80 max **Mo** 0.40-0.60 **Ni** 0.25 max **P** 0.04 max **S** 0.04 max **Si** 0.15-0.40	**MIL SPEC** MIL-S-872 (B)
K13047	Alloy Steel	**C** 0.26-0.34 **Cr** 0.80-1.15 **Mn** 0.40-0.70 **Mo** 0.15-0.25 **P** 0.035 max **S** 0.04 max **Si** 0.15-0.35	**ASTM** A372 (V-A); A649 (3)
K13048	Alloy Steel	**C** 0.28-0.33 **Cr** 0.40-0.60 **Cu** 0.35 max **Mn** 0.70-0.90 **Mo** 0.15-0.25 **Ni** 0.40-0.70 **P** 0.025 max **S** 0.025 max **Si** 0.20-0.35	**AMS** 6550
K13049	Alloy Steel	**C** 0.26-0.34 **Cr** 0.40-0.65 **Mn** 0.70-1.00 **Mo** 0.15-0.25 **Ni** 0.40-0.70 **P** 0.035 max **S** 0.04 max **Si** 0.15-0.35	**ASTM** A372
K13050	Alloy Steel	**C** 0.30 max **Mn** 1.35 max **Ni** 1.0-2.0 **P** 0.035 max **S** 0.040 max **Si** 0.20-0.35	**ASTM** A350 (LF5)
K13051	Alloy Steel, Superstrength	**C** 0.27-0.33 **Cr** 0.80-1.10 **Mn** 0.40-0.60 **Mo** 0.15-0.25 **P** 0.025 max **S** 0.025 max **Si** 0.20-0.35 **V** 0.05-0.10	**ASTM** A579 (13)
K13147	Alloy Steel Welding Wire	**C** 0.28-0.33 **Cr** 0.80-1.10 **Cu** 0.10 max **H** 0.0010 max **Mn** 0.40-0.60 **Mo** 0.15-0.25 **N** 0.005 max **Ni** 0.25 max **O** 0.0025 max **P** 0.008 max **S** 0.008 max **Si** 0.15-0.35 **V** 0.06 max **Other** S + P 0.012 max	**AMS** 6457
K13148	Alloy Steel Welding Wire	**C** 0.28-0.33 **Cr** 0.80-1.10 **Cu** 0.35 max **H** 0.0025 max **Mn** 0.60-0.90 **Mo** 0.06 max **N** 0.005 max **Ni** 0.25 max **O** 0.0025 max **P** 0.008 max **S** 0.008 max **Si** 0.15-0.35 **V** 0.15-0.25 **Other** S + P 0.012 max	**AMS** 6461
K13149	Alloy Steel Welding Wire	**C** 0.28-0.33 **Cr** 0.80-1.10 **Cu** 0.35 max **Mn** 0.70-0.90 **Mo** 0.06 max **Ni** 0.25 max **P** 0.025 max **S** 0.025 max **Si** 0.15-0.35 **V** 0.15-0.25	**AMS** 6462
K13247	Alloy Steel	**C** 0.30-0.35 **Cr** 0.80-1.10 **Cu** 0.35 max **Mn** 0.40-0.60 **Mo** 0.15-0.25 **Ni** 0.25 max **P** 0.025 max **S** 0.025 max **Si** 0.20-0.50	**AMS** 6356
K13262	Alloy Steel	**C** 0.32 max **Cr** 0.40 min **Mn** 1.00 max **Mo** 0.15 min **Ni** 0.70 min **P** 0.025 max **S** 0.025 max **Si** 0.75 max	

The chemical compositions listed are for identification purposes and should not be used in lieu of the cross referenced specifications.

MISCELLANEOUS STEELS AND FERROUS ALLOYS

UNIFIED NUMBER	DESCRIPTION	CHEMICAL COMPOSITION	CROSS REFERENCE SPECIFICATIONS
K13502	Alloy Steel	C 0.35 max Cr 0.25 max Mn 0.40-1.05 Mo 0.10 max Ni 0.40 max P 0.025 max S 0.025 max Si 0.15-0.40 V 0.05 max	ASTM A508 (1)
K13521	Alloy Steel	C 0.31-0.40 Cr 0.22-0.43 Mn 0.91-1.29 Mo 0.04-0.11 Ni 0.17-0.43 P 0.040 max S 0.045 max Si 0.18-0.37	ASTM A687 (I)
K13547	Alloy Steel	C 0.31-0.39 Cr 0.40-0.65 Mn 0.75-1.05 Mo 0.15-0.25 P 0.035 max S 0.04 max Si 0.15-0.35	ASTM A372
K13548	Alloy Steel	C 0.31-0.39 Cr 0.80-1.15 Mn 0.70-1.00 Mo 0.15-0.25 P 0.035 max S 0.04 max Si 0.15-0.35	ASTM A372
K13550	Alloy Steel	C 0.48-0.53 Cr 0.40-0.60 Cu 0.35 max Mn 0.75-1.00 Mo 0.20-0.30 Ni 0.40-1.00 P 0.025 max S 0.025 max Si 0.15-0.35	AMS 6328
K13586	Alloy Steel	C 0.35 max Cr 0.70 max Mn 0.60-0.90 Mo 0.20 min Ni 1.50-3.50 P 0.025 max S 0.025 max Si 0.15-0.35 V 0.03-0.12	ASTM A294 (C)
K13643	Alloy Steel	C 0.31-0.42 Cr 0.42-0.68 Cu 0.22-0.48 Mn 0.86-1.24 Ni 0.22-0.48 P 0.045 max S 0.055 max Si 0.13-0.32	ASTM A325 (A); A563 (C3A)
K13650	Alloy Steel	C 0.20-0.53 Cr 0.45 min Cu 0.20 min Mn 0.40 min Mo 0.15 min Ni 0.20 min P 0.046 max S 0.050 max	ASTM A563 (DH3)
K14044	Alloy Steel	C 0.36-0.45 Cr 0.75-1.15 Mn 0.71-1.04 Mo 0.13-0.27 P 0.040 max S 0.045 max Si 0.18-0.37	ASTM A687 (II)
K14047	Alloy Steel	C 0.38-0.43 Cr 0.80-1.10 Mn 0.75-1.00 Mo 0.15-0.25 P 0.025 max S 0.025 max Si 0.20-0.35	ASTM A646 (12)
K14048	Alloy Steel	C 0.35-0.45 Cr 0.80-1.15 Mn 0.70-1.00 Mo 0.15-0.25 P 0.040 max S 0.040 max Si 0.20-0.35 V 0.06 max	ASTM A290 (E, F)
K14072	Alloy Steel	C 0.36-0.44 Cr 0.80-1.15 Mn 0.45-0.70 Mo 0.50-0.65 P 0.04 max S 0.04 max Si 0.20-0.35 V 0.25-0.35	ASTM A193 (B16); A437 (B4D)
K14073	Alloy Steel	C 0.36-0.44 Cr 0.80-1.15 Mn 0.45-0.70 Mo 0.50-0.65 P 0.025 max S 0.025 max Si 0.15-0.35 V 0.25-0.35	ASTM A540 (B21)
K14185	Alloy Steel	C 0.38-0.45 Cr 0.95-1.35 Mn 0.75-1.00 Mo 0.55-0.70 P 0.025 max S 0.020 max Si 0.20-0.35 V 0.20-0.30	MIL SPEC MIL-S-46047
K14245	Alloy Steel	C 0.35-0.50 Cr 0.75-1.50 Mn 0.60-1.00 Mo 0.15 min P 0.025 max S 0.025 max Si 0.15-0.35	ASTM A294 (A)
K14247	Alloy Steel	C 0.35-0.55 Cr 0.80-1.15 Mn 0.55-1.05 Mo 0.15-0.50 P 0.035 max S 0.040 max Si 0.15-0.35	ASTM A649 (1A)
K14248	Alloy Steel	C 0.35-0.50 Cr 0.80-1.15 Mn 0.75-1.05 Mo 0.15-0.25 P 0.035 max S 0.04 max Si 0.15-0.35	ASME A372 (5E, 8)
K14358	Alloy Steel	C 0.36-0.50 Cr 0.47-0.83 Cu 0.17-0.43 Mn 0.67-0.93 Mo 0.07 max Ni 0.47-0.83 P 0.06-0.125 S 0.055 max Si 0.25-0.55	ASTM A325 (B); A563 (C3B)

The chemical compositions listed are for identification purposes and should not be used in lieu of the cross referenced specifications.

UNIFIED NUMBER	DESCRIPTION	CHEMICAL COMPOSITION	CROSS REFERENCE SPECIFICATIONS
K14394	Alloy Steel	**C** 0.41-0.46 **Mn** 0.75-1.00 **Mo** 0.45-0.60 **P** 0.025 max **S** 0.025 max **Si** 1.40-1.75 **V** 0.03-0.08	**ASTM** A579 (33)
K14501	Alloy Steel	**C** 0.45 max **Mn** 0.90 max **P** 0.035 max **S** 0.035 max **Si** 0.15-0.35 **V** 0.03-0.12	**ASTM** A293 (1); A469 (1); A470 (1)
K14507	Alloy Steel	**C** 0.45 max **Cr** 1.25 max **Cu** 0.35 **Mn** 0.40-0.90 **Mo** 0.15 min **Ni** 0.50 **P** 0.040 max **S** 0.040 max **Si** 0.15-0.30 **V** 0.10 max	**ASTM** A291 (3)
K14508	Alloy Steel	**C** 0.40-0.50 **Mn** 1.40-1.80 **Mo** 0.17-0.27 **P** 0.035 max **S** 0.04 max **Si** 0.15-0.35	**ASTM** A372 (4)
K14510	Alloy Steel	**C** 0.40-0.50 **Mn** 0.70-0.90 **Mo** 0.20-0.30 **P** 0.035 max **S** 0.040 max **Si** 0.15-0.35	**ASTM** A194 (4)
K14520	Alloy Steel	**C** 0.45 max **Mn** 0.50-0.90 **Mo** 0.30-0.60 **P** 0.040 max **S** 0.040 max **Si** 0.15-0.45 **V** 0.10-0.25	**ASTM** A336 (F30)
K14542	Alloy Steel	**C** 0.45 max **Cr** 0.70-1.25 **Mn** 0.60-1.00 **Mo** 0.15 min **P** 0.025 max **S** 0.020 max **Si** 0.15-0.35	**ASTM** A288 (2, 3)
K14557	Alloy Steel	**C** 0.45 max **Cr** 1.50 max **Cu** 0.35 **Mn** 0.40-0.90 **Mo** 0.15 min **Ni** 1.00-3.00 **P** 0.040 max **S** 0.040 max **Si** 0.15-0.30 **V** 0.10 max	**ASTM** A291 (3A)
K14675	Alloy Steel (17-22A)	**C** 0.42-0.50 **Cr** 0.80-1.10 **Cu** 0.35 max **Mn** 0.40-0.70 **Mo** 0.45-0.65 **Ni** 0.25 max **P** 0.025 max **S** 0.025 max **Si** 0.20-0.35 **V** 0.25-0.35	**AMS** 6304; 6305 **MIL SPEC** MIL-S-24502
K15047	Alloy Steel	**C** 0.48-0.53 **Cr** 0.80-1.10 **Mn** 0.70-0.90 **P** 0.020 max **S** 0.035 max **Si** 0.20-0.35 **V** 0.15 min	**ASTM** A232 **SAE** J132
K15048	Alloy Steel	**C** 0.48-0.53 **Cr** 0.80-1.10 **Mn** 0.70-0.90 **P** 0.040 max **S** 0.040 max **Si** 0.20-0.35 **V** 0.15 min	**ASTM** A231
K15590	Alloy Steel	**C** 0.51-0.59 **Cr** 0.60-0.80 **Mn** 0.60-0.80 **P** 0.035 max **S** 0.040 max **Si** 1.20-1.60	**ASTM** A401 **SAE** J157
K15747	Alloy Steel	**C** 0.52-0.62 **Cr** 0.80-1.15 **Mn** 0.75-1.05 **Mo** 0.15-0.25 **P** 0.04 max **S** 0.04 max **Si** 0.20-0.35	**MIL SPEC** MIL-S-12504 (Mn-Cr-Mo)
K17145	Alloy Steel	**C** 0.65-0.77 **Mn** 0.75-1.05 **Mo** 0.90-1.10 **P** 0.04 max **S** 0.04 max **Si** 0.20-0.35	**MIL SPEC** MIL-S-12504 (Mn-Mo)
K18597	Alloy Steel	**C** 0.80-0.90 **Cr** 0.85-1.15 **Cu** 0.15 max **Mn** 0.20-0.50 **Mo** 0.50-0.65 **Ni** 0.15 max **P** 0.015 max **S** 0.015 max **Si** 0.60-0.90	**AMS** 6426
K19195	Alloy Steel (Bearing Steel)	**C** 0.85-1.00 **Cr** 1.40-1.80 **Cu** 0.35 max **Mn** 1.40-1.70 **Mo** 0.10 max **Ni** 0.25 max **P** 0.025 max **S** 0.025 max **Si** 0.50-0.80	**ASTM** A485 (2)
K19667	Alloy Steel (Bearing Steel)	**C** 0.90-1.05 **Cr** 0.90-1.20 **Cu** 0.35 max **Mn** 0.95-1.25 **Mo** 0.10 max **Ni** 0.25 max **P** 0.025 max **S** 0.025 max **Si** 0.45-0.75	**ASTM** A485 (1)

The chemical compositions listed are for identification purposes and should not be used in lieu of the cross referenced specifications.

UNIFIED NUMBER	DESCRIPTION	CHEMICAL COMPOSITION	CROSS REFERENCE SPECIFICATIONS
K19964	Alloy Steel (Bearing Steel)	C 0.95-1.10 Cr 1.30-1.60 Cu 0.25 max Mn 0.25-0.45 Mo 0.08 max Ni 0.35 max P 0.025 max S 0.025 max Si 0.20-0.35	**MIL SPEC** MIL-S-980
K19965	Alloy Steel (Bearing Steel)	C 0.95-1.0 Cr 1.10-1.50 Cu 0.35 max Mn 0.65-0.90 Mo 0.20-0.30 Ni 0.25 max P 0.025 max S 0.025 max Si 0.15-0.35	**ASTM** A485 (3)
K19990	Alloy Steel (Bearing Steel)	C 0.95-1.10 Cr 1.10-1.50 Cu 0.35 max Mn 1.05-1.35 Mo 0.45-0.60 Ni 0.25 max P 0.025 max S 0.025 max Si 0.15-0.35	**ASTM** A485 (4)
K20500	Alloy Steel Electrode (ER80S-BL2)	C 0.05 max Cr 1.20-1.50 Cu 0.35 max Mn 0.40-0.70 Mo 0.40-0.65 Ni 0.20 max P 0.025 max S 0.025 max Si 0.40-0.70 Other total other elements (except iron) 0.50 percent max	**ASME** SFA5.28 (ER80S-B2L) **AWS** A5.28 (ER80S-B2L)
K20622	Alloy Steel	C 0.06 max Cb 0.02 min Cu 1.00-1.30 Mn 0.40-0.65 Ni 1.20-1.50 P 0.025 max S 0.025 max Si 0.15-0.40	**ASTM** A710 (B)
K20747	Alloy Steel	C 0.07 max Cb 0.02 min Cr 0.60-0.90 Cu 1.00-1.30 Mn 0.40-0.70 Mo 0.15-0.25 Ni 0.70-1.00 P 0.025 max S 0.025 max Si 0.40 max	**ASTM** A710 (A); A736
K20900	Alloy Steel Electrode (ER80S-B2)	C 0.07-0.12 Cr 1.20-1.50 Cu 0.35 max Mn 0.40-0.70 Mo 0.40-0.65 Ni 0.20 max P 0.025 max S 0.025 max Si 0.40-0.70 Other total other elements (except iron) 0.50 max	**ASME** SFA5.28 (ER80S-B2); SFA5.30 (IN515) **AWS** A5.28 (ER80S-B2); A5.30 (IN515) **MIL SPEC** MIL-I-23413 (MIL-515)
K20902	Alloy Steel	C 0.09 max Cb 0.05-0.11 Mn 1.75-2.30 Mo 0.22-0.38 P 0.030 max S 0.035 max Si 0.17 max	**ASTM** A707 (L6)
K20910	Alloy Steel Electrode (ENi2N)	C 0.12 max Cu 0.08 max Mn 0.75-1.25 Ni 2.10-2.90 P 0.012 max S 0.020 max Si 0.05-0.25 V 0.05 max Other total other elements (except iron) 0.50 max	**ASME** SFA5.23 (ENi2N) **AWS** A5.23 (ENi2N)
K20915	Alloy Steel Electrode (EM3N)	Al 0.10 max C 0.10 max Cr 0.55 max Cu 0.08 max Mn 1.40-1.80 Mo 0.25-0.65 Ni 1.90-2.60 P 0.012 max S 0.010 max Si 0.20-0.60 Ti 0.10 max V 0.05 max Zr 0.10 max Other total other elements (except iron) 0.50 percent max	**ASME** SFA5.23 (EM3N) **AWS** A5.23 (EM3N)
K20930	Alloy Steel Electrode (EM4N)	Al 0.10 max C 0.10 max Cr 0.60 max Cu 0.08 max Mn 1.40-1.80 Mo 0.30-0.65 Ni 2.00-2.80 P 0.012 max S 0.010 max Si 0.20-0.60 Ti 0.10 max V 0.05 max Zr 0.10 max Other total other elements (except iron) 0.50 max	**ASME** SFA5.23 (EM4N) **AWS** A5.23 (EM4N)
K20934	Alloy Steel	C 0.09 max Cb 0.02 min Cr 0.56-0.94 Cu 0.95-1.35 Mn 0.35-0.75 Mo 0.14-0.28 Ni 0.67-1.03 P 0.030 max S 0.035 max Si 0.37 max	**ASTM** A707 (L5)
K21010	Alloy Steel Electrode (ENi2)	C 0.12 max Cu 0.35 max Mn 0.75-1.25 Ni 2.10-2.90 P 0.020 max S 0.020 max Si 0.05-0.30 Other total other elements (except iron) 0.50 max	**ASME** SFA5.23 (ENi2) **AWS** A5.23 (ENi2)

The chemical compositions listed are for identification purposes and should not be used in lieu of the cross referenced specifications.

UNIFIED NUMBER	DESCRIPTION	CHEMICAL COMPOSITION	CROSS REFERENCE SPECIFICATIONS
K21015	Alloy Steel Electrode (EM3)	**Al** 0.10 max **C** 0.10 max **Cr** 0.55 max **Cu** 0.25 max **Mn** 1.40-1.80 **Mo** 0.25-0.65 **Ni** 1.90-2.60 **P** 0.010 max **S** 0.010 max **Si** 0.20-0.60 **Ti** 0.10 max **V** 0.04 max **Zr** 0.10 max **Other** total other elements (except iron) 0.50 max	**ASME** SFA5.23 (EM3); SFA5.28 (ER110S-1) **AWS** A5.23 (EM3); A5.28 (ER110S-1) **MIL SPEC** MIL-E-23765 (MIL-110S-1)
K21028	Alloy Steel	**B** 0.007 max **C** 0.08-0.13 **Cr** 0.40-0.60 **Cu** 0.35 max **Mn** 0.75-1.00 **Mo** 0.20-0.30 **Ni** 1.65-2.00 **P** 0.025 max **S** 0.025 max **Si** 0.20-0.40 **V** 0.03-0.08	**AMS** 6266
K21030	Alloy Steel Electrode (EM4)	**Al** 0.10 max **C** 0.10 max **Cr** 0.60 max **Cu** 0.25 max **Mn** 1.40-1.80 **Mo** 0.30-0.65 **Ni** 2.00-2.80 **P** 0.010 max **S** 0.010 max **Si** 0.20-0.60 **Ti** 0.10 max **V** 0.03 max **Zr** 0.10 max **Other** total other elements (except iron) 0.50 max	**ASME** SFA5.23 (EM4); SFA5.28 (ER120S-1) **AWS** A5.23 (EM4); A5.28 (ER120S-1) **MIL SPEC** MIL-E-23765 (MIL-120S-1)
K21035	Alloy Steel Electrode (EF6N)	**C** 0.07-0.15 **Cr** 0.20-0.55 **Cu** 0.08 max **Mn** 1.45-1.90 **Mo** 0.40-0.65 **Ni** 1.75-2.25 **P** 0.012 max **S** 0.015 max **Si** 0.10-0.30 **V** 0.05 max **Other** total other elements (except iron) 0.50 max	**ASME** SFA5.23 (EF6N) **AWS** A5.23 (EF6N)
K21135	Alloy Steel Electrode (EF6)	**C** 0.07-0.15 **Cr** 0.20-0.55 **Cu** 0.35 max **Mn** 1.45-1.90 **Mo** 0.40-0.65 **Ni** 1.75-2.25 **P** 0.015 max **S** 0.015 max **Si** 0.10-0.30 **Other** total other elements (except iron) 0.50 max	**ASME** SFA5.23 (EF6) **AWS** A5.23 (EF6)
K21205	Alloy Steel	**Al** 0.06 max **C** 0.12 max **Cr** 0.90-1.20 **Mn** 0.45-0.75 **Mo** 0.25-0.40 **Ni** 0.90-1.20 **P** 0.035 max **S** 0.015 max **Si** 0.40 max	**ASTM** A734/A734M (A)
K21240	Alloy Steel Electrode (ER80S-Ni2)	**C** 0.12 max **Cu** 0.35 max **Mn** 1.25 max **Ni** 2.00-2.75 **P** 0.025 max **S** 0.025 max **Si** 0.40-0.80 **Other** total other elements (except iron) 0.50 max	**ASME** SFA5.28 (ER80S-Ni2) **AWS** A5.28 (ER80S-Ni2)
K21350	Alloy Steel Electrode (EF2N)	**C** 0.10-0.18 **Cu** 0.08 max **Mn** 1.70-2.40 **Mo** 0.40-0.65 **Ni** 0.40-0.80 **P** 0.012 max **S** 0.025 max **Si** 0.20 max **V** 0.05 max **Other** total other elements (except iron) 0.50 max	**ASME** SFA5.23 (EF2N) **AWS** A5.23 (EF2N)
K21385	Alloy Steel Electrode (EF3N)	**C** 0.10-0.18 **Cu** 0.08 max **Mn** 1.70-2.40 **Mo** 0.40-0.65 **Ni** 0.70-1.10 **P** 0.012 max **S** 0.025 max **Si** 0.30 max **V** 0.05 max **Other** total other elements (except iron) 0.50 max	**ASME** SFA5.23 (EF3N) **AWS** A5.23 (EF3N)
K21390	Alloy Steel	**C** 0.15 max **Cr** 2.00-2.50 **Mn** 0.30-0.60 **Mo** 0.90-1.10 **P** 0.035 max **S** 0.040 max **Si** 0.50 max	**ASTM** A739 (B22)
K21450	Alloy Steel Electrode (EF2)	**C** 0.10-0.18 **Cu** 0.35 max **Mn** 1.70-2.40 **Mo** 0.40-0.65 **Ni** 0.40-0.80 **P** 0.025 max **S** 0.025 max **Si** 0.20 max **Other** total other elements (except iron) 0.50 max	**ASME** SFA5.23 (EF2) **AWS** A5.23 (EF2)
K21485	Alloy Steel Electrode (EF3)	**C** 0.10-0.18 **Cu** 0.35 max **Mn** 1.70-2.40 **Mo** 0.40-0.65 **Ni** 0.70-1.10 **P** 0.025 max **S** 0.025 max **Si** 0.30 max **Other** total other elements (except iron) 0.50 max	**ASME** SFA5.23 (EF3) **AWS** A5.23 (EF3) **MIL SPEC** MIL-E-23765 (MIL-80S-1)
K21509	Alloy Steel	**C** 0.15 max **Cr** 1.65-2.35 **Mn** 0.30-0.60 **Mo** 0.44-0.65 **P** 0.030 max **S** 0.030 max **Si** 0.50 max	**ASTM** A199 (T3b); A200 (T3b); A213 (T3b); A369 (FP3b)

The chemical compositions listed are for identification purposes and should not be used in lieu of the cross referenced specifications.

UNIFIED NUMBER	DESCRIPTION	CHEMICAL COMPOSITION	CROSS REFERENCE SPECIFICATIONS
K21590	Alloy Steel	**C** 0.15 max **Cr** 2.00-2.50 **Mn** 0.30-0.60 **Mo** 0.90-1.10 **P** 0.030 max **S** 0.030 max **Si** 0.50 max	**ASTM** A182 (F22); A199 (T22); A200 (T22); A213 (T22); A234 (WP22); A335 (P22); A336 (F22, F22a); A369 (FP22); A387 (22); A541 (6, 6A); A542 **MIL SPEC** MIL-S-18410
K21604	Alloy Steel	**B** 0.0015-0.005 **C** 0.12-0.20 **Cr** 1.40-2.00 **Cu** 0.20-0.40 **Mn** 0.40-0.70 **Mo** 0.40-0.60 **P** 0.035 max **S** 0.04 max **Si** 0.20-0.35 **Ti** 0.04-0.10	**ASTM** A514 (E); A517 (E)
K21650	Alloy Steel	**B** 0.001-0.005 **C** 0.12-0.21 **Cr** 0.85-1.20 **Mn** 0.45-0.70 **Mo** 0.45-0.60 **Ni** 1.20-1.50 **P** 0.035 max **S** 0.04 max **Si** 0.20-0.35	**ASTM** A514 (P); A517 (P)
K21703	Alloy Steel	**C** 0.17 max **Mn** 0.70 max **Ni** 2.03-2.57 **P** 0.035 max **S** 0.040 max **Si** 0.13-0.32	**ASTM** A203/A203M (A)
K21903	Alloy Steel	**C** 0.19 max **Mn** 0.90 max **Ni** 2.03-2.57 **P** 0.04 max **S** 0.05 max **Si** 0.13-0.32	**ASTM** A333 (7); A334 (7)
K21940	Alloy Steel (CBS600)	**Al** 0.03-0.12 **C** 0.16-0.22 **Cr** 1.25-1.65 **Cu** 0.35 max **Mn** 0.45-0.75 **Mo** 0.90-1.10 **Ni** 0.25 max **P** 0.010 max **S** 0.010 max **Si** 0.90-1.25	**AMS** 6255
K22033	Alloy Steel	**C** 0.33-0.38 **Cr** 0.55-0.75 **Cu** 0.35 max **Mn** 0.60-0.80 **Mo** 0.06 max **Ni** 1.10-1.40 **P** 0.025 max **S** 0.025 max **Si** 0.20-0.35	**AMS** 6330
K22035	Alloy Steel	**C** 0.20 max **Cu** 0.75-1.25 **Mn** 0.40-1.06 **Ni** 1.60-2.24 **P** 0.045 max **S** 0.050 max	**ASTM** A182 (FR); A234 (WPR); A333 (9); A334 (9); A714 (V)
K22036	Alloy Steel	**C** 0.20 max **Cu** 0.75-1.25 **Mn** 0.40-1.06 **Ni** 1.60-2.24 **P** 0.035 max **S** 0.040 max	**ASTM** A350 (LF9)
K22094	Alloy Steel	**C** 0.20 max **Cr** 1.00-1.50 **Mn** 0.30-0.80 **Mo** 0.44-0.65 **P** 0.040 max **S** 0.040 max **Si** 0.50-1.00	
K22097	Alloy Steel	**C** 0.92-1.02 **Cr** 0.90-1.15 **Cu** 0.35 max **Mn** 0.95-1.25 **Mo** 0.08 max **Ni** 0.25 max **P** 0.015 max **S** 0.015 max **Si** 0.50-0.70	**AMS** 6445
K22103	Alloy Steel	**C** 0.21 max **Mn** 0.70 max **Ni** 2.03-2.57 **P** 0.035 max **S** 0.040 max **Si** 0.13-0.45	**ASTM** A203/A203M (B)
K22375	Alloy Steel	**C** 0.23 max **Cr** 1.50-2.00 **Mn** 0.20-0.40 **Mo** 0.40-0.60 **Ni** 2.75-3.90 **P** 0.020 max **S** 0.020 max **Si** 0.15-0.40 **V** 0.03 max	**ASTM** A508 (4, 4a, 4b)
K22440	Alloy Steel	**C** 0.38-0.43 **Cr** 0.20 max **Cu** 0.35 max **Mn** 0.60-0.80 **Mo** 0.20-0.30 **Ni** 1.65-2.00 **P** 0.025 max **S** 0.025 max **Si** 0.15-0.35	**AMS** 6312; 6317
K22573	Alloy Steel	**C** 0.25 max **Cr** 0.50 max **Mn** 0.60 max **Mo** 0.20-0.50 **Ni** 2.50 min **P** 0.015 max **S** 0.018 max **Si** 0.15-0.30 **V** 0.03 min	**ASTM** A469 (2)
K22578	Alloy Steel	**C** 0.25 max **Cr** 0.75 max **Mn** 0.020-0.60 **Mo** 0.25 min **Ni** 2.50 min **P** 0.015 max **S** 0.018 max **Si** 0.15-0.30 **V** 0.03 min	**ASTM** A470 (2)
K22720	Alloy Steel	**C** 0.18-0.23 **Cr** 0.80-1.20 **Cu** 0.50 max **H** 0.0010 max **Mn** 0.40-0.60 **Mo** 0.80-1.20 **N** 0.005 max **O** 0.0025 max **P** 0.015 max **S** 0.008 max **Si** 0.60-0.90 **V** 0.08-0.15	**AMS** 6459

The chemical compositions listed are for identification purposes and should not be used in lieu of the cross referenced specifications.

UNIFIED NUMBER	DESCRIPTION	CHEMICAL COMPOSITION	CROSS REFERENCE SPECIFICATIONS
K22770	Alloy Steel	C 0.25-0.30 Cr 1.00-1.50 Cu 0.50 max Mn 0.60-0.90 Mo 0.40-0.60 Ni 0.50 max P 0.025 max S 0.025 max Si 0.55-0.75 V 0.75-0.95	AMS 6303; 6436
K22773	Alloy Steel	C 0.27 max Cr 0.50 max Mn 0.60 max Mo 0.20-0.50 Ni 2.50 min P 0.015 max S 0.018 max Si 0.15-0.30 V 0.03 min	ASTM A469 (3)
K22878	Alloy Steel	C 0.28 max Cr 0.75 max Mn 0.20-0.60 Mo 0.25 min Ni 2.50 min P 0.015 max S 0.018 max Si 0.15-0.30 V 0.03 min	ASTM A470 (3, 4)
K22925	Alloy Steel Welding Wire	C 0.26-0.32 Cr 0.90-1.20 Cu 0.35 max H 0.0010 max Mn 0.60-0.90 Mo 0.90-1.10 N 0.0050 max Ni 0.40-0.70 O 0.0025 max P 0.010 max S 0.010 max Si 0.10-0.30 V 0.05-0.10	AMS 5027
K22950	Alloy Steel	C 0.49-0.55 Cr 0.70-0.90 Cu 0.35 max Mn 0.65-0.85 Mo 0.20-0.30 Ni 1.65-2.00 P 0.025 max S 0.025 max Si 0.15-0.35	AMS 6396; 6424
K23010	Alloy Steel	C 0.25-0.35 Cr 0.90-1.50 Mn 1.00 max Mo 1.00-1.50 Ni 0.75 max P 0.015 max S 0.018 max Si 0.15-0.35 V 0.20-0.30	ASTM A470 (8)
K23015	Alloy Steel	C 0.27-0.33 Cr 1.00-1.50 Cu 0.35 max Mn 0.45-0.65 Mo 0.40-0.60 Ni 0.25 max P 0.025 max S 0.025 max Si 0.55-0.75 V 0.20-0.30	AMS 6302; 6385; 6458
K23016	Alloy Steel Electrode (EB2H)	C 0.28-0.33 Cr 1.00-1.50 Cu 0.30 max Mn 0.45-0.65 Mo 0.40-0.65 P 0.015 max S 0.015 max Si 0.55-0.75 V 0.20-0.30 Other total other elements (except iron) 0.50 percent max	ASME SFA5.23 (EB2H) AWS A5.23 (EB2H)
K23028	Alloy Steel	C 0.30 max Cr 0.75 max Mn 0.70 max Mo 0.25 min Ni 2.00 min P 0.035 max S 0.035 max Si 0.15-0.35 V 0.03-0.12	ASTM A293 (2, 3)
K23080	Alloy Steel	C 0.28-0.33 Cr 0.75-1.00 Cu 0.35 max Mn 0.75-1.00 Mo 0.35-0.50 Ni 1.65-2.00 P 0.015 max S 0.015 max Si 0.20-0.35 V 0.05-0.10	AMS 6411; 6427 MIL SPEC MIL-S-8699; MIL-S-46128
K23116	Alloy Steel Welding Wire	C 0.28-0.33 Cr 1.00-1.50 Cu 0.08 max Mn 0.45-0.65 Mo 0.40-0.65 P 0.012 max S 0.015 max Si 0.55-0.75 V 0.20-0.30 Other total other elements (except iron) 0.50 percent max	ASME SFA5.23 (EB2HN) AWS A5.23 (EB2HN)
K23205	Alloy Steel	C 0.27-0.37 Cr 0.85-1.25 Mn 0.70-1.00 Mo 1.00-1.50 Ni 0.50 max P 0.015 max S 0.015 max Si 0.20 min V 0.20-0.30	ASTM A471 (10)
K23477	Alloy Steel	C 0.31-0.38 Cr 0.65-0.90 Mn 0.60-0.90 Mo 0.30-0.60 Ni 1.65-2.00 P 0.025 max S 0.025 max Si 0.20-0.35 V 0.17-0.23	ASTM A579 (21)
K23505	Tool Steel, Non-Tempering	C 0.32-0.39 Cr 0.55-0.75 Cu 0.50-0.70 Mn 0.35-0.50 Mo 0.50-0.70 P 0.025 max S 0.025 max Si 0.50-0.70	
K23510	Nitriding Steel	Al 0.95-1.30 C 0.33-0.38 Cr 1.00-1.30 Mn 0.50-0.70 Mo 0.15-0.25 P 0.035 max S 0.040 max Si 0.15-0.35	ASTM A355 (D)

The chemical compositions listed are for identification purposes and should not be used in lieu of the cross referenced specifications.

UNIFIED NUMBER	DESCRIPTION	CHEMICAL COMPOSITION	CROSS REFERENCE SPECIFICATIONS
K23545	Alloy Steel	**C** 0.35 max **Mn** 0.50-0.90 **Mo** 0.20-0.50 **Ni** 2.25-3.00 **P** 0.040 max **S** 0.040 max **Si** 0.10-0.40 **V** 0.15 max	**ASTM** A336 (F31)
K23550	Alloy Steel	**C** 0.35 max **Cr** 0.80-2.00 **Mn** 0.90 max **Mo** 0.20-0.40 **Ni** 1.5-2.25 **P** 0.015 max **S** 0.015 max **Si** 0.35 max **V** 0.20 max	**ASTM** A723 (1)
K23577	Alloy Steel Welding Wire	**C** 0.33-0.38 **Cr** 0.65-0.90 **H** 0.0010 max **Mn** 0.60-0.90 **Mo** 0.30-0.40 **N** 0.0050 max **Ni** 1.65-2.00 **O** 0.0025 max **P** 0.008 max **S** 0.008 max **Si** 0.25 max **V** 0.17-0.23	**AMS** 5029
K23578	Alloy Steel	**C** 0.35 max **Cr** 0.75 max **Mn** 0.70 max **Mo** 0.25 min **Ni** 2.50 min **P** 0.035 max **S** 0.035 max **Si** 0.15-0.35 **V** 0.03-0.12	**ASTM** A293 (4)
K23579	Alloy Steel	**C** 0.35 max **Cr** 1.25 max **Mn** 0.70 max **Mo** 0.25 min **Ni** 2.50 min **P** 0.035 max **S** 0.035 max **Si** 0.15-0.35 **V** 0.03 min	**ASTM** A293 (5)
K23705	Alloy Steel	**C** 0.37 max **Cr** 0.85-1.25 **Mn** 1.00 max **Mo** 1.00-1.50 **Ni** 0.50 max **P** 0.035 max **S** 0.035 max **Si** 0.15-0.35 **V** 0.20-0.30	**ASTM** A293 (6)
K23725	Alloy Steel Welding Wire	**C** 0.34-0.40 **Cr** 0.90-1.20 **Cu** 0.35 max **H** 0.0010 max **Mn** 0.60-0.90 **Mo** 0.90-1.10 **N** 0.0050 max **Ni** 0.40-0.70 **O** 0.0025 max **P** 0.008 max **S** 0.008 max **Si** 0.15-0.30 **V** 0.05-0.10	**AMS** 5028
K23745	Nitriding Steel	**Al** 0.95-1.30 **C** 0.35-0.40 **Cr** 1.20-1.50 **Mn** 0.70-0.95 **Mo** 0.15-0.25 **P** 0.035 max **S** 0.060 max **Se** 0.15-0.25 **Si** 0.15-0.35	**ASTM** A355 (B)
K24040	Alloy Steel	**C** 0.37-0.44 **Cr** 0.65-0.95 **Mn** 0.60-0.95 **Mo** 0.20-0.30 **Ni** 1.55-2.00 **P** 0.025 max **S** 0.025 max **Si** 0.15-0.35	**ASTM** A649 (1B)
K24045	Alloy Steel	**C** 0.35-0.45 **Cr** 0.60-0.90 **Mn** 0.60-0.90 **Mo** 0.20-0.50 **Ni** 1.65-2.00 **P** 0.040 max **S** 0.040 max **Si** 0.20-0.35 **V** 0.10 max	**ASTM** A290 (G, H, I, J, K, L)
K24055	Alloy Steel	**C** 0.38-0.43 **Cr** 0.70-0.90 **Mn** 0.60-0.80 **Mo** 0.20-0.30 **Ni** 1.65-2.00 **P** 0.035 max **S** 0.04 max **Si** 0.15-0.35	**ASTM** A372 (7)
K24064	Alloy Steel	**C** 0.37-0.44 **Cr** 0.70-0.95 **Mn** 0.70-0.90 **Mo** 0.30-0.40 **Ni** 1.65-2.00 **P** 0.025 max **S** 0.025 max **Si** 0.20-0.35	**ASTM** A540 (B24)
K24065	Nitriding Steel	**Al** 0.95-1.30 **C** 0.38-0.43 **Cr** 1.40-1.80 **Cu** 0.35 max **Mn** 0.50-0.80 **Mo** 0.30-0.40 **Ni** 0.25 max **P** 0.025 max **S** 0.025 max **Si** 0.20-0.40	**AMS** 6470; 6471; 6472 **ASTM** A355 (A) **MIL SPEC** MIL-S-6709
K24070	Alloy Steel	**C** 0.38-0.43 **Cr** 0.70-0.90 **Mn** 0.60-0.90 **Mo** 0.30-0.60 **Ni** 1.65-2.00 **P** 0.025 max **S** 0.025 max **Si** 0.20-0.35 **V** 0.05-0.10	**ASTM** A540 (B24V); A579 (22)
K24245	Alloy Steel	**C** 0.35-0.50 **Cr** 0.60 min **Cu** 0.35 **Mn** 0.40-0.90 **Mo** 0.20-0.60 **Ni** 1.65 min **P** 0.040 max **S** 0.040 max **Si** 0.15-0.35 **V** 0.10 max	**ASTM** A291 (4 to 7)
K24336	Alloy Steel	**B** 0.0005-0.005 **C** 0.40-0.46 **Cr** 0.80-1.05 **Cu** 0.35 max **Mn** 0.75-1.00 **Mo** 0.45-0.60 **Ni** 0.60-0.90 **P** 0.025 max **S** 0.025 max **Si** 0.50-0.80 **V** 0.01-0.06	**AMS** 6423

The chemical compositions listed are for identification purposes and should not be used in lieu of the cross referenced specifications.

UNIFIED NUMBER	DESCRIPTION	CHEMICAL COMPOSITION	CROSS REFERENCE SPECIFICATIONS
K24535	Alloy Steel	C 0.45 max Cr 0.50-1.25 Mn 0.60-1.00 Mo 0.20 min Ni 1.65-3.50 P 0.025 max S 0.025 max Si 0.15-0.35	ASTM A294 (B)
K24562	Alloy Steel	C 0.45 max Cr 0.70-1.25 Mn 1.00 max Mo 0.20 min Ni 1.65-3.50 P 0.025 max S 0.020 max Si 0.15-0.35 V 0.07-0.12	ASTM A288 (4, 5, 6, 7, 8)
K24628	Alloy Steel combined with K24728		
K24728	Alloy Steel (D-6)	C 0.45-0.50 Cr 0.90-1.20 Mn 0.60-0.90 Mo 0.90-1.10 Ni 0.40-0.70 P 0.010 max S 0.010 max Si 0.15-0.30 V 0.08-0.15	AMS 6431; 6432; 6438; 6439 ASTM A355 (A); A579 (23) MIL SPEC MIL-S-8949; MIL-S-47036
K24729	Alloy Steel combined with K24728		
K30560	Alloy Steel Electrode (ER90S-B3L)	C 0.05 max Cr 2.30-2.70 Cu 0.35 max Mn 0.40-0.70 Mo 0.90-1.20 Ni 0.20 max P 0.025 max S 0.025 max Si 0.40-0.70 Other total other elements (except iron) 0.50 max	ASME SFA5.28 (ER90S-B3L) AWS A5.28 (ER90S-B3L)
K30960	Alloy Steel Electrode (ER-90S-B3)	C 0.07-0.12 Cr 2.30-2.70 Cu 0.35 max Mn 0.40-0.70 Mo 0.90-1.20 Ni 0.20 max P 0.025 max S 0.025 max Si 0.40-0.70 Other total other elements (except iron) 0.50 max	ASME SFA5.28 (ER90S-B3); SFA5.30 (IN521) AWS A5.28 (ER90S-B3); A5.30 (IN521) MIL SPEC MIL-I-23413B (MIL-521)
K31015	Alloy Steel Electrode (EB3N)	C 0.06-0.15 Cr 2.25-3.00 Cu 0.08 max Mn 0.40-0.80 Mo 0.90-1.10 P 0.012 max S 0.025 max Si 0.05-0.30 V 0.05 max Other total other elements (except iron) 0.50 max	ASME SFA5.23 (EB3N) AWS A5.23 (EB3N)
K31115	Alloy Steel Electrode (EB3)	C 0.06-0.15 Cr 2.25-3.00 Cu 0.35 max Mn 0.40-0.80 Mo 0.90-1.10 P 0.025 max S 0.025 max Si 0.05-0.30 Other total other elements (except iron) 0.50 percent max	ASME SFA5.23 (EB3) AWS A5.23 (EB3)
K31210	Alloy Steel Electrode (ENi-3N)	C 0.13 max Cr 0.15 max Cu 0.08 max Mn 0.60-1.20 Ni 3.10-3.80 P 0.012 max S 0.020 max Si 0.05-0.30 V 0.05 max Other total other elements (except iron) 0.50 max	ASME SFA5.23 (ENi3N) AWS A5.23 (ENi3N)
K31240	Alloy Steel Electrode (ER80S-Ni3)	C 0.12 max Cu 0.35 max Mn 1.25 max Ni 3.00-3.75 P 0.025 max S 0.025 max Si 0.40-0.80 Other total other elements (except iron) 0.50 max	ASME SFA5.28 (ER80S-Ni3) AWS A5.28 (ER80S-Ni3)
K31310	Alloy Steel Electrode (ENi3)	C 0.13 max Cr 0.15 max Cu 0.35 max Mn 0.60-1.20 Ni 3.10-3.80 P 0.012 max S 0.020 max Si 0.05-0.30 Other total other elements (except iron) 0.50 max	ASME SFA5.23 (ENi3) AWS A5.23 (ENi3)
K31509	Alloy Steel	C 0.15 max Cr 2.15-2.85 Mn 0.30-0.60 Mo 0.44-0.65 P 0.030 max S 0.030 max Si 0.50-1.00	ASTM A199 (T4); A200 (T4)
K31545	Alloy Steel	C 0.15 max Cr 2.65-3.35 Mn 0.30-0.60 Mo 0.80-1.06 P 0.030 max S 0.030 max Si 0.50 max	ASTM A182 (F21); A199 (T21); A200 (T21); A213 (T21); A335 (P21); A336 (F21, F21a); A369 (FP21); A387 (21)
K31718	Alloy Steel	C 0.17 max Mn 0.70 max Ni 3.18-3.82 P 0.035 max S 0.040 max Si 0.13-0.32	ASTM A203/A203M (D)
K31820	Alloy Steel (HY80)	C 0.18 max Cr 1.00-1.80 Mn 0.10-0.40 Mo 0.20-0.60 Ni 2.00-3.25 P 0.025 max S 0.025 max Si 0.15-0.35	ASTM A372 (6) MIL SPEC MIL-S-16216; MIL-S-21952; MIL-S-22664; MIL-S-23009; MIL-S-24451

The chemical compositions listed are for identification purposes and should not be used in lieu of the cross referenced specifications.

UNIFIED NUMBER	DESCRIPTION	CHEMICAL COMPOSITION	CROSS REFERENCE SPECIFICATIONS
K31830	Alloy Steel	**B** 0.001-0.003 **C** 0.18 max **Cr** 2.75-3.25 **Mn** 0.30-0.60 **Mo** 0.90-1.10 **P** 0.02 max **S** 0.02 max **Si** 0.12 max **Ti** 0.015-0.035 **V** 0.20-0.30	**ASTM** A182; A182M
K31918	Alloy Steel	**C** 0.19 max **Mn** 0.31-0.64 **Ni** 3.18-3.82 **P** 0.05 max **S** 0.05 max **Si** 0.18-0.37	**ASTM** A333 (3); A334 (3)
K32018	Alloy Steel	**C** 0.20 max **Mn** 0.70 max **Ni** 3.18-3.82 **P** 0.035 max **S** 0.040 max **Si** 0.13-0.32	**ASTM** A203/A203M (E)
K32025	Alloy Steel	**C** 0.20 max **Mn** 0.90 max **Ni** 3.25-3.75 **P** 0.035 max **S** 0.040 max **Si** 0.20-0.35	**ASTM** A350 (LF3)
K32026	Alloy Steel	**Al** 0.05 max **C** 0.20 max **Cr** 0.20 max **Cu** 0.35 max **Mn** 0.90 max **Mo** 0.06 max **Ni** 3.25-3.75 **P** 0.020 max **S** 0.020 max **Si** 0.15-0.35 **V** 0.05 max	**ASTM** A765 (III)
K32045	Alloy Steel (HY100)	**C** 0.20 max **Cr** 1.00-1.80 **Mn** 0.10-0.40 **Mo** 0.20-0.60 **Ni** 2.25-3.50 **P** 0.25 max **S** 0.25 max **Si** 0.15-0.35	**MIL SPEC** MIL-S-16216; MIL-S-21952; MIL-S-22664; MIL-S-23009; MIL-S-24451
K32218	Alloy Steel	**C** 0.22 max **Mn** 1.00 max **Ni** 3.18-3.82 **P** 0.030 max **S** 0.035 max **Si** 0.37 max	**ASTM** A707 (L7)
K32550	Alloy Steel	**C** 0.23-0.28 **Cr** 0.20-0.40 **Mn** 1.20-1.50 **Mo** 0.35-0.45 **Ni** 1.65-2.00 **P** 0.025 max **S** 0.025 max **Si** 1.30-1.70	**AMS** 6418 **ASTM** A579 (31) **MIL SPEC** MIL-S-7108
K32723	Alloy Steel	**C** 0.27 max **Cr** 0.50 max **Mn** 0.70 max **Mo** 0.20-0.60 **Ni** 3.00 min **P** 0.015 max **S** 0.018 max **Si** 0.15-0.30 **V** 0.03 min	**ASTM** A469 (4)
K32800	Alloy Steel	**C** 0.28 max **Cr** 0.75-2.00 **Mn** 0.70 max **Mo** 0.20-0.70 **Ni** 2.00-4.00 **P** 0.015 max **S** 0.015 max **Si** 0.15-0.35 **V** 0.05 min	**ASTM** A471 (1-9)
K33020	Alloy Steel	**C** 0.27-0.33 **Cr** 1.00-1.35 **Cu** 0.35 max **Mn** 0.60-0.80 **Mo** 0.35-0.55 **Ni** 1.85-2.25 **P** 0.025 max **S** 0.025 max **Si** 0.40-0.70	**AMS** 6407
K33125	Alloy Steel	**C** 0.31 max **Cr** 0.50 max **Mn** 0.70 max **Mo** 0.20-0.70 **Ni** 3.00 min **P** 0.015 max **S** 0.018 max **Si** 0.15-0.30 **V** 0.05-0.15	**ASTM** A469 (5)
K33370	Alloy Steel	**C** 0.31-0.35 **Cr** 0.65-0.90 **Mn** 0.60-0.90 **Mo** 0.30-0.50 **Ni** 2.75 min **O** 25 ppm max **P** 0.005 max **S** 0.005 max **Si** 0.20-0.30	**MIL SPEC** MIL-A-46173
K33517	Alloy Steel	**C** 0.33-0.38 **Cr** 0.65-0.90 **Cu** 0.35 max **Mn** 0.60-0.90 **Mo** 0.30-0.40 **Ni** 1.65-2.00 **P** 0.010 max **S** 0.010 max **Si** 0.40-0.60 **V** 0.17-0.23	**AMS** 6433; 6434; 6435
K33585	Alloy Steel	**C** 0.35 max **Cr** 3.00-3.60 **Mn** 0.50-0.90 **Mo** 0.30-0.50 **Ni** 0.50-1.00 **P** 0.040 max **S** 0.040 max **Si** 0.15-0.45 **V** 0.05-0.15	**ASTM** A336 (F32)
K34035	Alloy Steel	**C** 0.40 max **Cr** 0.80-2.00 **Mn** 0.90 max **Mo** 0.30-0.50 **Ni** 2.25-3.25 **P** 0.015 max **S** 0.015 max **Si** 0.35 max **V** 0.20 max	**ASTM** A723 (2)
K34378	Alloy Steel	**C** 0.41-0.46 **Cr** 1.90-2.25 **Cu** 0.35 max **Mn** 0.75-1.00 **Mo** 0.45-0.60 **Ni** 0.25 max **P** 0.015 max **S** 0.015 max **Si** 1.40-1.75 **V** 0.03-0.08	**AMS** 6406

The chemical compositions listed are for identification purposes and should not be used in lieu of the cross referenced specifications.

UNIFIED NUMBER	DESCRIPTION	CHEMICAL COMPOSITION	CROSS REFERENCE SPECIFICATIONS
K41245	Alloy Steel	**C** 0.12 max **Cr** 4.00-6.00 **Mn** 0.30-0.60 **Mo** 0.45-0.65 **P** 0.03 max **S** 0.03 max **Si** 0.50 max	**ASTM** A213 (T5c); A335 (P5c)
K41370	Alloy Steel Electrode (EF5)	**C** 0.10-0.17 **Cr** 0.25-0.50 **Cu** 0.50 max **Mn** 1.70-2.20 **Mo** 0.45-0.65 **Ni** 2.30-2.80 **P** 0.010 max **S** 0.010 max **Si** 0.20 max **Other** total other elements (except iron) 0.50 max	**ASME** SFA5.23 (EF5) **AWS** A5.23 (EF5)
K41545	Alloy Steel	**C** 0.15 max **Cr** 4.00-6.00 **Mn** 0.30-0.60 **Mo** 0.45-0.65 **P** 0.030 max **S** 0.030 max **Si** 0.50 max	**ASTM** A199 (T5); A200 (T5); A213 (T5); A234 (WP5); A335 (P5); A336 (F5); A369 (FP5); A387 (5)
K41583	Alloy Steel	**Al** 0.01-0.16 **C** 0.15 max **Mn** 0.25-0.66 **Mo** 0.17-0.38 **N** 0.025 max **Ni** 4.65-5.35 **P** 0.035 max **S** 0.035 max **Si** 0.18-0.37	**ASTM** A645/A645M
K42247	Alloy Steel	**C** 0.22 max **Cr** 1.44-2.06 **Mn** 0.15-0.45 **Mo** 0.35-0.65 **Ni** 2.68-3.97 **P** 0.025 max **S** 0.025 max **Si** 0.37 max **V** 0.05 max	**ASTM** A707 (L8)
K42338	Alloy Steel	**C** 0.23 max **Cr** 1.44-2.06 **Mn** 0.40 max **Mo** 0.41-0.64 **Ni** 2.53-3.32 **P** 0.035 max **S** 0.18-0.37 **V** 0.03 max	**ASTM** A543/A543M (B,C)
K42339	Alloy Steel	**C** 0.23 max **Cr** 1.44-2.06 **Mn** 0.40 max **Mo** 0.41-0.64 **Ni** 2.53-3.32 **P** 0.020 max **S** 0.020 max **V** 0.03 max	**ASTM** A543/A543M(B)
K42343	Alloy Steel	**C** 0.23 max **Cr** 1.25-2.00 **Mn** 0.20-0.40 **Mo** 0.40-0.60 **Ni** 2.75-3.90 **P** 0.035 max **S** 0.040 max **Si** 0.30 max **V** 0.03 max	**ASTM** A541 (7, 7A, 7B)
K42348	Alloy Steel	**C** 0.23 max **Cr** 1.25-2.00 **Mn** 0.20-0.40 **Mo** 0.40-0.60 **Ni** 2.75-3.90 **P** 0.035 max **S** 0.040 max **Si** 0.30 max **V** 0.08 max	**ASTM** A541 (8, 8A)
K42365	Alloy Steel	**C** 0.23 max **Cr** 1.50-2.00 **Mn** 0.20-0.40 **Mo** 0.40-0.60 **Ni** 2.75-3.90 **P** 0.020 max **S** 0.020 max **Si** 0.30 max **V** 0.08 max	**ASTM** A508 (5, 5a)
K42544	Alloy Steel	**C** 0.25 max **Cr** 4.0-6.0 **Mn** 0.60 max **Mo** 0.44-0.65 **Ni** 0.50 max **P** 0.040 max **S** 0.030 max **Si** 0.50 max	**ASTM** A182 (F5a); A336 (F5a)
K42570	Alloy Steel	**C** 0.23-0.27 **Cr** 1.25-1.75 **Mn** 0.70-1.00 **Mo** 0.20-0.30 **Ni** 3.25-3.75 **P** 0.020 max **S** 0.020 **Si** 0.20-0.35	
K42598	Alloy Steel	**C** 0.23-0.28 **Cb** 0.03-0.07 **Cr** 1.40-1.65 **Mn** 0.20 max **Mo** 0.8-1.0 **Ni** 2.75-3.25 **P** 0.01 max **S** 0.01 max **Si** 0.10 max	**ASTM** A579 (11)
K42885	Alloy Steel	**C** 0.28 max **Cr** 1.25-2.00 **Mn** 0.60 max **Mo** 0.30-0.60 **Ni** 3.25-4.00 **P** 0.015 max **S** 0.018 max **Si** 0.15-0.30 **V** 0.05-0.15	**ASTM** A469 (6, 7, 8); A470 (5, 6, 7)
K43170	Alloy Steel	**C** 0.29-0.33 **Mn** 0.70-1.00 **Mo** 0.20-0.30 **Ni** 3.25-3.75 **P** 0.020 max **S** 0.020 max **Si** 0.20-0.35	
K44045	Alloy Steel	**C** 0.40 max **Cr** 0.80-2.00 **Mn** 0.90 max **Mo** 0.40-0.80 **Ni** 3.25-4.50 **P** 0.015 max **S** 0.015 max **Si** 0.35 max **V** 0.20 max	**ASTM** A723 (3)
K44210	Alloy Steel replaced by K44220		

The chemical compositions listed are for identification purposes and should not be used in lieu of the cross referenced specifications.

MISCELLANEOUS STEELS AND FERROUS ALLOYS

UNIFIED NUMBER	DESCRIPTION	CHEMICAL COMPOSITION	CROSS REFERENCE SPECIFICATIONS
K44220	Alloy Steel (300M)	C 0.38-0.46 Cr 0.70-0.95 Mn 0.60-0.90 Mo 0.30-0.65 Ni 1.65-2.00 P 0.010 max S 0.010 max Si 1.45-1.80 V 0.05 min	AMS 6419 ASTM A579 (32) MIL SPEC MIL-S-8844 (3)
K44315	Alloy Steel (300M) combined with K44220		
K44414	Alloy Steel	C 0.11-0.17 Cr 1.00-1.40 Cu 0.35 max Mn 0.40-0.70 Mo 0.08-0.15 Ni 3.00-3.50 P 0.025 max S 0.025 max Si 0.15-0.35	AMS 6263; 6264
K44910	Alloy Steel	C 0.07-0.13 Cr 1.25-1.75 Cu 0.35 max Mn 0.40-0.70 Mo 0.06 max Ni 3.25-3.75 P 0.025 max S 0.025 max Si 0.20-0.35	AMS 6250
K51210	Alloy Steel, (Krupp)	C 0.10-0.15 Cr 1.35-1.75 Mn 0.45-0.65 Ni 3.75-4.25 Si 0.15-0.35	ASTM A534 (Krupp)
K51255	Alloy Steel (HY30)	C 0.12 max Cr 0.40-0.70 Mn 0.60-0.90 Mo 0.30-0.65 Ni 4.75-5.25 P 0.010 max S 0.010 max Si 0.20-0.35 V 0.05-0.10	ASTM A579 (12) MIL SPEC MIL-S-24371; MIL-S-24512
K51545	Alloy Steel	C 0.15 max Cr 4.00-6.00 Mn 0.30-0.60 Mo 0.45-0.65 P 0.03 max S 0.03 max Si 1.00-2.00	ASTM A213 (T5b); A335 (P5b)
K52355	Alloy Steel	Al 1.10-1.40 C 0.21-0.26 Cr 1.00-1.25 Cu 0.35 max Mn 0.50-0.70 Mo 0.20-0.30 Ni 3.25-3.75 P 0.025 max S 0.025 max Si 0.20-0.40	AMS 6475
K52440	Nitriding Steel	Al 0.95-1.30 C 0.22-0.27 Cr 1.00-1.35 Mn 0.50-0.70 Mo 0.20-0.30 Ni 3.25-3.75 P 0.035 max S 0.040 max Si 0.15-0.35	ASTM A355 (C)
K61595	Alloy Steel replaced by S50300		
K63005	Valve Steel (CNS) replaced by S63005		
K63007	Valve Steel (21-55 N) replaced by S63007		
K63008	Valve Steel (21-4 N) replaced by S63008		
K63011	Valve Steel (746) replaced by S63011		
K63012	Valve Steel (21-2 N) replaced by S63012		
K63013	Valve Steel (Gaman H) replaced by S63013		
K63014	Valve Steel (10) replaced by S63014		
K63015	Valve Steel (10 N) replaced by S63015		
K63016	Valve Steel (21-12) replaced by S63016		
K63017	Valve Steel (21-12 N) replaced by S63017		

The chemical compositions listed are for identification purposes and should not be used in lieu of the cross referenced specifications.

UNIFIED NUMBER	DESCRIPTION	CHEMICAL COMPOSITION	CROSS REFERENCE SPECIFICATIONS
K63198	Iron Base Superalloy (19-9-DL) replaced by S63198		
K63199	Iron Base Superalloy (19-9-DX or 19-9-W-Mo) replaced by S63199		
K64005	Valve Steel (2) replaced by S64005		
K64006	Valve Steel (F) replaced by S64006		
K64152	High-Strength Alloy Steel (M152) replaced by S64152		
K64299	Iron Base Superalloy (29-9) replaced by S64299		
K65006	Valve Steel (XB) replaced by S65006		
K65007	Valve Steel (1) replaced by S65007		
K65150	Iron Base Superalloy (Pyromet X-15) replaced by S65150		
K65770	Iron Base Superalloy (AFC77) replaced by S65770		
K66009	Valve Steel (TPA) replaced by S66009		
K66220	Iron Base Superalloy (Discaloy) replaced by S66220		
K66286	Iron Base Superalloy (A286) replaced by S66286		
K66545	Iron Base Superalloy (W545) replaced by S66545		
K66979	Iron-Nickel Base Superalloy (D-979) replaced by N09979		
K70640	Iron, Chromium, Nickel, Titanium Alloy replaced by N09902		
K71040	Steel (Pyrowear Alloy 53)	**C** 0.07-0.13 **Cr** 0.75-1.25 **Cu** 1.80-2.30 **Mn** 0.25-0.50 **Mo** 3.00-3.50 **Ni** 1.60-2.40 **P** 0.015 max **S** 0.010 max **Si** 0.60-1.20 **V** 0.05-0.15	
K71340	Alloy Steel	**C** 0.13 max **Mn** 0.90 max **Ni** 7.40-8.60 **P** 0.035 max **S** 0.040 max **Si** 0.13-0.32	**ASTM** A522 (II); A553 (II)
K71350	Cr-Ni-Mo Alloy Steel	**Al** 0.03-0.12 **C** 0.10-0.16 **Cr** 0.90-1.20 **Cu** 0.35 max **Mn** 0.40-0.70 **Mo** 4.00-5.00 **Ni** 2.75-3.25 **P** 0.010 max **S** 0.010 max **Si** 0.40-0.60 **V** 0.25-0.50	**AMS** 6256
K74015	Alloy Steel replaced by T20811		
K81340	Alloy Steel	**C** 0.13 max **Mn** 0.90 max **Ni** 8.40-9.60 **P** 0.045 max **S** 0.045 max **Si** 0.13-0.32	**ASTM** A333 (8); A334 (8); A353M; A522 (I); A553M (I)
K81590	Alloy Steel replaced by S50400		

The chemical compositions listed are for identification purposes and should not be used in lieu of the cross referenced specifications.

MISCELLANEOUS STEELS AND FERROUS ALLOYS

UNIFIED NUMBER	DESCRIPTION	CHEMICAL COMPOSITION	CROSS REFERENCE SPECIFICATIONS
K88165	Bearing Steel replaced by T11350		
K90941	Alloy Steel	C 0.15 max Cr 8.00-10.00 Mn 0.30-0.60 Mo 0.90-1.10 P 0.030 max S 0.030 max Si 0.50-1.00	ASTM A182 (F9); A234 (WP9); A369 (FP9)
K90987	High-Speed Steel Welding Rod (RFE5-B)	C 0.5-0.9 Cr 3.0-5.0 Mn 0.50 max Mo 5.0-9.5 Si 0.50 max V 0.8-1.3 W 1.0-2.5 Other total other elements (except iron) 1.0 max	ASME SFA5.13 (RFe5-B) AWS A5.13 (RFe5-B)
K91094	Alloy Steel, Superstrength	C 0.42-0.47 Co 3.5-4.5 Cr 0.20-0.35 Mn 0.10-0.35 Mo 0.20-0.35 Ni 7.0-8.5 P 0.01 max S 0.01 max Si 0.10 max V 0.06-0.12	ASTM A579 (83)
K91122	Alloy Steel, Superstrength (HP9-4-25)	C 0.24-0.30 Co 3.50-4.50 Cr 0.35-0.60 Mn 0.10-0.35 Mo 0.35-0.60 Ni 7.00-9.00 P 0.010 max S 0.010 max Si 0.10 max V 0.06-0.12	AMS 6546 ASTM A579 (81)
K91151	Alloy Steel replaced by S41000		
K91161	Alloy Steel, Superstrength replaced by S41001		
K91209	Manganese Steel, Non-magnetic	C 1.20-1.50 Cr 0.60 max Mn 12.00-15.00 Mo 0.10 max Ni 0.75 max P 0.08 max S 0.04 max Si 0.55 max	MIL SPEC MIL-A-13259
K91283	Alloy Steel, Superstrength (HP9-4-30)	C 0.28-0.34 Co 4.0-5.0 Cr 0.90-1.10 Mn 0.10-0.35 Mo 0.90-1.10 Ni 7.0-8.5 P 0.01 max S 0.01 max Si 0.10 max V 0.06-0.12	AMS 6524; 6526 ASTM A579 (82)
K91308	High-Speed Steel Welding Rod (RFe5-A)	C 0.7-1.0 Cr 3.0-5.0 Fe rem Mn 0.50 max Mo 4.0-6.0 Si 0.50 max V 1.0-2.5 W 5.0-7.0 Other total other elements (except iron) 1.0 max	ASME SFA5.13 (RFe5-A) AWS A5.13 (RFe5-A)
* K91313	Alloy Steel, Superstrength	C 0.29-0.34 Co 4.25-4.75 Cr 0.90-1.10 Mn 0.10-0.35 Mo 0.90-1.10 Ni 7.00-8.00 P 0.010 max S 0.010 max Si 0.10 max V 0.06-0.12	
K91342	Alloy Steel, Superstrength replaced by S42201		
K91352	Alloy Steel, Superstrength	Al 0.05 max C 0.20-0.25 Cr 11.00-12.50 Mn 0.50-1.00 Mo 0.90-1.25 Ni 0.50-1.00 P 0.025 max S 0.025 max Si 0.20-0.50 Sn 0.04 max Ti 0.05 max V 0.20-0.30 W 0.90-1.25	ASTM A437 (B4B, B4C)
K91401	Alloy Steel	C 0.16-0.23 Co 4.15-5.10 Cr 0.61-0.89 Mn 0.20-0.40 Mo 0.86-1.14 Ni 8.40-9.60 P 0.010 max S 0.010 max Si 0.12 max V 0.04-0.14	ASTM A605/A605M
K91456	Steel, High Expansion	C 0.55-0.65 Mn 5.00-6.00 Ni 8.50-10.50 P 0.040 max S 0.030 max Si 1.00 max	AMS 5623; 5625
K91461	Nickel-Cobalt Steel Welding Wire	C 0.14-0.17 Co 3.50-4.00 Cr 0.90-1.05 Cu 0.10 max H 0.0010 max Mn 0.40-0.55 Mo 0.40-0.50 N 0.0080 max Ni 9.75-10.25 O 0.0050 max P 0.008 max S 0.008 max Si 0.15-0.25 V 0.06-0.10	AMS 6468

*Boxed entries are no longer active and are retained for reference purposes only.

The chemical compositions listed are for identification purposes and should not be used in lieu of the cross referenced specifications.

UNIFIED NUMBER	DESCRIPTION	CHEMICAL COMPOSITION	CROSS REFERENCE SPECIFICATIONS
K91470	Iron, Electrical Heating Element Alloy	Al 2.75-3.75 Cr 12-15	ASTM B603 (IV)
K91472	Alloy Steel (HP 9-4-20)	C 0.17-0.23 Co 4.25-4.75 Cr 0.65-0.85 Cu 0.35 max Mn 0.20-0.40 Mo 0.90-1.10 Ni 8.50-9.50 P 0.010 max S 0.010 max Si 0.20 max V 0.06-0.12	AMS 6523; 6525
K91505	Steel, High Expansion	C 0.50-0.60 Cr 3.00-5.00 Cu 0.50 max Mn 3.50-5.50 Mo 0.50 max Ni 11.0-14.0 P 0.040 max S 0.030 max Si 0.50 max	AMS 5624
K91555	Alloy Steel, Non-magnetic	C 0.40-0.75 Cr 3.50-6.00 Mn 6.00-10.00 Ni 6.00-10.00 P 0.05 max S 0.045 max Si 0.20-0.65	ASTM A289 (A)
K91670	Iron, Electrical Heating Element Alloy	Al 3.75-4.75 Cr 13-16	ASTM B603 (III)
K91800	Iron, Chromium Sealing Alloy	C 0.08 max Cr 18 nom Fe rem Mn 1.00 max Ni 0.50 max P 0.04 max S 0.03 max Si 0.75 max Other Ti 5×C-0.60	ASTM F256 (I)
K91870	Iron, Electrical Heating Element Alloy	Al 4.75-5.75 Cr 14-17	ASTM B603 (IIA)
K91890	Alloy Steel	Al 0.40 max C 0.03 max Cr 4.50-5.50 Mn 0.10 max Mo 2.75-3.25 Ni 11.50-12.50 P 0.010 max S 0.010 max Si 0.10 max Ti 0.20-0.35	ASTM A590
K91930	Alloy Steel, Superstrength	Al 0.25-0.40 C 0.03 max Cr 4.75-5.25 Mn 0.10 max Mo 2.75-3.25 Ni 11.5-12.5 P 0.01 max S 0.01 max Si 0.12 max Ti 0.05-0.15	ASTM A579 (74)
K91940	Alloy Steel, Superstrength	Al 0.35-0.50 C 0.03 max Cr 4.75-5.25 Mn 0.10 max Mo 2.75-3.25 Ni 11.5-12.5 P 0.01 max S 0.01 max Si 0.12 max Ti 0.10-0.25	ASTM A579 (75)
K91955	Alloy Steel, Non-magnetic	C 0.40-0.60 Cr 3.50-6.00 Mn 16.00-20.00 Ni 2.00 max P 0.08 max S 0.025 max Si 0.20-0.65	ASTM A289 (B)
K91970	Alloy Steel	Al 0.025 max C 0.10-0.14 Co 7.50-8.50 Cr 1.80-2.20 Mn 0.05-0.25 Mo 0.90-1.10 N 0.0075 max Ni 9.50-10.50 O 0.0025 max P 0.010 max S 0.006 max Si 0.10 max Ti 0.015 max	AMS 6543
K91971	Nickel-Cobalt Steel Welding Wire	Al 0.01-0.03 C 0.10-0.14 Co 7.50-8.50 Cr 1.80-2.20 H 0.0003 max Mn 0.07-0.17 Mo 0.90-1.10 N 0.005 max Ni 9.50-10.50 O 0.0025 max P 0.006 max S 0.006 max Si 0.15-0.25 Ti 0.02 max V 0.04-0.09	AMS 6465
K92100	Iron, Thermostat Alloy	Cr 2 Fe 79 Ni 19	ASTM B388
K92350	Iron, Thermostat Alloy	Cr 8.5 Fe 66.5 Ni 25	ASTM B388
K92400	Iron, Electrical Heating Element Alloy	Al 4.00-5.25 Cr 20-24	ASTM B603 (IIB)
K92500	Iron, Electrical Heating Element Alloy	Al 5.00-6.00 Cr 20-24	ASTM B603 (I)
K92510	Iron, Thermostat Alloy	Cr 3 Fe 75 Ni 22	ASTM B388

The chemical compositions listed are for identification purposes and should not be used in lieu of the cross referenced specifications.

UNIFIED NUMBER	DESCRIPTION	CHEMICAL COMPOSITION	CROSS REFERENCE SPECIFICATIONS
K92571	Alloy Steel (AF 1410)	**Al** 0.015 max **C** 0.13-0.17 **Co** 13.5-14.5 **Cr** 1.80-2.20 **Mn** 0.10 max **Mo** 0.90-1.10 **N** 0.0015 max **Ni** 9.50-10.50 **O** 0.0015 max **P** 0.008 max **S** 0.005 max **Si** 0.10 max **Ti** 0.015 max	**AMS** 6522; 6527; 6544
K92801	Iron, Chromium Sealing Alloy	**C** 0.12 max **Cr** 28 nom **Fe** rem **Mn** 1.00 max **N** 0.20 max **Ni** 0.50 max **P** 0.04 max **S** 0.03 max **Si** 0.75 max	**ASTM** F256 (II)
K92810	Alloy Steel, Maraging	**Al** 0.05-0.15 **C** 0.03 max **Co** 7.0-8.5 **Mn** 0.10 max **Mo** 4.0-4.5 **Ni** 17.0-19.0 **P** 0.010 max **S** 0.010 max **Si** 0.10 max **Ti** 0.10-0.25	**ASTM** A538/A538M (A) **MIL SPEC** MIL-S-46850; MIL-S-47139
K92820	Alloy Steel, Superstrength	**Al** 0.05-0.15 **C** 0.03 max **Co** 8.0-9.0 **Mn** 0.10 max **Mo** 3.0-3.5 **Ni** 17.0-19.0 **P** 0.01 max **S** 0.01 max **Si** 0.10 max **Ti** 0.15-0.25	**ASTM** A579 (71) **MIL SPEC** MIL-S-46850 **SAE** J1099 (A538A)
K92850	Iron, Thermostat Alloy	**Al** 5 **Fe** 71.5 **Mn** 9.5 **Ni** 14	**ASTM** B388
K92890	Alloy Steel, Maraging	**Al** 0.05-0.15 **C** 0.03 max **Co** 7.0-8.5 **Mn** 0.10 max **Mo** 4.6-5.1 **Ni** 17.0-19.0 **P** 0.010 max **S** 0.010 max **Si** 0.10 max **Ti** 0.30-0.50	**AMS** 6501; 6512; 6520 **ASTM** A538/A538M (B) **MIL SPEC** MIL-S-46850; MIL-S-47139 **SAE** J1099 (A538B)
K92940	Alloy Steel, Superstrength	**Al** 0.05-0.15 **C** 0.03 max **Co** 7.5-8.5 **Mn** 0.10 max **Mo** 4.6-5.2 **Ni** 17.0-19.0 **P** 0.01 max **S** 0.01 max **Si** 0.10 max **Ti** 0.30-0.50	**ASTM** A579 (72)
K93120	Alloy Steel, Maraging	**Al** 0.05-0.15 **C** 0.03 max **Co** 8.0-9.5 **Mn** 0.10 max **Mo** 4.6-5.2 **Ni** 18.0-19.0 **P** 0.010 max **S** 0.010 max **Si** 0.10 max **Ti** 0.55-0.80	**AMS** 6514; 6521 **ASTM** A538/A538M (C) **MIL SPEC** MIL-S-46850; MIL-S-47139 **SAE** J1099 (A538C)
K93130	Nickel-Cobalt Steel Welding Wire	**Al** 0.05-0.15 **C** 0.010 max **Co** 8.00-9.00 **H** 0.0025 max **Mn** 0.10 max **Mo** 4.50-6.00 **N** 0.005 max **Ni** 18.00-19.00 **O** 0.0025 max **P** 0.010 max **S** 0.010 max **Si** 0.10 max **Ti** 0.65-0.80	**AMS** 6463
K93160	Alloy Steel, Superstrength	**Al** 0.05-0.15 **C** 0.03 max **Co** 8.5-9.5 **Mn** 0.10 max **Mo** 4.6-5.2 **Ni** 18.0-19.0 **P** 0.01 max **S** 0.01 max **Si** 0.10 max **Ti** 0.50-0.80	**ASTM** A579 (73)
K93600	Iron, Thermostat Alloy (Invar)	**Fe** 64 **Ni** 36	**ASTM** B388
K93601	Nickel Steel, 36% Ni	**C** 0.10 max **Co** 0.50 max **Cr** 0.50 max **Mn** 0.50 max **Mo** 0.50 max **Ni** 35.0-37.0 **P** 0.025 max **S** 0.025 max **Si** 0.45 max	**ASTM** A658/A658M
K93602	Nickel Steel, Free Cutting, 36% Ni	**C** 0.15 max **Mn** 0.50 max **Ni** 35.0-36.5 **Se** 0.15-0.25 **Si** 0.35 max	**MIL SPEC** MIL-S-16598
K93800	Iron, Thermostat Alloy	**Co** 7 **Fe** 62 **Ni** 31	**ASTM** B388
K94000	Iron, Thermostat Alloy	**Fe** 60 **Ni** 40	**ASTM** B388
K94100	Iron, Nickel Sealing Alloy	**Al** 0.10 max **C** 0.05 max **Cr** 0.25 max **Fe** rem **Mn** 0.80 max **Ni** 41 nom **P** 0.025 max **S** 0.025 max **Si** 0.30 max	**ASTM** F30 (42)
K94101	Iron, Nickel Sealing Alloy (DUMET)	**C** 0.10 max **Fe** rem **Mn** 0.75-1.25 **Ni** 41-43 **P** 0.02 max **S** 0.02 max **Si** 0.30 max	**ASTM** F29
K94200	Iron, Thermostat Alloy	**Fe** 58 **Ni** 42	**ASTM** B388
K94500	Iron, Thermostat Alloy	**Fe** 55 **Ni** 45	**ASTM** B388

The chemical compositions listed are for identification purposes and should not be used in lieu of the cross referenced specifications.

UNIFIED NUMBER	DESCRIPTION	CHEMICAL COMPOSITION	CROSS REFERENCE SPECIFICATIONS
K94600	Iron, Nickel Sealing Alloy	**Al** 0.10 max **C** 0.05 max **Cr** 0.25 max **Fe** rem **Mn** 0.80 max **Ni** 46 nom **P** 0.025 max **S** 0.025 max **Si** 0.30 max	**ASTM** F30 (46)
K94610	Iron, Nickel-Cobalt Sealing Alloy (KOVAR)	**Al** 0.10 max **C** 0.04 max **Co** 17 nom **Cr** 0.20 max **Cu** 0.20 max **Fe** 53 nom **Mg** 0.10 max **Mn** 0.50 max **Mo** 0.20 max **Ni** 29 nom **Si** 0.20 max **Ti** 0.10 max **Zr** 0.10 max **Other** Al+Mg+Zr+Ti 0.20 max	**AMS** 7726; 7727; 7728 **ASTM** F15
K94760	Iron, Nickel-Chromium Sealing Alloy	**Al** 0.20 max **C** 0.07 max **Cr** 5.6 nom **Fe** rem **Mn** 0.25 max **Ni** 42 nom **P** 0.025 max **S** 0.025 max **Si** 0.30 max	**ASTM** F31
K94800	Iron, Nickel Sealing Alloy	**Al** 0.10 max **C** 0.05 max **Cr** 0.25 max **Fe** rem **Mn** 0.80 max **Ni** 48 nom **P** 0.025 max **S** 0.025 max **Si** 0.30 max	**ASTM** F30 (48)
K95000	Iron, Nickel Thermostat Alloy	**Fe** 50 **Ni** 50	**AMS** 7717; 7718; 7719 **ASTM** B388
K95050	Iron, Nickel Sealing Alloy replaced by N14052		
K95100	Iron, Cobalt, Vanadium Ferromagnetic Alloy replaced by R30005		

The chemical compositions listed are for identification purposes and should not be used in lieu of the cross referenced specifications.

UNS NUMBERS ASSIGNED TO DATE

With Description of Each Material Covered and References To Documents In Which The Same or Similar Materials are Described

Lxxxxx Number Series
Low Melting Metals and Alloys

LOW MELTING METALS AND ALLOYS

UNIFIED NUMBER	DESCRIPTION	CHEMICAL COMPOSITION	CROSS REFERENCE SPECIFICATIONS
L01900	Cadmium Metal	**Ag** 0.01 max **As** 0.003 max **Cd** 99.90 min **Cu** 0.015 max **Pb** 0.025 max **Sb** 0.001 max **Sn** 0.01 max **Zn** 0.035 max **Other** Tl 0.003 max	**ASTM** B440 **FED** QQ-A-671
L01950	Cadmium Metal	**Cd** 99.95 min **Zn** 0.020 max	**ASTM** B440
L05020	Lead-Tin Solder replaced by L54210		
L05021	Lead-Tin Solder replaced by L54211		
L05025	Lead-Silver Solder replaced by L50150		
L05050	Lead-Tin Solder replaced by L54320		
L05051	Lead-Tin Solder replaced by L54321		
L05055	Lead-Silver Solder replaced by L50180		
L05070	Lead Alloy replaced by L53131		
L05090	Lead Die Casting Alloy replaced by L53340		
L05100	Lead-Tin Solder replaced by L54520		
L05110	Lead-Tin Solder replaced by L50151		
L05120	White Metal Bearing Alloy (Babbitt Metal)	**Al** 0.005 max **As** 0.20 max **Cu** 0.50 max **Fe** 0.10 max **Pb** 83.0-88.0 **Sb** 8.0-10.0 **Sn** 4.0-6.0 **Zn** 0.005 max **Other** total 0.75 max	**FED** QQ-T-390
L05150	Lead-Tin Solder replaced by L54560		
L05153	White Metal Bearing Alloy (Babbitt Metal) replaced by L53346		
L05155	White Metal Bearing Alloy (Babbitt Metal) replaced by L53620		
L05180	Lead Die Casting Alloy replaced by L53560		
L05188	White Metal Bearing Alloy (Babbitt Metal) replaced by L53565		
L05201	Lead-Tin Solder replaced by L54711		
L05202	Lead-Tin Solder replaced by L54712		
L05237	White Metal Bearing Alloy (Babbitt Metal) replaced by L53585		
L05250	Lead-Tin Solder replaced by L54720		
L05251	Lead-Tin Solder replaced by L54721		

The chemical compositions listed are for identification purposes and should not be used in lieu of the cross referenced specifications.

UNIFIED NUMBER	DESCRIPTION	CHEMICAL COMPOSITION	CROSS REFERENCE SPECIFICATIONS
L05252	Lead-Tin Solder replaced by L54722		
L05300	Lead-Tin Solder replaced by L54820		
L05301	Lead-Tin Solder replaced by L54821		
L05302	Lead-Tin Solder replaced by L54822		
L05350	Lead-Tin Solder replaced by L54850		
L05351	Lead-Tin Solder replaced by L54851		
L05352	Lead Tin Solder replaced by L54852		
L05400	Lead-Tin Solder replaced by L54915		
L05401	Lead-Tin Solder replaced by L54916		
L05402	Lead-Tin Solder replaced by L54918		
L05450	Lead-Tin Solder replaced by L54950		
L05451	Lead-Tin Solder replaced by L54951		
L05500	Lead-Tin (50-50) Solder replaced by L55030		
L05501	Lead-Tin (50-50) Solder replaced by L55031		
L06990	Lithium Metal	**Fe** 0.005 max **K** 0.025 max **Li** rem **Na** 0.800 max **Other** Chlorides 0.04 max	**ASTM** B357
L13002	Pig Tin	**As** 0.0005 max **Bi** 0.001 max **Cd** 0.001 max **Cu** 0.002 max **Fe** 0.005 max **Pb** 0.010 max **S** 0.002 max **Sb** 0.008 max **Sn** 99.98 max **Zn** 0.001 max **Other** Ni + Co 0.005 max	**ASTM** B339 (AAA)
L13004	Pig Tin	**As** 0.01 max **Bi** 0.01 max **Cd** 0.001 max **Cu** 0.02 max **Fe** 0.01 max **Pb** 0.02 max **S** 0.01 max **Sb** 0.02 max **Sn** 99.95 max **Zn** 0.005 max **Other** Ni + Co 0.01 max	**ASTM** B339 (AA)
L13006	Pig Tin	**As** 0.05 max **Sn** 99.80 min	**ASTM** B339 (B)
L13007	Pig Tin	**Ag** 0.01 max **Al** 0.001 max **As** 0.04 max **Bi** 0.015 max **Cd** 0.001 max **Cu** 0.03 max **Fe** 0.015 max **Pb** 0.05 max **S** 0.003 max **Sb** 0.04 max **Sn** 99.80 min **Zn** 0.001 max **Other** Ni+Co 0.015 max	**FED** QQ-T-371
L13008	Pig Tin	**As** 0.05 max **Bi** 0.015 max **Cd** 0.001 max **Cu** 0.04 max **Fe** 0.015 max **Pb** 0.05 max **S** 0.01 max **Sb** 0.04 max **Sn** 99.80 min **Zn** 0.005 max **Other** Ni + Co 0.01 max	**ASTM** B339 (A)
L13010	Pig Tin	**Sn** 99.65 min	**ASTM** B339 (C)
L13012	Pig Tin	**Sn** 99.50 min	**ASTM** B339 (D)

The chemical compositions listed are for identification purposes and should not be used in lieu of the cross referenced specifications.

UNIFIED NUMBER	DESCRIPTION	CHEMICAL COMPOSITION	CROSS REFERENCE SPECIFICATIONS
L13014	Pig Tin	**Sn** 99.00 min	**ASTM** B339 (E)
L13600	Tin-Lead Solder	**Al** 0.005 max **As** 0.03 max **Bi** 0.25 max **Co** 0.08 max **Fe** 0.02 max **Pb** 40 nom **Sb** 0.12 max **Sn** 60 nom **Zn** 0.005 max	**ASTM** B32 (60A); B486 (60A) **FED** QQ-S-571 (Sn60)
L13601	Tin-Lead Solder	**Al** 0.005 max **As** 0.03 max **Bi** 0.25 max **Cu** 0.08 max **Fe** 0.02 max **Pb** 40 nom **Sb** 0.20-0.50 **Sn** 60 nom **Zn** 0.005 max	**ASTM** B32 (60B); B486 (60B)
L13630	Tin-Lead Solder	**Al** 0.005 max **As** 0.03 max **Bi** 0.25 max **Cu** 0.08 max **Fe** 0.02 max **Pb** 37 nom **Sb** 0.12 max **Sn** 63 nom **Zn** 0.005 max	**AMS** 4751 **ASTM** B32 (grade 63A); B486 (grade 63A) **FED** QQ-S-571 (Sn63)
L13631	Tin-Lead Solder	**Al** 0.005 max **As** 0.03 max **Bi** 0.25 max **Cu** 0.08 max **Fe** 0.02 max **Pb** 37 nom **Sb** 0.20-0.50 **Sn** 63 nom **Zn** 0.005 max	**ASTM** B32 (63B); B486 (63B)
L13650	Tin Die Casting Alloy	**Al** 0.01 max **As** 0.15 max **Cu** 1.5-2.5 **Fe** 0.08 max **Pb** 17-19 **Sb** 14-16 **Sn** 64-66 **Zn** 0.01 max	**ASTM** B102 (PY1815A)
L13700	Tin-Lead Solder	**Al** 0.005 max **As** 0.03 max **Bi** 0.25 max **Cu** 0.08 max **Fe** 0.02 max **Pb** 30 nom **Sb** 0.20 nom **Sn** 70 nom **Zn** 0.005 max	**ASTM** B32 (70A); B486 (70A) **FED** QQ-S-571 (Sn70)
L13701	Tin-Lead Solder	**Al** 0.005 max **As** 0.03 max **Bi** 0.25 max **Cu** 0.08 max **Fe** 0.02 max **Pb** 30 nom **Sb** 0.20-0.50 **Sn** 70 nom **Zn** 0.005 max	**ASTM** B32 (70B); B486 (70B)
L13820	Tin Die Casting Alloy	**Al** 0.01 max **As** 0.08 max **Cu** 4-6 **Fe** 0.08 max **Pb** 0.35 max **Sb** 12-14 **Sn** 80-84 **Zn** 0.01 max	**ASTM** B102 (YC135A)
L13840	White Metal Bearing Alloy (Babbitt Metal)	**Al** 0.005 max **As** 0.10 max **Bi** 0.08 max **Cd** 0.05 max **Cu** 7.5-8.5 **Fe** 0.08 max **Pb** 0.35 max **Sb** 7.5-8.5 **Sn** 83.0-85.0 **Zn** 0.005 max	**ASTM** B23 (3) **FED** QQ-T-390
L13870	White Metal Bearing Alloy (Babbitt Metal)	**Al** 0.005 max **As** 0.10 max **Bi** 0.08 max **Cd** 0.05 max **Cu** 5.0-6.5 **Fe** 0.08 max **Pb** 0.50 max **Sb** 6.0-7.5 **Sn** 86.0-89.0 **Zn** 0.005 max	**ASTM** B23 (11)
L13890	White Metal Bearing Alloy (Babbitt Metal)	**Al** 0.005 max **As** 0.10 max **Bi** 0.08 max **Cd** 0.05 max **Cu** 3.0-4.0 **Fe** 0.08 max **Pb** 0.35 max **Sb** 7.0-8.0 **Sn** 88.0-90.0 **Zn** 0.005 max	**ASTM** B23 (2) **FED** QQ-T-390
L13910	White Metal Bearing Alloy (Babbitt Metal)	**Al** 0.005 max **As** 0.10 max **Bi** 0.08 max **Cd** 0.05 max **Cu** 4.0-5.0 **Fe** 0.08 max **Pb** 0.35 max **Sb** 4.0-5.0 **Sn** 90.0-92.0 **Zn** 0.005 max	**AMS** 4800 **ASTM** B23 (1) **FED** QQ-T-390
L13911	Modern Pewter Casting Alloys	**As** 0.05 max **Cu** 0.25-2.0 **Fe** 0.015 max **Pb** 0.05 max **Sb** 6-8 **Sn** 90-93 **Zn** 0.005 max	**ASTM** B560 (1)
L13912	Modern Pewter Sheet Alloy	**As** 0.05 max **Cu** 1.5-3.0 **Fe** 0.015 max **Pb** 0.05 max **Sb** 5-7.5 **Sn** 90-93 **Zn** 0.005 max	**ASTM** B560 (2)
L13913	Tin Die-Casting Alloy	**Al** 0.01 max **As** 0.08 max **Cu** 4-5 **Fe** 0.08 max **Pb** 0.35 max **Sb** 4-5 **Sn** 90-92 **Zn** 0.01 max	**ASTM** B102 (CY44A)

The chemical compositions listed are for identification purposes and should not be used in lieu of the cross referenced specifications.

UNIFIED NUMBER	DESCRIPTION	CHEMICAL COMPOSITION	CROSS REFERENCE SPECIFICATIONS
L13940	Tin-Antimony Solder	**Al** 0.03 max **As** 0.06 max **Cd** 0.03 max **Cu** 0.08 max **Fe** 0.08 max **Pb** 0.20 max **Sb** 4.0-6.0 **Sn** 94 min **Zn** 0.03 max **Other** total 0.03 max	**FED** QQ-S-571 (Sb5)
L13950	Tin-Antimony Solder	**Al** 0.005 max **As** 0.05 max **Bi** 0.15 max **Cu** 0.08 max **Fe** 0.04 max **Pb** 0.20 max **Sb** 4.5-5.5 **Sn** 95 nom **Zn** 0.005 max	**ASTM** B32 (95TA); B486 (95TA)
L13960	Tin-Silver Solder	**Ag** 3.6-4.4 **Al** 0.005 max **As** 0.05 max **Bi** 0.15 max **Cu** 0.08 max **Fe** 0.02 max **Pb** 0.20 max **Sb** 0.20-0.50 **Sn** 96 nom **Zn** 0.005 max	**ASTM** B486 (96TS)
L13961	Tin-Silver Solder	**Ag** 3.6-4.4 **As** 0.05 max **Cd** 0.005 max **Cu** 0.20 max **Pb** 0.10 maX **Sn** rem **Zn** 0.005 max	**FED** QQ-S-571 (Sn96)
L13963	Modern Pewter	**As** 0.05 max **Cu** 1.0-2.0 **Fe** 0.015 max **Pb** 0.05 max **Sb** 1.0-3.0 **Sn** 95-98 **Zn** 0.005 max	**ASTM** B560 (3)
L13965	Tin-Silver Solder	**Ag** 3.3-3.7 **Al** 0.005 max **As** 0.05 max **Bi** 0.15 max **Cu** 0.08 max **Fe** 0.02 max **Pb** 0.20 max **Sb** 0.20-0.50 **Sn** 96.5 nom **Zn** 0.005 max	**ASTM** B32 (96.5 TS)
L50001	Zone Refined Lead	**Pb** 99.9999 min	
L50005	Refined Soft Lead	**Pb** 99.999 min	
L50010	Refined Soft Lead	**Pb** 99.99 min	
L50020	Refined Soft Lead	**Pb** 99.985	
L50025	LME Grade Pure Lead	**Pb** 99.97	
L50035	Refined Soft Lead	**Pb** 99.95 min	
L50040	Refined Soft Lead	**Pb** 99.94 min	
L50042	Corroding Lead	**Ag** 0.0015 max **Bi** 0.050 max **Cu** 0.0015 max **Fe** 0.002 max **Pb** 99.94 min **Zn** 0.001 max **Other** copper + silver Cu + Ag 0.0025 max, arsenic, antimony and tin As + Sb + Sn 0.002 max	**ASTM** B29 (Corroding Lead)
L50045	Common Lead	**Ag** 0.005 max **Bi** 0.050 max **Cu** 0.0015 max **Fe** 0.002 max **Pb** 99.94 min **Zn** 0.001 max **Other** arsenic, antimony and tin together As + Sb + Sn 0.002 max	**ASTM** B29 (Common Lead)
L50050	Grade A Lead	**Pb** 99.90 min **Other** total other elements 0.10 max	**FED** QQ-L-171
L50060	Type III Lead	**Pb** 99.85 min	
L50065	Grade AA Lead, Grade C Lead	**Pb** 99.7 min **Sb** 0.02 max	**FED** QQ-C-40 (Grade AA and Grade C)
L50070	Grade B Remelted Lead	**Bi** 0.025 max (Optional) **Pb** 39.5 min	**FED** QQ-L-201
L50080	Grade B Lead	**Pb** 95.0 min **Other** total other elements 5.0 max	**FED** QQ-L-171
L50090	Chemical B Grade Lead	**Pb** N.S.	
L50101	Cable Sheathing Alloy	**Ag** 0.2 nom **Pb** 99.8 nom	
L50110	Electrowinning Anode Alloy	**Ag** 0.5 nom **Pb** 99.5 nom	
L50113	Solder Alloy	**Ag** 0.5 nom **Pb** 97.0 nom **Sn** 2.5 nom	

The chemical compositions listed are for identification purposes and should not be used in lieu of the cross referenced specifications.

UNIFIED NUMBER	DESCRIPTION	CHEMICAL COMPOSITION	CROSS REFERENCE SPECIFICATIONS
L50115	Electrowinning Anode Alloy	**Ag** 0.75 nom **Pb** 99.25 nom	
L50120	Electrowinning Anode Alloy	**Ag** 1.0 nom **Pb** 99.0 nom	
L50121	Solder Alloy	**Ag** 1.0 nom **Pb** 98. nom **Sn** 1.0 nom	
L50122	Electrowinning Anode Alloy	**Ag** 1.0 nom **As** 1.0 nom **Pb** 98. nom	
L50131	Solder Alloy-Grade 1.5S	**Ag** 1.3-1.7 **Al** 0.005 max **As** 0.02 max **Bi** 0.25 max **Cu** 0.08 max **Fe** 0.02 max **Pb** 97.5 nom **Sn** 0.75-1.25 **Zn** 0.005 max	**AMS** 4756 **ASTM** B32 (1.5S)
L50132	Solder Alloy-Grade Ag1.5	**Ag** 1.3-1.7 **Al** 0.005 max **As** 0.02 max **Bi** 0.25 max **Cd** 0.001 max **Cu** 0.30 max **Fe** 0.02 max **Pb** rem **Sb** 0.40 max **Sn** 0.75-1.25 **Zn** 0.005 max **Other** total of all others 0.08 max	**FED** QQ-S-571 (Ag1.5)
L50134	Solder Alloy-Grade 5S	**Ag** 1.5 nom **Pb** 93.5 nom **Sn** 5.0 nom	
L50140	Cathodic Protection Anode Alloy	**Ag** 2.0 nom **Pb** 98.0 nom	
L50150	Solder Alloy-Grade 2.5S	**Ag** 2.3-2.7 **Al** 0.005 max **As** 0.02 max **Bi** 0.25 max **Cu** 0.08 max **Fe** 0.02 max **Pb** 97.5 nom **Sn** 0.25 max **Zn** 0.005 max	**ASTM** B32 (2.5S)
L50151	Solder Alloy-Grade Ag2.5	**Ag** 2.3-2.7 **Al** 0.005 max **As** 0.02 max **Bi** 0.25 max **Cd** 0.001 max **Cu** 0.30 max **Fe** 0.02 max **Pb** rem **Sb** 0.40 max **Sn** 0.25 max **Zn** 0.005 max **Other** total of all others 0.03 max	**FED** QQ-S-571 (Ag2.5)
L50152	Solder Alloy	**Ag** 2.5 nom **Pb** 95.5 nom **Sn** 2.0 nom	
L50170	Solder Alloy	**Ag** 5.0 nom **Pb** 95.0 nom	
L50172	Solder Alloy	**Ag** 5.0 nom **In** 5.0 nom **Pb** 90.0 nom	
L50180	Solder Alloy-Grade Ag5.5	**Ag** 5.0-6.0 **Al** 0.005 max **As** 0.02 max **Bi** 0.25 max **Cd** 0.001 max **Cu** 0.30 max **Fe** 0.002 max **Pb** rem **Sb** 0.40 max **Sn** 0.25 max **Zn** 0.005 max **Other** total of all others 0.03 max	**AMS** 4755 **FED** QQ-S-571 (Ag5.5)
L50310	Arsenical Lead Cable Sheathing Alloy	**As** 0.15 nom **Bi** 0.10 nom **Pb** 99.6 nom **Sn** 0.10 nom	
L50510	Lead-Barium Alloy	**Pb** 99.9 nom **Other** barium Ba 0.05 nom	
L50520	Lead-Tin-Barium Alloy	**Pb** 99.0 nom **Sn** 1.0 nom **Other** barium Ba 0.05 nom	
L50521	Lead-Tin-Barium Alloy	**Pb** 98.5 nom **Sn** 1.5 nom **Other** barium Ba 0.05 nom	
L50522	Lead-Tin-Barium Alloy	**Pb** 98.0 nom **Sn** 2.0 nom **Other** barium Ba 0.05 nom	
L50530	Lead-Tin-Barium Alloy	**Pb** 98.9 nom **Sn** 1.0 nom **Other** barium Ba 0.10 nom	
L50535	Lead-Tin-Barium Alloy	**Pb** 97.9 nom **Sn** 2.0 nom **Other** barium Ba 0.10 nom	
L50540	Frary Metal	**Pb** 98.8 nom **Other** barium Ba 0.4 nom, calcium Ca 0.8 nom	
L50541	Frary Metal	**Pb** 98.5 nom **Other** barium Ba 1.0 nom, calcium Ca 0.5 nom	

The chemical compositions listed are for identification purposes and should not be used in lieu of the cross referenced specifications.

LOW MELTING METALS AND ALLOYS

UNIFIED NUMBER	DESCRIPTION	CHEMICAL COMPOSITION	CROSS REFERENCE SPECIFICATIONS
L50542	Frary Metal	**Pb** 98.0 nom **Other** barium Ba 1.2 nom, calcium Ca 0.8 nom	
L50543	Frary Metal	**Pb** 97.2 nom **Other** barium Ba 2.0 nom, calcium Ca 0.8 nom	
L50710	Lead-Calcium Alloy	**Pb** 99.9 nom **Other** calcium Ca 0.008 nom	
L50712	Cable Sheathing Alloy	**Pb** 99.9 nom **Other** calcium Ca 0.025 nom	
L50713	Cable Sheathing Alloy	**Pb** 99.7 nom **Sn** 0.3 nom **Other** calcium Ca 0.025 nom	
L50720	Lead-Calcium Alloy	**Pb** 99.9 nom **Other** calcium Ca 0.03 nom	
L50722	Lead-Copper-Calcium Alloy	**Cu** 0.06 nom **Pb** 99.9 nom **Other** calcium Ca 0.03 nom	
L50725	Cable Sheathing Alloy	**Cu** 0.06 nom **Pb** 99.9 nom **Other** calcium Ca 0.035 nom	
L50728	Battery Grid Alloy, Lead-Calcium-Tin	**Pb** 99.5 nom **Sn** 0.5 nom **Other** calcium Ca 0.04 nom	
L50730	Electrowinning Anode Alloy	**Ag** 0.5 nom **Pb** 99.4 nom **Other** calcium Ca 0.06 nom	
L50735	Battery Grid Alloy, Lead-Calcium	**Pb** 99.9 nom **Other** calcium Ca 0.06 nom	
L50736	Battery Grid Alloy, Lead-Calcium-Tin	**Pb** 99.7 nom **Sn** 0.2 nom **Other** calcium Ca 0.065 nom	
L50737	Battery Grid Alloy, Lead-Calcium-Tin	**Pb** 99.4 nom **Sn** 0.5 nom **Other** calcium Ca 0.065 nom	
L50740	Battery Grid Alloy, Lead-Calcium-Tin	**Pb** 99.2 nom **Sn** 0.7 nom **Other** calcium Ca 0.065 nom	
L50745	Battery Grid Alloy, Lead-Calcium-Tin	**Pb** 98.9 nom **Sn** 1.0 nom **Other** calcium Ca 0.065 nom	
L50750	Battery Grid Alloy, Lead-Calcium-Tin	**Pb** 98.6 nom **Sn** 1.3 nom **Other** calcium Ca 0.065 nom	
L50755	Battery Grid Alloy, Lead-Calcium-Tin	**Pb** 98.4 nom **Sn** 1.5 nom **Other** calcium Ca 0.065 nom	
L50760	Battery Grid Alloy, Lead-Calcium	**Pb** 99.9 nom **Other** calcium Ca 0.07 nom	
L50765	Battery Grid Alloy, Lead-Calcium-Tin	**Pb** 99.2 nom **Sn** 0.7 nom **Other** calcium Ca 0.07 nom	
L50770	Battery Grid Alloy, Lead-Calcium	**Pb** 99.9 nom **Other** calcium Ca 0.10 nom	
L50775	Battery Grid Alloy, Lead-Calcium-Tin	**Pb** 99.6 nom **Sn** 0.3 nom **Other** calcium Ca 0.10 nom	
L50780	Battery Grid Alloy, Lead-Calcium-Tin	**Pb** 99.4 nom **Sn** 0.5 nom **Other** calcium Ca 0.10 nom	
L50790	Battery Grid Alloy, Lead-Calcium-Tin	**Pb** 98.9 nom **Sn** 1.0 nom **Other** calcium Ca 0.10 nom	
L50795	Battery Grid Alloy, Lead-Calcium-Tin	**Pb** 99.6 nom **Sn** 0.3 nom **Other** calcium Ca 0.12 nom	
L50800	Battery Grid Alloy, Lead-Calcium-Tin	**Pb** 99.5 nom **Sn** 0.3 nom **Other** calcium Ca 0.21 nom	

The chemical compositions listed are for identification purposes and should not be used in lieu of the cross referenced specifications.

UNIFIED NUMBER	DESCRIPTION	CHEMICAL COMPOSITION	CROSS REFERENCE SPECIFICATIONS
L50810	Bahnmetal, Bearing Metal	**Al** 0.02 nom **Li** 0.04 nom **Pb** 98.7 nom **Other** calcium Ca 0.7 nom, sodium Na 0.6 nom	
L50820	Bahnmetal, Bearing Metal	**Al** 0.02 nom **Li** 0.04 nom **Pb** 98.7 nom **Other** calcium Ca 0.7 nom, sodium Na 0.2 nom, barium Ba 0.4 nom	
L50840	Lead-Calcium Alloy	**Pb** 99.0 nom **Other** calcium Ca 1.0 nom	
L50850	Lead-Calcium Master Alloy	**Pb** 98 nom **Other** calcium Ca 2.0 nom	
L50880	Lead-Calcium Alloy	**Pb** 94.0 nom **Other** calcium Ca 6.0 nom	
L50940	Lead-Cadmium Alloy-Eutectic	**Cd** 17.0 nom **Pb** 83.0 nom	
L51110	Copperized Lead	**Cu** 0.05 nom **Pb** 99.9 nom	
L51120	Chemical Lead	**Ag** 0.002-0.02 **Bi** 0.005 max **Cu** 0.04-0.08 **Fe** 0.002 max **Pb** 99.90 min **Zn** 0.001 max **Other** As + Sb + Sn 0.002 max	**ASTM** B29 (Chemical Lead)
L51121	Copper Bearing Lead	**Ag** 0.020 max **Bi** 0.025 max **Cu** 0.04-0.08 **Fe** 0.002 max **Pb** 99.90 min **Zn** 0.001 max **Other** As + Sb + Sn 0.002 max	**ASTM** B29 (Copper Bearing Lead) **FED** QQ-L-171 (Grade C); QQ-L-201 (Grade C)
L51123	Lead Alloy Grade D	**Ag** 0.020 max **Bi** 0.025 max **Cu** 0.04-0.08 **Fe** 0.002 max **Pb** 99.85 min **Te** 0.035-0.060 **Zn** 0.001 max **Other** As + Sb + Sn 0.002 max	**FED** QQ-L-201 (Grade D)
L51124	Lead Alloy Grade D	**Ag** 0.020 max **Bi** 0.025 max **Cu** 0.04-0.08 **Fe** 0.002 max **Pb** 99.82 min **Sn** 0.016 max **Te** 0.035-0.055 **Zn** 0.001 max **Other** Sb + As 0.002 max	**FED** QQ-C-40 (Grade D)
L51125	Pure Copper Lead, Copperized Soft Lead	**Cu** 0.06 **Pb** 99.9 min	
L51180	Copper-Lead Bearing Alloy	**Cu** rem **Fe** 0.35 max **Pb** 44.0-58.0 **Sn** 1.0-5.0 **Other** others, total 0.45 max, others, each 0.15 max	**SAE** J460, No.485
L51510	Lead-Indium-Silver Solder Alloy	**Ag** 2.38 nom **In** 4.76 nom **Pb** 92.8 nom	
L51511	Lead-Indium Solder Alloy	**In** 5.0 nom **Pb** 95.0 nom	
L51512	Lead-Indium-Silver Solder Alloy	**Ag** 2.5 nom **In** 5.0 nom **Pb** 92.5 nom	
L51530	Lead-Indium Alloy	**In** 19.0 nom **Pb** 81.0 nom	
L51532	Lead-Indium Alloy	**In** 20.0 nom **Pb** 80.0 nom	
L51535	Lead-Indium Alloy	**In** 25.0 nom **Pb** 75.0 nom	
L51540	Lead-Indium Alloy	**In** 40.0 nom **Pb** 60.0 nom	
L51550	Lead-Indium Alloy	**In** 50.0 nom **Pb** 50.0 nom	
L51705	Lead-Lithium Alloy	**Li** 0.01 nom **Pb** 99.9 nom	
L51708	Lead-Lithium Alloy	**Li** 0.02 nom **Pb** 99.9 nom	
L51710	Lead-Lithium Alloy	**Li** 0.03 nom **Pb** 99.9 nom	
L51720	Lead-Lithium Alloy	**Li** 0.06 nom **Pb** 99.9 nom	
L51730	Lead-Lithium Alloy	**Li** 0.7 nom **Pb** 99.3 nom	

The chemical compositions listed are for identification purposes and should not be used in lieu of the cross referenced specifications.

LOW MELTING METALS AND ALLOYS

UNIFIED NUMBER	DESCRIPTION	CHEMICAL COMPOSITION	CROSS REFERENCE SPECIFICATIONS
L51740	Lead-Tin-Lithium Alloy	**Li** 0.02 nom **Pb** 99.6 nom **Sn** 0.35 nom	
L51748	Lead-Tin-Lithium Alloy	**Li** 0.04 nom **Pb** 99.2 nom **Sn** 0.7 nom	
L51770	Lead-Tin-Lithium-Calcium Alloy	**Li** 0.08 nom **Pb** 98.8 **Sn** 1.0 nom **Other** calcium Ca 0.03 nom	
L51775	Lead-Tin-Lithium-Calcium Alloy	**Li** 0.12 nom **Pb** 98.8 nom **Sn** 1.0 nom **Other** calcium Ca 0.03 nom	
L51778	Lead-Tin-Lithium-Calcium Alloy	**Li** 0.15 nom **Pb** 97.8 nom **Sn** 2.0 nom **Other** calcium Ca 0.04 nom	
L51780	Lead-Tin-Lithium-Calcium Alloy	**Li** 0.15 nom **Pb** 99.6 nom **Sn** 0.1 nom **Other** calcium Ca 0.15 nom	
L51790	Lead-Tin-Lithium-Calcium Alloy	**Li** 0.65 nom **Pb** 98.3 nom **Sn** 1.0 nom **Other** calcium Ca 0.02 nom	
L52505	Lead-Antimony Alloy	**Pb** 99.9 nom **Sb** 0.1 nom	
L52510	99.8 Pb	**Cu** 0.06 **Pb** 99.8 nom **Sb** 0.1 **Sn** 0.1	
L52515	Cable Sheathing Alloy	**As** 0.015 nom **Pb** 99.8 nom **Sb** 0.2 nom	
L52520	Cable Sheathing Alloy	**Pb** 99.4 **Sb** 0.2 **Sn** 0.4	
L52525	Cable Sheathing Alloy	**As** 0.035 nom **Pb** 99.7 nom **Sb** 0.3 nom **Te** 0.035 nom	
L52530	Lead-Antimony Alloy	**Pb** 99.6 nom **Sb** 0.35 nom	
L52535	Cable Sheathing Alloy	**As** 0.03 nom **Pb** 99.6 nom **Sb** 0.4 nom	
L52540	Cable Sheathing Alloy	**Cd** 0.25 nom **Pb** 99.2 nom **Sb** 0.5 nom	
L52545	Lead-Antimony Alloy	**Pb** 99.4 nom **Sb** 0.6 nom	
L52550	Cable Sheathing Alloy	**Cu** 0.04 nom **Pb** 99.3 nom **Sb** 0.6 nom **Te** 0.04 nom	
L52555	Alloy B per DIN 17640	**Pb** 99.3 **Sb** 0.7	
L52560	Bullet Alloy	**Pb** 99.2 nom **Sb** 0.75 nom	
L52565	Overhead Cable Alloy	**Pb** 99.2 nom **Sb** 0.75 nom	
L52570	Cable Sheathing Alloy, Alloy B	**Cu** 0.06 max **Pb** 99.1 **Sb** 0.85	
L52595	Cable Sheathing Alloy	**Pb** 99.0 nom **Sb** 0.95 nom	
L52605	1% Antimonial Lead	**Pb** 99.0 nom **Sb** 1.0 nom	
L52615	Lead-Base Die Casting Alloy	**As** 0.1 nom **Pb** 98.6 nom **S** 0.003 nom **Sb** 1.0 nom **Sn** 0.3 nom	
L52618	Lead-Antimony-Gallium Alloy	**Pb** 98.0 nom **Sb** 1.2 nom **Other** gallium Ga 0.8 nom	
L52620	Battery Alloy	**Cd** 1.45 nom **Pb** 97.0 nom **Sb** 1.5 nom	
L52625	Shot Alloy	**As** 0.45 nom **Pb** 98.0 nom **Sb** 1.55 nom **Sn** 0.0005 max	
L52630	Battery Alloy	**As** 0.3 **Pb** 98 nom **S** 0.005 **Sb** 1.6 **Se** 0.02 **Sn** 0.1	
L52705	2% Antimonial Lead	**Pb** 98.0 nom **Sb** 2.0 nom	
L52710	Battery Alloy	**As** 0.15 **Pb** 97.5 nom **S** 0.003 **Sb** 2.0 **Sn** 0.3	

The chemical compositions listed are for identification purposes and should not be used in lieu of the cross referenced specifications.

UNIFIED NUMBER	DESCRIPTION	CHEMICAL COMPOSITION	CROSS REFERENCE SPECIFICATIONS
L52715	Battery Alloy	**As** 0.25 **Pb** 97.4 nom **Sb** 2.25 **Se** 0.02 **Sn** 0.1	
L52720	Battery Alloy	**As** 0.3 **Pb** 97.2 nom **S** 0.008 **Sb** 2.25 **Se** 0.02 **Sn** 0.2	
L52725	Bullet Alloy	**Pb** 97.5 nom **Sb** 2.5 nom	
L52730	Electrotype-General	**Pb** 95 nom **Sb** 2.5 nom **Sn** 2.5 nom	
L52750	Battery Alloy	**As** 0.4 **Pb** 96.5 nom **S** 0.005 **Sb** 2.75 **Sn** 0.3 **Other** calcium Ca 0.075	
L52755	Battery Alloy	**As** 0.5 **Pb** 96.4 nom **S** 0.007 **Sb** 2.75 **Sn** 0.3 **Other** calcium Ca 0.075	
L52760	Battery Alloy	**As** 0.18 **Cu** 0.075 **Pb** 96.8 nom **S** 0.008 **Sb** 2.75 **Sn** 0.2	
L52765	Battery Alloy	**As** 0.3 **Cu** 0.075 **Pb** 96.6 nom **S** 0.008 **Sb** 2.75 **Sn** 0.3	
L52770	Battery Alloy	**As** 0.15 **Cu** 0.04 max **Pb** 96.6 nom **S** 0.001 **Sb** 2.9 **Sn** 0.3	
L52775	Battery Alloy	**As** 0.15 **Cu** 0.05 **Pb** 96.6 nom **S** 0.004 **Sb** 2.9 **Sn** 0.3	
L52805	3% Antimonial Lead	**Pb** 97 nom **Sb** 3.0 nom	
L52810	Battery Alloy	**As** 0.15 **Pb** 96.5 nom **S** 0.003 **Sb** 3.0 **Sn** 0.3	
L52815	Shot Alloy	**As** 0.6 **Pb** 96.4 nom **Sb** 3.0 **Sn** 0.0005 max	
L52830	Electrotype-General	**Pb** 94 nom **Sb** 3 nom **Sn** 3 nom	
L52840	Battery Alloy	**As** 0.5 **Cu** 0.12 **Pb** 95.7 nom **Sb** 3.25 **Sn** 0.4 **Other** calcium Ca 0.06	
L52860	Lead-Base Bearing Alloy No. 16 (Overlay)	**Al** 0.005 max **As** 0.05 max **Bi** 0.10 max **Cd** 0.005 max **Cu** 0.10 max **Pb** rem **Sb** 3.0-4.0 **Sn** 3.5-4.7 **Zn** 0.005 max **Other** total of others 0.40 max	
L52901	4% Antimonial Lead	**Pb** 96 nom **Sb** 4 nom	
L52905	Battery Alloy	**As** 0.15 **Pb** 95.5 **S** 0.003 **Sb** 4.0 **Sn** 0.3	
L52910	Lead Antimony Tin Alloy	**Pb** 95 nom **Sb** 4 nom **Sn** 1 nom	
L52915	Type Metal Alloy	**Pb** 93 nom **Sb** 4 nom **Sn** 3 nom	
L52920	4.5% Antimonial Lead	**Pb** 95.5 nom **Sb** 4.5 nom	
L52922	Anode Alloy	**Pb** 95 nom **S** 0.5 nom **Sb** 4.5 nom	
L52930	Battery Alloy	**As** 0.3 **Cu** 0.05 **Pb** 94.6 nom **S** 0.007 **Sb** 4.75 **Sn** 0.3	
L52940	Battery Alloy	**As** 0.15 **Cu** 0.05 **Pb** 94.8 nom **S** 0.004 **Sb** 4.75 **Sn** 0.3	
L53020	Lyman's 2 Alloy-Bullet Alloy	**Pb** 90.0 nom **Sb** 5.0 nom **Sn** 5.0 nom	
L53105	6% Antimonial Lead	**Pb** 94 nom **Sb** 6.0 nom	
L53110	Rolled Sheet alloy	**Pb** 94 nom **Sb** 6 nom	
L53115	Lead Alloy	**Pb** 93.7 nom **Sb** 6.0 nom **Sn** 0.3 nom	
L53120	Electrowinning Anode Alloy	**As** 0.4 nom **Pb** 93.6 nom **Sb** 6.0 nom	

The chemical compositions listed are for identification purposes and should not be used in lieu of the cross referenced specifications.

LOW MELTING METALS AND ALLOYS

UNIFIED NUMBER	DESCRIPTION	CHEMICAL COMPOSITION	CROSS REFERENCE SPECIFICATIONS
L53122	High-Strength Sheet Lead	As 0.4 nom Pb 93.6 nom Sb 6.0 nom	
L53125	Creep Resistant Pipe and Sheet Alloy	As 0.65 nom Pb 93.3 nom Sb 6.0 nom	
L53130	Lead Alloy	As 0.6 nom Pb 92.8 nom Sb 6.0 nom Sn 0.6 nom	
L53131	Lead Alloy	Pb rem Sb 6.0-7.0 Sn 0.25-0.75	AMS 7720; 7721
L53135	Battery Alloy	As 0.3 Cu 0.07 Pb 93.4 nom S 0.006 Sb 6.0 Sn 0.3	
L53140	Hard Shot Alloy	As 1.2 nom Pb 92.6 nom Sb 6.2 nom Sn 0.0005 max	
L53210	Type Metal Alloy	Pb 89 nom Sb 7 nom Sn 4 nom	
L53220	Lead Alloy	As 0.6 Pb 91.8 nom Sb 7.0 Sn 0.6	
L53230	8% Antimonial Lead	Pb 92 nom Sb 8 nom	
L53235	Hard Shot Alloy	As 1.25 nom Pb 90.7 nom Sb 8.0 nom Sn 0.0005 max	
L53238	Hard Shot Alloy	As 2 nom Pb 90.0 nom Sb 8 nom Sn 0.0005 max	
L53260	Spin Casting Alloy	Pb 88.9 nom Sb 8.0 nom Sn 3.1 nom	
L53265	Type Metal Alloy	Pb 88 nom Sb 8.0 nom Sn 4.0 nom	
L53305	9% Antimonial Lead	Pb 91 nom Sb 9.0 nom	
L53310	Lead Alloy	Pb 90 nom Sb 9 nom Sn 1 nom	
L53320	Whitemetal Bearing Alloy	Pb 86 nom Sb 9 nom Sn 5 nom	
L53340	Lead-Base Die Casting Alloy	As 0.15 max Cu 0.50 max Pb 89-91 Sb 9.25-10.75 Zn 0.01 max	ASTM B102 (Y10A)
L53343	Lead-Base Bearing Alloy	Al 0.005 max As 0.25 max Bi 0.10 max Cd 0.05 max Cu 0.50 max Fe 0.1 max Pb rem Sb 9.5-10.5 Sn 5.5-6.5 Zn 0.005 max	
L53345	Lead-Base Bearing Alloy	Al 0.005 max As 0.25 max Bi 0.10 max Cd 0.05 max Cu 0.50 max Pb rem Sb 9.0-11.0 Sn 5.0-7.0 Zn 0.005 max Other total of others 0.20 max	SAE J460, Alloy 13
L53346	Lead-Base Bearing Alloy	Al 0.005 max As 0.25 max Bi 0.10 max Cd 0.05 max Cu 0.50 max Fe 0.10 max Pb rem Sb 9.5-10.5 Sn 5.5-6.5 Zn 0.005 max	ASTM B23 (Alloy 13)
L53405	11% Antimonial Lead	Pb 89 nom Sb 11 nom	
L53420	Linotype Alloy	Pb 86 nom Sb 11 nom Sn 3 nom	
L53425	Linotype-Special Alloy	Pb 84 nom Sb 11 nom Sn 5 nom	
L53454	Type Metal Alloy	Pb 85 nom Sb 12 nom Sn 3 nom	
L53455	Linotype B (Eutectic) Alloy	Pb 84 nom Sb 12 nom Sn 4 nom	
L53456	Type Metal Alloy	Pb 83 nom Sb 12 nom Sn 5 nom	
L53460	Type Metal Alloy	Pb 82 nom Sb 12 nom Sn 6 nom	
L53465	Lead Alloy	Cu 0.05 Pb 77.5 nom Sb 12.5 nom Sn 10.0 nom	
L53470	CT Metal	As 0.4 nom Pb 86 nom Sb 12.75 nom Sn 0.75 nom	

The chemical compositions listed are for identification purposes and should not be used in lieu of the cross referenced specifications.

UNIFIED NUMBER	DESCRIPTION	CHEMICAL COMPOSITION	CROSS REFERENCE SPECIFICATIONS
L53480	Arsenical Lead-G Babbitt	**As** 3.0 nom **Pb** 83.5 nom **Sb** 12.75 nom **Sn** 0.75 nom	
L53505	CT Metal	**As** 1 nom **Pb** 85 nom **Sb** 13 nom **Sn** 1 nom	
L53510	Stereotype-General Alloy	**Pb** 80.5 nom **Sb** 13 nom **Sn** 6.5 nom	
L53530	Stereotype-Flat Alloy	**Pb** 80 nom **Sb** 14 nom **Sn** 6 nom	
L53550	15% Antimonial Lead	**Pb** 85 nom **Sb** 15 nom	
L53555	Lead Alloy	**Pb** 83 nom **Sb** 15 nom **Sn** 2 nom	
L53558	Type Metal Alloy	**Pb** 81 nom **Sb** 15 nom **Sn** 4 nom	
L53560	Lead-Base Die Casting Alloy	**Al** 0.01 max **As** 0.15 max **Cu** 0.50 max **Pb** 79-81 **Sb** 14-16 **Sn** 4-6 **Zn** 0.01 max	**ASTM** B102 (YT155A)
L53565	Lead-Base White Metal Bearing Alloy	**Al** 0.005 max **As** 0.30-0.60 **Bi** 0.10 max **Cd** 0.05 max **Cu** 0.50 max **Fe** 0.10 max **Pb** rem **Sb** 14.0-16.0 **Sn** 4.5-5.5 **Zn** 0.005 max	**ASTM** B23 (Alloy 8)
L53570	Monotype-Ordinary Alloy	**Pb** 78 nom **Sb** 15 nom **Sn** 7 nom	
L53575	Stereotype-Curved Alloy	**Pb** 77 nom **Sb** 15 nom **Sn** 8 nom	
L53580	Rules Monotype Alloy	**Pb** 75 nom **Sb** 15 nom **Sn** 10 nom	
L53585	Lead-Base Whitemetal Bearing Alloy	**Al** 0.005 max **As** 0.30-0.60 **Bi** 0.10 max **Cd** 0.05 max **Cu** 0.50 max **Fe** 0.10 max **Pb** rem **Sb** 14.0-16.0 **Sn** 9.3-10.7 **Zn** 0.005 max	**ASTM** B23 (Alloy 7)
L53620	Lead-Base Bearing Alloy	**Al** 0.005 max **As** 0.8-1.4 **Bi** 0.10 max **Cd** 0.05 max **Cu** 0.6 max **Fe** 0.10 max **Pb** rem **Sb** 14.5-17.5 **Sn** 0.8-1.2 **Zn** 0.005 max	**ASTM** B23 (Alloy 15)
L53650	Display Monotype Alloy	**Pb** 75 nom **Sb** 17 nom **Sn** 8 nom	
L53655	Type Metal Alloy	**Pb** 74 nom **Sb** 17 nom **Sn** 9 nom	
L53685	Lanston Standard Case Type Monotype Alloy	**Pb** 72 nom **Sb** 19 nom **Sn** 9 nom	
L53710	Hard Foundry Type Alloy	**Cu** 1.5 nom **Pb** 58.5 nom **Sb** 20 nom **Sn** 20 nom	
L53740	Monotype Case Type Alloy	**Pb** 67 nom **Sb** 24 nom **Sn** 9 nom	
L53750	Monotype Case Type Alloy	**Pb** 64 nom **Sb** 24 nom **Sn** 12 nom	
L53780	Hard Foundry Type Alloy	**Cu** 1.5 nom **Pb** 60.5 nom **Sb** 25 nom **Sn** 13 nom	
L53790	Type Metal Alloy	**Pb** 67 nom **Sb** 28 nom **Sn** 5 nom	
L53795	Type Metal Alloy	**Pb** 65.5 nom **Sb** 29 nom **Sn** 5.5 nom	
L54030	Alloy 1/2C-Cable Sheathing Alloy	**Cd** 0.075 nom **Pb** 99.7 nom **Sn** 0.2 nom	
L54050	Alloy C-Cable Sheathing Alloy	**Cd** 0.15 nom **Pb** 99.4 nom **Sn** 0.4 nom	
L54210	2% Tin Solder	**Al** 0.005 max **As** 0.02 max **Bi** 0.25 max **Cu** 0.08 max **Fe** 0.02 max **Pb** 98 nom **Sb** 0.12 max **Sn** 1.5-2.5 **Zn** 0.005 max	**ASTM** B32 (Alloy 2A)

The chemical compositions listed are for identification purposes and should not be used in lieu of the cross referenced specifications.

LOW MELTING METALS AND ALLOYS

UNIFIED NUMBER	DESCRIPTION	CHEMICAL COMPOSITION	CROSS REFERENCE SPECIFICATIONS
L54211	2% Tin Antimonial Solder	**Al** 0.005 max **As** 0.02 max **Bi** 0.25 max **Cu** 0.08 max **Fe** 0.02 max **Pb** 98 nom **Sb** 0.20-0.50 **Sn** 1.5-2.5 **Zn** 0.005 max	**ASTM** B32 (Alloy 2B)
L54250	Solder Alloy	**Al** 0.005 max **As** 0.40-0.60 **Bi** 0.25 max **Cu** 0.08 max **Fe** 0.02 max **Pb** rem **Sb** 4.90-5.40 **Sn** 2.50-2.75 **Zn** 0.005 max **Other** other elements, total 0.08 max	**SAE** J473 (Alloy 9B)
L54280	Solder Alloy	**Pb** 91.9 nom **Sb** 5.1 nom **Sn** 3.0 nom	
L54310	Electrotype Curved Plate Alloy	**Pb** 93 nom **Sb** 3 nom **Sn** 4 nom	
L54320	5/95 Solder	**Al** 0.005 max **As** 0.02 max **Bi** 0.25 max **Cu** 0.08 max **Fe** 0.02 max **Pb** 95 nom **Sb** 0.12 max **Sn** 4.5-5.5 **Zn** 0.005 max	**ASTM** B32 (Alloy 5A)
L54321	5% Tin Antimonial Solder	**Al** 0.005 max **As** 0.02 max **Bi** 0.25 max **Cu** 0.08 max **Fe** 0.02 max **Pb** 95 nom **Sb** 0.20-0.50 **Sn** 4.5-5.5 **Zn** 0.005 max	**ASTM** B32 (Alloy 5B)
L54322	SN5 Solder	**Ag** 0.015 max **Al** 0.005 max **As** 0.02 max **Bi** 0.25 max **Cd** 0.001 max **Cu** 0.08 max **Fe** 0.02 max **Pb** rem **Sb** 0.50 max **Sn** 4.5-5.5 **Zn** 0.005 max **Other** total of all others 0.08 max	**FED** QQ-S-571 (Sn5)
L54360	Solder Alloy	**As** 0.5 nom **Pb** 90.5 nom **Sb** 4.0 nom **Sn** 5.0 nom	
L54370	Plated Overlay for Bearings	**Pb** rem **Sn** 5.0-9.0 **Other** other elements, total 3.5 max	**SAE** J460, No. 190
L54410	8% Tin Solder	**Pb** 92 nom **Sb** 0.3 nom **Sn** 8 nom	
L54510	Plated Overlay for Bearings	**Pb** rem **Sn** 8.0-12.0 **Other** other elements, total 3.5 max	**SAE** J460, No. 19
L54520	10/90 Solder	**Al** 0.005 max **As** 0.02 max **Bi** 0.25 max **Cu** 0.08 max **Fe** 0.02 max **Pb** 90 nom **Sb** 0.20-0.50 **Sn** 10 nom **Zn** 0.005 max	**ASTM** B32 (Alloy 10B)
L54525	88-10-2 Solder	**Ag** 1.7-2.4 **Al** 0.005 max **As** 0.02 max **Bi** 0.03 max **Cd** 0.001 max **Cu** 0.08 max **Pb** rem **Sb** 0.20 max **Sn** 9.0-11.0 **Zn** 0.005 max **Other** total of all others 0.10 max	**FED** QQ-S-571 (Sn10)
L54530	Solder Alloy	**As** 0.5 nom **Pb** 85.5 nom **Sb** 4 nom **Sn** 10 nom	
L54540	Solder Alloy	**Pb** 87.5 nom **Sb** 0.45 nom **Sn** 12 nom	
L54555	Solder Alloy	**Al** 0.005 max **Bi** 0.25 max **Cu** 0.08 max **Fe** 0.02 max **Pb** rem **Sb** as specified 2.75 max **Sn** 14.0-15.0 **Zn** 0.005 max **Other** other elements, total 0.08 max	**SAE** J473, No. 6B
L54560	15/85 Solder	**Al** 0.005 max **As** 0.02 max **Bi** 0.25 max **Cu** 0.08 max **Fe** 0.02 max **Pb** 85 nom **Sb** 0.20-0.50 **Sn** 15 desired **Zn** 0.005 max	**ASTM** B32 (Alloy 15B)
L54570	Solder Alloy	**Pb** 82.5 nom **Sb** 2.5 nom **Sn** 15 nom	
L54580	Type Metal Alloy	**Pb** 80.5 nom **Sb** 4.5 nom **Sn** 15 nom	

The chemical compositions listed are for identification purposes and should not be used in lieu of the cross referenced specifications.

UNIFIED NUMBER	DESCRIPTION	CHEMICAL COMPOSITION	CROSS REFERENCE SPECIFICATIONS
L54610	Solder Alloy	**Al** 0.005 max **Bi** 0.25 max **Cu** 0.08 max **Fe** 0.02 max **Pb** rem **Sb** 1.25-1.75 **Sn** 19.0-20.0 **Zn** 0.005 max **Other** other elements, total 0.08 max	**SAE** J473, No. 5B
L54710	20% Tin Solder	**Pb** 80 nom **Sb** 0.2 max **Sn** 20 nom	
L54711	Solder Alloy 20B	**Al** 0.005 max **As** 0.02 max **Bi** 0.25 max **Cu** 0.08 max **Fe** 0.02 max **Pb** 80 nom **Sb** 0.20-0.50 **Sn** 20 desired **Zn** 0.005 max	**ASTM** B32 (Alloy 20B) **FED** QQ-S-571 (Pb80)
L54712	Solder Alloy 20C	**Al** 0.005 max **As** 0.02 max **Bi** 0.25 max **Cu** 0.08 max **Fe** 0.02 max **Pb** 79 nom **Sb** 0.8-1.2 **Sn** 20 desired **Zn** 0.005 max	**ASTM** B32 (Alloy 20C) **FED** QQ-S-571 (Sn20)
L54713	Solder Alloy	**Pb** 77 nom **Sb** 1.2 nom **Sn** 22 nom	
L54715	Solder Alloy	**Pb** 74 nom **Sb** 3 nom **Sn** 23 nom	
L54720	25/75 Solder	**Al** 0.005 max **As** 0.02 max **Bi** 0.25 max **Cu** 0.08 max **Fe** 0.02 max **Pb** 75 nom **Sb** 0.25 max **Sn** 25 nom **Zn** 0.005 max	**ASTM** B32 (Alloy 25A)
L54721	Solder Alloy 25B	**Al** 0.005 max **As** 0.02 max **Bi** 0.25 max **Cu** 0.08 max **Fe** 0.02 max **Pb** 75 nom **Sb** 0.20-0.50 **Sn** 25 desired **Zn** 0.005 max	**ASTM** B32 (Alloy 25B)
L54722	Solder Alloy 25C	**Al** 0.005 max **As** 0.02 max **Bi** 0.25 max **Cu** 0.08 max **Fe** 0.02 max **Pb** 73.7 nom **Sb** 1.1-1.5 **Sn** 25 desired **Zn** 0.005 max	**ASTM** B32 (Alloy 25C)
L54727	Lead-Base Bearing Alloy	**Cu** 3 nom **Pb** 59 nom **Sb** 13 nom **Sn** 25 nom	
L54750	Silver-Loaded Solder	**Ag** 3 nom **Pb** 70 nom **Sn** 27 nom	
L54755	Fusible Alloy	**Bi** 21.5 nom **Pb** 51.5 nom **Sn** 27 nom	
L54805	Solder Alloy	**Pb** 70.5 nom **Sb** 1.5 nom **Sn** 28 nom	
L54810	Fusible Alloy	**Bi** 28.5 nom **Pb** 43 nom **Sn** 28.5 nom	
L54815	Solder Alloy	**Al** 0.005 max **Bi** 0.25 max **Cu** 0.08 max **Fe** 0.02 max **Pb** rem **Sb** 0.75-1.25 **Sn** 29.0-30.0 **Zn** 0.005 max **Other** other elements, total 0.08 max	**SAE** J473, No. 3B
L54820	30/70 Solder	**Al** 0.005 max **As** 0.02 max **Bi** 0.25 max **Cu** 0.08 max **Fe** 0.02 max **Pb** 70 nom **Sb** 0.25 max **Sn** 30 desired **Zn** 0.005 max	**ASTM** B32 (Alloy 30A)
L54821	Solder Alloy	**Al** 0.005 max **As** 0.02 max **Bi** 0.25 max **Cu** 0.08 max **Fe** 0.02 max **Pb** 70 nom **Sb** 0.20-0.50 **Sn** 30 desired **Zn** 0.005 max	**ASTM** B32 (Alloy 30B) **FED** QQ-S-571 (Pb70)
L54822	Solder Alloy	**Al** 0.005 max **As** 0.02 max **Bi** 0.25 max **Cu** 0.08 max **Fe** 0.02 max **Pb** 68.4 nom **Sb** 1.4-1.8 **Sn** 30 desired **Zn** 0.005 max	**ASTM** B32 (Alloy 30C) **FED** QQ-S-571 (Sn30)
L54827	Type Metal Alloy	**Pb** 64 nom **Sb** 6 nom **Sn** 30 nom	
L54829	Fusible Alloy	**Bi** 20 nom **Pb** 50 nom **Sn** 30 nom	
L54830	Fusible Alloy	**Bi** 30.8 nom **Pb** 38.4 nom **Sn** 30.8 nom	
L54832	Solder Alloy	**Pb** 66.7 nom **Sb** 1.8 nom **Sn** 31.5 nom	

The chemical compositions listed are for identification purposes and should not be used in lieu of the cross referenced specifications.

LOW MELTING METALS AND ALLOYS

UNIFIED NUMBER	DESCRIPTION	CHEMICAL COMPOSITION	CROSS REFERENCE SPECIFICATIONS
L54833	Solder Alloy	Pb 65 nom Sb 3 nom Sn 32 nom	
L54835	Fusible Alloy	Bi 33.3 nom Pb 33.4 nom Sn 33.3 nom	
L54840	Fusible Alloy	Bi 32 nom Pb 34 nom Sn 34 nom	
L54850	35/65 Solder	Al 0.005 max As 0.02 max Bi 0.25 max Cu 0.08 max Fe 0.02 max Pb 65 nom Sb 0.25 max Sn 35 nom Zn 0.005 max	ASTM B32 (Alloy 35A)
L54851	Solder Alloy	Al 0.005 max As 0.02 max Bi 0.25 max Cu 0.08 max Fe 0.02 max Pb 65 nom Sb 0.20-0.50 Sn 35 nom Zn 0.005 max	ASTM B32 (Alloy 35B) FED QQ-S-571 (Pb65)
L54852	Solder Alloy	Al 0.005 max As 0.02 max Bi 0.25 max Cu 0.08 max Fe 0.02 max Pb 63.2 nom Sb 1.6-2.0 Sn 35 nom Zn 0.005 max	ASTM B32 (Alloy 35C) FED QQ-S-571 (Sn35)
L54855	Silver-Loaded Solder Alloy	Ag 3.0 nom Pb 61.5 nom Sn 35.5 nom	
L54860	Fusible Alloy	Bi 21 nom Pb 43 nom Sn 36 nom	
L54865	Fusible Alloy	Bi 21 nom Pb 42 nom Sn 37 nom	
L54905	Antimonial Solder Alloy	Al 0.005 max Bi 0.25 max Cu 0.08 max Fe 0.02 max Pb rem Sb 1.5-2.00 Sn 38.0-38.5 Zn 0.005 max Other other elements, total 0.08 max	SAE J473, No. 2B
L54910	Fusible Alloy	Bi 12.6 nom Pb 47.5 nom Sn 39.9 nom	
L54915	40/60 Solder	Al 0.005 max As 0.02 max Bi 0.25 max Cu 0.08 max Fe 0.02 max Pb 60 nom Sb 0.12 max Sn 40 nom Zn 0.005 max	ASTM B32 (Alloy 40A)
L54916	Solder Alloy	Al 0.005 max As 0.02 max Bi 0.25 max Cu 0.08 max Fe 0.02 max Pb 60 nom Sb 0.20-0.50 Sn 40 nom Zn 0.005 max	ASTM B32 (Alloy 40B) FED QQ-S-571 (Sn40)
L54918	Solder Alloy	Al 0.005 max As 0.02 max Bi 0.25 max Cu 0.08 max Fe 0.02 max Pb 58 nom Sb 1.8-2.4 Sn 40 nom Zn 0.005 max	ASTM B32 (Alloy 40C)
L54925	Lead-Base Bearing Alloy	Cu 2 nom Pb 46 nom Sb 12 nom Sn 40 nom	
L54930	Fusible Alloy	Bi 4 nom Pb 55.5 nom Sn 40.5 nom	
L54935	Fusible Alloy	Bi 14 nom Pb 43 nom Sn 43 nom	
L54940	Solder Alloy	Al 0.005 max Bi 0.25 max Cu 0.08 max Fe 0.02 max Pb rem Sb 1.5-2.00 Sn 43.0-43.5 Zn 0.005 max Other other elements, total 0.08 max	SAE J473, Grade 1B
L54945	Silver-Loaded Solder	Ag 1 nom Pb 55 nom Sn 44 nom	
L54950	45/55 Solder	Al 0.005 max As 0.03 max Bi 0.25 max Cu 0.08 max Fe 0.02 max Pb 55 nom Sb 0.12 max Sn 45 nom Zn 0.005 max	AMS 4750 ASTM B32 (Alloy 45A)
L54951	Solder Alloy	Al 0.005 max As 0.03 max Bi 0.25 max Cu 0.08 max Fe 0.02 max Pb 55 nom Sb 0.20-0.50 Sn 45 nom Zn 0.005 max	ASTM B32 (Alloy 45B)

The chemical compositions listed are for identification purposes and should not be used in lieu of the cross referenced specifications.

UNIFIED NUMBER	DESCRIPTION	CHEMICAL COMPOSITION	CROSS REFERENCE SPECIFICATIONS
L54955	Solder Alloy	**Pb** 52.5 nom **Sb** 2.5 nom **Sn** 45 nom	
L55005	Solder Alloy	**Pb** 52 nom **Sn** 48 nom	
L55030	50/50 Solder	**Al** 0.005 max **As** 0.03 max **Bi** 0.25 max **Cu** 0.08 max **Fe** 0.02 max **Pb** 50 nom **Sb** 0.12 max **Sn** 50 nom **Zn** 0.005 max	**ASTM** B32 (Alloy 50A)
L55031	Solder Alloy	**Al** 0.005 max **As** 0.03 max **Bi** 0.25 max **Cu** 0.08 max **Fe** 0.02 max **Pb** 50 nom **Sb** 0.20-0.50 **Sn** 50 nom **Zn** 0.005 max	**ASTM** B32 (Alloy 50B) **FED** QQ-S-571 (Sn50)
L55210	Battery Alloy	**Pb** 99.6 nom **Sn** 0.3 nom **Other** calcium Ca 0.06 nom; strontium Sr 0.06 nom	
L55230	Battery Alloy	**Al** 0.03 nom **Pb** 99 nom **Sn** 0.8 nom **Other** strontium Sr 0.16 nom	
L55260	Battery Alloy	**Al** 0.03 nom **Pb** 99.8 nom **Other** strontium Sr 0.2 nom	
L55290	Lead-Strontium Alloy	**Pb** 98. nom **Other** strontium Sr 2. nom	

The chemical compositions listed are for identification purposes and should not be used in lieu of the cross referenced specifications.

UNS NUMBERS ASSIGNED TO DATE

With Description of Each Material Covered and References To Documents In Which The Same or Similar Materials are Described

Mxxxxx Number Series
Miscellaneous Nonferrous Metals and Alloys

MISCELLANEOUS NONFERROUS METALS AND ALLOYS

UNIFIED NUMBER	DESCRIPTION	CHEMICAL COMPOSITION	CROSS REFERENCE SPECIFICATIONS
M00995	Metallic Antimony	**Ag** 0.10 max **As** 0.10 max **Cu** 0.10 max **Fe** 0.10 max **Ni** 0.10 max **Pb** 0.20 max **S** 0.10 max **Sb** 99.50 min **Sn** 0.10 max	**ASTM** B237 (B) **MIL SPEC** MIL-A-10841
M00998	Metallic Antimony	**Ag** 0.05 max **As** 0.05 max **Cu** 0.05 max **Fe** 0.05 max **Ni** 0.05 max **Pb** 0.15 max **S** 0.10 max **Sb** 99.80 min **Sn** 0.05 max	**ASTM** B237 (A) **MIL SPEC** MIL-A-10841
* M05995	Plutonium Nuclear Grade	**Fe** 500 ppm max **Other** plutonium Pu 99.5 min, F+Cl 200 ppm max; Mg+Ca 500 ppm max	
M08990	Depleted Uranium	**C** 0.07 max **U** 99.00 min	**AMS** 7730
M10030	Magnesium Alloy-Casting	**Ag** 0.001 max **Al** 2.5-3.5 **Cd** 0.001 max **Cu** 0.005 max **Fe** 0.005 max **Mg** rem **Mn** 0.005 max **Ni** 0.001 max **Zn** 0.10 max **Other** total 0.30 max, Rare Earths 0.001 max	**ASTM** B275 (A3A)
M10100	Magnesium Alloy-Casting	**Al** 9.3-10.7 **Cu** 0.10 max **Mg** rem **Mn** 0.10 min **Ni** 0.01 max **Si** 0.30 max **Zn** 0.30 max **Other** total 0.30 max	**AMS** 4455; 4483 **ASTM** B80 (AM100A); B93 (AM100A); B149 (AM100A); B275 (AM100A); B403 (AM100A) **FED** QQ-M-55 **MIL SPEC** MIL-M-46062B **SAE** J465 (AM100-A)
M10102	Magnesium-Die Alloy Casting	**Al** 9.4-10.6 **Cu** 0.08 max **Mg** rem **Mn** 0.13 min **Ni** 0.01 max **Si** 1.0 max **Other** total 0.30 max	**ASTM** B275 (AM100B)
M10410	Magnesium-Die Alloy Casting	**Al** 3.5-5.0 **Cu** 0.06 max **Mg** rem **Mn** 0.20-0.50 **Ni** 0.03 max **Si** 0.50-1.5 **Zn** 0.12 max **Other** total 0.30 max	**ASTM** B93 (AS41A); B94 (AS41A); B275 (AS41A) **SAE** J465 (AS41-A)
M10600	Magnesium-Die-Mold Alloy Casting	**Al** 5.5-6.5 **Cu** 0.35 max **Mg** rem **Mn** 0.13 min **Ni** 0.03 max **Si** 0.50 max **Zn** 0.22 max **Other** total 0.30 max	**ASTM** B93 (AM60A); B94 (AM60A); B275 (AM60A) **SAE** J465 (AM60-A)
M10800	Magnesium Alloy-Casting	**Al** 8.0-9.0 **Cu** 0.08 max **Mg** rem **Mn** 0.18 min **Ni** 0.01 max **Si** 0.20 max **Zn** 0.20 max **Other** total 0.30 max	**ASTM** B275 (AM80A)
M10900	Magnesium Alloy-Casting	**Ag** 0.008 max **Al** 8.5-9.5 **Cu** 0.10 max **Fe** 0.008 max **Mg** rem **Mn** 0.15 min **Ni** 0.005 max **Si** 0.15 max **Zn** 0.20 max **Other** total 0.10 max; each 0.02 max	**ASTM** B275 (AM90A)
M11100	Magnesium Alloy-Casting	**Al** 1.0-1.5 **Cu** 0.10 max **Fe** 0.005 max **Mg** rem **Mn** 0.20 min **N** 0.005 max **Si** 0.10 max **Zn** 0.20-0.6	**ASTM** B275 (AZ10A) **FED** QQ-M-31
M11101	Magnesium Alloy-Casting	**Al** 9.5-10.5 **Be** 0.0002-0.0008 **Cu** 0.05 max **Fe** 0.005 max **Mg** rem **Mn** 0.13 min **Ni** 0.005 max **Si** 0.05 max **Zn** 0.75-1.25 **Other** total 0.30 max	**ASME** SFA5.19 ER AZ101A **ASTM** B275 (AZ101A) **AWS** A5.19 ER AZ101A
M11125	Magnesium Alloy-Casting	**Al** 11.0-13.0 **Be** 0.0002-0.0008 **Mg** rem **Zn** 4.5-5.5 **Other** total 0.30 max	**ASTM** B275 (AZ125A)
M11210	Magnesium Alloy-Casting	**Al** 1.6-2.5 **Cu** 0.05 max **Fe** 0.005 max **Mg** rem **Mn** 0.15 max **Ni** 0.002 max **Si** 0.05 max **Zn** 0.8-1.6 **Other** total 0.30 max, Ca 0.10-0.25	**ASTM** B275 (AZ21A)
M11310	Magnesium Alloy-Sheet and Plate	**Al** 2.5-3.5 **Cu** 0.05 max **Fe** 0.005 max **Mg** rem **Mn** 0.20 min **Ni** 0.005 max **Si** 0.30 max **Zn** 0.6-1.4 **Other** total 0.30 max, Ca 0.30 max	**ASTM** B90 (AZ31A); B275 (AZ31A)

*Boxed entries are no longer active and are retained for reference purposes only.

The chemical compositions listed are for identification purposes and should not be used in lieu of the cross referenced specifications.

UNIFIED NUMBER	DESCRIPTION	CHEMICAL COMPOSITION	CROSS REFERENCE SPECIFICATIONS
M11311	Magnesium Alloy-Wrought-Extruded Shapes-Forgings	Al 2.5-3.5 Cu 0.05 max Fe 0.005 max Mg rem Mn 0.20 min Ni 0.005 max Si 0.10 max Zn 0.6-1.4 Other total 0.30 max, Ca 0.04 max	**AMS** 4375; 4376; 4377; 4382 **ASTM** B90 (AZ31B); B91 (AZ31B); B107 (AZ31B); B275 (AZ31B) **FED** QQ-M-31; QQ-M-40; QQ-M-44 **SAE** J466 (AZ31B)
M11312	Magnesium Alloy-Wrought-Extruded Shapes	Al 2.4-3.6 Cu 0.10 max Mg rem Mn 0.15 min Ni 0.03 max Si 0.10 max Zn 0.50-1.5 Other total 0.30 max	**ASTM** B90 (AZ31C); B107 (AZ31C); B275 (AZ31C)
M11610	Magnesium Alloy-Forgings-Extruded Shapes	Al 5.8-7.2 Cu 0.05 max Fe 0.005 max Mg rem Mn 0.15 min Ni 0.005 max Si 0.10 max Zn 0.40-1.5 Other total 0.30 max	**AMS** 4350 **ASTM** B91 (AZ61A); B107 (AZ61A); B275 (AZ61A) **FED** QQ-M-31; QQ-M-40 **SAE** J466 (AZ61A)
M11611	Magnesium Welding Wire	Al 5.8-7.2 Be 0.0002-0.0008 Cu 0.05 max Fe 0.005 max Mg rem Mn 0.15 min Ni 0.005 max Si 0.05 max Zn 0.40-1.5 Other total 0.30 max	**ASME** SFA5.19 (ER AZ61A) **AWS** A5.19 (ER AZ61A)
M11630	Magnesium Alloy-Casting	Al 5.3-6.7 Cu 0.15 nom Mg rem Mn 0.15 min Ni 0.01 max Si 0.25 nom Zn 2.5-3.5 Other total 0.30 max	**AMS** 4420; 4422; 4424 **ASTM** B80 (AZ63A); B93 (AZ63A); B275 (AZ63A) **FED** QQ-M-55; QQ-M-56 **SAE** J465 (AZ63A)
M11800	Magnesium Alloy-Forgings-Extrusions	Al 7.8-9.2 Cu 0.05 max Fe 0.005 max Mg rem Mn 0.12 min Ni 0.005 max Si 0.10 max Zn 0.20-0.8 Other total 0.30 max	**AMS** 4360 **ASTM** B91 (AZ80A); B107 (AZ80A); B275 (AZ80A) **FED** QQ-M-31; QQ-M-40 **SAE** J466 (AZ80A)
M11810	Magnesium Alloy-Casting	Al 7.0-8.1 Cu 0.10 max Mg rem Mn 0.13 min Ni 0.01 max Si 0.30 max Zn 0.40-1.0 Other total 0.30 max	**ASTM** B80 (AZ81A); B93 (AZ81A); B199 (AZ81A); B275 (AZ81A); B403 (AZ81A) **FED** QQ-M-55; QQ-M-56
M11900	Magnesium Alloy-Casting	Al 8.5-9.5 Cu 0.02 max Fe 0.015 max Mg rem Mn 0.15 min Ni 0.005 max Si 0.20 max Zn 0.20 max Other each 0.07 max; total 0.30 max	**ASTM** B275 (AZ90A)
M11910	Magnesium Alloy-Casting	Al 8.3-9.7 Cu 0.10 max Mg rem Mn 0.13 min Mo 0.03 max Si 0.50 max Zn 0.35-1.0 Other total 0.30 max	**AMS** 4490 **ASTM** B93 (AZ91A); B94 (AZ91A); B275 (AZ91A) **FED** QQ-M-38 **SAE** J465 (AZ91A)
M11912	Magnesium Alloy-Casting	Al 8.5-9.5 Cu 0.25 max Mg rem Mn 0.15 min Ni 0.01 min Si 0.20 max Zn 0.45-0.9 Other total 0.30 max	**ASTM** B93 (AZ91B); B94 (AZ91B); B275 (AZ91B) **SAE** J465 (AZ91B)
M11914	Magnesium Alloy-Casting	Al 8.1-9.3 Cu 0.10 max Mg rem Mn 0.13 min Ni 0.01 max Si 0.30 max Zn 0.40-1.0 Other total 0.30 max	**AMS** 4437; 4452 **ASTM** B80 (AZ91C); B93 (AZ91C); B94 (AZ91C); B199 (AZ91C); B275 (AZ91C); B403 (AZ91C) **FED** QQ-M-55; QQ-M-56 **MIL SPEC** MIL-M-46062B **SAE** J465 (AZ91C)
M11920	Magnesium Alloy-Casting	Al 8.3-9.7 Cu 0.25 max Mg rem Mn 0.10 min Ni 0.01 max Si 0.30 max Zn 1.6-2.4 Other total 0.30 max	**AMS** 4434; 4453; 4484 **ASTM** B80 (AZ92A); B93 (AZ92A); B94 (AZ92A); B199 (AZ92A); B275 (AZ92A); B403 (AZ92A) **FED** QQ-M-55; QQ-M-56 **MIL SPEC** MIL-M-46062B **SAE** J465 (AZ92A)
M11922	Magnesium Alloy Welding Wire	Al 8.3-9.7 Be 0.0002-0.0008 Cu 0.05 max Fe 0.005 max Mg rem Mn 0.15 min Ni 0.005 max Si 0.05 max Zn 1.7-2.3 Other total 0.30 max	**AMS** 4395 **ASME** SFA5.19 (ER AZ92A) **AWS** A5.19 (ER AZ92A)
M12300	Magnesium Alloy-Casting	Cu 0.10 max Mg rem Ni 0.01 max Zn 0.30 max Zr 0.20 min Other total 0.30 max, Rare Earths 2.5-4.0	**ASTM** B80 (EK30A); B275 (EK30A)
M12330	Magnesium Alloy-Sand-Permanent Mold-Investment Casting	Cu 0.10 max Mg rem Ni 0.01 max Zn 2.0-3.1 Zr 0.50-1.0 Other total 0.30 max, Rare Earths 2.5-4.0	**AMS** 4396; 4442 **ASTM** B80 (EZ33A); B199 (EZ33A); B275 (EZ33A); B403 (EZ33A) **FED** QQ-M-55; QQ-M-56 **SAE** J465 (EZ33A)
M12331	Magnesium Alloy Welding Wire	Mg rem Re 2.5-4.0 Zn 2.0-3.1 Zr 0.45-1.0 Other total 0.30 max	**AMS** 4396 **ASME** SFA5.19 (ER EZ33A) **AWS** A5.19 (ER EZ33A)

The chemical compositions listed are for identification purposes and should not be used in lieu of the cross referenced specifications.

UNIFIED NUMBER	DESCRIPTION	CHEMICAL COMPOSITION	CROSS REFERENCE SPECIFICATIONS
M12350	Magnesium Alloy-Casting (QH-21)	**Ag** 2.0-3.0 **Cu** 0.10 max **Mg** rem **Ni** 0.01 max **Th** 0.6-1.6 **Zn** 0.20 max **Zr** 0.40-1.0 (total) 0.40 min (soluble) **Other** total Rare Earths 0.6-1.5, Th + Di 1.5-2.4, Other impurities 0.30 total	AMS 4419
M12410	Magnesium Alloy-Sand Casting	**Cu** 0.10 max **Mg** rem **Ni** 0.01 max **Zn** 0.30 max **Zr** 0.40-1.0 **Other** total 0.30 max, Rare Earths 3.0-5.0	AMS 4440; 4441 **ASTM** B80 (EK41A); B275 (EK41A)
M13210	Magnesium Alloy-Wrought Products-Forgings	**Mg** rem **Mn** 0.45-1.1 **Th** 1.5-2.5 **Other** total 0.30 max	AMS 4363; 4383; 4390 **ASTM** B90 (HM21A); B91 (HM21A); B275 (HM21A) **FED** QQ-M-40 **MIL SPEC** MIL-M-8917 **SAE** J466 (HM21A)
M13310	Magnesium Alloy-Casting	**Cu** 0.10 max **Mg** rem **Ni** 0.01 max **Th** 2.5-4.0 **Zn** 0.30 max **Zr** 0.40-1.0 **Other** total 0.30 max	AMS 4384; 4385; 4445 **ASTM** B80 (HK31A); B90 (HK31A); B199 (HK31A); B275 (HK31A); B403 (HK31A) **FED** QQ-M-55; QQ-M-56 **MIL SPEC** MIL-M-26075; MIL-M-46062 **SAE** J465 (HK31A); J466 (HK31A)
M13312	Magnesium Alloy-Casting	**Mg** rem **Mn** 1.2 min **Th** 2.5-3.5 **Other** 0.30 max total	AMS 4388; 4389 **ASTM** B275 (HM31A) **MIL SPEC** MIL-M-8916 **SAE** J466 (HM31A)
M13320	Magnesium Alloy-Casting	**Cu** 0.10 max **Mg** rem **Ni** 0.01 max **Th** 2.5-4.0 **Zn** 1.7-2.5 **Zr** 0.50-1.0 **Other** total 0.30 max, Rare Earths 0.10 max	AMS 4447 **ASTM** B80 (HZ32A); B275 (HZ32A) **FED** QQ-M-55; QQ-M-56 **SAE** J465 (HZ32A)
M14141	Magnesium Alloy-Wrought Products	**Al** 1.0-1.5 **Cu** 0.005 max **Li** 13.0-15.0 **Mg** rem **Mn** 0.15 min **Ni** 0.005 max **Si** 0.004 max **Other** total 0.20 max, Na 0.005 max	AMS 4386; 4397 **ASTM** B90 (LA141A) **MIL SPEC** MIL-M-46130 **SAE** J466 (LA141A)
M14142	Magnesium Alloy-Wrought Products	**Al** 0.05 max **Cu** 0.05 max **Fe** 0.005 max **Li** 12.0-15.0 **Mg** rem **Mn** 0.15 max **Ni** 0.005 max **Si** 0.50-0.6 **Other** Na 0.005 max	ASTM (LS141A) **MIL SPEC** MIL-M-46130; MIL-M-46143
M14145	Magnesium Alloy-Wrought Products	**Ag** 2.0-3.0 **Al** 0.05 max **Cu** 0.05 max **Fe** 0.005 max **Li** 12.0-15.0 **Mg** rem **Mn** 0.15 max **Ni** 0.005 max **Si** 1.5-2.0 **Zn** 4.5-5.0 **Other** total 0.20 max, Na 0.005 max	ASTM (LZ145A) **MIL SPEC** MIL-M-46130; MIL-M-46143
M15100	Magnesium Alloy-Extruded Shapes-Wire	**Cu** 0.05 max **Mg** rem **Mn** 1.2 min **Ni** 0.01 max **Si** 0.10 nom **Other** total 0.30 max, Ca 0.15 nom	ASTM B107 (M1A); B275 (M1A) **FED** QQ-M-31; QQ-M-40 **SAE** J466 (M1A)
M15101	Magnesium Alloy- Casting	**Cu** 0.08 max **Mg** rem **Mn** 1.3 min **Ni** 0.01 max **Si** 0.10 max **Other** total 0.20 max	ASTM B93 (M1B); B275 (M1B)
M15102	Magnesium Alloy- Forgings	**Al** 0.01 max **Cu** 0.02 max **Fe** 0.03 max **Mg** rem **Mn** 0.9-1.2 **Ni** 0.001 max **Other** each 0.05 max; total 0.30 max	ASTM B91 (MIC); B275 (MIC)
M16100	Magnesium Alloy- Wrought Products	**Mg** rem **Zn** 1.0-1.5 **Other** total 0.30 max, Rare Earths 0.12-0.22	ASTM B90 (ZE10A); B275 (ZE10A) **MIL SPEC** MIL-M-46037 **SAE** J466 (ZE10A)
M16210	Magnesium Alloy	**Mg** rem **Zn** 2.0-2.6 **Zr** 0.45-0.8 **Other** total 0.30 max	AMS 4387 **ASTM** B275 (ZK21A) **MIL SPEC** MIL-M-46039
M16400	Magnesium Alloy- Extruded Shapes	**Cu** 0.10 max **Mg** rem **Ni** 0.01 max **Zn** 3.5-4.5 **Zr** 0.45 min **Other** total 0.30 max	ASTM B107 (ZK40A); B275 (ZK40A) **SAE** J466 (ZK40A)
M16410	Magnesium Alloy- Casting	**Cu** 0.10 max **Mg** rem **Mn** 0.15 max **Ni** 0.01 max **Zn** 3.5-5.0 **Zr** 0.40-1.0 **Other** total 0.30 max, Rare Earths 0.75-1.75	AMS 4439 **ASTM** B80 (ZE41A); B275 (ZE41A) **SAE** J465 (ZE41A)
M16510	Magnesium Alloy- Casting	**Cu** 0.10 max **Mg** rem **Ni** 0.01 max **Zn** 3.6-5.5 **Zr** 0.50-1.0 **Other** total 0.30 max	AMS 4443 **ASTM** B80 (ZK51A); B275 (ZK51A) **FED** QQ-M-56 **MIL SPEC** MIL-M-46062 **SAE** J465 (ZK51A)
M16600	Magnesium Alloy-Wrought Products-Forgings	**Mg** rem **Zn** 4.8-6.2 **Zr** 0.45 min **Other** total 0.30 max	AMS 4352; 4362 **ASTM** B91 (ZK60A); B107 (ZK60A); B275 (ZK60A) **FED** QQ-M-31; QQ-M-40 **SAE** J466 (ZK60A)

The chemical compositions listed are for identification purposes and should not be used in lieu of the cross referenced specifications.

UNIFIED NUMBER	DESCRIPTION	CHEMICAL COMPOSITION	CROSS REFERENCE SPECIFICATIONS
M16601	Magnesium Alloy	**Mg** rem **Zn** 4.8-6.8 **Zr** 0.45 min **Other** total 0.30 max	**ASTM** B275 (ZK60B) **MIL SPEC** MIL-M-26696
M16610	Magnesium Alloy- Casting	**Mg** rem **Zn** 5.5-6.5 **Zr** 0.6-1.0 **Other** total 0.30 max	**AMS** 4444 **ASTM** B80 (ZK61A); B275 (ZK61A); B403 (ZK61A) **FED** QQ-M-56 **MIL SPEC** MIL-M-46062 **SAE** J465 (ZK61A)
M16620	Magnesium Alloy- Casting	**Cu** 0.10 max **Mg** rem **Ni** 0.01 max **Th** 1.4-2.2 **Zn** 5.2-6.2 **Zr** 0.50-1.0 **Other** total 0.30 max	**AMS** 4438 **ASTM** B80 (ZH62A); B275 (ZH62A) **FED** QQ-M-56 **MIL SPEC** MIL-M-46062 **SAE** J465 (ZH62A)
M16630	Magnesium Alloy- Casting	**Mg** rem **Zn** 5.5-6.0 **Zr** 0.40-1.0 **Other** total 0.30 max, Rare Earths 2.1-3.0	**AMS** 4425 **ASTM** B80 (ZE63A); B275 (ZE63A) **MIL SPEC** MIL-M-46062 **SAE** J465 (ZE63A)
M18010	Magnesium Alloy- Casting	**Mg** rem **Zr** 0.40-1.0 **Other** total 0.30 max	**ASTM** B80 (K1A); B275 (K1A); B403 (K1A)
M18220	Magnesium Alloy- Casting	**Ag** 2.0-3.0 **Cu** 0.10 max **Mg** rem **Ni** 0.01 max **Zr** 0.40-1.0 **Other** total 0.30 max, Rare Earths 1.8-2.5	**AMS** 4418 **ASTM** B80 (QE22A); B199 (QE22A); B403 (QE22A) **FED** QQ-M-55; QQ-M-56 **MIL SPEC** MIL-M-46062 **SAE** J465 (QE22A)
M18540	Magnesium Alloy	**Al** 3.0-4.0 **Cu** 0.05 max **Mg** rem **Mn** 0.20 min **Ni** 0.01 max **Si** 0.30 max **Sn** 4.0-6.0 **Zn** 0.30 max **Other** total 0.30 max	**ASTM** B275 (TA54A) **FED** QQ-M-40
M19001	Magnesium Brazing Filler Metal	**Al** 8.3-9.7 **Be** 0.0002-0.0008 **Cu** 0.05 **Fe** 0.005 max **Mg** rem **Mn** 0.15 min **Ni** 0.005 max **Si** 0.05 max **Zn** 1.7-2.3 **Other** 0.30	**ASME** SFA5.8 (BMg-1) **AWS** A5.8 (BMg-1) **FED** QQ-B-655
M19980	Magnesium Remelt Alloy	**Cu** 0.02 max **Mg** 99.80 min **Mn** 0.10 max **Ni** 0.001 max **Pb** 0.01 max **Sn** 0.01 max **Other** each 0.05 max	**ASTM** B92 (9980A); B275 (9980A)
M19990	Magnesium Remelt Alloy	**Al** 0.003 max **B** 0.00007 max or 0.00003 **Cd** 0.0001 max or 0.00005 **Fe** 0.04 max **Mg** 99.90 min **Mn** 0.004 max **Ni** 0.001 max **Si** 0.005 max **Other** each 0.01 max	**ASTM** B92 (9990A); B275 (9990A)
M19991	Magnesium Remelt Alloy	**Al** 0.005 max **Fe** 0.011 max **Mg** 99.90 min **Mn** 0.004 max **Ni** 0.001 max **Si** 0.005 max **Other** each 0.01 max	**ASTM** B92 (9990B); B275 (9990B)
M19995	Magnesium Remelt Alloy	**Al** 0.01 max **B** 0.00003 max **Cd** 0.00005 max **Fe** 0.003 max **Mg** 99.95 min **Mn** 0.004 max **Ni** 0.001 max **Si** 0.005 max **Ti** 0.01 max **Other** each 0.005 max	**ASTM** B92 (9995A); B275 (9995A)
M19998	Magnesium Remelt Alloy	**Al** 0.004 max **B** 0.00003 max **Cd** 0.00005 max **Cu** 0.0005 max **Fe** 0.002 max **Mg** 99.98 min **Mn** 0.002 max **Ni** 0.0005 max **Pb** 0.001 max **Si** 0.003 max **Ti** 0.001 max **Other** each 0.005 max	**ASTM** B92 (9998A); B275 (9998A)
M26800	Manganese Brazing Alloy	**B** 0.5-1.1 **C** 0.06 max **Co** 14.0-18.0 **Mn** rem **Ni** 14.0-18.0 **Si** 1.00 max	**AMS** 4780
M27200	Manganese Thermostat Alloy	**Cu** 18 **Mn** 72 **Ni** 10	**ASTM** B388
M29350	Electrolytic Manganese	**Mn** 93-94 **N** 5.5-6.5 **S** 0.035 max	**ASTM** A601 (E)
M29450	Electrolytic Manganese	**Mn** 94-95 **N** 4.0-5.4 **S** 0.035 max	**ASTM** A601 (D)
M29951	Electrolytic Manganese	**H** 0.015 max **Mn** 99.5 min **S** 0.030 max	**ASTM** A601 (A)
M29952	Electrolytic Manganese	**H** 0.005 max **Mn** 99.5 min **S** 0.030 max	**ASTM** A601 (B)
M29953	Electrolytic Manganese	**H** 0.0010 max **Mn** 99.5 min **S** 0.030 max	**ASTM** A601 (C)
M29954	Electrolytic Manganese	**H** 0.0030 max **Mn** 99.5 max **S** 0.035 max	**ASTM** A601 (F)

The chemical compositions listed are for identification purposes and should not be used in lieu of the cross referenced specifications.

UNS NUMBERS ASSIGNED TO DATE

With Description of Each Material Covered and References To Documents In Which The Same or Similar Materials are Described

Nxxxxx Number Series
Nickel and Nickel Alloys

NICKEL AND NICKEL ALLOYS

UNIFIED NUMBER	DESCRIPTION	CHEMICAL COMPOSITION	CROSS REFERENCE SPECIFICATIONS
N02016	Ni-Mn-Al Thermocouple Alloy (Chromel)	**Al** 1.75-2.25 **C** 0.15 max **Fe** 0.50 max **Mn** 2.0-3.0 **Ni** rem **Si** 1.6 max	
N02061	Ni Welding Filler Metal (Nickel FM61)	**Al** 1.5 max **C** 0.15 max **Cu** 0.25 max **Fe** 1.0 max **Mn** 1.0 max **Ni** 93.0 min **P** 0.03 max **S** 0.015 max **Si** 0.75 max **Ti** 2.0-3.5	**ASME** SFA5.14 (ERNi-1); SFA5.30 (IN61) **AWS** A5.14 (ERNi-1); A5.30 (IN61) **MIL SPEC** MIL-E-21562 (EN61, RN61); MIL-I-23413 (61)
N02100	Nickel Alloy (ACI CZ-100)	**C** 1.00 max **Cu** 1.25 max **Fe** 3.00 max **Mn** 1.50 max **Ni** rem **Si** 2.00 max	
N02200	Commercially Pure Ni Alloy (Nickel 200)	**C** 0.15 max **Cu** 0.25 max **Fe** 0.40 max **Mn** 0.35 max **Ni** 99.0 min **S** 0.010 max **Si** 0.35 max	**ASME** SB160; SB161; SB162; SB163 **ASTM** B160; B161; B162; B163; B366; B725; B730
N02201	Commercially Pure Ni Alloy (Nickel 201)	**C** 0.02 max **Cu** 0.25 max **Fe** 0.40 max **Mn** 0.35 max **Ni** 99.0 min **S** 0.010 max **Si** 0.35 max	**AMS** 5553 **ASME** SB160; SB161; SB162; SB163 **ASTM** B160; B161; B162; B163; B366; B725; B730
N02205	Commercially Pure Ni Alloy (Nickel 205)	**C** 0.15 max **Cu** 0.15 max **Fe** 0.20 max **Mg** 0.01-0.08 **Mn** 0.35 max **Ni** 99.0 min **S** 0.008 max **Si** 0.15 max **Ti** 0.01-0.05	**AMS** 5555 **ASTM** F1; F2; F3; F9 **MIL SPEC** MIL-N-46025
N02211	Nickel Alloy, Solution Strengthened (Nickel 211)	**C** 0.20 max **Cu** 0.25 max **Fe** 0.75 max **Mn** 4.25-5.25 **Ni** 93.7 min **S** 0.015 max **Si** 0.15 max	**ASTM** F290
N02220	Commercially Pure Ni Alloy (Nickel 220)	**C** 0.15 max **Cu** 0.10 max **Fe** 0.10 max **Mg** 0.01-0.08 **Mn** 0.20 max **Ni** 99.00 min **S** 0.008 max **Si** 0.01-0.05 **Ti** 0.01-0.05	**ASTM** F239
N02225	Commercially Pure Ni Alloy (Nickel 225)	**C** 0.15 max **Cu** 0.10 max **Fe** 0.10 max **Mg** 0.01-0.08 **Mn** 0.20 max **Ni** 99.00 min **S** 0.008 max **Si** 0.15-0.25 **Ti** 0.01-0.05	**ASTM** F239
N02230	Commercially Pure Ni Alloy (Nickel 230)	**C** 0.15 max **Cu** 0.10 max **Fe** 0.10 max **Mg** 0.04-0.08 **Mn** 0.15 max **Ni** 99.00 min **S** 0.008 max **Si** 0.010-0.035 **Ti** 0.005 max	**ASTM** F239
N02233	Commercially Pure Ni Alloy (Nickel 233)	**C** 0.15 max **Cu** 0.10 max **Fe** 0.10 max **Mg** 0.01-0.10 **Mn** 0.30 max **Ni** 99.00 min **S** 0.008 max **Si** 0.10 max **Ti** 0.005 max	**ASTM** F1; F2; F3; F4
N02270	Commercially Pure Ni Alloy (Nickel 270)	**C** 0.02 max **Co** 0.001 max **Cr** 0.001 max **Cu** 0.001 max **Fe** 0.005 max **Mg** 0.001 max **Mn** 0.001 max **Ni** 99.97 min **S** 0.001 max **Si** 0.001 max **Ti** 0.001 max	**ASTM** F1; F2; F3; F239
N02290	Commercially Pure Nickel	**Al** 0.001 max **C** 0.006 max **Cr** 0.001 max **Cu** 0.02 max **Fe** 0.015 max **Mg** 0.001 max **Mn** 0.001 max **N** 0.001 max **Ni** rem **O** 0.025 max **S** 0.0008 max **Si** 0.001 max	
N03220	Beryllium Nickel, Precipitation Hardenable Castings (M220C)	**Be** 1.80-2.30 **C** 0.30-0.50 **Ni** rem	
N03260	Nickel - Thoria Dispersion Strengthened	**C** 0.02 max **Co** 0.20 max **Cr** 0.05 max **Cu** 0.15 max **Fe** 0.05 max **Ni** rem **S** 0.0025 max **Ti** 0.05 max **Other** ThO_2 1.80-2.60	**AMS** 5865; 5890
N03300	Nickel Alloy, Precipitation Hardenable (Permanickel 300)	**C** 0.40 max **Cu** 0.25 max **Fe** 0.60 max **Mg** 0.20-0.50 **Mn** 0.50 max **Ni** 97.0 min **S** 0.01 max **Si** 0.35 max **Ti** 0.20-0.60	**ASTM** F290
N03301	Nickel Alloy, Precipitation Hardenable (Duranickel 301)	**Al** 4.00-4.75 **C** 0.30 max **Cu** 0.25 max **Fe** 0.60 max **Mn** 0.50 max **Ni** 93.0 min **S** 0.01 max **Si** 1.00 max **Ti** 0.25-1.00	

The chemical compositions listed are for identification purposes and should not be used in lieu of the cross referenced specifications.

UNIFIED NUMBER	DESCRIPTION	CHEMICAL COMPOSITION	CROSS REFERENCE SPECIFICATIONS
N03360	Commercially Pure Be-Ni Alloy Precipitation Hardenable	**Be** 1.85-2.05 **Ni** rem **Ti** 0.4-0.6	**MIL SPEC** MIL-B-63573
N04019	Nickel Base Castings	**C** 0.25 max **Cu** 27.0-31.0 **Fe** 2.50 max **Mn** 1.50 max **Ni** 60.0 min **S** 0.015 max **Si** 3.50-4.50 max	**AMS** 4892; 4893 **FED** QQ-N-288
N04060	Ni-Cu Welding Filler Metal (Monel FM60)	**Al** 1.25 max **C** 0.15 max **Cu** bal **Fe** 2.5 max **Mn** 4.0 max **Ni** 62.0-69.0 **P** 0.02 max **S** 0.015 max **Si** 1.25 max **Ti** 1.5-3.0	**ASME** SFA5.14 (ERNiCu-7); SFA5.30 (IN60) **AWS** A5.14 (ERNiCu-7); A5.30 (IN60) **MIL SPEC** MIL-E-21562 (EN60, RN60); MIL-I-23413 (60)
N04400	Ni-Cu Alloy Solid Solution Strengthened (Monel 400)	**C** 0.3 max **Cu** bal **Fe** 2.50 max **Mn** 2.00 max **Ni** 63.00-70.00 **S** 0.024 max **Si** 0.50 max	**AMS** 4544; 4574; 4575; 4675; 4730; 4731; 7233; **ASME** SB127; SB163; SB164; SB165; SB564 **ASTM** B127; B163; B164; B165; B366; B564 **FED** QQ-N-281 **MIL SPEC** MIL-T-1368; MIL-T-23520; MIL-N-24106
N04401	Ni-Cu Alloy, Solid Solution Strengthened (Monel 401)	**C** 0.10 max **Co** 0.25 max **Cu** rem **Fe** 0.75 max **Mn** 2.25 max **Ni** 40.0-45.0 **S** 0.015 max **Si** 0.25 max	
N04404	Ni-Cu Alloy Solid Solution Strengthened (Monel 404)	**Al** 0.05 max **C** 0.15 max **Cu** bal **Fe** 0.50 max **Mn** 0.10 max **Ni** 52.0-57.0 **S** 0.024 max **Si** 0.10 max	**ASTM** F96
N04405	Ni-Cu Alloy Solid Solution Strengthened (Monel R405)	**C** 0.30 max **Cu** bal **Fe** 2.5 max **Mn** 2.0 max **Ni** 63.0-70.0 **S** 0.025-0.060 **Si** 0.50 max	**AMS** 4674; 7234 **ASME** SB164 **ASTM** B164 **FED** QQ-N-281
N05500	Ni-Cu Alloy Precipitation Hardenable (Monel K500)	**Al** 2.30-3.15 **C** 0.25 max **Cu** bal **Fe** 2.00 max **Mn** 1.50 max **Ni** 63.0-70.0 **S** 0.01 max **Si** 0.50 max **Ti** 0.35-0.85	**AMS** 4676 **FED** QQ-N-286 **MIL SPEC** MIL-N-24549
N05502	Ni-Cu Alloy Precipitation Hardenable (Monel 502)	**Al** 2.50-3.50 **C** 0.10 max **Cu** bal **Fe** 2.00 max **Mn** 1.50 max **Ni** 63.0-70.0 **S** 0.010 max **Si** 0.5 max **Ti** 0.50 max	**AMS** 4677
N06001	Ni-Cr Alloy Solid Solution Strengthened (Hastelloy F)	**C** 0.05 max **Cb** 1.8-2.5 **Co** 2.5 max **Cr** 21.0-23.0 **Fe** bal **Mn** 1.00-2.00 **Mo** 5.5-7.5 **Ni** 44.0-47.0 **P** 0.04 max **S** 0.03 max **Si** 1.00 max **W** 1.00 max	**ASME** SB436 **ASTM** B366; B436
N06002	Ni-Cr Alloy Solid Solution Strengthened (Hastelloy X)	**C** 0.05-0.15 **Co** 0.5-2.5 **Cr** 20.5-23.0 **Fe** 17.0-20.0 **Mn** 1.00 max **Mo** 8.0-10.0 **Ni** bal **P** 0.040 max **S** 0.030 max **Si** 1.00 max **W** 0.20-1.0	**AMS** 5390; 5536; 5587; 5588; 5754; 5798; 7237 **ASME** SB435; SB572; SB619; SB622; SB626; SFA5.14 (ERNiCrMo-2) **ASTM** B366; B435; A567; B572; B619, B622; B626 **AWS** A5.14 (ERNiCrMo-2)
N06003	Ni-Cr Alloy Solid Solution Strengthened (Nichrome V)	**C** 0.15 max **Cr** 19-21 **Fe** 1.0 max **Mn** 2.5 max **Ni** bal **S** 0.01 max **Si** 0.75-1.6	**AMS** 5676; 5677; 5682 **ASTM** B344 **MIL SPEC** MIL-C-5031 (Class 7)
N06004	Ni-Cr Alloy Solid Solution Strengthened (Nichrome)	**C** 0.15 max **Cr** 14-18 **Fe** bal **Mn** 1.0 max **Ni** 57 min **S** 0.01 max **Si** 0.75-1.6	**ASTM** B344
N06005	Ni-Cr Alloy Solid Solution Strengthened (Eatonite)	**C** 2.40 nom **Co** 10.00 nom **Cr** 29.00 nom **Fe** 6.5 max **Ni** 39.00 nom **Si** 0.70 nom **W** 15.00 nom	**SAE** J775 (VF3)
N06006	Ni-Cr Alloy Solid Solution Strengthened (ACI HX)	**C** 0.35-0.75 **Cr** 15.0-19.0 **Fe** rem **Mn** 2.00 max **Mo** 0.50 max **Ni** 64.0-68.0 **P** 0.04 max **S** 0.04 max **Si** 2.50 max	**ASTM** A297
N06007	Ni-Cr Alloy (Hastelloy G)	**C** 0.05 max **Cb** 1.75-2.5 **Co** 2.5 max **Cr** 21.0-23.5 **Cu** 1.5-2.5 **Fe** 18.0-21.0 **Mn** 1.0-2.0 **Mo** 5.5-7.5 **Ni** Rem **P** 0.04 max **S** 0.03 max **Si** 1.0 max **W** 1.0 max	**ASME** SB366; SB619; SFA5.14 (ERNiCrMo-1) **ASTM** B366; B581; B582; B619 **AWS** A5.14 (ERNiCrMo-1)
N06008	Ni-Cr 30 Alloy Solid Solution Strengthened	**Al** 0.20 max **C** 0.15 max **Cr** 29.0-31.0 **Fe** 1.0 max **Mn** 0.10 max **Ni** rem **P** 0.030 max **S** 0.010 max **Si** 0.75-1.60	
N06009	Ni-20 Cr, Columbium Stabilized	**C** 0.15 max **Cb** 0.75-1.50 **Cr** 19.0-21.0 **Fe** 1.00 max **Mn** 2.5 max **Ni** rem **S** 0.010 max **Si** 0.75-1.60	

The chemical compositions listed are for identification purposes and should not be used in lieu of the cross referenced specifications.

UNIFIED NUMBER	DESCRIPTION	CHEMICAL COMPOSITION	CROSS REFERENCE SPECIFICATIONS
N06010	Ni-Cr 10 Thermocouple Alloy (Alumel)	**Al** 0.20 max **C** 0.15 max **Cr** 9.0-11.0 **Fe** 0.50 max **Mn** 0.10 max **Ni** rem **Si** 1.60 max	
N06017	Ni-Cr Alloy ERNiCrMo-9 replaced by N06985		
N06022	Ni-Cr-Mo Alloy (Hastelloy C-22)	**C** 0.015 max **Co** 2.5 max **Cr** 20.0-22.5 **Fe** 2.0-6.0 **Mn** 0.50 max **Mo** 12.5-14.5 **Ni** rem **P** 0.02 max **S** 0.02 max **Si** 0.08 max **V** 0.35 max **W** 2.5-3.5	**ASME** SB366; SB574; SB575; SB619; SB622; SB626 **ASTM** B366; B574; B575; B619; B622; B626
N06030	Wrought Ni-Cr-Fe-Mo-Cu Alloy Solid Solution Strengthened (Hastelloy G-30)	**C** 0.03 max **Cb** 0.30-1.50 **Co** 5.0 max **Cr** 28.0-31.5 **Cu** 1.0-2.4 **Fe** 13.0-17.0 **Mn** 1.5 **Mo** 4.0-6.0 **P** 0.04 max **S** 0.02 max **Si** 0.8 max **W** 1.5-4.0	**ASME** SB366; SB581; SB582; SB619; SB622; SB626 **ASTM** B366; B581; B582; B619; B622; B626
N06040	Ni-Cr Alloy Solid Solution Strengthened (ACI CY-40)	**C** 0.40 max **Cr** 14.0-17.0 **Fe** 11.0 max **Mn** 1.50 max **Ni** rem **Si** 3.00 max	**ASTM** A296
N06050	Ni-Cr Alloy Solid Solution Strengthened (ACI HX-50)	**C** 0.40-0.60 **Cr** 15.0-19.0 **Fe** rem **Mn** 1.50 max **Mo** 0.50 max **Ni** 64.0-68.0 **P** 0.04 max **S** 0.04 max **Si** 0.50-2.00	**ASTM** A608
N06062	Ni-Cr Alloy Solid Solution Strengthened (Inconel FM62)	**C** 0.08 max **Cb** 1.50-3.00 **Cr** 14.00-17.00 **Cu** 0.50 max **Fe** 6.00-10.00 **Mn** 1.00 max **Ni** 70.00 min **P** 0.030 max **S** 0.015 max **Si** 0.35 max	**AMS** 5679 **ASME** SFA5.14 (ERNiCrFe-5) **AWS** A5.14 (ERNiCrFe-5) **MIL SPEC** MIL-E-21562 (EN62, RN62); MIL-I-23413 (62)
N06075	Ni-Cr Alloy Solid Solution Strengthened (Nimonic 75)	**C** 0.08-0.15 **Cr** 18.0-21.0 **Cu** 0.50 max **Fe** 5.00 max **Mn** 1.00 max **Ni** rem **Si** 1.00 max **Ti** 0.20-0.60	
N06082	Ni-Cr Welding Filler Metal (Inconel FM82)	**C** 0.10 max **Cb** 2.0-3.0 includes Ta **Cr** 18.0-22.0 **Cu** 0.50 max **Fe** 3.0 max **Mn** 2.5-3.5 **Ni** 67.0 min **P** 0.03 max **S** 0.015 max **Si** 0.50 max **Ti** 0.75 max	**ASME** SFA5.14 (ERNiCr-3); SFA5.30 (IN82) **AWS** A5.14 (ERNiCr-3); A5.30 (IN82) **MIL SPEC** MIL-R-5031 (Class 8A); MIL-E-21562 (EN82, RN82); MIL-I-23413 (82)
N06102	Ni-Cr Alloy Solid Solution Strengthened (IN-102)	**Al** 0.30-0.60 **B** 0.003-0.008 **C** 0.08 max **Cb** 2.75-3.25 **Cr** 14.0-16.0 **Fe** 5.0-9.0 **Mg** 0.01-0.05 **Mn** 0.75 max **Mo** 2.75-3.25 **Ni** bal **P** 0.010 max **S** 0.010 max **Si** 0.40 max **Ti** 0.40-0.70 **W** 2.75-3.25 **Zr** 0.01-0.05	**ASTM** B445; B518; B519
N06110	Austenitic Corrosion Resistant Nickel Alloy (Allcorr)	**Al** 1.50 max **C** 0.15 max **Cb** 2.00 max **Co** 12.0 max **Cr** 27.0-33.0 **Mo** 8.00-12.0 **Ni** rem **Ti** 1.50 max **W** 4.00 max	
N06132	Ni-Cr Alloy Solid Solution Strengthened (Inconel WE132) replaced by W86132		
N06333	Ni-Cr Alloy Solid Solution Strengthened (RA333)	**C** 0.08 max **Co** 2.50-4.00 **Cr** 24.00-27.00 **Cu** 0.50 max **Fe** bal **Mn** 2.00 max **Mo** 2.50-4.00 **Ni** 44.00-47.00 **P** 0.030 max **Pb** 0.025 max **S** 0.030 max **Si** 0.75-1.50 **Sn** 0.025 max **W** 2.50-4.00	**AMS** 5717 **ASTM** B718; B719; B722; B723; B726
N06455	Ni-Cr Alloy (Hastelloy C-4)	**C** 0.015 max **Co** 2.0 max **Cr** 14.0-18.0 **Fe** 3.0 max **Mn** 1.0 max **Mo** 14.0-17.0 **Ni** bal **P** 0.04 max **S** 0.03 max **Si** 0.08 max **Ti** 0.70 max	**ASME** SB366; SB574; SB575; SB619; SB622; SB626; SFA5.14 (ERNiCrMo-7) **ASTM** B574; B575; B619; B622; B626 **AWS** A5.14 (ERNiCrMo-7)
N06600	Ni-Cr Alloy Solid Solution Strengthened (Inconel 600)	**C** 0.15 max **Cr** 14.00-17.00 **Cu** 0.50 max **Fe** 6.00-10.00 **Mn** 1.00 max **Ni** 72.0 min **S** 0.015 max **Si** 0.50 max	**AMS** 5540; 5580; 5665; 5687; 7232 **ASME** SB163; SB166; SB167; SB168; SB564 **ASTM** B163; B166; B167; B168; B366; B516; B517; B564 **FED** QQ-W-390 **MIL SPEC** MIL-R-5031 (Class 8); MIL-T-23227; MIL-N-23228; MIL-N-23229
N06601	Ni-Cr Alloy Solid Solution Strengthened (Inconel 601)	**Al** 1.0-1.7 **C** 0.1 max **Cr** 21.0-25.0 **Cu** 1.0 max **Fe** bal **Mn** 1.0 max **Ni** 58.0-63.0 **S** 0.015 max **Si** 0.50 max	**AMS** 5715; 5870

The chemical compositions listed are for identification purposes and should not be used in lieu of the cross referenced specifications.

UNIFIED NUMBER	DESCRIPTION	CHEMICAL COMPOSITION	CROSS REFERENCE SPECIFICATIONS
N06617	Ni-Cr Alloy Solid Solution Strengthened (Inconel 617)	Al 0.80-1.50 B 0.006 max C 0.05-0.15 Co 10.0-15.0 Cr 20.0-24.0 Cu 0.50 max Fe 3.00 max Mn 1.00 max Mo 8.00-10.0 Ni 44.5 min S 0.015 max Si 1.00 max Ti 0.60 max	
N06625	Ni-Cr Alloy Solid Solution Strengthened (Inconel 625)	Al 0.40 max C 0.10 max Cb 3.15-4.15 Cr 20.0-23.0 Fe 5.0 max Mn 0.50 max Mo 8.0-10.0 Ni bal P 0.015 max S 0.015 max Si 0.50 max Ti 0.40 max	AMS 5401; 5402; 5581; 5599; 5666; 5837 ASME SB443; SB444; SB446; SFA5.14 (ERNiCrMo-3) ASTM B366; B443; B444; B446; B704; B705 AWS A5.14 (ERNiCrMo-3) MIL SPEC MIL-E-21562 (RN625) (EN625)
N06635	Ni-Cr-Mo Alloy (Hastelloy S)	Al 0.10-0.50 B 0.015 max C 0.02 max Co 2.00 max Cr 14.5-17.0 Cu 0.35 max Fe 3.00 max Mn 0.30-1.00 Mo 14.0-16.5 Ni rem P 0.020 max S 0.015 max Si 0.20-0.75 W 1.00 max Other La 0.01-0.10	AMS 5711; 5838; 5873
N06690	Ni-Cr Alloy Solid Solution Strengthened (Inconel 690)	C 0.05 max Cr 27.0-31.0 Cu 0.50 max Fe 7.0-11.0 Mn 0.50 max Ni 58.0 min S 0.015 max Si 0.50 max	ASME Code Case N-20 (1484); SB163; SB166; SB167; SB168 ASTM B163; B166; B167; B168; B366
N06782	Ni-Cr Alloy Solid Solution Strengthened (X-782)	C 2.00 nom Co 0.50 nom Cr 26.00 nom Fe 4.0 max Mn 0.30 nom Ni bal Si 0.30 nom W 8.75 nom	SAE J775 (VF4)
N06804	Ni-Cr Alloy Solid Solution Strengthened (Incoloy 804)	Al 0.60 max C 0.10 max Cr 28.0-31.0 Cu 0.50 max Fe rem Mn 1.50 max Ni 39.0-43.0 S 0.015 max Si 0.75 max Ti 1.20 max	
N06975	Ni-Cr Alloy (Hastelloy G-2)	C 0.03 max Cr 23.0-26.0 Cu 0.70-1.20 Fe bal Mn 1.0 max Mo 5.0-7.0 Ni 47.0-52.0 P 0.03 max S 0.03 max Si 1.0 max Ti 0.70-1.50	ASME SB366; SB581; SB582; SB619; SFA5.14 (ERNiCrMo-8) ASTM B366; B581; B582; B619 AWS A5.14 (ERNiCrMo-8)
N06985	Ni-Cr Alloy ERNiCrMo-9 (Hastelloy G-3)	C 0.15 max Co 5.0 Cr 21.0-23.5 Cu 1.5-2.5 Fe 18.0-21.0 Mn 1.0 max Mo 6.0-8.0 Ni rem P 0.04 max S 0.03 max Si 1.0 max W 1.5 max Other Cb + Ta 0.50 max	ASME SB581; SB582; SB619; SB622; SB626; SFA5.14 (ERNiCrMo-9) ASTM B581; B582; B619; B622; B626 AWS A5.14 (ERNiCrMo-9)
N07001	Ni-Cr Alloy Precipitation Hardenable (Waspaloy)	Al 1.20-1.60 B 0.003-0.01 C 0.03-0.10 Co 12.00-15.00 Cr 18.00-21.00 Cu 0.50 max Fe 2.00 max Mn 1.00 max Mo 3.50-5.00 Ni bal P 0.030 max S 0.030 max Si 0.75 max Ti 2.75-3.25 Zr 0.02-0.12	AMS 5544; 5586; 5704; 5706; 5707; 5708; 5709 ASTM B637
N07002	Ni-Cr Alloy Precipitation Hardenable (TPM)	Al 0.05 nom C 0.05 nom Co 0.50 nom Cr 16.00 nom Mn 2.30 nom Ni bal Si 0.05 nom Ti 3.10 nom	SAE J775 (HEV2)
N07012	Cr-Co-Mo-W Nickel Alloy (AF2-1DA)	Al 4.20-4.80 B 0.01-0.02 Bi 0.00005 max C 0.30-0.35 Co 9.50-10.50 Cr 11.5-12.5 Fe 1.00 max Mn 0.10 max Mo 2.50-3.50 N 0.005 max Ni rem O 0.010 max P 0.015 max Pb 0.002 max S 0.015 max Si 0.10 max Ta 1.00-2.00 Ti 2.75-3.25 W 5.50-6.50 Zr 0.05-0.15	AMS 5855; 5856; 5881
N07013	Cr-Co-Mo Nickel Alloy Castings	Al 3.20-3.60 B 0.010-0.020 C 0.07-0.20 Cb 0.10 max Co 8.50-9.50 Cr 12.2-13.0 Fe 0.50 max Hf 0.75-1.05 Mn 0.10 max Mo 1.70-2.10 Ni rem P 0.015 max S 0.015 max Si 0.10 max Ta 3.85-4.50 Ti 3.85-4.15 W 3.85-4.50 Zr 0.05-0.14 Other Al + Ti 7.30-7.70	AMS 5404
N07031	Ni-Cr Alloy Precipitation Hardenable (Pyromet 31)	Al 1.00-1.70 B 0.003-0.007 C 0.03-0.06 Cr 22.0-23.0 Cu 0.60-1.20 Fe rem Mn 0.20 max Mo 1.70-2.30 Ni 55.0-58.0 P 0.015 max S 0.015 max Si 0.20 max Ti 2.10-2.60	

The chemical compositions listed are for identification purposes and should not be used in lieu of the cross referenced specifications.

UNIFIED NUMBER	DESCRIPTION	CHEMICAL COMPOSITION	CROSS REFERENCE SPECIFICATIONS
N07041	Ni-Cr Alloy Precipitation Hardenable (Rene 41)	**Al** 1.40-1.80 **B** 0.0030-0.010 **C** 0.12 max **Co** 10.00-12.00 **Cr** 18.00-20.00 **Fe** 5.00 max **Mn** 0.10 max **Mo** 9.00-10.50 **Ni** bal **S** 0.015 max **Si** 0.50 max **Ti** 3.00-3.30	**AMS** 5399; 5545; 5713; 5800; 7469
N07069	Ni-Cr Welding Filler Metal (Inconel FM69)	**Al** 0.40-1.00 **C** 0.08 **Cb** 0.70-1.20 includes Ta **Cr** 14.0-17.0 **Cu** 0.50 max **Fe** 5.0-9.0 **Mn** 1.0 max **Ni** 70.0 min **P** 0.03 max **S** 0.015 max **Si** 0.50 max **Ti** 2.00-2.75	**ASME** SFA5.14 (ERNiCrFe-7) **AWS** A5.14 (ERNiCrFe-7) **MIL SPEC** MIL-E-21562 (RN69)
N07080	Ni-Cr Alloy Precipitation Hardenable (Nimonic 80A)	**Al** 1.0-1.8 **B** 0.008 max **C** 0.10 max **Co** 2.0 max **Cr** 18.0-21.0 **Cu** 0.2 max **Fe** 3.0 max **Mn** 1.0 max **Ni** bal **S** 0.015 max **Si** 1.00 max **Ti** 1.8-2.7	**ASTM** B637
N07090	Ni-Cr Alloy Precipitation Hardenable (Nimonic 90)	**Al** 0.8-2.0 **C** 0.13 max **Co** 15.0-21.0 **Cr** 18.0-21.0 **Fe** 3.0 max **Mn** 1.0 max **Ni** bal **Si** 1.5 max **Ti** 1.8-3.0	**AMS** 5829
N07092	Ni-Cr Alloy Precipitation Hardenable (Inconel FM92)	**C** 0.08 max **Cr** 14.00-17.00 **Cu** 0.50 max **Fe** 8.0 max **Mn** 2.00-2.75 **Ni** 67.00 min **P** 0.030 max **S** 0.015 max **Si** 0.35 max **Ti** 2.50-3.50	**AMS** 5675 **ASME** SFA5.14 (ERNiCrFe-6); SFA5.30 (IN6A) **AWS** A5.14 (ERNiCrFe-6); A5.30 (IN6A) **MIL SPEC** MIL-E-21562 (EN6A, RN6A); MIL-I-23413 (6A)
N07252	Ni-Cr Alloy Precipitation Hardenable (M252)	**Al** 0.75-1.25 **B** 0.003-0.01 **C** 0.10-0.20 **Co** 9.00-11.00 **Cr** 18.00-20.00 **Fe** 5.00 max **Mn** 0.50 max **Mo** 9.00-10.50 **Ni** bal **P** 0.015 max **S** 0.015 max **Si** 0.50 max **Ti** 2.25-2.75	**AMS** 5551; 5756; 5757 **ASTM** B637
N07263	Ni-Cr Alloy Precipitation Hardenable (Nimonic Alloy 263)	**Al** 0.3-0.6 **C** 0.04-0.08 **Co** 19.0-21.0 **Cr** 19.0-21.0 **Cu** 0.20 max **Fe** 0.7 max **Mn** 0.60 max **Mo** 5.6-6.1 **Ni** rem **P** 0.015 max **S** 0.007 max **Si** 0.40 max **Ti** 1.9-2.4 **Other** Al+Ti 2.4-2.8	**AMS** 5872
N07500	Ni-Cr Alloy Precipitation Hardenable (Udimet 500)	**Al** 2.50-3.25 **B** 0.003-0.01 **C** 0.15 max **Co** 13.00-20.00 **Cr** 15.00-20.00 **Cu** 0.15 max **Fe** 4.00 max **Mn** 0.75 max **Mo** 3.00-5.00 **Ni** bal **P** 0.015 max **S** 0.015 max **Si** 0.75 max **Ti** 2.50-3.25	**AMS** 5384; 5751; 5753 **ASTM** A567; B637
N07702	Ni-Cr Alloy Precipitation Hardenable (Inconel 702)	**Al** 2.75-3.75 **C** 0.10 max **Cr** 14.0-17.0 **Cu** 0.5 max **Fe** 2.0 max **Mn** 1.0 max **Ni** bal **S** 0.01 max **Si** 0.7 max **Ti** 0.25-1.00	**AMS** 5550
N07713	Ni-Cr Alloy Precipitation Hardenable (IN-713)	**Al** 5.5-6.5 **B** 0.005-0.015 **C** 0.08-0.20 **Cb** 1.8-2.8 **Cr** 12.00-14.00 **Fe** 2.50 max **Mn** 0.25 max **Mo** 3.8-5.2 **Ni** bal **Si** 0.50 max **Ti** 0.5-1.0 **Zr** 0.05-0.15	**AMS** 5377; 5391 **ASTM** A567
N07718	Ni-Cr Alloy Precipitation Hardenable (Inconel 718)	**Al** 0.20-0.80 **B** 0.006 max **C** 0.08 max **Cb** 4.75-5.50 **Co** 1.00 max **Cr** 17.0-21.0 **Cu** 0.30 max **Fe** bal **Mn** 0.35 max **Mo** 2.80-3.30 **Ni** 50.0-55.0 **P** 0.015 max **S** 0.015 max **Si** 0.35 max **Ti** 0.65-1.15	**AMS** 5383; 5589; 5590; 5596; 5597; 5662; 5663; 5664; 5832 **MIL SPEC** MIL-N-24469 **ASTM** B637; B670
N07721	Ni-Cr Alloy Precipitation Hardenable (Inconel 721)	**Al** 0.10 max **C** 0.07 max **Cr** 15.0-17.0 **Cu** 0.20 max **Fe** 8.00 max **Mn** 2.00-2.50 **Ni** rem **S** 0.01 max **Si** 0.15 max **Ti** 2.75-3.35	
N07722	Ni-Cr Alloy Precipitation Hardenable (Inconel 722)	**Al** 0.4-1.0 **C** 0.08 max **Cr** 14.0-17.0 **Cu** 0.5 max **Fe** 5.0-9.0 **Mn** 1.0 max **Ni** 70.0 min **S** 0.01 max **Si** 0.07 max **Ti** 2.00-2.75	**AMS** 5541; 5714
N07750	Ni-Cr Alloy Precipitation Hardenable (Inconel X750)	**Al** 0.40-1.0 **C** 0.08 max **Cb** 0.70-1.20 **Cr** 14.0-17.0 **Cu** 0.5 max **Fe** 5.0-9.0 **Mn** 1.0 max **Ni** 70.0 min **S** 0.01 max **Si** 0.50 max **Ti** 2.25-2.75	**AMS** 5384; 5542; 5582; 5583; 5598; 5667; 5668; 5669; 5670; 5671; 5698; 5699; 5747; 5749; 5779; 7246 **ASME** SA637 **ASTM** B637 **MIL SPEC** MIL-N-7786; MIL-N-8550; MIL-S-23192; MIL-N-24114

The chemical compositions listed are for identification purposes and should not be used in lieu of the cross referenced specifications.

UNIFIED NUMBER	DESCRIPTION	CHEMICAL COMPOSITION	CROSS REFERENCE SPECIFICATIONS
N07751	Ni-Cr Alloy Precipitation Hardenable (Inconel 751)	**Al** 0.90-1.50 **C** 0.10 max **Cb** 0.70-1.20 **Cr** 14.0-17.0 **Cu** 0.50 max **Fe** 5.00-9.00 **Mn** 1.00 max **Ni** 70.0 min **S** 0.01 max **Si** 0.50 max **Ti** 2.00-2.60	
N07754	Ni-Cr Alloy Y Dispersion Strengthened (Inconel MA754)	**Al** 0.2-0.5 **C** 0.05 max **Cr** 19.0-23.0 **Fe** 2.5 max **Ni** rem **Ti** 0.3-0.6 **Other** Y_2O_3 0.5-0.7	
N08001	Ni-Fe-Cr Alloy Solid Solution Strengthened (ACI HW)	**C** 0.35-0.75 **Cr** 10.0-14.0 **Fe** rem **Mn** 2.00 max **Mo** 0.50 max **Ni** 58.0-62.0 **P** 0.04 max **S** 0.04 max **Si** 2.50 max	**ASTM** A297
N08002	Ni-Fe-Cr Alloy Solid Solution Strengthened (ACI HT)	**C** 0.35-0.75 **Cr** 13.0-17.0 **Fe** rem **Mn** 2.00 max **Mo** 0.50 max **Ni** 33.0-37.0 **P** 0.04 max **S** 0.04 max **Si** 2.50 max	**ASTM** A297
N08004	Ni-Fe-Cr Alloy Solid Solution Strengthened (ACI HU)	**C** 0.35-0.75 **Cr** 17.0-21.0 **Fe** rem **Mn** 2.00 max **Mo** 0.50 max **Ni** 37.0-41.0 **P** 0.04 max **S** 0.04 max **Si** 2.50 max	**ASTM** A297
N08005	Ni-Fe-Cr Alloy Solid Solution Strengthened (ACI HU-50)	**C** 0.40-0.60 **Cr** 17.0-21.0 **Fe** rem **Mn** 1.50 max **Mo** 0.50 max **Ni** 37.0-41.0 **P** 0.04 max **S** 0.04 max **Si** 0.50-2.00	**ASTM** A608
N08006	Ni-Fe-Cr Alloy Solid Solution Strengthened (ACI HW-50)	**C** 0.40-0.60 **Cr** 10.0-14.0 **Fe** rem **Mn** 1.50 max **Mo** 0.50 max **Ni** 58.0-62.0 **P** 0.04 max **S** 0.04 max **Si** 0.50-2.00	**ASTM** A608
N08007	Ni-Fe-Cr Alloy Solid Solution Strengthened (ACI CN-7M)	**C** 0.07 max **Cr** 19.0-22.0 **Cu** 3.00-4.00 **Fe** rem **Mn** 1.50 max **Mo** 2.00-3.00 **Ni** 27.5-30.5 **Si** 1.50 max	**ASTM** A296; A351
N08008	Ni-Fe-Cr Alloy Solid Solution Strengthened (ACI HT-50C)	**C** 0.40-0.60 **Cb** 0.75-1.25 **Cr** 13.0-17.0 **Fe** rem **Mo** 0.50 max **Ni** 33.0-37.0	**ASTM** A567
N08020	Ni-Fe-Cr Alloy Solid Solution Strengthened (Carpenter 20Cb3)	**C** 0.07 max **Cb** 8xC-1.00 **Cr** 19.00-21.00 **Cu** 3.00-4.00 **Fe** bal **Mn** 2.00 max **Mo** 2.00-3.00 **Ni** 32.00-38.00 **P** 0.045 max **S** 0.035 max **Si** 1.00 max	**ASME** SB462; SB463; SB464 **ASTM** B366; B462; B463; B464; B468; B471; B472; B473; B474; B475
N08021	Austenitic Cr-Ni-Cu-Mo Alloy Welding Filler Metal	**C** 0.07 max **Cr** 19.0-21.0 **Cu** 3.0-4.0 **Mn** 2.5 max **Mo** 2.0-3.0 **Ni** 32.0-36.0 **P** 0.03 max **S** 0.03 max **Si** 0.60 max **Other** Cb + Ta 8xC-1.0	**ASME** SFA5.9 (ER320) **AWS** A5.9 (ER320)
N08022	Austenitic Cr-Ni-Cu-Mo Alloy Welding Filler Metal	**C** 0.025 max **Cr** 19.0-21.0 **Cu** 3.0-4.0 **Mn** 1.5-2.0 **Mo** 2.0-3.0 **Ni** 32.0-36.0 **P** 0.015 max **S** 0.020 max **Si** 0.15 max **Other** Cb + Ta 8xC-0.40	**ASME** SFA5.9 (ER320LR) **AWS** A5.9 (ER320LR)
N08024	Cr-Ni-Fe-Mo-Cu Alloy Columbium Stabilized	**C** 0.03 max **Cb** 0.15-0.35 **Cr** 22.5-25.0 **Cu** 0.50-1.50 **Fe** rem **Mn** 1.00 max **Mo** 3.50-5.00 **Ni** 35.0-40.0 **P** 0.035 max **S** 0.035 max **Si** 0.50 max	**ASTM** B463; B464; B468
N08026	Ni-Cr Wrought Alloy (Carpenter 20Mo6)	**C** 0.03 max **Cr** 22.00-26.00 **Cu** 2.00-4.00 **Fe** rem **Mn** 1.00 max **Mo** 5.00-6.70 **Ni** 33.00-37.20 **P** 0.03 max **S** 0.03 max **Si** 0.50 max	**ASME** SB463; SB464; SB468 **ASTM** B463; B464; B468; B474
N08028	Ni-Fe-Cr Alloy Solid Solution Strengthened (Sanicro 28)	**C** 0.03 max **Cr** 26.0-28.0 **Cu** 0.6-1.4 **Fe** rem **Mn** 2.50 max **Mo** 3.0-4.0 **Ni** 29.5-32.5 **P** 0.030 max **S** 0.030 max **Si** 1.00 max	**ASME** SB668; SB709 **ASTM** B668; B709
N08030	Ni-Fe-Cr Alloy Solid Solution Strengthened (ACI HT-30)	**C** 0.25-0.35 **Cr** 13.0-17.0 **Fe** rem **Mn** 2.00 max **Mo** 0.50 max **Ni** 33.0-37.0 **P** 0.040 max **S** 0.040 max **Si** 2.50 max	**ASTM** A351
N08050	Ni-Fe-Cr Alloy Solid Solution Strengthened (ACI HT-50)	**C** 0.40-0.60 **Cr** 15.0-19.0 **Fe** rem **Mn** 1.50 max **Mo** 0.50 max **Ni** 33.0-37.0 **P** 0.04 max **S** 0.04 max **Si** 0.50-2.00	**ASTM** A608

The chemical compositions listed are for identification purposes and should not be used in lieu of the cross referenced specifications.

UNIFIED NUMBER	DESCRIPTION	CHEMICAL COMPOSITION	CROSS REFERENCE SPECIFICATIONS
N08065	Ni-Fe-Cr Welding Filler Metal (Incoloy FM65)	**Al** 0.20 max **C** 0.05 max **Cr** 19.5-23.5 **Cu** 1.50-3.0 **Fe** 22.0 min **Mn** 1.0 max **Mo** 2.5-3.5 **Ni** 38.0-46.0 **P** 0.03 max **S** 0.03 max **Si** 0.50 max **Ti** 0.60-1.2	**ASME** SFA5.14 (ERNiFeCr-1) **AWS** A5.14 (ERNiFeCr-1) **MIL SPEC** MIL-E-21562 (RN65)
N08221	Ni-Fe-Cr-Cu Alloy	**Al** 0.20 max **C** 0.025 max **Cr** 20.0-22.0 **Cu** 1.50-3.00 **Fe** rem **Mn** 1.00 max **Mo** 5.00-6.50 **Ni** 36.0-46.0 **S** 0.03 max **Si** 0.50 max **Ti** 0.60-1.00	**ASTM** B366; B423; B424; B425
* N08245	Ni-Fe-Cr Alloy (ARMCO 20-45-5)	**C** 0.08 max **Cb** 0.40 max **Cr** 18.0-22.0 **Fe** rem **Mn** 3.0-7.0 **Mo** 1.5-3.0 **Ni** 43.0-49.0 **P** 0.045 max **S** 0.030 max **Si** 1.00 max	**ASTM** B590
N08310	Ni-Fe-Cr-Mo Alloy	**C** 0.020 max **Cr** 24.0-26.0 **Fe** rem **Mn** 2.00-4.00 **Mo** 2.00-4.00 **N** 0.20-0.40 **Ni** 18.0-22.0 **P** 0.035 max **S** 0.015 max **Si** 0.050 max	**ASTM** B720
N08320	Ni-Fe-Cr Alloy (Haynes No. 20 Mod)	**C** 0.05 max **Cr** 21.0-23.0 **Fe** bal **Mn** 2.5 max **Mo** 4.0-6.0 **Ni** 25.0-27.0 **P** 0.04 max **S** 0.03 max **Si** 1.0 max **Ti** 4xC min	**ASME** SB366; SB620; SB621; SB622 **ASTM** B366; B620; B621; B622
N08321	Ni-Fe-Cr Alloy Solution Strengthened (20 Cb3 Modified)	**C** 0.035 max **Cb** 8xC-0.4 max **Cr** 19.0-21.0 **Cu** 3.0-4.0 **Fe** rem **Mn** 1.5-2.5 **Mo** 2.0-3.0 **Ni** 32.0-36.0 **P** 0.02 max **S** 0.015 max **Si** 0.3 max	**ASME** SFA5.9 (ER320) **AWS** A5.9 (ER320)
N08330	Ni-Fe-Cr Alloy Solid Solution Strengthened (RA-330)	**C** 0.08 max **Cr** 17.0-20.0 **Cu** 1.00 max **Fe** bal **Mn** 2.00 max **Ni** 34.0-37.0 **P** 0.03 max **Pb** 0.005 max **S** 0.03 max **Si** 0.75-1.50 **Sn** 0.025 max	**AMS** 5592; 5716 **ASME** SB366; SB511; SB710 **ASTM** B366; B511; B512; B535; B536; B546; B710 **SAE** J405 (30330); J412 (30330)
N08331	Ni-Fe-Cr Alloy Solution Strengthened (RA-330 Modified)	**C** 0.10-0.25 **Cr** 15.0-17.0 **Cu** 0.5 max **Fe** rem **Mn** 1.0-2.5 **Mo** 0.5 max **Ni** 34.0-37.0 **P** 0.03 max **S** 0.03 max **Si** 0.30-0.65	**AWS** ER330
N08332	Ni-Fe-Cr-Si Alloy (RA 330TX)	**Al** 0.10-0.50 **C** 0.05-0.10 **Cr** 17.0-20.0 **Cu** 1.00 max **Fe** rem **Mn** 2.00 max **Ni** 34.0-37.0 **P** 0.03 max **Pb** 0.005 max **S** 0.03 **Si** 0.75-1.50 **Sn** 0.025 max **Ti** 0.20-0.60 **Other** note: iron shall be determined arithmetically by difference	**ASME** SB366; SB511; SB535; SB536 **ASTM** B366; B511; B512; B535; B536; B546; B710
N08366	Ni-Fe-Cr Alloy Solid Solution Strengthened (AL-6X)	**C** 0.035 max **Cr** 20.0-22.0 **Fe** bal **Mn** 2.00 max **Mo** 6.00-7.00 **Ni** 23.5-25.5 **P** 0.030 max **S** 0.030 max **Si** 1.00 max	**ASME** SB675 **ASTM** B376; B675; B676; B688; B690; B691
N08367	Cr-Ni-Mo Alloy (AL 6XN)	**C** 0.030 max **Cr** 20.0-22.0 **Fe** rem **Mn** 2.00 max **Mo** 6.00-7.00 **N** 0.18-0.25 **Ni** 23.50-25.50 **P** 0.040 max **S** 0.030 max **Si** 1.00 max	**ASTM** B675; B676; B688; B690; B691
N08421	Ni-Fe-Cr-Mo-Cu Alloy	**Al** 0.2 max **C** 0.025 max **Cr** 20.0-22.0 **Cu** 1.5-2.0 **Fe** rem **Mn** 1.00 max **Mo** 5.0-6.5 **Ni** 39.0-41.0 **S** 0.03 max **Si** 0.5 max **Ti** 0.6-1.0	
N08700	Fe-Ni-Cr Alloy (JS 700)	**C** 0.04 max **Cb** 8 x C to 0.50 **Cr** 19.0-23.0 **Cu** 0.50 max **Fe** bal **Mn** 2.00 max **Mo** 4.3-5.0 **Ni** 24.0-26.0 **P** 0.04 max **Pb** 0.005 max **S** 0.03 max **Si** 1.00 max **Sn** 0.035 max	**ASME** SB599; SB672 **ASTM** B599; B672
N08800	Fe-Ni-Cr Alloy Solid Solution Strengthened (Incoloy 800)	**Al** 0.15-0.60 **C** 0.10 max **Cr** 19.0-23.0 **Cu** 0.75 max **Fe** bal **Mn** 1.5 max **Ni** 30.0-35.0 **S** 0.015 max **Si** 1.0 max **Ti** 0.15-0.60	**AMS** 5766; 5871 **ASME** SB163; SB407; SB408; SB409; SB564 **ASTM** B163; B366; B407; B408; B409; B514; B515; B564; B574

*Boxed entries are no longer active and are retained for reference purposes only.

The chemical compositions listed are for identification purposes and should not be used in lieu of the cross referenced specifications.

UNIFIED NUMBER	DESCRIPTION	CHEMICAL COMPOSITION	CROSS REFERENCE SPECIFICATIONS
N08801	Fe-Ni-Cr Alloy Solid Solution Strengthened (Incoloy 801)	C 0.10 max Cr 19.0-22.0 Cu 0.5 max Fe bal Mn 1.5 max Ni 30.0-34.0 S 0.015 max Si 1.0 max Ti 0.75-1.5	AMS 5552; 5742
N08802	Ni-Fe-Cr Alloy Solid Solution Strengthened (Incoloy 802)	Al 0.15-1.00 C 0.20-0.50 Cr 19.0-23.0 Cu 0.75 max Fe rem Mn 1.50 max Ni 30.0-35.0 S 0.015 max Si 0.75 max Ti 0.25-1.25	
N08810	Fe-Ni-Cr Alloy (Incoloy 800H)	Al 0.15-0.60 C 0.05-0.10 Cr 19.0-23.0 Cu 0.75 max Fe bal Mn 1.5 max Ni 30.0-35.0 S 0.015 max Si 1.0 max Ti 0.15-0.60	ASME SB163; SB407; SB408; SB409; SB514; SB564 ASTM B163; B366; B407; B408; B409; B514; B515; B564
N08811	Wrought Ni-Fe-Cr Alloy (Incoloy 800HT)	Al 0.15-0.60 C 0.06-0.10 Cr 19.0-23.0 Cu 0.75 max Fe 39.5 min Mn 1.5 max Ni 30.0-35.0 S 0.015 max Si 1.0 max Ti 0.15-0.60	ASTM B163; B366; B407; B408; B409
N08825	Ni-Fe-Cr Alloy Solid Solution Strengthened (Incoloy 825)	Al 0.2 max C 0.05 max Cr 19.5-23.5 Cu 1.5-3.0 Fe bal Mn 1.0 max Mo 2.5-3.5 Ni 38.0-46.0 S 0.03 max Si 0.5 max Ti 0.6-1.2	ASME SB163; SB423; SB424; SB425 ASTM B163; B423; B424; B425; B704; B705
N08904	Ni-Cr-Mo Alloy	C 0.020 max Cr 19.0-23.0 Cu 1.00-2.00 Fe rem Mn 2.00 max Mo 4.00-5.00 Ni 23.0-28.0 P 0.045 max S 0.035 max Si 1.00 max	ASME SB625; SB649; SB673; SB674; SB677 ASTM B625; B649; B673; B674; B677
N08925	Ni-Fe-Cr-Mo-Cu Alloy, Low Carbon	C 0.020 max Cr 19.0-21.0 Cu 0.50-1.50 Fe rem Mn 1.00 max Mo 6.00-7.00 N 0.10-0.20 Ni 24.0-26.0 P 0.045 max S 0.030 max Si 0.50 max	ASME SB625; SB649; SB673; SB674 ASTM B625; B649; B673; B674; B677
N09027	Ni-Fe-Cr Alloy Precipitation Hardenable (CG27)	Al 1.45-1.75 B 0.003-0.015 C 0.02-0.08 Cb 0.60-1.10 Cr 12.50-14.00 Fe bal Mn 0.25 max Mo 5.00-6.00 Ni 36.50-39.50 P 0.015 max S 0.015 max Si 0.25 max Ti 2.30-2.70	AMS 5633; 5634
N09706	Ni-Fe-Cr Alloy Precipitation Hardenable (Inconel 706)	Al 0.40 max B 0.006 max C 0.06 max Cb 2.5-3.3 Cr 14.5-17.5 Cu 0.30 max Fe bal Mn 0.35 max Ni 39.0-44.0 P 0.020 max S 0.015 max Si 0.35 max Ti 1.5-2.0	AMS 5605; 5606; 5701; 5702; 5703
N09901	Ni-Fe-Cr Alloy Precipitation Hardenable (Incoloy 901)	Al 0.35 max B 0.010-0.020 C 0.10 max Cr 11.00-14.00 Cu 0.50 max Fe bal Mn 1.00 max Mo 5.00-7.00 Ni 40.00-45.00 S 0.030 max Si 0.60 max Ti 2.35-3.10	AMS 5660; 5661
N09902	Ni-Fe-Cr Alloy Precipitation Hardenable (Ni-Span-C 902)	Al 0.30-0.80 C 0.06 max Cr 4.90-5.75 Fe bal Mn 0.80 max Ni 41.0-43.5 P 0.04 max S 0.04 max Si 1.0 max Ti 2.20-2.75	AMS 5210; 5221; 5223; 5225
N09925	Ni-Fe-Cr Alloy Precipitation Hardenable	Al 0.10-0.50 C 0.03 max Cb 0.50 max Cr 19.5-23.5 Cu 1.50-3.00 Fe 22.0 min Mn 1.00 max Mo 2.50-3.50 Ni 38.0-46.0 S 0.03 max Si 0.50 max Ti 1.90-2.40	
N09926	Ni-Fe-Cr-Cu Alloy Precipitation Hardenable (Incoloy 926)	Al 0.3 max C 0.04 max Cr 14.0-18.0 Cu 3.5-5.5 Fe 39.0 min Mn 1.5 max Mo 2.5-3.5 Ni 26.0-30.0 S 0.015 max Si 0.75 max Ti 1.5-2.3	
N09979	Ni-Fe-Cr Alloy Precipitation Hardenable (D979)	Al 0.75-1.30 B 0.008-0.016 C 0.08 max Cr 14.00-16.00 Fe bal Mn 0.75 max Mo 3.75-4.50 Ni 42.00-48.00 P 0.015 max S 0.015 max Si 0.75 max Ti 2.70-3.30 W 3.75-4.50 Zr 0.050 max	AMS 5509; 5746 SAE J467 (D-979)

The chemical compositions listed are for identification purposes and should not be used in lieu of the cross referenced specifications.

UNIFIED NUMBER	DESCRIPTION	CHEMICAL COMPOSITION	CROSS REFERENCE SPECIFICATIONS
N10001	Ni-Mo Alloy Solid Solution Strengthened (Hastelloy B)	**C** 0.12 max **Co** 2.50 max **Cr** 1.00 max **Fe** 6.00 max **Mn** 1.00 max **Mo** 26.0-33.0 **Ni** bal **P** 0.040 max **S** 0.030 max **Si** 1.00 max **V** 0.60 max	**AMS** 5396 **ASME** SB333; SB335; SB366; SB619; SFA5.14 (ERNiMo-1) **ASTM** A296; A494; B333; B335; B366; B619 **AWS** A5.14 (ErNiMo-1) **MIL SPEC** MIL-R-5031 (Class 10)
N10002	Ni-Mo Alloy Solid Solution Strengthened (Hastelloy C)	**C** 0.08 max **Co** 2.5 max **Cr** 14.5-16.5 **Fe** 4.0-7.0 **Mn** 1.00 max **Mo** 15.0-17.0 **Ni** bal **P** 0.040 max **S** 0.030 max **Si** 1.00 max **V** 0.35 max **W** 3.0-4.5	**AMS** 5388; 5389; 5530; 5750 **ASTM** A494; A567 **MIL SPEC** MIL-R-5031 (Class 11)
N10003	Ni-Mo Alloy Solid Solution Strengthened (Hastelloy N)	**Al** 0.50 max **B** 0.010 max **C** 0.04-0.08 **Co** 0.20 max **Cr** 6.0-8.0 **Cu** 0.35 max **Fe** 5.0 max **Mn** 1.00 max **Mo** 15.0-18.0 **Ni** bal **P** 0.015 max **S** 0.020 max **Si** 1.00 max **V** 0.50 max **W** 0.50 max	**AMS** 5607; 5771 **ASME** SB434; SFA5.14 (ERNiMo-2) **ASTM** B366; B434; B573 **AWS** A5.14 (ERNiMo-2)
N10004	Ni-Mo Alloy Solid Solution Strengthened (Hastelloy W)	**C** 0.12 max **Cr** 4.00-6.00 **Fe** 4.00-7.00 **Mn** 1.00 max **Mo** 23.00-26.00 **Ni** bal **P** 0.050 max **S** 0.050 max **Si** 1.00 max **V** 0.60 max	**AMS** 5755; 5786; 5787 **ASME** SFA5.14 (ERNiMo-3) **AWS** A5.14 (ERNiMo-3) **MIL SPEC** MIL-R-5031 (Class 12)
N10276	Ni-Mo Alloy Solid Solution Strengthened (Hastelloy C276)	**C** 0.02 max **Co** 2.5 max **Cr** 14.5-16.5 **Fe** 4.0-7.0 **Mn** 1.0 max **Mo** 15.0-17.0 **Ni** bal **P** 0.030 max **S** 0.030 max **Si** 0.08 max **V** 0.35 max **W** 3.0-4.5	**ASME** SB366; SB574; SB575; SB619; SB622; SB626; SFA5.14 (NiCrMo-4) **ASTM** B366; B574; B575; B619; B622; B626 **AWS** A5.14 (NiCrMo-4)
N10665	Ni-Mo Alloy (Hastelloy B-2)	**C** 0.02 max **Co** 1.0 max **Cr** 1.0 max **Fe** 2.0 max **Mn** 1.0 max **Mo** 26.0-30.0 **Ni** bal **P** 0.04 **S** 0.03 max **Si** 0.10 max	**ASME** SB333; SB335; SB366; SB619; SB622; SB626; SFA5.14 (ERNiMo-7) **ASTM** B333; B335; B366; B619; B622; B626 **AWS** A5.14 (ERNiMo-7)
N13009	Ni-W-Co-Cr Alloy Castings	**Al** 4.75-5.25 **B** 0.010-0.020 **Bi** 0.00005 max **C** 0.12-0.17 **Cb** 0.75-1.25 **Co** 9.00-11.00 **Cr** 8.00-10.00 **Cu** 0.10 max **Fe** 1.50 max **Mn** 0.20 max **Ni** rem **Pb** 0.0010 max **S** 0.015 max **Si** 0.20 max **Ti** 1.75-2.25 **W** 11.5-13.5 **Zr** 0.03-0.08	**AMS** 5407
N13010	Ni-Co-Cr-Mo Alloy Castings	**Al** 5.75-6.25 **B** 0.010-0.020 **Bi** 0.00005 max **C** 0.08-0.13 **Cb** 0.10 max **Co** 9.50-10.50 **Cr** 7.50-8.50 **Fe** 0.35 max **Mn** 0.20 max **Mo** 5.75-6.25 **Ni** rem **P** 0.015 max **Pb** 0.0005 max **S** 0.015 max **Si** 0.25 max **Ta** 4.00-4.50 **Ti** 0.80-1.20 **W** 0.10 max **Zr** 0.05-0.10	**AMS** 5405; 5406
N13017	Co-Cr-Mo-Ni Alloy (Astroloy M)	**Al** 3.85-4.15 **B** 0.020-0.030 **Bi** 0.00005 max **C** 0.02-0.06 **Co** 16.0-18.0 **Cr** 14.0-16.0 **Cu** 0.10 max **Fe** 0.50 max **Mn** 0.15 max **Mo** 4.50-5.50 **N** 0.0050 max **Ni** rem **O** 0.010 max **P** 0.015 max **Pb** 0.0002 max **S** 0.015 max **Si** 0.20 max **Ti** 3.35-3.65 **W** 0.05 max **Zr** 0.06 max	**AMS** 5851; 5852; 5882
N13020	Co-Cr-Mo-Ni Alloy	**Al** 3.75-4.75 **B** 0.025-0.035 **Bi** 0.00005 max **C** 0.03-0.10 **Co** 17.0-20.0 **Cr** 14.0-16.0 **Cu** 0.10 max **Fe** 2.00 max **Mn** 0.15 max **Mo** 4.50-5.50 **Ni** rem **Ti** 2.75-3.75 **Zr** 0.06 max	**AMS** 5846
N13100	Ni-Co Alloy Precipitation Hardenable (IN-100)	**Al** 5.00-6.00 **B** 0.01-0.02 **C** 0.15-0.20 **Co** 13.0-17.0 **Cr** 8.0-11.0 **Fe** 1.0 max **Mn** 0.20 max **Mo** 2.0-4.0 **Ni** bal **S** 0.015 max **Si** 0.20 max **Ti** 4.50-5.00 **V** 0.70-1.20 **Zr** 0.03-0.09	**AMS** 5397
N14052	Ni-Fe Alloy (Alloy No. 52)	**Al** 0.10 max **C** 0.05 max **Co** 0.50 max **Cr** 0.25 max **Fe** bal **Mn** 0.60 max **Ni** 50.5 nom **P** 0.025 max **S** 0.025 max **Si** 0.30 max	**ASTM** ASTM F30 (52)

The chemical compositions listed are for identification purposes and should not be used in lieu of the cross referenced specifications.

UNIFIED NUMBER	DESCRIPTION	CHEMICAL COMPOSITION	CROSS REFERENCE SPECIFICATIONS
N19903	Ni-Fe-Co Alloy Precipitation Hardenable (Incoloy 903)	**Al** 0.30-1.15 **B** 0.012 max **C** 0.06 max **Cb** 2.40-3.50 **Co** 13.0-17.0 **Cr** 1.00 max **Cu** 0.50 max **Fe** rem **Mn** 1.00 max **Ni** 36.0-40.0 **S** 0.015 max **Si** 0.35 max **Ti** 1.00-1.25	
N19907	Ni-Fe-Co Alloy Precipitation Hardenable	**Al** 0.20 max **B** 0.012 max **C** 0.06 max **Cb** 4.3-5.2 **Co** 12.0-16.0 **Cr** 1.0 max **Cu** 0.5 max **Fe** rem **Mn** 1.0 max **Ni** 35.0-40.0 **P** 0.015 max **S** 0.015 max **Si** 0.35 max **Ti** 1.2-1.8	
N19909	Ni-Fe-Co Alloy Precipitation Hardenable	**Al** 0.15 max **B** 0.012 max **C** 0.06 max **Cb** 4.3-5.2 **Co** 12.0-16.0 **Cr** 1.0 max **Cu** 0.5 max **Fe** rem **Mn** 1.0 max **Ni** 35.0-40.0 **P** 0.015 max **S** 0.015 max **Si** 0.25-0.50 **Ti** 1.3-1.8	
N99600	Ni-Cr-B Brazing Filler Metal (BNi-1)	**Al** 0.05 **B** 2.75-3.50 **C** 0.6-0.9 **Cr** 13.0-15.0 **Fe** 4.0-5.0 **Ni** rem **P** 0.02 **S** 0.02 **Si** 4.0-5.0 **Ti** 0.05 **Zr** 0.05 **Other** 0.50	**AMS** 4775 **ASME** SFA-5.8 (BNi-1) **AWS** A5.8 (BNi-1) **MIL SPEC** MIL-B-7883 (BNi-1)
N99610	Ni-Cr-B Brazing Filler Metal (BNi-1a)	**Al** 0.05 **B** 2.75-3.50 **C** 0.06 **Cr** 13.0-15.0 **Fe** 4.0-5.0 **Ni** rem **P** 0.02 **S** 0.02 **Si** 4.0-5.0 **Ti** 0.05 **Zr** 0.05 **Other** 0.50	**AMS** 4776 **ASME** SFA5.8 (BNi-1a) **AWS** A5.8 (BNi-1a) **MIL SPEC** MIL-B-7883 (BNi-1a)
N99612	Ni-Cr-B Brazing Filler Metal (BNi-9)	**Al** 0.05 max **B** 3.25-4.0 **C** 0.06 max **Co** 0.10 max **Cr** 13.5-16.5 **Fe** 1.5 max **Ni** rem **P** 0.02 max **S** 0.02 max **Ti** 0.05 max **Zr** 0.05 max **Other** total 0.50 max	**ASME** SFA5.8 (BNi-9) **AWS** A5.8 (BNi-9)
N99620	Ni-Cr-B Brazing Filler Metal (BNi-2)	**Al** 0.05 **B** 2.75-3.50 **C** 0.06 **Cr** 6.0-8.0 **Fe** 2.5-3.5 **Ni** rem **P** 0.02 **S** 0.02 **Si** 4.0-5.0 **Ti** 0.05 **Zr** 0.05 **Other** 0.50	**AMS** 4777 **ASME** SFA-5.8 (BNi-2) **AWS** A5.8 (BNi-2) **MIL SPEC** MIL-B-7883 (BNi-2)
N99622	Ni-W-Cr-B Brazing Filler Metal (BNi-10)	**Al** 0.05 max **B** 2.0-3.0 **C** 0.40-0.55 **Co** 0.10 max **Cr** 10.0-13.0 **Fe** 2.5-4.5 **Ni** rem **P** 0.02 max **S** 0.02 max **Si** 3.0-4.0 **Ti** 0.05 max **W** 15.0-17.0 **Zr** 0.05 max **Other** total 0.50 max	**ASME** SFA5.8 (BNi-10) **AWS** A5.8 (BNi-10)
N99624	Ni-W-Cr-B Brazing Filler Metal (BNi-11)	**Al** 0.02 max **B** 2.2-3.1 **C** 0.30-0.42 **Cr** 9.0-11.75 **Fe** 2.5-4.0 **Ni** rem **P** 0.02 max **S** 0.02 max **Si** 3.25-4.25 **Ti** 0.05 max **W** 11.50-12.75 **Zr** 0.05 max **Other** total 0.50 max	**ASME** SFA 5.8 (BNi-10) **AWS** A5.8 (BNi-10)
N99630	Ni-Si-B Brazing Filler Metal (BNi-3)	**Al** 0.05 **B** 2.75-3.50 **C** 0.06 **Fe** 0.5 **Ni** rem **P** 0.02 **S** 0.02 **Si** 4.0-5.0 **Ti** 0.05 **Zr** 0.05 **Other** 0.50	**AMS** 4778 **ASME** SFA5.8 (BNi-3) **AWS** A5.8 (BNi-3) **MIL SPEC** MIL-B-7883B (BNi-3)
N99640	Ni-Si-B Brazing Filler Metal (BNi-4)	**Al** 0.05 **B** 1.5-2.2 **C** 0.06 **Fe** 1.5 **Ni** rem **P** 0.02 **S** 0.02 **Si** 3.0-4.0 **Ti** 0.05 **Zr** 0.05 **Other** 0.50	**AMS** 4779 **ASME** SFA5.8 (BNi-4) **AWS** A5.8 (BNi-4) **MIL SPEC** MIL-B-7883 (BNi-4)
N99644	Ni-Cr-B Weld Filler Metal for Hard Surfacing (RNiCr-A) (Colmonoy 44)	**B** 2.00-3.00 **C** 0.30-0.60 **Co** 1.50 max **Cr** 8.0-14.0 **Fe** 1.25-3.25 **Ni** bal **Si** 1.25-3.25	**ASME** SFA5.13 (RNiCr-A) **AWS** A5.13 (RNiCr-A)
N99645	Ni-Cr-B Weld Filler Metal for Hard Surfacing (RNiCr-B) (Colmonoy 45)	**B** 2.00-4.00 **C** 0.40-0.80 **Co** 1.25 max **Cr** 10.0-16.0 **Fe** 3.00-5.00 **Ni** bal **Si** 3.00-5.00	**ASME** SFA5.13 (RNiCr-B) **AWS** A5.13 (RNiCr-B) **MIL SPEC** MIL-R-17131 (MIL-RNiCr-B)
N99646	Ni-Cr-B Weld Filler Metal for Hard Surfacing (RNiCr-C) (Colmonoy 46)	**B** 2.50-4.50 **C** 0.50-1.00 **Co** 1.00 max **Cr** 12.0-18.0 **Fe** 3.50-5.50 **Ni** bal **Si** 3.50-5.50	**ASME** SFA5.13 (RNiCr-C) **AWS** A5.13 (RNiCr-C) **MIL SPEC** MIL-R-17131 (MIL-RNiCr-C)
N99650	Ni-Cr-Si Brazing Filler Metal (BNi-5)	**Al** 0.05 **B** 0.03 **C** 0.10 **Cr** 18.5-19.5 **Ni** rem **P** 0.02 **S** 0.02 **Si** 9.75-10.50 **Ti** 0.05 **Zr** 0.05 **Other** 0.50	**AMS** 4782 **ASME** SFA5.8 (BNi-5) **AWS** A5.8 (BNi-5) **MIL SPEC** MIL-B-7883 (BNi-5)
N99700	Ni-P Brazing Filler Metal (BNi-6)	**Al** 0.05 **C** 0.10 **Ni** rem **P** 10.0-12.0 **S** 0.02 **Ti** 0.05 **Zr** 0.05 **Other** 0.50	**ASME** SFA5.8 (BNi-6) **AWS** A5.8 (BNi-6)

The chemical compositions listed are for identification purposes and should not be used in lieu of the cross referenced specifications.

UNIFIED NUMBER	DESCRIPTION	CHEMICAL COMPOSITION	CROSS REFERENCE SPECIFICATIONS
N99710	Ni-Cr-P Brazing Filler Metal (BNi-7)	**Al** 0.05 **B** 0.01 **C** 0.08 **Cr** 13.0-15.0 **Fe** 0.2 **Mn** 0.04 **Ni** rem **P** 9.7-10.5 **S** 0.02 **Si** 0.10 **Ti** 0.05 **Zr** 0.05 **Other** 0.50	**ASME** SFA5.8 (BNi-7) **AWS** A5.8 (BNi-7)
N99800	Ni-Mn-Si-Cu Brazing Filler Metal (BNi-8)	**Al** 0.05 **C** 0.10 **Cu** 4.0-5.0 **Mn** 21.5-24.5 **Ni** rem **P** 0.02 **S** 0.02 **Si** 6.0-8.0 **Ti** 0.05 **Zr** 0.05 **Other** 0.50	**ASME** SFA5.8 (BNi-8) **AWS** A5.8 (BNi-8)

The chemical compositions listed are for identification purposes and should not be used in lieu of the cross referenced specifications.

UNS NUMBERS ASSIGNED TO DATE

With Description of Each Material Covered and References To Documents In Which The Same or Similar Materials are Described

Pxxxxx Number Series
Precious Metals and Alloys

PRECIOUS METALS AND ALLOYS

UNIFIED NUMBER	DESCRIPTION	CHEMICAL COMPOSITION	CROSS REFERENCE SPECIFICATIONS
P00010	Refined Gold	**Ag** 0.001 max **Au** 99.995 min **Bi** 0.001 max **Cr** 0.0003 max **Cu** 0.001 max **Fe** 0.001 max **Mg** 0.001 max **Mn** 0.0003 max **Pb** 0.001 max **Pd** 0.001 max **Si** 0.001 max **Sn** 0.001 max	**ASTM** B562 (99.995)
P00015	Refined Gold	**As** 0.003 max **Au** 99.99 min **Bi** 0.002 max **Cr** 0.0003 max **Cu** 0.005 max **Fe** 0.002 max **Mg** 0.003 max **Mn** 0.0003 max **Ni** 0.0003 max **Pb** 0.002 max **Pd** 0.005 max **Si** 0.005 max **Sn** 0.001 max	**ASTM** B562 (Grade 99.99) **FED** QQ-G-00545B
P00016	Gold	**Au** 99.99 min **Other** Ag+Cu 0.009 max, each, other impurity 0.003 max	**ASTM** F72
P00020	Refined Gold	**Ag** 0.035 max **Au** 99.95 min **Cu** 0.02 max **Fe** 0.005 max **Pb** 0.005 max **Pd** 0.02 max **Other** Ag+Cu 0.04 max	**AMS** 7731 **ASTM** B562 (99.95)
P00025	Refined Gold	**Au** 99.5 min	**ASTM** B562 (99.5)
P00300	Gold-Palladium-Nickel Brazing Alloy	**Au** 29.5-30.5 **Ni** 35.5-36.5 **Pd** 33.5-34.5	**AMS** 4785 **ASME** SFA5.8 (BAu-5) **AWS** A5.8 (BAu-5)
P00350	Gold Brazing Alloy	**Au** 34.5-35.5 **Cu** rem **Ni** 2.5-3.5	**ASME** SFA5.8 (BAu-3) **AWS** A5.8 (BAu-3)
P00375	Gold Brazing Alloy	**Au** 37.0-38.0 **Cu** rem	**ASME** SFA5.8 (BAu-1) **AWS** A5.8 (BAu-1)
P00500	Gold Brazing Alloy	**Au** 49.5-50.5 **Ni** 24.5-25.5 **Pd** 24.5-25.5	**AMS** 4784
P00507	Gold-Palladium-Nickel Brazing Alloy	**Au** 49.5-50.5 **Co** 0.06 max **Ni** 24.5-25.5 **Pd** rem	**ASME** SFA5.8 (BVAu-7) **AWS** A5.8 (BVAu-7)
P00560	Gold: 14 Karat Yellow Gold - Old U.S. (Prior to October 1981)	**Ag** 3.70-4.70 **Au** 56.25-56.55 **Cu** 32.30-33.30 **Fe** 0.05 max **Ni** 0.10 max **Si** 0.01 max **Zn** 6.00-7.50 **Other** all elements except Au, Ag, Cu, Zn: max 0.15	
P00580	Gold: 14 Karat Yellow Gold - New U.S. (After September 1981)	**Ag** 3.0-4.0 **Au** 58.03-58.63 **Cu** 30.74-31.74 **Fe** 0.05 max **Ni** 0.10 max **Si** 0.01 max **Zn** 5.68-7.18 **Other** all elements except Au, Ag, Cu, Zn: max 0.15	
P00691	Gold Electrical Contact Alloy	**Ag** 23.5-26.5 **Au** 68.0-70.0 **Pt** 5.0-7.0	**ASTM** B522 (I)
P00692	Gold Electrical Contact Alloy	**Ag** 24.5-25.5 **Au** 68.5-69.5 **Pt** 5.5-6.5 **S** 0.01 max	**ASTM** B522 (II)
P00700	Gold Brazing Alloy	**Au** 69.5-70.5 **Ni** 21.5-22.5 **Pd** 7.5-8.5	**AMS** 4786 **ASME** SFA5.8 (BAu-6) **AWS** A5.8 (BAu-6)
P00707	Gold Brazing Alloy replaced by P00807		
P00710	Gold Electrical Contact Alloy	**Ag** 4.0-5.0 **Au** 70.5-72.5 **Cu** 13.5-15.5 **Pt** 8.0-9.0 **Zn** 0.7-1.3	**ASTM** B541
P00750	Gold Electrical Contact Alloy	**Ag** 21.4-22.6 **Au** 74.2-75.8 **Ni** 2.6-3.4	**ASTM** B477
P00800	Gold Brazing Alloy	**Au** 79.5-80.5 **Cu** rem	**ASME** SFA5.8 (BAu-2) **AWS** A5.8 (BAu-2)
P00807	Gold Brazing Alloy (Vacuum Grade)	**Au** 79.5-80.5 **Cu** rem	**ASME** SFA5.8 (BVAu-2) **AWS** A5.8 (BVAu-2)
P00820	Gold Brazing Alloy	**Au** 81.5-82.5 **Ni** 17.5-18.5	**AMS** 4787 **ASME** SFA5.8 (BAu-4) **AWS** A5.8 (BAu-4)
P00827	Gold Brazing Alloy (Vacuum Grade)	**Au** 81.5-82.5 **Ni** rem	**ASME** SFA5.8 (BVAu-4) **AWS** A5.8 (BVAu-4)
P00900	Gold-Silver Alloy	**Ag** 3.0-10.0 **Au** 90.0 min	**FED** QQ-G-540
P00901	Gold Electrical Contact Alloy	**Au** 89.0-91.0 **Cu** 9.0-11.0	**ASTM** B596

The chemical compositions listed are for identification purposes and should not be used in lieu of the cross referenced specifications.

UNIFIED NUMBER	DESCRIPTION	CHEMICAL COMPOSITION	CROSS REFERENCE SPECIFICATIONS
P00927	Gold-Palladium Brazing Alloy (Vacuum Grade)	**Au** 91.0-93.0 **Pd** rem	**ASME** SFA5.8 (BVAu-8) **AWS** A5.8 (BVAu-8)
P03300	Palladium Alloy (Paliney No. 7, "W" Alloy 3350, No. 226 Alloy)	**Ag** 29.0-31.0 **Au** 9.5-10.5 **Cu** 13.5-14.5 **Pd** 34.0-36.0 **Pt** 9.5-10.5 **Other** elements 0.01 max total	**AMS** 7735
P03350	Palladium Electrical Contact Alloy	**Ag** 29-31 **Au** 9.5-10.5 **Cu** 13.5-14.5 **Pd** 34.0-36.0 **Pt** 9.5-10.5 **Zn** 0.6-1.2 **Other** total 0.1 max	**ASTM** B540
P03440	Palladium Electrical Contact Alloy	**Ag** 37.0-39.0 **Cu** 15.5-16.5 **Ni** 0.8-1.2 **Pd** 43.0-45.0 **Pt** 0.8-1.2	**ASTM** B563
* P03590	Palladium Alloy	**Cu** 39-41 **Pd** 59 min	
P03657	Palladium-Cobalt Brazing Alloy (Vacuum Grade)	**Co** rem **Ni** 0.06 max **Pd** 64.0-66.0	**ASME** SFA5.8 (BVPd-1) **AWS** A5.8 (BVPd-1)
P03980	Refined Palladium	**Ir** 0.05 max **Pd** 99.80 min **Pt** 0.15 max **Rh** 0.10 max **Other** Ru 0.05 max	**ASTM** B589 (99.80)
* P03990	Palladium	**Other** Pd 99.07 min	
P03995	Refined Palladium	**Ag** 0.005 max **Al** 0.005 max **Au** 0.01 max **Ca** 0.005 max **Co** 0.001 max **Cr** 0.001 max **Cu** 0.005 max **Fe** 0.005 max **Mg** 0.005 max **Mn** 0.001 max **Ni** 0.005 max **Pb** 0.005 max **Pd** 99.95 min **Sb** 0.002 max **Si** 0.005 max **Sn** 0.005 max **Zn** 0.0025 max **Other** Pt + Rh + Ru + Ir 0.03 max	**ASTM** B589 (99.95)
* P04840	Platinum Iridium Alloy	**Ir** 13-17 **Pt** 84 min	
P04980	Refined Platinum	**Ir** 0.05 max **Pt** 99.80 min **Rh** 0.10 max **Other** Pd 0.15 max, Ru 0.05 max	**ASTM** B561 (99.80)
P04995	Refined Platinum	**Ag** 0.005 max **Al** 0.005 max **As** 0.005 max **Au** 0.01 max **Bi** 0.005 max **Cd** 0.005 max **Cr** 0.005 max **Cu** 0.01 max **Ir** 0.015 max **Fe** 0.01 max **Mg** 0.005 max **Mn** 0.005 max **Mo** 0.01 max **Ni** 0.005 max **Pb** 0.005 max **Pt** 99.95 min **Rh** 0.03 max **Sb** 0.005 max **Si** 0.01 max **Sn** 0.005 max **Te** 0.005 max **Zn** 0.005 max **Other** Ca 0.005 max, Ru 0.01 max, Pd 0.02 max	**ASTM** B561 (99.95) **FED** QQ-P-00428
P05980	Rhodium	**As** 0.01 max **Bi** 0.01 max **Cd** 0.01 max **Ir** 0.01 max **Fe** 0.01 max **Pb** 0.01 max **Pd** 0.05 max **Pt** 0.01 max **Rh** 99.8 min **Ru** 0.05 max **Sn** 0.01 max **Zn** 0.01 max	**ASTM** B616 (99.80)
P05981	Refined Rhodium	**Ag** 0.02 max **Al** 0.01 max **As** 0.005 max **Au** 0.01 max **B** 0.005 max **Bi** 0.005 max **Cd** 0.005 max **Co** 0.005 max **Cr** 0.01 max **Cu** 0.01 max **Fe** 0.01 max **Ir** 0.05 max **Mg** 0.01 max **Mn** 0.005 max **Ni** 0.01 max **Pb** 0.01 max **Pt** 0.05 max **Rh** 99.90 min **Sb** 0.005 max **Si** 0.01 max **Sn** 0.01 max **Te** 0.01 max **Zn** 0.01 max **Other** Pd 0.05 max, Ru 0.05 max, Ca 0.01 max	**ASTM** B616-78 (99.90)

*Boxed entries are no longer active and are retained for reference purposes only.

The chemical compositions listed are for identification purposes and should not be used in lieu of the cross referenced specifications.

UNIFIED NUMBER	DESCRIPTION	CHEMICAL COMPOSITION	CROSS REFERENCE SPECIFICATIONS
P05982	Refined Rhodium	**Ag** 0.005 max **Al** 0.005 max **As** 0.003 max **Au** 0.003 max **B** 0.001 max **Bi** 0.005 max **Cd** 0.005 max **Co** 0.001 max **Cr** 0.005 max **Cu** 0.005 max **Fe** 0.003 max **Ir** 0.02 max **Mg** 0.005 max **Mn** 0.005 max **Ni** 0.003 max **Pb** 0.005 max **Pt** 0.02 max **Rh** 99.95 min **Sb** 0.003 max **Si** 0.005 max **Sn** 0.003 max **Te** 0.005 max **Zn** 0.003 max **Other** Pd 0.005 max, Ru 0.01 max, Ca 0.005 max	**ASTM** B616 (99.95)
P05990	Rhodium	**Ag** 0.02 max **Al** 0.01 max **As** 0.005 max **Au** 0.01 max **B** 0.005 max **Bi** 0.005 max **Ca** 0.01 max **Cd** 0.005 max **Co** 0.005 max **Cr** 0.01 max **Cu** 0.01 max **Fe** 0.01 max **Ir** 0.05 max **Mg** 0.01 max **Mn** 0.005 max **Ni** 0.01 max **Pb** 0.01 max **Pd** 0.05 max **Pt** 0.05 max **Rn** 99.90 min **Ru** 0.05 max **Sb** 0.005 max **Si** 0.01 max **Sn** 0.01 max **Te** 0.01 max **Zn** 0.01 max	**ASTM** B616 (99.90)
P05995	Refined Rhodium	**Ag** 0.005 max **Al** 0.005 max **As** 0.003 max **Au** 0.003 max **B** 0.001 max **Bi** 0.005 max **Ca** 0.005 max **Cd** 0.005 max **Co** 0.001 max **Cr** 0.005 max **Cu** 0.005 max **Fe** 0.003 max **Ir** 0.02 max **Mg** 0.005 max **Mn** 0.005 max **Ni** 0.003 max **Pb** 0.005 max **Pd** 0.005 max **Pt** 0.02 max **Rh** 99.95 min **Ru** 0.01 max **Sb** 0.003 max **Si** 0.005 max **Sn** 0.003 max **Te** 0.005 max **Zn** 0.003 max	**ASTM** B616 (99.95)
P06100	Refined Iridium	**As** 0.01 max **Bi** 0.01 max **Cd** 0.01 max **Fe** 0.01 max **Ir** 99.80 min **Pb** 0.02 max **Pt** 0.10 max **Rh** 0.15 max **Si** 0.02 max **Sn** 0.01 max **Other** Pd 0.05 max, Ru 0.05 max	**ASTM** B671 (99.80)
P07010	Refined Silver	**Ag** 99.99 min **Bi** 0.0005 max **Cu** 0.010 max **Fe** 0.001 max **Pb** 0.001 max **Se** 0.0005 max **Te** 0.0005 max **Other** Pd 0.001 max	**ASTM** B413 (Grade 99.99)
P07015	Refined Silver	**Ag** 99.95 min **Bi** 0.001 max **Cu** 0.04 max **Fe** 0.002 max **Pb** 0.015 max	**ASTM** B413 (99.95) **MIL SPEC** MIL-S-13282
P07016	Silver	**Ag** 99.95 min **Cu** 0.05 max **P** 0.002 max **Other** total volatile impurities 0.010 max	**ASTM** F106 (TB Ag)
P07017	Silver Brazing Alloy	**Ag** 99.95 min **Cu** 0.05 max	**ASME** SFA5.8 (BVAg-0) **AWS** A5.8 (BVAg-0)
P07020	Refined Silver	**Ag** 99.90 min **Bi** 0.001 max **Cu** 0.08 max **Fe** 0.002 max **Pb** 0.025 max **Other** Ag+Cu 99.95 min	**ASTM** B413 (99.90)
P07200	Silver Brazing Alloy	**Ag** 19.0-21.0 **Cu** 39.0-41.0 **Mn** 4.5-5.5 **Zn** 33.0-37.0	**ASME** SFA5.8 (BAg-25) **AWS** A5.8 (BAg-25)
P07250	Silver Brazing Alloy	**Ag** 24.0-26.0 **Cu** 37.0-39.0 **Mn** 1.5-2.5 **Ni** 1.5-2.5 **Zn** 31.0-35.0	**ASME** SFA5.8 (BAg-26) **AWS** A5.8 (BAg-26)
P07251	Silver Brazing Alloy	**Ag** 24.0-26.0 **Cd** 12.5-14.5 **Cu** 34.0-36.0 **Zn** 24.5-28.5	**ASME** SFA5.8 (BAg-27) **AWS** A5.8 (BAg-27)
P07300	Silver Brazing Alloy	**Ag** 29.0-31.0 **Cd** 19.0-21.0 **Cu** 26.0-28.0 **Zn** 21.0-25.0	**ASME** SFA5.8 (BAg-2a) **AWS** A5.8 (BAg-2a)
P07301	Silver Brazing Alloy	**Ag** 29.0-31.0 **Cu** 37.0-39.0 **Zn** 30.0-34.0	**ASME** SFA5.8 (BAg-20) **AWS** A5.8 (BAg-20)
P07350	Silver Brazing Alloy	**Ag** 34.0-36.0 **Cd** 17.0-19.0 **Cu** 25.0-27.0 **Zn** 19.0-23.0	**AMS** 4768 **ASME** SFA5.8 (BAg-2) **AWS** A5.8 (BAg-2) **FED** QQ-B-654 (VIII) **MIL SPEC** MIL-B-7883 (BAg-2)

The chemical compositions listed are for identification purposes and should not be used in lieu of the cross referenced specifications.

UNIFIED NUMBER	DESCRIPTION	CHEMICAL COMPOSITION	CROSS REFERENCE SPECIFICATIONS
P07400	Silver Brazing Alloy	**Ag** 39.0-41.0 **Cu** 29.0-31.0 **Ni** 1.5-2.5 **Zn** 26.0-30.0	**ASME** SFA5.8 (BAg-4) **AWS** A5.8 (BAg-4)
P07401	Silver Brazing Alloy	**Ag** 39.0-41.0 **Cu** 29.0-31.0 **Sn** 1.5-2.5 **Zn** 26.0-30.0	**ASME** SFA5.8 (BAg-28) **AWS** A5.8 (BAg-28)
P07450	Silver Brazing Alloy	**Ag** 44.0-46.0 **Cd** 23.0-25.0 **Cu** 14.0-16.0 **Zn** 14.0-18.0	**AMS** 4769 **ASME** SFA5.8 (BAg-1) **AWS** A5.8 (BAg-1) **FED** QQ-B-654 (VII) **MIL SPEC** MIL-B-7883 (BAg-1)
P07453	Silver Brazing Alloy	**Ag** 44.0-46.0 **Cu** 29.0-31.0 **Zn** 23.0-27.0	**ASME** SFA5.8 (BAg-5) **AWS** A5.8 (BAg-5) **FED** QQ-B-654 (I)
P07490	Silver Brazing Alloy	**Ag** 48.0-50.0 **Cu** 15.0-17.0 **Mn** 7.0-8.0 **Ni** 4.0-5.0 **Zn** 21.0-25.0	**ASME** SFA5.8 (BAg-22) **AWS** A5.8 (BAg-22)
P07500	Silver Brazing Alloy	**Ag** 49.0-51.0 **Cd** 17.0-19.0 **Cu** 14.5-16.5 **Zn** 14.5-18.5	**AMS** 4770 **ASME** SFA5.8 (BAg-1a) **AWS** A5.8 (BAg-1a) **FED** QQ-B-654 (IV) **MIL SPEC** MIL-B-7883 (BAg-1a)
P07501	Silver Brazing Alloy	**Ag** 49.0-51.0 **Cd** 15.0-17.0 **Cu** 14.5-16.5 **Ni** 2.5-3.5 **Zn** 13.5-17.5	**AMS** 4771 **ASME** SFA5.8 (BAg-3) **AWS** A5.8 (BAg-3) **FED** QQ-B-654 (V) **MIL SPEC** MIL-B-7883 (BAg-3)
P07503	Silver Brazing Alloy	**Ag** 49.0-51.0 **Cu** 33.0-35.0 **Zn** 14.0-18.0	**ASME** SFA5.8 (BAg-6) **AWS** A5.8 (BAg-6)
P07505	Silver Brazing Alloy	**Ag** 49.0-51.0 **Cu** 19.0-21.0 **Ni** 1.5-2.5 **Zn** 26.0-30.0	**ASME** SFA5.8 (BAg-24) **AWS** A5.8 (BAg-24)
P07507	Silver Brazing Alloy	**Ag** 49.0-51.0 **Cu** rem	**ASME** SFA5.8 (BVAg-6b) **AWS** A5.8 (BVAg-6b)
P07540	Silver Brazing Alloy	**Ag** 53.0-55.0 **Cu** rem **Ni** 0.5-1.5 **Zn** 4.0-6.0	**AMS** 4772 **ASME** SFA5.8 (BAg-13) **AWS** A5.8 (BAg-13) **MIL SPEC** MIL-B-7883 (BAg-13)
P07547	Silver-Palladium Brazing Alloy	**Ag** 53.0-55.0 **Cu** 20.0-22.0 **Pd** rem	**ASME** SFA5.8 (BVAg-32) **AWS** A5.8 (BVAg-32)
P07560	Silver Brazing Alloy	**Ag** 55.0-57.0 **Cu** rem **Ni** 1.5-2.5	**AMS** 4765 **ASME** SFA5.8 (BAg-13a) **AWS** A5.8 (BAg-13a)
P07563	Silver Brazing Alloy	**Ag** 55.0-57.0 **Cu** 21.0-23.0 **Sn** 4.5-5.5 **Zn** 15.0-19.0	**ASME** SFA5.8 (BAg-7) **AWS** A5.8 (BAg-7)
P07587	Silver Brazing Alloy	**Ag** 57.0-59.0 **Cu** 31.0-33.0 **Pd** rem	**ASME** SFA5.8 (BVAg-31) **AWS** A5.8 (BVAg-31)
P07600	Silver Brazing Alloy	**Ag** 59.0-61.0 **Cu** rem **Sn** 9.5-10.5	**AMS** 4773 **ASME** SFA5.8 (BAg-18) **AWS** A5.8 (BAg-18)
P07607	Silver Brazing Alloy (Vacuum Grade)	**Ag** 59.0-61.0 **Cu** rem **Sn** 9.5-10.5	**ASME** SFA5.8 (BVAg-18) **AWS** A5.8 (BVAg-18)
P07627	Silver Brazing Alloy	**Ag** 60.5-62.5 **Cu** rem **In** 14.0-15.0	**ASME** SFA5.8 (BVAg-29) **AWS** A5.8 (BVAg-29)
P07630	Silver Brazing Alloy	**Ag** 62.0-64.0 **Cu** 27.5-29.5 **Ni** 2.0-3.0 **Ti** 5.0-7.0	**AMS** 4774 **ASME** SFA5.8 (BAg-21) **AWS** A5.8 (BAg-21)
P07650	Silver Brazing Alloy	**Ag** 64.0-66.0 **Cu** 19.0-21.0 **Zn** 13.0-17.0	**ASME** SFA5.8 (BAg-9) **AWS** A5.8 (BAg-9) **FED** QQ-B-654 (II)
P07687	Silver Brazing Alloy	**Ag** 67.0-69.0 **Cu** rem **Pd** 4.5-5.5	**ASME** SFA5.8 (BVAg-30) **AWS** A5.8 (BVAg-30)
P07700	Silver Brazing Alloy	**Ag** 69.0-71.0 **Cu** 19.0-21.0 **Zn** 8.0-12.0	**ASME** SFA5.8 (BAg-10) **AWS** A5.8 (BAg-10)
P07720	Silver Brazing Alloy	**Ag** 71.0-73.0 **Cu** rem	**ASME** SFA5.8 (BAg-8) **AWS** A5.8 (BAg-8)
P07723	Silver Brazing Alloy	**Ag** 71.0-73.0 **Cu** rem **Li** 0.25-0.50	**ASME** SFA5.8 (BAg-8a) **AWS** A5.8 (BAg-8a)
P07727	Silver Brazing Alloy	**Ag** 71.0-73.0 **Cu** rem	**ASME** SFA5.8 (BVAg-8) **AWS** A5.8 (BVAg-8)
P07728	Silver Brazing Alloy	**Ag** 70.5-72.5 **Cu** rem **Ni** 0.3-0.7	**ASME** SFA5.8 (BVAg-8b) **AWS** A5.8 (BVAg-8b)
P07850	Silver Brazing Alloy	**Ag** 84.0-86.0 **Mn** rem	**AMS** 4766 **ASME** SFA5.8 (BAg-23) **AWS** A5.8 (BAg-23) **MIL SPEC** MIL-B-7883 (BAg-23)
P07900	Silver Electrical Contact Alloy	**Ag** 89.6-91.0 **Al** 0.005 max **Cd** 0.05 max **Cu** 9.0-10.4 **Fe** 0.05 max **Ni** 0.01 max **P** 0.02 max **Pb** 0.03 max **Zn** 0.06 max	**ASTM** B617

The chemical compositions listed are for identification purposes and should not be used in lieu of the cross referenced specifications.

UNIFIED NUMBER	DESCRIPTION	CHEMICAL COMPOSITION	CROSS REFERENCE SPECIFICATIONS
P07925	Silver Brazing Alloy	**Ag** 92.0-93.0 **Cu** rem **Li** 0.15-0.30	**AMS** 4767 **ASME** SFA5.8 (BAg-19) **AWS** A5.8 (BAg-19) **MIL SPEC** MIL-B-7883 (BAg-19)
P07931	Silver, Sterling Silver-Standard Grade	**Ag** 92.10-93.50 **Cd** 0.05 max **Cu** 6.50-7.90 **Fe** 0.05 max **Pb** 0.03 max **Zn** 0.06 max **Other** 0.06 max	
P07932	Silver, Sterling Silver, Silversmiths Grade	**Ag** 92.50-93.50 **Cd** 0.05 max **Cu** 6.50-7.50 **Fe** 0.05 max **Pb** 0.03 max **Zn** 0.06 max **Other** 0.06 max	

The chemical compositions listed are for identification purposes and should not be used in lieu of the cross referenced specifications.

UNS NUMBERS ASSIGNED TO DATE

With Description of Each Material Covered and References To Documents In Which The Same or Similar Materials are Described

Rxxxxx Number Series
Reactive and Refractory Metals and Alloys

REACTIVE AND REFRACTORY METALS AND ALLOYS

UNIFIED NUMBER	DESCRIPTION	CHEMICAL COMPOSITION	CROSS REFERENCE SPECIFICATIONS
R03600	Molybdenum, Unalloyed	C 0.010-0.040 Fe 0.010 max Mo bal N 0.0010 max Ni 0.005 max O 0.0030 max Si 0.010 max	ASTM B384 (360); B385 (360); B386 (360); B387 (360)
R03601	Molybdenum Sealing Alloy	C 0.04 max Fe 0.01 max Mo 99.90 min N 0.001 max Ni 0.01 max O 0.005 max Si 0.01 max W 0.02 max Other each 0.005 max	ASTM F49 (Arc-Casting Grade)
R03602	Molybdenum Sealing Alloy	C 0.005 max Fe 0.01 max Mo 99.90 min N 0.002 max Ni 0.01 max O 0.008 max Si 0.01 max W 0.02 max Other each 0.005 max	ASTM F49 (Powder Grade)
R03603	Molybdenum	Al 0.015 max C 0.0015 max Fe 0.001 max H 0.001 max Mo 99.90 min N 0.001 max O 0.0175 max Si 0.035 max Sn 0.0025 max W 0.02 max Other each 0.005 max, Ca 0.005 max, K 0.015 max	ASTM F364 (II)
R03604	Molybdenum	Al 0.015 max C 0.005 max Fe 0.01 max H 0.001 max Mo 99.90 min N 0.002 max O 0.008 max Si 0.01 max Sn 0.0025 max W 0.02 max Other each 0.005 max, Ca 0.005 max, K 0.015 max	ASTM F364 (1)
R03605	Molybdenum Metal	C 0.030 max Fe 0.020 max Mo 99.90 min N 0.0010 max Ni 0.010 max O 0.0030 max Si 0.010 max	AMS 7801
R03606	Molybdenum Metal	C 0.030 max Fe 0.008 max H 0.0005 max Mo 99.95 min N 0.002 max Ni 0.002 max O 0.0015 max Si 0.008 max	AMS 7805
R03610	Molybdenum Unalloyed	C 0.010 max Fe 0.010 max Mo bal N 0.0020 max Ni 0.005 max O 0.0070 max Si 0.010 max	AMS 7800 ASTM B384 (361); B385 (361); B386 (361); B387 (361)
R03620	Molybdenum Alloy	C 0.010-0.040 Fe 0.010 max Mo bal N 0.0010 max Ni 0.005 max O 0.0030 max Si 0.010 max Ti 0.40-0.55	ASTM B384 (362); B385 (362); B386 (362); B387 (362)
R03630	Molybdenum Alloy	C 0.010-0.040 Fe 0.010 max Mo bal N 0.0010 max Ni 0.005 max O 0.0030 max Si 0.010 max Ti 0.40-0.55 Zr 0.06-0.12	AMS 7817; 7819 ASTM B384 (363); B385 (363); B386 (363); B387 (363)
R03640	Molybdenum Alloy	C 0.010-0.040 Fe 0.010 max Mo bal N 0.0020 max Ni 0.005 max O 0.030 max Si 0.005 max Ti 0.40-0.55 Zr 0.06-0.12	ASTM B384 (364); B385 (364); B386 (364); B387 (364)
R03650	Molybdenum Unalloyed, Low Carbon	C 0.010 max Fe 0.010 max Mo bal N 0.0010 max Ni 0.005 max O 0.0030 max Si 0.010 max	ASTM B384 (365); B385 (365); B386 (365); B387 (365)
R04200	Niobium (Columbium), Unalloyed, Reactor Grade	C 0.01 max Fe 0.005 max H 0.001 max Hf 0.01 max Mo 0.005 max N 0.01 max Nb rem Ni 0.005 max O 0.015 max Si 0.005 max Ta 0.1 max W 0.03 max Zr 0.01 max	ASTM B391; B392; B393; B394
R04210	Niobium (Columbium), Unalloyed, Commercial Grade	C 0.01 max Fe 0.01 max H 0.001 max Hf 0.01 max Mo 0.005 max N 0.01 max Nb rem Ni 0.005 max O 0.025 max Si 0.005 max Ta 0.2 max W 0.05 max Zr 0.01 max	ASTM B391; B392; B393; B394
R04211	Niobium (Columbium), Unalloyed, Commercial Grade	C 0.005 max Cb rem Fe 0.010 max H 0.002 max N 0.010 max O 0.030 max Si 0.005 max Ta 0.10 max Ti 0.005 max Zr 0.010 max Other each 0.010 max, total 0.15 max	AMS 7850

The chemical compositions listed are for identification purposes and should not be used in lieu of the cross referenced specifications.

REACTIVE AND REFRACTORY METALS AND ALLOYS

UNIFIED NUMBER	DESCRIPTION	CHEMICAL COMPOSITION	CROSS REFERENCE SPECIFICATIONS
R04251	Niobium (Columbium)- 1% Zirconium, Reactor Grade	C 0.01 max Fe 0.005 max H 0.001 max Hf 0.01 max Mo 0.005 max N 0.01 max Nb rem Ni 0.005 max O 0.015 max Si 0.005 max Ta 0.1 max W 0.03 max Zr 0.8-1.2	ASTM B391; B392; B393; B394
R04261	Niobium (Columbium)- 1% Zirconium, Commercial Grade	C 0.01 max Fe 0.01 max H 0.001 max Hf 0.01 max Mo 0.005 max N 0.01 max Nb rem Ni 0.005 max O 0.025 max Si 0.005 max Ta 0.2 max W 0.05 max Zr 0.8-1.2	ASTM B391; B393; B394
R04271	Niobium (Columbium) Alloy, 10W-2.5Zr	C 0.030 max Cb rem Fe 0.02 max H 0.001 max N 0.010 max O 0.020 max Si 0.02 max Ta 0.15 max Ti 0.01 max W 9.00-11.00 Zr 2.00-3.00	AMS 7851; 7855
R04295	Niobium (Columbium) Alloy	C 0.015 max Cb rem H 0.0015 max Hf 9-11 N 0.010 max O 0.020 max Ta 0.500 max Ti 0.7-1.3 W 0.500 max Zr 0.700 max	AMS 7857 ASTM B652; B654; B655
R05200	Tantalum, Unalloyed, Cast	C 0.01 max Cb 0.05 max Fe 0.01 max H 0.001 max Mo 0.01 max N 0.01 max Ni 0.01 max O 0.015 max Si 0.005 max Ta rem Ti 0.01 max W 0.03 max	ASTM B364; B365
R05210	Tantalum Metal	C 0.0075 max Cb 0.100 max Co 0.002 max Fe 0.01 max H 0.0010 max Mo 0.01 max N 0.0075 max Ni 0.005 max O 0.015 max Si 0.005 max Ta 99.85 min Ti 0.005 max W 0.03 max Zr 0.01 max Other total 0.15 max	AMS 7849
R05255	90 Tantalum-10 Tungsten	C 0.01 max Cb 0.10 max Fe 0.01 max H 0.001 max Mo 0.01 max N 0.01 max Ni 0.01 max O 0.015 max Si 0.005 max Ta rem Ti 0.01 max W 9.0-11.0	AMS 7847; 7848 ASTM B364; B365
R05400	Tantalum, Unalloyed, Sintered	C 0.01 max Cb 0.05 max Fe 0.01 max H 0.001 max Mo 0.01 max N 0.015 max Ni 0.01 max O 0.03 max Si 0.005 max Ta rem Ti 0.01 max W 0.03 max	ASTM B364; B365
* R07004	Tungsten	Al 0.003 max C 0.010 max Fe 0.005 max H 0.002 max Mo 0.005 max N 0.005 max Ni 0.008 max O 0.006 max Si 0.003 max W 99.95 min	ASTM B410
R07005	Tungsten	W 99.95 min Other 0.01 max each, 0.05 max total	AMS 7897 ASTM F288 (1A and 1B); F290
R07006	Tungsten Metal	Al 0.005 max C 0.008 max Fe 0.005 max H 0.001 max Mo 0.020 max N 0.002 max Ni 0.005 max O 0.005 max Si 0.005 max W rem	AMS 7898
R07030	Tungsten	W 96-98	ASTM B459 (4)
R07031	Tungsten-Rhenium	Re 2.5-3.5 W rem Other each 0.05 max, total 0.01 max	ASTM F73 (Electronic Grade)
R07050	Tungsten	W 94-96	ASTM B459 (3)
R07080	Tungsten	W 91-94	ASTM B459 (2)
R07100	Tungsten	W 89-91	ASTM B459 (1)
R07900	Tungsten Arc Welding Electrode	W 99.5 min	ASME SFA5.12 (EWP) AWS A5.12 (EWP)

*Boxed entries are no longer active and are retained for reference purposes only.

The chemical compositions listed are for identification purposes and should not be used in lieu of the cross referenced specifications.

UNIFIED NUMBER	DESCRIPTION	CHEMICAL COMPOSITION	CROSS REFERENCE SPECIFICATIONS
R07911	Tungsten-Thorium Alloy Arc Welding Electrode	Th 0.8-1.2 W 98.5 min	ASME SFA5.12 (EWTh-1) AWS A5.12 (EWTh-1)
R07912	Tungsten-Thorium Alloy Arc Welding Electrode	Th 1.7-2.2 W 97.5 min	ASME SFA5.12 (EWTh-2) AWS A5.12 (EWTh-2)
R07913	Tungsten-Thorium Alloy Arc Welding Electrode	Th 0.35-0.55 W 98.95 min	ASME SFA5.12 (EWTh-3) AWS A5.12(EWTh-3)
R07920	Tungsten-Zirconium Alloy Arc Welding Electrode	W 99.2 min Zr 0.15-0.40	ASME SFA5.12 (EWZr) AWS A5.12 (EWZr)
R19800	Beryllium	Al 0.20 max Be 98.00 min C 0.15 max Fe 0.20 max Mg 0.08 max Si 0.12 max	AMS 7901
R19801	Beryllium	Al 0.20 max Be 98.00 C 0.15 max Fe 0.20 max Mg 0.08 max Si 0.12 max	AMS 7902
R19920	Beryllium	Al 0.18 max Be 99.20 min C 0.15 max Fe 0.20 max Mg 0.08 max Si 0.10 max	AMS 7900
R20500	Chromium-Nickel Alloy	Al 0.25 max C 0.10 max Cr 48.0-52.0 Fe 1.00 max Mn 0.30 max N 0.30 max Ni bal P 0.02 max S 0.02 max Si 1.00 max Ti 0.50 max	ASTM A560 (50 Cr-50 Ni)
R20600	Chromium-Nickel Alloy	Al 0.25 max Cr 58.0-62.0 Fe 1.00 max Mn 0.30 max N 0.30 max Ni bal P 0.02 max S 0.02 max Si 1.00 max Ti 0.50 max	ASTM A560 (60 Cr-40 Ni)
R20990	Chromium Metal	C 0.050 max Cr 99.0 min P 0.010 max S 0.030 max Si 0.15 max	ASTM A481 (A)
R20994	Chromium Metal	C 0.050 max Cr 99.4 min P 0.010 max S 0.010 max Si 0.10 max	ASTM A481 (B)
R30001	Co-Cr-W Alloy (Stellite 1)	C 2.5 nom Co bal Cr 30.0 nom Fe 3.0 min Mn 0.5 nom Mo 0.5 nom Ni 1.5 nom Si 1.3 nom W 13.0 nom	ASME SFA5.13 (RCoCr-C) AWS A5.13 (RCoCr-C) SAE J775 (VF6) MIL SPEC MIL-R-17131 (MIL-RCoCr-C)
R30002	Co-Cr-Ni-W Alloy (Stellite F)	C 1.75 nom Co bal Cr 25.00 nom Fe 2.0 max Mn 0.30 nom Ni 22.00 nom Si 1.00 nom W 12.00 nom	SAE J775 (VF5)
R30003	Co-Cr-Ni-Fe-Mo (Elgiloy)	Be 1.0 max C 0.15 max Co 39.0-41.0 Cr 19.0-21.0 Fe bal Mn 1.5-2.5 Mo 6.0-8.0 Ni 15.0-16.0	AMS 5833; 5834; 5875; 5876
R30004	Co-Cr-Ni-Fe Alloy (Havar)	Be 0.02-0.06 C 0.17-0.23 Co 41.0-44.0 Cr 19.0-21.0 Fe bal Mn 1.35-1.80 Mo 2.0-2.8 Ni 12.0-14.0 W 2.3-3.3	
R30005	Co-Fe-V Ferromagnetic Alloy	Co 49.0 Fe 49.0 V 2.0	MIL SPEC MIL-A-47182
R30006	Co-Cr-W Alloy (Stellite 6)	C 0.9-1.4 Co bal Cr 27.0-31.0 Fe 3.0 max Mn 1.0 max Mo 1.5 max Ni 3.0 max Si 1.5 max W 3.5-5.5	AMS 5373; 5387; 5788 ASME SFA5.13 (RCoCr-A) AWS A5.13 (RCoCr-A) MIL SPEC MIL-R-17131 (MIL-RCoCr-A)
R30012	Co-Cr-W Alloy (Stellite 12)	C 1.4 nom Co bal Cr 30.0 nom Fe 3.0 min Mn 2.5 nom Ni 1.5 nom Si 0.7 nom W 8.3 nom	ASME SFA5.13 (RCoCr-B) AWS A5.13 (RCoCr-B) SAE J775 (VF7)
R30021	Co-Cr-Mo-Ni Alloy (Stellite 21)	B 0.007 max C 0.20-0.30 Co bal Cr 25.00-29.00 Fe 3.00 max Mn 1.00 max Mo 5.00-6.00 Ni 1.75-3.75 Si 1.00 max	AMS 5385 ASTM A567 (1)
R30023	Co-Cr-W-Ni Alloy (Stellite 23)	C 0.40 nom Co bal Cr 24 nom Fe 1.0 nom Mn 0.3 nom Ni 2 nom Si 0.6 nom W 5 nom	AMS 5375
R30027	Co-Ni-Cr Alloy (Stellite 27)	C 0.40 nom Co bal Cr 25 nom Fe 1.0 nom Mn 0.3 nom Mo 5.5 nom Ni 32 nom Si 0.6 nom	AMS 5378

The chemical compositions listed are for identification purposes and should not be used in lieu of the cross referenced specifications.

UNIFIED NUMBER	DESCRIPTION	CHEMICAL COMPOSITION	CROSS REFERENCE SPECIFICATIONS
R30030	Co-Cr-Ni-Mo Alloy (Stellite 30)	C 0.45 nom Co bal Cr 26 nom Fe 1.0 nom Mn 0.6 nom Mo 6 nom Ni 15 nom Si 0.6 nom	AMS 5380
R30031	Co-Cr-Ni-Mo Alloy (Stellite 31)	C 0.45-0.55 Co bal Cr 24.50-26.50 Fe 2.00 max Mn 1.00 max Ni 9.50-11.50 Si 1.00 max W 7.00-8.00	AMS 5382; 5789 ASTM A567 (2)
R30035	Co-Ni-Cr-Mo Alloy (MP-35-N)	C 0.025 max Co bal Cr 19.00-21.00 Fe 1.00 max Mn 0.15 max Mo 9.00-10.50 Ni 33.00-37.00 P 0.015 max S 0.010 max Si 0.15 max Ti 1.00 max	AMS 5844; 5845; 5758; 7468
R30036	Co-Cr-W-Ni Alloy Solid Solution Strengthened (Haynes 36)	B 0.01-0.05 C 0.35-0.45 Co rem Cr 17.5-19.5 Fe 2.00 max Mn 1.50 max Ni 9.00-11.0 Si 1.00 max W 14.0-16.0	
R30040	Co-Cr-Ni-Si-W Brazing Filler Metal	Al 0.05 max B 0.70-0.90 C 0.35-0.45 Co rem Cr 18.0-20.0 Fe 1.00 max Ni 16.0-18.0 P 0.02 max S 0.02 max Si 7.50-8.50 Ti 0.05 max W 3.50-4.50 Zr 0.05 max Other each 0.03 max, total 0.15 max	AMS 4783
R30100	Fe-Cr-Ni-Co Alloy	C 0.10 max Co 18.5-21.0 Cr 20.0-22.5 Fe rem Mn 1.00-2.00 Mo 2.50-3.50 N 0.10-0.20 Ni 19.0-21.0 P 0.040 max S 0.030 max Si 1.00 max W 2.00-3.00	AMS 7724
R30155	Fe-Cr-Ni-Co Alloy (N-155)	C 0.08-0.16 Cb 0.75-1.25 Co 18.50-21.00 Cr 20.00-22.50 Fe bal Mn 1.00-2.00 Mo 2.50-3.50 N 0.20 max Ni 19.00-21.00 P 0.040 max S 0.030 max Si 1.00 max W 2.00-3.00	AMS 5376; 5531; 5532; 5585; 5768; 5769; 5794 ASTM A639 (661); A567 (661) MIL SPEC MIL-R-5031 (Class 9)
R30159	Co-Ni-Cr-Fe-Mo Alloy (MP-159)	Al 0.10-0.30 B 0.03 max C 0.04 max Cb 0.25-0.75 Co 34.0-38.0 Cr 18.0-20.0 Fe 8.00-10.00 Mn 0.20 max Mo 6.00-8.00 Ni rem P 0.020 max S 0.010 max Si 0.20 max Ti 2.50-3.25	AMS 5841; 5842; 5843
R30188	Co-Cr-Ni-W Alloy (HS188)	C 0.05-0.15 Co bal Cr 20.00-24.00 Fe 3.00 max La 0.03-0.15 Mn 1.25 max Ni 20.00-24.00 Si 0.20-0.50 W 13.00-16.00	AMS 5608; 5772; 5801
R30260	Co-Ni-Cr-Mo-W Alloy Age-Hardenable (Duratherm 2602)	Be 0.20-0.30 C 0.05 max Cb 0.10 max Co 41.0-42.0 Cr 11.7-12.3 Cu 0.30 max Fe 9.80-10.4 Mn 0.40-1.10 Mo 3.70-4.30 Ni rem Si 0.20-0.60 Ti 0.80-1.20 W 3.60-4.20	
R30477	Co-Ni-Cr-Mo-W Alloy Age-Hardenable (Duratherm 477)	Al 0.15 max Be 0.05 max Cb 0.10 max Co 39.0-46.0 Cr 17.7-18.3 Cu 0.30 max Fe 4.50-5.50 Mn 0.70 max Mo 3.80-4.20 Ni rem Si 0.50 max Ti 0.80-1.20 W 3.80-4.20	ASTM F563 ISO 5832/8
R30556	Fe-Cr-Ni-Co Alloy (HS 556)	Al 0.10-0.50 B 0.02 max C 0.05-0.15 Cb 0.30 max Co 16.0-21.0 Cr 21.0-23.0 Fe rem La 0.005-0.10 Mn 0.50-2.00 Mo 2.50-4.00 N 0.10-0.30 Ni 19.0-22.5 P 0.04 max S 0.015 max Si 0.20-0.80 Ta 0.30-1.25 W 2.00-3.50 Zr 0.001-0.10	AMS 5874
R30590	Fe-Cr-Co-Ni Alloy (S-590)	C 0.38-0.48 Cb 3.50-4.50 Co 18.50-21.50 Cr 19.00-22.00 Cu 0.50 max Fe bal Mn 2.00 max Mo 3.50-4.50 Ni 18.50-21.50 P 0.040 max S 0.030 max Si 1.00 max W 3.50-4.50	AMS 5770

The chemical compositions listed are for identification purposes and should not be used in lieu of the cross referenced specifications.

UNIFIED NUMBER	DESCRIPTION	CHEMICAL COMPOSITION	CROSS REFERENCE SPECIFICATIONS
R30600	Wrought Co-Ni-Cr-Mo-W Alloy Age-Hardenable (Duratherm 600)	Al 0.6-0.8 Be 0.05 max C 0.05 max Cb 0.1 max Co 41.0-42.0 Cr 11.7-12.3 Cu 0.3 max Fe 8.5-8.9 Mn 0.4-1.1 Mo 3.7-4.3 Ni bal Si 0.2-0.6 Ti 1.8-2.2 W 3.6-4.2	
R30605	Co-Cr-Ni-W Alloy (L-605)	C 0.05-0.15 Co bal Cr 19.0-21.0 Fe 3.0 max Mn 2.00 max Ni 9.0-11.0 Si 1.00 max W 14.0-16.0	AMS 5537; 5759; 5796; 7236 ASTM F90 MIL SPEC MIL-R-5031 (Class 13)
R30700	Co-Ni-Cr-Mo-W Alloy Age-Hardenable (Duratherm 700)	Al 1.00-1.40 Be 0.05 max C 0.05 max Cb 0.10 max Co 41.0-42.0 Cr 11.7-12.3 Cu 0.30 max Fe 6.40-7.00 Mn 0.40-1.10 Mo 3.70-4.30 Ni rem Si 0.20-0.60 Ti 3.30-3.70 W 3.60-4.20	
R30816	Co-Cr-Ni-Cb-Mo-W Alloy (S-816)	C 0.32-0.42 Cb 3.50-4.50 Co 40.00 min Cr 19.00-21.00 Fe 5.00 max Mn 1.00-2.00 Mo 3.50-4.50 Ni 19.00-21.00 P 0.040 max S 0.030 max Si 1.00 max W 3.50-4.50	AMS 5534; 5765 ASTM A461; A639 (671)
R39001	Cobalt Brazing Filler Metal	Al 0.05 B 0.7-0.9 C 0.35-0.45 Co rem Cr 18.0-20.0 Fe 1.0 Ni 16.0-18.0 P 0.02 S 0.02 Si 7.5-8.5 Ti 0.05 W 3.5-4.5 Zr 0.05 Other 0.50	ASME SFA5.8 (BCo-1) AWS A5.8 (BCo-1)
R50100	Titanium Welding Filler Metal, Unalloyed	C 0.03 max Fe 0.10 max H 0.005 max N 0.012 max O 0.10 max Ti rem	AWS A5.16 (ERTi-1) MIL SPEC MIL-R-81586 (CP-E)
R50120	Titanium Welding Filler Metal	C 0.05 max Fe 0.20 max H 0.008 max N 0.020 max O 0.10 max Ti rem	AWS A5.16 (ERTi-2)
R50125	Titanium Welding Filler Metal	C 0.05 max Fe 0.20 max H 0.008 max N 0.020 max O 0.10-0.15 Ti rem	AMS 4951 AWS A5.16 (ERTi-3) MIL SPEC MIL-R-81588 (CP-F)
R50130	Titanium Welding Filler Metal	C 0.05 max Fe 0.30 max H 0.008 max N 0.020 max O 0.15-0.25 Ti rem	AWS A5.16 (ERTi-4)
R50250	Titanium, Unalloyed	C 0.10 max Fe 0.20 max H 0.015 max N 0.03 max O 0.18 max Ti rem	ASTM F67 (1); B265 (1); B337 (1); B338 (1); B348 (1); B367 (C-1); B381 (F-1); F467 (1); F468 (1) MIL SPEC MIL-T-81556; MIL-T-81915
R50400	Titanium, Unalloyed	C 0.10 max Fe 0.30 max H 0.015 max N 0.03 max O 0.25 max Ti rem	AMS 4902; 4941; 4942 ASTM F67 (2); B265 (2); B337 (2); B338 (2); B348 (2); B367 (C-2); B381 (F-2); F467 (2); F468 (2) MIL SPEC MIL-T-9046; MIL-T-81556
R50550	Titanium, Unalloyed	C 0.10 max Fe 0.30 max H 0.015 max N 0.05 max O 0.35 max Ti rem Other each 0.1 max, total 0.4 max	AMS 4900; 4951 ASTM F67 (3); B265 (3); B337 (3); B338 (3); B348 (3); B367 (C-3); B381 (F-3) MIL SPEC MIL-T-9046; MIL-T-81556
R50700	Titanium, Unalloyed	C 0.10 max Fe 0.50 max H 0.015 max N 0.05 max O 0.40 max Ti rem Other each 0.1 max, total 0.4 max	AMS 4901; 4921 ASTM F67 (4); B265 (4); B348 (4); B367 (C-4); B381 (F-4); F467 (4); F468 (4) MIL SPEC MIL-T-9046; MIL-T-9047; MIL-T-81556; MIL-F-83142
R52250	Titanium, Low Alloyed	C 0.10 max Fe 0.20 max H 0.015 max N 0.03 max O 0.18 max Ti rem Other Pd 0.12-0.25	ASTM B265 (11); B337 (11); B338 (11); B348 (11); B367 (C-7A); B381 (F-11)
R52400	Titanium, Low Alloyed	C 0.10 max Fe 0.30 max H 0.015 max N 0.03 max O 0.25 max Ti rem Other Pd 0.12-0.25	ASTM B265 (7); B337 (7); B338 (7); B348 (7); B367 (C-7B); B381 (F-7); F467 (7); F468 (7)
R52401	Titanium Welding Filler Metal	C 0.05 max Fe 0.25 max H 0.008 max N 0.020 max O 0.15 max Pd 0.15-0.25 Ti rem	AWS A5.16 (ERTi-0.2Pd)
R52550	Titanium, Low Alloyed	C 0.10 max Fe 0.30 max H 0.015 max N 0.05 max O 0.35 max Ti rem Other Pd 0.12 min	ASTM B367 (C-8A)
R52700	Titanium Alloy	C 0.10 max Fe 0.50 max H 0.0100 max N 0.05 max O 0.40 max Pd 0.12 min Ti rem	ASTM B367 (C-8B)

The chemical compositions listed are for identification purposes and should not be used in lieu of the cross referenced specifications.

UNIFIED NUMBER	DESCRIPTION	CHEMICAL COMPOSITION	CROSS REFERENCE SPECIFICATIONS
R53400	Titanium Alloy	C 0.08 max Fe 0.30 max H 0.015 max Mo 0.2-0.4 N 0.03 max Ni 0.6-0.9 O 0.25 max Ti rem	ASTM B265 (12); B337 (12); B338 (12); B348 (12); B381 (F-12)
R53401	Titanium Alloy Welding Filler Metal	C 0.03 max Fe 0.30 max H 0.008 max Mo 0.2-0.4 N 0.020 max Ni 0.6-0.9 O 0.25 max Ti rem	AWS A5.16 (ERTi-12)
R54520	Titanium Alloy	Al 4.0-6.0 C 0.10 max Fe 0.50 max H 0.020 max N 0.05 max O 0.20 max Sn 2.0-3.0 Ti rem	AMS 4910; 4926; 4966 ASTM B265 (6); B348 (6); B367 (C-6); B381 (F-6) MIL SPEC MIL-T-9046; MIL-T-9047; MIL-T-81556; MIL-T-81915; MIL-F-83142
R54521	Titanium Alloy	Al 5 Sn 2.5 Ti rem	AMS 4909; 4924 MIL SPEC MIL-T-9046; MIL-T-9047; MIL-T-81556; MIL-F-83142
R54522	Titanium Alloy Welding Filler Metal	Al 4.7-5.6 C 0.05 max Fe 0.40 max H 0.008 max N 0.030 max O 0.12 max Sn 2.0-3.0 Ti rem	AMS 4953 AWS A5.16 (ERTi-5Al-2.5Sn)
R54523	Titanium Alloy Welding Filler Metal	Al 4.7-5.6 C 0.04 max Fe 0.25 max H 0.005 max N 0.012 max O 0.10 max Sn 2.0-3.0 Ti rem	AWS A5.16 (ERTi-5Al-2.5Sn-1) MIL SPEC MIL-R-81588 (5Al-2.5Sn ELI)
R54550	Titanium Alloy	Al 5 Sn 5 Ti rem Zr 5	MIL SPEC MIL-F-83142
R54560	Titanium Alloy	Al 5 Mo 1 Si 0.2 Sn 6 Ti rem Zr 2	MIL SPEC MIL-T-9046; MIL-T-9047
R54620	Titanium Alloy	Al 6 Mo 2 Sn 2 Ti rem Zr 4	AMS 4919; 4975; 4976 MIL SPEC MIL-T-9046; MIL-T-9047; MIL-T-81915; MIL-F-83142
R54621	Titanium Alloy Welding Filler Metal	Al 5.50-6.50 C 0.04 max Cr 0.25 max Fe 0.05 H 0.15 max Mo 1.80-2.20 N 0.015 max O 0.30 max Sn 1.80-2.20 Ti rem Zr 3.60-4.40	MIL SPEC MIL-R-81588 (6Al-2Sn-4Zr-2Mo)
R54790	Titanium Alloy	Al 2 Mo 1 Si 0.2 Sn 11 Ti rem Zr 5	AMS 4974 MIL SPEC MIL-T-9047; MIL-F-83142
R54810	Titanium Alloy	Al 8 Mo 1 Ti rem V 1	AMS 4915; 4916; 4933; 4955; 4972; 4973 AWS A5.16 (ERTi-8Al-1Mo-1V) MIL SPEC MIL-T-9046; MIL-T-9047; MIL-T-81556; MIL-R-81588 (8Al-1Mo-1VELI); MIL-F-83142
R56080	Titanium Alloy	Mn 8 Ti rem	AMS 4908 MIL SPEC MIL-T-9046
R56210	Titanium Alloy	Al 6 Cb 2 Mo 0.8 Ta 1 Ti rem	AWS A5.16 (ERTi-8Al-2Cb-1Ta-1Mo) MIL SPEC MIL-T-9046; MIL-T-9047; MIL-R-81588 (6Al-2Cb-1Ta-0.8Mo)
R56260	Titanium Alloy	Al 6 Mo 6 Sn 2 Ti rem Zr 4	AMS 4981 MIL SPEC MIL-T-9047
R56320	Titanium Alloy	Al 2.5-3.5 C 0.05 max Fe 0.25 max H 0.013 max N 0.02 max O 0.12 max Ti rem V 2.0-3.0	AMS 4943; 4944 ASTM B337 (9); B338 (9) AWS A5.16 (ERTi-3Al-2.5V) MIL SPEC MIL-T-9046; MIL-T-9047
R56321	Titanium Welding Filler Metal	Al 2.5-3.5 C 0.04 max Fe 0.25 max H 0.005 max N 0.012 max O 0.10 max Ti rem V 2.0-3.0	AWS A5.16 (ERTi-3Al-2.5V-1)
R56400	Titanium Alloy	Al 5.5-6.75 C 0.10 max Fe 0.40 max H 0.015 max N 0.05 max O 0.20 max Ti rem V 3.5-4.5	AMS 4905; 4906; 4911; 4920; 4928; 4934; 4935; 4954; 4965; 4967 ASTM B265 (5); B348 (5); B367 (C-5); B381 (F-5) AWS A5.16 (ERTi-6Al-4V) MIL SPEC MIL-T-9046; MIL-T-9047; MIL-T-81556; MIL-T-81915; MIL-F-83142
R56401	Titanium Alloy	Al 6 Ti rem V 4	AMS 4907; 4930; 4985; 4991; 4996; 4998 ASTM F136 (5); F467 (5); F468 (5); F468 (S) MIL SPEC MIL-T-9046; MIL-T-9047; MIL-T-81556; MIL-F-83142
R56402	Titanium Alloy Welding Filler Metal	Al 5.5-6.75 C 0.04 max Fe 0.15 max H 0.005 max N 0.012 max O 0.10 max Ti rem V 3.5-4.5	AMS 4956 AWS A5.16 (ERTi-6Al-4V-1) MIL SPEC MIL-R-81588 (6Al-4V-ELI)
R56430	Titanium Alloy	Al 4 Mo 3 Ti rem V 1	AMS 4912; 4913 MIL SPEC MIL-T-9046
* R56440	Titanium Alloy	Al 4 Mn 4 Ti rem	

*Boxed entries are no longer active and are retained for reference purposes only.

The chemical compositions listed are for identification purposes and should not be used in lieu of the cross referenced specifications.

UNIFIED NUMBER	DESCRIPTION	CHEMICAL COMPOSITION	CROSS REFERENCE SPECIFICATIONS
R56620	Titanium Alloy	**Al** 5.5 **Sn** 2 **Ti** rem **V** 5.5	**AMS** 4918; 4936; 4971; 4978; 4979 **MIL SPEC** MIL-T-9046; MIL-T-9047; MIL-T-81556; MIL-F-83142
R56740	Titanium Alloy	**Al** 7 **Mo** 4 **Ti** rem	**AMS** 4970 **MIL SPEC** MIL-T-9047; MIL-T-81556; MIL-F-83142
R58010	Titanium Alloy	**Al** 3 **Cr** 11 **Ti** rem **V** 13	**AMS** 4917; 4959 **AWS** A5.16 (ERTi-13V-11Cr-3Al) **MIL SPEC** MIL-T-9046; MIL-T-9047; MIL-R-81588 (13V-11Cr-3Al); MIL-F-83142
R58030	Titanium Alloy	**C** 0.10 max **Fe** 0.35 max **H** 0.020 max **Mo** 10.0-13.0 **N** 0.05 max **O** 0.18 max **Sn** 3.75-5.25 **Ti** rem **Zr** 4.50-7.50	**AMS** 4977; 4980 **ASTM** B265 (10); B337 (10); B338 (10); B348 (10) **MIL SPEC** MIL-T-9046; MIL-T-9047; MIL-F-83142
R58450	Titanium Alloy	**Cb** 45 **Ti** rem	**AMS** 4982
R58640	Titanium Alloy	**Al** 3 **Cr** 6 **Mo** 4 **Ti** rem **V** 8 **Zr** 4	**MIL SPEC** MIL-T-9046; MIL-T-9047; MIL-F-83142
R58650	Titanium Alloy, 5Al-2Sn-2Zr-4Cr-4Mo	**Al** 4.50-5.50 **C** 0.05 max **Cr** 3.50-4.50 **Cu** 0.10 max **Fe** 0.30 max **H** 0.0125 max **Mn** 0.10 max **Mo** 3.50-4.50 **N** 0.04 max **O** 0.08-0.13 **Sn** 1.50-2.50 **Ti** rem **Y** 0.0050 max **Zr** 1.50-2.50 **Other** each 0.10 max, total 0.30 max	**AMS** 4995; 4997
R58820	Titanium Alloy	**Al** 3 **Fe** 2 **Mo** 8 **Ti** rem **V** 8	**MIL SPEC** MIL-T-9046; MIL-T-9047; MIL-F-83142
R60001	Zirconium Unalloyed Reactor Grade	**Al** 0.0075 max **B** 0.00005 max **C** 0.027 max **Cd** 0.00005 max **Co** 0.0020 max **Cr** 0.020 max **Cu** 0.0050 max **Fe** 0.150 max **H** 0.0025 max **Hf** 0.010 max **Mn** 0.0050 max **N** 0.0065 max **Ni** 0.0070 max **Si** 0.0120 max **Ti** 0.0050 max **U** 0.00055 max **W** 0.010 max	**ASTM** B349; B350; B351; B352; B353
R60701	Zirconium, Unalloyed	**C** 0.05 max **H** 0.005 max **Hf** 4.5 max **N** 0.025 max **Other** Zr+Hf 99.5 min, Fe+Cr 0.05 max	**ASTM** B493; B494; B495; B523; B550; B551
R60702	Zirconium, Unalloyed	**C** 0.05 max **H** 0.005 max **Hf** 4.5 max **N** 0.025 max **Other** Zr+Hf 99.2 min, Fe+Cr 0.2 max	**ASME** SB493; SB494; SB495; SB523; SB550; SB551; SFA5.24 (ERZr2) **ASTM** B493; B494; B495; B523; B550; B551; B653; B658 **AWS** A5.24 (ERZr2)
R60703	Zirconium Alloy	**Hf** 4.5 max **Other** Zr+Hf 98.0 min	**ASTM** B494; B495
R60704	Zirconium Alloy	**C** 0.05 max **H** 0.005 max **Hf** 4.5 max **N** 0.025 max **Sn** 1.00-2.00 **Other** Zr+Hf 97.5 min, Fe+Cr 0.20-0.40	**ASME** SFA5.24 (ERZr3) **ASTM** B493; B495; B523; B550; B551; B653; B658 **AWS** A5.24 (ERZr3)
R60705	Zirconium Alloy	**C** 0.05 max **Cb** 2.0-3.0 **H** 0.005 max **Hf** 4.5 max **N** 0.025 max **O** 0.18 max **Other** Zr+Hf 95.5 min	**ASTM** B493; B495; B523; B550; B551; B653
R60706	Zirconium Alloy	**C** 0.050 max **Cb** 2.0-3.0 **H** 0.005 max **Hf** 4.5 max **N** 0.025 max **O** 0.16 max **Other** Zr+Hf 95.5 min Fe+Cr 0.2 max	**ASTM** B495; B551
R60707	Zirconium Alloy (2.5 Cb) Welding Filler Metal	**C** 0.05 max **Cb** 2.0-3.0 **Fe** 0.20 max includes Cr **H** 0.005 max **Hf** 4.5 max **N** 0.025 max **Zr** 95.5 min includes Hf	**ASME** SFA5.24 (ERZr4) **AWS** A5.24 (ERZr4)
R60802	Zirconium Alloy Reactor Grade	**Al** 0.0075 max **B** 0.00005 max **C** 0.027 max **Cd** 0.00005 max **Co** 0.0020 max **Cr** 0.05-0.15 **Cu** 0.0050 max **Fe** 0.07-0.20 **H** 0.0025 max **Hf** 0.010 max **Mn** 0.0050 max **N** 0.0065 max **Ni** 0.05-0.08 **Si** 0.020 max **Sn** 1.20-1.70 **Ti** 0.0050 max **U** 0.00035 max **W** 0.010 max **Other** Fe+Cr+Ni 0.18-0.38	**ASTM** B350; B351; B352; B353

The chemical compositions listed are for identification purposes and should not be used in lieu of the cross referenced specifications.

UNIFIED NUMBER	DESCRIPTION	CHEMICAL COMPOSITION	CROSS REFERENCE SPECIFICATIONS
R60804	Zirconium Alloy Reactor Grade	**Al** 0.0075 max **B** 0.00005 max **C** 0.027 max **Cd** 0.00005 max **Co** 0.0020 max **Cr** 0.07-0.13 **Cu** 0.0050 max **Fe** 0.18-0.24 **H** 0.0025 max **Hf** 0.010 max **Mn** 0.0050 max **N** 0.0065 max **Ni** 0.0070 max **Si** 0.0120 max **Sn** 1.20-1.70 **Ti** 0.0050 max **U** 0.00035 max **W** 0.010 max **Other** Fe+Cr 0.28-0.37	**ASTM** B350; B351; B352; B353
R60901	Zirconium Alloy Reactor Grade	**Al** 0.0075 max **B** 0.00005 max **C** 0.027 max **Cb** 2.40-2.80 **Cd** 0.00005 max **Co** 0.0020 max **Cr** 0.020 max **Cu** 0.0050 max **Fe** 0.150 max **H** 0.0025 max **Hf** 0.010 max **Mn** 0.0050 max **N** 0.0065 max **Ni** 0.0070 max **O** 0.09-0.13 **Si** 0.0120 max **Ti** 0.0050 max **U** 0.00035 max **W** 0.010 max	**ASTM** B350; B351; B352; B353
R60902	Zirconium Alloy Reactor Grade	**Al** 0.0075 max **B** 0.00005 max **C** 0.030 max **Cb** 2.4-2.8 **Cd** 0.00005 max **Co** 0.0020 max **Cr** 0.020 max **Cu** 0.3-0.7 **Fe** 0.15 max **H** 0.0025 max **Hf** 0.010 max **Mg** 0.0020 max **Mn** 0.0050 max **N** 0.0065 max **Ni** 0.0070 max **O** 0.08-0.12 **Pb** 0.013 max **Si** 0.012 max **Sn** 0.0050 max **Ta** 0.020 max **Ti** 0.0050 max **U** 0.00035 max **V** 0.0050 max **W** 0.010 max **Zr** bal	

The chemical compositions listed are for identification purposes and should not be used in lieu of the cross referenced specifications.

UNS NUMBERS ASSIGNED TO DATE

With Description of Each Material Covered and References To Documents In Which The Same or Similar Materials are Described

Sxxxxx Number Series
Heat and Corrosion Resistant Steels (Including Stainless), Valve Steels, and Iron-Base "Superalloys"

HEAT AND CORROSION RESISTANT STEELS (INCLUDING STAINLESS), VALVE STEELS, AND IRON-BASE "SUPERALLOYS"

UNIFIED NUMBER	DESCRIPTION	CHEMICAL COMPOSITION	CROSS REFERENCE SPECIFICATIONS
S13800	Precipitation Hardenable Cr-Ni-Al-Mo Stainless Steel (PH 13-8 Mo)	Al 0.90-1.35 C 0.05 max Cr 12.25-13.25 Mn 0.20 max Mo 2.00-2.50 N 0.01 max Ni 7.50-8.50 P 0.01 max S 0.008 max Si 0.10 max	AISI S13800 AMS 5629; 5864 ASME SA705 (XM-13) ASTM A564 (XM-13); A693 (XM-13); A705 (XM-13)
S13889	Precipitation Hardenable Cr-Ni-Al-Mo Stainless Steel Bare Filler Metal (PH 13-8 Mo)	Al 0.90-1.35 C 0.05 max Cr 12.25-13.25 H 0.0025 max Mn 0.10 max Mo 2.00-2.50 N 0.01 max Ni 7.50-8.50 O 0.005 max P 0.008 max S 0.010 max Si 0.10 max	AMS 5840
S14800	Precipitation Hardenable Cr-Ni-Mo-Al Stainless Steel (PH 14-8 Mo)	Al 0.75-1.50 C 0.05 max Cr 13.75-15.00 Mn 1.00 max Mo 2.00-3.00 Ni 7.75-8.75 P 0.015 max S 0.010 max Si 1.00 max	AMS 5601; 5603
S15500	Precipitation Hardenable Cr-Ni-Cu Stainless Steel (15-5 PH)	C 0.07 max Cb 0.15-0.45 Cr 14.00-15.50 Cu 2.50-4.50 Mn 1.00 max Ni 3.50-5.50 P 0.040 max S 0.030 max Si 1.0 max	AISI S15500 AMS 5658; 5659; 5826; 5862 ASME SA705 (XM-12) ASTM A564 (XM-12); A693 (XM-12); A705 (XM-12)
S15700	Precipitation Hardenable Cr-Ni-Mo-Al Stainless Steel (PH 15-7 Mo)	Al 0.75-1.50 C 0.09 max Cr 14.00-16.00 Mn 1.00 max Mo 2.00-3.00 Ni 6.50-7.75 P 0.04 max S 0.03 max Si 1.00 max	AMS 5520; 5627; 5863 ASTM A564 (632); A579 (632); A693 (632); A705 (632) MIL SPEC MIL-S-8955 SAE J467 (PH15-7-Mo)
S15780	Precipitation Hardenable Cr-Ni-Mo-Al Stainless Steel Bare Filler Metal (PH 15-7 Mo)	Al 0.75-1.25 C 0.09 max Cr 14.00-15.25 Mn 1.00 max Mo 2.00-2.75 Ni 6.50-7.75 P 0.025 max S 0.025 max Si 0.50 max	AMS 5813
S15789	Precipitation Hardenable Cr-Ni-Mo-Al Stainless Steel Bare Filler Metal (PH 15-7 Mo)	Al 0.75-1.25 C 0.09 max Cr 14.00-15.25 H 0.0025 max Mn 1.00 max Mo 2.00-2.75 Ni 6.50-7.75 O 0.005 max P 0.010 max S 0.010 max Si 0.50 max	AMS 5812
S16600	Precipitation Hardenable Cr-Ni-Ti-Al Stainless Steel (Croloy 16-6 PH)	Al 0.25-0.40 C 0.025-0.045 Cr 15.00-16.00 Mn 0.70-0.90 Ni 7.00-8.00 P 0.025 max S 0.025 max Si 0.50 max Ti 0.30-0.50	
S16800	Cr-Ni-Mo Stainless Steel (16-8-2-H)	C 0.05-0.10 Cr 14.5-16.5 Mn 2.00 max Mo 1.50-2.00 Ni 7.50-9.50 P 0.040 max S 0.030 max Si 0.75 max	ASTM A376 (16-8-2-H); A430 (16-8-2-H)
S16880	Cr-Ni-Mo Stainless Steel Bare Filler Metal (16-8-2)	C 0.10 max Cr 14.5-16.5 Cu 0.75 max Mn 1.00-2.50 Mo 1.00-2.00 Ni 7.50-9.50 P 0.03 max S 0.03 max Si 0.30-0.65	ASME SFA5.9 (ER16-8-2) AWS A5.9 (ER16-8-2)
S17400	Precipitation Hardenable Cr-Ni-Cu Stainless Steel (17-4 PH)	C 0.07 max Cb 0.15-0.45 Cr 15.50-17.50 Cu 3.00-5.00 Mn 1.00 max Ni 3.00-5.00 P 0.040 max S 0.030 max Si 1.00 max	AISI S17400 AMS 5604; 5622; 5643 ASME SA564 (630); SA705 (630) ASTM A564 (630); A693 (630); A705 (630) MIL SPEC MIL-C-24111; MIL-S-81591 SAE J467 (17-4PH)
S17480	Precipitation Hardenable Cr-Ni-Cu Stainless Steel Bare Filler Metal (17-4 PH)	C 0.05 max Cr 16.00-16.75 Cu 3.25-4.00 Mn 0.25-0.75 Mo 0.75 max Ni 4.50-5.00 P 0.04 max S 0.03 max Si 0.75 max Other Cb+Ta 0.15-0.30	AMS 5825 ASME SFA5.9 (ER630) AWS A5.9 (ER630)
S17600	Precipitation Hardenable Cr-Ni-Al-Ti Stainless Steel (Stainless W)	Al 0.40 max C 0.08 max Cr 15.00-17.50 Mn 1.00 max Ni 6.00-7.50 P 0.040 max S 0.030 max Si 1.00 max Ti 0.40-1.20	ASTM A564 (635); A693 (635); A705 (635)
S17700	Precipitation Hardenable Cr-Ni-Al Stainless Steel (17-7 PH)	Al 0.75-1.50 C 0.09 max Cr 16.00-18.00 Mn 1.00 max Ni 6.50-7.75 P 0.040 max S 0.040 max Si 1.00 max	AISI S17700 AMS 5528; 5529; 5568; 5644; 5673; 5678 ASME SA705 (631) ASTM A313 (631); A564 (631); A579 (62); A693 (631); A705 (631) MIL SPEC MIL-S-25043 SAE J217; J467 (17-7PH)
S17780	Precipitation Hardenable Cr-Ni-Al Stainless Steel Bare Filler Metal (17-7-PH)	Al 0.75-1.25 C 0.09 max Cr 16.00-17.25 Mn 1.00 max Ni 6.50-7.75 P 0.025 max S 0.025 max Si 0.50 max	AMS 5824

The chemical compositions listed are for identification purposes and should not be used in lieu of the cross referenced specifications.

UNIFIED NUMBER	DESCRIPTION	CHEMICAL COMPOSITION	CROSS REFERENCE SPECIFICATIONS
S18200	Free Machining Ferritic Cr-Mo Stainless Steel	C 0.08 max Cr 17.50-19.50 Mn 1.25-2.50 Mo 1.50-2.50 P 0.04 max S 0.15 min Si 1.00 max	ASTM A581(XM-34); A582(XM-34)
S20100	Austenitic Cr-Mn-Ni Stainless Steel	C 0.15 max Cr 16.00-18.00 Mn 5.50-7.50 N 0.25 max Ni 3.50-5.50 P 0.060 max S 0.030 max Si 1.00 max	AISI 201 ASME SA412 (201) ASTM A412 (201); A429 (201); A666 (201) FED QQ-S-766 (201) SAE J405 (30201)
S20200	Austenitic Cr-Mn-Ni Stainless Steel	C 0.15 max Cr 17.00-19.00 Mn 7.50-10.00 N 0.25 max Ni 4.00-6.00 P 0.060 max S 0.030 max Si 1.00 max	AISI 202 ASTM A314 (202); A412 (202); A429 (202); A473 (202); A666 (202) FED QQ-S-763 (202); QQ-S-766 (202) SAE J405 (30202)
S20300	Austenitic Cr-Mn-Ni-Cu Free Machining Stainless Steel (203 EZ)	C 0.08 max Cr 16.00-18.00 Cu 1.75-2.25 Mn 5.00-6.50 Mo 0.50 max Ni 5.00-6.50 P 0.040 max S 0.18-0.35 Si 1.0 max	AMS 5762 ASTM A581 (XM-1); A582 (XM-1) SAE J405 (203 EZ)
S20500	Austenitic Cr-Mn-Ni Stainless Steel	C 0.12-0.25 Cr 16.00-18.00 Mn 14.00-15.50 N 0.32-0.40 Ni 1.00-1.75 P 0.060 max S 0.030 max Si 1.00 max	AISI 205
S20910	Austenitic Cr-Ni-Mn-Mo Stainless Steel (22-13-5)	C 0.06 max Cb 0.10-0.30 Cr 20.50-23.50 Mn 4.00-6.00 Mo 1.50-3.00 N 0.20-0.40 Ni 11.50-13.50 P 0.040 max S 0.030 max Si 1.00 max V 0.10-0.30	AMS 5764 ASME SA182 (XM-19); SA240 (XM-19); SA249 (XM-19); SA312 (XM-19); SA403 (XM-19); SA412 (XM-19); SA479 (XM-19) ASTM A182(XM-19); A240 (XM-19); A249 (XM-19); A269 (XM-19); A312 (XM-19); A403 (XM-19); A412 (XM-19); A429 (XM-19); A479 (XM-19); A580 (XM-19)
S20980	Austenitic Cr-Ni-Mn-Mo Stainless Steel Bare Filler Metal (22-10-5)	C 0.05 max Cr 20.5-24.0 Cu 0.75 max Mn 4.00-7.00 Mo 1.50-3.00 N 0.10-0.30 Ni 9.50-12.00 P 0.03 max S 0.03 max Si 0.90 max	ASME SFA5.9 (ER209) AWS A5.9 (ER209)
S21400	Austenitic Cr-Mn Stainless Steel (Tenelon)	C 0.12 max Cr 17.00-18.50 Mn 14.50-16.00 N 0.35 max Ni 0.75 max P 0.045 max S 0.030 max Si 0.30-1.00	ASTM A240 (XM-31)
S21460	Austenitic Cr-Mn-Ni-Mo Stainless Steel (Cryogenic Tenelon)	C 0.12 max Cr 17.00-19.00 Mn 14.00-16.00 N 0.35-0.50 Ni 5.00-6.00 P 0.060 max S 0.030 max Si 1.00 max	ASTM A412 (XM-14)
S21500	Austenitic Cr-Mn-Ni-Mo Stainless Steel (Esshete 1250)	B 0.003-0.009 C 0.06-0.15 Cb 0.75-1.25 Cr 14.0-16.0 Mn 5.50-7.00 Mo 0.80-1.20 Ni 9.0-11.0 P 0.040 max S 0.030 max Si 0.20-1.20 V 0.15-0.40	ASTM A213
S21600	Austenitic Cr-Mn-Ni-Mo Stainless Steel (216)	C 0.08 max Cr 17.50-22.00 Mn 7.50-9.00 Mo 2.00-3.00 N 0.25-0.50 Ni 5.00-7.00 P 0.045 max S 0.030 max Si 1.00 max	ASME SA240 (XM-17); SA479 (XM-17) ASTM A240 (XM-17); A479 (XM-17); A492 (XM-17)
S21603	Austenitic Cr-Mn-Ni-Mo Stainless Steel, Low Carbon (216L)	C 0.03 max Cr 17.50-22.00 Mn 7.50-9.00 Mo 2.00-3.00 N 0.25-0.50 Ni 5.00-7.00 P 0.045 max S 0.030 max Si 1.00 max	ASME SA240 (XM-18); SA479 (XM-18) ASTM A240 (XM-18); A479 (XM-18); A492 (XM-18)
S21800	Austenitic Cr-Ni-Mn Stainless Steel (Nitronic 60)	C 0.10 max Cr 16.00-18.00 Mn 7.00-9.00 N 0.08-0.18 Ni 8.00-9.00 P 0.040 max S 0.030 max Si 3.50-4.50	ASME Code Case 1817 ASTM A193; A194; A276; A479; A555
S21880	Austenitic Cr-Ni-Mn Stainless Steel Bare Filler Metal (Nitronic 60W)	C 0.10 max Cr 16.0-18.0 Cu 0.75 max Mn 7.00-9.00 Mo 0.75 max N 0.08-0.18 Ni 8.00-9.00 P 0.03 max S 0.03 max Si 3.50-4.50	ASME SFA5.9 (ER218) AWS A5.9 (ER218)
S21900	Austenitic Cr-Mn-Ni Stainless Steel (21-6-9)	C 0.08 max Cr 19.00-21.50 Mn 8.00-10.00 N 0.15-0.40 Ni 5.50-7.50 P 0.060 max S 0.030 max Si 1.00 max	AMS 5561 ASTM A276 (XM-10); A314 (XM-10); A412 (XM-10); A429 (XM-10); A473 (XM-10); A580 (XM-10)
S21904	Austenitic Cr-Mn-Ni Stainless Steel, Low Carbon (21-6-9 LC)	C 0.04 max Cr 19.00-21.50 Mn 8.00-10.00 N 0.15-0.40 Ni 5.50-7.50 P 0.060 max S 0.030 max Si 1.00 max	AMS 5562; 5595; 5656 ASME SA412 (21904) ASTM A276 (XM-11); A314 (XM-11); A412 (XM-11); A429 (XM-11); A473 (XM-11); A580 (XM-11)

The chemical compositions listed are for identification purposes and should not be used in lieu of the cross referenced specifications.

UNIFIED NUMBER	DESCRIPTION	CHEMICAL COMPOSITION	CROSS REFERENCE SPECIFICATIONS
S21980	Austenitic Cr-Mn-Ni Stainless Steel Bare Filler Metal (21-6-9)	**C** 0.05 max **Cr** 19.0-21.5 **Cu** 0.75 max **Mn** 8.00-10.00 **Mo** 0.75 max **N** 0.10-0.30 **Ni** 5.50-7.00 **P** 0.03 max **S** 0.03 max **Si** 1.00 max	**ASME** SFA5.9 (ER219) **AWS** A5.9 (ER219)
S23980	Austenitic Cr-Mn-Ni Stainless Steel Bare Filler Metal (18-4-Mn)	**C** 0.05 max **Cr** 17.0-19.0 **Cu** 0.75 max **Mn** 10.5-13.5 **Mo** 0.75 max **N** 0.10-0.20 **Ni** 4.00-6.00 **P** 0.03 max **S** 0.03 max **Si** 1.00 max	**ASME** SFA5.9 (ER240) **AWS** A5.9 (ER240)
S24000	Austenitic Cr-Mn-Ni Stainless Steel (18-3-Mn)	**C** 0.08 max **Cr** 17.00-19.00 **Mn** 11.50-14.50 **N** 0.20-0.40 **Ni** 2.50-3.75 **P** 0.060 max **S** 0.030 max **Si** 1.00 max	**ASME** SA240 (XM-29); SA249 (XM-29); SA312 (XM-29); SA688 (XM-29) **ASTM** A240 (XM-29); A249 (XM-29); A269 (XM-29); A312 (XM-29); A412 (XM-29); A688 (XM-29)
S24100	Austenitic Cr-Mn-Ni Stainless Steel (18-2-Mn)	**C** 0.15 max **Cr** 16.50-19.50 **Mn** 11.00-14.00 **N** 0.20-0.45 **Ni** 0.50-2.50 **P** 0.060 max **S** 0.030 max **Si** 1.00 max	**ASTM** A580 (XM-28)
S28200	Austenitic Cr-Mn-Cu-Mo Stainless Steel (18-18-Plus)	**C** 0.15 max **Cr** 17.00-19.00 **Cu** 0.50-1.50 **Mn** 17.00-19.00 **Mo** 0.50-1.50 **N** 0.40-0.60 **P** 0.045 max **S** 0.030 max **Si** 1.00 max	**ASTM** A276; A314; A473; A493; A580
S30100	Austenitic Cr-Ni Stainless Steel	**C** 0.15 max **Cr** 16.00-18.00 **Mn** 2.00 max **Ni** 6.00-8.00 **P** 0.045 max **S** 0.030 max **Si** 1.00 max	**AISI** 301 **AMS** 5517; 5518; 5519 **ASTM** A167 (301); A177 (301); A554 (301); A666 (301) **FED** QQ-S-766 (301) **MIL SPEC** MIL-S-5059 (301) **SAE** J405 (30301)
S30115	Austenitic Cr-Ni Stainless Steel	**C** 0.07-0.11 **Cr** 16.50-17.50 **Mn** 1.00-1.50 **Mo** 0.60-0.80 **Ni** 7.70-8.30 **P** 0.03 max **S** 0.03 max **Si** 0.90-1.40	**MIL SPEC** MIL-S-46889
S30200	Austenitic Cr-Ni Stainless Steel	**C** 0.15 max **Cr** 17.00-19.00 **Mn** 2.00 max **Ni** 8.00-10.00 **P** 0.045 max **S** 0.030 max **Si** 1.00 max	**AISI** 302 **AMS** 5515; 5516; 5600; 5636; 5637; 5688; 5693; 7210; 7241 **ASME** SA240 (302); SA479 (302) **ASTM** A167 (302); A240 (302); A276 (302); A313 (302); A314 (302); A368 (302); A473 (302); A478 (302); A479 (302); A492 (302); A493 (302); A511 (302); A554 (302); A666 (302) **FED** QQ-S-763 (302); QQ-S-766 (302); QQ-W-423 (302) **MIL SPEC** MIL-S-862 (302); MIL-S-5059; MIL-S-7720; MIL-S-46044 **SAE** J230; J405 (30302)
S30210	Austenitic Cr-Ni Stainless Steel, Low Permeability	**C** 0.26-0.33 **Cr** 17.0-19.0 **Mn** 3.00-4.00 **Ni** 8.0-10.0 **P** 0.18-0.33 **S** 0.035 max **Si** 1.00 max	**MIL SPEC** MIL-S-17759 (Ships)
S30215	Austenitic Cr-Ni Stainless Steel	**C** 0.15 max **Cr** 17.00-19.00 **Mn** 2.00 max **Ni** 8.00-10.00 **P** 0.045 max **S** 0.030 max **Si** 2.00-3.00	**AISI** 302B **ASTM** A167 (302 B); A276 (302 B); A314 (302 B); A473 (302 B); A580 (302 B) **SAE** J405 (30302 B)
S30260	Austenitic Cr-Ni Stainless Steel, Low Permeability	**C** 0.15 max **Cr** 16.0-18.0 **Mn** 1.00 max **Ni** 9.5-12.0 **P** 0.20-0.40 **S** 0.040 max **Si** 1.00 max	**MIL SPEC** MIL-S-17759 (Ships)
S30300	Free Machining Austenitic Cr-Ni Stainless Steel	**C** 0.15 max **Cr** 17.00-19.00 **Mn** 2.00 max **Mo** 0.60 max (Optional) **Ni** 8.00-10.00 **P** 0.20 max **S** 0.15 min **Si** 1.00 max	**AISI** 303 **AMS** 5640 (Type 1) **ASME** SA194 (303); SA320 (303) **ASTM** A194 (303); A314 (303); A320 (303); A473 (303); A581 (303); A582 (303) **MIL SPEC** MIL-S-862 (303) **SAE** J405 (30303)
S30310	Austenitic Cr-Ni-Mn Free Machining Stainless Steel	**C** 0.15 max **Cr** 17.00-19.00 **Mn** 2.50-4.50 **Mo** 0.75 max **Ni** 7.00-10.00 **P** 0.20 max **S** 0.25 min **Si** 1.00 max	**AMS** 5640 (Type 2) **ASTM** A581 (XM-5); A582 (XM-5) **SAE** J405 (303 plus X); J412 (303 plus X)
S30323	Free Machining Austenitic Cr-Ni Stainless Steel (Se Bearing)	**C** 0.15 max **Cr** 17.00-19.00 **Mn** 2.00 max **Ni** 8.00-10.00 **P** 0.20 max **S** 0.060 max **Se** 0.15 min **Si** 1.00 max	**AISI** 303 Se **AMS** 5640 (Type 2); 5641; 5738 **ASME** SA194 (303 Se); SA320 (303 Se) **ASTM** A194 (303 Se); A314 (303 Se); A320 (303 Se); A473 (303 Se); A581 (303 Se); A582 (303 Se) **MIL SPEC** MIL-S-862 (303 Se) **SAE** J405 (30303 Se)
S30330	Free Machining Copper Bearing Cr-Ni Stainless Steel	**C** 0.15 max **Cr** 17.00-19.00 **Cu** 2.50-4.00 **Mn** 2.00 max **Ni** 6.00-10.00 **P** 0.15 max **S** 0.10 max **Se** 0.10 max **Si** 1.00 max	**SAE** J405 (303 Cu)
S30345	Austenitic Cr-Ni-Al Free Machining Stainless Steel (303 MA)	**Al** 0.60-1.00 **C** 0.15 max **Cr** 17.00-19.00 **Mn** 2.00 max **Mo** 0.40-0.60 **Ni** 8.00-10.00 **P** 0.05 max **S** 0.11-0.16 **Si** 1.00 max	**AMS** 5638 **ASTM** A581 (XM-2); A582 (XM-2)

The chemical compositions listed are for identification purposes and should not be used in lieu of the cross referenced specifications.

UNIFIED NUMBER	DESCRIPTION	CHEMICAL COMPOSITION	CROSS REFERENCE SPECIFICATIONS
S30360	Austenitic Cr-Ni-Pb Free Machining Stainless Steel (303 Pb)	**C** 0.15 max **Cr** 17.00-19.00 **Mn** 2.00 max **Mo** 0.75 max **Ni** 8.00-10.00 **P** 0.040 max **Pb** 0.12-0.30 **S** 0.12-0.30 **Si** 1.00 max	**AMS** 5635 **ASTM** A581 (XM-3); A582 (XM-3)
S30400	Austenitic Cr-Ni Stainless Steel	**C** 0.08 max **Cr** 18.00-20.00 **Mn** 2.00 max **Ni** 8.00-10.50 **P** 0.045 max **S** 0.030 max **Si** 1.00 max	**AISI** 304 **AMS** 5501; 5513; 5560; 5563; 5564; 5565; 5566; 5567; 5639; 5697; 7228; 7245 **ASME** SA182 (304); SA194 (8); SA213 (304); SA240 (304); SA249 (304); SA312 (304); SA320 (B8); SA358 (304); SA376 (304); SA403 (304); SA409 (304); SA430 (304); SA479 (304); SA688 (304) **ASTM** A167 (304); A182 (304); A193 (304); A194 (304); A213 (304); A240 (304); A249 (304); A269 (304); A270 (304); A271 (304); A276 (304); A312 (304); A313 (304); A314 (304); A320 (304); A358 (304); A368 (304); A376 (304); A409 (304); A430 (304); A473 (304); A478 (304); A479 (304); A492 (304); A493 (304); A511 (304); A554 (304); A580 (304); A632 (304); A651 (304); A666 (304); A688 (304) **FED** QQ-S-763 (304); QQ-S-766 (304); QQ-W-423 (304) **MIL SPEC** MIL-S-862 (304); MIL-S-5059 (304); MIL-T-5695; MIL-T-6845 (304); MIL-T-8504 (304); MIL-T-8506 (304); MIL-F-20138; MIL-S-23195 (304); MIL-S-23196 (304); MIL-S-27419 **SAE** J405 (30304)
S30403	Austenitic Cr-Ni Stainless Steel (Low Carbon)	**C** 0.03 max **Cr** 18.00-20.00 **Mn** 2.00 max **Ni** 8.00-12.00 **P** 0.045 max **S** 0.030 max **Si** 1.00 max	**AISI** 304 L **AMS** 5511; 5647 **ASME** SA182 (304 L); SA213 (304 L); SA240 (304 L); SA249 (304 L); SA312 (304 L); SA403 (304 L); SA479 (304 L); SA688 (304 L) **ASTM** A167 (304 L); A182 (304 L); A213 (304 L); A240 (304 L); A249 (304 L); A269 (304 L); A276 (304 L); A312 (304 L); A314 (304 L); A403 (304 L); A473 (304 L); A478 (304 L); A479 (304 L); A511 (304 L); A554 (304 L); A580 (304 L); A632 (304 L); A688 (304 L) **FED** QQ-S-763 (304 L); QQ-S-766 (304 L) **MIL SPEC** MIL-S-862 (304 L); MIL-S-4043; MIL-S-23195 (304 L); MIL-S-23196 (304 L) **SAE** J405 (30304 L)
S30409	Austenitic Cr-Ni Stainless Steel, High Carbon (304 H)	**C** 0.04-0.10 **Cr** 18.00-20.00 **Mn** 2.00 max **Ni** 8.00-11.00 **P** 0.040 max **S** 0.030 max **Si** 1.00 max	**ASME** SA182 (304 H); SA213 (304 H); SA240 (304 H); SA249 (304 H); SA312 (304 H); SA376 (304 H); SA403 (304 H); SA430 (304 H); SA479 (304 H) **ASTM** A182 (304 H); A213 (304 H); A240 (304 H); A249 (304 H); A271 (304 H); A312 (304 H); A358 (304 L); A376 (304 H); A403 (304 L); A430 (304 H); A479 (304 H)
S30430	Austenitic Cr-Ni-Cu Low Work Hardening Stainless Steel (18-9-LW)	**C** 0.10 max **Cr** 17.00-19.00 **Cu** 3.00-4.00 **Mn** 2.00 max **Ni** 8.00-10.00 **P** 0.045 max **S** 0.030 max **Si** 1.00 max	**AISI** S30430 **ASTM** A493 (XM-7)
S30451	Austenitic Nitrogen Bearing Cr-Ni Stainless Steel (304 N)	**C** 0.08 max **Cr** 18.00-20.00 **Mn** 2.00 max **N** 0.10-0.16 **Ni** 8.00-10.50 **P** 0.045 max **S** 0.030 max **Si** 1.00 max	**AISI** 304 N **ASME** SA182 (304 N); SA213 (304 N); SA240 (304 N); SA249 (304 N); SA312 (304 N); SA358 (304 N); SA376 (304 N); SA430 (304 N); SA479 (304 N) **ASTM** A182 (304 N); A213 (304 N); A240 (304 N); A249 (304 N); A312 (304 N); A358 (304 N); A376 (304 N); A403 (304 N); A430 (304 N); A479 (304 N)
S30452	Austenitic Cr-Ni Stainless Steel, High Nitrogen	**C** 0.08 max **Cr** 18.00-20.00 **Mn** 2.00 max **N** 0.16-0.30 **Ni** 8.00-10.50 **P** 0.045 max **S** 0.030 max **Si** 1.00 max	**ASME** SA240 (XM-21) **ASTM** A240 (XM-21); A276 (XM-21)
S30453	Austenitic Nitrogen Bearing Cr-Ni Stainless Steel (304 LN)	**C** 0.030 max **Cr** 18.00-20.00 **Mn** 2.00 max **N** 0.10-0.16 **Ni** 8.00-12.00 **P** 0.045 max **S** 0.030 max **Si** 1.00 max	**ASTM** A276
S30454	Austenitic Nitrogen Bearing Cr-Ni Stainless Steel (304L (Hi)N)	**C** 0.03 max **Cr** 18.00-20.00 **Mn** 2.00 max **N** 0.16-0.30 **Ni** 8.00-12.00 **P** 0.045 max **S** 0.030 max **Si** 1.00 max	**ASTM** A276
S30483	Austenitic Cr-Ni Stainless Steel Bare Filler Metal. Low Carbon	**C** 0.03 max **Cr** 18.0 min **Mn** 2.00 max **Ni** 8.00 min **P** 0.030 max **S** 0.030 max **Si** 1.00 max **Other** others (total) 1.00 max	
S30500	Austenitic Cr-Ni Stainless Steel (Low Work Hardening)	**C** 0.12 max **Cr** 17.00-19.00 **Mn** 2.00 max **Ni** 10.00-13.00 **P** 0.045 max **S** 0.030 max **Si** 1.00 max	**AISI** 305 **AMS** 5514; 5685; 5686 **ASME** SA193 (305); SA194 (305); SA240 (305) **ASTM** A167 (305); A240 (305); A249 (305); A276 (305); A313 (305); A314 (305); A368 (305); A473 (305); A478 (305); A492 (305); A493 (305); A511 (305); A554 (305); A580 (305) **FED** QQ-S-763 (305); QQ-S-766 (305); QQ-W-423 (305) **SAE** J405 (30305)

The chemical compositions listed are for identification purposes and should not be used in lieu of the cross referenced specifications.

UNIFIED NUMBER	DESCRIPTION	CHEMICAL COMPOSITION	CROSS REFERENCE SPECIFICATIONS
S30560	Austenitic Cr-Ni Stainless Steel Low Permeability	C 0.10 max Cr 17.0 min Cu 0.50 max Mn 0.2-2.5 Ni 8.5-14.0 P 0.035 max S 0.035 max Si 0.2-1.5	MIL SPEC MIL-W-30508
S30600	Austenitic Cr-Ni Stainless Steel (18-15)	C 0.18 max Cr 17.0-18.5 Cu 0.50 max Mn 2.00 max Mo 0.20 max Ni 14.0-15.5 P 0.02 max S 0.02 max Si 3.70-4.30	ASTM A240; A269; A312; A358; A479
S30780	Austenitic Cr-Ni-Mn Stainless Steel Bare Filler Metal	C 0.04-0.14 Cr 19.5-22.0 Cu 0.75 max Mn 3.30-4.75 Mo 0.50-1.50 Ni 8.00-10.70 P 0.03 max S 0.03 max Si 0.30-0.65	ASME SFA5.9 (ER307) AWS A5.9 (ER307)
S30800	Austenitic Cr-Ni Heat Resisting Steel	C 0.08 max Cr 19.00-21.00 Mn 2.00 max Ni 10.00-12.00 P 0.045 max S 0.030 max Si 1.00 max	AISI 308 ASTM A167 (308); A276 (308); A314 (308); A473 (308); A580 (308) SAE J405 (30308)
S30815	Austenitic Cr-Ni Heat Resisting Steel (253 MA)	C 0.10 max Cr 20.00-22.00 Mn 0.80 max N 0.14-0.20 Ni 10.00-12.00 P 0.040 max S 0.030 max Si 1.40-2.00 Other cerium 0.03-0.08	ASTM A276; A473; A479
S30880	Austenitic Cr-Ni Heat Resisting Steel Bare Filler Metal	C 0.08 max Cr 19.50-22.00 Cu 0.75 max Mn 1.00-2.50 Mo 0.75 max Ni 9.00-11.00 P 0.03 max S 0.03 max Si 0.30-0.65	ASME SFA5.9 (ER308) (ER308H); SFA5.30 (IN308) AWS A5.9 (ER308) (ER308H); A5.30 (IN308) MIL SPEC MIL-R-5031 (Class 1); MIL-E-19933 (MIL-308) (MIL-308Co); MIL-I-23413 (MIL-308) (MIL-308Co)
S30881	Austenitic Cr-Ni Heat Resisting Steel Bare Filler Metal	C 0.08 max Cr 19.50-22.00 Cu 0.5 max Mn 1.00-2.50 Mo 0.50 max Ni 9.00-11.00 P 0.03 max S 0.03 max Si 0.65-1.00	ASME SFA5.9 (ER308Si) AWS A5.9 (ER308Si)
S30882	Austenitic Cr-Ni-Mo Heat Resisting Steel Bare Filler Metal	C 0.08 max Cr 18.00-21.00 Cu 0.75 max Mn 1.00-2.50 Mo 2.00-3.00 Ni 9.00-12.00 P 0.03 max S 0.03 max Si 0.30-0.65	ASME SFA5.9 (ER308Mo) AWS A5.9 (ER308Mo)
S30883	Austenitic Cr-Ni Heat Resisting Steel Bare Filler Metal (Low Carbon)	C 0.03 max Cr 19.50-22.00 Cu 0.75 max Mn 1.00-2.50 Mo 0.75 max Ni 9.00-11.00 P 0.03 max S 0.03 max Si 0.30-0.65	ASME SFA5.9 (ER308L); SFA5.30 (IN308L) AWS A5.9 (ER308L); A5.30 (IN308L) MIL SPEC MIL-R-5031 (Class 16); MIL-E-19933 (MIL-308L) (MIL-308CoL); MIL-I-23413 (MIL-308L) (MIL-308CoL)
S30884	Austenitic Cr-Ni Heat Resisting Steel Bare Filler Metal (19-9 High Carbon)	C 0.08-0.15 Cr 18.0-20.0 Mn 1.00-2.50 Ni 9.00-11.00 P 0.03 max S 0.03 max Si 0.25-0.60	MIL SPEC MIL-E-19933D, (MIL-308HC)
S30886	Austenitic Cr-Ni-Mo Heat Resisting Steel Bare Filler Metal (19-9-Mo Low Carbon)	C 0.04 max Cr 18.00-21.00 Cu 0.75 max Mn 1.00-2.50 Mo 2.00-3.00 Ni 9.00-12.00 P 0.03 max S 0.03 max Si 0.30-0.65	ASME SFA5.9 (ER308MoL) AWS A5.9 (ER308MoL)
S30888	Austenitic Cr-Ni Heat Resisting Steel Bare Filler Metal (308LSi)	C 0.03 max Cr 19.5-22.0 Cu 0.5 max Mn 1.00-2.50 Mo 0.5 max Ni 9.00-11.00 P 0.03 max S 0.03 max Si 0.65-1.00	ASME SFA5.9 (ER308LSi) AWS A5.9 (ER308LSi)
S30900	Austenitic Cr-Ni Heat Resisting Steel	C 0.20 max Cr 22.00-24.00 Mn 2.00 max Ni 12.00-15.00 P 0.045 max S 0.030 max Si 1.00 max	AISI 309 ASME SA249 (309); SA312 (309); SA358 (309); SA403 (309); SA409 (309) ASTM A167 (309); A249 (309); A276 (309); A312 (309); A314 (309); A358 (309); A403 (309); A409 (309); A473 (309); A511 (309); A554 (309); A580 (309) FED QQ-S-763 (309); QQ-S-766 (309) MIL SPEC MIL-S-862 (309) SAE J405 (30309)
S30908	Austenitic Cr-Ni Heat Resisting Stainless Steel (309 S)	C 0.08 max Cr 22.00-24.00 Mn 2.00 max Ni 12.00-15.00 P 0.045 max S 0.030 max Si 1.00 max	AISI 309 S AMS 5523; 5574; 5650 ASME SA240 (309 S) ASTM A167 (309 S); A240 (309 S); A276 (309 S); A314 (309 S); A473 (309 S); A511 (309 S); A554 (309 S); A580 (309 S) SAE J405 (30309 S)
S30940	Austenitic Cr-Ni-Cb Heat Resisting Stainless Steel (309 S Cb)	C 0.08 max Cb 10xC-1.00 Cr 22.00-24.00 Mn 2.00 max Ni 12.00-15.00 P 0.040 max S 0.030 max Si 1.00 max	ASTM A313 (309-Cb); A554 (309-S-Cb)

The chemical compositions listed are for identification purposes and should not be used in lieu of the cross referenced specifications.

UNIFIED NUMBER	DESCRIPTION	CHEMICAL COMPOSITION	CROSS REFERENCE SPECIFICATIONS
S30980	Austenitic Cr-Ni Heat Resisting Steel Bare Filler Metal	**C** 0.12 max **Cr** 23.0-25.0 **Cu** 0.75 max **Mn** 1.00-2.50 **Mo** 0.75 max **Ni** 12.0-14.0 **P** 0.03 max **S** 0.03 max **Si** 0.30-0.65	**ASME** SFA5.9 (ER309) **AWS** A5.9 (ER309) **MIL SPEC** MIL-R-5031 (Class 2); MIL-E-19933 (MIL-309) (MIL-309Co)
S30981	Austenitic Cr-Ni Heat Resisting Steel Bare Filler Metal (309Si)	**C** 0.12 max **Cr** 23.0-25.0 **Cu** 0.5 max **Mn** 1.00-2.50 **Mo** 0.5 max **Ni** 12.0-14.0 **P** 0.03 max **S** 0.03 max **Si** 0.65-1.00	**ASME** SFA5.9 (ER309Si) **AWS** A5.9 (ER309Si)
S30983	Austenitic Cr-Ni Heat Resisting Steel Bare Filler Metal (309L)	**C** 0.03 max **Cr** 23.0-25.0 **Cu** 0.75 max **Mn** 1.00-2.50 **Mo** 0.75 max **Ni** 12.0-14.0 **P** 0.03 max **S** 0.03 max **Si** 0.30-0.65	**ASME** SFA5.9 (ER309L) **AWS** A5.9 (ER309L)
S31000	Austenitic Cr-Ni Heat Resisting Steel	**C** 0.25 max **Cr** 24.00-26.00 **Mn** 2.00 max **Ni** 19.00-22.00 **P** 0.045 max **S** 0.030 max **Si** 1.50 max	**AISI** 310 **ASME** SA182 (310); SA213 (310); SA249 (310); SA312 (310); SA358 (310); SA403 (310); SA409 (310) **ASTM** A167 (310); A182 (310); A213 (310); A249 (310); A276 (310); A312 (310); A314 (310); A358 (310); A403 (310); A409 (310); A473 (310); A511 (310); A554 (310); A632 (310) **FED** QQ-S-763 (310); QQ-S-766 (310); QQ-W-423 (310) **MIL SPEC** MIL-S-862 (310) **SAE** J405 (30310)
S31008	Austenitic Cr-Ni Heat Resisting Steel (310 S)	**C** 0.08 max **Cr** 24.00-26.00 **Mn** 2.00 max **Ni** 19.00-22.00 **P** 0.045 max **S** 0.030 max **Si** 1.50 max	**AISI** 310 S **AMS** 5521; 5572; 5577; 5651 **ASME** SA240 (310 S); SA479 (310 S) **ASTM** A167 (310 S); A240 (310 S); A276 (310 S); A314 (310 S); A473 (310 S); A479 (310 S); A511 (310 S); A554 (310 S); A580 (310 S) **FED** QQ-S-763 **SAE** J405 (30310 S)
S31040	Austenitic Cr-Ni Heat Resisting Steel (310Cb)	**C** 0.08 max **Cr** 24.0-26.00 **Mn** 2.00 max **N** 0.10 max **Ni** 19.00-22.00 **P** 0.045 max **S** 0.030 max **Si** 1.50 max **Other** Cb+Ta 10xC min-1.10 max	**ASTM** A313 (310Cb)
S31080	Austenitic Cr-Ni Heat Resisting Steel Bare Filler Metal	**C** 0.08-0.15 **Cr** 25.0-28.0 **Cu** 0.75 max **Mn** 1.00-2.50 **Mo** 0.75 max **Ni** 20.0-22.5 **P** 0.03 max **S** 0.03 max **Si** 0.30-0.65	**AMS** 5694 **ASME** SFA5.9 (ER310); SFA5.30 (IN310) **AWS** A5.9 (ER310); A5.30 (IN310) **MIL SPEC** MIL-R-5031 (Class 3); MIL-E-19933 (MIL-310); MIL-I-23413 (MIL-310)
S31100	Stainless Steel (744X)	**C** 0.05 max **Cr** 25.00-27.00 **Mn** 1.00 max **Ni** 6.00-7.00 **P** 0.030 max **S** 0.030 max **Si** 0.60 max **Ti** 0.25 max	**ASTM** A276 (XM-26)
S31200	Ferritic-Austenitic Cr-Ni-Mo-N Stainless Steel (44LN)	**C** 0.030 max **Cr** 24.0-26.0 **Mn** 2.00 max **Mo** 1.20-2.00 **N** 0.14-0.20 **Ni** 5.50-6.50 **P** 0.045 max **S** 0.030 max **Si** 1.00 max	**ASTM** A182 (F50)
S31254	Austenitic Cr-Ni-Mo-Cu-N Stainless Steel (254 SMO)	**C** 0.020 max **Cr** 19.50-20.50 **Cu** 0.50-1.00 **Mn** 1.00 max **Mo** 6.00-6.50 **N** 0.180-0.220 **Ni** 17.50-18.50 **P** 0.030 max **S** 0.010 max **Si** 0.80 max	
S31260	Ferritic-Austenitic Cr-Ni-Mo-W Stainless Steel	**C** 0.030 max **Cr** 24.0-26.0 **Cu** 0.20-0.80 **Mn** 1.00 max **Mo** 2.50-3.50 **N** 0.10-0.30 **Ni** 5.50-7.50 **P** 0.030 max **S** 0.030 max **Si** 0.75 max **W** 0.10-0.50	**ASTM** A240; A789; A790
S31380	Austenitic Cr-Ni Stainless Steel Bare Filler Metal	**C** 0.15 max **Cr** 28.0-32.0 **Cu** 0.75 max **Mn** 1.00-2.50 **Mo** 0.75 max **Ni** 8.00-10.5 **P** 0.03 max **S** 0.03 max **Si** 0.30-0.65	**ASME** SFA5.9 (ER312); SFA5.30 (IN312) **AWS** A5.9 (ER312); A5.30 (IN312) **MIL SPEC** MIL-E-19933 (MIL-312) (MIL-312Co)
S31400	Austenitic Cr-Ni Heat Resisting Steel	**C** 0.25 max **Cr** 23.00-26.00 **Mn** 2.00 max **Ni** 19.00-22.00 **P** 0.045 max **S** 0.030 max **Si** 1.50-3.00	**AISI** 314 **AMS** 5522; 5652 **ASTM** A276 (314); A314 (314); A473 (314); A580 (314) **SAE** J405 (30314)
S31500	Ferritic- Austenitic Stainless Steel	**C** 0.030 max **Cr** 18.00-19.00 **Mn** 1.20-2.00 **Mo** 2.50-3.00 **Ni** 4.25-5.25 **P** 0.030 max **S** 0.030 max **Si** 1.40-2.00	**ASTM** A669

The chemical compositions listed are for identification purposes and should not be used in lieu of the cross referenced specifications.

UNIFIED NUMBER	DESCRIPTION	CHEMICAL COMPOSITION	CROSS REFERENCE SPECIFICATIONS
S31600	Austenitic Cr-Ni-Mo Stainless Steel	**C** 0.08 max **Cr** 16.00-18.00 **Mn** 2.00 max **Mo** 2.00-3.00 **Ni** 10.00-14.00 **P** 0.045 max **S** 0.030 max **Si** 1.00 max	**AISI** 316 **AMS** 5524; 5573; 5648; 5690 **ASME** SA182 (316); SA193 (316); SA194 (316); SA213 (316); SA240 (316); SA249 (316); SA312 (316); SA320 (316); SA358 (316); SA376 (316); SA403 (316); SA409 (316); SA430 (316); SA479 (316); SA688 (316) **ASTM** A167 (316); A182 (316); A193 (316); A194 (316); A213 (316); A240 (316); A249 (316); A269 (316); A276 (316); A312 (316); A313 (316); A314 (316); A320 (316); A358 (316); A368 (316); A376 (316); A403 (316); A409 (316); A430 (316); A473 (316); A478 (316); A479 (316); A511 (316); A554 (316); A651 (316) **FED** QQ-S-763 (316); QQ-S-766 (316); QQ-W-423 (316) **MIL SPEC** MIL-S-862 (316); MIL-S-5059 (316); MIL-S-7720 (316) **SAE** J405 (30316)
S31603	Austenitic Cr-Ni-Mo Stainless Steel (Low Carbon)	**C** 0.030 max **Cr** 16.00-18.00 **Mn** 2.00 max **Mo** 2.00-3.00 **Ni** 10.00-14.00 **P** 0.045 max **S** 0.030 max **Si** 1.00 max	**AISI** 316 L **AMS** 5507; 5653 **ASME** SA182 (316 L); SA213 (316 L); SA240 (316 L); SA249 (316 L); SA312 (316 L); SA403 (316 L); SA479 (316 L); SA688 (316 L) **ASTM** A167 (316 L); A182 (316 L); A213 (316 L); A240 (316 L); A249 (316 L); A269 (316 L); A276 (316 L); A312 (316 L); A314 (316 L); A403 (316 L); A473 (316 L); A478 (316 L); A479 (316 L); A511 (316 L); A554 (316 L); A580 (316 L); A632 (316 L); A688 (316 L) **FED** QQ-S-763 (316 L); QQ-S-766 (316 L) **MIL SPEC** MIL-S-862 (316 L) **SAE** J405 (30316 L)
S31609	Austenitic Cr-Ni-Mo Stainless Steel (High Carbon)	**C** 0.04-0.10 **Cr** 16.00-18.00 **Mn** 2.00 max **Mo** 2.00-3.00 **Ni** 10.00-14.00 **P** 0.040 max **S** 0.030 max **Si** 1.00 max	**ASME** SA182 (316 H); SA213 (316 H); SA240 (316 H); SA249 (316 H); SA 312 (316 H); SA376 (316 H); SA403 (316 H); SA430 (316 H); SA479 (316 H) **ASTM** A182 (316 H); A213 (316 H); A240 (316 H); A249 (316 H); A312 (316 H); A358 (316 H); A376 (316 H); A403 (316 H); A430 (316 H); A479 (316 H)
S31620	Austenitic Cr-Ni-Mo Free Machining Stainless Steel	**C** 0.08 max **Cr** 17.00-19.00 **Mn** 2.00 max **Mo** 1.75-2.50 **Ni** 12.00-14.00 **P** 0.20 max **S** 0.10 min **Si** 1.00 max	**AISI** 316 F **AMS** 5649
S31635	Austenitic Cr-Ni-Mo Stainless Steel (Titanium Stabilized) (316 Ti)	**C** 0.08 max **Cr** 16.00-18.00 **Mn** 2.00 max **Mo** 2.00-3.00 **N** 0.10 max **Ni** 10.00-14.00 **P** 0.045 max **S** 0.030 max **Si** 1.00 max **Ti** 5x(C + N) min-0.70 max	**ASTM** A313 (316 Ti)
S31640	Austenitic Cr-Ni-Mo Stainless Steel (Columbium Stabilized) (316Cb)	**C** 0.08 max **Cr** 16.00-18.00 **Mn** 2.00 max **Mo** 2.00-3.00 **N** 0.10 max **Ni** 10.00-14.00 **P** 0.045 max **S** 0.030 max **Si** 1.50 max **Other** Cb+Ta 10xC min-1.10 max	**ASTM** A313 (316Cb)
S31651	Austenitic Nitrogen Bearing Cr-Ni-Mo Stainless Steel (316 N)	**C** 0.08 max **Cr** 16.00-18.00 **Mn** 2.00 max **Mo** 2.00-3.00 **N** 0.10-0.16 **Ni** 10.00-14.00 **P** 0.045 max **S** 0.030 max **Si** 1.00 max	**AISI** 316 N **ASME** SA182 (316 N); SA213 (316 N); SA240 (316 N); SA249 (316 N); SA312 (316 N); SA358 (316 N); SA376 (316 N); SA403 (316 N); SA430 (316 N); SA479 (316 N) **ASTM** A182 (316 N); A213 (316 N); A240 (316 N); A249 (316 N); A276 (316 N); A312 (316 N); A358 (316 N); A376 (316 N); A403 (316 N); A430 (316 N); A479 (316 N)
S31653	Austenitic Nitrogen Bearing Cr-Ni-Mo Stainless Steel (316LN)	**C** 0.030 max **Cr** 16.00-18.00 **Mn** 2.00 max **Mo** 2.00-3.00 **N** 0.10-0.16 **Ni** 10.00-14.00 **P** 0.045 max **S** 0.030 max **Si** 1.00 max	**ASTM** A276
S31654	Austenitic Nitrogen Bearing Cr-Ni-Mo Stainless Steel (316 L(Hi)N)	**C** 0.03 max **Cr** 16.00-18.00 **Mn** 2.00 max **Mo** 2.00-3.00 **N** 0.16-0.30 **Ni** 10.00-14.00 **P** 0.045 max **S** 0.030 max **Si** 1.00 max	**ASTM** A276
S31680	Austenitic Cr-Ni-Mo Stainless Steel Bare Filler Metal	**C** 0.08 max **Cr** 18.0-20.0 **Cu** 0.75 max **Mn** 1.00-2.50 **Mo** 2.00-3.00 **Ni** 11.0-14.0 **P** 0.03 max **S** 0.03 max **Si** 0.30-0.65	**AMS** 5692 **ASME** SFA5.9 (ER316) (316H); SFA5.30 (IN316) **AWS** A5.9 (ER316) (316H); A5.30 (IN316) **MIL SPEC** MIL-R-5031 (Class 4); MIL-E-19933 (MIL-316); MIL-I-23413 (MIL-316)
S31681	Austenitic Cr-Ni-Mo Stainless Steel Bare Filler Metal (316 Si)	**C** 0.08 max **Cr** 18.0-20.0 **Cu** 0.5 max **Mn** 1.00-2.50 **Mo** 2.00-3.00 **Ni** 11.0-14.0 **P** 0.03 max **S** 0.03 max **Si** 0.65-1.00	**ASME** SFA5.9 (ER316Si) **AWS** A5.9 (ER316Si)

The chemical compositions listed are for identification purposes and should not be used in lieu of the cross referenced specifications.

UNIFIED NUMBER	DESCRIPTION	CHEMICAL COMPOSITION	CROSS REFERENCE SPECIFICATIONS
S31683	Austenitic Cr-Ni-Mo Stainless Steel Bare Filler Metal (316L)	**C** 0.03 max **Cr** 18.0-22.0 **Cu** 0.75 max **Mn** 1.00-2.50 **Mo** 2.00-3.00 **Ni** 11.0-14.0 **P** 0.03 max **S** 0.03 max **Si** 0.30-0.65	**ASME** SFA5.9 (ER316L); SFA5.30 (IN316L) **AWS** A5.9 (ER316L); A5.30 (ER316L) **MIL SPEC** MIL-R-5031 (Class 17); MIL-E-19933 (MIL-316L); MIL-I-23413 (MIL-316L)
S31688	Austenitic Cr-Ni-Mo Stainless Steel Bare Filler Metal (316LSi)	**C** 0.03 max **Cr** 18.0-20.0 **Cu** 0.5 max **Mn** 1.00-2.50 **Mo** 2.00-3.00 **Ni** 11.0-14.0 **P** 0.03 max **S** 0.03 max **Si** 0.65-1.00	**ASME** SFA5.9 (ER316LSi) **AWS** A5.9 (ER316LSi)
S31700	Austenitic Cr-Ni-Mo Stainless Steel	**C** 0.08 max **Cr** 18.00-20.00 **Mn** 2.00 max **Mo** 3.00-4.00 **Ni** 11.00-15.00 **P** 0.045 max **S** 0.030 max **Si** 1.00 max	**AISI** 317 **ASME** SA240 (317); SA249 (317); SA312 (317); SA403 (317); SA409 (317) **ASTM** A167 (317); A240 (317); A249 (317); A269 (317); A276 (317); A312 (317); A314 (317); A403 (317); A409 (317); A473 (317); A478 (317); A511 (317); A554 (317); A580 (317); A632 (317) **FED** QQ-S-763 (317) **MIL SPEC** MIL-S-862 (317) **SAE** J405 (30317)
S31703	Austenitic Cr-Ni-Mo Stainless Steel (Low Carbon)	**C** 0.030 max **Cr** 18.00-20.00 **Mn** 2.00 max **Mo** 3.00-4.00 **Ni** 11.00-15.00 **P** 0.045 max **S** 0.030 max **Si** 1.00 max	**AISI** 317 L **ASME** SA240 (317 L) **ASTM** A167 (317 L); A240 (317 L)
S31725	Austenitic Cr-Ni-Mo Stainless Steel (317LM)	**C** 0.03 max **Cr** 18.0-20.0 **Cu** 0.75 max **Mn** 2.00 max **Mo** 4.00-5.00 **N** 0.10 max **Ni** 13.0-17.0 **P** 0.045 max **S** 0.030 max **Si** 0.075 max	**ASTM** A167; A213; A240; A249; A269; A276; A312; A358; A376; A409; A479
S31726	Austenitic Cr-Ni-Mo Stainless Steel (317L4)	**C** 0.03 max **Cr** 17.0-20.0 **Cu** 0.75 max **Mn** 2.00 max **Mo** 4.00-5.00 **N** 0.10-0.20 **Ni** 13.5-17.5 **P** 0.045 max **S** 0.030 max **Si** 0.75 max	**ASTM** A167; A213; A240; A249; A269; A276; A312; A358; A376; A409; A479
S31753	Austenitic Nitrogen-Bearing Stainless Steel (317LN)	**C** 0.030 max **Cr** 18.0-20.0 **Mn** 2.00 max **Mo** 3.00-4.00 **N** 0.10-0.20 **Ni** 11.0-15.0 **P** 0.045 max **S** 0.30 max **Si** 1.00 max	
S31780	Austenitic Cr-Ni-Mo Stainless Steel Bare Filler Metal	**C** 0.08 max **Cr** 18.5-20.5 **Cu** 0.75 max **Mn** 1.00-2.50 **Mo** 3.00-4.00 **Ni** 13.0-15.0 **P** 0.03 max **S** 0.03 max **Si** 0.30-0.65	**ASME** SFA5.9 (ER317) **AWS** A5.9 (ER317) **MIL SPEC** MIL-E-19933 (MIL-317)
S31783	Austenitic Cr-Ni-Mo Stainless Steel Bare Filler Metal (317L)	**C** 0.03 max **Cr** 18.5-20.5 **Cu** 0.75 max **Mn** 1.00-2.50 **Mo** 3.00-4.00 **Ni** 13.0-15.0 **P** 0.04 max **S** 0.03 max **Si** 0.30-0.65	**ASME** SFA5.9 (ER 317L) **AWS** A5.9 (ER 317L)
S31803	Ferritic-Austenitic Low Carbon Cr-Ni-Mo-N Stainless Steel (2205)	**C** 0.030 max **Cr** 21.0-23.0 **Mn** 2.00 max **Mo** 2.50-3.50 **N** 0.08-0.20 **Ni** 4.50-6.50 **P** 0.030 max **S** 0.020 max **Si** 1.00 max	**ASTM** A276; A473; A479
S31980	Austenitic Cr-Ni-Mo Stainless Steel (Columbium Stabilized) Bare Filler Metal	**C** 0.08 max **Cr** 18.0-20.0 **Cu** 0.75 max **Mn** 1.00-2.50 **Mo** 2.00-3.00 **Ni** 11.0-14.0 **P** 0.03 max **S** 0.03 max **Si** 0.30-0.65 **Other** Cb+Ta 8xC-1.0 max	**ASME** SFA5.9 (ER318) **AWS** A5.9 (ER318) **MIL SPEC** MIL-E-19933 (MIL-318)
S32100	Austenitic Cr-Ni Stainless Steel (Titanium Stabilized)	**C** 0.08 max **Cr** 17.00-19.00 **Mn** 2.00 max **Ni** 9.00-12.00 **P** 0.045 max **S** 0.030 max **Si** 1.00 max **Ti** 5XC min	**AISI** 321 **AMS** 5510; 5557; 5559; 5570; 5576; 5645; 5689; 7211 **ASME** SA182 (321); SA193 (321); SA194 (321); SA213 (321); SA240 (321); SA249 (321); SA312 (321); SA320 (321); SA358 (321); SA376 (321); SA403 (321); SA409 (321); SA430 (321); SA479 (321) **ASTM** A167 (321); A182 (321); A193 (321); A194 (321); A213 (321); A240 (321); A249 (321); A269 (321); A271 (321); A276 (321); A312 (321); A314 (321); A320 (321); A358 (321); A376 (321); A403 (321); A409 (321); A430 (321); A473 (321); A479 (321); A493 (321); A511 (321); A554 (321); A632 (321) **FED** QQ-S-763 (321); QQ-S-766 (321); QQ-W-423 (321) **MIL SPEC** MIL-S-862 (321); MIL-S-27419 **SAE** J405 (30321)
S32109	Austenitic Cr-Ni Stainless Steel (Titanium Stabilized), High Carbon (321H)	**C** 0.04-0.10 **Cr** 17.00-20.00 **Mn** 2.00 max **Ni** 9.00-12.00 **P** 0.040 max **S** 0.030 max **Si** 1.00 max **Ti** 4xC-0.60	**ASME** SA182 (321 H); SA213 (321 H); SA240 (321 H); SA249 (321 H); SA312 (321 H); SA376 (321 H); SA403 (321 H); SA430 (321 H); SA479 (321 H) **ASTM** A182 (321 H); A213 (321 H); A240 (321 H); A249 (321 H); A271 (321 H); A312 (321 H); A376 (321 H); A403 (321 H); A430 (321 H); A479 (321 H)

The chemical compositions listed are for identification purposes and should not be used in lieu of the cross referenced specifications.

UNIFIED NUMBER	DESCRIPTION	CHEMICAL COMPOSITION	CROSS REFERENCE SPECIFICATIONS
S32180	Austenitic Cr-Ni Stainless Steel (Titanium Stabilized) Bare Filler Metal	**C** 0.08 max **Cr** 18.5-20.5 **Cu** 0.75 max **Mn** 1.00-2.50 **Mo** 0.75 max **Ni** 9.00-10.50 **P** 0.03 max **S** 0.03 max **Si** 0.30-0.65 **Ti** 9xC-1.0	**ASME** SFA5.9 (ER321) **AWS** A5.9 (ER321) **MIL SPEC** MIL-E-19933 (MIL-321)
S32304	Ferritic-Austenitic Cr-Ni Stainless Steel, Low Carbon with Nitrogen, Mo, and Cu (SAF 2304)	**C** 0.030 max **Cr** 21.5-24.5 **Cu** 0.05-0.60 **Mn** 2.50 max **Mo** 0.05-0.60 **N** 0.05-0.20 **Ni** 3.0-5.5 **P** 0.040 max **S** 0.040 max **Si** 1.0 max	**ASTM** A789; A790
S32404	Ferritic-Austenitic Cr-Ni-Mo Stainless Steel (Uranus 50)	**C** 0.04 max **Cr** 20.5-22.5 **Cu** 1.0-2.0 **Mn** 2.0 max **Mo** 2.0-3.0 **N** 0.20 max **Ni** 5.5-8.5 **P** 0.030 max **S** 0.010 max **Si** 1.0 max	
S32550	Ferritic-Austenitic Cr-Ni-Mo-Cu Stainless Steel (Ferralium 255)	**C** 0.04 max **Cr** 24.0-27.0 **Cu** 1.50-2.50 **Mn** 1.50 max **Mo** 2.00-4.00 **N** 0.10-0.25 **Ni** 4.50-6.50 **P** 0.04 max **S** 0.03 max **Si** 1.00 max	**ASTM** A240; A479
S32900	Ferritic-Austenitic Cr-Ni Stainless Steel	**C** 0.20 max **Cr** 23.00-28.00 **Mn** 1.00 max **Mo** 1.00-2.00 **Ni** 2.50-5.00 **P** 0.040 max **S** 0.030 max **Si** 0.75 max	**AISI** 329 **ASME** SA268 (329) **ASTM** A268 (329); A511 (329)
S32950	Ferritic-Austenitic Cr-Ni-Mo Stainless Steel (7 Mo Plus)	**C** 0.03 max **Cr** 26.0-29.0 **Mn** 2.00 max **Mo** 1.00-2.50 **N** 0.15-0.35 **Ni** 3.50-5.20 **P** 0.035 max **S** 0.010 max **Si** 0.60 max	**ASTM** A240; A789; A790
S33100	Ni-Cr Stainless Steel (F-10)	**C** 0.10-0.20 **Cr** 7.00-9.00 **Mn** 0.50-0.80 **Ni** 19.00-22.00 **P** 0.030 max **S** 0.030 max **Si** 1.00-1.40	**ASTM** A182 (F-10)
S34700	Austenitic Cr-Ni Stainless Steel (Columbium Stabilized)	**C** 0.08 max **Cb** 10XC min **Cr** 17.00-19.00 **Mn** 2.00 max **Ni** 9.00-13.00 **P** 0.045 max **S** 0.030 max **Si** 1.00 max	**AISI** 347 **AMS** 5512; 5556; 5558; 5571; 5575; 5646; 5654; 5674; 7229 **ASME** SA182 (347); SA193 (347); SA194 (347); SA213 (347); SA240 (347); SA249 (347); SA312 (347); SA320 (347); SA358 (347); SA376 (347); SA403 (347); SA409 (347); SA430 (347); SA479 (347) **ASTM** A167 (347); A182 (347); A193 (347); A194 (347); A213 (347); A249 (347); A269 (347); A271 (347); A276 (347); A312 (347); A314 (347); A320 (347); A358 (347); A376 (347); A403 (347); A409 (347); A430 (347); A473 (347); A479 (347); A493 (347); A511 (347); A554 (347); A580 (347); A632 (347) **FED** QQ-S-763 (347); QQ-S-766 (347); QQ-W-423 (347) **MIL SPEC** MIL-S-862 (347); MIL-S-23195 (347); MIL-S-23196 (347) **SAE** J405 (30347)
S34709	Austenitic Cr-Ni Stainless Steel (Columbium Stabilized), High Carbon (347-H)	**C** 0.04-0.10 **Cb** 8xC-1.00 **Cr** 17.00-20.00 **Mn** 2.00 max **Ni** 9.00-13.00 **P** 0.040 max **S** 0.030 max **Si** 1.00 max	**ASME** SA182 (347 H); SA213 (347 H); SA240 (347 H); SA249 (347 H); SA312 (347 H); SA376 (347 H); SA403 (347 H); SA430 (347 H); SA479 (347 H) **ASTM** A182 (347 H); A213 (347 H); A240 (347 H); A249 (347 H); A271 (347 H); A312 (347 H); A376 (347 H); A403 (347 H); A430 (347 H); A479 (347 H)
S34720	Austenitic Cr-Ni Free Machining Stainless Steel (Columbium Stabilized)	**C** 0.08 max **Cb** 10xC-1.10 **Cr** 17.00-19.00 **Mn** 2.00 max **Ni** 9.00-12.00 **P** 0.040 max **S** 0.18-0.35 **Si** 1.00 max	**AMS** 5642 (1)
S34723	Austenitic Cr-Ni Selenium Bearing Free Machining Stainless Steel (Columbium Stabilized)	**C** 0.08 max **Cb** 10xC-1.10 **Cr** 17.00-19.00 **Mn** 2.00 max **Ni** 9.00-12.00 **P** 0.11-0.17 **S** 0.030 max **Se** 0.15-0.35 **Si** 1.00 max	**AMS** 5642 (2)
* S34740	Austenitic Cr-Ni Free Machining Stainless Steel (Columbium Stabilized)	**C** 0.08 max **Cb** 10xC-1.10 **Cr** 17.00-19.00 **Mn** 2.00 max **Ni** 9.00-12.00 **P** 0.040 max **S** 0.18-0.35 **Si** 1.00 max	
* S34741	Austenitic Cr-Ni Selenium Bearing Free Machining Stainless Steel (Columbium Stabilized)	**C** 0.08 max **Cb** 10xC-1.10 **Cr** 17.00-19.00 **Mn** 2.00 max **Ni** 9.00-12.00 **P** 0.11-0.17 **S** 0.030 max **Se** 0.15-0.35 **Si** 1.00 max	

*Boxed entries are no longer active and are retained for reference purposes only.

The chemical compositions listed are for identification purposes and should not be used in lieu of the cross referenced specifications.

UNIFIED NUMBER	DESCRIPTION	CHEMICAL COMPOSITION	CROSS REFERENCE SPECIFICATIONS
S34780	Austenitic Cr-Ni Stainless Steel (Columbium Stabilized) Bare Filler Metal	**C** 0.08 max **Cb** 10xC-1.0 **Cr** 19.0-21.5 **Cu** 0.75 max **Mn** 1.00-2.50 **Mo** 0.75 max **Ni** 9.00-11.00 **P** 0.03 max **S** 0.03 max **Si** 0.30-0.65	**AMS** 5790 **ASME** SFA5.9 (ER347); SFA5.30 (IN347) **AWS** A5.9 (ER347); A5.30 (IN347) **MIL SPEC** MIL-R-5031 (Classes 5 and 5a); MIL-E-19933 (MIL-347) (MIL-347Co); MIL-I-23413 (MIL-348) (MIL-348Co)
S34781	Austenitic Cr-Ni Stainless Steel (Columbium Stabilized) Bare Filler Metal	**C** 0.07 max **Cr** 17.0-19.0 **Cu** 0.50 max **Mn** 2.00 max **Mo** 0.75 max **Ni** 9.00-13.00 **P** 0.040 max **S** 0.030 max **Si** 0.50-1.00 **Other** Cb+Ta, 12xC-0.50	**AMS** 5680
S34788	Austenitic Cr-Ni Stainless Steel (Columbium Stabilized) Bare Filler Metal (347Si)	**C** 0.08 max **Cr** 19.0-21.5 **Cu** 0.5 max **Mn** 1.00-2.50 **Mo** 0.5 max **Ni** 9.00-11.00 **P** 0.03 max **S** 0.03 max **Si** 0.65-1.00 **Other** Cb+Ta, 10xC-1.0	**ASME** SFA5.9 (ER347Si) **AWS** A5.9 (ER347Si)
S34800	Austenitic Cr-Ni Stainless Steel (Columbium Stabilized) (Ta and Co Restricted)	**C** 0.08 max **Cb** 10XC min **Co** 0.20 max **Cr** 17.00-19.00 max **Mn** 2.00 max **Ni** 9.00-13.00 **P** 0.045 max **S** 0.030 max **Si** 1.00 max **Ta** 0.10 max	**AISI** 348 **ASME** SA182 (348); SA213 (348); SA240 (348); SA249 (348); SA312 (348); SA358 (348); SA376 (348); SA403 (348); SA409 (348); SA479 (348) **ASTM** A167 (348); A182 (348); A213 (348); A240 (348); A249 (348); A269 (348); A276 (348); A312 (348); A314 (348); A358 (348); A376 (348); A403 (348); A409 (348); A479 (348); A580 (348); A632 (348) **FED** QQ-S-766 (348) **MIL SPEC** MIL-S-23195 (348); MIL-S-23196 (348) **SAE** J405 (30348)
S34809	Austenitic Cr-Ni Stainless Steel (Columbium Stabilized), High Carbon (Ta and Co Restricted) (348-H)	**C** 0.04-0.10 **Cb** 8xC-1.00 **Co** 0.20 max **Cr** 17.00-20.00 **Mn** 2.00 max **Ni** 9.00-1300 **P** 0.045 max **S** 0.030 max **Si** 1.00 max **Ta** 0.10 max	**ASME** SA182 (348 H); SA213 (348 H); SA240 (348 H); SA249 (348 H); SA312 (348 H); SA4479 (348 H) **ASTM** A182 (348 H); A213 (348 H); A240 (348 H); A249 (348 H); A312 (348 H); A479 (348 H)
S35000	Precipitation Hardenable Cr-Ni-Mo Stainless Steel (AM 350)	**C** 0.07-0.11 **Cr** 16.00-17.00 **Mn** 0.50-1.25 **Mo** 2.50-3.25 **N** 0.07-0.13 **Ni** 4.00-5.00 **P** 0.040 max **S** 0.030 max **Si** 0.50 max	**AMS** 5546; 5548; 5554; 5745 **ASTM** A579 (61); A693 (350) **MIL SPEC** MIL-S-8840 **SAE** J467 (AM-350)
S35080	Precipitation Hardenable Cr-Ni-Mo Stainless Steel Bare Filler Metal (AM350)	**C** 0.08-0.12 **Cr** 16.0-17.0 **Mn** 0.50-1.25 **Mo** 2.50-3.25 **N** 0.07-0.13 **Ni** 4.00-5.00 **P** 0.040 max **S** 0.030 max **Si** 0.50 max	**AMS** 5774
S35500	Precipitation Hardenable Cr-Ni-Mo Stainless Steel (AM 355)	**C** 0.10-0.15 **Cr** 15.00-16.00 **Mn** 0.50-1.25 **Mo** 2.50-3.25 **N** 0.07-0.13 **Ni** 4.00-5.00 **P** 0.040 max **S** 0.030 max **Si** 0.50 max	**AMS** 5547; 5549; 5594; 5743; 5744 **ASTM** A564 (634); A579 (634); A693 (634); A705 (634) **MIL SPEC** MIL-S-8840 **SAE** J467 (AM-355)
S35580	Precipitation Hardenable Cr-Ni-Mo Stainless Steel Bare Filler Metal (AM355)	**C** 0.10-0.15 **Cr** 15.0-16.0 **Cu** 0.50 max **Mn** 0.50-1.25 **Mo** 2.50-3.25 **N** 0.07-0.13 **Ni** 4.00-5.00 **P** 0.04 max **S** 0.03 max **Si** 0.50 max	**AMS** 5780
S36200	Precipitation Hardenable Cr-Ni-Ti Stainless Steel (Almar 362)	**Al** 0.10 max **C** 0.05 max **Cr** 14.00-15.00 **Mn** 0.50 max **Ni** 6.00-7.00 **P** 0.030 max **S** 0.030 max **Si** 0.30 max **Ti** 0.55-0.90	**AMS** 5739; 5740 **ASTM** A564 (XM-9); A693 (XM-9); A705 (XM-9)
S37000	Austenitic Cr-Ni-Mo-Ti Stainless Steel	**Al** 0.050 max **As** 0.030 max **B** 0.0020 max **C** 0.030-0.050 **Cb** 0.050 max **Co** 0.050 max **Cr** 12.5-14.5 **Cu** 0.04 max **Mn** 1.65-2.35 **Mo** 1.50-2.50 **N** 0.005 max **Ni** 14.5-16.5 **P** 0.040 max **S** 0.010 max **Si** 0.50-1.00 **Ta** 0.020 max **Ti** 0.10-0.40 **V** 0.05 max **Other** aim for 0.25	
S38100	Austenitic Cr-Ni-Si Stainless Steel (18-18-2)	**C** 0.08 max **Cr** 17.00-19.00 **Mn** 2.00 max **Ni** 17.50-18.50 **P** 0.030 max **S** 0.030 max **Si** 1.50-2.50	**ASME** SA213 (XM-15); SA240 (XM-15); SA249 (XM-15); SA312 (XM-15) **ASTM** A167 (XM-15); A213 (XM-15); A240 (XM-15); A249 (XM-15); A269 (XM-15); A312 (XM-15)
S38400	Austenitic Cr-Ni Stainless Steel (Low Work Hardening)	**C** 0.08 max **Cr** 15.00-17.00 **Mn** 2.00 max **Ni** 17.00-19.00 **P** 0.045 max **S** 0.030 max **Si** 1.00 max	**AISI** 384 **ASTM** A493 (384) **SAE** J405 (30384)

The chemical compositions listed are for identification purposes and should not be used in lieu of the cross referenced specifications.

UNIFIED NUMBER	DESCRIPTION	CHEMICAL COMPOSITION	CROSS REFERENCE SPECIFICATIONS
* S38500	Austenitic Cr-Ni Stainless Steel (Low Work Hardening)	C 0.08 max Cr 11.50-13.50 Mn 2.00 max Ni 14.00-16.00 P 0.045 max S 0.030 max Si 1.00 max	AISI 385 ASTM A493 (385) SAE J405 (30385)
S38660	Austenitic Cr-Ni-Mo Stainless Steel-Reactor Grade	Al 0.050 max As 0.030 max B 0.0020 max C 0.03-0.05 Cb 0.050 max Co 0.050 max Cr 12.5-14.5 Cu 0.04 max Mn 1.65-2.35 Mo 1.50-2.50 N 0.005 max Ni 14.5-16.5 P 0.040 max S 0.010 max Si 0.50-1.00 Ta 0.020 max Ti 0.10-0.40 V 0.05 max	ASTM A771
S40300	Cr Stainless Steel (Hardenable by Heat Treatment)	C 0.15 max Cr 11.50-13.00 Mn 1.00 max P 0.040 max S 0.030 max Si 0.50 max	AISI 403 ASTM A176 (403); A276 (403); A314 (403); A473 (403); A479 (403); A511 (403); A580 (403) FED QQ-S-763 (403) MIL SPEC MIL-S-861; MIL-S-862 (403) SAE J405 (51403)
S40500	Cr Stainless Steel (Not Hardenable by Heat Treatment)	Al 0.10-0.30 C 0.08 max Cr 11.50-14.50 Mn 1.00 max P 0.040 max S 0.030 max Si 1.00 max	AISI 405 ASME SA240 (405); SA268 (405); SA479 (405) ASTM A176 (405); A240 (405); A268 (405); A276 (405); A314 (405); A473 (405); A479 (405); A511 (405); A580 (405) FED QQ-S-763 (405) MIL SPEC MIL-S-861; MIL-S-862 (405) SAE J405 (51405)
S40800	Ferritic Cr Stainless Steel	C 0.08 max Cr 11.5-13.0 Mn 1.00 max Ni 0.80 max P 0.045 max S 0.045 max Si 1.00 max Ti 12xC-1.10	ASTM A268
S40900	Ferritic Cr Stainless Steel	C 0.08 max Cr 10.50-11.75 Mn 1.00 max Ni 0.50 max P 0.045 max S 0.045 max Si 1.00 max Ti 6xC-0.75	AISI 409 ASME SA268 (409) ASTM A176 (409); A268 (409); A651 (409) SAE J405 (51409)
S41000	Cr Stainless Steel (Hardenable by Heat Treatment)	C 0.15 max Cr 11.50-13.50 Mn 1.00 max P 0.040 max S 0.030 max Si 1.00 max	AISI 410 AMS 5504; 5505; 5591; 5613; 5776 ASME SA194 (6); SA240 (410); SA268 (410); SA479 (410) ASTM A176 (410); A182 (F62); A193 (410); A194 (410); A240 (410); A268 (410); A276 (410); A314 (410); A336 (F6); A473 (410); A479 (410); A493 (410); A511 (410); A580 (410) FED QQ-S-763 (410); QQ-S-766 (410); QQ-W-423 (410) MIL SPEC MIL-S-861; MIL-S-862 (410) SAE J405 (51410); J412 (51410)
S41001	Cr Super Strength Steel	Al 0.05 max C 0.15 max Cr 11.5-13.5 Cu 0.50 max Mn 1.00 max Mo 0.50 max Ni 0.75 max P 0.025 max S 0.025 max Si 1.00 max Sn 0.05 max	ASTM A579 (51)
S41008	Cr Stainless Steel (Hardenable by Heat Treatment), Low Carbon (410 S)	C 0.08 max Cr 11.50-13.50 Mn 1.00 max Ni 0.60 max P 0.040 max S 0.030 max Si 1.00 max	ASME SA240 (410 S) ASTM A176 (410 S); A240 (410 S); A473 (410 S)
S41025	Martensitic Cr Stainless Steel with Mo	C 0.15 max Cr 11.50-13.50 Mn 1.00 max Mo 0.40-0.60 Ni 0.60 max P 0.040 max S 0.030 max Si 1.00 max	AMS 5614
S41026	Martensitic Cr Stainless Steel with Mo	C 0.15 max Cr 11.5-13.5 Cu 0.50 max Mn 1.00 max Mo 0.40-0.60 Ni 1.0-2.0 P 0.02 max S 0.02 max Si 1.00 max	ASTM A182 (F6b)
S41040	Martensitic Cr Stainless Steel with Cb	C 0.15 max Cb 0.05-0.20 Cr 11.50-13.50 Mn 1.00 max P 0.040 max S 0.030 max Si 1.00 max	AMS 5609; 5611; 5612 ASME SA479 (XM-30) ASTM A479 (XM-30)
S41041	Martensitic Cr Stainless Steel with Cb	Al 0.05 max C 0.13-0.18 Cb 0.15-0.45 Cr 11.50-13.00 Mn 0.40-0.60 Mo 0.20 max Ni 0.50 max P 0.030 max S 0.030 max Si 0.50 max Sn 0.05 max	
S41050	Ferritic Cr Stainless Steel with Ni (E-4)	C 0.040 max Cr 10.50-12.50 Mn 1.00 max N 0.10 max Ni 0.60-1.10 P 0.045 max S 0.030 max Si 1.00 max	ASTM A176; A240
S41080	Martensitic Cr Stainless Steel Bare Filler Metal	C 0.12 max Cr 11.5-13.5 Cu 0.75 max Mn 0.6 max Mo 0.75 max Ni 0.6 max P 0.03 max S 0.03 max Si 0.50 max	AMS 5778 ASME SFA5.9 (ER410) AWS A5.9 (ER410) MIL SPEC MIL-E-19933 (MIL-410)

*Boxed entries are no longer active and are retained for reference purposes only.

The chemical compositions listed are for identification purposes and should not be used in lieu of the cross referenced specifications.

UNIFIED NUMBER	DESCRIPTION	CHEMICAL COMPOSITION	CROSS REFERENCE SPECIFICATIONS
S41081	Martensitic Cr Stainless Steel Bare Filler Metal	**Al** 0.05 max **C** 0.11-0.15 **Cr** 11.5-12.5 **Cu** 0.50 max **Mn** 0.60 max **Mo** 0.20 max **N** 0.08 max **Ni** 0.75 max **P** 0.025 max **S** 0.025 max **Si** 0.50 max **Ti** 0.05 max	**AMS** 5821
S41086	Martensitic Cr-Ni-Mo Stainless Steel Bare Filler Metal (410NiMo)	**C** 0.06 max **Cr** 11.0-12.5 **Cu** 0.75 max **Mn** 0.60 max **Mo** 0.40-0.70 **Ni** 4.00-5.00 **P** 0.03 max **S** 0.03 max **Si** 0.50 max	**ASME** SFA5.9 (ER410NiMo) **AWS** A5.9 (ER410NiMo)
S41400	Cr Stainless Steel with Low Nickel Content	**C** 0.15 max **Cr** 11.50-13.50 **Mn** 1.00 max **Ni** 1.25-2.50 **P** 0.040 max **S** 0.030 max **Si** 1.00 max	**AISI** 414 **AMS** 5615 **ASTM** A276 (414); A314 (414); A473 (414); A511 (414); A580 (414) **FED** QQ-S-763 (414) **SAE** J405 (51414)
S41500	Ferritic Cr Stainless Steel with Nickel	**C** 0.05 max **Cr** 11.5-14.0 **Mn** 0.50-1.00 **Mo** 0.50-1.00 **Ni** 3.50-5.50 **P** 0.030 max **S** 0.030 max **Si** 0.60 max	**ASTM** A182 (F6NM)
S41600	Cr Stainless Steel (Free Machining)	**C** 0.15 max **Cr** 12.00-14.00 **Mn** 1.25 max **Mo** 0.60 max (Optional) **P** 0.060 max **S** 0.15 min **Si** 1.00 max	**AISI** 416 **AMS** 5610 **ASME** SA194 (416) **ASTM** A194 (416); A314 (416); A473 (416); A581 (416); A582 (416) **MIL SPEC** MIL-S-862 (416) **SAE** J405 (51416)
S41610	Martensitic Cr Free Machining Stainless Steel	**C** 0.15 max **Cr** 12.00-14.00 **Mn** 1.50-2.50 **Mo** 0.60 max **P** 0.06 max **S** 0.15 min **Si** 1.00 max	**ASTM** A581 (XM-6); A582 (XM-6) **SAE** J405 (416 plus X)
S41623	Martensitic Cr Free Machining Stainless Steel (Se Bearing)	**C** 0.15 max **Cr** 12.00-14.00 **Mn** 1.25 max **P** 0.060 max **S** 0.060 **Se** 0.15 min **Si** 1.00 max	**AISI** 416 Se **AMS** 5610 **ASME** SA194 (416 Se) **ASTM** A194 (416 Se); A314 (416 Se); A473 (416 Se); A511 (416 Se); A581 (416 Se); A582 (416 Se) **FED** QQ-S-763 **MIL SPEC** MIL-S-763; MIL-S-862 (416 Se) **SAE** J405 (51416 Se)
S41780	Ferritic Cr-Ni-Co-Mo Stainless Steel Bare Filler Metal	**C** 0.10-0.15 **Co** 1.30-2.00 **Cr** 11.0-12.5 **Cu** 0.75 max **Mn** 0.40-1.30 **Mo** 1.50-2.00 **N** 0.045 **Ni** 2.50-3.00 **P** 0.030 max **S** 0.030 max **Si** 0.40 max **V** 0.25-0.40	**AMS** 5823
S41800	Martensitic Cr-W-Ni Stainless and Heat Resisting Steel (Greek Ascoloy)	**C** 0.15-0.20 **Cr** 12.00-14.00 **Mn** 0.50 max **Ni** 1.80-2.20 **P** 0.040 max **S** 0.030 max **Si** 0.50 max **W** 2.50-3.50	**AMS** 5508; 5616 **ASTM** A565 (615) **SAE** J467 (Greek Ascoloy)
S41880	Martensitic Cr-W-Ni Stainless and Heat Resisting Steel Bare Filler Metal (Greek Ascoloy)	**Al** 0.15 max **C** 0.15-0.20 **Cr** 12.0-14.0 **Cu** 0.50 max **Mn** 0.50 max **Mo** 0.50 max **Ni** 1.80-2.20 **P** 0.040 max **S** 0.030 max **Si** 0.50 max **Sn** 0.05 max **W** 2.50-3.50	**AMS** 5817
S42000	Hardenable Cr Stainless Steel	**C** over 0.15 **Cr** 12.00-14.00 **Mn** 1.00 max **P** 0.040 max **S** 0.030 max **Si** 1.00 max	**AISI** 420 **AMS** 5506; 5621; 7207 **ASTM** A276 (420); A314 (420); A473 (420); A580 (420) **FED** QQ-S-763 (420); QQ-S-766 (420); QQ-W-423 (420) **MIL SPEC** MIL-S-862 (420) **SAE** J405 (51420)
S42010	Hardenable Cr Stainless Steel (Low Ni - Low Mo)	**C** 0.15-0.30 **Cr** 13.5-15.0 **Mn** 1.00 max **Mo** 0.40-1.00 **Ni** 0.25-1.00 **P** 0.040 max **S** 0.030 max **Si** 1.00 max	**ASTM** A276; A314
S42020	Hardenable Cr Free Machining Stainless Steel	**C** over 0.15 **Cr** 12.00-14.00 **Mn** 1.25 max **Mo** 0.60 max (Optional) **P** 0.060 max **S** 0.15 min **Si** 1.00 max	**AISI** 420F **AMS** 5620 **SAE** J405 (51420F)
S42023	Hardenable Cr Free Machining Stainless Steel (Se Bearing)	**C** 0.30-0.40 **Cr** 12.00-14.00 **Mn** 1.25 max **Mo** 0.60 max **P** 0.06 max **S** 0.06 max **Se** 0.15 min **Si** 1.00 max **Other** Zr or Cu 0.60 max	**AMS** 5620 **ASTM** A582 (420 F Se) **SAE** J405 (51420F Se)
S42080	Hardenable Cr Stainless Steel Bare Filler Metal	**C** 0.25-0.40 **Cr** 12.0-14.0 **Cu** 0.75 max **Mn** 0.60 max **Mo** 0.75 max **Ni** 0.60 max **P** 0.03 max **S** 0.03 max **Si** 0.50 max	**ASME** SFA5.9 (ER420) **AWS** A5.9 (ER420)
S42100	Martensitic Cr Stainless Steel	**Al** 0.05 max **C** 0.17-0.23 **Cb** 0.050 max **Cr** 11.0-12.5 **Mn** 0.40-0.70 **Mo** 0.80-1.20 **Ni** 0.30-0.80 **P** 0.040 max **S** 0.010 max **Si** 0.20-0.30 **V** 0.25-0.35 **W** 0.40-0.60	

The chemical compositions listed are for identification purposes and should not be used in lieu of the cross referenced specifications.

UNIFIED NUMBER	DESCRIPTION	CHEMICAL COMPOSITION	CROSS REFERENCE SPECIFICATIONS
S42200	Martensitic Cr Stainless Steel	C 0.20-0.25 Cr 11.50-13.50 Mn 1.00 max Mo 0.75-1.25 Ni 0.50-1.00 P 0.040 max S 0.030 max Si 0.75 max V 0.15-0.30 W 0.75-1.25	AISI 422 AMS 5655 ASTM A565 (616) MIL SPEC MIL-S-861 SAE J467 (422)
S42201	Martensitic Cr, Superstrength Steel	Al 0.05 max C 0.20-0.25 Co 0.25 max Cr 11.0-13.5 Mn 0.50-1.00 Mo 0.75-1.25 Ni 0.75-1.25 P 0.025 max S 0.025 max Si 0.20-0.60 Sn 0.04 max Ti 0.05 max V 0.20-0.30 W 0.75-1.25	ASTM A579 (52)
S42300	Martensitic Cr-Mo-V Stainless Steel (Lapelloy)	C 0.27-0.32 Cr 11.00-12.00 Mn 0.95-1.35 Mo 2.50-3.00 Ni 0.50 max P 0.025 max S 0.025 max Si 0.50 max V 0.20-0.30	ASTM A565 (619)
S42400	Ferritic Stainless Steel	C 0.06 max Cr 12.00-14.00 Mn 0.50-1.00 Mo 0.30-0.70 Ni 3.50-4.50 P 0.03 max S 0.03 max Si 0.30-0.60	ASTM A182 (F6NM)
S42700	Martensitic Cr-Mo-V Stainless Steel (for Bearings)	C 1.10-1.20 Cr 14.0-15.0 Cu 0.35 max Mn 0.30-0.60 Mo 3.75-4.25 Ni 0.40 max P 0.015 max S 0.010 max Si 0.20-0.40 V 1.10-1.30	AMS 5749
S42800	Martensitic Cr-Mo-V Stainless Steel (For Bearings)	C 1.05-1.15 Cb 0.25-0.35 Cr 13.7-14.8 Cu 0.35 max Mn 0.25-0.50 Mo 1.90-2.25 Ni 0.35 max P 0.015 max S 0.010 max Si 0.20-0.40 V 0.90-1.15	AMS 5900
S42900	Cr Stainless Steel	C 0.12 max Cr 14.00-16.00 Mn 1.00 max P 0.040 max S 0.030 max Si 1.00 max	AISI 429 ASME SA182 (429); SA240 (429); SA268 (429) ASTM A176 (429); A182 (429); A240 (429); A268 (429); A276 (429); A314 (429); A473 (429); A493 (429); A511 (429); A554 (429) FED QQ-S-763 (429); QQ-S-766 (429) MIL SPEC MIL-S-862 (430) SAE J405 (51429)
S43000	Cr Stainless Steel	C 0.12 max Cr 16.00-18.00 Mn 1.00 max P 0.040 max S 0.030 max Si 1.00 max	AISI 430 AMS 5503; 5627 ASME SA182 (430); SA240 (430); SA268 (430); SA479 (430) ASTM A176 (430); A182 (430); A240 (430); A268 (430); A276 (430); A314 (430); A473 (430); A479 (430); A493 (430); A511 (430); A554 (430); A580 (430); A651 (430) FED QQ-S-763 (430); QQ-S-766 (430); QQ-W-423 (430) MIL SPEC MIL-S-862 (430) SAE J405 (51430)
S43020	Free Machining Cr Stainless Steel (S Bearing)	C 0.12 max Cr 16.00-18.00 Mn 1.25 max Mo 0.60 max (Optional) P 0.060 max S 0.15 min Si 1.00 max	AISI 430 F ASTM A314 (430 F); A473 (430 F); A581 (430 F); A582 (430 F) MIL SPEC MIL-S-862 (430 F) SAE J405 (51430 F)
S43023	Free Machining Cr Stainless Steel (Se Bearing)	C 0.12 max Cr 16.00-18.00 Mn 1.25 max P 0.060 max S 0.060 max Se 0.15 min Si 1.00 max	AISI 430F Se ASTM A314 (430F Se); A473 (430F Se); A581 (430F Se); A582 (430F Se) MIL SPEC MIL-S-862 (430F Se) SAE J405 (51430F Se)
S43035	Ferritic Cr-Ti Stainless Steel (HWT, Aqualloy)	Al 0.15 max C 0.07 max Cr 17.00-19.00 Mn 1.00 max Ni 0.50 max P 0.040 max S 0.030 max Si 1.00 max Ti 12xC-1.10	AISI 439 ASME SA240 (439); SA268 (439); SA479 (439) ASTM A240 (439); A268 (439); A479 (439); A651 (439)
S43036	Ferritic Cr Stainless Steel with Titanium (430 Ti)	C 0.10 max Cr 16.00-19.50 Mn 1.00 max Ni 0.75 max P 0.040 max S 0.030 max Si 1.00 max Other Ti, 5×C-0.75	ASTM A268 (430 Ti); A554 (430 Ti); A651 (430 Ti)
S43080	Ferritic Cr Stainless Steel Bare Filler Metal	C 0.10 max Cr 15.5-17.0 Cu 0.75 max Mn 0.60 max Mo 0.75 max Ni 0.60 max P 0.03 max S 0.03 max Si 0.50 max	ASME SFA5.9 (ER430) AWS A5.9 (ER430)
S43100	Ni Bearing Hardenable Cr Stainless Steel	C 0.20 max Cr 15.00-17.00 Mn 1.00 max Ni 1.25-2.50 P 0.040 max S 0.030 max Si 1.00 max	AISI 431 AMS 5628 ASTM A276 (431); A314 (431); A473 (431); A493 (431); A579 (63); A580 (431) MIL SPEC MIL-S-862 (431); MIL-S-8967; MIL-S-18732 SAE J405 (51431)
S43400	Cr-Mo Stainless Steel	C 0.12 max Cr 16.00-18.00 Mn 1.00 max Mo 0.75-1.25 P 0.040 max S 0.030 max Si 1.00 max	AISI 434 ASTM A651 (434) SAE J405 (51434)

The chemical compositions listed are for identification purposes and should not be used in lieu of the cross referenced specifications.

UNIFIED NUMBER	DESCRIPTION	CHEMICAL COMPOSITION	CROSS REFERENCE SPECIFICATIONS
S43600	Cr-Mo Stainless Steel	C 0.12 max Cr 16.00-18.00 Mn 1.00 max Mo 0.75-1.25 P 0.040 max S 0.030 max Si 1.00 max Other Cb+Ta, 5XC-0.70	AISI 436 SAE J405 (51436)
S44002	Hardenable Cr Stainless Steel (440 A)	C 0.60-0.75 Cr 16.00-18.00 Mn 1.00 max Mo 0.75 max P 0.040 max S 0.030 max Si 1.00 max	AISI 440 A AMS 5631; 5632; 7445 ASTM A276 (440 A); A314 (440 A); A473 (440 A); A511 (440 A); A580 (440 A) FED QQ-S-763 (440 A) MIL SPEC MIL-S-862 (440 A) SAE J405 (51440 A)
S44003	Hardenable Cr Stainless Steel (440 B)	C 0.75-0.95 Cr 16.00-18.00 Mn 1.00 max Mo 0.75 max P 0.040 max S 0.030 max Si 1.00 max	AISI 440 B AMS 7445 ASTM A276 (440 B); A314 (440 B); A473 (440 B); A580 (440 B) FED QQ-S-763 (440 B) MIL SPEC MIL-S-862 (440 B) SAE J405 (51440 B)
S44004	Hardenable Cr Stainless Steel (440 C)	C 0.95-1.20 Cr 16.00-18.00 Mn 1.00 max Mo 0.75 max P 0.040 max S 0.030 max Si 1.00 max	AISI 440 C AMS 5618; 5630; 5880; 7445 ASTM A276 (440 C); A314 (440 C); A473 (440 C); A493 (440 C); A580 (440 C) FED QQ-S-763 (440 C) MIL SPEC MIL-S-862 (440 C) SAE J405 (51440 C)
S44020	Martensitic Cr Free Machining Stainless Steel (440 F)	C 0.95-1.20 Cr 16.00-18.00 Mn 1.25 max Mo 0.40-0.60 N 0.08 max Ni 0.75 max P 0.040 max S 0.10-0.35 Si 1.00 max	AMS 5632 (1) MIL SPEC MIL-S-862 (440 F) SAE J405 (51440 F)
S44023	Martensitic CrSe Selenium Bearing Free Machining Stainless Steel (440 F Se)	C 0.95-1.20 Cr 16.00-18.00 Mn 1.25 max Mo 0.60 max N 0.08 max Ni 0.75 max P 0.040 max S 0.030 max Se 0.15 min Si 1.00 max	AMS 5632 (2) MIL SPEC MIL-S-862 (440 F Se) SAE J405 (51440 F Se)
S44025	Hardenable Cr Stainless Steel with Mo	C 0.95-1.10 Cr 16.00-18.00 Cu 0.50 max Mn 1.00 max Mo 0.40-0.65 Ni 0.75 max P 0.025 max S 0.025 max Si 1.00 max	ASTM A756
S44100	Ferritic Cr Stainless Steel (Columbium- Titanium Stabilized)	C 0.03 max Cb 0.3 + (9xC) min - 0.9 max Cr 17.5-19.5 Mn 1.00 max N 0.03 max Ni 1.00 max P 0.04 max S 0.03 max Si 1.00 max Ti 0.10-0.50	
S44200	Cr Heat Resisting Steel	C 0.20 max Cr 18.00-23.00 Mn 1.00 max P 0.040 max S 0.030 max Si 1.00 max	AISI 442 ASTM A176 (442) SAE J405 (51442)
S44300	High Cr-Cu Heat Resisting Steel	C 0.20 max Cr 18.00-23.00 Cu 0.90-1.25 Mn 1.00 max Ni 0.50 max P 0.040 max S 0.030 max Si 1.00 max	ASTM A176 (443); A268 (443); A511 (443) FED QQ-S-763
S44400	Cr-Mo Stainless Steel (Stabilized Low-Interstitial Ferritic)	C 0.025 max Cr 17.5-19.5 Mn 1.00 max Mo 1.75-2.50 N 0.025 max Ni 1.00 max P 0.040 max S 0.030 max Si 1.00 max Other Ti+Cb [0.20+4(C+N)] min - 0.80 max	ASTM A176; A240; A268; A276
S44600	Cr Heat Resisting Steel	C 0.20 max Cr 23.00-27.00 Mn 1.50 max N 0.25 max P 0.040 max S 0.030 max Si 1.00 max	AISI 446 ASME SA268 (446) ASTM A176 (446); A268 (446); A276 (446); A314 (446); A473 (446); A511 (446); A580 (446) FED QQ-S-763 (446); QQ-S-766 (446) MIL SPEC MIL-S-862 (446) SAE J405 (51446)
* S44625	Low Interstitial Ferritic Cr Stainless Steel	C 0.01 max Cr 25.00-27.50 Cu 0.20 max Mn 0.40 max Mo 0.75-1.50 N 0.015 max Ni 0.50 max P 0.020 max S 0.020 max Si 0.40 max Other Ni+Cu 0.50 max	ASME SA182 (XM-27); SA240 (XM-27); SA268 (XM-27); SA479 (XM-27) ASTM A176 (XM-27); A182 (XM-27); A240 (XM-27); A268 (XM-27); A276 (XM-27); A314 (XM-27); A479 (XM-27); A493 (XM-27)
S44626	High Cr+Mo Stainless Steel (26-1)	C 0.06 max Cr 25.00-27.00 Cu 0.20 max Mn 0.75 max Mo 0.75-1.50 N 0.04 max Ni 0.50 max P 0.040 max S 0.020 max Si 0.75 max Ti 0.20-1.00 Other Ti 7x(C+N) min	ASME SA240 (XM-33); SA268 (XM-33) ASTM A176 (XM-33); A240 (XM-33); A268 (XM-33)

*Boxed entries are no longer active and are retained for reference purposes only.

The chemical compositions listed are for identification purposes and should not be used in lieu of the cross referenced specifications.

UNIFIED NUMBER	DESCRIPTION	CHEMICAL COMPOSITION	CROSS REFERENCE SPECIFICATIONS
S44627	Low Interstitial Ferritic Cr Stainless Steel with Columbium	**C** 0.010 max **Cb** 0.05-0.20 **Cr** 25.0-27.0 **Cu** 0.20 max **Mn** 0.40 max **Mo** 0.75-1.50 **N** 0.015 max **Ni** 0.50 max **P** 0.020 max **S** 0.020 max **Si** 0.40 max	**ASME** SA240; SA268; SA479; SA731; SFA5.9 (Er26-1) **ASTM** A176; A240; A268; A276; A314; A479; A731
S44635	Ferritic Cr-Mo-Ni Stainless Steel (25-4-4)	**C** 0.025 max **Cr** 24.5-26.0 **Mn** 1.00 max **Mo** 3.50-4.50 **N** 0.035 max **Ni** 3.50-4.50 **P** 0.040 max **S** 0.030 max **Si** 0.75 max **Other** Cb+Ti 0.20+4(C+N) - 0.80 max	**ASTM** A176; A240; A268
S44660	Ferritic Cr-Mo-Ni Stainless Steel (Columbium plus Titanium Stabilized) (SC-1)	**C** 0.025 max **Cr** 25.0-27.0 **Mn** 1.00 max **Mo** 2.50-3.50 **N** 0.035 max **Ni** 1.50-3.50 **P** 0.040 max **S** 0.030 max **Si** 1.00 max **Other** Cb+Ti 0.20+4(C+N) - 0.80 max	**ASTM** A176; A240; A268
S44687	Ferritic Cr-Mo Stainless Steel Bare Filler Metal	**C** 0.01 max **Cr** 25.0-27.5 **Cu** 0.20 max **Mn** 0.40 max **Mo** 0.75-1.50 **N** 0.015 max **P** 0.02 max **S** 0.02 max **Si** 0.40 max **Other** Ni+Cu 0.50 max	**ASME** SFA5.9 (ER26-1) **AWS** A5.9 (ER26-1)
S44700	Ferritic High Cr+Mo Stainless Steel (28-4)	**C** 0.010 max **Cr** 28.0-30.0 **Cu** 0.15 max **Mn** 0.30 max **Mo** 3.5-4.2 **N** 0.020 max **Ni** 0.15 max **P** 0.025 max **S** 0.020 max **Si** 0.20 max **Other** C+N, 0.025 max	**ASTM** A176; A240; A268; A276; A479
S44735	Ferritic High Cr+Mo Stainless Steel (29-4C)	**C** 0.030 max **Cr** 28.0-30.0 **Mn** 1.00 max **Mo** 3.60-4.20 **N** 0.045 max **Ni** 1.00 max **P** 0.040 max **S** 0.030 max **Si** 1.00 max **Other** Cb+Ti 6(C+N) 0.20 min-1.00 max	**ASTM** A176; A240; A268; A511
S44800	Ferritic High Cr Stainless Steel (28-4-2)	**C** 0.010 max **Cr** 28.0-30.0 **Cu** 0.15 max **Mn** 0.30 max **Mo** 3.5-4.2 **N** 0.020 max **Ni** 2.0-2.5 **P** 0.025 max **S** 0.020 max **Si** 0.20 max **Other** C+N, 0.025 max	**ASTM** A176; A240; A268; A276; A479
S45000	Stainless Steel (Custom 450)	**C** 0.05 max **Cb** 8xC min **Cr** 14.00-16.00 **Cu** 1.25-1.75 **Mn** 1.00 max **Mo** 0.50-1.00 **Ni** 5.00-7.00 **P** 0.030 max **S** 0.030 max **Si** 1.00 max	**AMS** 5763; 5773; 5859 **ASME** SA564 (XM-25); SA705 (XM-25) **ASTM** A564 (XM-25); A693 (XM-25); A705 (XM-25)
S45500	Stainless Steel (Custom 455)	**C** 0.05 max **Cb** 0.10-0.50 **Cr** 11.00-12.50 **Cu** 1.50-2.50 **Mn** 0.50 max **Mo** 0.50 max **Ni** 7.50-9.50 **P** 0.040 max **S** 0.030 max **Si** 0.50 max **Ti** 0.80-1.40	**AMS** 5578; 5617; 5672; 5860 **ASTM** A313 (XM-16); A564 (XM-16); A693 (XM-16); A705 (XM-16) **MIL SPEC** MIL-S-83311
S50100	Cr Heat Resisting Steel	**C** over 0.10 **Cr** 4.00-6.00 **Mn** 1.00 max **Mo** 0.40-0.65 **P** 0.040 max **S** 0.030 max **Si** 1.00 max	**AISI** 501 **AMS** 5502; 5602 **ASME** SA194 (3); SA387 (5) **ASTM** A193 (501); A194 (501); A314 (501); A387 (5); A473 (501) **SAE** J405 (51501)
S50180	Cr Heat Resisting Steel Bare Filler Metal	**C** 0.25-0.40 **Cr** 4.80-6.00 **Cu** 0.35 max **Mn** 0.75-1.00 **Mo** 0.45-0.65 **P** 0.025 max **S** 0.030 max **Si** 0.25-0.50	**ASME** SFA5.23 (EB6H) **AWS** A5.23 (EB6H)
S50181	Cr Heat Resisting Steel Bare Filler Metal (Nuclear Grade)	**C** 0.25-0.40 **Cr** 4.80-6.00 **Cu** 0.08 max **Mn** 0.75-1.00 **Mo** 0.45-0.65 **P** 0.012 max **S** 0.030 max **Si** 0.25-0.50 **V** 0.05 max	**ASME** SFA5.23 (EB6HN) **AWS** A5.23 (EB6HN)
S50200	Cr Heat Resisting Steel	**C** 0.10 max **Cr** 4.00-6.00 **Mn** 1.00 max **Mo** 0.40-0.65 **P** 0.040 max **S** 0.030 max **Si** 1.00 max	**AISI** 502 **ASME** SA387 (5) **ASTM** A199; A200; A213; A314 (502); A335; A387 (5); A473 (502) **MIL SPEC** MIL-T-16286; MIL-S-20146 **SAE** J405 (51502)
S50280	Cr Heat Resisting Steel Base Filler Metal	**C** 0.10 max **Cr** 4.60-6.00 **Cu** 0.75 max **Mn** 0.60 max **Mo** 0.45-0.65 **Ni** 0.60 max **P** 0.03 max **S** 0.03 max **Si** 0.50 max	**AMS** 6466 **ASME** SFA5.9 (ER502); SFA5.23 (EB6); SFA5.30 (IN502) **AWS** A5.9 (ER502); A5.23 (EB6); A5.30 (IN502) **MIL SPEC** MIL-I-23431 (MIL-505)

The chemical compositions listed are for identification purposes and should not be used in lieu of the cross referenced specifications.

UNIFIED NUMBER	DESCRIPTION	CHEMICAL COMPOSITION	CROSS REFERENCE SPECIFICATIONS
S50281	Cr Heat Resisting Steel Bare Filler Metal (Nuclear Grade)	C 0.10 max Cr 4.50-6.00 Cu 0.08 max Mn 0.40-0.65 Mo 0.45-0.65 P 0.012 max S 0.025 max Si 0.20-0.50 V 0.05 max	ASME SFA5.23 (EB6N) AWS A5.23 (EB6N)
S50300	Cr Heat Resisting Steel	C 0.15 max Cr 6.00-8.00 Mn 1.00 max Mo 0.45-0.65 P 0.040 max S 0.040 max Si 1.00 max	ASTM A182 (F7); A199 (T7); A200 (T7); A213 (T7); A335 (P7); A369 (FP7); A387; A473 (501A)
S50400	Cr Heat Resisting Steel	C 0.15 max Cr 8.00-10.00 Mn 1.00 max Mo 0.90-1.10 P 0.040 max S 0.040 max Si 1.00 max	ASTM A199 (T9); A200 (T9); A213 (T9); A276; A335 (P9); A387; A473 (501B)
S50480	Cr Heat Resisting Steel Bare Filler Metal	C 0.10 max Cr 8.00-10.50 Cu 0.75 max Mn 0.60 max Mo 0.80-1.20 Ni 0.50 max P 0.04 max S 0.03 max Si 0.50 max	ASME SFA5.9 (ER505); SFA5.23 (EB8) AWS A5.9 (ER505); A5.23 (EB8)
S50481	Cr Heat Resisting Steel Bare Filler Metal, Nuclear Grade	C 0.10 max Cr 8.00-10.5 Cu 0.08 max Mn 0.30-0.65 Mo 0.80-1.20 P 0.012 max S 0.030 max Si 0.05-0.50 V 0.05 max	ASME SFA5.23 (EB8N) AWS A5.23 (EB8N)
S63005	Valve Steel (CNS)	C 0.25-0.35 Cr 12.00-13.50 Mn 0.50 max Mo 0.50 max Ni 7.00-8.50 P 0.030 max S 0.030 max Si 2.00-3.00	AMS 5705 SAE J775 (HNV-5)
S63007	Valve Steel (21-55N)	C 0.15-0.25 Cr 20.0-22.0 Mn 5.00-6.50 N 0.20-0.35 Ni 4.50-6.00 P 0.04 max S 0.03 max Si 1.00 max	SAE J775 (EV-7)
S63008	Valve Steel (21-4N)	C 0.48-0.58 Cr 20.0-22.0 Mn 8.00-10.00 N 0.28-0.50 Ni 3.25-4.50 P 0.030 max S 0.040-0.09 Si 0.25 max	SAE J775 (EV-8)
S63011	Valve Steel (746)	C 0.65-0.75 Cr 20.50-22.00 Mn 5.50-6.90 N 0.18-0.28 Ni 1.40-1.90 P 0.040 max S 0.025-0.055 Si 0.45-0.85	SAE J775 (EV-11)
S63012	Valve Steel (21-2 N)	C 0.50-0.60 Cr 19.25-21.50 Mn 7.00-9.50 N 0.20-0.40 Ni 1.50-2.75 P 0.050 max S 0.040-0.090 Si 0.25 max	SAE J775 (EV-12)
S63013	Valve Steel (Gaman H)	C 0.45-0.55 Cr 20.00-22.00 Mn 11.00-13.00 N 0.40-0.50 P 0.03 max S 0.05 max Si 2.00-3.00 SAE J775 (EV-13)	
S63014	Valve Steel (10)	C 0.30-0.45 Cr 18.0-20.0 Mn 0.80-1.30 Ni 7.75-8.25 P 0.030 max S 0.030 max Si 2.75-3.25	SAE J775 (EV-5)
S63015	Valve Steel (10 N)	C 0.35-0.45 Cr 18.0-20.0 Mn 0.80-1.30 N 0.15-0.25 Ni 7.75-8.15 P 0.030 max S 0.030 max Si 2.75-3.25	SAE J775 (EV-6)
S63016	Valve Steel (21-12)	C 0.15-0.25 Cr 20.50-22.00 Mn 1.00-1.40 Ni 10.50-12.00 P 0.030 max S 0.030 max Si 0.70 max	SAE J775 (EV-3)
S63017	Valve Steel (21-12N)	C 0.15-0.25 Cr 20.0-22.0 Mn 1.00-1.50 N 0.15-0.25 Ni 10.50-12.50 P 0.030 max S 0.030 max Si 0.70-1.25	SAE J775 (EV-4)
S63197	Iron Base Superalloy, Welding Filler Metal (ER349)	C 0.07-0.13 Cb + Ta 1.00-1.40 Cr 19.0-21.5 Cu 0.75 max Mn 1.00-2.50 Mo 0.35-0.65 Ni 8.00-9.50 P 0.03 max S 0.03 max Si 0.30-0.65 Ti 0.10-0.30 W 1.25-1.75	ASME SFA5.9 (ER349) AWS A5.9 (ER349) MIL SPEC MIL-R-5031 (Class 6); MIL-E-19933 (MIL-349)
S63198	Iron Base Superalloy (19-9-DL)	C 0.28-0.35 Cb 0.25-0.60 Cr 18.0-21.0 Cu 0.50 max Mn 0.75-1.50 Mo 1.00-1.75 Ni 8.00-11.00 P 0.040 max S 0.030 max Si 0.30-0.80 Ti 0.10-0.35 W 1.00-1.75	AMS 5526; 5527; 5579; 5721 ASTM A453 (651); A457 (651); A458 (651); A477 (651) SAE J467 (19-9DL)

The chemical compositions listed are for identification purposes and should not be used in lieu of the cross referenced specifications.

UNIFIED NUMBER	DESCRIPTION	CHEMICAL COMPOSITION	CROSS REFERENCE SPECIFICATIONS
S63199	Iron Base Superalloy (19-9-DX or 19-9-W-Mo)	C 0.28-0.35 Cr 18.0-21.0 Cu 0.50 max Mn 0.75-1.50 Mo 1.25-2.00 Ni 8.00-11.00 P 0.040 max S 0.030 max Si 0.30-0.80 Ti 0.40-0.75 W 1.00-1.75	AMS 5782 SAE J467 (19-9DX)
S64005	Valve Steel (2)	C 0.50-0.60 Cr 7.50-8.50 Mn 0.20-0.60 Mo 0.60-0.90 P 0.020 max S 0.020 max Si 1.25-1.75	SAE J775 (HNV-1)
S64006	Valve Steel (F)	C 0.35-0.50 Cr 1.50-2.50 Mn 0.20-0.60 P 0.030 max S 0.030 max Si 3.50-4.50	SAE J775 (HNV-2)
S64152	High Strength Alloy Steel (M152)	C 0.08-0.15 Cr 11.00-12.50 Mn 0.50-0.90 Mo 1.50-2.00 N 0.01-0.05 Ni 2.00-3.00 P 0.025 max S 0.025 max Si 0.35 max V 0.25-0.40	AMS 5718; 5719 ASTM A565 (XM-32)
S64299	Iron Base Superalloy (29-9)	C 0.08-0.15 Cr 27.0-31.0 Cu 0.50 max Mn 1.00-2.00 Mo 0.50 max Ni 8.50-10.50 P 0.040 max S 0.030 max Si 0.75 max	AMS 5784
S65006	Valve Steel (XB)	C 0.75-0.90 Cr 19.0-21.0 Mn 0.80 max Ni 1.00-1.70 P 0.040 max S 0.040 max Si 1.75-2.60	AMS 5710 SAE J775 (HNV-6)
S65007	Valve Steel (1)	C 0.40-0.50 Cr 8.00-9.00 Mn 0.20-0.60 P 0.025 max S 0.025 max Si 3.00-3.50	SAE J775 (HNV-3)
S65150	Iron Base Superalloy (Pyromet X-15)	C 0.03 max Co 19.0-21.0 Cr 14.50-16.00 Mn 0.10 max Mo 2.50-3.00 Ni 0.20 max P 0.015 max S 0.015 max Si 0.10 max	AMS 5761
S65770	Iron Base Superalloy (AFC 77)	C 0.12-0.17 Co 13.0-14.0 Cr 13.50-14.50 Mn 0.30 max Mo 4.50-5.50 Ni 0.30-0.70 P 0.015 max S 0.015 max Si 0.25 max V 0.10-0.30	AMS 5748
S66009	Valve Steel (TPA)	C 0.35-0.50 Cr 12.0-15.0 Mn 1.00 max Mo 0.20-0.50 Ni 12.0-15.0 P 0.045 max S 0.030 max Si 0.30-0.80 W 1.50-3.00 Other if Mo not used, W 2.00-3.00	AMS 5700 SAE J775 (EV-9)
S66220	Iron Base Superalloy (Discaloy)	Al 0.35 max B 0.0010-0.010 C 0.08 max Cr 12.0-15.0 Cu 0.50 max Mn 1.50 max Mo 2.50-3.50 Ni 24.0-28.0 P 0.040 max S 0.030 max Si 1.00 max Ti 1.55-2.00	AMS 5733 ASTM A453 (662); A638 (662) SAE J467 (Discaloy)
S66286	Iron Base Superalloy (A286)	Al 0.35 max B 0.0010-0.010 C 0.08 max Cr 13.50-16.00 Mn 2.00 max Mo 1.00-1.50 Ni 24.0-27.0 P 0.040 max S 0.030 max Si 1.00 max Ti 1.90-2.35 V 0.10-0.50	AMS 5525; 5726; 5731; 5732; 5734; 5737; 5804; 5805; 5858; 5895; 7235 ASME SA638 (660) ASTM A453 (660); A638 (660) SAE J467 (A286)
S66545	Iron Base Superalloy (W545)	Al 0.25 max B 0.01-0.07 C 0.08 max Cr 12.0-15.0 Cu 0.25 max Mn 1.25-2.00 Mo 1.25-2.25 Ni 24.0-28.0 P 0.040 max S 0.030 max Si 0.10-0.80 Ti 2.70-3.30	AMS 5543 ASTM A453 (665) SAE J467 (W545)

The chemical compositions listed are for identification purposes and should not be used in lieu of the cross referenced specifications.

UNS NUMBERS ASSIGNED TO DATE

With Description of Each Material Covered and References To Documents In Which The Same or Similar Materials are Described

Txxxxx Number Series
Tool Steels, Wrought and Cast

TOOL STEELS, WROUGHT AND CAST

UNIFIED NUMBER	DESCRIPTION	CHEMICAL COMPOSITION	CROSS REFERENCE SPECIFICATIONS
T11301	High-Speed Tool Steel (M-1)	**C** 0.78-0.88 **Cr** 3.50-4.00 **Mn** 0.15-0.40 **Mo** 8.20-9.20 **P** 0.03 max **S** 0.03 max **Si** 0.20-0.50 **V** 1.00-1.35 **W** 1.40-2.10	**AISI** M-1 **ASTM** A600 (M-1) **SAE** J437 (M1); J438 (M1)
T11302	High-Speed Tool Steel (M-2)	**C** 0.78-1.05 **Cr** 3.75-4.50 **Mn** 0.15-0.40 **Mo** 4.50-5.50 **P** 0.03 max **S** 0.03 max **Si** 0.20-0.45 **V** 1.75-2.20 **W** 5.50-6.75	**AISI** M-2 **ASTM** A600 (M-2) **SAE** J437 (M2); J438 (M2)
T11304	High-Speed Tool Steel (M-4)	**C** 1.25-1.40 **Cr** 3.75-4.75 **Mn** 0.15-0.40 **Mo** 4.25-5.50 **P** 0.03 max **S** 0.03 max **Si** 0.20-0.45 **V** 3.75-4.50 **W** 5.25-6.50	**AISI** M-4 **ASTM** A600 (M-4) **SAE** J437 (M4); J438 (M4)
T11306	High-Speed Tool Steel (M-6)	**C** 0.75-0.85 **Co** 11.00-13.00 **Cr** 3.75-4.50 **Mn** 0.15-0.40 **Mo** 4.50-5.50 **P** 0.03 max **S** 0.03 max **Si** 0.20-0.45 **V** 1.30-1.70 **W** 3.75-4.75	**AISI** M-6 **ASTM** A600 (M-6)
T11307	High-Speed Tool Steel (M-7)	**C** 0.97-1.05 **Cr** 3.50-4.00 **Mn** 0.15-0.40 **Mo** 8.20-9.20 **P** 0.03 max **S** 0.03 max **Si** 0.20-0.55 **V** 1.75-2.25 **W** 1.40-2.10	**AISI** M-7 **ASTM** A600 (M-7)
T11310	High-Speed Tool Steel (M-10)	**C** 0.84-1.05 **Cr** 3.75-4.50 **Mn** 0.10-0.40 **Mo** 7.75-8.50 **P** 0.03 max **S** 0.03 max **Si** 0.20-0.45 **V** 1.80-2.20	**AISI** M-10 **ASTM** A600 (M-10)
T11313	High-Speed Tool Steel (M-3, Class 1)	**C** 1.00-1.10 **Cr** 3.75-4.50 **Mn** 0.15-0.40 **Mo** 4.75-6.50 **P** 0.03 max **S** 0.03 max **Si** 0.20-0.45 **V** 2.25-2.75 **W** 5.00-6.75	**AISI** M-3 (Cl.1) **ASTM** A600 (M-3, Cl.1) **SAE** J437 (M3); J438 (M3)
T11323	High-Speed Tool Steel (M-3, Class 2)	**C** 1.15-1.25 **Cr** 3.75-4.50 **Mn** 0.15-0.40 **Mo** 4.75-6.50 **P** 0.03 max **S** 0.03 max **Si** 0.20-0.45 **V** 2.75-3.25 **W** 5.00-6.75	**AISI** M-3 (Cl.2) **ASTM** A600 (M-3, Cl.2) **SAE** J437 (M3); J438 (M3)
T11330	High-Speed Tool Steel (M-30)	**C** 0.75-0.85 **Co** 4.50-5.50 **Cr** 3.50-4.25 **Mn** 0.15-0.40 **Mo** 7.75-9.00 **P** 0.03 max **S** 0.03 max **Si** 0.20-0.45 **V** 1.00-1.40 **W** 1.30-2.30	**AISI** M-30 **ASTM** A600 (M-30)
T11333	High-Speed Tool Steel (M-33)	**C** 0.85-0.92 **Co** 7.75-8.75 **Cr** 3.50-4.00 **Mn** 0.15-0.40 **Mo** 9.00-10.00 **P** 0.03 max **S** 0.03 max **Si** 0.25-0.55 **V** 1.00-1.35 **W** 1.30-2.10	**AISI** M-33 **ASTM** A600 (M-33)
T11334	High-Speed Tool Steel (M-34)	**C** 0.85-0.92 **Co** 7.75-8.75 **Cr** 3.50-4.00 **Mn** 0.15-0.40 **Mo** 7.75-9.20 **P** 0.03 max **S** 0.03 max **Si** 0.20-0.45 **V** 1.90-2.30 **W** 1.40-2.10	**AISI** M-34 **ASTM** A600 (M-34)
T11336	High-Speed Tool Steel (M-36)	**C** 0.80-0.90 **Co** 7.75-8.75 **Cr** 3.75-4.50 **Mn** 0.15-0.40 **Mo** 4.50-5.50 **P** 0.03 max **S** 0.03 max **Si** 0.20-0.45 **V** 1.75-2.25 **W** 5.50-6.50	**AISI** M-36 **ASTM** A600 (M-36)
T11341	High-Speed Tool Steel (M-41)	**C** 1.05-1.15 **Co** 4.75-5.75 **Cr** 3.75-4.50 **Mn** 0.20-0.60 **Mo** 3.25-4.25 **P** 0.03 max **S** 0.03 max **Si** 0.15-0.50 **V** 1.75-2.25 **W** 6.25-7.00	**AISI** M-41 **ASTM** A600 (M-41)
T11342	High-Speed Tool Steel (M-42)	**C** 1.05-1.15 **Co** 7.75-8.75 **Cr** 3.50-4.25 **Mn** 0.15-0.40 **Mo** 9.00-10.00 **P** 0.03 max **S** 0.03 max **Si** 0.15-0.65 **V** 0.95-1.35 **W** 1.15-1.85	**AISI** M-42 **ASTM** A600 (M-42)
* T11343	High-Speed Tool Steel (M-43)	**C** 1.15-1.25 **Co** 7.75-8.75 **Cr** 3.50-4.25 **Mn** 0.20-0.40 **Mo** 7.50-8.50 **P** 0.03 max **S** 0.03 max **Si** 0.15-0.65 **V** 1.50-1.75 **W** 2.25-3.00	**AISI** M-43 **ASTM** A600 (M-43)

*Boxed entries are no longer active and are retained for reference purposes only.

The chemical compositions listed are for identification purposes and should not be used in lieu of the cross referenced specifications.

UNIFIED NUMBER	DESCRIPTION	CHEMICAL COMPOSITION	CROSS REFERENCE SPECIFICATIONS
*T11344	High-Speed Tool Steel (M-44)	C 1.10-1.20 Co 11.00-12.25 Cr 4.00-4.75 Mn 0.20-0.40 Mo 6.00-7.00 P 0.03 max S 0.03 max Si 0.30-0.55 V 1.85-2.20 W 5.00-5.75	AISI M-44 ASTM A600 (M-44)
T11346	High-Speed Tool Steel (M-46)	C 1.22-1.30 Co 7.80-8.80 Cr 3.70-4.20 Mn 0.20-0.40 Mo 8.00-8.50 P 0.03 max S 0.03 max Si 0.40-0.65 V 3.00-3.30 W 1.90-2.20	AISI M-46 ASTM A600 (M-46)
*T11347	High-Speed Tool Steel (M-47)	C 1.05-1.15 Co 4.75-5.25 Cr 3.50-4.00 Mn 0.15-0.40 Mo 9.25-10.00 P 0.03 max S 0.03 max Si 0.20-0.45 V 1.15-1.35 W 1.30-1.80	AISI M-47 ASTM A600 (M-47)
T11348	High-Speed Tool Steel, Powder Metallurgy (M-48)	C 1.50 Co 9.00 Cr 3.75 Mo 5.25 V 3.10 W 10.0	AISI M-48
T11350	Bearing Steel	C 0.77-0.85 Co 0.25 max Cr 3.75-4.25 Cu 0.10 max Mn 0.35 max Mo 4.00-4.50 Ni 0.15 max P 0.015 max S 0.015 max Si 0.25 max V 0.90-1.10 W 0.25 max	AMS 6490; 6491
T11352	High-Speed Tool Steel, Powder Metallurgy (M-52)	C 0.90 Cr 4.00 Mo 4.00 V 2.00 W 1.25	AISI M-52
T11361	High-Speed Tool Steel, Powder Metallurgy (M-61)	C 1.60 Cr 4.00 Mo 6.50 V 5.00 W 12.0	AISI M-61
T11362	High-Speed Tool Steel, Powder Metallurgy (M-62)	C 1.30 Cr 3.75 Mo 10.5 V 2.00 W 6.25	AISI M-62
T12001	High-Speed Tool Steel (T-1)	C 0.65-0.80 Cr 3.75-4.50 Mn 0.10-0.40 P 0.03 max S 0.03 max Si 0.20-0.40 V 0.90-1.30 W 17.25-18.75	AISI T-1 AMS 5626 ASTM A600 (T-1) SAE J437 (T1); J438 (T1)
*T12002	High-Speed Tool Steel (T-2)	C 0.80-0.90 Cr 3.75-4.50 Mn 0.20-0.40 Mo 1.00 max P 0.03 max S 0.03 max Si 0.20-0.40 V 1.80-2.40 W 17.50-19.00	AISI T-2 ASTM A600 (T-2) SAE J437 (T2); J438 (T2)
T12004	High-Speed Tool Steel (T-4)	C 0.70-0.80 Co 4.25-5.75 Cr 3.75-4.50 Mn 0.10-0.40 Mo 0.40-1.00 P 0.03 max S 0.03 max Si 0.20-0.40 V 0.80-1.20 W 17.50-19.00	AISI T-4 ASTM A600 (T-4) SAE J437 (T4); J438 (T4)
T12005	High-Speed Tool Steel (T-5)	C 0.75-0.85 Co 7.00-9.50 Cr 3.75-5.00 Mn 0.20-0.40 Mo 0.50-1.25 P 0.03 max S 0.03 max Si 0.20-0.40 V 1.80-2.40 W 17.50-19.00	AISI T-5 ASTM A600 (T-5) SAE J437 (T5); J438 (T5)
T12006	High-Speed Tool Steel (T-6)	C 0.75-0.85 Co 11.00-13.00 Cr 4.00-4.75 Mn 0.20-0.40 Mo 0.40-1.00 P 0.03 max S 0.03 max Si 0.20-0.40 V 1.50-2.10 W 18.50-21.00	AISI T-6 ASTM A600 (T-6)
T12008	High-Speed Tool Steel (T-8)	C 0.75-0.85 Co 4.25-5.75 Cr 3.75-4.50 Mn 0.20-0.40 Mo 0.40-1.00 P 0.03 max S 0.03 max Si 0.20-0.40 V 1.80-2.40 W 13.25-14.75	AISI T-8 ASTM A600 (T-8) SAE J437 (T8); J438 (T8)
T12015	High-Speed Tool Steel (T-15)	C 1.50-1.60 Co 4.75-5.25 Cr 3.75-5.00 Mn 0.15-0.40 Mo 1.00 max P 0.03 max S 0.03 max Si 0.15-0.40 V 4.50-5.25 W 11.75-13.00	AISI T-15 ASTM A600 (T-15)

*Boxed entries are no longer active and are retained for reference purposes only.

The chemical compositions listed are for identification purposes and should not be used in lieu of the cross referenced specifications.

UNIFIED NUMBER	DESCRIPTION	CHEMICAL COMPOSITION	CROSS REFERENCE SPECIFICATIONS
T20810	Hot-Work Tool Steel (H-10)	C 0.35-0.45 Cr 3.00-3.75 Mn 0.25-0.70 Mo 2.00-3.00 P 0.030 max S 0.030 max Si 0.80-1.20 V 0.25-0.75	AISI H-10 ASTM A681 (H-10) FED QQ-T-570 (H-10)
T20811	Hot-Work Tool Steel (H-11)	C 0.33-0.43 Cr 4.75-5.50 Mn 0.20-0.50 Mo 1.10-1.60 P 0.030 max S 0.030 max Si 0.80-1.20 V 0.30-0.60	AISI H-11 AMS 6437; 6485; 6487; 6488 ASTM A579 (H-11); A681 (H-11) FED QQ-T-570 (H-11) MIL SPEC MIL-S-47086; MIL-S-47262 SAE J437 (H-11); J438 (H-11); J467 (H-11)
T20812	Hot-Work Tool Steel (H-12)	C 0.30-0.40 Cr 4.75-5.50 Mn 0.20-0.50 Mo 1.25-1.75 P 0.030 max S 0.030 max Si 0.80-1.20 V 0.50 max W 1.00-1.70	AISI H-12 ASTM A681 (H-12) FED QQ-T-570 (H-12) SAE J437 (H12); J438 (H12); J467 (H12)
T20813	Hot-Work Tool Steel (H-13)	C 0.32-0.45 Cr 4.75-5.50 Mn 0.20-0.50 Mo 1.10-1.75 P 0.030 max S 0.030 max Si 0.80-1.20 V 0.80-1.20	AISI H-13 ASTM A681 (H-13) FED QQ-T-570 (H-13) SAE J437 (H13); J438 (H13); J467 (H13)
T20814	Hot-Work Tool Steel (H-14)	C 0.35-0.45 Cr 4.75-5.50 Mn 0.20-0.50 P 0.030 max S 0.030 max Si 0.80-1.20 W 4.00-5.25	AISI H-14 ASTM A681 (H-14) FED QQ-T-570 (H-14)
T20819	Hot-Work Tool Steel (H-19)	C 0.32-0.45 Co 4.00-4.50 Cr 4.00-4.75 Mn 0.20-0.50 Mo 0.30-0.55 P 0.030 max S 0.030 max Si 0.20-0.50 V 1.75-2.20 W 3.75-4.50	AISI H-19 ASTM A681 (H-19) FED QQ-T-570 (H-19)
T20821	Hot-Work Tool Steel (H-21)	C 0.26-0.36 Cr 3.00-3.75 Mn 0.15-0.40 P 0.030 max S 0.030 max Si 0.15-0.50 V 0.30-0.60 W 8.50-10.00	AISI H-21 ASTM A681 (H-21) FED QQ-T-570 (H-21) SAE J437 (H21); J438 (H21)
T20822	Hot-Work Tool Steel (H-22)	C 0.30-0.40 Cr 1.75-3.75 Mn 0.15-0.40 P 0.030 max S 0.030 max Si 0.15-0.40 V 0.25-0.50 W 10.00-11.75	AISI H-22 ASTM A681 (H-22) FED QQ-T-570 (H-22)
T20823	Hot-Work Tool Steel (H-23)	C 0.25-0.35 Cr 11.00-12.75 Mn 0.15-0.40 P 0.030 max S 0.030 max Si 0.15-0.60 V 0.75-1.25 W 11.00-12.75	AISI H-23 ASTM A681 (H-23) FED QQ-T-570 (H-23)
T20824	Hot-Work Tool Steel (H-24)	C 0.42-0.53 Cr 2.50-3.50 Mn 0.15-0.40 P 0.030 max S 0.030 max Si 0.15-0.40 V 0.40-0.60 W 14.00-16.00	AISI H-24 ASTM A681 (H-24) FED QQ-T-570 (H-24)
* T20825	Hot-Work Tool Steel (H-25)	C 0.22-0.32 Cr 3.75-4.50 Mn 0.15-0.40 P 0.030 max S 0.030 max Si 0.15-0.40 V 0.40-0.60 W 14.00-16.00	AISI H-25 ASTM A681 (H-25) FED QQ-T-570 (H-25)
T20826	Hot-Work Tool Steel (H-26)	C 0.45-0.55 Cr 3.75-4.50 Mn 0.15-0.40 P 0.030 max S 0.030 max Si 0.15-0.40 V 0.75-1.25 W 17.25-19.00	AISI H-26 ASTM A681 (H-26) FED QQ-T-570 (H-26)
* T20841	Hot-Work Tool Steel (H-41)	C 0.60-0.75 Cr 3.50-4.00 Mn 0.15-0.40 Mo 8.20-9.20 P 0.030 max S 0.030 max Si 0.20-0.45 V 1.00-1.30 W 1.40-2.10	AISI H-41 ASTM A681 (H-41) FED QQ-T-570 (H-41)
T20842	Hot-Work Tool Steel (H-42)	C 0.55-0.70 Cr 3.75-4.50 Mn 0.15-0.40 Mo 4.50-5.50 P 0.030 max S 0.030 max Si 0.20-0.45 V 1.75-2.20 W 5.50-6.75	AISI H-42 ASTM A681 (H-42) FED QQ-T-570 (H-42)
* T20843	Hot-Work Tool Steel (H-43)	C 0.50-0.65 Cr 3.75-4.50 Mn 0.15-0.40 Mo 7.75-8.50 P 0.030 max S 0.030 max Si 0.20-0.45 V 1.80-2.20	AISI H-43 ASTM A681 (H-43) FED QQ-T-570 (H-43)
T30102	Cold-Work Tool Steel (A-2)	C 0.95-1.05 Cr 4.75-5.50 Mn 1.00 max Mo 0.90-1.40 P 0.030 max S 0.030 max Si 0.50 max V 0.15-0.50	AISI A-2 ASTM A681 (A-2) FED QQ-T-570 (A-2) SAE J437 (A2); J438 (A2)

*Boxed entries are no longer active and are retained for reference purposes only.

The chemical compositions listed are for identification purposes and should not be used in lieu of the cross referenced specifications.

UNIFIED NUMBER	DESCRIPTION	CHEMICAL COMPOSITION	CROSS REFERENCE SPECIFICATIONS
* T30103	Cold-Work Tool Steel (A-3)	C 1.20-1.30 Cr 4.75-5.50 Mn 0.40-0.60 Mo 0.90-1.40 P 0.030 max S 0.030 max Si 0.50 max V 0.80-1.40	AISI A-3 ASTM A681 (A-3) FED QQ-T-570 (A-3)
T30104	Cold-Work Tool Steel (A-4)	C 0.95-1.05 Cr 0.90-2.20 Mn 1.80-2.20 Mo 0.90-1.40 P 0.030 max S 0.030 max Si 0.50 max	AISI A-4 ASTM A681 (A-4) FED QQ-T-570 (A-4)
* T30105	Cold-Work Tool Steel (A-5)	C 0.95-1.05 Cr 0.90-1.20 Mn 2.80-3.20 Mo 0.90-1.40 P 0.030 max S 0.030 max Si 0.50 max	AISI A-5 ASTM A681 (A-5) FED QQ-T-570 (A-5)
T30106	Cold-Work Tool Steel (A-6)	C 0.65-0.75 Cr 0.90-1.20 Mn 1.80-2.50 Mo 0.90-1.40 P 0.030 max S 0.030 max Si 0.50 max	AISI A-6 ASTM A681 (A-6) FED QQ-T-570 (A-6)
T30107	Cold-Work Tool Steel (A-7)	C 2.00-2.85 Cr 5.00-5.75 Mn 0.80 max Mo 0.90-1.40 P 0.030 max S 0.030 max Si 0.50 max V 3.90-5.15 W 0.50-1.50	AISI A-7 ASTM A681 (A-7) FED QQ-T-570 (A-7)
T30108	Cold-Work Tool Steel (A-8)	C 0.50-0.60 Cr 4.75-5.50 Mn 0.50 max Mo 1.15-1.65 P 0.030 max S 0.030 max Si 0.75-1.10 W 1.00-1.50	AISI A-8 ASTM A681 (A-8) FED QQ-T-570 (A-8)
T30109	Cold-Work Tool Steel (A-9)	C 0.45-0.55 Cr 4.75-5.50 Mn 0.50 max Mo 1.30-1.80 Ni 1.25-1.75 P 0.030 max S 0.030 max Si 0.95-1.15 V 0.80-1.40	AISI A-9 ASTM A681 (A-9) FED QQ-T-570 (A-9)
T30110	Cold-Work Tool Steel (A-10)	C 1.25-1.50 Mn 1.60-2.10 Mo 1.25-1.75 Ni 1.55-2.05 P 0.030 max S 0.030 max Si 1.00-1.50	AISI A-10 ASTM A681 (A-10) FED QQ-T-570 (A-10)
T30111	Cold-Work Tool Steel (A-11)	C 2.45 Cr 5.25 Mn 0.50 Mo 1.30 Si 0.90 V 9.75	AISI A-11
T30402	Cold-Work Tool Steel (D-2)	C 1.40-1.60 Co 1.00 max Cr 11.00-13.00 Mn 0.60 max Mo 0.70-1.20 P 0.030 max S 0.030 max Si 0.60 max V 1.10 max	AISI D-2 ASTM A681 (D-2) FED QQ-T-570 (D-2) SAE J437 (D2); J438 (D2)
T30403	Cold-Work Tool Steel (D-3)	C 2.00-2.35 Cr 11.00-13.50 Mn 0.60 max P 0.030 max S 0.030 max Si 0.60 max V 1.00 max W 1.00 max	AISI D-3 ASTM A681 (D-3) FED QQ-T-570 (D-3) SAE J437 (D3); J438 (D3)
T30404	Cold-Work Tool Steel (D-4)	C 2.05-2.40 Cr 11.00-13.00 Mn 0.60 max Mo 0.70-1.20 P 0.030 max S 0.030 max Si 0.60 max V 1.00 max	AISI D-4 ASTM A681 (D-4) FED QQ-T-570 (D-4)
T30405	Cold-Work Tool Steel (D-5)	C 1.40-1.60 Co 2.50-3.50 Cr 11.0-13.00 Mn 0.60 max Mo 0.70-1.20 P 0.030 max S 0.030 max Si 0.60 max V 1.00 max	AISI D-5 ASTM A681 (D-5) FED QQ-T-570 (D-5) SAE J437 (D5); J438 (D5)
T30407	Cold-Work Tool Steel (D-7)	C 2.15-2.50 Cr 11.50-13.50 Mn 0.60 max Mo 0.70-1.20 P 0.030 max S 0.030 max Si 0.60 max V 3.80-4.40	AISI D-7 ASTM A681 (D-7) FED QQ-T-570 (D-7) SAE J437 (D7); J438 (D7)
T31501	Oil-Hardening Tool Steel (O-1)	C 0.85-1.00 Cr 0.40-0.60 Mn 1.00-1.40 P 0.030 max S 0.030 max Si 0.50 max V 0.30 max W 0.40-0.60	AISI O-1 ASTM A681 (O-1) FED QQ-T-570 (O-1) SAE J437 (O1); J438 (O1)
T31502	Oil-Hardening Tool Steel (O-2)	C 0.85-0.95 Cr 0.35 max Mn 1.40-1.80 Mo 0.30 max P 0.030 max S 0.030 max Si 0.50 max V 0.30 max	AISI O-2 ASTM A681 (O-2) FED QQ-T-570 (O-2) SAE J437 (O2); J438 (O2)
T31506	Oil-Hardening Tool Steel (O-6)	C 1.25-1.55 Cr 0.30 max Mn 0.30-1.10 Mo 0.20-0.30 P 0.030 max S 0.030 max Si 0.55-1.50	AISI O-6 ASTM A681 (O-6) FED QQ-T-570 (O-6) SAE J437 (06); J438 (06)

*Boxed entries are no longer active and are retained for reference purposes only.

The chemical compositions listed are for identification purposes and should not be used in lieu of the cross referenced specifications.

UNIFIED NUMBER	DESCRIPTION	CHEMICAL COMPOSITION	CROSS REFERENCE SPECIFICATIONS
T31507	Oil Hardening Tool Steel (O-7)	**C** 1.10-1.30 **Cr** 0.35-0.85 **Mn** 1.00 max **Mo** 0.30 max **P** 0.030 max **S** 0.030 max **Si** 0.60 max **V** 0.40 max **W** 1.00-2.00	**AISI** O-7 **ASTM** A681 (O-7) **FED** QQ-T-570 (O-7)
T41901	Shock Resisting Tool Steel (S-1)	**C** 0.40-0.55 **Cr** 1.00-1.80 **Mn** 0.10-0.40 **Mo** 0.50 max **P** 0.030 max **S** 0.030 max **Si** 0.15-1.20 **V** 0.15-0.30 **W** 1.50-3.00	**AISI** S-1 **ASTM** A681 (S-1) **FED** QQ-T-570 (S-1) **SAE** J437; J438 (S-1)
T41902	Shock-Resisting Tool Steel (S-2)	**C** 0.40-0.55 **Mn** 0.30-0.50 **Mo** 0.30-0.60 **P** 0.030 max **S** 0.030 max **Si** 0.90-1.20 **V** 0.50 max	**AISI** S-2 **ASTM** A681 (S-2) **FED** QQ-T-570 (S-2) **SAE** J437 (S2); J438 (S2)
T41904	Shock-Resisting Tool Steel (S-4)	**C** 0.50-0.65 **Cr** 0.35 max **Mn** 0.60-0.95 **P** 0.030 max **S** 0.030 max **Si** 1.75-2.25 **V** 0.35 max	**AISI** S-4 **ASTM** A681 (S-4) **FED** QQ-T-570 (S-4)
T41905	Shock-Resisting Tool Steel (S-5)	**C** 0.50-0.65 **Cr** 0.35 max **Mn** 0.60-1.00 **Mo** 0.20-1.35 **P** 0.030 max **S** 0.030 max **Si** 1.75-2.25 **V** 0.35 max	**AISI** S-5 **ASTM** A681 (S-5) **FED** QQ-T-570 (S-5) **SAE** J437 (S5); J438 (S5)
T41906	Shock-Resisting Tool Steel (S-6)	**C** 0.40-0.50 **Cr** 1.20-1.50 **Mn** 1.20-1.50 **Mo** 0.30-0.50 **P** 0.030 max **S** 0.030 max **Si** 2.00-2.50 **V** 0.20-0.40	**AISI** S-6 **ASTM** A681 (S-6) **FED** QQ-T-570 (S-6)
T41907	Shock-Resisting Tool Steel (S-7)	**C** 0.45-0.55 **Cr** 3.00-3.50 **Mn** 0.20-0.80 **Mo** 1.30-1.80 **P** 0.030 max **S** 0.030 max **Si** 0.20-1.00 **Other** V 0.20-0.30 (Optional)	**AISI** S7 **ASTM** A681 (S-7) **FED** QQ-T-570 (S-7)
* T51602	Mold Steel (P-2)	**C** 0.10 max **Cr** 0.75-1.25 **Mn** 0.10-0.40 **Mo** 0.15-0.40 **Ni** 0.10-0.50 **P** 0.030 max **S** 0.030 max **Si** 0.10-0.40	**AISI** P-2 **ASTM** A681 (P-2)
* T51603	Mold Steel (P-3)	**C** 0.10 max **Cr** 0.40-0.75 **Mn** 0.20-0.60 **Ni** 1.00-1.50 **P** 0.030 max **S** 0.030 max **Si** 0.40 max	**AISI** P-3 **ASTM** A681 (P-3)
* T51604	Mold Steel (P-4)	**C** 0.12 max **Cr** 4.00-5.25 **Mn** 0.20-0.60 **Mo** 0.40-1.00 **P** 0.030 max **S** 0.030 max **Si** 0.10-0.40	**AISI** P-4 **ASTM** A681 (P-4)
* T51605	Mold Steel (P-5)	**C** 0.10 max **Cr** 2.00-2.50 **Mn** 0.20-0.60 **Ni** 0.35 max **P** 0.030 max **S** 0.030 max **Si** 0.40 max	**AISI** P-5 **ASTM** A681 (P-5)
T51606	Mold Steel (P-6)	**C** 0.05-0.15 **Cr** 1.25-1.75 **Mn** 0.35-0.70 **Ni** 3.25-3.75 **P** 0.030 max **S** 0.030 max **Si** 0.10-0.40	**AISI** P-6 **ASTM** A681 (P-6)
T51620	Mold Steel (P-20)	**C** 0.28-0.40 **Cr** 1.40-2.00 **Mn** 0.60-1.00 **Mo** 0.30-0.55 **P** 0.030 max **S** 0.030 max **Si** 0.20-0.80	**AISI** P-20 **ASTM** A681 (P-20)
T51621	Mold Steel (P-21)	**Al** 1.05-1.25 **C** 0.18-0.22 **Cr** 0.20-0.30 **Mn** 0.20-0.40 **Ni** 4.00-4.25 **P** 0.030 max **S** 0.030 max **Si** 0.20-0.40 **V** 0.15-0.25	**AISI** P-21 **ASTM** A681 (P-21)
* T60601	C-W Tool Steel (F-1)	**C** 0.95-1.25 **Mn** 0.50 max **P** 0.030 max **S** 0.030 max **Si** 0.50 max **W** 1.00-1.75	**AISI** F-1 **ASTM** A681 (F-1) **FED** QQ-T-570 (F-1)

*Boxed entries are no longer active and are retained for reference purposes only.

The chemical compositions listed are for identification purposes and should not be used in lieu of the cross referenced specifications.

UNIFIED NUMBER	DESCRIPTION	CHEMICAL COMPOSITION	CROSS REFERENCE SPECIFICATIONS
* T60602	C-W Tool Steel (F-2)	C 1.20-1.40 Cr 0.20-0.40 Mn 0.50 max P 0.030 max S 0.030 max Si 0.50 max W 3.00-4.50	AISI F-2 ASTM A681 (F-2) FED QQ-T-570 (F-2)
T61202	Low-Alloy Tool Steel (L-2)	C 0.45-1.00 Cr 0.70-1.20 Mn 0.10-0.90 Mo 0.25 max P 0.030 max S 0.030 max Si 0.50 max V 0.10-0.30	AISI L-2 ASTM A681 (L-2) FED QQ-T-570 (L-2)
* T61203	Low-Alloy Tool Steel (L-3)	C 0.95-1.10 Cr 1.30-1.70 Mn 0.25-0.50 P 0.030 max S 0.030 max Si 0.50 max V 0.10-0.30	AISI L-3 ASTM A681 (L-3) FED QQ-T-570 (L-3)
T61206	Low-Alloy Tool Steel (L-6)	C 0.65-0.75 Cr 0.60-1.20 Mn 0.25-0.80 Mo 0.50 max Ni 1.25-2.00 P 0.030 max S 0.030 max Si 0.50 max	AISI L-6 ASTM A681 (L-6) FED QQ-T-570 (L-6) SAE J437 (L6); J438 (L6)
T72301	Water Hardening Tool Steel (W-1)	C 0.70-1.50 Cr 0.15 max Cu 0.20 max Mn 0.10-0.40 Mo 0.10 max Ni 0.20 max P 0.025 max S 0.025 max Si 0.10-0.40 V 0.10 max W 0.15 max	AISI W-1 ASTM A686 (W-1) SAE J437 (W108), (W109), (W110), (W112); J438 (W108), (W109), (W110), (W112)
T72302	Water Hardening Tool Steel (W-2)	C 0.85-1.50 Cr 0.15 max Cu 0.20 max Mn 0.10-0.40 Mo 0.10 max Ni 0.20 max P 0.030 max S 0.030 max Si 0.10-0.40 V 0.15-0.35 W 0.15 max	AISI W-2 ASTM A686 (W-2) SAE J437 (W209), (W210); J438 (W209), (W210)
T72305	Water-Hardening Tool Steel (W-5)	C 1.05-1.15 Cr 0.40-0.60 Cu 0.20 max Mn 0.10-0.40 Mo 0.10 max Ni 0.20 max P 0.030 max S 0.030 max Si 0.10-0.40 V 0.10 max W 0.15 max	AISI W-5 ASTM A686 (W-5)
T90102	Cast Cold Work Tool Steel (CA-2)	C 0.95-1.05 Cr 4.75-5.50 Mn 0.75 max Mo 0.90-1.40 P 0.030 max S 0.030 max Si 1.50 max Other vanadium optional; if present, 0.20-0.50	ASTM A597 (CA-2)
T90402	Cast Cold Work Tool Steel (CD-2)	C 1.40-1.60 Cr 11.00-13.00 Mn 1.00 max Mo 0.70-1.20 P 0.030 max S 0.030 max Si 1.50 max Other cobalt optional; if present 0.70-1.00. vanadium optional; if present, 0.40-1.00	ASTM A597 (CD-2)
T90405	Cast Cold Work Tool Steel (CD-5)	C 1.35-1.60 Co 2.50-3.50 Cr 11.00-13.00 Mn 0.75 max Mo 0.70-1.20 P 0.030 max S 0.030 max Si 1.50 max V 0.35-0.55 Other nickel optional; if present 0.40-0.60	ASTM A597 (CD-5)
T90812	Cast Hot Work Tool Steel (CH-12)	C 0.30-0.40 Co 0.20-0.50 Cr 4.75-5.75 Mn 0.75 max Mo 1.25-1.75 P 0.030 max S 0.030 max Si 1.50 max V 0.20-0.50 W 1.00-1.70	ASTM A597 (CH-12)
T90813	Cast Hot Work Tool Steel (CH-13)	C 0.30-0.42 Cr 4.75-5.75 Mn 0.75 max Mo 1.25-1.75 P 0.030 max S 0.030 max Si 1.50 max V 0.75-1.20	ASTM A597 (CH-13)
T91501	Cast Oil Hardening Tool Steel (CO-1)	C 0.85-1.00 Cr 0.40-1.00 Mn 1.00-1.30 P 0.030 max S 0.030 max Si 1.50 max V 0.30 max W 0.40-0.60	ASTM A597 (CO-1)
T91905	Cast Shock Resisting Tool Steel (CS-5)	C 0.50-0.65 Cr 0.35 max Mn 0.60-1.00 Mo 0.20-0.80 P 0.030 max S 0.030 max Si 1.75-2.25 V 0.35 max	ASTM A597 (CS-5)

*Boxed entries are no longer active and are retained for reference purposes only.

The chemical compositions listed are for identification purposes and should not be used in lieu of the cross referenced specifications.

UNS NUMBERS ASSIGNED TO DATE

With Description of Each Material Covered and References To Documents In Which The Same or Similar Materials are Described

**Wxxxxx Number Series
Welding Filler Metals**

WELDING FILLER METALS

UNIFIED NUMBER	DESCRIPTION	CHEMICAL COMPOSITION	CROSS REFERENCE SPECIFICATIONS
W06010	Mild Steel Arc Welding Electrode (E6010)		**ASME** SFA5.1 (E6010) **AWS** A5.1 (E6010)
W06011	Mild Steel Arc Welding Electrode (E6011)		**ASME** SFA5.1 (E6011) **AWS** A5.1 (E6011)
W06012	Mild Steel Arc Welding Electrode (E6012)		**ASME** SFA5.1 (E6012) **AWS** A5.1 (E6012)
W06013	Mild Steel Arc Welding Electrode (E6013)		**AMS** 5031 **ASME** SFA5.1 (E6013) **AWS** A5.1 (E6013) **FED** QQ-E-450 (6013)
W06020	Mild Steel Arc Welding Electrode (E6020)		**ASME** SFA5.1 (E6020) **AWS** A5.1 (E6020)
W06022	Mild Steel Arc Welding Electrode (E6022)		**ASME** SFA5.1 (E6022) **AWS** A5.1 (E6022)
W06027	Mild Steel Arc Welding Electrode (E6027)		**ASME** SFA5.1 (E6027) **AWS** A5.1 (E6027)
W06040	Mild Steel Composite Wire for Electroslag Welding (EWT1)	**C** 0.13 max **Mn** 2.00 max **P** 0.03 max **S** 0.03 max **Si** 0.60 max	**ASME** SFA5.25 (EWT1) **AWS** A5.25 (EWT1)
W06301	Mild Steel Flux Cored Wire for Electrogas Welding (EG6xT1)	**Cr** 0.20 max **Mn** 1.50 max **Mo** 0.35 max **Ni** 0.30 max **P** 0.03 max **S** 0.03 max **Si** 0.50 max **V** 0.08 max	**ASME** SFA5.26 (EG6xT1) **AWS** A5.26 (EG6xT1)
W06302	Mild Steel Flux Cored Wire for Electrogas Welding (EG6xT2)	**Cr** 0.20 max **Mn** 2.00 max **Mo** 0.35 max **Ni** 0.30 max **P** 0.03 max **S** 0.04 max **Si** 0.90 max **V** 0.08 max	**ASME** SFA5.26 (EG6xT2) **AWS** A5.26 (EG6xT2)
W06601	Mild Steel Flux Cored Wire (E6xT-1)	**Al** 1.8 max **Cr** 0.20 max **Mn** 1.75 max **Mo** 0.30 max **Ni** 0.50 max **P** 0.04 max **S** 0.03 max **Si** 0.90 max **V** 0.08 max	**ASME** SFA5.20 (E6xT-1) **AWS** A5.20 (E6xT-1)
W06604	Mild Steel Flux Cored Wire (E6xT-4)	**Al** 1.8 max **Cr** 0.20 max **Mn** 1.75 max **Mo** 0.30 max **Ni** 0.50 max **P** 0.04 max **S** 0.03 max **Si** 0.90 max **V** 0.08 max	**ASME** SFA5.20 (E6xT-4) **AWS** A5.20 (E6xT-4)
W06605	Mild Steel Flux Cored Wire (E6xT-5)	**Al** 1.8 max **Cr** 0.20 max **Mn** 1.75 max **Mo** 0.30 max **Ni** 0.50 max **P** 0.04 max **S** 0.03 max **Si** 0.90 max **V** 0.08 max	**ASME** SFA5.20 (E6xT-5) **AWS** A5.20 (E6xT-5)
W06606	Mild Steel Flux Cored Wire (E6xT-6)	**Al** 1.8 max **Cr** 0.20 max **Mn** 1.75 max **Mo** 0.30 max **Ni** 0.50 max **P** 0.04 max **S** 0.03 max **Si** 0.90 max **V** 0.08 max	**ASME** SFA5.20 (E6xT-6) **AWS** A5.20 (E6xT-6)
W06607	Mild Steel Flux Cored Wire (E6xT-7)	**Al** 1.8 max **Cr** 0.20 max **Mn** 1.75 max **Mo** 0.30 max **Ni** 0.50 max **P** 0.04 max **S** 0.03 max **Si** 0.90 max **V** 0.08 max	**ASME** SFA5.20 (E6xT-7) **AWS** A5.20 (E6xT-7)
W06608	Mild Steel Flux Cored Wire (E6xT-8)	**Al** 1.8 max **Cr** 0.20 max **Mn** 1.75 max **Mo** 0.30 max **Ni** 0.50 max **P** 0.04 max **S** 0.03 max **Si** 0.90 max **V** 0.08 max	**ASME** SFA5.20 (E6xT-8) **AWS** A5.20 (E6xT-8)
W06611	Mild Steel Flux Cored Wire (E6xT-11)	**Al** 1.8 max **Cr** 0.20 max **Mn** 1.75 max **Mo** 0.30 max **Ni** 0.50 max **P** 0.04 max **S** 0.03 max **Si** 0.90 max **V** 0.08 max	**ASME** SFA5.20 (E6xT-11) **AWS** A5.20 (E6xT-11)
W07014	Mild Steel Arc Welding Electrode (E7014)	**Cr** 0.20 max **Mn** 1.25 max **Mo** 0.30 max **Ni** 0.30 max **Si** 0.90 max **V** 0.08 max	**ASME** SFA5.1 (E7014) **AWS** A5.1 (E7014)
W07015	Mild Steel Arc Welding Electrode (E7015)	**Cr** 0.20 max **Mn** 1.25 max **Mo** 0.30 max **Ni** 0.30 max **Si** 0.90 max **V** 0.08 max	**ASME** SFA5.1 (E7015) **AWS** A5.1 (E7015) **MIL SPEC** MIL-E-22200/6 (MIL-7015)
W07016	Mild Steel Arc Welding Electrode (E7016)	**Cr** 0.20 max **Mn** 1.60 max **Mo** 0.30 max **Ni** 0.30 max **Si** 0.75 max **V** 0.08 max	**ASME** SFA5.1 (7016) **AWS** A5.1 (7016) **MIL SPEC** MIL-E-22200/6 (MIL-7016)
W07018	Mild Steel Arc Welding Electrode (E7018)	**Cr** 0.20 max **Mn** 1.60 max **Mo** 0.30 max **Ni** 0.30 max **Si** 0.75 max **V** 0.08 max	**ASME** SFA5.1 (7018) **AWS** A5.1 (7018) **MIL SPEC** MIL-E-22200 (MIL-7018)

The chemical compositions listed are for identification purposes and should not be used in lieu of the cross referenced specifications.

UNIFIED NUMBER	DESCRIPTION	CHEMICAL COMPOSITION	CROSS REFERENCE SPECIFICATIONS
W07024	Mild Steel Arc Welding Electrode (E7024)	**Cr** 0.20 max **Mn** 1.25 max **Mo** 0.30 max **Ni** 0.30 max **Si** 0.90 max **V** 0.08 max	**ASME** SFA5.1 (E7024) **AWS** A5.1 (E7024)
W07027	Mild Steel Arc Welding Electrode (E7027)	**Cr** 0.20 max **Mn** 1.60 max **Mo** 0.30 max **Ni** 0.30 max **Si** 0.75 max **V** 0.08 max	**ASME** SFA5.1 (E7027) **AWS** A5.1 (E7027)
W07028	Mild Steel Arc Welding Electrode (E7028)	**Cr** 0.20 max **Mn** 1.25 max **Mo** 0.30 max **Ni** 0.30 max **Si** 0.90 max **V** 0.08 max	**ASME** SFA5.1 (E7028) **AWS** A5.1 (E7028)
W07048	Mild Steel Arc Welding Electrode (E7048)	**Cr** 0.20 max **Mn** 1.25 max **Mo** 0.30 max **Ni** 0.30 max **Si** 0.90 max **V** 0.08 max	**ASME** SFA5.1 (E7048) **AWS** A5.1 (E7048)
W07116	Mild Steel Arc Welding Electrode (E7016-1)	**Cr** 0.20 max **Mn** 1.60 max **Mo** 0.30 max **Ni** 0.30 max **Si** 0.75 max **V** 0.08 max	**ASME** SFA5.1 (E7016-1) **AWS** A5.1 (E7016-1)
W07118	Mild Steel Arc Welding Electrode (E7018-1)	**Cr** 0.20 max **Mn** 1.60 max **Mo** 0.30 max **Ni** 0.30 max **Si** 0.75 max **V** 0.08 max	**ASME** SFA5.1 (E7018-1) **AWS** A5.1 (E7018-1)
W07124	Mild Steel Arc Welding Electrode (E7024-1)	**Cr** 0.20 max **Mn** 1.25 max **Mo** 0.30 max **Ni** 0.30 max **Si** 0.90 max **V** 0.08 max	**ASME** SFA5.1 (E7024-1) **AWS** A5.1 (E7024-1)
W07301	Mild Steel Flux Cored Wire for Electrogas Welding (EG7xT1)	**Cr** 0.20 max **Mn** 1.50 max **Mo** 0.35 max **Ni** 0.30 max **P** 0.03 max **S** 0.03 max **Si** 0.50 max **V** 0.08 max	**ASME** SFA5.26 (EG7xT1) **AWS** A5.26 (EG7xT1)
W07302	Mild Steel Flux Cored Wire for Electrogas Welding (EG7xT2)	**Cr** 0.20 max **Mn** 2.00 max **Mo** 0.35 max **Ni** 0.30 max **P** 0.03 max **S** 0.04 max **Si** 0.90 max **V** 0.08 max	**ASME** SFA5.26 (EG7xT2) **AWS** A5.26 (EG7xT2)
W07601	Mild Steel Flux Cored Wire (E7xT-1)	**Al** 1.8 max **Cr** 0.20 max **Mn** 1.75 max **Mo** 0.30 max **Ni** 0.50 max **P** 0.04 max **S** 0.03 max **Si** 0.90 max **V** 0.08 max	**ASME** SFA5.20 (E7xT-1) **AWS** A5.20 (E7xT-1)
W07602	Mild Steel Flux Cored Wire (E7xT-2)		**ASME** SFA5.20 (E7xT-2) **AWS** A5.20 (E7xT-2)
W07603	Carbon Steel Bare Electrode (or Rod) for Flux Cored Arc Welding (E7xT-3)		**ASME** SFA5.20 (E7xT-3) **AWS** A5.20 (E7xT-3)
W07604	Mild Steel Flux Cored Wire (E7xT-4)	**Al** 1.8 max **Cr** 0.20 max **Mn** 1.75 max **Mo** 0.30 max **Ni** 0.50 max **P** 0.04 max **S** 0.03 max **Si** 0.90 max **V** 0.08 max	**ASME** SFA5.20 (E7xT-4) **AWS** A5.20 (E7xT-4)
W07605	Mild Steel Flux Cored Wire (E7xT-5)	**Al** 1.8 max **Cr** 0.20 max **Mn** 1.75 max **Mo** 0.30 max **Ni** 0.50 max **P** 0.04 max **S** 0.03 max **Si** 0.90 max **V** 0.08 max	**ASME** SFA5.20 (E7xT-5) **AWS** A5.20 (E7xT-5) **MIL SPEC** MIL-E-24403/1 (MIL-70T-5)
W07606	Mild Steel Flux Cored Wire (E7xT-6)	**Al** 1.8 max **Cr** 0.20 max **Mn** 1.75 max **Mo** 0.30 max **Ni** 0.50 max **P** 0.04 max **S** 0.03 max **Si** 0.90 max **V** 0.08 max	**ASME** SFA5.20 (E7xT-6) **AWS** A5.20 (E7xT-6)
W07607	Mild Steel Flux Cored Wire (E7xT-7)	**Al** 1.8 max **Cr** 0.20 max **Mn** 1.75 max **Mo** 0.30 max **Ni** 0.50 max **P** 0.04 max **S** 0.03 max **Si** 0.90 max **V** 0.08 max	**ASME** SFA5.20 (E7xT-7) **AWS** A5.20 (E7xT-7)
W07608	Mild Steel Flux Cored Wire (E7xT-8)	**Al** 1.8 max **Cr** 0.20 max **Mn** 1.75 max **Mo** 0.30 max **Ni** 0.50 max **P** 0.04 max **S** 0.03 max **Si** 0.90 max **V** 0.08 max	**ASME** SFA5.20 (E7xT-8) **AWS** A5.20 (E7xT-8)
W07610	Mild Steel Flux Cored Wire (E7xT-10)		**ASME** SFA5.20 (E7xT-10) **AWS** A5.20 (E7xT-10)
W07611	Mild Steel Flux Cored Wire (E7xT-11)	**Al** 1.8 max **Cr** 0.20 max **Mn** 1.75 max **Mo** 0.30 max **Ni** 0.50 max **P** 0.04 max **S** 0.03 max **Si** 0.90 max **V** 0.08 max	**ASME** SFA5.20 (E7xT-11) **AWS** A5.20 (E7xT-11)

The chemical compositions listed are for identification purposes and should not be used in lieu of the cross referenced specifications.

UNIFIED NUMBER	DESCRIPTION	CHEMICAL COMPOSITION	CROSS REFERENCE SPECIFICATIONS
W10013	Low Alloy Steel Arc Welding Electrode (10013)	C 0.06-0.12 **Cr** 0.20 max **Cu** 0.35 max **Mn** 0.35-0.70 **Mo** 0.90-1.20 **Ni** 0.25 max **P** 0.025 max **S** 0.025 max **Si** 0.30-0.60 **V** 0.10-0.30	**AMS** 6464 **MIL SPEC** MIL-E-6843 (10013)
W10015	Low Alloy Steel Arc Welding Electrode (E10015-D2)	C 0.15 max **Mn** 1.65-2.00 **Mo** 0.25-0.45 **P** 0.03 max **S** 0.04 max **Si** 0.60 max	**ASME** SFA5.5 (E10015-D2) **AWS** A5.5 (E10015-D2)
W10016	Low Alloy Steel Arc Welding Electrode (E10016-D2)	C 0.15 max **Mn** 1.65-2.00 **Mo** 0.25-0.45 **P** 0.03 max **S** 0.04 max **Si** 0.60 max	**ASME** SFA5.5 (E10016-D2) **AWS** A5.5 (E10016-D2)
W10018	Low Alloy Steel Arc Welding Electrode (E10018-D2)	C 0.15 max **Mn** 1.65-2.00 **Mo** 0.25-0.45 **P** 0.03 max **S** 0.04 max **Si** 0.80 max	**ASME** SFA5.5 (E10018-D2) **AWS** A5.5 (E10018-D2)
W10235	Low Alloy Steel Flux Cored Wire (E100T5-D2)	C 0.15 max **Mn** 1.65-2.25 **Mo** 0.25-0.55 **P** 0.03 max **S** 0.03 max **Si** 0.80 max	**ASME** SFA5.29 (E100T5-D2) **AWS** A5.29 (E100T5-D2)
W17010	Low Alloy Steel Arc Welding Electrode (E7010-A1)	C 0.12 max **Mn** 0.60 max **Mo** 0.40-0.65 **P** 0.03 max **S** 0.04 max **Si** 0.40 max	**ASME** SFA5.5 (E7010-A1) **AWS** A5.5 (E7010-A1) **MIL SPEC** MIL-E-22200/7 (MIL-7010-A1)
W17011	Low Alloy Steel Arc Welding Electrode (E7011-A1)	C 0.12 max **Mn** 0.60 max **Mo** 0.40-0.65 **P** 0.03 max **S** 0.04 max **Si** 0.40 max	**ASME** SFA5.5 (E7011-A1) **AWS** A5.5 (E7011-A1) **MIL SPEC** MIL-E-22200/7 (MIL-7011-A1)
W17015	Low Alloy Steel Arc Welding Electrode (E7015-A1)	C 0.12 max **Mn** 0.90 max **Mo** 0.40-0.65 **P** 0.03 max **S** 0.04 max **Si** 0.60 max	**ASME** SFA5.5 (E7015-A1) **AWS** A5.5 (E7015-A1)
W17016	Low Alloy Steel Arc Welding Electrode (E7016-A1)	C 0.12 **Mn** 0.90 **Mo** 0.40-0.65 **P** 0.03 **S** 0.04 **Si** 0.60	**ASME** SFA5.5 (E7016-A1) **AWS** A5.5 (E7016-A1)
W17018	Low Alloy Steel Arc Welding Electrode (E7018-A1)	C 0.12 max **Mn** 0.90 max **Mo** 0.40-0.65 **P** 0.03 max **S** 0.04 max **Si** 0.80 max	**ASME** SFA5.5 (E7018-A1) **AWS** A5.5 (E7018-A1) **MIL SPEC** MIL-E-22200/7 (MIL-7018-A1)
W17020	Low Alloy Steel Arc Welding Electrode (E7020-A1)	C 0.12 max **Mn** 0.60 max **Mo** 0.40-0.65 **P** 0.03 max **S** 0.04 max **Si** 0.40 max	**ASME** SFA5.5 (E7020-A1) **AWS** A5.5 (E7020-A1) **MIL SPEC** MIL-E-22200/7 (MIL-7020-A1)
W17027	Low Alloy Steel Arc Welding Electrode (E7027-A1)	C 0.12 max **Mn** 1.00 max **Mo** 0.40-0.65 **P** 0.03 max **S** 0.04 max **Si** 0.40 max	**ASME** SFA5.5 (E7027-A1) **AWS** A5.5 (E7027-A1)
W17031	Low Alloy Steel Flux Cored Wire (E8xT1-A1)	C 0.12 max **Mn** 1.25 max **Mo** 0.40-0.65 **P** 0.03 max **S** 0.03 max **Si** 0.80 max	**ASME** SFA5.29 (E8xT1-A1) **AWS** A5.29 (E8xT1-A1)
W17035	Low Alloy Steel Flux Cored Wire (E70T5-A1)	C 0.12 max **Mn** 1.25 max **Mo** 0.40-0.65 **P** 0.03 max **S** 0.03 max **Si** 0.80 max	**ASME** SFA5.29 (E70T5-A1) **AWS** A5.29 (E70T5-A1)
W17041	Low Alloy Steel Composite Wire for Submerged Arc Welding (ECA1)	C 0.12 max **Cu** 0.35 max **Mn** 1.00 max **Mo** 0.40-0.65 **P** 0.030 max **S** 0.040 max **Si** 0.80 max	**ASME** SFA5.23 (ECA1) **AWS** A5.23 (ECA1)
W17042	Low Alloy Steel Composite Wire for Submerged Arc Welding (ECA2)	C 0.12 max **Cu** 0.35 max **Mn** 1.40 max **Mo** 0.40-0.65 **P** 0.030 max **S** 0.040 max **Si** 0.80 max	**ASME** SFA5.23 (ECA2) **AWS** A5.23 (ECA2)
W17043	Low Alloy Steel Composite Wire for Submerged Arc Welding (ECA3)	C 0.15 max **Cu** 0.35 max **Mn** 2.10 max **Mo** 0.40-0.65 **P** 0.030 max **S** 0.040 max **Si** 0.80 max	**ASME** SFA5.23 (ECA3) **AWS** A5.23 (ECA3)
W17044	Low Alloy Steel Composite Wire for Submerged Arc Welding (ECA4)	C 0.15 max **Cu** 0.35 max **Mn** 1.60 max **Mo** 0.40-0.65 **P** 0.030 max **S** 0.040 max **Si** 0.80 max	**ASME** SFA5.23 (ECA4) **AWS** A5.23 (ECA4)

The chemical compositions listed are for identification purposes and should not be used in lieu of the cross referenced specifications.

UNIFIED NUMBER	DESCRIPTION	CHEMICAL COMPOSITION	CROSS REFERENCE SPECIFICATIONS
W17141	Low Alloy Steel Composite Wire for Submerged Arc Welding (ECA1N) Nuclear Grade	**C** 0.12 max **Cu** 0.08 max **Mn** 1.00 max **Mo** 0.40-0.65 **P** 0.012 max **S** 0.040 max **Si** 0.80 max **V** 0.05 max	**ASME** SFA5.23 (ECA1N) **AWS** A5.23 (ECA1N)
W17142	Low Alloy Steel Composite Wire for Submerged Arc Welding (ECA2N) Nuclear Grade	**C** 0.12 max **Cu** 0.08 max **Mn** 1.40 max **Mo** 0.40-0.65 **P** 0.012 max **S** 0.040 max **Si** 0.80 max **V** 0.05 max	**ASME** SFA5.23 (ECA2N) **AWS** A5.23 (ECA2N)
W17143	Low Alloy Steel Composite Wire for Submerged Arc Welding (ECA3N) Nuclear Grade	**C** 0.15 max **Cu** 0.08 max **Mn** 2.10 max **Mo** 0.40-0.65 **P** 0.012 max **S** 0.040 max **Si** 0.80 max **V** 0.05 max	**ASME** SFA5.23 (ECA3N) **AWS** A5.23 (ECA3N)
W17144	Low Alloy Steel Composite Wire for Submerged Arc Welding (ECA4N) Nuclear Grade	**C** 0.15 max **Cu** 0.08 max **Mn** 1.60 max **Mo** 0.40-0.65 **P** 0.012 max **S** 0.040 max **Si** 0.80 max **V** 0.05 max	**ASME** SFA5.23 (ECA4N) **AWS** A5.23 (ECA4N)
W18016	Low Alloy Steel Arc Welding Electrode (E8016-D3)	**C** 0.12 max **Mn** 1.00-1.75 **Mo** 0.40-0.65 **P** 0.03 max **S** 0.04 max **Si** 0.60 max	**ASME** SFA5.5 (E8016-D3) **AWS** A5.5 (E8016-D3)
W18018	Low Alloy Steel Arc Welding Electrode (E8018-D3)	**C** 0.12 max **Mn** 1.00-1.75 **Mo** 0.40-0.65 **P** 0.03 max **S** 0.04 max **Si** 0.80 max	**ASME** SFA5.5 (E8018-D3) **AWS** A5.5 (E8018-D3)
W19015	Low Alloy Steel Arc Welding Electrode (E9015-D1)	**C** 0.12 max **Mn** 1.25-1.75 **Mo** 0.25-0.45 **P** 0.03 max **S** 0.04 max **Si** 0.60 max	**ASME** SFA5.5 (E9015-D1) **AWS** A5.5 (E9015-D1)
W19016	Low Alloy Steel Arc Welding Electrode (9016A)	**C** 0.12 max **Cr** 0.15 max **Mn** 1.25-1.75 **Mo** 0.30-0.60 **Ni** 0.10 max **P** 0.03 max **S** 0.04 max **Si** 0.05-0.60 **V** 0.05 max	
W19018	Low Alloy Steel Arc Welding Electrode (E9018-D1)	**C** 0.12 max **Mn** 1.25-1.75 **Mo** 0.25-0.45 **P** 0.03 max **S** 0.04 max **Si** 0.80 max	**ASME** SFA5.5 (E9018-D1) **AWS** A5.5 (E9018-D1)
W19131	Low Alloy Steel Flux Cored Wire (E91T1-D1)	**C** 0.12 max **Mn** 1.25-2.00 **Mo** 0.25-0.55 **P** 0.03 max **S** 0.03 max **Si** 0.80 max	**ASME** SFA5.29 (E91T1-D1) **AWS** A5.29 (E91T1-D1)
W19235	Low Alloy Steel Flux Cored Wire (E90T5-D2)	**C** 0.15 max **Mn** 1.65-2.25 **Mo** 0.25-0.55 **P** 0.03 max **S** 0.03 max **Si** 0.80 max	**ASME** SFA5.29 (E90T5-D2) **AWS** A5.29 (E90T5-D2)
W19331	Low Alloy Steel Flux Cored Wire (E90T1-D3)	**C** 0.12 max **Mn** 1.00-1.75 **Mo** 0.40-0.65 **P** 0.03 max **S** 0.03 max **Si** 0.80 max	**ASME** SFA5.29 (E90T1-D3) **AWS** A5.29 (E90T1-D3)
W20018	Low Alloy Steel Arc Welding Electrode (E7018-W)	**C** 0.12 max **Cr** 0.15-0.30 **Cu** 0.30-0.60 **Mn** 0.40-0.70 **Ni** 0.20-0.40 **P** 0.025 max **S** 0.025 max **Si** 0.40-0.70 **V** 0.08 max	**ASME** SFA5.5 (E7018-W) **AWS** A5.5 (E7018-W)
W20118	Low Alloy Steel Arc Welding Electrode (E8018-W)	**C** 0.12 max **Cr** 0.45-0.70 **Cu** 0.30-0.75 **Mn** 0.50-1.30 **Ni** 0.40-0.80 **P** 0.03 max **S** 0.04 max **Si** 0.35-0.80	**ASME** SFA5.5 (E8018-W) **AWS** A5.5 (E8018-W)
W20131	Low Alloy Steel Flux Cored Wire (E80T1-W) (EGxxT-5)	**C** 0.12 max **Cr** 0.45-0.70 **Cu** 0.30-0.75 **Mn** 0.50-1.30 **Ni** 0.40-0.80 **P** 0.03 max **S** 0.03 max **Si** 0.30-0.80	**ASME** SFA5.26 (EGxxT5); SFA5.29 (E80T1-W) **AWS** A5.26 (EGxxT5); A5.29 (E80T1-W)
W20140	Low Alloy Steel Composite Wire for Submerged Arc and Electroslag Welding (ECW) (EWT2)	**C** 0.12 max **Cr** 0.45-0.70 **Cu** 0.30-0.75 **Mn** 0.50-1.60 **Ni** 0.40-0.80 **P** 0.030 max **S** 0.040 max **Si** 0.80 max	**ASME** SFA5.23 (ECW); SFA5.25 (EWT2) (EWT4) **AWS** A5.23 (ECW); A5.25 (EWT2) (EWT4)
W20240	Low Alloy Steel Composite Wire for Submerged Arc Welding (ECF2)	**C** 0.17 max **Cu** 0.35 max **Mn** 1.25-2.25 **Mo** 0.40-0.65 **Ni** 0.40-0.80 **P** 0.030 max **S** 0.040 max **Si** 0.80 max	**ASME** SFA5.23 (ECF2) **AWS** A5.23 (ECF2)

The chemical compositions listed are for identification purposes and should not be used in lieu of the cross referenced specifications.

UNIFIED NUMBER	DESCRIPTION	CHEMICAL COMPOSITION	CROSS REFERENCE SPECIFICATIONS
W20241	Low Alloy Steel Composite Wire for Submerged Arc Welding (ECF2N) Nuclear Grade	**C** 0.17 max **Cu** 0.08 max **Mn** 1.25-2.25 **Mo** 0.40-0.65 **Ni** 0.40-0.80 **P** 0.012 max **S** 0.040 max **Si** 0.80 max **V** 0.05 max	**ASME** SFA5.23 (ECF2N) **AWS** A5.23 (ECF2N)
W20440	Low Alloy Steel Composite Wire for Submerged Arc Welding (ECF4)	**C** 0.17 max **Cr** 0.60 max **Cu** 0.35 max **Mn** 1.60 max **Mo** 0.25 max **Ni** 0.40-0.80 **P** 0.030 max **S** 0.040 max **Si** 0.80 max	**ASME** SFA5.23 (ECF4) **AWS** A5.23 (ECF4)
W20441	Low Alloy Steel Composite Wire for Submerged Arc Welding (ECF4N) Nuclear Grade	**C** 0.17 max **Cr** 0.60 max **Cu** 0.08 max **Mn** 1.60 max **Mo** 0.25 max **Ni** 0.40-0.80 **P** 0.012 max **S** 0.040 max **Si** 0.80 max **V** 0.05 max	**ASME** SFA5.23 (ECF4N) **AWS** A5.23 (ECF4N)
W21015	Low Alloy Arc Welding Electrode (8015)	**C** 0.12 max **Cr** 0.15 max **Mn** 0.40-1.10 **Mo** 0.35 max **Ni** 0.80-1.10 **P** 0.03 max **S** 0.04 max **Si** 0.05-.60 **V** 0.05 max	**MIL SPEC** MIL-E-22200/6 (MIL-8015-C3)
W21016	Low Alloy Arc Welding Electrode (E8016-C3)	**C** 0.12 max **Cr** 0.15 max **Mn** 0.40-1.25 **Mo** 0.35 max **Ni** 0.80-1.10 **P** 0.03 max **S** 0.04 max **Si** 0.80 max **V** 0.05 max	**ASME** SFA5.5 (E8016-C3) **AWS** A5.5 (E8016-C3) **MIL SPEC** MIL-E-22200/6 (MIL-8016-C3); MIL-E-18038 (MIL-8016)
W21018	Low Alloy Steel Arc Welding Electrode (E8018-C3)	**C** 0.12 max **Cr** 0.15 max **Mn** 0.40-1.25 **Mo** 0.35 max **Ni** 0.80-1.10 **P** 0.03 max **S** 0.04 max **Si** 0.80 max **V** 0.05 max	**ASME** SFA5.5 (E8018-C3) **AWS** A5.5 (E8018-C3) **MIL SPEC** MIL-E-22200/1 (MIL-8018-C3) (MIL-8018-C3SR)
W21030	Low Alloy Steel Metal Cored Wire for Gas Shielded Arc Welding (E80C-Ni1)	**C** 0.12 max **Cu** 0.35 max **Mn** 1.25 max **Mo** 0.65 max **Ni** 0.80-1.10 **P** 0.025 max **S** 0.030 max **Si** 0.60 max **V** 0.05 max	**ASME** SFA5.28 (E80C-Ni1) **AWS** A5.28 (E80C-Ni1)
W21031	Low Alloy Steel Flux Cored Wire (E8xT1-Ni1)	**Al** 1.8 max **C** 0.12 max **Cr** 0.15 max **Mn** 1.50 max **Mo** 0.35 max **Ni** 0.80-1.10 **P** 0.03 max **Si** 0.80 max **V** 0.05 max	**ASME** SFA5.29 (E8xT1-Ni1) **AWS** A5.29 (E8xT1-Ni1)
W21033	Low Alloy Steel Flux Cored Wire (EGxxT-3)	**C** 0.10 max **Mn** 1.0-1.8 **Mo** 0.30 max **Ni** 0.7-1.1 **P** 0.03 max **S** 0.030 max **Si** 0.50 max	**ASME** SFA5.26 (EGxxT-3) **AWS** A5.26 (EGxxT-3)
W21035	Low Alloy Steel Flux Cored Wire (E80T5-Ni1)	**Al** 1.8 max **C** 0.12 max **Cr** 0.15 max **Mn** 1.50 max **Mo** 0.35 max **Ni** 0.80-1.10 **P** 0.03 max **Si** 0.80 max **V** 0.05 max	**ASME** SFA5.29 (E80T5-Ni1) **AWS** A5.29 (E80T5-Ni1)
W21038	Low Alloy Steel Flux Cored Wire (E71T8-Ni1)	**Al** 1.8 max **C** 0.12 max **Cr** 0.15 max **Mn** 1.50 max **Mo** 0.35 max **Ni** 0.80-1.10 **P** 0.03 max **S** 0.03 max **Si** 0.80 max **V** 0.05	**ASME** SFA5.29 (E71T8-Ni1) **AWS** A5.29 (E71T8-Ni1)
W21040	Low Alloy Steel Composite Wire for Submerged Arc Welding (ECNi1)	**C** 0.12 max **Cr** 0.15 max **Cu** 0.35 max **Mn** 1.60 max **Mo** 0.35 max **Ni** 0.80-1.10 **P** 0.030 max **S** 0.030 max **Si** 0.80 max **V** 0.05 max	**ASME** SFA5.23 (ECNi1) **AWS** A5.23 (ECNi1)
W21041	Low Alloy Steel Composite Wire for Submerged Arc Welding (ECNi1N) Nuclear Grade	**C** 0.12 max **Cr** 0.15 max **Cu** 0.08 max **Mn** 1.60 max **Mo** 0.35 max **Ni** 0.80-1.10 **P** 0.012 max **S** 0.030 max **Si** 0.80 max	**ASME** SFA5.23 (ECNi1N) **AWS** A5.23 (ECNi1N)
W21048	Low Alloy Steel Flux Cored Wire (ExxT8-K6)	**Al** 1.8 max **C** 0.15 max **Cr** 0.15 max **Mn** 0.50-1.50 **Mo** 0.15 max **Ni** 0.40-1.10 **P** 0.03 max **S** 0.03 max **Si** 0.80 max **V** 0.05 max	**ASME** SFA5.29 (ExxT8-K6) **AWS** A5.29 (ExxT8-K6)
W21118	Low Alloy Steel Arc Welding Electrode (E8018-NM)	**Al** 0.05 max **C** 0.10 max **Cr** 0.05 max **Cu** 0.10 max **Mn** 0.80-1.25 **Mo** 0.40-0.65 **Ni** 0.80-1.10 **P** 0.020 max **S** 0.030 max **Si** 0.60 max **V** 0.02 max	**ASME** SFA5.5 (E8018-NM) **AWS** A5.5 (E8018-NM)
W21135	Low Alloy Steel Flux Cored Wire (E80T5-K1)	**C** 0.15 **Cr** 0.15 **Mn** 0.80-1.40 **Mo** 0.20-0.65 **Ni** 0.80-1.10 **P** 0.03 **S** 0.03 **Si** 0.80 **V** 0.05	**ASME** SFA5.29 (E80T5-K1) **AWS** A5.29 (E80T5-K1)

The chemical compositions listed are for identification purposes and should not be used in lieu of the cross referenced specifications.

UNIFIED NUMBER	DESCRIPTION	CHEMICAL COMPOSITION	CROSS REFERENCE SPECIFICATIONS
W21140	Low Alloy Steel Composite Wire for Submerged Arc Welding (ECF3)	C 0.17 max Cu 0.35 max Mn 1.25-2.25 Mo 0.40-0.65 Ni 0.70-1.10 P 0.030 max S 0.040 max Si 0.80 max	ASME SFA5.23 (ECF3) AWS A5.23 (ECF3)
W21141	Low Alloy Steel Composite Wire for Submerged Arc Welding (ECF3N) Nuclear Grade	C 0.17 max Cu 0.08 max Mn 1.25-2.25 Mo 0.40-0.65 Ni 0.70-1.10 P 0.012 max S 0.040 max Si 0.80 max V 0.05 max	ASME SFA5.23 (ECF3N) AWS A5.23 (ECF3N)
W21150	Low Alloy Steel Composite Wire for Submerged Arc Welding (ECF1)	C 0.12 max Cr 0.15 max Cu 0.35 max Mn 0.70-1.50 Mo 0.55 max Ni 0.90-1.70 P 0.030 max S 0.040 max Si 0.80 max	ASME SFA5.23 (ECF1) AWS A5.23 (ECF1)
W21151	Low Alloy Steel Composite Wire for Submerged Arc Welding (ECF1N) Nuclear Grade	C 0.12 max Cr 0.15 max Cu 0.08 max Mn 0.70-1.50 Mo 0.55 max Ni 0.90-1.70 P 0.012 max S 0.040 max Si 0.80 max V 0.05 max	ASME SFA5.23 (ECF1N) AWS A5.23 (ECF1N)
W21215	Low Alloy Steel Arc Welding Electrode (9015) (10015)	C 0.10 max Cr 0.15 max Mn 0.70-1.10 Mo 0.35 max Ni 1.40-1.80 P 0.025 max S 0.030 max Si 0.05-0.60 V 0.20 max	MIL SPEC MIL-E-22200/6 (MIL-10015)
W21216	Low Alloy Steel Arc Welding Electrode (9015) (10016)	C 0.10 max Cr 0.15 max Mn 0.70-1.10 Mo 0.35 max Ni 1.40-1.80 P 0.025 S 0.030 Si 0.05-0.60 V 0.20 max	MIL SPEC MIL-E-22200/6 (MIL-10016)
W21218	Low Alloy Steel Arc Welding Electrode (E9018-M)	C 0.10 max Cr 0.15 max Mn 0.60-1.25 Mo 0.35 max Ni 1.40-1.80 P 0.030 max S 0.030 max Si 0.80 max V 0.05 max	ASME SFA5.5 (E9018-M) AWS A5.5 (E9018-M) MIL SPEC MIL-E-22200/1 (MIL-9018-M)
W21231	Low Alloy Steel Flux Cored Wire (ExxT1-K2)	Al 1.8 max C 0.15 max Cr 0.15 max Mn 0.50-1.75 Mo 0.35 max Ni 1.00-2.00 P 0.03 max S 0.03 max Si 0.80 max V 0.05 max	ASME SFA5.29 (ExxT1-K2) AWS A5.29 (ExxT1-K2)
W21234	Low Alloy Steel Flux Cored Wire (E70T4-K2)	Al 1.8 max C 0.15 max Cr 0.15 max Mn 0.50-1.75 Mo 0.35 max Ni 1.00-2.00 P 0.03 max S 0.03 max Si 0.80 max V 0.05 max	ASME SFA5.29 (E70T4-K2) AWS A5.29 (E70T4-K2)
W21235	Low Alloy Steel Flux Cored Wire (E80T5-K2)	Al 1.8 max C 0.15 max Cr 0.15 max Mn 0.50-1.75 Mo 0.35 max Ni 1.00-2.00 P 0.03 max S 0.03 max Si 0.80 max V 0.05 max	ASME SFA5.29 (E80T5-K2) AWS A5.29 (E80T5-K2)
W21238	Low Alloy Steel Flux Cored Wire (E71T8-K2)	Al 1.8 max C 0.15 max Cr 0.15 max Mn 0.50-1.75 Mo 0.35 max Ni 1.00-2.00 P 0.03 max S 0.03 max Si 0.80 max V 0.05 max	ASME SFA5.29 (E71T8-K2) AWS A5.29 (E71T8-K2)
W21240	Low Alloy Steel Composite Wire for Submerged Arc Welding (ECM1)	C 0.10 max Cr 0.15 max Cu 0.30 max Mn 0.60-1.60 Mo 0.35 max Ni 1.25-2.00 P 0.030 max S 0.040 max Si 0.80 max	ASME SFA5.23 (ECM1) AWS A5.23 (ECM1)
W21241	Low Alloy Steel Composite Wire for Submerged Arc Welding (ECM1N) Nuclear Grade	C 0.10 Cr 0.15 Cu 0.08 Mn 0.60-1.60 Mo 0.35 Ni 1.25-2.00 P 0.012 S 0.040 Si 0.80 V 0.05	ASME SFA5.23 (ECM1N) AWS A5.23 (ECM1N)
W21250	Low Alloy Steel Composite Wire for Submerged Arc Welding (ECNi4)	C 0.14 max Cu 0.35 max Mn 1.60 max Mo 0.35 max Ni 1.40-2.10 P 0.030 max S 0.030 max Si 0.80 max	ASME SFA5.23 (ECNi4) AWS A5.23 (ECNi4)
W21251	Low Alloy Steel Composite Wire for Submerged Arc Welding (ECNi4N) Nuclear Grade	C 0.14 max Cu 0.08 max Mn 1.60 max Mo 0.35 max Ni 1.40-2.10 P 0.012 max S 0.030 max Si 0.80 max V 0.05 max	ASME SFA5.23 (ECNi4N) AWS A5.23 (ECNi4N)

The chemical compositions listed are for identification purposes and should not be used in lieu of the cross referenced specifications.

UNIFIED NUMBER	DESCRIPTION	CHEMICAL COMPOSITION	CROSS REFERENCE SPECIFICATIONS
W21318	Low Alloy Steel Arc Welding Electrode (E10018-M)	**C** 0.10 max **Cr** 0.35 max **Mn** 0.75-1.70 **Mo** 0.25-0.50 **Ni** 1.40-2.10 **P** 0.030 max **S** 0.030 max **Si** 0.60 max **V** 0.05 max	**ASME** SFA5.5 (E10018-M) **AWS** A5.5 (E10018-M) **MIL SPEC** MIL-E-22200/1 (MIL-10018-M)
W21331	Low Alloy Steel Flux Cored Wire (ExxxT1-K3)	**C** 0.15 max **Cr** 0.15 max **Mn** 0.75-2.25 **Mo** 0.25-0.65 **Ni** 1.25-2.60 **P** 0.03 max **S** 0.03 max **Si** 0.80 max **V** 0.05 max	**ASME** SFA5.29 (ExxxT1-K3) **AWS** A5.29 (ExxxT1-K3)
W21335	Low Alloy Steel Flux Cored Wire (ExxxT5-K3)	**C** 0.15 max **Cr** 0.15 max **Mn** 0.75-2.25 **Mo** 0.25-0.65 **Ni** 1.25-2.60 **P** 0.03 max **S** 0.03 max **Si** 0.80 max **V** 0.05 max	**ASME** SFA5.29 (ExxxT5-K3) **AWS** A5.29 (ExxxT5-K3)
W21340	Low Alloy Steel Composite Wire for Submerged Arc Welding (ECM2)	**C** 0.10 max **Cr** 0.35 max **Cu** 0.30 max **Mn** 0.90-1.80 **Mo** 0.25-0.65 **Ni** 1.40-2.10 **P** 0.030 max **S** 0.040 max **Si** 0.80 max	**ASME** SFA5.23 (ECM2) **AWS** A5.23 (ECM2)
W21341	Low Alloy Steel Composite Wire for Submerged Arc Welding (ECM2N) Nuclear Grade	**C** 0.10 max **Cr** 0.35 max **Cu** 0.08 max **Mn** 0.90-1.80 **Mo** 0.25-0.65 **Ni** 1.40-2.10 **P** 0.012 max **S** 0.040 max **Si** 0.80 max	**ASME** SFA5.23 (ECM2N) **AWS** A5.23 (ECM2N)
W21380	Low Alloy Steel Solid Wire for Submerged Arc Welding (MI-88)	**Al** 0.10 max **C** 0.06 max **Cr** 0.10-0.30 **Cu** 0.10-0.30 **Mn** 1.00-1.50 **Mo** 0.20-0.40 **Ni** 1.40-1.90 **P** 0.010 max **S** 0.010 max **Si** 0.50 max **Ti** 0.10 max **V** 0.05 max **Zr** 0.10 max	
W21418	Low Alloy Steel Arc Welding Electrode (E11018-M)	**C** 0.10 max **Cr** 0.40 max **Mn** 1.30-1.80 **Mo** 0.25-0.50 **Ni** 1.25-2.50 **P** 0.030 max **S** 0.030 max **Si** 0.60 max **V** 0.05 max	**ASME** SFA5.5 (E11018-M) **AWS** A5.5 (E11018-M) **MIL SPEC** MIL-E-22200/1 (MIL-11018-M)
W21631	Low Alloy Steel Flux Cored Wire (E120T1-K5)	**C** 0.10-0.25 **Cr** 0.20-0.70 **Mn** 0.60-1.60 **Mo** 0.15-0.55 **Ni** 0.75-2.00 **P** 0.03 max **S** 0.03 max **Si** 0.80 max **V** 0.05 max	**ASME** SFA5.29 (E120T1-K5) **AWS** A5.29 (E120T1-K5)
W21780	Low Alloy Steel Solid Wire for Submerged Arc Welding (EB82)	**C** 0.12 max **Cr** 0.30 max **Cu** 0.40-1.10 **Mn** 0.80-1.25 **Mo** 0.15-0.60 **Ni** 0.80-1.25 **P** 0.020 max **S** 0.020 max **Si** 0.80 max **V** 0.05 max	
W22016	Low Alloy Steel Arc Welding Electrode (E8016-C1)	**C** 0.12 max **Mn** 1.25 **Ni** 2.00-2.75 **P** 0.03 max **S** 0.04 max **Si** 0.60 max	**ASME** SFA5.5 (E8016-C1) **AWS** A5.5 (E8016-C1)
W22018	Low Alloy Steel Arc Welding Electrode (E8018-C1)	**C** 0.12 max **Mn** 1.25 max **Ni** 2.00-2.75 **P** 0.03 max **S** 0.04 max **Si** 0.80 max	**ASME** SFA5.5 (E8018-C1) **AWS** A5.5 (E8018-C1)
W22030	Low Alloy Steel Metal Cored Wire for Gas Shielded Arc Welding (E80C-Ni2)	**C** 0.12 max **Cu** 0.35 max **Mn** 1.25 max **Ni** 2.00-2.75 **P** 0.025 max **S** 0.030 max **Si** 0.60 max	**ASME** SFA5.28 (E80C-Ni2) **AWS** A5.28 (E80C-Ni2)
W22031	Low Alloy Steel Flux Cored Wire (ExxT1-Ni2)	**Al** 1.8 max **C** 0.12 max **Mn** 1.50 max **Ni** 1.75-2.75 **P** 0.03 max **S** 0.03 max **Si** 0.80 max	**ASME** SFA5.29 (ExxT1-Ni2) **AWS** A5.29 (ExxT1-Ni2)
W22035	Low Alloy Steel Flux Cored Wire (E80T5-Ni2)	**Al** 1.8 max **C** 0.12 max **Mn** 1.50 max **Ni** 1.75-2.75 **P** 0.03 max **S** 0.03 max **Si** 0.80 max	**ASME** SFA5.29 (E80T5-Ni2) **AWS** A5.29 (E80T5-Ni2)
W22038	Low Alloy Steel Flux Cored Wire (E71T8-Ni2)	**Al** 1.8 max **C** 0.12 max **Mn** 1.50 max **Ni** 1.75-2.75 **P** 0.03 max **S** 0.03 max **Si** 0.80 max	**ASME** SFA5.29 (E71T8-Ni2) **AWS** A5.29 (E71T8-Ni2)
W22040	Low Alloy Steel Composite Wire for Submerged Arc Welding (ECNi2)	**C** 0.12 max **Cu** 0.35 max **Mn** 1.80 max **Ni** 2.00-2.90 **P** 0.030 max **S** 0.030 max **Si** 0.80 max	**ASME** SFA5.23 (ECNi2) **AWS** A5.23 (ECNi2)
W22041	Low Alloy Steel Composite Wire for Submerged Arc Welding (ECNi2N) Nuclear Grade	**C** 0.12 max **Cu** 0.08 max **Mn** 1.80 max **Ni** 2.00-2.90 **P** 0.012 max **S** 0.030 max **Si** 0.80 max **V** 0.05 max	**ASME** SFA5.23 (ECNi2N) **AWS** A5.23 (ECNi2N)

The chemical compositions listed are for identification purposes and should not be used in lieu of the cross referenced specifications.

UNIFIED NUMBER	DESCRIPTION	CHEMICAL COMPOSITION	CROSS REFERENCE SPECIFICATIONS
W22051	Low Alloy Steel Flux Cored Wire (E101T1-K7)	**C** 0.15 max **Mn** 1.00-1.75 **Ni** 2.00-2.75 **P** 0.03 max **S** 0.03 max **Si** 0.80 max	**ASME** SFA5.29 (E101T1-K7) **AWS** A5.29 (E101T1-K7)
W22115	Low Alloy Steel Arc Welding Electrode (E7015-C1L)	**C** 0.05 max **Mn** 1.25 max **Ni** 2.00-2.75 **P** 0.03 max **S** 0.04 max **Si** 0.50 max	**ASME** SFA5.5 (E7015-C1L) **AWS** A5.5 (E7015-C1L)
W22116	Low Alloy Steel Arc Welding Electrode (E7016-C1L)	**C** 0.05 max **Mn** 1.25 max **Ni** 2.00-2.75 **P** 0.03 max **S** 0.04 max **Si** 0.50 max	**ASME** SFA5.5 (E7016-C1L) **AWS** A5.5 (E7016-C1L)
W22118	Low Alloy Steel Arc Welding Electrode (E7018-C1L)	**C** 0.05 max **Mn** 1.25 max **Ni** 2.00-2.75 **P** 0.03 max **S** 0.04 max **Si** 0.50 max	**ASME** SFA5.5 (E7018-C1L) **AWS** A5.5 (E7018-C1L)
W22218	Low Alloy Steel Arc Welding Electrode (E12018-M)	**C** 0.10 max **Cr** 0.30-1.50 **Mn** 1.30-2.25 **Mo** 0.30-0.55 **Ni** 1.75-2.50 **P** 0.030 max **S** 0.030 max **Si** 0.60 max **V** 0.05 max	**ASME** SFA5.5 (E12018-M) **AWS** A5.5 (E12018-M)
W22231	Low Alloy Steel Flux Cored Wire (E111T1-K4)	**C** 0.15 max **Cr** 0.20-0.60 **Mn** 1.20-2.25 **Mo** 0.30-0.65 **Ni** 1.75-2.60 **P** 0.03 max **S** 0.03 max **Si** 0.80 max **V** 0.05 max	**ASME** SFA5.29 (E111T1-K4) **AWS** A5.29 (E111T1-K4)
W22235	Low Alloy Steel Flux Cored Wire (E1xxT5-K4)	**C** 0.15 max **Cr** 0.20-0.60 **Mn** 1.20-2.25 **Mo** 0.30-0.65 **Ni** 1.75-2.60 **P** 0.03 max **S** 0.03 max **Si** 0.80 max **V** 0.05 max	**ASME** SFA5.29 (E1xxT5-K4) **AWS** A5.29 (E1xxT5-K4)
W22240	Low Alloy Steel Composite Wire for Submerged Arc Welding (ECM3)	**C** 0.10 max **Cr** 0.65 max **Cu** 0.30 max **Mn** 0.90-1.80 **Mo** 0.20-0.70 **Ni** 1.80-2.60 **P** 0.030 max **S** 0.030 max **Si** 0.80 max	**ASME** SFA5.23 (ECM3) **AWS** A5.23 (ECM3)
W22241	Low Alloy Steel Composite Wire for Submerged Arc Welding (ECM3N) Nuclear Grade	**C** 0.10 max **Cr** 0.65 max **Cu** 0.08 max **Mn** 0.90-1.80 **Mo** 0.20-0.70 **Ni** 1.80-2.60 **P** 0.012 max **S** 0.030 max **Si** 0.80 max	**ASME** SFA5.23 (ECM3N) **AWS** A5.23 (ECM3N)
W22334	Low Alloy Steel Flux Cored Wire (EGxxT4)	**C** 0.12 max **Cr** 0.20 max **Mn** 1.00-2.00 **Mo** 0.40-0.65 **Ni** 1.50-2.00 **P** 0.02 max **S** 0.03 max **Si** 0.15-0.50 **V** 0.05 max	**ASME** SFA5.26 (EGxxT4) **AWS** A5.26 (EGxxT4)
W22340	Low Alloy Steel Composite Wire for Electroslag Welding (EWT3)	**C** 0.12 max **Cr** 0.20 max **Mn** 1.00-2.00 **Mo** 0.40-0.65 **Ni** 1.50-2.50 **P** 0.02 max **S** 0.03 max **Si** 0.15-0.50 **V** 0.05 max	**ASME** SFA5.25 (EWT3) **AWS** A5.25 (EWT3)
W22440	Low Alloy Steel Tubular Wire for Submerged Arc Welding (ECM4)	**C** 0.10 max **Cr** 0.80 max **Cu** 0.30 max **Mn** 1.30-2.25 **Mo** 0.30-0.80 **Ni** 2.00-2.80 **P** 0.030 max **S** 0.030 max **Si** 0.80 max	**ASME** SFA5.23 (ECM4) **AWS** A5.23 (ECM4)
W22441	Low Alloy Steel Composite Wire for Submerged Arc Welding (ECM4N) Nuclear Grade	**C** 0.10 max **Cr** 0.80 max **Cu** 0.08 max **Mn** 1.30-2.25 **Mo** 0.30-0.80 **Ni** 2.00-2.80 **P** 0.012 max **S** 0.030 max **Si** 0.80 max	**ASME** SFA5.23 (ECM4N) **AWS** A5.23 (ECM4N)
W22540	Low Alloy Steel Composite Wire for Submerged Arc Welding (ECF5)	**C** 0.17 max **Cr** 0.65 max **Cu** 0.50 max **Mn** 1.20-1.80 **Mo** 0.30-0.80 **Ni** 2.00-2.80 **P** 0.030 max **S** 0.030 max **Si** 0.80 max	**ASME** SFA5.23 (ECF5) **AWS** A5.23 (ECF5)
W22640	Low Alloy Steel Composite Wire for Submerged Arc Welding (ECF6)	**C** 0.14 max **Cr** 0.65 max **Cu** 0.40 max **Mn** 0.80-1.85 **Mo** 0.60 max **Ni** 1.50-2.25 **P** 0.030 max **S** 0.030 max **Si** 0.80 max	**ASME** SFA5.23 (ECF6) **AWS** A5.23 (ECF6)
W22641	Low Alloy Steel Composite Wire for Submerged Arc Welding (ECF6N) Nuclear Grade	**C** 0.14 max **Cr** 0.65 max **Cu** 0.08 max **Mn** 0.80-1.85 **Mo** 0.60 max **Ni** 1.50-2.25 **P** 0.012 max **S** 0.030 max **Si** 0.80 max **V** 0.05 max	**ASME** SFA5.23 (ECF6N) **AWS** A5.23 (ECF6N)
W22718	Low Alloy Steel Arc Welding Electrode (10018-N1)	**C** 0.15 max **Cr** 0.90-1.20 **Mn** 0.80-1.15 **Mo** 0.45-0.75 **Ni** 1.50-2.00 **P** 0.030 max **S** 0.030 max **Si** 0.30-0.60 **V** 0.02 max	**MIL SPEC** MIL-E-22200/5 (MIL-10018-N1)

The chemical compositions listed are for identification purposes and should not be used in lieu of the cross referenced specifications.

UNIFIED NUMBER	DESCRIPTION	CHEMICAL COMPOSITION	CROSS REFERENCE SPECIFICATIONS
W22815	Low Alloy Steel Arc Welding Electrode (11015) (12015)	**C** 0.10 max **Cr** 1.00-1.40 **Mn** 1.00-1.50 **Mo** 0.20-0.40 **Ni** 1.75-2.25 **P** 0.03 max **S** 0.04 max **Si** 0.05-0.60 **V** 0.20 max	
W22816	Low Alloy Steel Arc Welding Electrode (11016) (12016)	**C** 0.10 max **Cr** 1.00-1.40 **Mn** 1.00-1.50 **Mo** 0.20-0.40 **Ni** 1.75-2.25 **P** 0.03 max **S** 0.04 max **Si** 0.05-0.60 **V** 0.20 max	
W23016	Low Alloy Steel Arc Welding Electrode (E8016-C2)	**C** 0.12 max **Mn** 1.25 max **Ni** 3.00-3.75 **P** 0.03 max **S** 0.04 max **Si** 0.60 max	**ASME** SFA5.5 (E8016-C2) **AWS** A5.5 (E8016-C2)
W23018	Low Alloy Steel Arc Welding Electrode (W23018)	**C** 0.12 max **Mn** 1.25 max **Ni** 3.00-3.75 **P** 0.03 max **S** 0.04 max **Si** 0.80 max	**ASME** SFA5.5 (E8018-C2) **AWS** A5.5 (E8018-C2)
W23030	Low Alloy Steel Metal Cored Wire for Gas Shielded Arc Welding (E80C-Ni3)	**C** 0.12 max **Cu** 0.35 max **Mn** 1.25 max **Ni** 3.00-3.75 **P** 0.025 max **S** 0.030 max **Si** 0.60 max	**ASME** SFA5.28 (E80C-Ni3) **AWS** A5.28 (E80C-Ni3)
W23035	Low Alloy Steel Flux Cored Wire (ExxT5-Ni3)	**C** 0.12 max **Mn** 1.50 max **Ni** 2.75-3.75 **P** 0.03 max **S** 0.03 max **Si** 0.80 max	**ASME** SFA5.29 (ExxT5-Ni3) **AWS** A5.29 (ExxT5-Ni3)
W23040	Low Alloy Steel Composite Wire for Submerged Arc Welding (ECNi3)	**C** 0.12 max **Cr** 0.15 max **Cu** 0.35 max **Mn** 1.60 max **Ni** 2.80-3.75 **P** 0.030 max **S** 0.030 max **Si** 0.80 max	**ASME** SFA5.23 (ECNi3) **AWS** A5.23 (ECNi3)
W23041	Low Alloy Steel Composite Wire for Submerged Arc Welding (ECNi3N) Nuclear Grade	**C** 0.12 max **Cr** 0.15 max **Cu** 0.08 max **Mn** 1.60 max **Ni** 2.80-3.75 **P** 0.012 max **S** 0.030 max **Si** 0.80 max **V** 0.05 max	**ASME** SFA5.23 (ECNi3N) **AWS** A5.23 (ECNi3N)
W23115	Low Alloy Arc Welding Electrode (E7015-C2L)	**C** 0.05 max **Mn** 1.25 max **Ni** 3.00-3.75 **P** 0.03 max **S** 0.04 max **Si** 0.50 max	**ASME** SFA5.5 (E7015-C2L) **AWS** A5.5 (E7015-C2L)
W23116	Low Alloy Steel Arc Welding Electrode (E7016-C2L)	**C** 0.05 max **Mn** 1.25 max **Ni** 3.00-3.75 **P** 0.03 max **S** 0.04 max **Si** 0.50 max	**ASME** SFA5.5 (E7016-C2L) **AWS** A5.5 (E7016-C2L)
W23118	Low Alloy Steel Arc Welding Electrode (E7018-C2L)	**C** 0.05 max **Mn** 1.25 max **Ni** 3.00-3.75 **P** 0.03 max **S** 0.04 max **Si** 0.50 max	**ASME** SFA5.5 (E7018-C2L) **AWS** A5.5 (E7018-C2L)
W23218	Low Alloy Steel Arc Welding Electrode (E12018-M1)	**C** 0.10 max **Cr** 0.65 max **Mn** 0.80-1.60 **Mo** 0.20-0.30 **Ni** 3.00-3.80 **P** 0.015 max **S** 0.012 max **Si** 0.65 max **V** 0.05 max	**ASME** SFA5.5 (E12018-M1) **AWS** A5.5 (E12018-M1) **MIL SPEC** MIL-E-22200/10, (MIL-12018-M1)
W23318	Low Alloy Steel Arc Welding Electrode (14018-M1)	**C** 0.10 max **Cr** 0.35-1.20 **Mn** 0.75-1.35 **Mo** 0.30-1.10 **Ni** 3.10-3.90 **P** 0.013 max **S** 0.010 max **Si** 0.65 max **V** 0.09 max	**MIL SPEC** MIL-E-22200/9 (MIL-14018-M1)
W26018	Nickel Alloy Steel Arc Welding Electrode (14018-HT)	**C** 0.10 max **Cr** 0.05-0.60 **Mn** 0.40-0.80 **Mo** 0.40-0.70 **Ni** 7.00-9.00 **P** 0.013 max **S** 0.010 max **Si** 0.50 max **V** 0.04-0.10	**MIL SPEC** MIL-E-22200/11 (MIL-14018-HT)
W30710	Austenitic Cr-Ni-Mn Stainless Steel Arc Welding Electrode (E307)	**C** 0.04-0.14 **Cr** 18.0-21.5 **Cu** 0.75 max **Mn** 3.3-4.75 **Mo** 0.5-1.5 **Ni** 9.0-10.7 **P** 0.04 max **S** 0.03 max **Si** 0.90 max	**ASME** SFA5.4 (E307) **AWS** A5.4 (E307) **MIL SPEC** MIL-E-13080 (MIL-307)
W30731	Austenitic Cr-Ni-Mn Stainless Steel Flux Cored Wire (E307T-X)	**C** 0.13 max **Cr** 18.0-20.5 **Cu** 0.5 max **Mn** 3.3-4.75 **Mo** 0.5-1.5 **Ni** 9.0-10.5 **P** 0.04 max **S** 0.03 max **Si** 1.0 max	**ASME** SFA5.22 (E307T-1) (E307T-2) **AWS** A5.22 (E307T-1) (E307T-2)
W30733	Austenitic Cr-Ni-Mn Stainless Steel Flux Cored Wire (E307T-3)	**C** 0.13 max **Cr** 19.5-22.0 **Cu** 0.5 max **Mn** 3.3-4.75 **Mo** 0.5-1.5 **Ni** 9.0-10.5 **P** 0.04 max **S** 0.03 max **Si** 1.0 max	**ASME** SFA5.22 (E307T-3) **AWS** A5.22 (E307T-3)
W30740	Austenitic Cr-Ni-Mn Stainless Steel Composite Welding Filler Metal (ER307)	**C** 0.04-0.14 **Cr** 19.5-22.0 **Cu** 0.75 max **Mn** 3.3-4.75 **Mo** 0.5-1.5 **Ni** 8.0-10.7 **P** 0.03 max **S** 0.03 max **Si** 0.30-0.65	**ASME** SFA5.9 (ER307) **AWS** A5.9 (ER307)
W30810	Austenitic Cr-Ni Stainless Steel Arc Welding Electrode (E308)	**C** 0.08 max **Cr** 18.0-21.0 **Cu** 0.75 max **Mn** 0.5-2.5 **Mo** 0.75 max **Ni** 9.0-11.0 **P** 0.04 max **S** 0.03 max **Si** 0.90 max	**ASME** SFA5.4 (E308) (E308H) **AWS** A5.4 (E308) (E308H) **MIL SPEC** MIL-E-22200/2 (MIL-308) (MIL-308Co)

The chemical compositions listed are for identification purposes and should not be used in lieu of the cross referenced specifications.

WELDING FILLER METALS

UNIFIED NUMBER	DESCRIPTION	CHEMICAL COMPOSITION	CROSS REFERENCE SPECIFICATIONS
W30813	Austenitic Cr-Ni Stainless Steel Arc Welding Electrode (E308L)	**C** 0.04 max **Cr** 18.0-21.0 **Cu** 0.75 max **Mn** 0.5-2.5 **Mo** 0.75 max **Ni** 9.0-11.0 **P** 0.04 max **S** 0.03 max **Si** 0.90 max	**ASME** SFA5.4 (E308L) **AWS** A5.4 (E308L) **MIL SPEC** MIL-E-22200/2 (MIL-308L) (MIL-308CoL)
W30815	Austenitic Cr-Ni Stainless Steel Arc Welding Electrode (308HC) High Carbon	**C** 0.08-0.15 **Cr** 18.0-21.0 **Mn** 2.5 max **Ni** 9.0-11.0 **P** 0.03 max **S** 0.03 max **Si** 5xC max	**MIL SPEC** MIL-E-22200/2, Type MIL-308HC
W30820	Austenitic Cr-Ni-Mo Stainless Steel Arc Welding Electrode (E308Mo)	**C** 0.08 max **Cr** 18.0-21.0 **Cu** 0.75 max **Mn** 0.5-2.5 **Mo** 2.0-3.0 **Ni** 9.0-12.0 **P** 0.04 max **S** 0.03 max **Si** 0.90 max	**ASME** SFA5.4 (E308Mo) **AWS** A5.4 (E308Mo)
W30821	Austenitic Cr-Ni-Mo Stainless Steel Arc Welding Electrode - for Armor	**C** 0.07-0.17 **Cr** 18.00-21.50 **Mn** 1.25-2.25 **Mo** 1.85-2.25 **Ni** 9.00-10.7 **P** 0.04 max **S** 0.03 max **Si** 0.80 max	**MIL SPEC** MIL-E-13080 (MIL-308)
W30823	Austenitic Cr-Ni-Mo Stainless Steel Arc Welding Electrode (E308MoL) - Low Carbon	**C** 0.04 max **Cr** 18.0-21.0 **Cu** 0.75 max **Mn** 0.5-2.5 **Mo** 2.0-3.0 **Ni** 9.0-12.0 **P** 0.04 max **S** 0.03 max **Si** 0.90 max	**ASME** SFA5.4 (E308MoL) **AWS** A5.4 (E308MoL)
W30831	Austenitic Cr-Ni Stainless Steel Flux Cored Wire (E308T-X)	**C** 0.08 max **Cr** 18.0-21.0 **Cu** 0.5 max **Mn** 0.5-2.5 **Mo** 0.5 max **Ni** 9.0-11.0 **P** 0.04 max **S** 0.03 max **Si** 1.0 max	**ASME** SFA5.22 (E308T-1) (E308T-2) **AWS** A5.22 (E308T-1) (E308T-2)
W30832	Austenitic Cr-Ni-Mo Stainless Steel Flux Cored Wire (E308MoT-X)	**C** 0.08 max **Cr** 18.0-21.0 **Cu** 0.5 max **Mn** 0.5-2.5 **Mo** 2.0-3.0 **Ni** 9.0-12.0 **P** 0.04 max **S** 0.03 max **Si** 1.0 max	**ASME** SFA5.22 (E308MoT-1) (E308MoT-2) (E308MoT-3) **AWS** A5.22 (E308MoT-1) (E308MoT-2) (E308MoT-3)
W30833	Austenitic Cr-Ni Stainless Steel Flux Cored Wire (E308T-3)	**C** 0.08 max **Cr** 19.5-22.0 **Cu** 0.5 max **Mn** 0.5-2.5 **Mo** 0.5 max **Ni** 9.0-11.0 **P** 0.04 max **S** 0.03 max **Si** 1.0 max	**ASME** SFA5.22 (E308T-3) **AWS** A5.22 (E308T-3)
W30835	Austenitic Cr-Ni Stainless Steel Flux Cored Wire (E308LT-X)	**C** 0.03 max **Cr** 18.0-21.0 **Cu** 0.5 max **Mn** 0.5-2.5 **Mo** 0.5 max **Ni** 9.0-11.0 **P** 0.04 max **S** 0.03 max **Si** 1.0 max	**ASME** SFA5.22 (E308LT-1) (E308LT-2) **AWS** A5.22 (E308LT-1) (E308LT-2)
W30837	Austenitic Cr-Ni Stainless Steel Flux Cored Wire (E308LT-3)	**C** 0.03 max **Cr** 19.5-22.0 **Cu** 0.5 max **Mn** 0.5-2.5 **Mo** 0.5 max **Ni** 9.0-11.0 **P** 0.04 max **S** 0.03 max **Si** 1.0 max	**ASME** SFA5.22 (E308LT-3) **AWS** A5.22 (E308LT-3)
W30838	Austenitic Cr-Ni-Mo Stainless Steel Flux Cored Wire (E308LT-X)	**C** 0.03 **Cr** 18.0-21.0 **Cu** 0.5 **Mn** 0.5-2.5 **Mo** 2.0-3.0 **Ni** 9.0-12.0 **P** 0.04 **S** 0.03 **Si** 1.0	**ASME** SFA5.22 (E308MoLT-1) (E308MoLT-2) (E308MoLT-3) **AWS** A5.22 (E308MoLT-1) (E308MoLT-2) (E308MoLT-3)
W30840	Austenitic Cr-Ni Stainless Steel Composite Welding Filler Metal ER308	**C** 0.08 max **Cr** 19.5-22.0 **Cu** 0.75 max **Mn** 1.0-2.5 **Mo** 0.75 max **Ni** 9.0-11.0 **P** 0.03 max **S** 0.03 max **Si** 0.30-0.65	**ASME** SFA5.9 (ER308) (ER308H) **AWS** A5.9 (ER308) (ER308H)
W30841	Austenitic Cr-Ni Stainless Steel Composite Welding Filler Metal ER308Si	**C** 0.08 max **Cr** 19.5-22.0 **Cu** 0.5 max **Fe** rem **Mn** 1.0-2.5 **Mo** 0.5 max **Ni** 9.0-11.0 **P** 0.03 max **S** 0.03 max **Si** 0.65-1.00	**ASME** SFA5.9 (ER308Si) **AWS** A5.9 (ER308Si)
W30843	Austenitic Cr-Ni Stainless Steel Composite Welding Filler Metal (ER308L)	**C** 0.03 max **Cr** 19.5-22.0 **Cu** 0.75 max **Mn** 1.0-2.5 **Mo** 0.75 max **Ni** 9.0-11.0 **P** 0.03 max **S** 0.03 max **Si** 0.30-0.65	**ASME** SFA5.9 (ER308L) **AWS** A5.9 (ER308L)
W30848	Austenitic Cr-Ni Stainless Steel Composite Welding Filler Metal ER308LSi	**C** 0.03 max **Cr** 19.5-22.0 **Cu** 0.5 max **Fe** rem **Mn** 1.0-2.5 **Mo** 0.5 max **Ni** 9.0-11.0 **P** 0.03 max **S** 0.03 max **Si** 0.65-1.00	**ASME** SFA5.9 (ER308LSi) **AWS** A5.9 (ER308LSi)
W30850	Austenitic Cr-Ni-Mo Stainless Steel Composite Welding Filler Metal (ER308Mo)	**C** 0.08 max **Cr** 18.0-21.0 **Cu** 0.75 max **Mn** 1.0-2.5 **Mo** 2.0-3.0 **Ni** 9.0-12.0 **P** 0.03 max **S** 0.03 max **Si** 0.30-0.65	**ASME** SFA5.9 (ER308Mo) **AWS** A5.9 (ER308Mo)
W30853	Austenitic Cr-Ni-Mo Stainless Steel Composite Welding Filler Metal (ER308MoL)	**C** 0.04 max **Cr** 18.0-21.0 **Cu** 0.75 max **Mn** 1.0-2.5 **Mo** 2.0-3.0 **Ni** 9.0-12.0 **P** 0.03 max **S** 0.03 max **Si** 0.30-0.65	**ASME** SFA5.9 (ER308MoL) **AWS** A5.9 (ER308MoL)
W30910	Austenitic Cr-Ni Stainless Steel Arc Welding Electrode (E309)	**C** 0.15 max **Cr** 22.0-25.0 **Cu** 0.75 max **Mn** 0.5-2.5 **Mo** 0.75 max **Ni** 12.0-14.0 **P** 0.04 max **S** 0.03 max **Si** 0.90 max	**ASME** SFA5.4 (E309) **AWS** A5.4 (E309) **MIL SPEC** MIL-E-22200/2 (MIL-309) (MIL-309Co)

The chemical compositions listed are for identification purposes and should not be used in lieu of the cross referenced specifications.

UNIFIED NUMBER	DESCRIPTION	CHEMICAL COMPOSITION	CROSS REFERENCE SPECIFICATIONS
W30913	Austenitic Cr-Ni Stainless Steel Arc Welding Electrode (E309L)	**C** 0.04 max **Cr** 22.0-25.0 **Cu** 0.75 max **Mn** 0.5-2.5 **Mo** 0.75 max **Ni** 12.0-14.0 **P** 0.04 max **S** 0.03 max **Si** 0.90 max	**ASME** SFA5.4 (E309L) **AWS** A5.4 (E309L) **MIL SPEC** MIL-E-22200/2, Types (MIL-309L) (MIL-309CoL)
W30917	Austenitic Cr-Ni-Cb Stainless Steel Arc Welding Electrode (E309Cb)	**C** 0.12 max **Cb** 0.70-1.00 includes Ta **Cr** 22.0-25.0 **Cu** 0.75 max **Mn** 0.5-2.5 **Mo** 0.75 max **Ni** 12.0-14.0 **P** 0.04 max **S** 0.03 max **Si** 0.90 max	**ASME** SFA5.4 (E309Cb) **AWS** A5.4 (E309Cb) **MIL SPEC** MIL-E-22200/2 (MIL-309Cb)
W30920	Austenitic Cr-Ni-Mo Stainless Steel Arc Welding Electrode (E309Mo)	**C** 0.12 max **Cr** 22.0-25.0 **Cu** 0.75 max **Mn** 0.5-2.5 **Mo** 2.0-3.0 **Ni** 12.0-14.0 **P** 0.04 max **S** 0.03 max **Si** 0.90 max	**ASME** SFA5.4 (E309Mo) **AWS** A5.4 (E309Mo)
W30931	Austenitic Cr-Ni Stainless Steel Flux Cored Wire (E309T-X)	**C** 0.10 max **Cr** 22.0-25.0 **Cu** 0.5 max **Mn** 0.5-2.5 **Mo** 0.5 max **Ni** 12.0-14.0 **P** 0.04 max **S** 0.03 max **Si** 1.0 max	**ASME** SFA5.22 (E309T-1) (E309T-2) **AWS** A5.22 (E309T-1) (E309T-2)
W30932	Austenitic Cr-Ni-Cb Stainless Steel Flux Cored Wire (E309CbLT-X)	**C** 0.03 max **Cb** 0.70-1.00 includes Ta **Cr** 22.0-25.0 **Cu** 0.5 max **Mn** 0.5-2.5 **Mo** 0.5 max **Ni** 12.0-14.0 **P** 0.04 max **S** 0.03 max **Si** 1.0 max	**AWS** A5.22 (E309CbLT-1) (E309CbLT-2)
W30933	Austenitic Cr-Ni Stainless Steel Flux Cored Wire (E309T-3)	**C** 0.10 max **Cr** 23.0-25.5 **Cu** 0.5 max **Mn** 0.5-2.5 **Mo** 0.5 max **Ni** 12.0-14.0 **P** 0.04 max **S** 0.03 max **Si** 1.0 max	**ASME** SFA5.22 (E309T-3) **AWS** A5.22 (E309T-3)
W30934	Austenitic Cr-Ni-Cb Stainless Steel Flux Cored Wire (E309CbLT-3)	**C** 0.03 max **Cb** 0.70-1.00 includes Ta **Cr** 23.0-25.5 **Cu** 0.5 max **Mn** 0.5-2.5 **Mo** 0.5 max **Ni** 12.0-14.0 **P** 0.04 max **S** 0.03 max **Si** 1.0 max	**ASME** SFA5.22 (E309CbLT-3) **AWS** A5.22 (E309CbLT-3)
W30935	Austenitic Cr-Ni Stainless Steel Flux Cored Wire (E309LT-X)	**C** 0.03 max **Cr** 22.0-25.0 **Cu** 0.5 max **Mn** 0.5-2.5 **Mo** 0.5 max **Ni** 12.0-14.0 **P** 0.04 max **S** 0.03 max **Si** 1.0 max	**ASME** SFA5.22 (E309LT-1) (E309LT-2) **AWS** A5.22 (E309LT-1) (E309LT-2)
W30937	Austenitic Cr-Ni Stainless Steel Flux Cored Wire (E309LT-3)	**C** 0.03 max **Cr** 23.0-25.5 **Cu** 0.5 max **Mn** 0.5-2.5 **Mo** 0.5 max **Ni** 12.0-14.0 **P** 0.04 max **S** 0.03 max **Si** 1.0 max	**ASME** SFA5.22 (E309LT-3) **AWS** A5.22 (E309LT-3)
W30940	Austenitic Cr-Ni Stainless Steel Composite Welding Filler Metal (ER309)	**C** 0.12 max **Cr** 23.0-25.0 **Cu** 0.75 max **Mn** 1.0-2.5 **Mo** 0.75 max **Ni** 12.0-14.0 **P** 0.03 max **S** 0.03 max **Si** 0.30-0.65	**ASME** SFA5.9 (ER309) **AWS** A5.9 (ER309)
W30941	Austenitic Cr-Ni Stainless Steel Composite Welding Filler Metal (ER309Si)	**C** 0.12 max **Cr** 23.0-25.0 **Cu** 0.5 max **Fe** rem **Mn** 1.0-2.5 **Mo** 0.5 max **Ni** 12.0-14.0 **P** 0.03 max **S** 0.03 max **Si** 0.65-1.00	**ASME** SFA5.9 (ER309Si) **AWS** A5.9 (ER309Si)
W30943	Austenitic Cr-Ni Stainless Steel Composite Welding Filler Metal (ER309L)	**C** 0.03 max **Cr** 23.0-25.0 **Cu** 0.75 max **Mn** 1.0-2.5 **Mo** 0.75 max **Ni** 12.0-14.0 **P** 0.03 max **S** 0.03 max **Si** 0.30-0.65	**ASME** SFA5.9 (ER309L) **AWS** A5.9 (ER309L)
W31010	Austenitic Cr-Ni Stainless Steel Arc Welding Electrode (E310)	**C** 0.08-0.20 **Cr** 25.0-28.0 **Cu** 0.75 max **Mn** 1.0-2.5 **Mo** 0.75 max **Ni** 20.0-22.5 **P** 0.03 max **S** 0.03 max **Si** 0.75 max	**AMS** 5695 **ASME** SFA5.4 (E310) **AWS** A5.4 (E310) **MIL SPEC** MIL-E-22200/2 (MIL-310)
W31015	Austenitic Cr-Ni Stainless Steel Arc Welding Electrode (E310H) High Carbon	**C** 0.35-0.45 **Cr** 25.0-28.0 **Cu** 0.75 max **Mn** 1.0-2.5 **Mo** 0.75 max **Ni** 20.0-22.5 **P** 0.03 max **S** 0.03 max **Si** 0.75 max	**ASME** SFA5.4 (E310H) **AWS** A5.4 (E310H)
W31017	Austenitic Cr-Ni-Cb Stainless Steel Arc Welding Electrode (E310Cb)	**C** 0.12 max **Cb** 0.70-1.00 **Cr** 25.0-28.0 **Cu** 0.75 max **Mn** 1.0-2.5 **Mo** 0.75 max **Ni** 20.0-22.0 **P** 0.03 max **S** 0.03 max **Si** 0.75 max	**ASME** SFA5.4 (E310Cb) **AWS** A5.4 (E310Cb)
W31020	Austenitic Cr-Ni-Mo Stainless Steel Arc Welding Electrode (E310Mo)	**C** 0.12 max **Cr** 25.0-28.0 **Cu** 0.75 max **Mn** 1.0-2.5 **Mo** 2.0-3.0 **Ni** 20.0-22.0 **P** 0.03 max **S** 0.03 max **Si** 0.75 max	**ASME** SFA5.4 (E310Mo) **AWS** A5.4 (E310Mo)
W31031	Austenitic Cr-Ni Stainless Steel Flux Cored Wire (E310T-X)	**C** 0.20 max **Cr** 25.0-28.0 **Cu** 0.5 max **Mn** 1.0-2.50 **Mo** 0.5 max **Ni** 20.0-22.5 **P** 0.03 max **S** 0.03 max **Si** 1.0 max	**ASME** SFA5.22 (E310T-1) (E310T-2) (E310T-3) **AWS** A5.22 (E310T-1) (E310T-2) (E310T-3)

The chemical compositions listed are for identification purposes and should not be used in lieu of the cross referenced specifications.

UNIFIED NUMBER	DESCRIPTION	CHEMICAL COMPOSITION	CROSS REFERENCE SPECIFICATIONS
W31040	Austenitic Cr-Ni Stainless Steel Composite Welding Filler Metal (ER310)	**C** 0.08-0.15 **Cr** 25.0-28.0 **Cu** 0.75 max **Mn** 1.0-2.5 **Mo** 0.75 max **Ni** 20.0-22.5 **P** 0.03 max **S** 0.03 max **Si** 0.30-0.65	**ASME** SFA5.9 (ER310) **AWS** A5.9 (ER310)
W31310	Austenitic-Ferritic Cr-Ni Stainless Steel Arc Welding Electrode (E312)	**C** 0.15 max **Cr** 28.0-32.0 **Cu** 0.75 max **Mn** 0.5-2.5 **Mo** 0.75 max **Ni** 8.0-10.5 **P** 0.04 max **S** 0.03 max **Si** 0.90 max	**AMS** 5785 **ASME** SFA5.4 (E312) **AWS** A5.4 (E312) **MIL SPEC** MIL-E-22200/2 (MIL-312)
W31331	Austenitic-Ferritic Cr-Ni Stainless Steel Arc Welding Electrode (E312T-X)	**C** 0.15 max **Cr** 28.0-32.0 **Cu** 0.5 max **Mn** 0.5-2.5 **Mo** 0.5 max **Ni** 8.0-10.5 **P** 0.04 max **S** 0.03 max **Si** 1.0 max	**ASME** SFA5.22 (E312T-1) (E312T-2) (E312T-3) **AWS** A5.22 (E312T-1) (E312T-2) (E312T-3)
W31340	Austenitic Cr-Ni Stainless Steel Composite Welding Filler Metal (ER312)	**C** 0.15 max **Cr** 28.0-32.0 **Mn** 1.00-2.50 **Mo** 0.75 max **Ni** 8.00-10.5 **P** 0.03 max **S** 0.03 max **Si** 0.30-0.65	**ASME** SFA5.9 (ER312) **AWS** A5.9 (ER312)
W31610	Austenitic Cr-Ni-Mo Stainless Steel Arc Welding Electrode (E316)	**C** 0.08 max **Cr** 17.0-20.0 **Cu** 0.75 max **Mn** 0.5-2.5 **Mo** 2.0-3.0 **Ni** 11.0-14.0 **P** 0.04 max **S** 0.03 max **Si** 0.90 max	**AMS** 5691 **ASME** SFA5.4 (E316) (E316H) **AWS** A5.4 (E316) (E316H) **MIL SPEC** MIL-E-22200/2 (MIL-316)
W31613	Austenitic Cr-Ni-Mo Stainless Steel Arc Welding Electrode (E316L)	**C** 0.04 max **Cr** 17.0-20.0 **Cu** 0.75 max **Mn** 0.5-2.5 **Mo** 2.0-3.0 **Ni** 11.0-14.0 **P** 0.04 max **S** 0.03 max **Si** 0.90 max	**ASME** SFA5.4 (E316L) **AWS** A5.4 (E316L) **MIL SPEC** MIL-E-22200/2 (MIL-316L)
W31631	Austenitic Cr-Ni-Mo Stainless Steel Flux Cored Wire (E316T-X)	**C** 0.08 max **Cr** 17.0-20.0 **Cu** 0.5 max **Mn** 0.5-2.5 **Mo** 2.0-3.0 **Ni** 11.0-14.0 **P** 0.04 max **S** 0.03 max **Si** 1.0 max	**ASME** SFA5.22 (E316T-1) (E316T-2) **AWS** A5.22 (E316T-1) (E316T-2)
W31633	Austenitic Cr-Ni-Mo Stainless Steel Flux Cored Wire (E316T-3)	**C** 0.08 max **Cr** 18.0-20.5 **Cu** 0.5 max **Mn** 0.5-2.5 **Mo** 2.0-3.0 **Ni** 11.0-14.0 **P** 0.04 max **S** 0.03 max **Si** 1.0 max	**ASME** SFA5.22 (E316T-3) **AWS** A5.22 (E316T-3)
W31635	Austenitic Cr-Ni-Mo Stainless Steel Flux Cored Wire (E316LT-X)	**C** 0.03 max **Cr** 17.0-20.0 **Cu** 0.5 max **Mn** 0.5-2.5 **Mo** 2.0-3.0 **Ni** 11.0-14.0 **P** 0.04 max **S** 0.03 max **Si** 1.0 max	**ASME** SFA5.22 (E316LT-1) (E316LT-2) **AWS** A5.22 (E316LT-1) (E316LT-2)
W31637	Austenitic Cr-Ni-Mo Stainless Steel Flux Cored Wire (E316LT-3)	**C** 0.03 max **Cr** 18.0-20.5 **Cu** 0.5 max **Mn** 0.5-2.5 **Mo** 2.0-3.0 **Ni** 11.0-14.0 **P** 0.04 max **S** 0.03 max **Si** 1.0 max	**ASME** SFA5.22 (E316LT-3) **AWS** A5.22 (E316LT-3)
W31640	Austenitic Cr-Ni-Mo Stainless Steel Composite Welding Filler Metal (ER316)	**C** 0.08 max **Cr** 18.0-20.0 **Cu** 0.75 max **Mn** 1.0-2.5 **Mo** 2.0-3.0 **Ni** 11.0-14.0 **P** 0.03 max **S** 0.03 max **Si** 0.30-0.65	**ASME** SFA5.9 (ER316) (ER316H) **AWS** A5.9 (ER316) (ER316H)
W31641	Austenitic Cr-Ni Stainless Steel Composite Welding Filler Metal (ER316Si)	**C** 0.08 max **Cr** 18.0-20.0 **Cu** 0.5 max **Mn** 1.0-2.5 **Mo** 2.0-3.0 **Ni** 11.0-14.0 **P** 0.03 max **S** 0.03 max **Si** 0.65-1.00	**ASME** SFA5.9 (ER316Si) **AWS** A5.9 (ER316Si)
W31643	Austenitic Cr-Ni-Mo Stainless Steel Composite Welding Filler Metal (ER316L)	**C** 0.03 max **Cr** 18.0-22.0 **Cu** 0.75 max **Mn** 1.0-2.5 **Mo** 2.0-3.0 **Ni** 11.0-14.0 **P** 0.03 max **S** 0.03 max **Si** 0.30-0.65	**ASME** SFA5.9 (ER316L) **AWS** A5.9 (ER316L)
W31648	Austenitic Cr-Ni Stainless Steel Composite Welding Filler Metal (ER316LSi)	**C** 0.03 max **Cr** 18.0-20.0 **Cu** 0.5 max **Mn** 1.0-2.5 **Mo** 2.0-3.0 **Ni** 11.0-14.0 **P** 0.03 max **S** 0.03 max **Si** 0.65-1.00	**ASME** SFA5.9 (ER316LSi) **AWS** A5.9 (ER316LSi)
W31710	Austenitic Cr-Ni-Mo Stainless Steel Arc Welding Electrode (E317)	**C** 0.08 max **Cr** 18.0-21.0 **Cu** 0.75 max **Mn** 0.5-2.5 **Mo** 3.0-4.0 **Ni** 12.0-14.0 **P** 0.04 max **S** 0.03 max **Si** 0.90 max	**ASME** SFA5.4 (E317) **AWS** A5.4 (E317) **MIL SPEC** MIL-E-22200/2 (MIL-317)
W31713	Austenitic Cr-Ni-Mo Stainless Steel Arc Welding Electrode (E317L)	**C** 0.04 max **Cr** 18.0-21.0 **Cu** 0.75 max **Mn** 0.5-2.5 **Mo** 3.0-4.0 **Ni** 12.0-14.0 **P** 0.04 max **S** 0.03 max **Si** 0.90 max	**ASME** SFA5.4 (E317L) **AWS** A5.4 (E317L)
W31735	Austenitic Cr-Ni-Mo Stainless Steel Flux Cored Wire (E317LT-X)	**C** 0.03 max **Cr** 18.0-21.0 **Cu** 0.5 max **Mn** 0.5-2.5 **Mo** 3.0-4.0 **Ni** 12.0-14.0 **P** 0.04 max **S** 0.03 max **Si** 1.0 max	**ASME** SFA5.22 (E317LT-1) (E317LT-2) **AWS** A5.22 (E317LT-1) (E317LT-2)
W31737	Austenitic Cr-Ni-Mo Stainless Steel Flux Cored Wire (E317LT-3)	**C** 0.03 max **Cr** 18.5-21.0 **Cu** 0.5 max **Mn** 0.5-2.5 **Mo** 3.0-4.0 **Ni** 13.5-15.0 **P** 0.04 max **S** 0.03 max **Si** 1.0 max	**ASME** SFA5.22 (E317LT-3) **AWS** A5.22 (E317LT-3)
W31740	Austenitic Cr-Ni-Mo Stainless Steel Composite Welding Filler Metal (ER317)	**C** 0.08 max **Cr** 18.5-20.5 **Cu** 0.75 max **Mn** 1.0-2.5 **Mo** 3.0-4.0 **Ni** 13.0-15.0 **P** 0.03 max **S** 0.03 max **Si** 0.30-0.65	**ASME** SFA5.9 (ER317) **AWS** A5.9 (ER317)

The chemical compositions listed are for identification purposes and should not be used in lieu of the cross referenced specifications.

UNIFIED NUMBER	DESCRIPTION	CHEMICAL COMPOSITION	CROSS REFERENCE SPECIFICATIONS
W31743	Austenitic Cr-Ni-Mo Stainless Steel Composite Welding Filler Metal (ER317L)	**C** 0.03 max **Cr** 18.5-20.5 **Cu** 0.75 max **Mn** 1.0-2.5 **Mo** 3.0-4.0 **Ni** 13.0-15.0 **P** 0.04 max **S** 0.03 max **Si** 0.30-0.65	**ASME** SFA5.9 (ER317L) **AWS** A5.9 (ER317L)
W31910	Austenitic Cr-Ni-Mo Stainless Steel Arc Welding Electrode (E318)	**C** 0.08 max **Cb** 6xC-1.00 includes Ta **Cr** 17.0-20.0 **Cu** 0.75 max **Mn** 0.5-2.5 **Mo** 2.0-2.5 **Ni** 11.0-14.0 **P** 0.04 max **S** 0.03 max **Si** 0.90 max	**ASME** SFA5.4 (E318) **AWS** A5.4 (E318) **MIL SPEC** MIL-E-22200/2 (MIL-318)
W31940	Austenitic Cr-Ni-Mo-Cb Stainless Steel Composite Welding Filler Metal (ER318)	**C** 0.08 max **Cb** 8xC-1.0 includes Ta **Cr** 18.0-20.0 **Cu** 0.75 max **Mn** 1.0-2.5 **Mo** 2.0-3.0 **Ni** 11.0-14.0 **P** 0.03 max **S** 0.03 max **Si** 0.30-0.65	**ASME** SFA5.9 (ER318) **AWS** A5.9 (ER318)
W32140	Austenitic Cr-Ni-Ti Stainless Steel Composite Welding Filler Metal (ER321)	**C** 0.08 max **Cr** 18.5-20.5 **Cu** 0.75 max **Mn** 1.0-2.5 **Mo** 0.75 max **Ni** 9.0-10.5 **P** 0.03 max **S** 0.03 max **Si** 0.30-0.65 **Ti** 9xC-1.0	**ASME** SFA5.9 (ER321) **AWS** A5.9 (ER321)
W32210	Nitrogen Strengthened Cr-Ni-Mn-Mo Stainless Steel Arc Welding Electrode (22-10-5)	**C** 0.06 max **Cr** 20.5-24.0 **Cu** 0.75 max **Mn** 4.0-7.0 **Mo** 1.5-3.0 **N** 0.10-0.30 **Ni** 9.5-12.0 **P** 0.03 max **S** 0.03 max **Si** 0.90 max	**ASME** SFA5.4 (E209) **AWS** A5.4 (E209)
W32240	Nitrogen Strengthened Austenitic Cr-Ni-Mn-Mo Stainless Steel Composite Welding Filler Metal (22-10-5)	**C** 0.05 max **Cr** 20.5-24.0 **Cu** 0.75 max **Mn** 4.0-7.0 **Mo** 1.5-3.0 **N** 0.10-0.30 **Ni** 9.5-12.0 **P** 0.03 max **S** 0.03 max **Si** 0.90 max	**ASME** SFA5.9 (ER209) **AWS** A5.9 (ER209)
W32310	Nitrogen Strengthened Austenitic Cr-Ni-Mn-Mo Stainless Steel Arc Welding Electrode (21-6-9)	**C** 0.06 max **Cr** 19.0-21.5 **Cu** 0.75 max **Mn** 8.0-10.0 **Mo** 0.75 max **N** 0.10-0.30 **Ni** 5.5-7.0 **P** 0.03 max **S** 0.03 max **Si** 1.00 max	**ASME** SFA5.4 (E219) **AWS** A5.4 (E219)
W32340	Nitrogen Strengthened Austenitic Cr-Mn-Ni Stainless Steel Composite Welding Filler Metal (21-6-9)	**C** 0.05 max **Cr** 19.0-21.5 **Cu** 0.75 max **Mn** 8.0-10.0 **Mo** 0.75 max **N** 0.10-0.30 **Ni** 5.5-7.0 **P** 0.03 max **S** 0.03 max **Si** 1.00 max	**ASME** SFA5.9 (ER219) **AWS** A5.9 (ER219)
W32410	Nitrogen Strengthened Cr-Ni-Mn-Mo Stainless Steel Arc Welding Electrode (18-5-11)	**C** 0.06 max **Cr** 17.0-19.0 **Cu** 0.75 max **Mn** 10.5-13.5 **Mo** 0.75 max **N** 0.10-0.20 **Ni** 4.0-6.0 **P** 0.03 max **S** 0.03 max **Si** 1.00 max	**ASME** SFA5.4 (E240) **AWS** A5.4 (E240)
W32440	Nitrogen Strengthened Austenitic Cr-Mn-Ni Stainless Steel Composite Welding Filler Metal (18-5-11)	**C** 0.05 max **Cr** 17.0-19.0 **Cu** 0.75 max **Mn** 10.5-13.5 **Mo** 0.75 max **N** 0.10-0.20 **Ni** 4.0-6.0 **P** 0.03 max **S** 0.03 max **Si** 1.00 max	**ASME** SFA5.9 (ER240) **AWS** A5.9 (ER240)
W32540	Nitrogen Strengthened Austenitic Cr-Ni-Mn Stainless Steel Composite Welding Filler Metal (Nitronic 60W)	**C** 0.10 max **Cr** 16.0-18.0 **Cu** 0.75 max **Mn** 7.0-9.0 **Mo** 0.75 max **N** 0.08-0.18 **Ni** 8.0-9.0 **P** 0.03 max **S** 0.03 max **Si** 3.5-4.5	**ASME** SFA5.9 (ER218) **AWS** A5.9 (ER218)
W34710	Austenitic Cr-Ni-Cb Stainless Steel Arc Welding Electrode (E347)	**C** 0.08 max **Cb** 8xC-1.00 includes Ta **Cr** 18.0-21.0 **Cu** 0.75 max **Mn** 0.5-2.5 **Ni** 9.0-11.0 **P** 0.04 max **S** 0.03 max **Si** 0.90 max	**AMS** 5681 **ASME** SFA5.4 (E347) **AWS** A5.4 (E347) **MIL SPEC** MIL-E-22200/2 (MIL-347) (MIL-347Co)
W34712	Austenitic Cr-Ni-Cb Stainless Steel Arc Welding Electrode (347HC) High Carbon	**C** 0.08-0.15 **Cb** 8xC-1.4 includes Ta **Cr** 18.0-21.0 **Mn** 2.5 max **Ni** 9.0-11.0 **P** 0.025 max **S** 0.025 max **Si** 5xC max	**MIL SPEC** MIL-E-22200/2 (MIL-347HC)
W34731	Austenitic Cr-Ni-Cb Stainless Steel Flux Cored Wire (E347T-X)	**C** 0.08 max **Cb** 8xC-1.0 **Cr** 18.0-21.0 **Cu** 0.5 max **Mn** 0.5-2.5 **Mo** 0.5 max **Ni** 9.0-11.0 **P** 0.04 max **S** 0.03 max **Si** 1.0 max	**ASME** SFA5.22 (E347T-1) (E347T-2) **AWS** A5.22 (E347T-1) (E347T-2)
W34733	Austenitic Cr-Ni-Cb Stainless Steel Flux Cored Wire (E347T-3)	**C** 0.08 max **Cb** 8xC-1.0 **Cr** 19.0-21.5 **Cu** 0.5 max **Mn** 0.5-2.50 **Mo** 0.5 max **Ni** 9.0-11.0 **P** 0.04 max **S** 0.03 max **Si** 1.0 max	**ASME** SFA5.22 (E347T-3) **AWS** A5.22 (E347T-3)

The chemical compositions listed are for identification purposes and should not be used in lieu of the cross referenced specifications.

UNIFIED NUMBER	DESCRIPTION	CHEMICAL COMPOSITION	CROSS REFERENCE SPECIFICATIONS
W34740	Austenitic Cr-Ni-Cb Stainless Steel Composite Welding Filler Metal (ER347)	C 0.08 max Cb 10xC-1.0 Cr 19.0-21.5 Cu 0.75 max Mn 1.0-2.5 max Mo 0.75 max Ni 9.0-11.0 P 0.03 max S 0.03 max Si 0.30-0.65	ASME SFA5.9 (ER347) AWS A5.9 (ER347)
W34748	Austenitic Cr-Ni Stainless Steel Composite Welding Filler Metal (ER347Si)	C 0.08 max Cb 10xC-1.0 includes Ta Cr 19.0-21.5 Cu 0.5 max Mn 1.0-2.5 Mo 0.5 max Ni 9.0-11.0 P 0.03 max S 0.03 max Si 0.65-1.00	ASME SFA5.9 (ER347Si) AWS A5.9 (ER347Si)
W34910	Austenitic Cr-Ni-W-Mo Heat Resisting Arc Welding Electrode (E349)	C 0.13 max Cb 0.75-1.2 Cr 18.0-21.0 Cu 0.75 max Mn 0.5-2.5 Mo 0.35-0.65 Ni 8.0-10.0 P 0.04 max S 0.03 max Si 0.90 max Ti 0.15 max W 1.25-1.75	AMS 5783 ASME SFA5.4 (E349) AWS A5.4 (E349) MIL SPEC MIL-E-22200/2 (MIL-349)
W34940	Austenitic Cr-Ni Stainless Steel Composite Welding Filler Metal (ER349)	C 0.07-0.13 Cb 1.00-1.40 Cr 19.0-21.5 Mn 1.00-2.50 Mo 0.35-0.65 Ni 8.00-9.50 P 0.03 max S 0.03 max Si 0.30-0.65	ASME SFA5.9 (ER349) AWS A5.9 (ER349)
W35010	Precipitation Hardenable Cr-Ni-Mo Stainless Steel Arc Welding Electrode (350)	C 0.08-0.12 Cr 16.00-17.00 Mn 0.50-1.25 Mo 2.50-3.25 N 0.07-0.13 Ni 4.00-5.00 P 0.040 max S 0.030 max Si 0.50 max	AMS 5775
W35510	Precipitation Hardenable Stainless Steel Arc Welding Electrode	C 0.10-0.15 Cr 15.0-16.0 Cu 0.50 max Mn 0.50-1.25 Mo 2.50-3.25 N 0.07-0.13 Ni 4.00-5.00 P 0.040 max S 0.030 max Si 0.50 max	AMS 5781
W36810	Austenitic Cr-Ni-Mo Stainless Steel Arc Welding Electrode (E16-8-2)	C 0.10 max Cr 14.5-16.5 Cu 0.75 max Mn 0.5-2.5 Mo 1.0-2.0 Ni 7.5-9.5 P 0.03 max S 0.03 max Si 0.60 max	ASME SFA5.4 (E16-8-2) AWS A5.4 (E16-8-2) MIL SPEC MIL-E-22200/2C Type MIL-16-8-2
W36840	Austenitic Cr-Ni-Mo Stainless Steel Composite Welding Filler Metal (16-8-2)	C 0.10 max Cr 14.5-16.5 Cu 0.75 max Mn 1.0-2.5 Mo 1.0-2.0 Ni 7.5-9.5 P 0.03 max S 0.03 max Si 0.30-0.65	ASME SFA5.9 (ER16-8-2) AWS A5.9 (ER16-8-2)
W37410	Precipitation Hardenable Cr-Ni-Cu Stainless Steel Arc Welding Electrode (17-4PH) (E630)	C 0.05 max Cb 0.15-0.30 includes Ta Cr 16.00-16.75 Cu 3.25-4.00 Mn 0.25-0.75 Mo 0.75 max Ni 4.50-5.00 P 0.04 max S 0.03 max Si 0.75 max	AMS 5827 ASME SFA5.4 (E630) AWS A5.4 (E630)
W37440	Precipitation Hardenable Cr-Ni-Cu Stainless Steel Composite Welding Filler Metal (17-4PH) (ER630)	C 0.05 max Cb 0.15-0.30 includes Ta Cr 16.00-16.75 Cu 3.25-4.00 Mn 0.25-0.75 Mo 0.75 max Ni 4.50-5.00 P 0.04 max S 0.03 max Si 0.75 max	ASME SFA5.9 (ER630) AWS A5.9 (ER630)
W40931	Ferritic Cr-Ti Stainless Steel Flux Cored Wire (E409T-X)	C 0.10 max Cr 10.5-13.0 Cu 0.5 max Mn 0.80 max Mo 0.5 max Ni 0.60 max P 0.04 max S 0.03 max Si 1.0 max Ti 10xC-1.5	ASME SFA5.22 (E409T-1) (E409T-2) (E409T-3) AWS A5.22 (E409T-1) (E409T-2) (E409T-3)
W41010	Martensitic Cr Stainless Steel Arc Welding Electrode (E410)	C 0.12 max Cr 11.0-13.5 Cu 0.75 max Mn 1.0 max Mo 0.75 max Ni 0.6 max P 0.04 max S 0.03 max Si 0.90 max	AMS 5777 ASME SFA5.4 (E410) AWS A5.4 (E410) MIL SPEC MIL-E-22200/8 (MIL-410)
W41016	Martensitic Cr-Ni-Mo Stainless Steel Arc Welding Electrode (E410NiMo)	C 0.06 max Cr 11.0-12.5 Cu 0.75 max Mn 1.0 max Mo 0.40-0.70 Ni 4.0-5.0 P 0.04 max S 0.03 max Si 0.90 max	ASME SFA5.4 (E410NiMo) AWS A5.4 (E410NiMo)
W41031	Martensitic Cr Stainless Steel Flux Cored Wire (E410T-X)	C 0.12 max Cr 11.0-13.5 Cu 0.5 max Mn 1.2 max Mo 0.5 max Ni 0.60 max P 0.04 max S 0.03 max Si 1.0 max	ASME SFA5.22 (E410T-1) (E410T-2) (E410T-3) AWS A5.22 (E410T-1) (E410T-2) (E410T-3)
W41036	Martensitic Cr-Ni-Mo Stainless Steel Flux Cored Wire (E410NiMoT-X)	C 0.06 max Cr 11.0-12.5 Cu 0.5 max Mn 1.0 max Mo 0.40-0.70 Ni 4.0-5.0 P 0.04 max S 0.03 max Si 1.0 max	ASME SFA5.22 (E410NiMoT-1) (E410NiMoT-2) (E410NiMoT-3) AWS A5.22 (E410NiMoT-1) (E410NiMoT-2) (E410NiMoT-3)
W41038	Martensitic Cr-Ni-Ti Stainless Steel Flux Cored Wire (E410NiTiT-X)	C 0.03 max Cr 11.0-12.0 Cu 0.5 max Mn 0.70 max Mo 0.05 max Ni 3.6-4.5 P 0.03 max S 0.03 max Si 0.50 max Ti 10xC-1.5	ASME SFA5.22 (E410NiTiT-1) (E410NiTiT-2) (E410NiTiT-3) AWS A5.22 (E410NiTiT-1) (E410NiTiT-2) (E410NiTiT-3)

The chemical compositions listed are for identification purposes and should not be used in lieu of the cross referenced specifications.

UNIFIED NUMBER	DESCRIPTION	CHEMICAL COMPOSITION	CROSS REFERENCE SPECIFICATIONS
W41040	Martensitic Cr Stainless Steel Composite Welding Filler Metal (ER410)	C 0.12 max Cr 11.5-13.5 Cu 0.75 max Mn 0.6 max Mo 0.75 max Ni 0.6 max P 0.03 max S 0.03 max Si 0.50 max	**ASME** SFA5.9 (ER410) **AWS** A5.9 (ER410)
W41046	Martensitic Cr-Ni-Mo Stainless Steel Composite Welding Filler Metal (ER410NiMo)	C 0.06 max Cr 11.0-12.5 Cu 0.75 max Mn 0.6 max Mo 0.4-0.7 Ni 4.0-5.0 P 0.03 max S 0.03 max Si 0.50 max	**ASME** SFA5.9 (ER410NiMo) **AWS** A5.9 (ER410NiMo)
W42040	Martensitic Cr Stainless Steel Composite Welding Filler Metal (ER420)	C 0.25-0.40 Cr 12.0-14.0 Cu 0.75 max Mn 0.6 max Mo 0.75 max Ni 0.6 max P 0.03 max S 0.03 max Si 0.50 max	**ASME** SFA5.9 (ER420) **AWS** A5.9 (ER420)
W43010	Ferritic Cr Stainless Steel Arc Welding Electrode (E430)	C 0.10 max Cr 15.0-18.0 Cu 0.75 max Mn 1.0 max Mo 0.75 max Ni 0.60 max P 0.04 max S 0.03 max Si 0.90 max	**ASME** SFA5.4 (E430) **AWS** A5.4 (E430)
W43031	Ferritic Cr Stainless Steel Flux Cored Wire (E430T-X)	C 0.10 max Cr 15.0-18.0 Cu 0.5 max Mn 1.2 max Mo 0.5 max Ni 0.60 max P 0.04 max S 0.03 Si 1.0 max	**ASME** SFA5.22 (E430T-1) (E430T-2) (E430T-3) **AWS** A5.22 (E430T-1) (E430T-2) (E430T-3)
W43040	Ferritic Cr Stainless Steel Composite Welding Filler Metal (ER430)	C 0.10 max Cr 15.5-17.0 Cu 0.75 max Mn 0.6 max Mo 0.75 max Ni 0.6 max P 0.03 max S 0.03 max Si 0.50 max	**ASME** SFA5.9 (ER430) **AWS** A5.9 (ER430)
W44647	Ferritic Cr-Mo Stainless Steel Composite Welding Filler Metal (ER26-1)	C 0.01 max Cr 25.0-27.5 Cu 0.20 max Mn 0.40 max Mo 0.75-1.50 N 0.015 max Ni 0.5 max includes Cu P 0.02 max S 0.02 max Si 0.40 max	**ASME** SFA5.9 (ER26-1) **AWS** A5.9 (ER26-1)
W50140	Martensitic 5% Cr-Mo Steel Composite Wire for Submerged Arc Welding (ECB6H) High Carbon	C 0.25 max Cr 4.50-6.00 Cu 0.35 max Mn 1.20 max Mo 0.40-0.65 P 0.030 max S 0.040 max Si 0.80 max	**ASME** SFA5.23 (ECB6H) **AWS** A5.23 (ECB6H)
W50141	Martensitic 5% Cr-Mo Steel Composite Wire for Submerged Arc Welding (ECB6HN) High Carbon-Nuclear Grade	C 0.25 max Cr 4.50-6.00 Cu 0.08 max Mn 1.20 max Mo 0.40-0.65 P 0.012 max S 0.040 max Si 0.80 max V 0.05 max	**ASME** SFA5.23 (ECB6HN) **AWS** A5.23 (ECB6HN)
W50210	Martensitic 5% Cr-Mo Steel Arc Welding Electrode (E502)	C 0.10 max Cr 4.0-6.0 Cu 0.75 max Mn 1.0 max Mo 0.45-0.65 Ni 0.40 max P 0.04 max S 0.03 max Si 0.90 max	**AMS** 6467 **ASME** SFA5.4 (E502) **AWS** A5.4 (E502) **MIL SPEC** MIL-E-22200/8 (MIL-502)
W50213	Martensitic 5% Cr-Mo Steel Arc Welding Electrode (502L)	C 0.05 max Cr 4.00-6.00 Mn 1.00 max Mo 0.45-0.65 P 0.04 max S 0.03 max Si 0.90 max	**MIL SPEC** MIL-E-22200/8 (MIL-502-1x-L)
W50230	Martensitic 5% Cr-Mo Steel, Flux Cored Welding Wire (B6L)	C 0.05 max Cr 4.00-6.00 Cu 0.50 max Mn 1.25 max Mo 0.45-0.65 Ni 0.40 max P 0.04 max S 0.03 max Si 1.00 max	**ASME** SFA5.29 (E6xT5-B6L) **AWS** A5.29 (E6xT5-B6L)
W50231	Martensitic 5% Cr-Mo Steel Flux Cored Wire (E502T-X)	C 0.10 max Cr 4.0-6.0 Cu 0.5 max Mn 1.2 max Mo 0.45-0.65 Ni 0.4 max P 0.04 max S 0.03 max Si 1.0 max	**ASME** SFA5.22 (E502T-1) (E502T-2) **AWS** A5.22 (E502T-1) (E502T-2)
W50240	Martensitic 5% Cr-Mo Steel Composite Wire for Submerged Arc Welding (ECB6)	C 0.12 max Cr 4.50-6.00 Cu 0.35 max Mn 1.20 max Mo 0.40-0.65 P 0.030 max S 0.040 max Si 0.80 max	**ASME** SFA5.23 (ECB6); SFA5.9 (ER502) **AWS** A5.23 (ECB6); A5.9 (ER502)
W50241	Martensitic 5% Cr-Mo Steel Composite Wire for Submerged Arc Welding (ECB6N) Nuclear Grade	C 0.12 max Cr 4.50-6.00 Cu 0.08 max Mn 1.20 max Mo 0.40-0.65 P 0.012 max S 0.040 max Si 0.80 max V 0.05 max	**ASME** SFA5.23 (ECB6N) **AWS** A5.23 (ECB6N)
W50310	Martensitic 7% Cr-Mo Steel Arc Welding Electrode (E7Cr)	C 0.10 max Cr 6.0-8.0 Cu 0.75 max Mn 1.0 max Mo 0.45-0.65 Ni 0.40 max P 0.04 max S 0.03 max Si 0.90 max	**ASME** SFA5.4 (E7Cr) **AWS** A5.4 (E7Cr)
W50410	Martensitic 9% Cr-Mo Steel Arc Welding Electrode (E505)	C 0.10 max Cr 8.0-10.5 Cu 0.75 max Mn 1.0 max Mo 0.85-1.20 Ni 0.40 max P 0.04 max S 0.03 max Si 0.90 max	**ASME** SFA5.4 (E505) **AWS** A5.4 (E505) **MIL SPEC** MIL-E-22200/8 (MIL-505)

The chemical compositions listed are for identification purposes and should not be used in lieu of the cross referenced specifications.

WELDING FILLER METALS

UNIFIED NUMBER	DESCRIPTION	CHEMICAL COMPOSITION	CROSS REFERENCE SPECIFICATIONS
W50411	Martensitic 9% Cr-Mo Steel Arc Welding Electrode (365)	C 0.10 max Cr 8.50-10.50 Mo 1.00-1.50 Si 0.60 max	
W50430	Martensitic 9% Cr-Mo Steel, Flux Cored Wire (B8L)	C 0.05 max Cr 8.00-10.5 Cu 0.50 max Mn 1.25 max Mo 0.85-1.20 Ni 0.40 max P 0.03 max S 0.03 max Si 1.00 max	ASME SFA5.29 (E6xT5-B8L) AWS A5.29 (E6xT5-B8L)
W50431	Martensitic 9% Cr-Mo Steel Flux Cored Wire (E505T-X)	C 0.10 max Cr 8.0-10.5 Cu 0.5 max Mn 1.2 max Mo 0.85-1.20 Ni 0.40 max P 0.04 max S 0.03 max Si 1.0 max	ASME SFA5.22 (E505T-1) (E505T-2) AWS A5.22 (E505T-1) (E505T-2)
W50440	Martensitic Cr-Mo Stainless Steel Composite Welding Filler Metal (ER-505)	C 0.10 max Cr 8.0-10.5 Cu 0.75 max Mn 0.6 max Mo 0.8-1.2 Ni 0.5 max P 0.04 max S 0.03 max Si 0.50 max	ASME SFA5.9 (ER505); SFA5.23 (EC-B8) AWS A5.9 (ER505); A5.23 (EC-B8)
W50441	Martensitic 9% Cr-Mo Steel Composite Wire for Submerged Arc Welding (ECB8N) Nuclear Grade	C 0.12 max Cr 8.00-10.00 Cu 0.08 max Mn 1.20 max Mo 0.80-1.20 P 0.012 max S 0.040 max Si 0.80 max V 0.05 max	ASME SFA5.23 (ECB8N) AWS A5.23 (ECB8N)
W51016	Cr-Mo Low Alloy Steel Arc Welding Electrode (E8016-B1)	C 0.12 max Cr 0.40-0.65 Mn 0.90 max Mo 0.40-0.65 P 0.03 max S 0.04 max Si 0.60 max	ASME SFA5.5 (E8016-B1) AWS A5.5 (E8016-B1)
W51018	Cr-Mo Low Alloy Steel Arc Welding Electrode (E8018-B1)	C 0.12 max Cr 0.40-0.65 Mn 0.90 max Mo 0.40-0.65 P 0.03 max S 0.04 max Si 0.80 max	ASME SFA5.5 (E8018-B1) AWS A5.5 (E8018-B1)
W51031	Cr-Mo Low Alloy Steel Flux Cored Wire (E81T1-B1)	C 0.12 max Cr 0.40-0.65 Mn 1.25 max Mo 0.40-0.65 P 0.03 max S 0.03 max Si 0.80 max	ASME SFA5.29 (E81T1-B1) AWS A5.29 (E81T1-B1)
W51040	Cr-Mo Low Alloy Steel Composite Wire for Submerged Arc Welding (ECB1)	C 0.12 max Cr 0.40-0.65 Cu 0.35 max Mn 1.60 max Mo 0.40-0.65 P 0.030 max S 0.040 max Si 0.80 max	ASME SFA5.23 (ECB1) AWS A5.23 (ECB1)
W51041	Cr-Mo Low Alloy Steel Composite Wire for Submerged Arc Welding (ECB1N) Nuclear Grade	C 0.12 max Cr 0.40-0.65 Cu 0.08 max Mn 1.60 max Mo 0.40-0.65 P 0.012 max S 0.040 max Si 0.80 max V 0.05 max	ASME SFA5.23 (ECB1N) AWS A5.23 (ECB1N)
W51131	Cr-Mo Low Alloy Steel, Flux Cored Wire (B1L)	C 0.05 max Cr 0.40-0.65 Mn 1.25 max Mo 0.40-0.65 P 0.03 max S 0.03 max Si 0.80 max	ASME SFA5.29 (E8xT1-B1L) AWS A5.29 (EBxT1-B1L)
W51316	Cr-Mo Low Alloy Steel Arc Welding Electrode (E8016-B5)	C 0.07-0.15 Cr 0.40-0.60 Mn 0.40-0.70 Mo 1.00-1.25 P 0.03 max S 0.04 max Si 0.30-0.60 V 0.05 max	ASME SFA5.5 (E8016-B5) AWS A5.5 (E8016-B5)
W51340	Cr-Mo Low Alloy Steel Composite Wire for Submerged Arc Welding (ECB5)	C 0.18 max Cr 0.40-0.65 Cu 0.35 max Mn 1.20 max Mo 0.90-1.20 P 0.030 max S 0.040 max Si 0.80 max	ASME SFA5.23 (ECB5) AWS A5.23 (ECB5)
W51341	Cr-Mo Low Alloy Steel Composite Wire for Submerged Arc Welding (EB5N) Nuclear Grade	C 0.18 max Cr 0.40-0.65 Cu 0.08 max Mn 1.20 max Mo 0.90-1.20 P 0.012 max S 0.040 max Si 0.80 max V 0.05 max	ASME SFA5.23 (ECB5N) AWS A5.23 (ECB5N)
W52015	Cr-Mo Low Alloy Steel Arc Welding Electrode (8015-B2)	C 0.10 max Cr 1.00-1.50 Mn 0.90 max Mo 0.40-0.65 P 0.03 max S 0.04 max Si 0.60 max	MIL SPEC MIL-E-22200/8 (MIL-8015-B2)
W52016	Cr-Mo Low Alloy Steel Arc Welding Electrode (E8016B-2)	C 0.12 max Cr 1.00-1.50 Mn 0.90 max Mo 0.40-0.65 P 0.03 max S 0.04 max Si 0.60 max	ASME SFA5.5 (E8016-B2) AWS A5.5 (E8016-B2) MIL SPEC MIL-E-22200/8 (MIL-8016-B2)
W52018	Cr-Mo Low Alloy Steel Arc Welding Electrode (E8018-B2)	C 0.12 max Cr 1.00-1.50 Mn 0.90 max Mo 0.40-0.65 P 0.03 max S 0.04 max Si 0.80 max	ASME SFA5.5 (E8018-B2) AWS A5.5 (E8018-B2) MIL SPEC MIL-E-22200/8 (MIL-8018-B2)

The chemical compositions listed are for identification purposes and should not be used in lieu of the cross referenced specifications.

UNIFIED NUMBER	DESCRIPTION	CHEMICAL COMPOSITION	CROSS REFERENCE SPECIFICATIONS
W52030	Cr-Mo Low Alloy Steel Composite Wire for Gas Shielded Arc Welding	**C** 0.07-0.12 **Cr** 1.00-1.50 **Cu** 0.35 max **Mn** 0.40-1.00 **Mo** 0.40-0.65 **Ni** 0.20 max **P** 0.025 max **S** 0.030 max **Si** 0.25-0.60	**ASME** SFA5.28 (E80C-B2) **AWS** A5.28 (E80C-B2)
W52031	Cr-Mo Low Alloy Steel Flux Cored Wire (E8XT1-B2)	**C** 0.12 max **Cr** 1.00-1.50 **Mn** 1.25 max **Mo** 0.40-0.65 **P** 0.03 max **S** 0.03 max **Si** 0.80 max	**ASME** SFA5.29 (E80T1-B2) (E81T1-B2) **AWS** A5.29 (E80T1-B2) (E81T1-B2)
W52035	Cr-Mo Low Alloy Steel Flux Cored Wire (E80T5-B2)	**C** 0.12 max **Cr** 1.00-1.50 **Mn** 1.25 max **Mo** 0.40-0.65 **P** 0.03 max **S** 0.03 max **Si** 0.80 max	**ASME** SFA5.29 (E80T5-B2) **AWS** A5.29 (E80T5-B2)
W52040	Cr-Mo Low Alloy Steel Composite Wire for Submerged Arc Welding (ECB2)	**C** 0.15 max **Cr** 1.00-1.50 **Cu** 0.35 max **Mn** 1.20 max **Mo** 0.40-0.65 **P** 0.030 max **S** 0.040 max **Si** 0.80 max	**ASME** SFA5.23 (ECB2) **AWS** A5.23 (ECB2)
W52041	Cr-Mo Low Alloy Steel Composite Wire for Submerged Arc Welding (ECB2N) Nuclear Grade	**C** 0.15 max **Cr** 1.00-1.50 **Cu** 0.08 max **Mn** 1.20 max **Mo** 0.40-0.65 **P** 0.012 max **S** 0.040 max **Si** 0.80 max **V** 0.05 max	**ASME** SFA5.23 (ECB2N) **AWS** A5.23 (ECB2N)
W52115	Cr-Mo Low Alloy Steel Arc Welding Electrode (E8015-B2L)	**C** 0.05 max **Cr** 1.00-1.50 **Mn** 0.90 max **Mo** 0.40-0.65 **P** 0.03 max **S** 0.04 max **Si** 1.00 max	**ASME** SFA5.5 (E8015-B2L) **AWS** A5.5 (E8015-B2L) **MIL SPEC** MIL-E-22200/8 (MIL-8015-B2L)
W52116	Cr-Mo Low Alloy Steel Arc Welding Electrode	**C** 0.05 max **Cr** 1.00-1.50 **Mn** 0.60-0.85 **Mo** 0.45-0.65 **Si** 0.50-0.90	
W52118	Cr-Mo Low Alloy Steel Arc Welding Electrode (E8018-B2L)	**C** 0.05 max **Cr** 1.00-1.50 **Mn** 0.90 max **Mo** 0.40-0.65 **P** 0.03 max **S** 0.04 max **Si** 0.80 max	**ASME** SFA5.5 (E8018-B2L) **AWS** A5.5 (E8018-B2L) **MIL SPEC** MIL-E-22200/8 (MIL-8018-B2L)
W52130	Cr-Mo Low Alloy Steel Composite Wire for Gas Shielded Arc Welding (E80C-B2L)	**C** 0.05 max **Cr** 1.00-1.50 **Cu** 0.35 max **Mn** 0.40-1.00 **Mo** 0.40-0.65 **Ni** 0.20 max **P** 0.025 max **S** 0.030 max **Si** 0.25-0.60	**ASME** SFA5.28 (E80C-B2L) **AWS** A5.28 (E80C-B2L)
W52135	Cr-Mo Low Alloy Steel Flux Cored Wire (E80T5-B2L)	**C** 0.05 max **Cr** 1.00-1.50 **Mn** 1.25 max **Mo** 0.40-0.65 **P** 0.03 max **S** 0.03 max **Si** 0.80 max	**ASME** SFA5.29 (E80T5-B2L) **AWS** A5.29 (E80T5-B2L)
W52231	Cr-Mo Low Alloy Steel Flux Cored Wire (E80T1-B2H)	**C** 0.10-0.15 **Cr** 1.00-1.50 **Mn** 1.25 max **Mo** 0.40-0.65 **P** 0.03 max **S** 0.03 max **Si** 0.80 max	**ASME** SFA5.29 (E80T1-B2H) **AWS** A5.29 (E80T1-B2H)
W52240	Cr-Mo Low Alloy Steel Composite Wire for Submerged Arc Welding (ECB2H)	**C** 0.25 max **Cr** 1.00-1.50 **Cu** 0.35 max **Mn** 1.20 max **Mo** 0.40-0.65 **P** 0.030 max **S** 0.040 max **Si** 0.80 max **V** 0.30 max	**ASME** SFA5.23 (ECB2H) **AWS** A5.23 (ECB2H)
W52241	Cr-Mo Low Alloy Steel Flux Cored Wire for Submerged Arc Welding (ECB2HN) Nuclear Grade	**C** 0.25 max **Cr** 1.00-1.50 **Cu** 0.08 max **Mn** 1.20 max **Mo** 0.40-0.65 **P** 0.012 max **S** 0.040 max **Si** 0.80 max **V** 0.05 max	**ASME** SFA5.23 (ECB2HN) **AWS** A5.23 (ECB2HN)
W53015	Cr-Mo Low Alloy Steel Arc Welding Electrode (E9015-B3)	**C** 0.12 max **Cr** 2.00-2.50 **Mn** 0.90 max **Mo** 0.90-1.20 **P** 0.03 max **S** 0.04 max **Si** 0.60 max	**ASME** SFA5.5 (E9015-B3) **AWS** A5.5 (E9015-B3) **MIL SPEC** MIL-E-22200/8 (MIL-9015-B3)
W53016	Cr-Mo Low Alloy Steel Arc Welding Electrode (E9016-B3)	**C** 0.12 max **Cr** 2.00-2.50 **Mn** 0.90 max **Mo** 0.90-1.20 **P** 0.03 max **S** 0.04 max **Si** 0.60 max	**ASME** SFA5.5 (E9016-B3) **AWS** A5.5 (E9016-B3) **MIL SPEC** MIL-E-22200/8 (MIL-9016-B3)
W53018	Cr-Mo Low Alloy Steel Arc Welding Electrode (E9018B-3)	**C** 0.12 max **Cr** 2.00-2.50 **Mn** 0.90 max **Mo** 0.90-1.20 **P** 0.03 max **S** 0.04 max **Si** 0.80 max	**ASME** SFA5.5 (E9018-B3) **AWS** A5.5 (E9018-B3) **MIL SPEC** MIL-E-22200/8 (MIL-9018-B3)
W53030	Cr-Mo Low Alloy Steel Composite Wire for Gas Shielded Arc Welding (E90C-B3)	**C** 0.07-0.12 **Cr** 2.00-2.50 **Cu** 0.35 max **Mn** 0.40-1.00 **Mo** 0.90-1.20 **Ni** 0.20 max **P** 0.025 max **S** 0.030 max **Si** 0.25-0.60	**ASME** SFA5.28 (E90C-B3) **AWS** A5.28 (E90C-B3)

The chemical compositions listed are for identification purposes and should not be used in lieu of the cross referenced specifications.

UNIFIED NUMBER	DESCRIPTION	CHEMICAL COMPOSITION	CROSS REFERENCE SPECIFICATIONS
W53031	Cr-Mo Low Alloy Steel Flux Cored Wire (ExxxT1-B3)	**C** 0.12 max **Cr** 2.00-2.50 **Mn** 1.25 max **Mo** 0.90-1.20 **P** 0.03 max **S** 0.03 max **Si** 0.80 max	**ASME** SFA5.29 (E90T1-B3) (E91T1-B3) (E100T1-B3) **AWS** A5.29 (E90T1-B3) (E91T1-B3) (E100T1-B3)
W53035	Cr-Mo Low Alloy Steel Flux Cored Wire (E90T5-B3)	**C** 0.12 max **Cr** 2.00-2.50 **Mn** 1.25 max **Mo** 0.90-1.20 **P** 0.03 max **S** 0.03 max **Si** 0.80 max	**ASME** SFA5.29 (E90T5-B3) **AWS** A5.29 (E90T5-B3)
W53040	Cr-Mo Low Alloy Steel Composite Wire for Submerged Arc Welding (ECB3)	**C** 0.15 max **Cr** 2.00-2.50 **Cu** 0.35 max **Mn** 1.20 max **Mo** 0.90-1.20 **P** 0.030 max **S** 0.040 max **Si** 0.80 max	**ASME** SFA5.23 (ECB3) **AWS** A5.23 (ECB3)
W53041	Cr-Mo Low Alloy Steel Composite Wire for Submerged Arc Welding (ECB3N) Nuclear Grade	**C** 0.15 max **Cr** 2.00-2.50 **Cu** 0.08 max **Mn** 1.20 max **Mo** 0.90-1.20 **P** 0.012 max **S** 0.040 max **Si** 0.80 max **V** 0.05 max	**ASME** SFA5.23 (ECB3N) **AWS** A5.23 (ECB3N)
W53115	Cr-Mo Low Alloy Steel Arc Welding Electrode (E9015-B3L)	**C** 0.05 max **Cr** 2.00-2.50 **Mn** 0.90 max **Mo** 0.90-1.20 **P** 0.03 max **S** 0.04 max **Si** 1.00 max	**ASME** SFA5.5 (E9015-B3L) **AWS** A5.5 (E9015-B3L) **MIL SPEC** MIL-E-22200/8 (MIL-9015-B3L)
W53116	Cr-Mo Low Alloy Steel Arc Welding Electrode	**C** 0.05 max **Cr** 2.0-2.50 **Mn** 0.60-0.85 **Mo** 0.90-1.20 **Si** 0.50-0.90	
W53118	Cr-Mo Low Alloy Steel Arc Welding Electrode (E9018-B3L)	**C** 0.05 max **Cr** 2.00-2.50 **Mn** 0.90 max **Mo** 0.90-1.20 **P** 0.03 max **S** 0.04 max **Si** 0.80 max	**ASME** SFA5.5 (E9018-B3L) **AWS** A5.5 (E9018-B3L) **MIL SPEC** MIL-E-22200/8 (MIL-9018-B3L)
W53130	Cr-Mo Low Alloy Steel Composite Wire for Gas Shielded Arc Welding (E90C-B3L)	**C** 0.05 max **Cr** 2.00-2.50 **Cu** 0.35 max **Mn** 0.40-1.00 **Mo** 0.90-1.20 **Ni** 0.20 max **P** 0.025 max **S** 0.030 max **Si** 0.25-0.60	**ASME** SFA5.28 (E90C-B3L) **AWS** A5.28 (E90C-B3L)
W53131	Cr-Mo Low Alloy Steel Flux Cored Wire (E90T1-B3L)	**C** 0.05 max **Cr** 2.00-2.50 **Mn** 1.25 max **Mo** 0.90-1.20 **P** 0.03 max **S** 0.03 max **Si** 0.80 max	**ASME** SFA5.29 (E90T1-B3L) **AWS** A5.29 (E90T1-B3L)
W53231	Cr-Mo Low Alloy Steel Flux Cored Wire (E90T1-B3H)	**C** 0.10-0.15 **Cr** 2.00-2.50 **Mn** 1.25 max **Mo** 0.90-1.20 **P** 0.03 max **S** 0.03 max **Si** 0.80 max	**ASME** SFA5.29 (E90T1-B3H) **AWS** A5.29 (E90T1-B3H)
W53315	Cr-Mo Low Alloy Steel Arc Welding Electrode	**C** 0.10 max **Cr** 1.75-2.25 **Mo** 0.40-0.60 **Si** 0.60 max	
W53316	Cr-Mo Low Alloy Steel Arc Welding Electrode	**C** 0.10 **Cr** 1.75-2.25 **Mo** 0.40-0.60 **Si** 0.60 max	
W53340	Cr-Mo Low Alloy Steel Tubular Wire for Submerged Arc Welding (ECB4)	**C** 0.12 max **Cr** 1.75-2.25 **Cu** 0.35 max **Mn** 1.20 max **Mo** 0.40-0.65 **P** 0.030 max **S** 0.040 max **Si** 0.80 max	**ASME** SFA5.23 (ECB4) **AWS** A5.23 (ECB4)
W53341	Cr-Mo Low Alloy Composite Wire for Submerged Arc Welding (ECB4N) Nuclear Grade	**C** 0.12 max **Cr** 1.75-2.25 **Cu** 0.08 max **Mn** 1.20 max **Mo** 0.40-0.65 **P** 0.012 max **S** 0.040 max **Si** 0.80 max **V** 0.05 max	**ASME** SFA5.23 (ECB4N) **AWS** A5.23 (ECB4N)
W53415	Cr-Mo Low Alloy Steel Arc Welding Electrode (E8015-B4L)	**C** 0.05 max **Cr** 1.75-2.25 **Mn** 0.90 max **Mo** 0.40-0.65 **P** 0.03 max **S** 0.04 max **Si** 1.00 max	**ASME** SFA5.5 (E8015-B4L) **AWS** A5.5 (E8015-B4L)
W53416	Cr-Mo Low Alloy Steel Arc Welding Electrode	**C** 0.05 max **Cr** 1.75-2.25 **Mn** 0.60-0.85 **Mo** 0.45-0.65 **Si** 0.50-0.90	
W60189	Copper Arc Welding Electrode (ECu)	**Al** 0.1 max **Cu** rem includes Ag **Fe** 0.2 max **Mn** 0.1 max **Si** 0.1 max	**ASME** SFA5.6 (ECu) **AWS** A5.6 (ECu)
W60518	Phosphor Bronze A Arc Welding Electrode (ECuSn-A)	**Al** 0.01 max **Cu** rem **Fe** 0.25 max **P** 0.05-0.35 **Pb** 0.02 max **Sn** 4.0-6.0	**ASME** SFA5.6 (ECuSn-A); SFA5.13 (ECuSn-A) **AWS** A5.6 (ECuSn-A); A5.13 (ECuSn-A)
W60521	Phosphor Bronze C Arc Welding Electrode (ECuSn-C)	**Al** 0.01 max **Cu** rem **Fe** 0.25 max **P** 0.05-0.35 **Pb** 0.02 max **Sn** 7.0-9.0	**ASME** SFA5.6 (ECuSn-C); SFA5.13 (ECuSn-C) **AWS** A5.6 (ECuSn-C); A5.13 (ECuSn-C)
W60614	Aluminum Bronze Arc Welding Electrode (ECuAl-A2)	**Al** 7.0-9.0 **Cu** rem **Fe** 0.5-5.0 **Pb** 0.02 max **Si** 1.0 max	**ASME** SFA5.6 (ECuAl-A2) **AWS** A5.6 (ECuAl-A2) **MIL SPEC** MIL-E-278 (E-CuAl-A)

The chemical compositions listed are for identification purposes and should not be used in lieu of the cross referenced specifications.

UNIFIED NUMBER	DESCRIPTION	CHEMICAL COMPOSITION	CROSS REFERENCE SPECIFICATIONS
W60619	Aluminum Bronze Arc Welding Electrode (ECuAl-B)	Al 8.4-10.0 Cu rem Fe 2.5-5.0 Pb 0.02 max Si 1.0 max	ASME SFA5.13 (ECuAl-B) AWS A5.13 (ECuAl-B) MIL SPEC MIL-E-278 (E-CuAl-B)
W60625	Aluminum Bronze Arc Welding Electrode (ECuAl-C)	Al 12.0-13.0 Cu rem Fe 3.0-5.0 Pb 0.02 max Si 0.04 max Zn 0.02 max	ASME SFA5.13 (ECuAl-C) AWS A5.13 (ECuAl-C)
W60632	Ni-Al Bronze Arc Welding Electrode (ECuNiAl)	Al 6.5-8.5 Cu rem Fe 3.0-6.0 Mn 0.5-3.5 Ni 4.0-6.0 Pb 0.02 max Si 1.0 max	ASME SFA5.6 (ECuNiAl) AWS A5.6 (ECuNiAl)
W60633	Mn-Al Bronze Arc Welding Electrode (ECuMnNiAl)	Al 5.5-7.5 Cu rem Fe 2.0-6.0 Mn 11.0-13.0 Ni 1.0-2.5 Pb 0.02 max Si 1.5 max	ASME SFA5.6 (ECuMnNiAl) AWS A5.6 (ECuMnNiAl)
W60656	Silicon Bronze Arc Welding Electrode (ECuSi)	Al 0.01 max Cu rem Fe 0.5 max Mn 1.5 max Pb 0.02 max Si 2.4-4.0 Sn 1.5 max	ASME SFA5.6 (ECuSi); SFA5.13 (ECuSi) AWS A5.6 (ECuSi); A5.13 (ECuSi)
W60715	Copper-Nickel Arc Welding Electrode (ECuNi)	Cu rem Fe 0.4-0.75 Mn 1.0-2.5 Ni 29.0-33.0 P 0.020 max Pb 0.02 max S 0.015 max	ASME SFA5.6 (ECuNi) AWS A5.6 (ECuNi) MIL SPEC MIL-E-22200/4 (MIL-CuNi)
W61625	Aluminum Bronze Arc Welding Electrode (ECuAl-D)	Al 13.0-14.0 Cu rem Fe 3.0-5.0 Pb 0.02 max Si 0.04 max Zn 0.02 max	ASME SFA5.13 (ECuAl-D) AWS A5.13 (ECuAl-D)
W62625	Aluminum Bronze Arc Welding Electrode (ECuAl-E)	Al 14.0-15.0 Fe 3.0-5.0 Pb 0.02 max Si 0.04 max Zn 0.02 max	ASME SFA5.13 (ECuAl-E) AWS A5.13 (ECuAl-E)
W73001	Cobalt-Base Hard Facing Electrode (ECoCr-C) (Stellite 1)	C 1.75-3.00 Co rem Cr 25.0-33.0 Fe 5.00 max Mn 2.00 max Mo 1.00 max Ni 3.00 max Si 2.00 max W 11.0-14.0	ASME SFA5.13 (ECoCr-C) AWS A5.13 (ECoCr-C)
W73006	Cobalt-Base Hard Facing Electrode (ECoCr-A) (Stellite 6)	C 0.70-1.40 Co rem Cr 25.0-32.0 Fe 5.00 max Mn 2.00 max Mo 1.00 max Ni 3.00 max Si 2.00 max W 3.00-6.00	ASME SFA5.13 (ECoCr-A) AWS A5.13 (ECoCr-A)
W73012	Cobalt-Base Hard Facing Electrode (ECoCr-B) (Stellite 12)	C 1.00-1.70 Co rem Cr 25.0-32.0 Fe 5.00 max Mn 2.00 max Mo 1.00 max Ni 3.00 max Si 2.00 max W 7.00-9.50	ASME SFA5.13 (ECoCr-B) AWS A5.13 (ECoCr-B)
W73155	Fe-Cr-Ni-Co Superalloy Arc Welding Electrode	C 0.15 max Cb 0.75-1.25 Co 18.5-21.0 Cr 20.0-22.5 Fe rem Mn 1.00-2.50 Mo 2.5-3.5 Ni 19.0-21.0 P 0.040 max S 0.030 max Si 1.0 max W 2.0-3.0	AMS 5795 MIL SPEC MIL-E-22200/3 (MIL-3N1N)
W73605	Cobalt-Base Superalloy Arc Welding Electrode (L605)	C 0.15 max Co rem Cr 19.0-21.0 Fe 3.0 max Mn 0.5-2.0 Ni 9.0-11.0 P 0.030 max S 0.030 max Si 1.0 max W 14.0-16.0	AMS 5797 MIL SPEC MIL-E-22200/3 (MIL-3N1L)
W74510	Austenitic Chromium Iron Arc Welding Electrode for Surfacing (EFeCr-A1)	C 3.0-5.0 Cr 26.0-32.0 Fe rem Mn 4.0-8.0 Mo 2.0 Si 1.0-2.5	ASME SFA5.13 (EFeCr-A1); SFA5.21 (EFeCr-A1) AWS A5.13 (EFeCr-A1); A5.21 (EFeCr-A1)
W74530	Austenitic Chromium Iron Composite Welding Rod for Surfacing (RFeCr-A1)	C 3.7-5.0 Cr 27.0-35.0 Fe rem Mn 2.0-6.0 Si 1.0-2.5	ASME SFA5.21 (RFeCr-A1) AWS A5.21 (RFeCr-A1)
W75010	Air Hardening Steel Arc Welding Electrode for Surfacing (A2a)	C 0.30 max Cr 0.65 max Mn 0.50-1.60 Mo 0.70 max P 0.04 max S 0.04 max Si 0.20-0.65	MIL SPEC MIL-E-19141 (MIL-I-A2a)
W75110	Air Hardening Steel Arc Welding Electrode for Surfacing (A2c)	C 0.80 max Cr 1.5-3.0 Mn 1.25-2.50 P 0.04 max S 0.04 max Si 0.90 max	MIL SPEC MIL-E-19141 (MIL-I-A2C)
W77310	High-Speed Steel Arc Welding Electrode for Surfacing (EFe5-C)	C 0.3-0.5 Cr 3.0-5.0 Fe rem Mn 0.60 max Mo 5.0-9.0 Si 0.80 max V 0.8-1.2 W 1.0-2.5	ASME SFA5.13 (EFe5-C); SFA5.21 (EFe5-C) AWS A5.13 (EFe5-C); A5.21 (EFe5-C)
W77510	High-Speed Steel Arc Welding Electrode for Surfacing (EFe5-B)	C 0.5-0.9 Cr 3.0-5.0 Fe rem Mn 0.60 max Mo 5.0-9.5 Si 0.80 max V 0.8-1.3 W 1.0-2.5	ASME SFA5.13 (EFe5-B); SFA5.21(EFe5-B) AWS A5.13 (EFe5-B); A5.21 (EFe5-B) MIL SPEC MIL-E-19141 (MIL-Fe5-B)

The chemical compositions listed are for identification purposes and should not be used in lieu of the cross referenced specifications.

UNIFIED NUMBER	DESCRIPTION	CHEMICAL COMPOSITION	CROSS REFERENCE SPECIFICATIONS
W77530	High-Speed Steel Composite Welding Rod for Surfacing (RFe5-B)	**C** 0.5-0.9 **Cr** 3.0-5.0 **Fe** rem **Mn** 0.50 max **Mo** 5.0-9.5 **Si** 0.50 max **V** 0.8-1.3 **W** 1.0-2.5	**ASME** SFA5.21 (RFe5-B) **AWS** A5.21 (RFe5-B)
W77710	High-Speed Steel Arc Welding Electrode for Surfacing (EFe5-A)	**C** 0.7-1.0 **Cr** 3.0-5.0 **Fe** rem **Mn** 0.60 max **Mo** 4.0-6.0 **Si** 0.80 max **V** 1.0-2.5 **W** 5.0-7.0	**ASME** SFA5.13 (EFe5-A); SFA5.21 (EFe5-A) **AWS** A5.13 (EFe5-A); A5.21 (EFe5-A)
W77730	High-Speed Steel Composite Welding Rod for Surfacing (RFe5-A)	**C** 0.7-1.0 **Cr** 3.0-5.0 **Fe** rem **Mn** 0.50 max **Mo** 4.0-6.0 **Si** 0.50 max **V** 1.0-2.5 **W** 5.0-7.0	**ASME** SFA5.21 (RFe5-A) **AWS** A5.21 (RFe5-A)
W79110	Austenitic Manganese Steel Arc Welding Electrode for Surfacing (EFe-Mn-A)	**C** 0.5-0.9 **Cr** 0.50 max **Fe** rem **Mn** 11.0-16.0 **Ni** 2.75-6.0 **P** 0.03 max **Si** 1.3 max	**ASME** SFA5.13 (EFe-Mn-A); SFA5.21 (EFeMn-A) **AWS** A5.13 (EFeMn-A); A5.21 (EFeMn-A) **MIL SPEC** MIL-E-19141 (MIL-FeMn-A)
W79310	Austenitic Manganese Steel Arc Welding Electrode for Surfacing (EFeMn-B)	**C** 0.5-0.9 **Cr** 0.50 max **Fe** rem **Mn** 11.0-16.0 **Mo** 0.6-1.4 **P** 0.03 max **Si** 0.3-1.3	**ASME** SFA5.13 (EFeMn-B); SFA5.21 (EFeMn-B) **AWS** A5.13 (EFeMn-B); A5.21 (EFeMn-B)
W80001	Ni-Mo Arc Welding Electrode 3N1B (Hastelloy B)	**C** 0.07 max **Co** 2.5 max **Cr** 1.0 max **Cu** 0.50 max **Fe** 4.0-7.0 **Mn** 1.0 max **Mo** 26.0-30.0 **Ni** rem **P** 0.04 max **S** 0.03 max **Si** 1.0 max **V** 0.60 max **W** 1.0 max	**ASME** SFA5.11 (ENiMo-1) **AWS** A5.11 (ENiMo-1) **MIL SPEC** MIL-E-22200/3 (MIL-3N1B)
W80002	Ni-Mo Arc Welding Electrode 3N1C (Hastelloy C)	**C** 0.12 max **Co** 2.50 max **Cr** 14.50-16.50 **Cu** 0.35 max **Fe** 4.0-7.0 **Mn** 1.0 max **Mo** 15.0-17.0 **Ni** rem **P** 0.040 max **S** 0.030 max **Si** 1.0 max **W** 3.00-4.50	**ASME** SFA5.11 (ENiCrMo-5) **AWS** A5.11 (ENiCrMo-5) **MIL SPEC** MIL-E-22200/3 (MIL-3N1C)
W80004	Ni-Mo Arc Welding Electrode ENiMo-3 (Hastelloy W)	**C** 0.12 max **Co** 2.5 max **Cr** 2.5-5.5 **Cu** 0.50 max **Fe** 4.0-7.0 **Mn** 1.0 max **Mo** 23.0-27.0 **Ni** rem **P** 0.04 max **S** 0.03 max **Si** 1.0 max **V** 0.60 max **W** 1.0 max	**AMS** 5787 **ASME** SFA5.11 (ENiMo-3) **AWS** A5.11 (ENiMo-3) **MIL SPEC** MIL-E-22200/3 (MIL-4N1W)
W80276	Ni-Mo Arc Welding Electrode ENiMo-7 (Hastelloy C-276)	**C** 0.02 max **Co** 2.5 max **Cr** 14.5-16.5 **Cu** 0.50 max **Fe** 4.0-7.0 **Mn** 1.0 max **Mo** 15.0-17.0 **Ni** rem **P** 0.04 max **S** 0.03 max **Si** 0.2 max **V** 0.35 max **W** 3.0-4.5	**ASME** SFA5.11 (ENiCrMo-4) **AWS** A5.11 (ENiCrMo-4)
W80665	Ni-Mo Arc Welding Electrode ENiMo-7 (Hastelloy B-2)	**C** 0.02 max **Co** 1.0 max **Cr** 1.0 max **Cu** 0.50 max **Fe** 2.0 max **Mn** 1.75 max **Mo** 26.0-30.0 **Ni** rem **P** 0.04 max **S** 0.03 max **Si** 0.2 max **W** 1.0 max	**ASME** SFA5.11 (ENiMo-7) **AWS** A5.11 (ENiMo-7)
W82001	Ni Arc Welding Electrode for Cast Iron (ENi-CI)	**C** 2.00 max **Cu** 2.50 max **Fe** 8.0 max **Mn** 1.00 max **Ni** 85.00 min **S** 0.03 max **Si** 4.00 max	**AWS** A5.15 (ENi-CI)
W82002	Ni-Fe Arc Welding Electrode for Cast Iron (ENiFe-CI)	**C** 2.00 max **Cu** 2.50 max **Fe** rem **Mn** 1.00 max **Ni** 45.0-60.0 **S** 0.03 max **Si** 4.00 max	**AWS** A5.15 (ENiFe-CI)
W82003	Ni Arc Welding Electrode for Cast Iron (ENi-CI-A)	**Al** 1.00-3.00 **C** 2.00 max **Cu** 2.50 max **Fe** 8.00 max **Mn** 1.00 max **Ni** 85.0 min **S** 0.03 max **Si** 4.00 max	**ASME** SFA5.15 (ENi-CI-A) **AWS** A5.15 (ENi-CI-A)
W82004	Ni-Fe Arc Welding Electrode for Cast Iron (ENiFe-CI-A)	**Al** 1.00-3.00 **C** 2.00 max **Cu** 2.50 max **Fe** rem **Mn** 1.00 max **Ni** 45.0-60.0 **S** 0.03 max **Si** 4.0 max	**ASME** SFA5.15 (ENiFe-CI-A) **AWS** A5.15 (ENiFe-CI-A)
W82141	Ni Arc Welding Electrode ENi-1 (Nickel W.E.141)	**Al** 1.0 max **C** 0.10 max **Cu** 0.25 max **Fe** 0.75 max **Mn** 0.75 max **Ni** 92.0 min **P** 0.03 max **S** 0.02 max **Si** 1.25 max **Ti** 1.0-4.0	**ASME** SFA5.11 (ENi-1) **AWS** A5.11 (ENi-1) **MIL SPEC** MIL-E-22200/3 (MIL-4N11)
W84001	Ni-Cu Arc Welding Electrode for Cast Iron (ENiCu-A)	**C** 0.35-0.55 **Cu** 35.0-45.0 **Fe** 3.0-6.0 **Mn** 2.25 max **Ni** 50.0-60.0 **S** 0.025 max **Si** 0.75 max	**AWS** A5.15 (ENiCu-A)

The chemical compositions listed are for identification purposes and should not be used in lieu of the cross referenced specifications.

UNIFIED NUMBER	DESCRIPTION	CHEMICAL COMPOSITION	CROSS REFERENCE SPECIFICATIONS
W84002	Ni-Cu Arc Welding Electrode for Cast Iron (ENiCu-B)	**C** 0.35-0.55 **Cu** 25.0-35.0 **Fe** 3.0-6.0 **Mn** 2.25 max **Ni** 60.0-70.0 **S** 0.025 max **Si** 0.75 max	**AWS** A5.15 (ENiCu-B)
W84190	Ni-Cu Arc Welding Electrode ENiCu-7 (Monel W.E.190)	**Al** 0.75 max **C** 0.15 max **Cu** rem **Fe** 2.5 max **Mn** 4.0 max **Ni** 62.0-68.0 **P** 0.02 max **S** 0.015 max **Si** 1.0 max **Ti** 1.0	**ASME** SFA5.11 (ENiCu-7) **AWS** A5.11 (ENiCu-7) **MIL SPEC** MIL-E-22200/3 (MIL-9N10)
W86002	Ni-Cr-Mo Arc Welding Electrode ENiCrMo-2 (Hastelloy X)	**C** 0.05-0.15 **Co** 0.50-2.50 **Cr** 20.5-23.0 **Cu** 0.50 max **Fe** 17.0-20.0 **Mn** 1.0 max **Mo** 8.0-10.0 **Ni** rem **P** 0.04 max **S** 0.03 max **Si** 1.0 max **W** 0.20-1.0	**AMS** 5799 **ASME** SFA5.11 (ENiCrMo-2) **AWS** A5.11 (ENiCrMo-2)
W86007	Ni-Cr Arc Welding Electrode ENiCrMo-1 (Hastelloy G)	**C** 0.05 max **Cb** 1.75-2.50 **Co** 2.5 max **Cr** 21.0-23.5 **Cu** 1.5-2.5 **Fe** 18.0-21.0 **Mn** 1.0-2.0 **Mo** 5.5-7.5 **Ni** rem **P** 0.04 max **S** 0.03 max **Si** 1.0 max **W** 1.0 max	**ASME** SFA5.11 (ENiCrMo-1) **AWS** A5.11 (ENiCrMo-1)
W86022	Ni-Cr Welding Electrode (ENiCrMo-10) (Hastelloy C-22)	**C** 0.02 max **Co** 2.5 max **Cr** 20.0-22.5 **Cu** 0.50 max **Fe** 2.0-6.0 **Mn** 1.0 max **Mo** 12.5-14.5 **Ni** rem **P** 0.03 max **S** 0.015 max **Si** 0.2 max **V** 0.35 **W** 2.5-3.5	
W86030	Ni-Cr Welding Electrode (ENiCrMo-11) (Hastelloy G-30)	**C** 0.03 max **Cb** 0.30-1.50 **Co** 5.0 max **Cr** 28.0-31.5 **Cu** 1.0-2.4 **Fe** 13.0-17.0 **Mn** 1.5 max **Mo** 4.0-6.0 **Ni** rem **P** 0.04 max **S** 0.02 max **Si** 1.0 max **Ti** 1.5-4.0	
W86112	Ni-Cr-Mo Arc Welding Electrode ENiCrMo-3 (Inconel W.E. 112)	**C** 0.10 max **Cb** 3.15-4.15 includes Ta **Co** 0.12 max **Cr** 20.0-23.0 **Cu** 0.50 max **Fe** 7.0 max **Mn** 1.0 max **Mo** 8.0-10.0 **Ni** 55.0 min **P** 0.03 max **S** 0.02 max **Si** 0.75 max	**ASME** SFA5.11 (ENiCrMo-3) **AWS** A5.11 (ENiCrMo-3) **MIL SPEC** MIL-E-22200/3 (MIL-1N12)
W86117	Ni-Cr-Co-Mo Arc Welding Electrode (ENiCrCoMo-1) (Inconel W. E. 117)	**C** 0.05-0.15 **Cb** + Ta 1.0 max **Co** 9.0-15.0 **Cr** 21.0-26.0 **Cu** 0.50 max **Fe** 5.0 max **Mn** 0.30-2.5 **Mo** 8.0-10.0 **Ni** rem **P** 0.03 max **S** 0.015 max **Si** 0.75 max	**ASME** SFA5.11 (ENiCrCoMo-1) **AWS** A5.11 (ENiCrCoMo-1)
W86132	Ni-Cr Arc Welding Electrode ENiCrFe-1 (Inconel W.E. 132)	**C** 0.08 max **Cb** 1.5-4.0 includes Ta **Cr** 13.0-17.0 **Cu** 0.50 max **Fe** 11.0 max **Mn** 3.5 max **Ni** 62.0 min **P** 0.03 max **S** 0.02 max **Si** 0.75 max	**AMS** 5684 **ASME** SFA5.11 (ENiCrFe-1) **AWS** A5.11 (ENiCrFe-1) **MIL SPEC** MIL-E-22200/3 (MIL-3N12)
W86133	Ni-Cr Arc Welding Electrode ENiCrFe-2 (Inco-Weld A)	**C** 0.10 max **Cb** 0.5-3.0 includes Ta **Cr** 13.0-17.0 **Cu** 0.50 max **Fe** 12.0 max **Mn** 1.0-3.5 **Mo** 0.50-2.50 **Ni** 62.0 min **P** 0.03 max **S** 0.02 max **Si** 0.75 max	**ASME** SFA5.11 (ENiCrFe-2) **AWS** A5.11 (ENiCrFe-2) **MIL SPEC** MIL-E-22200/3 (MIL-4N1A)
W86134	Ni-Cr Arc Welding Electrode ENiCrFe-4 (Inco-Weld B)	**C** 0.20 max **Cb** 1.0-3.5 includes Ta **Cr** 13.0-17.0 **Cu** 0.50 max **Fe** 12.0 max **Mn** 1.0-3.5 **Mo** 1.0-3.5 **Ni** 60.0 min **P** 0.03 max **S** 0.02 max **Si** 1.0 max	**ASME** SFA5.11 (ENiCrFe-4) **AWS** A5.11 (ENiCrFe-4)
W86182	Ni-Cr Arc Welding Electrode ENiCrFe-3 (Inconel W.E. 182)	**C** 0.10 max **Cb** 1.0-3.5 includes Ta **Cr** 13.0-17.0 **Cu** 0.50 **Fe** 10.0 **Mn** 5.0-9.5 **Ni** 59.0 min **P** 0.03 max **S** 0.015 **Si** 1.0 max **Ti** 1.0 max	**ASME** SFA5.11 (ENiCrFe-3) **AWS** A5.11 (ENiCrFe-3) **MIL SPEC** MIL-E-22200/3 (MIL-8N12)
W86455	Ni-Mo-Cr Arc Welding Electrode (ENiCrMo-7)	**C** 0.015 max **Co** 2.0 max **Cr** 14.0-18.0 **Cu** 0.50 max **Fe** 3.0 max **Mn** 1.5 max **Mo** 14.0-17.0 **Ni** rem **P** 0.04 max **S** 0.03 max **Si** 0.2 max **Ti** 0.7 max **W** 0.5 max	**ASME** SFA5.11 (ENiCrMo-7) **AWS** A5.11 (ENiCrMo-7)
W86620	Ni-Cr-Mo Arc Welding Electrode (ENiCrMo-6)	**C** 0.10 max **Cb** 0.5-2.0 includes Ta **Cr** 12.0-17.0 **Cu** 0.50 max **Fe** 10.0 max **Mn** 2.0-4.0 **Mo** 5.0-9.0 **Ni** 55.0 min **P** 0.03 max **S** 0.02 max **Si** 1.0 max **W** 1.0-2.0	**ASME** SFA5.11 (ENiCrMo-6) **AWS** A5.11 (ENiCrMo-6)

The chemical compositions listed are for identification purposes and should not be used in lieu of the cross referenced specifications.

WELDING FILLER METALS

UNIFIED NUMBER	DESCRIPTION	CHEMICAL COMPOSITION	CROSS REFERENCE SPECIFICATIONS
W86985	Ni-Cr-Mo Arc Welding Electrode (ENiCrMo-9)	**C** 0.02 max **Cb** 0.5 max includes Ta **Co** 5.0 max **Cr** 21.0-23.5 **Cu** 1.5-2.5 **Fe** 18.0-21.0 **Mn** 1.0 max **Mo** 6.0-8.0 **Ni** rem **P** 0.04 max **S** 0.03 max **Si** 1.0 max **W** 1.5 max	**ASME** SFA5.11 (ENiCrMo-9) **AWS** A5.11 (ENiCrMo-9)
W88021	Ni-Fe-Cr Arc Welding Electrode (E320) (Carpenter 20Cb3)	**C** 0.07 **Cb** 8xC-1.00 includes Ta **Cr** 19.0-21.0 **Cu** 3.0-4.0 **Mn** 0.5-2.5 **Mo** 2.0-3.0 **Ni** 32.0-36.0 **P** 0.04 **S** 0.03 **Si** 0.60	**ASME** SFA5.4 (E320) **AWS** A5.4 (E320)
W88022	Ni-Fe-Cr Arc Welding Electrode (E320LR) Low Residuals (Carpenter 20Cb3L.R.)	**C** 0.035 max **Cb** 8xC-0.40 includes Ta **Cr** 19.0-21.0 **Cu** 3.0-4.0 **Mn** 1.50-2.50 **Mo** 2.0-3.0 **Ni** 32.0-36.0 **P** 0.020 max **S** 0.015 max **Si** 0.30 max	**ASME** SFA5.4 (E320LR) **AWS** A5.4 (E320LR)
W88028	Ni-Fe-Cr Welding Electrode (E028L)	**C** 0.03 max **Cr** 26.5-29.0 **Cu** 0.60-1.50 **Mn** 0.50-2.50 **Mo** 3.20-4.20 **Ni** 30.0-33.0 **P** 0.02 max **S** 0.02 max **Si** 0.90 max	**ASME** SFA5.4 (E028L) **AWS** A5.4 (E028L)
W88331	Ni-Fe-Cr Arc Welding Electrode (E330)	**C** 0.18-0.25 **Cr** 14.0-17.0 **Cu** 0.75 max **Mn** 1.0-2.5 **Mo** 0.75 max **Ni** 33.0-37.0 **P** 0.04 max **S** 0.03 max **Si** 0.90 max	**ASME** SFA5.4 (E330) **AWS** A5.4 (E330) **MIL SPEC** MIL-E-22200/2 (MIL-330)
W88335	Ni-Fe-Cr Arc Welding Electrode (E330H) High Carbon	**C** 0.35-0.45 **Cr** 14.0-17.0 **Cu** 0.75 max **Mn** 1.0-2.5 **Mo** 0.75 max **Ni** 33.0-37.0 **P** 0.04 max **S** 0.03 max **Si** 0.90 max	**ASME** SFA5.4 (E330H) **AWS** A5.4 (E330H)
W88904	Ni-Fe-Cr Arc Welding Electrode (E904L)	**C** 0.03 **Cr** 19.0-21.5 **Cu** 1.0-2.0 **Mn** 1.0-2.5 **Mo** 4.0-5.2 **Ni** 24.0-26.0 **P** 0.03 **S** 0.02 **Si** 0.90	**ASME** SFA5.4 (E904L) **AWS** A5.4 (E904L)
W89604	Ni-Cr-B Arc Welding Electrode for Hard Surfacing (ENiCr-A)	**B** 2.00-3.00 **C** 0.30-0.60 **Co** 1.50 max **Cr** 8.0-14.0 **Fe** 1.25-3.25 **Ni** rem **Si** 1.25-3.25	**AWS** A5.13 (ENiCr-A)
W89605	Ni-Cr-B Arc Welding Electrode for Hard Surfacing (ENiCr-B)	**B** 2.00-4.00 **C** 0.40-0.80 **Co** 1.25 max **Cr** 10.0-16.0 **Fe** 3.00-5.00 **Ni** rem **Si** 3.00-5.00	**AWS** A5.13 (ENiCr-B)
W89606	Ni-Cr-B Arc Welding Electrode for Hard Surfaciang (ENiCr-C)	**B** 2.50-4.50 **C** 0.50-1.00 **Co** 1.00 max **Cr** 12.0-18.0 **Fe** 3.50-5.50 **Ni** rem **Si** 3.50-5.50	**AWS** A5.13 (ENiCr-C)

The chemical compositions listed are for identification purposes and should not be used in lieu of the cross referenced specifications.

UNS NUMBERS ASSIGNED TO DATE

With Description of Each Material Covered and References To Documents In Which The Same or Similar Materials are Described

Zxxxxx Number Series
Zinc and Zinc Alloys

ZINC AND ZINC ALLOYS

UNIFIED NUMBER	DESCRIPTION	CHEMICAL COMPOSITION	CROSS REFERENCE SPECIFICATIONS
Z13000	Zinc Anodes Type II	Al 0.005 max Cd 0.003 max Fe 0.0014 max	ASTM B418 (Zinc Anodes Type II)
Z13001	Zinc Metal	Cd 0.003 max Fe 0.003 max Pb 0.003 max Sn 0.001 max Zn 99.990 min	ASTM B6 (Special High Grade)
Z15001	Zinc Metal	Cd 0.02 max Fe 0.02 max Pb 0.03 max Zn 99.90 min	ASTM B6 (High Grade)
* Z16001	Zinc Metal	Cd 0.40 max Fe 0.03 max Pb 0.20 max Zn 99.5 min	ASTM B6 (Intermediate)
* Z17001	Zinc Metal	Cd 0.05 max Fe 0.03 max Pb 0.6 max Zn 99.0 min	ASTM B6 (Brass Special)
Z19001	Zinc Metal	Cd 0.20 max Fe 0.05 max Pb 1.4 max Zn 98.0 min	ASTM B6 (Prime Western)
Z21210	Zinc Rolled	Cd 0.005 max Cu 0.001 max Fe 0.010 max Pb 0.05 max Zn rem	ASTM B69 Rolled Zinc
Z21310	Zinc Rolled	Cd 0.005 max Cu 0.001 max Fe 0.012 max Pb 0.05-0.12 Zn rem	ASTM B69 Rolled Zinc
Z21540	Zinc Rolled	Cd 0.20-0.35 Cu 0.005 max Fe 0.020 max Pb 0.30-0.65 Zn rem	ASTM B69 (Rolled Zinc)
Z32120	Zinc Anodes Type I	Al 0.10-0.4 Cd 0.03-0.10 Fe 0.005 max	ASTM B418 (Zinc Anodes Type I)
Z32121	Zinc Anodes Type III	Al 0.1-0.5 Cd 0.025-0.15 Cu 0.005 max Fe 0.005 max Pb 0.006 max Si 0.125 max	ASTM B418 (Type III) MIL SPEC MIL-A-18001
Z33520	Zinc Alloy (AG40A)	Al 3.5-4.3 Cd 0.004 max Cu 0.25 max Fe 0.100 max Mg 0.02-0.05 Pb 0.005 max Sn 0.003 max Zn rem	AMS 4803 ASTM B86 (AG40A); B240 (AG40A) FED QQ-Z-363 SAE J468 (903) (Castings) (Ingot)
Z33522	Zinc Alloy (AG40B)	Al 3.9-4.3 Cd 0.0020 max Cr 0.02 max Cu 0.10 max Fe 0.075 max Mg 0.010-0.020 Mn 0.5 max Ni 0.005-0.020 Pb 0.0020 max Si 0.035 max Sn 0.0010 max Zn rem	ASTM B240 (AG40B)
Z33523	Zinc Alloy, Casting (AG40B)	Al 3.5-4.3 Cd 0.0020 max Cr 0.02 max Cu 0.25 max Fe 0.075 max Mg 0.005-0.020 Mn 0.5 max Ni 0.005-0.020 Pb 0.0030 max Si 0.035 max Sn 0.0010 max Zn rem	ASTM B86 (AG40B)
Z35531	Zinc Alloy (AG41A)	Al 3.5-4.3 Cd 0.004 max Cu 0.75-1.25 Fe 0.100 max Mg 0.03-0.08 Pb 0.005 max Sn 0.003 max Zn rem	ASTM B86 (AG41A); B240 (AG41A) FED QQ-Z-363 SAE J468 (925) (Castings) (Ingot)
Z35541	Zinc Alloy, Die Casting (AC43A)	Al 3.5-4.3 Cd 0.004 max Cu 2.5-3.0 Fe 0.100 max Mg 0.020-0.050 Pb 0.005 max Sn 0.003 max Zn rem	ASTM B240 (AC43A)
Z35542	Zinc Forming Die Alloy (Kirksite II or B)	Al 3.9-4.3 Cd 0.003 max Cu 2.5-2.9 Fe 0.075 max Mg 0.02-0.05 Pb 0.003 max Sn 0.001 max Zn rem	MIL SPEC MIL-Z-7068 (Class II); MIL-Z-16460 (A) SAE J469 (921) Other ISO 301-1981 (E) (ZnAl4Cu3)
Z35543	Zinc Forming Die Alloy (Kirksite I or A)	Al 3.5-4.5 Cd 0.005 max Cu 2.5-3.5 Fe 0.100 max Mg 0.02-0.10 Pb 0.007 max Sn 0.005 max Zn rem	MIL SPEC MIL-Z-7068 (Class I)
Z35630	Zinc Foundry Alloy, Ingot	Al 10.5-11.5 Cd 0.003 max Cu 0.50-1.25 Fe 0.075 max Mg 0.015-0.030 Pb 0.004 max Sn 0.002 max Zn rem	ASTM B669

*Boxed entries are no longer active and are retained for reference purposes only.

The chemical compositions listed are for identification purposes and should not be used in lieu of the cross referenced specifications.

ZINC AND ZINC ALLOYS

UNIFIED NUMBER	DESCRIPTION	CHEMICAL COMPOSITION	CROSS REFERENCE SPECIFICATIONS
Z35635	Zinc Alloy (ZA-8)	**Al** 8.0-8.8 **Cd** 0.003 max **Cu** 0.8-1.3 **Fe** 0.10 max **Mg** 0.015-0.030 **Pb** 0.004 max **Sn** 0.002 max **Zn** rem	**ASTM** B669 (ZA-8); ZI ZA-8
Z35840	Zinc Alloy (ZA-27)	**Al** 25.0-28.0 **Cd** 0.003 max **Cu** 2.0-2.5 **Fe** 0.10 max **Mg** 0.010-0.020 **Pb** 0.004 max **Sn** 0.002 max **Zn** rem	**ASTM** B669 (ZA-27); ZI ZA-27
Z38510	Zinc Coating Alloy	**Al** 4.7-6.2 **Cd** 0.005 max **Cu** 0.1 max **Fe** 0.075 max **Mg** 0.05 max **Pb** 0.005 max **Sb** 0.002 max **Si** 0.015 max **Sn** 0.002 max **Ti** 0.02 max **Zn** rem **Zr** 0.02 max **Other** Ce + La 0.03-0.10 **Other** each 0.02 max, total 0.04 max	**ASTM** B750 (Zn-5Al-MM)
Z44330	Zinc Rolled	**Cd** 0.005 max **Cu** 0.65-1.25 **Fe** 0.012 max **Pb** 0.05-0.12 **Zn** rem	**ASTM** B69 (Rolled Zinc)
Z45330	Zinc Rolled	**Cd** 0.005 max **Cu** 0.75-1.25 **Fe** 0.015 max **Mg** 0.007-0.02 **Pb** 0.05-0.12 **Zn** rem	**ASTM** B69 (Rolled Zinc)

The chemical compositions listed are for identification purposes and should not be used in lieu of the cross referenced specifications.

Cross Index of Commonly Known Documents Which Describe Materials Same as or Similar to Those Covered by UNS Numbers

AA (Aluminum Association) NUMBERS
(The Symbol **ob** Indicates an Obsolete UNS Number.)

Cross Reference Specifications	Unified Number	Cross Reference Specifications	Unified Number	Cross Reference Specifications	Unified Number
AA		343.1	A03431	518.2	A05182
		354.0	A03540	520.0	A05200
100.1	A01001	354.1	A03541	520.2	A05202
130.1	A01301	355.0	A03550	535.0	A05350
150.1	A01501	355.1	A03551	535.2	A05352
160.1	A01601	355.2	A03552	705.0	A07050
170.1	A01701	356.0	A03560	705.1	A07051
201.0	A02010	356.1	A03561	707.0	A07070
201.2	A02012	356.2	A03562	707.1	A07071
202.0	A02020	357.0	A03570	710.0	A07100
202.2	A02022	357.1	A03571	710.1	A07101
203.0	A02030	358.0	A03580	711.0	A07110
203A	A92034	358.2	A03582	711.1	A07111
203.2	A02032	359.0	A03590	712.0	A07120
204.0	A02040	359.2	A03592	712.2	A07122
204.2	A02042	360.0	A03600	713.0	A07130
206.0	A02060	360.2	A03602	713.1	A07131
206.2	A02062	361.0	A03610	771.0	A07710
208.0	A02080	361.1	A03611	771.2	A07712
208.1	A02081	363.0	A03630	772.0	A07720
208.2	A02082	363.1	A03631	772.2	A07722
213.0	A02130	364.0	A03640	850.0	A08500
213.1	A02131	364.2	A03642	850.1	A08501
222.0	A02220	369.0	A03690	851.0	A08510
222.1	A02221	369.1	A03691	851.1	A08511
224.0	A02240	380.0	A03800	852.0	A08520
224.2	A02242	380.2	A03802	852.1	A08521
238.0	A02380	383.0	A03830	853.0	A08530
238.1	A02381	383.1	A03831	853.2	A08532
238.2	A02382	383.2	A03832	1030	A91030
240.0	A02400	384.0	A03840	1035	A91035
240.1	A02401	384.1	A03841	1040	A91040
242.0	A02420	384.2	A03842	1045	A91045
242.1	A02421	385.0	A03850	1050	A91050
242.2	A02422	385.1	A03851	1055	A91055
243.0	A02430	390.0	A03900	1060	A91060
243.1	A02431	390.2	A03902	1065	A91065
249.0	A02490	392.0	A03920	1070	A91070
249.2	A02492	392.1	A03921	1075	A91075
295.0	A02950	393.0	A03930	1080	A91080
295.1	A02951	393.1	A03931	1085	A91085
295.2	A02952	393.2	A03932	1090	A91090
296.0	A02960	408.2	A04082	1095	A91095
296.1	A02961	409.2	A04092	1100	A91100
296.2	A02962	411.2	A04112	1135	A91135
305.0	A03050	413.0	A04130	1145	A91145
305.2	A03052	413.2	A04132	1170	A91170
308.0	A03080	435.2	A04352	1175	A91175
308.1	A03081	443.0	A04430	1180	A91180
308.2	A03082	443.1	A04431	1185	A91185
319.0	A03190	443.2	A04432	1188	A91188
319.1	A03191	444.0	A04440	1193	A91193 ob
319.2	A03192	444.2	A04442	1199	A91199
320.0	A03200	445.2	A04452	1200	A91200
320.1	A03201	511.0	A05110	1230	A91230
324.0	A03240	511.1	A05111	1235	A91235
324.1	A03241	511.2	A05112	1250	A91250
324.2	A03242	512.0	A05120 ob	1260	A91260 ob
328.0	A03280	512.2	A05122	1285	A91285
328.1	A03281	513.0	A05130	1345	A91345
332.0	A03320	513.2	A05132	1350	A91350
332.1	A03321	514.0	A05140	1435	A91435
332.2	A03322	514.1	A05141	2011	A92011
333.0	A03330	514.2	A05142	2014	A92014
333.1	A03331	515.0	A05150	2017	A92017
336.0	A03360	515.2	A05152	2018	A92018
336.1	A03361	516.0	A05160	2020	A92020 ob
336.2	A03362	516.1	A05161	2021	A92021 ob
339.0	A03390	518.0	A05180	2024	A92024
343.0	A03430	518.1	A05181	2025	A92025

AA (Aluminum Association) Numbers

Cross Reference Specifications	Unified Number	Cross Reference Specifications	Unified Number	Cross Reference Specifications	Unified Number
2036	A92036	5356	A95356	7150	A97150
2037	A92037	5357	A95357	7175	A97175
2038	A92038	5451	A95451	7178	A97178
2048	A92048	5451	A95454	7179	A97179
2090	A92090	5456	A95456	7277	A97277
2117	A92117	5457	A95457	7472	A97472
2124	A92124	5554	A95554	7475	A97475
2214	A92214	5556	A95556	8001	A98001
2218	A92218	5557	A95557	8006	A98006
2219	A92219	5652	A95652	8007	A98007
2224	A92224	5654	A95654	8013	A98013 ob
2319	A92319	5657	A95657	8014	A98014
2324	A92324	6003	A96003	8017	A98017
2419	A92419	6004	A96004	8020	A98020
2519	A92519	6005	A96005	8030	A98030
2618	A92618	6005A	A96005	8040	A98040
3002	A93002	6006	A96006	8076	A98076
3003	A93003	6007	A96007	8077	A98077
3004	A93004	6009	A96009	8079	A98079
3005	A93005	6010	A96010	8081	A98081
3006	A93006	6011	A96011	8111	A98111
3007	A93007	6017	A96017	8112	A98112
3009	A93009	6053	A96053	8130	A98130
3010	A93010	6060	A96060	8176	A98176
3011	A93011	6061	A96061	8177	A98177
3102	A93102	6063	A96063	8280	A98280
3104	A93104	6066	A96066	A201.0	A12010
3105	A93105	6070	A96070	A201.1	A12011
3107	A93107	6101	A96101	A201.2	A12012
3303	A93303	6105	A96105	A206.0	A12060
4002	A94002 ob	6110	A96110	A206.2	A12062
4004	A94004	6111	A96111	A242.0	A12420
4008	A94008	6151	A96151	A242.1	A12421
4032	A94032	6162	A96162	A242.2	A12422
4043	A94043	6201	A96201	A305.0	A13050
4044	A94044	6205	A96205	A305.1	A13051
4045	A94045	6253	A96253	A305.2	A13052
4047	A94047	6261	A96261	A319.0	A13190
4104	A94104	6262	A96262	A319.1	A13191
4145	A94145	6301	A96301	A333.0	A13330
4343	A94343	6351	A96351	A333.1	A13331
4543	A94543	6463	A96463	A355.0	A13550
4643	A94643	6763	A96763	A355.2	A13552
5005	A95005	6951	A96951	A356.0	A13560
5006	A95006	7001	A97001	A356.1	A13561
5010	A95010	7004	A97004	A356.2	A13562
5016	A95016	7005	A97005	A357.0	A13570
5034	A95034 ob	7008	A97008	A357.2	A13572
5039	A95039 ob	7011	A97011 ob	A360.0	A13600
5040	A95040	7013	A97013	A360.1	A13601
5042	A95042	7016	A97016	A360.2	A13602
5043	A95043	7021	A97021	A380.0	A13800
5050	A95050	7029	A97029	A380.1	A13801
5051	A95051	7039	A97039	A380.2	A13802
5052	A95052	7046	A97046	A384.0	A13840
5056	A95056	7049	A97049	A384.1	A13841
5082	A95082	7050	A97050	A390.0	A13900
5083	A95083	7070	A97070	A390.1	A13901
5086	A95086	7072	A97072	A413.0	A14130
5151	A95151	7075	A97075	A413.1	A14131
5154	A95154	7076	A97076	A413.2	A14132
5182	A95182	7079	A97079	A443.0	A14430
5183	A95183	7090	A97090	A443.1	A14431
5205	A95205	7091	A97091	A444.0	A14440
5250	A95250	7104	A97104	A444.1	A14441
5252	A95252	7108	A97108	A444.2	A14442
5254	A95254	7116	A97116	A535.0	A15350
5351	A95351	7129	A97129	A535.1	A15351
5352	A95352	7146	A97146	B201.0	A22010
		7149	A97149	B237.0	A23570

Cross Reference Specifications	Unified Number
B319.0	A23190
B319.1	A23191
B356.0	A23560
B356.2	A23562
B357.2	A23572
B380.0	A23800
B380.1	A23801
B390.0	A23900
B390.1	A23901
B413.0	A24130
B413.1	A24131
B443.0	A24430
B443.1	A24431
B535.0	A25350
B535.2	A25352
C355.0	A33550
C355.1	A33551
C355.2	A33552
C356.0	A33560
C356.2	A33562
C357.0	A33570
C357.2	A33572
C443.0	A34430
C443.1	A34431
C443.2	A34432
D357.0	A43570
F356.0	A63560
F356.2	A63562

Cross Index of Commonly Known Documents Which Describe Materials Same as or Similar to Those Covered by UNS Numbers

ACI (Steel Founders Society of America) NUMBERS

Cross Reference Specifications	Unified Number

ACI

CA-6NM	J91540
CA-15	J91150
CA-15M	J91151
CA-40	J91153
CB-30	J91803
CC-50	J92615
CE-30	J93423
CF-3	J92500
CF-3	J92700
CF-3M	J92800
CF-8	J92600
CF-8C	J92710
CF-8M	J92900
CF-10MC	J92971
CF-16F	J92701
CF-20	J92602
CG-12	J93001
CH-8	J93400
CH-10	J93401
CH-20	J93402
CK-20	J94202
CN-7M	J95150
HC	J92605
HC-30	J92613
HD	J93005
HE	J93403
HE-35	J93413
HF	J92603
HH	J93503
HH-30	J93513
HI	J94003
HK	J94224
HK-30	J94203
HK-40	J94204
HL	J94604
HL-30	J94613
HN	J94213
HT	J94605
HT-30	J94603
HT50	J94805
HU50	J95404

Cross Index of Commonly Known Documents Which
Describe Materials Same as or Similar to Those
Covered by UNS Numbers

AISI (American Iron and Steel Institute) NUMBERS
(Carbon and Low Alloy Steels)
including
SAE (Society of Automotive Engineers) NUMBERS
(Carbon and Low Alloy Steels)

(The Symbol **ob** Indicates an Obsolete UNS Number.)

AISI & SAE

Cross Reference Specifications	Unified Number	Cross Reference Specifications	Unified Number	Cross Reference Specifications	Unified Number
12L13	G12134	416 Se	S41623	1069	G10690
12L14	G12144	420	S42000	1070	G10700
15B21 H	H15211	420F	S42020	1072	G15720
15B35 H	H15351	422	S42200	1074	G10740
15B37 H	H15371	429	S42900	1075	G10750
15B41 H	H15411	430	S43000	1078	G10780
15B48 H	H15481	430 F	S43020	1080	G10800
15B62 H	H15621	430F Se	S43023	1084	G10840
50B40	G50401	431	S43100	1085	G10850
50B40 H	H50401	434	S43400	1086	G10860
50B44	G50441	436	S43600	1090	G10900
50B44 H	H50441	439	S43035	1095	G10950
50B46	G50461	440 A	S44002	1108	G11080
50B46 H	H50461	440 B	S44003	1109	G11090
50B50	G50501	440 C	S44004	1110	G11100
50B50 H	H50501	442	S44200	1116	G11160
50B60	G50601	446	S44600	1117	G11170
50B60 H	H50601	501	S50100	1118	G11180
51B60	G51601	502	S50200	1119	G11190
51B60 H	H51601	1005	G10050	1123	G11230
81B45	G81451	1006	G10060	1132	G11320
81B45 H	H81451	1008	G10080	1137	G11370
86B30 H	H86301	1009	G10090	1139	G11390
86B45	G86451	1010	G10100 ob	1140	G11400
86B45 H	H86451	1011	G10110	1141	G11410
94B15	G94151	1012	G10120	1144	G11440
94B15 H	H94151	1015	G10150	1145	G11450
94B17	G94171	1016	G10160	1146	G11460
94B17 H	H94171	1017	G10170	1151	G11510
94B30	G94301	1018	G10180	1211	G12110
94B30 H	H94301	1019	G10190	1212	G12120
94B40	G94401	1020	G10200	1213	G12130
201	S20100	1021	G10210	1215	G12150
202	S20200	1022	G10220	1330	G13300
205	S20500	1023	G10230	1330 H	H13300
301	S30100	1024	G15240	1335	G13350
302	S30200	1025	G10250	1335 H	H13350
302B	S30215	1026	G10260	1340	G13400
303	S30300	1027	G15270	1340 H	H13400
303 Se	S30323	1029	G10290	1345	G13450
304	S30400	1030	G10300	1345 H	H13450
304 L	S30403	1033	G10330	1513	G15130
304 N	S30451	1034	G10340	1518	G15180
305	S30500	1035	G10350	1522	G15220
308	S30800	1036	G15360	1522 H	H15220
309	S30900	1037	G10370	1524	G15240
309 S	S30908	1038	G10380	1524 H	H15240
310	S31000	1038 H	H10380	1525	G15250
310 S	S31008	1039	G10390	1526	G15260
314	S31400	1040	G10400	1526 H	H15260
316	S31600	1041	G15410	1527	G15270
316 F	S31620	1042	G10420	1533	G15330
316 L	S31603	1043	G10430	1534	G15340
316 N	S31651	1044	G10440	1536	G15360
317	S31700	1045	G10450	1541	G15410
317 L	S31703	1045 H	H10450	1541 H	H15410
321	S32100	1046	G10460	1547	G15470
329	S32900	1048	G15480	1548	G15480
347	S34700	1049	G10490	1551	G15510
348	S34800	1050	G10500	1552	G15520
384	S38400	1051	G15510	1553	G15530
385	S38500 ob	1052	G15520	1561	G15610
403	S40300	1053	G10530	1566	G15660
405	S40500	1055	G10550	1570	G15700
409	S40900	1059	G10590	1572	G15720
410	S41000	1060	G10600	1580	G15800
414	S41400	1061	G15610	1590	G15900
416	S41600	1064	G10640	3140	G31400
		1065	G10650	4012	G40120
		1066	G15660	4023	G40230

AISI (American Iron and Steel Institute) and SAE (Society of Automotive Engineers) Numbers

Cross Reference Specifications	Unified Number	Cross Reference Specifications	Unified Number	Cross Reference Specifications	Unified Number
4024	G40240	5120	G51200	A-3	T30103 ob
4027	G40270	5120 H	H51200	A-4	T30104
4027 H	H40270	5130	G51300	A-5	T30105 ob
4028	G40280	5130 H	H51300	A-6	T30106
4028 H	H40280	5132	G51320	A-7	T30107
4032	G40320	5132 H	H51320	A-8	T30108
4032H	H40320	5135	G51350	A-9	T30109
4037	G40370	5135 H	H51350	A-10	T30110
4037 H	H40370	5140	G51400	A-11	T30111
4042	G40420	5140 H	H51400	D-2	T30402
4042 H	H40420	5145	G51450	D-3	T30403
4047	G40470	5145 H	H51450	D-4	T30404
4047 H	H40470	5147	G51470	D-5	T30405
4063	G40630	5147 H	H51470	D-7	T30407
4118	G41180	5150	G51500	E3310	G33106
4118 H	H41180	5150 H	H51500	E4337	G43376
4120	G41200	5155	G51550	E4340	G43406
4121	G41210	5155 H	H51550	E4340 H	H43406
4130	G41300	5160	G51600	E9310	G93106
4130 H	H41300	5160 H	H51600	E50100	G50986
4135	G41350	6118	G61180	E51100	G51986
4135 H	H41350	6118 H	H61180	E52100	G52986
4137	G41370	6120	G61200	E71400	G71406
4137 H	H41370	6150	G61500	F-1	T60601 ob
4140	G41400	6150 H	H61500	F-2	T60602 ob
4140 H	H41400	8115	G81150	H-10	T20810
4142	G41420	8615	G86150	H-11	T20811
4142 H	H41420	8617	G86170	H-12	T20812
4145	G41450	8617 H	H86170	H-13	T20813
4145 H	H41450	8620	G86200	H-14	T20814
4147	G41470	8620 H	H86200	H-19	T20819
4147 H	H41470	8622	G86220	H-21	T20821
4150	G41500	8622 H	H86220	H-22	T20822
4150 H	H41500	8625	G86250	H-23	T20823
4161	G41610	8625 H	H86250	H-24	T20824
4161 H	H41610	8627	G86270	H-25	T20825 ob
4320	G43200	8627 H	H86270	H-26	T20826
4320 H	H43200	8630	G86300	H-41	T20841 ob
4337	G43370	8630 H	H86300	H-42	T20842
4340	G43400	8637	G86370	H-43	T20843 ob
4419	G44190	8637 H	H86370	L-2	T61202
4419 H	H44190	8640	G86400	L-3	T61203 ob
4422	G44220	8640 H	H86400	L-6	T61206
4427	G44270	8642	G86420	M-1	T11301
4520	G45200	8642 H	H86420	M-2	T11302
4615	G46150	8645	G86450	M-3 (Cl.1)	T11313
4617	G46170	8645 H	H86450	M-3 (Cl.2)	T11323
4620	G46200	8650	G86500	M-4	T11304
4620 H	H46200	8650 H	H86500	M-6	T11306
4621	G46210	8655	G86550	M-7	T11307
4621 H	H46210	8655 H	H86550	M-10	T11310
4626	G46260	8660	G86600	M-30	T11330
4626 H	H46260	8660 H	H86600	M-33	T11333
4715	G47150	8720	G87200	M-34	T11334
4718	G47180	8720 H	H87200	M-36	T11336
4718 H	H47180	8735	G87350	M-41	T11341
4720	G47200	8740	G87400	M-42	T11342
4720 H	H47200	8740 H	H87400	M-43	T11343 ob
4815	G48150	8742	G87420	M-44	T11344 ob
4815 H	H48150	8822	G88220	M-46	T11346
4817	G48170	8822 H	H88220	M-47	T11347 ob
4817 H	H48170	9254	G92540	M-48	T11348
4820	G48200	9255	G92550	M-52	T11352
4820 H	H48200	9260	G92600	M-61	T11361
5015	G50150	9260 H	H92600	M-62	T11362
5046	G50460	9262	G92620	O-1	T31501
5046 H	H50460	9310 H	H93100	O-2	T31502
5060	G50600	9840	G98400	O-6	T31506
5115	G51150	9850	G98500	O-7	T31507
5117	G51170	A-2	T30102	P-2	T51602 ob

Cross Reference Specifications	Unified Number
P-3	T51603 ob
P-4	T51604 ob
P-5	T51605 ob
P-6	T51606
P-20	T51620
P-21	T51621
S-1	T41901
S-2	T41902
S-4	T41904
S-5	T41905
S-6	T41906
S7	T41907
S13800	S13800
S15500	S15500
S17400	S17400
S17700	S17700
S30430	S30430
T-1	T12001
T-2	T12002 ob
T-4	T12004
T-5	T12005
T-6	T12006
T-8	T12008
T-15	T12015
W-1	T72301
W-2	T72302
W-5	T72305

Cross Index of Commonly Known Documents Which Describe Materials Same as or Similar to Those Covered by UNS Numbers

AMS (SAE/Aerospace Materials Specification) NUMBERS

(The Symbol **ob** Indicates an Obsolete UNS Number.)

Cross Reference Specifications	Unified Number	Cross Reference Specifications	Unified Number	Cross Reference Specifications	Unified Number
AMS					
		4121	A92014	4201	A97050
		4122	A97075	4202	A97475
4000	A91060	4123	A97075	4210	A03550
4001	A91100	4124	A97075	4212	A03550
4002	A91100	4125	A96151	4214	A03550
4004	A95052	4126	A97075	4215	A33550
4005	A95056	4127	A96061	4217	A03560
4006	A93003	4128	A96061	4218	A13560
4007	A92024	4130	A92025	4219	A13570
4008	A93003	4131	A97075	4222	A02420
4009	A96061	4132	A92618	4222	A12420
4010	A93003	4133	A92014	4223	A12010
4011	A91145	4134	A92014	4224	A02430
4015	A95052	4135	A92014	4225	A02030
4016	A95052	4136	A97079	4226	A02240
4017	A95052	4138	A97079	4227	A02400
4018	A95154	4139	A97079	4228	A12010
4019	A95154	4140	A92018	4229	A12010
4024	A97079	4141	A97075	4231	A02950
4025	A96061	4142	A92218	4235	A12060
4026	A96061	4143	A92219	4236	A12060
4027	A96061	4144	A92219	4237	A02060
4028	A92014	4145	A94032	4238	A03570
4029	A92014	4146	A96061	4238	A05350
4031	A92219	4147	A97075	4239	A03570
4035	A92024	4147	A97175	4239	A05350
4037	A92024	4148	A97175	4240	A05200
4044	A97075	4149	A97175	4260	A03560
4045	A97075	4150	A96061	4261	A03560
4050	A97050	4152	A92024	4275	A08500
4056	A95083	4153	A92014	4280	A03550
4057	A95083	4154	A97075	4281	A03550
4058	A95083	4156	A96063	4282	A02961
4059	A95083	4157	A97049	4284	A03560
4062	A91100	4159	A97049	4285	A03560
4065	A93003	4160	A96061	4286	A03560
4066	A92219	4161	A96061	4290	A03600
4067	A93003	4162	A92219	4290	A13600
4068	A92219	4163	A92219	4291	A13800
4069	A95052	4164	A92024	4310	A97075
4070	A95052	4165	A92024	4311	A97075
4071	A95052	4166	A97075	4312	A96061
4078	A97075	4167	A97075	4313	A92219
4079	A96061	4168	A97075	4314	A92014
4080	A96061	4169	A97075	4320	A97149
4081	A96061	4172	A96061	4340	A97050
4082	A96061	4173	A96061	4341	A97050
4083	A96061	4174	A97075	4342	A97050
4084	A97475	4175	A95052	4343	A97149
4085	A97475	4176	A95056	4344	A97175
4086	A92024	4177	A95056	4349	A95056
4087	A92024	4178	A95052	4350	M11610
4088	A92024	4179	A97175	4352	M16600
4089	A97475	4180	A91100	4360	M11800
4090	A97475	4182	A95056	4362	M16600
4100	A97475	4184	A94145	4363	M13210
4101	A92124	4185	A94047	4375	M11311
4102	A91100	4186	A97075	4376	M11311
4107	A97050	4187	A97075	4377	M11311
4108	A97050	4188/1	A02010	4382	M11311
4111	A97049	4188/2	A02060	4383	M13210
4112	A92024	4188/3	A03550	4384	M13310
4113	A96061	4188/4	A03560	4385	M13310
4114	A95052	4188/5	A03570	4386	M14141
4115	A96061	4189	A94643	4387	M16210
4116	A96061	4190	A94043	4388	M13312
4117	A96061	4191	A92319	4389	M13312
4118	A92017	4192	A92024	4390	M13210
4119	A92024	4193	A92024	4395	M11922
4120	A92024	4200	A97049	4396	M12330

AMS (SAE/Aerospace Materials Specification) Numbers

Cross Reference Specifications	Unified Number	Cross Reference Specifications	Unified Number	Cross Reference Specifications	Unified Number
4396	M12331	4765	P07560	4953	R54522
4397	M14141	4766	P07850	4954	R56400
4418	M18220	4767	P07925	4955	R54810
4419	M12350	4768	P07350	4956	R56402
4420	M11630	4769	P07450	4959	R58010
4422	M11630	4770	P07500	4965	R56400
4424	M11630	4771	P07501	4966	R54520
4425	M16630	4772	P07540	4967	R56400
4434	M11920	4773	P07600	4970	R56740
4437	M11914	4774	P07630	4971	R56620
4438	M16620	4775	N99600	4972	R54810
4439	M16410	4776	N99610	4973	R54810
4440	M12410	4777	N99620	4974	R54790
4441	M12410	4778	N99630	4975	R54620
4442	M12330	4779	N99640	4976	R54620
4443	M16510	4780	M26800	4977	R58030
4444	M16610	4782	N99650	4978	R56620
4445	M13310	4783	R30040	4979	R56620
4447	M13320	4784	P00500	4980	R58030
4452	M11914	4785	P00300	4981	R56260
4453	M11920	4786	P00700	4982	R58450
4455	M10100	4787	P00820	4985	R56401
4483	M10100	4800	L13910	4991	R56401
4484	M11920	4803	Z33520	4995	R58650
4490	M11910	4842	C93700	4996	R56401
4500	C11000	4845	C90500	4997	R58650
4501	C10200	4855	C83600	4998	R56401
4505	C26000	4860	C86500	5010	G12120
4507	C26000	4862	C86300	5020	G11374
4510	C51000	4870	C95420	5022	G11170
4520	C54400	4871	C95420	5024	G11370
4530	C17200	4872	C95420	5027	K22925
4532	C17200	4873	C95420	5028	K23725
4544	N04400	4881	C95520	5029	K23577
4555	C26000	4890	C82500	5030	K00606
4555	C33000	4892	N04019	5031	W06013
4558	C33200	4893	N04019	5032	G10200
4574	N04400	4900	R50550	5040	G10100 ob
4575	N04400	4901	R50700	5041	G10060
4602	C10200	4902	R50400	5042	G10100 ob
4610	C36000	4905	R56400	5044	G10100 ob
4611	C46400	4906	R56400	5045	G10200
4612	C46400	4907	R56401	5047	G10100 ob
4614	C37700	4908	R56080	5050	G10100 ob
4615	C65500	4909	R54521	5053	G10100 ob
4625	C51000	4910	R54520	5060	G10150
4635	C62300	4911	R56400	5061	K00802
4640	C63000	4912	R56430	5062	K02508
4650	C17200	4913	R56430	5069	G10180
4651	C17200	4915	R54810	5070	G10220
4665	C65500	4916	R54810	5075	G10250
4674	N04405	4917	R58010	5077	G10250
4675	N04400	4918	R56620	5080	G10350
4676	N05500	4919	R54620	5082	G10350
4677	N05502	4920	R56400	5085	G10500
4700	C14180	4921	R50700	5110	G10800
4701	C10200	4924	R54521	5112	G10860
4710	C27000	4926	R54520	5112	G10900
4712	C27000	4928	R56400	5115	G10700
4713	C27000	4930	R56401	5120	G10740
4720	C51000	4933	R54810	5121	G10950
4725	C17200	4934	R56400	5122	G10950
4730	N04400	4935	R56400	5132	G10950
4731	N04400	4936	R56620	5210	N09902
4732	C71110	4941	R50400	5221	N09902
4750	L54950	4942	R50400	5223	N09902
4751	L13630	4943	R56320	5225	N09902
4755	L50180	4944	R56320	5310	F23330
4756	L50131	4951	R50125	5315	F33101
4764	C69950	4951	R50550	5316	F34100

Cross Reference Specifications	Unified Number	Cross Reference Specifications	Unified Number	Cross Reference Specifications	Unified Number
5328	J23260	5404	N07013	5572	S31008
5329	J23260	5405	N13010	5573	S31600
5330	J24060	5406	N13010	5574	S30908
5331	J24060	5407	N13009	5575	S34700
5333	J11442	5501	S30400	5576	S32100
5334	J13042	5502	S50100	5577	S31008
5335	J13050	5503	S43000	5578	S45500
5336	J13046	5504	S41000	5579	S63198
5337	J93150	5505	S41000	5580	N06600
5338	J14046	5506	S42000	5581	N06625
5339	J93010	5507	S31603	5582	N07750
5340	J92240	5508	S41800	5583	N07750
5341	J92711	5509	N09979	5585	R30155
5342	J92200	5510	S32100	5586	N07001
5343	J92200	5511	S30403	5587	N06002
5344	J92200	5512	S34700	5588	N06002
5346	J92110	5513	S30400	5589	N07718
5347	J92110	5514	S30500	5590	N07718
5348	J92110	5515	S30200	5591	S41000
5349	J91161	5516	S30200	5592	N08330
5350	J91152	5517	S30100	5594	S35500
5351	J91150	5518	S30100	5595	S21904
5352	J91639	5519	S30100	5596	N07718
5353	J91651	5520	S15700	5597	N07718
5354	J91631	5521	S31008	5598	N07750
5355	J92200	5522	S31400	5599	N06625
5356	J92110	5523	S30908	5600	S30200
5357	J92110	5524	S31600	5601	S14800
5358	J92512	5525	S66286	5602	S50100
5358	J92602	5526	S63198	5603	S14800
5359	J92001	5527	S63198	5604	S17400
5360	J92951	5528	S17700	5605	N09706
5361	J93072	5529	S17700	5606	N09706
5362	J92811	5530	N10002	5607	N10003
5363	J92641	5531	R30155	5608	R30188
5364	J92641	5532	R30155	5609	S41040
5365	J94211	5534	R30816	5610	S41600
5366	J94211	5536	N06002	5610	S41623
5368	J92001	5537	R30605	5611	S41040
5369	J92843	5540	N06600	5612	S41040
5370	J92620	5541	N07722	5613	S41000
5371	J92620	5542	N07750	5614	S41025
5372	J91601	5543	S66545	5615	S41400
5373	R30006	5544	N07001	5616	S41800
5375	R30023	5545	N07041	5617	S45500
5376	R30155	5546	S35000	5618	S44004
5377	N07713	5547	S35500	5620	S42020
5378	R30027	5548	S35000	5620	S42023
5380	R30030	5549	S35500	5621	S42000
5382	R30031	5550	N07702	5622	S17400
5383	N07718	5551	N07252	5623	K91456
5384	N07500	5552	N08801	5624	K91505
5384	N07750	5553	N02201	5625	K91456
5385	R30021	5554	S35000	5626	T12001
5387	R30006	5555	N02205	5627	S15700
5388	N10002	5556	S34700	5627	S43000
5389	N10002	5557	S32100	5628	S43100
5390	N06002	5558	S34700	5629	S13800
5391	N07713	5559	S32100	5630	S44004
5392	F47004	5560	S30400	5631	S44002
5393	F47005	5561	S21900	5632	S44002
5394	F47006	5562	S21904	5632 (1)	S44020
5395	F43030	5563	S30400	5632 (2)	S44023
5396	N10001	5564	S30400	5633	N09027
5397	N13100	5565	S30400	5634	N09027
5398	J92200	5566	S30400	5635	S30360
5399	N07041	5567	S30200	5636	S30200
5400	J92110	5568	S17700	5637	S30200
5401	N06625	5570	S32100	5638	S30345
5402	N06625	5571	S34700	5639	S30400

AMS (SAE/Aerospace Materials Specification) Numbers

Cross Reference Specifications	Unified Number	Cross Reference Specifications	Unified Number	Cross Reference Specifications	Unified Number
5640 (Type 1)	S30300	5710	S65006	5797	W73605
5640 (Type 2)	S30310	5711	N06635	5798	N06002
5640 (Type 2)	S30323	5713	N07041	5799	W86002
5641	S30323	5714	N07722	5800	N07041
5642 (1)	S34720	5715	N06601	5801	R30188
5642 (2)	S34723	5716	N08330	5804	S66286
5643	S17400	5717	N06333	5805	S66286
5644	S17700	5718	S64152	5812	S15789
5645	S32100	5719	S64152	5813	S15780
5646	S34700	5721	S63198	5817	S41880
5647	S30403	5726	S66286	5821	S41081
5648	S31600	5731	S66286	5823	S41780
5649	S31620	5732	S66286	5824	S17780
5650	S30908	5733	S66220	5825	S17480
5651	S31008	5734	S66286	5826	S15500
5652	S31400	5737	S66286	5827	W37410
5653	S31603	5738	S30323	5829	N07090
5654	S34700	5739	S36200	5832	N07718
5655	S42200	5740	S36200	5833	R30003
5656	S21904	5742	N08801	5834	R30003
5658	S15500	5743	S35500	5837	N06625
5659	S15500	5744	S35500	5838	N06635
5660	N09901	5745	S35000	5840	S13889
5661	N09901	5746	N09979	5841	R30159
5662	N07718	5747	N07750	5842	R30159
5663	N07718	5748	S65770	5843	R30159
5664	N07718	5749	N07750	5844	R30035
5665	N06600	5749	S42700	5845	R30035
5666	N06625	5750	N10002	5846	N13020
5667	N07750	5751	N07500	5851	N13017
5668	N07750	5753	N07500	5852	N13017
5669	N07750	5754	N06002	5855	N07012
5670	N07750	5755	N10004	5856	N07012
5671	N07750	5756	N07252	5858	S66286
5672	S45500	5757	N07252	5859	S45000
5673	S17700	5758	R30035	5860	S45500
5674	S34700	5759	R30605	5862	S15500
5675	N07092	5761	S65150	5863	S15700
5676	N06003	5762	S20300	5864	S13800
5677	N06003	5763	S45000	5865	N03260
5678	S17700	5764	S20910	5870	N06601
5679	N06062	5765	R30816	5871	N08800
5680	S34781	5766	N08800	5872	N07263
5681	W34710	5768	R30155	5873	N06635
5682	N06003	5769	R30155	5874	R30556
5684	W86132	5770	R30590	5875	R30003
5685	S30500	5771	N10003	5876	R30003
5686	S30500	5772	R30188	5880	S44004
5687	N06600	5773	S45000	5881	N07012
5688	S30200	5774	S35080	5882	N13017
5689	S32100	5775	W35010	5890	N03260
5690	S31600	5776	S41000	5895	S66286
5691	W31610	5777	W41010	5900	S42800
5692	S31680	5778	S41080	6250	K44910
5693	S30200	5779	N07750	6255	K21940
5694	S31080	5780	S35580	6256	K71350
5695	W31010	5781	W35510	6260	G93106
5697	S30400	5782	S63199	6263	K44414
5698	N07750	5783	W34910	6264	K44414
5699	N07750	5784	S64299	6265	G93106
5700	S66009	5785	W31310	6266	K21028
5701	N09706	5786	N10004	6267	G93106
5702	N09706	5787	N10004	6270	G86150
5703	N09706	5787	W80004	6272	G86170
5704	N07001	5788	R30006	6274	G86200
5705	S63005	5789	R30031	6275	G94171
5706	N07001	5790	S34780	6276	G86200
5707	N07001	5794	R30155	6277	G86200
5708	N07001	5795	W73155	6280	G86300
5709	N07001	5796	R30605	6281	G86300

Cross Reference Specifications	Unified Number	Cross Reference Specifications	Unified Number	Cross Reference Specifications	Unified Number
6282	G87350	6431	K24728	7232	N06600
6290	G46150	6432	K24728	7233	N04400
6292	G46170	6433	K33517	7234	N04405
6294	G46200	6434	K33517	7235	S66286
6299	H43200	6435	K33517	7236	R30605
6300	G40370	6436	K22770	7237	N06002
6302	K23015	6437	T20811	7240	G10600
6303	K22770	6438	K24728	7241	S30200
6304	K14675	6439	K24728	7245	S30400
6305	K14675	6440	G52986	7246	N07750
6312	K22440	6442	G50986	7301	G61500
6317	K22440	6443	G51986	7304	G10950
6320	G87350	6444	G52986	7320	C92800
6321	K03810	6445	K22097	7322	C91300
6322	G87400	6446	G51986	7445	S44002
6323	G87400	6447	G51986	7445	S44003
6324	K11640	6447	G52986	7445	S44004
6325	G87400	6448	G61500	7468	R30035
6327	G87400	6449	G51986	7469	N07041
6328	K13550	6450	G61500	7707	K00095
6330	K22033	6451	G92540	7717	K95000
6342	G98400	6454	G43400	7718	K95000
6348	G41300	6455	G61500	7719	K95000
6349	G41400	6457	K13147	7720	L53131
6350	G41300	6458	K23015	7721	L53131
6351	G41300	6459	K22720	7724	R30100
6352	G41350	6460	K11365	7726	K94610
6354	K11914	6461	K13148	7727	K94610
6355	G86300	6462	K13149	7728	K94610
6356	K13247	6463	K93130	7730	M08990
6357	G87350	6464	W10013	7731	P00020
6358	G87400	6465	K91971	7735	P03300
6359	G43400	6466	S50280	7800	R03610
6360	G41300	6467	W50210	7801	R03605
6361	G41300	6468	K91461	7805	R03606
6362	G41300	6470	K24065	7817	R03630
6365	G41350	6471	K24065	7819	R03630
6370	G41300	6472	K24065	7847	R05255
6371	G41300	6475	K52355	7848	R05255
6372	G41350	6485	T20811	7849	R05210
6373	G41300	6487	T20811	7850	R04211
6374	G41300	6488	T20811	7851	R04271
6378	K11542	6490	T11350	7855	R04271
6379	K11546	6491	T11350	7857	R04295
6381	G41400	6501	K92890	7897	R07005
6382	G41400	6512	K92890	7898	R07006
6385	K23015	6514	K93120	7900	R19920
6386 (1)	K11856	6520	K92890	7901	R19800
6386 (2)	K11630	6521	K93120	7902	R19801
6386 (3)	K11511	6522	K92571		
6386 (4)	K11662	6523	K91472		
6386 (5)	K11625	6524	K91283		
6390	G41400	6525	K91472		
6395	G41400	6526	K91283		
6396	K22950	6527	K92571		
6406	K34378	6530	G86300		
6407	K33020	6535	G87350		
6411	K23080	6543	K91970		
6412	G43370	6544	K92571		
6413	G43370	6546	K91122		
6414	G43400	6550	K13048		
6415	G43400	7207	S42000		
6415	G43406	7210	S30200		
6418	K32550	7211	S32100		
6419	K44220	7220	A91100		
6422	K11940	7222	A92117		
6423	K24336	7223	A92024		
6424	K22950	7225	G10100 ob		
6426	K18597	7228	S30400		
6427	K23080	7229	S34700		

Cross Index of Commonly Known Documents Which Describe Materials Same as or Similar to Those Covered by UNS Numbers

ASME (American Society of Mechanical Engineers) NUMBERS

(The Symbol **ob** Indicates an Obsolete UNS Number.)

Cross Reference Specifications	Unified Number	Cross Reference Specifications	Unified Number	Cross Reference Specifications	Unified Number
ASME		SA240 (317 L)	S31703	SA312 (316 L)	S31603
		SA240 (321)	S32100	SA312 (316 N)	S31651
A372 (5E, 8)	K14248	SA240 (321 H)	S32109	SA312 (317)	S31700
Code Case 1817	S21800	SA240 (347)	S34700	SA312 (321)	S32100
Code Case N-20 (1484)	N06690	SA240 (347 H)	S34709	SA312 (321 H)	S32109
SA 312 (316 H)	S31609	SA240 (348)	S34800	SA312 (347)	S34700
SA182 (304)	S30400	SA240 (348 H)	S34809	SA312 (347 H)	S34709
SA182 (304 H)	S30409	SA240 (405)	S40500	SA312 (348)	S34800
SA182 (304 L)	S30403	SA240 (410)	S41000	SA312 (348 H)	S34809
SA182 (304 N)	S30451	SA240 (410 S)	S41008	SA312 (XM-15)	S38100
SA182 (310)	S31000	SA240 (429)	S42900	SA312 (XM-19)	S20910
SA182 (316)	S31600	SA240 (430)	S43000	SA312 (XM-29)	S24000
SA182 (316 H)	S31609	SA240 (439)	S43035	SA320 (303)	S30300
SA182 (316 L)	S31603	SA240 (XM-15)	S38100	SA320 (303 Se)	S30323
SA182 (316 N)	S31651	SA240 (XM-17)	S21600	SA320 (316)	S31600
SA182 (321)	S32100	SA240 (XM-18)	S21603	SA320 (321)	S32100
SA182 (321 H)	S32109	SA240 (XM-19)	S20910	SA320 (347)	S34700
SA182 (347)	S34700	SA240 (XM-21)	S30452	SA320 (B8)	S30400
SA182 (347 H)	S34709	SA240 (XM-27)	S44625 ob	SA351	J95151
SA182 (348)	S34800	SA240 (XM-29)	S24000	SA358 (304)	S30400
SA182 (348 H)	S34809	SA240 (XM-33)	S44626	SA358 (304 N)	S30451
SA182 (429)	S42900	SA249 (304)	S30400	SA358 (309)	S30900
SA182 (430)	S43000	SA249 (304 H)	S30409	SA358 (310)	S31000
SA182 (XM-19)	S20910	SA249 (304 L)	S30403	SA358 (316)	S31600
SA182 (XM-27)	S44625 ob	SA249 (304 N)	S30451	SA358 (316 N)	S31651
SA193 (305)	S30500	SA249 (309)	S30900	SA358 (321)	S32100
SA193 (316)	S31600	SA249 (310)	S31000	SA358 (347)	S34700
SA193 (321)	S32100	SA249 (316)	S31600	SA358 (348)	S34800
SA193 (347)	S34700	SA249 (316 H)	S31609	SA376 (304)	S30400
SA194 (3)	S50100	SA249 (316 L)	S31603	SA376 (304 H)	S30409
SA194 (6)	S41000	SA249 (316 N)	S31651	SA376 (304 N)	S30451
SA194 (8)	S30400	SA249 (317)	S31700	SA376 (316)	S31600
SA194 (303)	S30300	SA249 (321)	S32100	SA376 (316 H)	S31609
SA194 (303 Se)	S30323	SA249 (321 H)	S32109	SA376 (316 N)	S31651
SA194 (305)	S30500	SA249 (347)	S34700	SA376 (321)	S32100
SA194 (316)	S31600	SA249 (347 H)	S34709	SA376 (321 H)	S32109
SA194 (321)	S32100	SA249 (348)	S34800	SA376 (347)	S34700
SA194 (347)	S34700	SA249 (348 H)	S34809	SA376 (347 H)	S34709
SA194 (416)	S41600	SA249 (XM-15)	S38100	SA376 (348)	S34800
SA194 (416 Se)	S41623	SA249 (XM-19)	S20910	SA387 (5)	S50100
SA213 (304)	S30400	SA249 (XM-29)	S24000	SA387 (5)	S50200
SA213 (304 H)	S30409	SA268	S44627	SA395 (60-40-18)	F32800
SA213 (304 L)	S30403	SA268 (329)	S32900	SA403 (304)	S30400
SA213 (304 N)	S30451	SA268 (405)	S40500	SA403 (304 H)	S30409
SA213 (310)	S31000	SA268 (409)	S40900	SA403 (304 L)	S30403
SA213 (316)	S31600	SA268 (410)	S41000	SA403 (309)	S30900
SA213 (316 H)	S31609	SA268 (429)	S42900	SA403 (310)	S31000
SA213 (316 L)	S31603	SA268 (430)	S43000	SA403 (316)	S31600
SA213 (316 N)	S31651	SA268 (439)	S43035	SA403 (316 H)	S31609
SA213 (321)	S32100	SA268 (446)	S44600	SA403 (316 L)	S31603
SA213 (321 H)	S32109	SA268 (XM-27)	S44625 ob	SA403 (316 N)	S31651
SA213 (347)	S34700	SA268 (XM-33)	S44626	SA403 (317)	S31700
SA213 (347 H)	S34709	SA278 (20)	F11401	SA403 (321)	S32100
SA213 (348)	S34800	SA278 (25)	F11701	SA403 (321 H)	S32109
SA213 (348 H)	S34809	SA278 (30)	F12101	SA403 (347)	S34700
SA213 (XM-15)	S38100	SA278 (35)	F12401	SA403 (347 H)	S34709
SA240	S44627	SA278 (40)	F12803	SA403 (348)	S34800
SA240 (302)	S30200	SA278 (45)	F13102	SA403 (XM-19)	S20910
SA240 (304)	S30400	SA278 (50)	F13502	SA409 (304)	S30400
SA240 (304 H)	S30409	SA278 (55)	F13802	SA409 (309)	S30900
SA240 (304 L)	S30403	SA278 (60)	F14102	SA409 (310)	S31000
SA240 (304 N)	S30451	SA278 (70)	F14801	SA409 (316)	S31600
SA240 (305)	S30500	SA278 (80)	F15501	SA409 (317)	S31700
SA240 (309 S)	S30908	SA312 (304)	S30400	SA409 (321)	S32100
SA240 (310 S)	S31008	SA312 (304 H)	S30409	SA409 (347)	S34700
SA240 (316)	S31600	SA312 (304 L)	S30403	SA409 (348)	S34800
SA240 (316 H)	S31609	SA312 (304 N)	S30451	SA412 (201)	S20100
SA240 (316 L)	S31603	SA312 (309)	S30900	SA412 (21904)	S21904
SA240 (316 N)	S31651	SA312 (310)	S31000	SA412 (XM-19)	S20910
SA240 (317)	S31700	SA312 (316)	S31600	SA430 (304)	S30400

Cross Reference Specifications	Unified Number	Cross Reference Specifications	Unified Number	Cross Reference Specifications	Unified Number
SA430 (304 H)	S30409	SB43	C23000	SB171	C44300
SA430 (304 N)	S30451	SB61	C92200	SB171	C44400
SA430 (316)	S31600	SB62	C83600	SB171	C44500
SA430 (316 H)	S31609	SB75	C10200	SB171	C46400
SA430 (316 N)	S31651	SB75	C12000	SB171	C46500
SA430 (321)	S32100	SB75	C12200	SB171	C46600
SA430 (321 H)	S32109	SB75	C14200	SB171	C46700
SA430 (347)	S34700	SB96	C65500	SB171	C61400
SA430 (347 H)	S34709	SB98	C65100	SB171	C63000
SA479	S44627	SB98	C65500	SB171	C70600
SA479 (302)	S30200	SB98	C66100	SB171	C71500
SA479 (304)	S30400	SB111	C10200	SB271	C83600
SA479 (304 H)	S30409	SB111	C12000	SB271	C92200
SA479 (304 L)	S30403	SB111	C12200	SB271	C93700
SA479 (304 N)	S30451	SB111	C14200	SB271	C95200
SA479 (310 S)	S31008	SB111	C19200	SB271	C95400
SA479 (316)	S31600	SB111	C23000	SB271	C97600
SA479 (316 H)	S31609	SB111	C28000	SB283	C37700
SA479 (316 L)	S31603	SB111	C44300	SB315	C65100
SA479 (316 N)	S31651	SB111	C44400	SB315	C65500
SA479 (321)	S32100	SB111	C44500	SB315	C65800
SA479 (321 H)	S32109	SB111	C60800	SB333	N10001
SA479 (347)	S34700	SB111	C68700	SB333	N10665
SA479 (347 H)	S34709	SB111	C70400	SB335	N10001
SA479 (348)	S34800	SB111	C70600	SB335	N10665
SA479 (405)	S40500	SB111	C71000	SB359	C10200
SA479 (410)	S41000	SB111	C71500	SB359	C12000
SA479 (430)	S43000	SB127	N04400	SB359	C12200
SA479 (439)	S43035	SB148	C95200	SB359	C14200
SA479 (XM-17)	S21600	SB148	C95400	SB359	C19200
SA479 (XM-18)	S21603	SB150	C61400	SB359	C23000
SA479 (XM-19)	S20910	SB150	C62300	SB359	C44300
SA479 (XM-27)	S44625 ob	SB150	C63000	SB359	C44400
SA479 (XM-30)	S41040	SB150	C64200	SB359	C44500
SA564 (630)	S17400	SB152	C10200	SB359	C60800
SA564 (XM-25)	S45000	SB152	C10400	SB359	C68700
SA637	N07750	SB152	C10500	SB359	C70400
SA638 (660)	S66286	SB152	C10700	SB359	C70600
SA688 (304)	S30400	SB152	C11300	SB359	C71000
SA688 (304 L)	S30403	SB152	C12200	SB359	C71500
SA688 (316)	S31600	SB152	C12300	SB366	N06007
SA688 (316 L)	S31603	SB160	N02200	SB366	N06022
SA688 (XM-29)	S24000	SB160	N02201	SB366	N06030
SA705 (630)	S17400	SB161	N02200	SB366	N06455
SA705 (631)	S17700	SB161	N02201	SB366	N06975
SA705 (XM-12)	S15500	SB162	N02200	SB366	N08320
SA705 (XM-13)	S13800	SB162	N02201	SB366	N08330
SA705 (XM-25)	S45000	SB163	N02200	SB366	N08332
SA731	S44627	SB163	N02201	SB366	N10001
SA4479 (348 H)	S34809	SB163	N04400	SB366	N10276
SB11	C11000	SB163	N06600	SB366	N10665
SB11	C12200	SB163	N06690	SB395	C10200
SB11	C12500	SB163	N08800	SB395	C12000
SB11	C14100 ob	SB163	N08810	SB395	C12200
SB11	C14200	SB163	N08825	SB395	C14200
SB12	C10200	SB164	N04400	SB395	C19200
SB12	C11000	SB164	N04405	SB395	C23000
SB12	C12000	SB165	N04400	SB395	C44300
SB12	C12100	SB166	N06600	SB395	C44400
SB12	C12200	SB166	N06690	SB395	C44500
SB12	C12300	SB167	N06600	SB395	C60800
SB12	C12500	SB167	N06690	SB395	C68700
SB12	C12700	SB168	N06600	SB395	C70400
SB12	C12800	SB168	N06690	SB395	C70600
SB12	C13000	SB169	C61000	SB395	C71000
SB12	C14100 ob	SB169	C61400	SB395	C71500
SB12	C14200	SB171	C36500	SB402	C70600
SB42	C10200	SB171	C36600	SB402	C71500
SB42	C12000	SB171	C36700	SB407	N08800
SB42	C12200	SB171	C36800	SB407	N08810

Cross Reference Specifications	Unified Number	Cross Reference Specifications	Unified Number	Cross Reference Specifications	Unified Number
SB408	N08800	SB619	N06022	SFA5.4 (E307)	W30710
SB408	N08810	SB619	N06030	SFA5.4 (E308) (E308H)	W30810
SB409	N08800	SB619	N06455	SFA5.4 (E308L)	W30813
SB409	N08810	SB619	N06975	SFA5.4 (E308Mo)	W30820
SB423	N08825	SB619	N06985	SFA5.4 (E308MoL)	W30823
SB424	N08825	SB619	N10001	SFA5.4 (E309)	W30910
SB425	N08825	SB619	N10276	SFA5.4 (E309Cb)	W30917
SB434	N10003	SB619	N10665	SFA5.4 (E309L)	W30913
SB435	N06002	SB620	N08320	SFA5.4 (E309Mo)	W30920
SB436	N06001	SB621	N08320	SFA5.4 (E310)	W31010
SB443	N06625	SB622	N06002	SFA5.4 (E310Cb)	W31017
SB444	N06625	SB622	N06022	SFA5.4 (E310H)	W31015
SB446	N06625	SB622	N06030	SFA5.4 (E310Mo)	W31020
SB462	N08020	SB622	N06455	SFA5.4 (E312)	W31310
SB463	N08020	SB622	N06985	SFA5.4 (E316) (E316H)	W31610
SB463	N08026	SB622	N08320	SFA5.4 (E316L)	W31613
SB464	N08020	SB622	N10276	SFA5.4 (E317)	W31710
SB464	N08026	SB622	N10665	SFA5.4 (E317L)	W31713
SB466	C70400	SB625	N08904	SFA5.4 (E318)	W31910
SB466	C70600	SB625	N08925	SFA5.4 (E320)	W88021
SB466	C71000	SB626	N06002	SFA5.4 (E320LR)	W88022
SB466	C71500	SB626	N06022	SFA5.4 (E330)	W88331
SB467	C70600	SB626	N06030	SFA5.4 (E330H)	W88335
SB467	C71000	SB626	N06455	SFA5.4 (E347)	W34710
SB467	C71500	SB626	N06985	SFA5.4 (E349)	W34910
SB468	N08026	SB626	N10276	SFA5.4 (E410)	W41010
SB493	R60702	SB626	N10665	SFA5.4 (E410NiMo)	W41016
SB494	R60702	SB649	N08904	SFA5.4 (E430)	W43010
SB495	R60702	SB649	N08925	SFA5.4 (E502)	W50210
SB511	N08330	SB668	N08028	SFA5.4 (E505)	W50410
SB511	N08332	SB672	N08700	SFA5.4 (E630)	W37410
SB514	N08810	SB673	N08904	SFA5.4 (E904L)	W88904
SB523	R60702	SB673	N08925	SFA5.5 (E7010-A1)	W17010
SB535	N08332	SB674	N08904	SFA5.5 (E7011-A1)	W17011
SB536	N08332	SB674	N08925	SFA5.5 (E7015-A1)	W17015
SB543	C12200	SB675	N08366	SFA5.5 (E7015-C1L)	W22115
SB543	C19400	SB677	N08904	SFA5.5 (E7015-C2L)	W23115
SB543	C23000	SB709	N08028	SFA5.5 (E7016-A1)	W17016
SB543	C44300	SB710	N08330	SFA5.5 (E7016-C1L)	W22116
SB543	C44400	SFA 5.8 (BNi-10)	N99624	SFA5.5 (E7016-C2L)	W23116
SB543	C44500	SFA-5.8 (BNi-1)	N99600	SFA5.5 (E7018-A1)	W17018
SB543	C68700	SFA-5.8 (BNi-2)	N99620	SFA5.5 (E7018-C1L)	W22118
SB543	C70400	SFA5.1 (7016)	W07016	SFA5.5 (E7018-C2L)	W23118
SB543	C70600	SFA5.1 (7018)	W07018	SFA5.5 (E7018-W)	W20018
SB543	C71500	SFA5.1 (E6010)	W06010	SFA5.5 (E7020-A1)	W17020
SB550	R60702	SFA5.1 (E6011)	W06011	SFA5.5 (E7027-A1)	W17027
SB551	R60702	SFA5.1 (E6012)	W06012	SFA5.5 (E8015-B2L)	W52115
SB564	N04400	SFA5.1 (E6013)	W06013	SFA5.5 (E8015-B4L)	W53415
SB564	N06600	SFA5.1 (E6020)	W06020	SFA5.5 (E8016-B1)	W51016
SB564	N08800	SFA5.1 (E6022)	W06022	SFA5.5 (E8016-B2)	W52016
SB564	N08810	SFA5.1 (E6027)	W06027	SFA5.5 (E8016-B5)	W51316
SB572	N06002	SFA5.1 (E7014)	W07014	SFA5.5 (E8016-C1)	W22016
SB574	N06022	SFA5.1 (E7015)	W07015	SFA5.5 (E8016-C2)	W23016
SB574	N06455	SFA5.1 (E7016-1)	W07116	SFA5.5 (E8016-C3)	W21016
SB574	N10276	SFA5.1 (E7018-1)	W07118	SFA5.5 (E8016-D3)	W18016
SB575	N06022	SFA5.1 (E7024)	W07024	SFA5.5 (E8018-B1)	W51018
SB575	N06455	SFA5.1 (E7024-1)	W07124	SFA5.5 (E8018-B2)	W52018
SB575	N10276	SFA5.1 (E7027)	W07027	SFA5.5 (E8018-B2L)	W52118
SB581	N06030	SFA5.1 (E7028)	W07028	SFA5.5 (E8018-C1)	W22018
SB581	N06975	SFA5.1 (E7048)	W07048	SFA5.5 (E8018-C2)	W23018
SB581	N06985	SFA5.2 (R100)	K12048	SFA5.5 (E8018-C3)	W21018
SB582	N06030	SFA5.2 (RG45)	K00045	SFA5.5 (E8018-D3)	W18018
SB582	N06975	SFA5.2 (RG60)	K00060	SFA5.5 (E8018-NM)	W21118
SB582	N06985	SFA5.2 (RG65)	K00065	SFA5.5 (E8018-W)	W20118
SB584	C92200	SFA5.4 (E7Cr)	W50310	SFA5.5 (E9015-B3)	W53015
SB584	C97600	SFA5.4 (E16-8-2)	W36810	SFA5.5 (E9015-B3L)	W53115
SB584	C93700	SFA5.4 (E028L)	W88028	SFA5.5 (E9015-D1)	W19015
SB599	N08700	SFA5.4 (E209)	W32210	SFA5.5 (E9016-B3)	W53016
SB619	N06002	SFA5.4 (E219)	W32310	SFA5.5 (E9018-B3)	W53018
SB619	N06007	SFA5.4 (E240)	W32410	SFA5.5 (E9018-B3L)	W53118

Cross Reference Specifications	Unified Number	Cross Reference Specifications	Unified Number	Cross Reference Specifications	Unified Number
SFA5.5 (E9018-D1)	W19018	SFA5.8 (BCuP-5)	C55284	SFA5.9 (ER310)	W31040
SFA5.5 (E9018-M)	W21218	SFA5.8 (BCuP-6)	C55280	SFA5.9 (ER312)	S31380
SFA5.5 (E10015-D2)	W10015	SFA5.8 (BCuP-7)	C55282	SFA5.9 (ER312)	W31340
SFA5.5 (E10016-D2)	W10016	SFA5.8 (BCuZn-E)	C28580	SFA5.9 (ER316) (316H)	S31680
SFA5.5 (E10018-D2)	W10018	SFA5.8 (BCuZn-F)	C49080	SFA5.9 (ER316) (ER316H)	W31640
SFA5.5 (E10018-M)	W21318	SFA5.8 (BCuZn-G)	C26380	SFA5.9 (ER316L)	S31683
SFA5.5 (E11018-M)	W21418	SFA5.8 (BCuZn-H)	C24080	SFA5.9 (ER316L)	W31643
SFA5.5 (E12018-M)	W22218	SFA5.8 (BCuZn-H)	C26000	SFA5.9 (ER316LSi)	S31688
SFA5.5 (E12018-M1)	W23218	SFA5.8 (BMg-1)	M19001	SFA5.9 (ER316LSi)	W31648
SFA5.6 (ECu)	W60189	SFA5.8 (BNi-1a)	N99610	SFA5.9 (ER316Si)	S31681
SFA5.6 (ECuAl-A2)	W60614	SFA5.8 (BNi-3)	N99630	SFA5.9 (ER316Si)	W31641
SFA5.6 (ECuMnNiAl)	W60633	SFA5.8 (BNi-4)	N99640	SFA5.9 (ER317)	S31780
SFA5.6 (ECuNi)	W60715	SFA5.8 (BNi-5)	N99650	SFA5.9 (ER317)	W31740
SFA5.6 (ECuNiAl)	W60632	SFA5.8 (BNi-6)	N99700	SFA5.9 (ER317L)	W31743
SFA5.6 (ECuSi)	W60656	SFA5.8 (BNi-7)	N99710	SFA5.9 (ER318)	S31980
SFA5.6 (ECuSn-A)	W60518	SFA5.8 (BNi-8)	N99800	SFA5.9 (ER318)	W31940
SFA5.6 (ECuSn-C)	W60521	SFA5.8 (BNi-9)	N99612	SFA5.9 (ER320)	N08021
SFA5.7 (ERCu)	C18980	SFA5.8 (BNi-10)	N99622	SFA5.9 (ER320)	N08321
SFA5.7 (ERCuAl-A1)	C61000	SFA5.8 (BVAg-0)	P07017	SFA5.9 (ER320LR)	N08022
SFA5.7 (ERCuAl-A2)	C61800	SFA5.8 (BVAg-6b)	P07507	SFA5.9 (ER321)	S32180
SFA5.7 (ERCuAl-A3)	C62400	SFA5.8 (BVAg-8)	P07727	SFA5.9 (ER321)	W32140
SFA5.7 (ERCuMnNiAl)	C63380	SFA5.8 (BVAg-8b)	P07728	SFA5.9 (ER347)	S34780
SFA5.7 (ERCuNi)	C71581	SFA5.8 (BVAg-18)	P07607	SFA5.9 (ER347)	W34740
SFA5.7 (ERCuNiAl)	C63280	SFA5.8 (BVAg-29)	P07627	SFA5.9 (ER347Si)	S34788
SFA5.7 (ERCuSi-A)	C65600	SFA5.8 (BVAg-30)	P07687	SFA5.9 (ER347Si)	W34748
SFA5.7 (ERCuSn-A)	C51800	SFA5.8 (BVAg-31)	P07587	SFA5.9 (ER349)	S63197
SFA5.8 (BAg-1)	P07450	SFA5.8 (BVAg-32)	P07547	SFA5.9 (ER349)	W34940
SFA5.8 (BAg-1a)	P07500	SFA5.8 (BVAu-2)	P00807	SFA5.9 (ER410)	S41080
SFA5.8 (BAg-2)	P07350	SFA5.8 (BVAu-4)	P00827	SFA5.9 (ER410)	W41040
SFA5.8 (BAg-2a)	P07300	SFA5.8 (BVAu-7)	P00507	SFA5.9 (ER410NiMo)	S41086
SFA5.8 (BAg-3)	P07501	SFA5.8 (BVAu-8)	P00927	SFA5.9 (ER410NiMo)	W41046
SFA5.8 (BAg-4)	P07400	SFA5.8 (BVCu-1x)	C14181	SFA5.9 (ER420)	S42080
SFA5.8 (BAg-5)	P07453	SFA5.8 (BVPd-1)	P03657	SFA5.9 (ER420)	W42040
SFA5.8 (BAg-6)	P07503	SFA5.8 (RBCuZn-A)	C47000	SFA5.9 (ER430)	S43080
SFA5.8 (BAg-7)	P07563	SFA5.8 (RBCuZn-C)	C68100	SFA5.9 (ER430)	W43040
SFA5.8 (BAg-8)	P07720	SFA5.8 (RBCuZn-D)	C77300	SFA5.9 (ER502)	S50280
SFA5.8 (BAg-8a)	P07723	SFA5.9 (ER 317L)	S31783	SFA5.9 (ER502)	W50240
SFA5.8 (BAg-9)	P07650	SFA5.9 (ER16-8-2)	S16880	SFA5.9 (ER505)	S50480
SFA5.8 (BAg-10)	P07700	SFA5.9 (ER16-8-2)	W36840	SFA5.9 (ER505)	W50440
SFA5.8 (BAg-13)	P07540	SFA5.9 (ER26-1)	S44687	SFA5.9 (ER630)	S17480
SFA5.8 (BAg-13a)	P07560	SFA5.9 (ER26-1)	W44647	SFA5.9 (ER630)	W37440
SFA5.8 (BAg-18)	P07600	SFA5.9 (ER209)	S20980	SFA5.9 (Er26-1)	S44627
SFA5.8 (BAg-19)	P07925	SFA5.9 (ER209)	W32240	SFA5.10 (ER1100)	A91100
SFA5.8 (BAg-20)	P07301	SFA5.9 (ER218)	S21880	SFA5.10 (ER1188)	A91188
SFA5.8 (BAg-21)	P07630	SFA5.9 (ER218)	W32540	SFA5.10 (ER2319)	A92319
SFA5.8 (BAg-22)	P07490	SFA5.9 (ER219)	S21980	SFA5.10 (ER4043)	A94043
SFA5.8 (BAg-23)	P07850	SFA5.9 (ER219)	W32340	SFA5.10 (ER4047)	A94047
SFA5.8 (BAg-24)	P07505	SFA5.9 (ER240)	S23980	SFA5.10 (ER4145)	A94145
SFA5.8 (BAg-25)	P07200	SFA5.9 (ER240)	W32440	SFA5.10 (ER4643)	A94643
SFA5.8 (BAg-26)	P07250	SFA5.9 (ER307)	S30780	SFA5.10 (ER5183)	A95183
SFA5.8 (BAg-27)	P07251	SFA5.9 (ER307)	W30740	SFA5.10 (ER5356)	A95356
SFA5.8 (BAg-28)	P07401	SFA5.9 (ER308) (ER308H)	S30880	SFA5.10 (ER5554)	A95554
SFA5.8 (BAlSi-2)	A94343	SFA5.9 (ER308) (ER308H)	W30840	SFA5.10 (ER5556)	A95556
SFA5.8 (BAlSi-3)	A94145	SFA5.9 (ER308L)	S30883	SFA5.10 (ER5654)	A95654
SFA5.8 (BAlSi-4)	A94047	SFA5.9 (ER308L)	W30843	SFA5.10 (R206.0)	A02060
SFA5.8 (BAlSi-5)	A94045	SFA5.9 (ER308LSi)	S30888	SFA5.10 (R242.0)	A02042
SFA5.8 (BAlSi-7)	A94004	SFA5.9 (ER308LSi)	W30848	SFA5.10 (R242.0)	A02420
SFA5.8 (BAlSi-11)	A94104	SFA5.9 (ER308Mo)	S30882	SFA5.10 (R295.0)	A02950
SFA5.8 (BAu-1)	P00375	SFA5.9 (ER308Mo)	W30850	SFA5.10 (R355.0)	A03550
SFA5.8 (BAu-2)	P00800	SFA5.9 (ER308MoL)	S30886	SFA5.10 (R356.0)	A03560
SFA5.8 (BAu-3)	P00350	SFA5.9 (ER308MoL)	W30853	SFA5.11 (ENi-1)	W82141
SFA5.8 (BAu-4)	P00820	SFA5.9 (ER308Si)	S30881	SFA5.11 (ENiCrCoMo-1)	W86117
SFA5.8 (BAu-5)	P00300	SFA5.9 (ER308Si)	W30841	SFA5.11 (ENiCrFe-1)	W86132
SFA5.8 (BAu-6)	P00700	SFA5.9 (ER309)	S30980	SFA5.11 (ENiCrFe-2)	W86133
SFA5.8 (BCo-1)	R39001	SFA5.9 (ER309)	W30940	SFA5.11 (ENiCrFe-3)	W86182
SFA5.8 (BCu-1)	C14180	SFA5.9 (ER309L)	S30983	SFA5.11 (ENiCrFe-4)	W86134
SFA5.8 (BCuP-1)	C55180	SFA5.9 (ER309L)	W30943	SFA5.11 (ENiCrMo-1)	W86007
SFA5.8 (BCuP-2)	C55181	SFA5.9 (ER309Si)	S30981	SFA5.11 (ENiCrMo-2)	W86002
SFA5.8 (BCuP-3)	C55281	SFA5.9 (ER309Si)	W30941	SFA5.11 (ENiCrMo-3)	W86112
SFA5.8 (BCuP-4)	C55283	SFA5.9 (ER310)	S31080	SFA5.11 (ENiCrMo-4)	W80276

Cross Reference Specifications	Unified Number
SFA5.11 (ENiCrMo-5)	W80002
SFA5.11 (ENiCrMo-6)	W86620
SFA5.11 (ENiCrMo-7)	W86455
SFA5.11 (ENiCrMo-9)	W86985
SFA5.11 (ENiCu-7)	W84190
SFA5.11 (ENiMo-1)	W80001
SFA5.11 (ENiMo-3)	W80004
SFA5.11 (ENiMo-7)	W80665
SFA5.12 (EWP)	R07900
SFA5.12 (EWTh-1)	R07911
SFA5.12 (EWTh-2)	R07912
SFA5.12 (EWTh-3)	R07913
SFA5.12 (EWZr)	R07920
SFA5.13 (ECoCr-A)	W73006
SFA5.13 (ECoCr-B)	W73012
SFA5.13 (ECoCr-C)	W73001
SFA5.13 (ECuAl-B)	W60619
SFA5.13 (ECuAl-C)	W60625
SFA5.13 (ECuAl-D)	W61625
SFA5.13 (ECuAl-E)	W62625
SFA5.13 (ECuSi)	W60656
SFA5.13 (ECuSn-A)	W60518
SFA5.13 (ECuSn-C)	W60521
SFA5.13 (EFe-Mn-A)	W79110
SFA5.13 (EFe5-A)	W77710
SFA5.13 (EFe5-B)	W77510
SFA5.13 (EFe5-C)	W77310
SFA5.13 (EFeCr-A1)	W74510
SFA5.13 (EFeMn-B)	W79310
SFA5.13 (ERCuSi-A)	C65600
SFA5.13 (RCoCr-A)	R30006
SFA5.13 (RCoCr-B)	R30012
SFA5.13 (RCoCr-C)	R30001
SFA5.13 (RCuAl-C)	C62580
SFA5.13 (RCuAl-D)	C62581
SFA5.13 (RCuAl-E)	C62582
SFA5.13 (RFe5-A)	K91308
SFA5.13 (RFe5-B)	K90987
SFA5.13 (RFeCr-A1)	F45100
SFA5.13 (RNiCr-A)	N99644
SFA5.13 (RNiCr-B)	N99645
SFA5.13 (RNiCr-C)	N99646
SFA5.14 (ERNi-1)	N02061
SFA5.14 (ERNiCr-3)	N06082
SFA5.14 (ERNiCrFe-5)	N06062
SFA5.14 (ERNiCrFe-6)	N07092
SFA5.14 (ERNiCrFe-7)	N07069
SFA5.14 (ERNiCrMo-1)	N06007
SFA5.14 (ERNiCrMo-2)	N06002
SFA5.14 (ERNiCrMo-3)	N06625
SFA5.14 (ERNiCrMo-7)	N06455
SFA5.14 (ERNiCrMo-8)	N06975
SFA5.14 (ERNiCrMo-9)	N06985
SFA5.14 (ERNiCu-7)	N04060
SFA5.14 (ERNiFeCr-1)	N08065
SFA5.14 (ERNiMo-1)	N10001
SFA5.14 (ERNiMo-2)	N10003
SFA5.14 (ERNiMo-3)	N10004
SFA5.14 (ERNiMo-7)	N10665
SFA5.14 (NiCrMo-4)	N10276
SFA5.15 (ENi-Cl-A)	W82003
SFA5.15 (ENiFe-Cl-A)	W82004
SFA5.17 (EH11K)	K11140
SFA5.17 (EH14)	K11585
SFA5.17 (EL8)	K01008
SFA5.17 (EL8K)	K01009
SFA5.17 (EL12)	K01012
SFA5.17 (EM12)	K01112
SFA5.17 (EM12K)	K01113

Cross Reference Specifications	Unified Number
SFA5.17 (EM13K)	K01313
SFA5.17 (EM14K)	K01314
SFA5.17 (EM15K)	K01515
SFA5.18 (ER70S-2)	K10726
SFA5.18 (ER70S-3)	K11022
SFA5.18 (ER70S-4)	K11132
SFA5.18 (ER70S-5)	K11357
SFA5.18 (ER70S-6)	K11140
SFA5.18 (ER70S-7)	K11125
SFA5.19 (ER AZ61A)	M11611
SFA5.19 (ER AZ92A)	M11922
SFA5.19 (ER AZ101A)	M11101
SFA5.19 (ER EZ33A)	M12331
SFA5.20 (E6xT-1)	W06601
SFA5.20 (E6xT-4)	W06604
SFA5.20 (E6xT-5)	W06605
SFA5.20 (E6xT-6)	W06606
SFA5.20 (E6xT-7)	W06607
SFA5.20 (E6xT-8)	W06608
SFA5.20 (E6xT-11)	W06611
SFA5.20 (E7xT-1)	W07601
SFA5.20 (E7xT-2)	W07602
SFA5.20 (E7xT-3)	W07603
SFA5.20 (E7xT-4)	W07604
SFA5.20 (E7xT-5)	W07605
SFA5.20 (E7xT-6)	W07606
SFA5.20 (E7xT-7)	W07607
SFA5.20 (E7xT-8)	W07608
SFA5.20 (E7xT-10)	W07610
SFA5.20 (E7xT-11)	W07611
SFA5.21 (EFe5-A)	W77710
SFA5.21 (EFe5-B)	W77510
SFA5.21 (EFe5-C)	W77310
SFA5.21 (EFeCr-A1)	W74510
SFA5.21 (EFeMn-A)	W79110
SFA5.21 (EFeMn-B)	W79310
SFA5.21 (RFe5-A)	W77730
SFA5.21 (RFe5-B)	W77530
SFA5.21 (RFeCr-A1)	W74530
SFA5.22 (E307T-1) (E307T-2)	W30731
SFA5.22 (E307T-3)	W30733
SFA5.22 (E308LT-1) (E308LT-2)	W30835
SFA5.22 (E308LT-3)	W30837
SFA5.22 (E308MoLT-1) (E308MoLT-2) (E308MoLT-3)	W30838
SFA5.22 (E308MoT-1) (E308MoT-2) (E308MoT-3)	W30832
SFA5.22 (E308T-1) (E308T-2)	W30831
SFA5.22 (E308T-3)	W30833
SFA5.22 (E309CbLT-3)	W30934
SFA5.22 (E309LT-1) (E309LT-2)	W30935
SFA5.22 (E309LT-3)	W30937
SFA5.22 (E309T-1) (E309T-2)	W30931
SFA5.22 (E309T-3)	W30933
SFA5.22 (E310T-1) (E310T-2) (E310T-3)	W31031
SFA5.22 (E312T-1) (E312T-2) (E312T-3)	W31331
SFA5.22 (E316LT-1) (E316LT-2)	W31635
SFA5.22 (E316LT-3)	W31637
SFA5.22 (E316T-1) (E316T-2)	W31631
SFA5.22 (E316T-3)	W31633
SFA5.22 (E317LT-1) (E317LT-2)	W31735
SFA5.22 (E317LT-3)	W31737
SFA5.22 (E347T-1) (E347T-2)	W34731
SFA5.22 (E347T-3)	W34733

Cross Reference Specifications	Unified Number
SFA5.22 (E409T-1) (E409T-2) (E409T-3)	W40931
SFA5.22 (E410NiMoT-1) (E410NiMoT-2) (E410NiMoT-3)	W41036
SFA5.22 (E410NiTiT-1) (E410NiTiT-2) (E410NiTiT-3)	W41038
SFA5.22 (E410T-1) (E410T-2) (E410T-3)	W41031
SFA5.22 (E430T-1) (E430T-2) (E430T-3)	W43031
SFA5.22 (E502T-1) (E502T-2)	W50231
SFA5.22 (E505T-1) (E505T-2)	W50431
SFA5.23 (EA1)	K11222
SFA5.23 (EA1N)	K11122
SFA5.23 (EA2)	K11223
SFA5.23 (EA2N)	K11123
SFA5.23 (EA3)	K11423
SFA5.23 (EA3N)	K11323
SFA5.23 (EA4)	K11424
SFA5.23 (EA4N)	K11324
SFA5.23 (EB1)	K11043
SFA5.23 (EB1N)	K10943
SFA5.23 (EB2)	K11172
SFA5.23 (EB2H)	K23016
SFA5.23 (EB2HN)	K23116
SFA5.23 (EB2N)	K11072
SFA5.23 (EB3)	K31115
SFA5.23 (EB3N)	K31015
SFA5.23 (EB5)	K12187
SFA5.23 (EB5N)	K12087
SFA5.23 (EB6)	S50280
SFA5.23 (EB6H)	S50180
SFA5.23 (EB6HN)	S50181
SFA5.23 (EB6N)	S50281
SFA5.23 (EB8)	S50480
SFA5.23 (EB8N)	S50481
SFA5.23 (EC-B8)	W50440
SFA5.23 (ECA1)	W17041
SFA5.23 (ECA1N)	W17141
SFA5.23 (ECA2)	W17042
SFA5.23 (ECA2N)	W17142
SFA5.23 (ECA3)	W17043
SFA5.23 (ECA3N)	W17143
SFA5.23 (ECA4)	W17044
SFA5.23 (ECA4N)	W17144
SFA5.23 (ECB1)	W51040
SFA5.23 (ECB1N)	W51041
SFA5.23 (ECB2)	W52040
SFA5.23 (ECB2H)	W52240
SFA5.23 (ECB2HN)	W52241
SFA5.23 (ECB2N)	W52041
SFA5.23 (ECB3)	W53040
SFA5.23 (ECB3N)	W53041
SFA5.23 (ECB4)	W53340
SFA5.23 (ECB4N)	W53341
SFA5.23 (ECB5)	W51340
SFA5.23 (ECB5N)	W51341
SFA5.23 (ECB6)	W50240
SFA5.23 (ECB6H)	W50140
SFA5.23 (ECB6HN)	W50141
SFA5.23 (ECB6N)	W50241
SFA5.23 (ECB8N)	W50441
SFA5.23 (ECF1)	W21150
SFA5.23 (ECF1N)	W21151
SFA5.23 (ECF2)	W20240
SFA5.23 (ECF2N)	W20241
SFA5.23 (ECF3)	W21140
SFA5.23 (ECF3N)	W21141
SFA5.23 (ECF4)	W20440
SFA5.23 (ECF4N)	W20441

ASME (American Society of Mechanical Engineers) Numbers

Cross Reference Specifications	Unified Number
SFA5.23 (ECF5)	W22540
SFA5.23 (ECF6)	W22640
SFA5.23 (ECF6N)	W22641
SFA5.23 (ECM1)	W21240
SFA5.23 (ECM1N)	W21241
SFA5.23 (ECM2)	W21340
SFA5.23 (ECM2N)	W21341
SFA5.23 (ECM3)	W22240
SFA5.23 (ECM3N)	W22241
SFA5.23 (ECM4)	W22440
SFA5.23 (ECM4N)	W22441
SFA5.23 (ECNi1)	W21040
SFA5.23 (ECNi1N)	W21041
SFA5.23 (ECNi2)	W22040
SFA5.23 (ECNi2N)	W22041
SFA5.23 (ECNi3)	W23040
SFA5.23 (ECNi3N)	W23041
SFA5.23 (ECNi4)	W21250
SFA5.23 (ECNi4N)	W21251
SFA5.23 (ECW)	W20140
SFA5.23 (EF1)	K11160
SFA5.23 (EF1N)	K11060
SFA5.23 (EF2)	K21450
SFA5.23 (EF2N)	K21350
SFA5.23 (EF3)	K21485
SFA5.23 (EF3N)	K21385
SFA5.23 (EF4)	K12048
SFA5.23 (EF4N)	K11948
SFA5.23 (EF5)	K41370
SFA5.23 (EF6)	K21135
SFA5.23 (EF6N)	K21035
SFA5.23 (EL12)	K01012
SFA5.23 (EL12N)	K00912
SFA5.23 (EM2)	K10882
SFA5.23 (EM2N)	K10982
SFA5.23 (EM3)	K21015
SFA5.23 (EM3N)	K20915
SFA5.23 (EM4)	K21030
SFA5.23 (EM4N)	K20930
SFA5.23 (EM12K)	K01113
SFA5.23 (EM12KN)	K01013
SFA5.23 (ENi1)	K11040
SFA5.23 (ENi1K)	K11058
SFA5.23 (ENi1KN)	K10958
SFA5.23 (ENi1N)	K10940
SFA5.23 (ENi2)	K21010
SFA5.23 (ENi2N)	K20910
SFA5.23 (ENi3)	K31310
SFA5.23 (ENi3N)	K31210
SFA5.23 (ENi4)	K11485
SFA5.23 (ENi4N)	K11385
SFA5.23 (EW)	K11245
SFA5.24 (ERZr2)	R60702
SFA5.24 (ERZr3)	R60704
SFA5.24 (ERZr4)	R60707
SFA5.25 (EH10K-EW)	K01010
SFA5.25 (EH10Mo-EW)	K10945
SFA5.25 (EH11K-EW)	K11140
SFA5.25 (EH14-EW)	K11585
SFA5.25 (EM5K-EW)	K10726
SFA5.25 (EM12-EW)	K01112
SFA5.25 (EM12K-EW)	K01113
SFA5.25 (EM13K-EW)	K01313
SFA5.25 (EM15K-EW)	K01515
SFA5.25 (EWS-EW)	K11245
SFA5.25 (EWT1)	W06040
SFA5.25 (EWT2) (EWT4)	W20140
SFA5.25 (EWT3)	W22340
SFA5.26 (EG6xT1)	W06301
SFA5.26 (EG6xT2)	W06302

Cross Reference Specifications	Unified Number
SFA5.26 (EG7xT1)	W07301
SFA5.26 (EG7xT2)	W07302
SFA5.26 (EGxxS-1)	K01313
SFA5.26 (EGxxS-1B)	K10945
SFA5.26 (EGxxS-2)	K10726
SFA5.26 (EGxxS-3)	K11022
SFA5.26 (EGxxS-5)	K11325
SFA5.26 (EGxxS-6)	K11140
SFA5.26 (EGxxT-3)	W21033
SFA5.26 (EGxxT4)	W22334
SFA5.26 (EGxxT5)	W20131
SFA5.27 (ERCu)	C18980
SFA5.27 (ERCuNi)	C71581
SFA5.27 (ERCuSi-A)	C65600
SFA5.27 (RBCuZn-A)	C47000
SFA5.27 (RBCuZn-D)	C77300
SFA5.27 (RCuZn-B)	C68000
SFA5.27 (RCuZn-C)	C68100
SFA5.28 (E80C-B2)	W52030
SFA5.28 (E80C-B2L)	W52130
SFA5.28 (E80C-Ni1)	W21030
SFA5.28 (E80C-Ni2)	W22030
SFA5.28 (E80C-Ni3)	W23030
SFA5.28 (E90C-B3)	W53030
SFA5.28 (E90C-B3L)	W53130
SFA5.28 (ER80S-B2)	K20900
SFA5.28 (ER80S-B2L)	K20500
SFA5.28 (ER80S-D2)	K10945
SFA5.28 (ER80S-Ni1)	K11260
SFA5.28 (ER80S-Ni2)	K21240
SFA5.28 (ER80S-Ni3)	K31240
SFA5.28 (ER90S-B3)	K30960
SFA5.28 (ER90S-B3L)	K30560
SFA5.28 (ER100S-1)	K10882
SFA5.28 (ER100S-2)	K11250
SFA5.28 (ER110S-1)	K21015
SFA5.28 (ER120S-1)	K21030
SFA5.29 (E1xxT5-K4)	W22235
SFA5.29 (E6xT5-B6L)	W50230
SFA5.29 (E6xT5-B8L)	W50430
SFA5.29 (E8xT1-A1)	W17031
SFA5.29 (E8xT1-B1L)	W51131
SFA5.29 (E8xT1-Ni1)	W21031
SFA5.29 (E70T4-K2)	W21234
SFA5.29 (E70T5-A1)	W17035
SFA5.29 (E71T8-K2)	W21238
SFA5.29 (E71T8-Ni1)	W21038
SFA5.29 (E71T8-Ni2)	W22038
SFA5.29 (E80T1-B2) (E81T1-B2)	W52031
SFA5.29 (E80T1-B2H)	W52231
SFA5.29 (E80T1-W)	W20131
SFA5.29 (E80T5-B2)	W52035
SFA5.29 (E80T5-B2L)	W52135
SFA5.29 (E80T5-K1)	W21135
SFA5.29 (E80T5-K2)	W21235
SFA5.29 (E80T5-Ni1)	W21035
SFA5.29 (E80T5-Ni2)	W22035
SFA5.29 (E81T1-B1)	W51031
SFA5.29 (E90T1-B3) (E91T1-B3) (E100T1-B3)	W53031
SFA5.29 (E90T1-B3H)	W53231
SFA5.29 (E90T1-B3L)	W53131
SFA5.29 (E90T1-D3)	W19331
SFA5.29 (E90T5-B3)	W53035
SFA5.29 (E90T5-D2)	W19235
SFA5.29 (E91T1-D1)	W19131
SFA5.29 (E100T5-D2)	W10235
SFA5.29 (E101T1-K7)	W22051
SFA5.29 (E111T1-K4)	W22231

Cross Reference Specifications	Unified Number
SFA5.29 (E120T1-K5)	W21631
SFA5.29 (ExxT1-K2)	W21231
SFA5.29 (ExxT1-Ni2)	W22031
SFA5.29 (ExxT5-Ni3)	W23035
SFA5.29 (ExxT8-K6)	W21048
SFA5.29 (ExxxT1-K3)	W21331
SFA5.29 (ExxxT5-K3)	W21335
SFA5.30 (IN-Ms3)	K11140
SFA5.30 (IN6A)	N07092
SFA5.30 (IN60)	N04060
SFA5.30 (IN61)	N02061
SFA5.30 (IN67)	C71581
SFA5.30 (IN82)	N06082
SFA5.30 (IN308)	S30880
SFA5.30 (IN308L)	S30883
SFA5.30 (IN310)	S31080
SFA5.30 (IN312)	S31380
SFA5.30 (IN316)	S31680
SFA5.30 (IN316L)	S31683
SFA5.30 (IN347)	S34780
SFA5.30 (IN502)	S50280
SFA5.30 (IN515)	K20900
SFA5.30 (IN521)	K30960
SFA5.30 (INMs1)	K10726
SFA5.30 (INMs2)	K01313

Cross Index of Commonly Known Documents Which Describe Materials Same as or Similar to Those Covered by UNS Numbers

ASTM (American Society for Testing and Materials) NUMBERS

(The Symbol **ob** Indicates an Obsolete UNS Number.)

Cross Reference Specifications	Unified Number	Cross Reference Specifications	Unified Number	Cross Reference Specifications	Unified Number
ASTM		A29 (1109)	G11090	A29 (5135)	G51350
		A29 (1110)	G11100	A29 (5140)	G51400
A1 (61-80)	K06100	A29 (1116)	G11160	A29 (5145)	G51450
A1 (81-90)	K07000	A29 (1117)	G11170	A29 (5147)	G51470
A1 (91-120)	K07301	A29 (1118)	G11180	A29 (5150)	G51500
A1 (121 and over)	K07500	A29 (1119)	G11190	A29 (5155)	G51550
A2 (A)	K06703	A29 (1132)	G11320	A29 (5160)	G51600
A2 (B)	K07700	A29 (1137)	G11370	A29 (5515)	G51150
A2 (C)	K08201	A29 (1139)	G11390	A29 (6118)	G61180
A21 (F)	K05200	A29 (1140)	G11400	A29 (6150)	G61500
A21 (U)	K04700	A29 (1141)	G11410	A29 (8115)	G81150
A25 (A)	K05700	A29 (1144)	G11440	A29 (8615)	G86150
A25 (B)	K06200	A29 (1145)	G11450	A29 (8617)	G86170
A25 (C)	K07201	A29 (1146)	G11460	A29 (8620)	G86200
A25 (U)	K07200	A29 (1151)	G11510	A29 (8622)	G86220
A26 (1064)	G10640	A29 (1211)	G12110	A29 (8622)	G88220
A27 (60-30)	J03000	A29 (1212)	G12120	A29 (8625)	G86250
A27 (65-35)	J03001	A29 (1213)	G12130	A29 (8627)	G86270
A27 (70-36)	J03501	A29 (1215)	G12150	A29 (8630)	G86300
A27 (70-40)	J02501	A29 (1513)	G15130	A29 (8637)	G86370
A27 (N1, U-60-30)	J02500	A29 (1518)	G15180	A29 (8640)	G86400
A27 (N2)	J03500	A29 (1522)	G15220	A29 (8642)	G86420
A29 (12L13)	G12134	A29 (1524)	G15240	A29 (8645)	G86450
A29 (1005)	G10050	A29 (1525)	G15250	A29 (8650)	G86500
A29 (1006)	G10060	A29 (1526)	G15260	A29 (8655)	G86550
A29 (1008)	G10080	A29 (1527)	G15270	A29 (8720)	G87200
A29 (1010)	G10100 ob	A29 (1536)	G15360	A29 (8740)	G87400
A29 (1011)	G10110	A29 (1541)	G15410	A29 (9254)	G92540
A29 (1012)	G10120	A29 (1547)	G15470	A29 (9255)	G92550
A29 (1013)	G10130	A29 (1548)	G15480	A29 (9260)	G92600
A29 (1015)	G10150	A29 (1551)	G15510	A29 (E9310)	G93106
A29 (1016)	G10160	A29 (1552)	G15520	A29 (E50100)	G50986
A29 (1017)	G10170	A29 (1561)	G15610	A29 (E51100)	G51986
A29 (1018)	G10180	A29 (1566)	G15660	A29 (E52100)	G52986
A29 (1019)	G10190	A29 (1572)	G15720	A31 (B)	K03100
A29 (1020)	G10200	A29 (4023)	G40230	A36 (SHAPES)	K02600
A29 (1021)	G10210	A29 (4024)	G40240	A47 (32510)	F22200
A29 (1022)	G10220	A29 (4027)	G40270	A47 (35018)	F22400
A29 (1023)	G10230	A29 (4028)	G40280	A48 (20)	F11401
A29 (1025)	G10250	A29 (4037)	G40370	A48 (25)	F11701
A29 (1026)	G10260	A29 (4042)	G40420	A48 (30)	F12101
A29 (1029)	G10290	A29 (4047)	G40470	A48 (35)	F12401
A29 (1030)	G10300	A29 (4118)	G41180	A48 (40)	F12801
A29 (1034)	G10340	A29 (4130)	G41300	A48 (45)	F13101
A29 (1035)	G10350	A29 (4135)	G41350	A48 (50)	F13501
A29 (1037)	G10370	A29 (4137)	G41370	A48 (55)	F13801
A29 (1038)	G10380	A29 (4140)	G41400	A48 (60)	F14101
A29 (1039)	G10390	A29 (4142)	G41420	A49	K04701
A29 (1040)	G10400	A29 (4145)	G41450	A53 (E-A) (S-A)	K02504
A29 (1042)	G10420	A29 (4147)	G41470	A53 (E-B) (S-B)	K03005
A29 (1043)	G10430	A29 (4150)	G41500	A57 (1064)	G10640
A29 (1045)	G10450	A29 (4161)	G41610	A59 (9260)	G92600
A29 (1046)	G10460	A29 (4320)	G43200	A65	K11210
A29 (1049)	G10490	A29 (4340)	G43400	A67	K01505
A29 (1050)	G10500	A29 (4419)	G44190	A67 (2)	K06002
A29 (1059)	G10590	A29 (4422)	G44220	A67 (1046)	G10450
A29 (1060)	G10600	A29 (4427)	G44270	A105	K03504
A29 (1064)	G10640	A29 (4615)	G46150	A106 (A)	K02501
A29 (1065)	G10650	A29 (4620)	G46200	A106 (B)	K03006
A29 (1069)	G10690	A29 (4626)	G46260	A106 (C)	K03501
A29 (1070)	G10700	A29 (4718)	G47180	A107 (1118)	G11180
A29 (1074)	G10740	A29 (4720)	G47200	A108	G15130
A29 (1075)	G10750	A29 (4815)	G48150	A108	G15180
A29 (1078)	G10780	A29 (4817)	G48170	A108	G15220
A29 (1080)	G10800	A29 (4820)	G48200	A108	G15240
A29 (1084)	G10840	A29 (5015)	G50150	A108	G15250
A29 (1086)	G10860	A29 (5046)	G50460	A108	G15260
A29 (1090)	G10900	A29 (5120)	G51200	A108	G15270
A29 (1095)	G10950	A29 (5130)	G51300	A108	G15360
A29 (1108)	G11080	A29 (5132)	G51320	A108	G15410

Cross Reference Specifications	Unified Number	Cross Reference Specifications	Unified Number	Cross Reference Specifications	Unified Number
A108	G15470	A167 (310)	S31000	A182(XM-19)	S20910
A108	G15480	A167 (310 S)	S31008	A182 (XM-27)	S44625 ob
A108	G15510	A167 (316)	S31600	A183	K03015
A108	G15520	A167 (316 L)	S31603	A192	K01201
A108	G15610	A167 (317)	S31700	A193	S21800
A108	G15660	A167 (317 L)	S31703	A193 (304)	S30400
A108	G15720	A167 (321)	S32100	A193 (316)	S31600
A108 (12L14)	G12144	A167 (347)	S34700	A193 (321)	S32100
A108 (1008)	G10080	A167 (348)	S34800	A193 (347)	S34700
A108 (1010)	G10100 ob	A167 (XM-15)	S38100	A193 (410)	S41000
A108 (1016)	G10160	A176	S41050	A193 (501)	S50100
A108 (1017)	G10170	A176	S44400	A193 (B16)	K14072
A108 (1018)	G10180	A176	S44627	A194	S21800
A108 (1117)	G11170	A176	S44635	A194 (1)	K01503
A108 (1118)	G11180	A176	S44660	A194 (2, 2HM, 2H)	K04002
A108 (1141)	G11410	A176	S44700	A194 (4)	K14510
A108 (1144)	G11440	A176	S44735	A194 (303)	S30300
A108 (1151)	G11510	A176	S44800	A194 (303 Se)	S30323
A108 (1211)	G12110	A176 (403)	S40300	A194 (304)	S30400
A108 (1212)	G12120	A176 (405)	S40500	A194 (316)	S31600
A108 (1213)	G12130	A176 (409)	S40900	A194 (321)	S32100
A108 (1215)	G12150	A176 (410)	S41000	A194 (347)	S34700
A109 (1, 2, 3)	K02500	A176 (410 S)	S41008	A194 (410)	S41000
A109 (4,5)	K01507	A176 (429)	S42900	A194 (416)	S41600
A109 (1137)	G11370	A176 (430)	S43000	A194 (416 Se)	S41623
A126 (A)	F11501	A176 (442)	S44200	A194 (501)	S50100
A126 (B)	F12102	A176 (443)	S44300	A197	F22000
A126 (C)	F12802	A176 (446)	S44600	A199	S50200
A128 (A)	J91109	A176 (XM-27)	S44625 ob	A199 (T3b)	K21509
A128 (B-1)	J91119	A176 (XM-33)	S44626	A199 (T4)	K31509
A128 (B-2)	J91129	A177 (301)	S30100	A199 (T5)	K41545
A128 (B-3)	J91139	A178 (A)	K01200	A199 (T7)	S50300
A128 (B-4)	J91149	A178 (C)	K03503	A199 (T9)	S50400
A128 (C)	J91309	A179	K01200	A199 (T11)	K11597
A128 (D)	J91459	A181	K03502	A199 (T21)	K31545
A128 (E-1)	J91249	A182	K31830	A199 (T22)	K21590
A128 (E-2)	J91339	A182 (304)	S30400	A200	S50200
A131 (A)	K02300	A182 (304 H)	S30409	A200 (T3b)	K21509
A131 (AH32, DH32, EH32)	K11846	A182 (304 L)	S30403	A200 (T4)	K31509
A131 (AH36, DH36, EH36)	K11852	A182 (304 N)	S30451	A200 (T5)	K41545
A131 (B)	K02102	A182 (310)	S31000	A200 (T7)	S50300
A131 (CS) (DS)	K01601	A182 (316)	S31600	A200 (T9)	S50400
A131 (D)	K02101	A182 (316 H)	S31609	A200 (T11)	K11597
A131 (E)	K01801	A182 (316 L)	S31603	A200 (T21)	K31545
A139 (B)	K03003	A182 (316 N)	S31651	A200 (T22)	K21590
A139 (C)	K03004	A182 (321)	S32100	A202 (A)	K11742
A139 (D)	K03010	A182 (321 H)	S32109	A202/A202M (B)	K12542
A139 (E)	K03012	A182 (347)	S34700	A203/A203M (A)	K21703
A159 (G1800)	F10004	A182 (347 H)	S34709	A203/A203M (B)	K22103
A159 (G2500)	F10005	A182 (348)	S34800	A203/A203M (D)	K31718
A159 (G2500a)	F10009	A182 (348 H)	S34809	A203/A203M (E)	K32018
A159 (G3000)	F10006	A182 (429)	S42900	A204/A204M (A)	K11820
A159 (G3500)	F10007	A182 (430)	S43000	A204/A204M (B)	K12020
A159 (G3500b)	F10010	A182 (F1)	K12822	A204/A204M (C)	K12320
A159 (G3500c)	F10011	A182 (F2)	K12122	A209 (T1)	K11522
A159 (G4000)	F10008	A182 (F5a)	K42544	A209 (T1a)	K12023
A159 (G4000d)	F10012	A182 (F6NM)	S41500	A209 (T1b)	K11422
A161 (LCST)	K01504	A182 (F6NM)	S42400	A210 (A-1)	K02707
A161 (T1)	K11522	A182 (F6b)	S41026	A210 (C)	K03501
A167	S31725	A182 (F7)	S50300	A213	S21500
A167	S31726	A182 (F9)	K90941	A213	S31725
A167 (301)	S30100	A182 (F-10)	S33100	A213	S31726
A167 (302)	S30200	A182 (F11)	K11572	A213	S50200
A167 (302 B)	S30215	A182 (F12)	K11564	A213 (304)	S30400
A167 (304)	S30400	A182 (F21)	K31545	A213 (304 H)	S30409
A167 (304 L)	S30403	A182 (F22)	K21590	A213 (304 L)	S30403
A167 (305)	S30500	A182 (F50)	S31200	A213 (304 N)	S30451
A167 (308)	S30800	A182 (F62)	S41000	A213 (310)	S31000
A167 (309)	S30900	A182 (FR)	K22035	A213 (316)	S31600
A167 (309 S)	S30908	A182M	K31830	A213 (316 H)	S31609

Cross Reference Specifications	Unified Number
A213 (316 L)	S31603
A213 (316 N)	S31651
A213 (321)	S32100
A213 (321 H)	S32109
A213 (347)	S34700
A213 (347 H)	S34709
A213 (348)	S34800
A213 (348 H)	S34809
A213 (T2)	K11547
A213 (T3b)	K21509
A213 (T5)	K41545
A213 (T5b)	K51545
A213 (T5c)	K41245
A213 (T7)	S50300
A213 (T9)	S50400
A213 (T11)	K11597
A213 (T12)	K11562
A213 (T17)	K12047
A213 (T21)	K31545
A213 (T22)	K21590
A213 (XM-15)	S38100
A214	K01807
A216 (WCA)	J02502
A216 (WCB)	J03002
A216 (WCC)	J02503
A217 (C5)	J42045
A217 (C12)	J82090
A217 (CA-15)	J91150
A217 (WC 4)	J12082
A217 (WC 6)	J12072
A217 (WC1)	J12524
A217 (WC5)	J22000
A217 (WC9)	J21890
A217 (WC11)	J11872
A217 (WCl)	J12520
A220 (40010)	F22830
A220 (45006)	F23131
A220 (45008)	F23130
A220 (50005)	F23530
A220 (60004)	F24130
A220 (70003)	F24830
A220 (80002)	F25530
A220 (90001)	F26230
A225/A225M	K11803
A225/A225M	K12003
A225/A225M (C)	K12524
A226	K01201
A227	K06501
A228	K08500
A228 (1086)	G10860
A229	K07001
A229 (1065)	G10650
A230	K06701
A230 (1064)	G10640
A231	K15048
A232	K15047
A234 (WP1)	K12821
A234 (WP5)	K41545
A234 (WP9)	K90941
A234 (WP12)	K12062
A234 (WP22)	K21590
A234 (WPB)	K03006
A234 (WPC)	K03501
A234 (WPR)	K22035
A236 (1046)	G10450
A240	S30600
A240	S31260
A240	S31725
A240	S31726
A240	S32550

Cross Reference Specifications	Unified Number
A240	S32950
A240	S41050
A240	S44400
A240	S44627
A240	S44635
A240	S44660
A240	S44700
A240	S44735
A240	S44800
A240 (302)	S30200
A240 (304)	S30400
A240 (304 H)	S30409
A240 (304 L)	S30403
A240 (304 N)	S30451
A240 (305)	S30500
A240 (309 S)	S30908
A240 (310 S)	S31008
A240 (316)	S31600
A240 (316 H)	S31609
A240 (316 L)	S31603
A240 (316 N)	S31651
A240 (317)	S31700
A240 (317 L)	S31703
A240 (321)	S32100
A240 (321 H)	S32109
A240 (347 H)	S34709
A240 (348)	S34800
A240 (348 H)	S34809
A240 (405)	S40500
A240 (410)	S41000
A240 (410 S)	S41008
A240 (429)	S42900
A240 (430)	S43000
A240 (439)	S43035
A240 (XM-15)	S38100
A240 (XM-17)	S21600
A240 (XM-18)	S21603
A240 (XM-19)	S20910
A240 (XM-21)	S30452
A240 (XM-27)	S44625 ob
A240 (XM-29)	S24000
A240 (XM-31)	S21400
A240 (XM-33)	S44626
A241	K05800
A242 (1)	K11510
A242 (2)	K12010
A249	S31725
A249	S31726
A249 (304)	S30400
A249 (304 H)	S30409
A249 (304 L)	S30403
A249 (304 N)	S30451
A249 (305)	S30500
A249 (309)	S30900
A249 (310)	S31000
A249 (316)	S31600
A249 (316 H)	S31609
A249 (316 L)	S31603
A249 (316 N)	S31651
A249 (317)	S31700
A249 (321)	S32100
A249 (321 H)	S32109
A249 (347)	S34700
A249 (347 H)	S34709
A249 (348)	S34800
A249 (348 H)	S34809
A249 (XM-15)	S38100
A249 (XM-19)	S20910
A249 (XM-29)	S24000
A250 (T1)	K11522

Cross Reference Specifications	Unified Number
A250 (T1b)	K11422
A254	K01001
A266 (1, 2)	K03506
A266 (3)	K05001
A266 (4)	K03017
A266 (1046)	G10450
A268	S40800
A268	S44400
A268	S44627
A268	S44635
A268	S44660
A268	S44700
A268	S44735
A268	S44800
A268 (329)	S32900
A268 (405)	S40500
A268 (409)	S40900
A268 (410)	S41000
A268 (429)	S42900
A268 (430)	S43000
A268 (430 Ti)	S43036
A268 (439)	S43035
A268 (443)	S44300
A268 (446)	S44600
A268 (XM-27)	S44625 ob
A268 (XM-33)	S44626
A269	S30600
A269	S31725
A269	S31726
A269 (304)	S30400
A269 (304 L)	S30403
A269 (316)	S31600
A269 (316 L)	S31603
A269 (317)	S31700
A269 (321)	S32100
A269 (347)	S34700
A269 (348)	S34800
A269 (XM-15)	S38100
A269 (XM-19)	S20910
A269 (XM-29)	S24000
A270 (304)	S30400
A271 (304)	S30400
A271 (304 H)	S30409
A271 (321)	S32100
A271 (321 H)	S32109
A271 (347)	S34700
A271 (347 H)	S34709
A273 (1026)	G10260
A273 (1029)	G10290
A273 (1042)	G10420
A274 (3140)	G31400
A274 (E9310)	G93106
A276	S21800
A276	S28200
A276	S30453
A276	S30454
A276	S30815
A276	S31653
A276	S31654
A276	S31725
A276	S31726
A276	S31803
A276	S42010
A276	S44400
A276	S44627
A276	S44700
A276	S44800
A276	S50400
A276 (302)	S30200
A276 (302 B)	S30215

Cross Reference Specifications	Unified Number	Cross Reference Specifications	Unified Number	Cross Reference Specifications	Unified Number
A276 (304)	S30400	A294 (A)	K14245	A304 (4147 H)	H41470
A276 (304 L)	S30403	A294 (B)	K24535	A304 (4150 H)	H41500
A276 (305)	S30500	A294 (C)	K13586	A304 (4161 H)	H41610
A276 (308)	S30800	A295 (50100)	G50986	A304 (4320 H)	H43200
A276 (309)	S30900	A295 (E51100)	G51986	A304 (4340 H)	H43400
A276 (309 S)	S30908	A296	N06040	A304 (4419 H)	H44190
A276 (310)	S31000	A296	N08007	A304 (4620 H)	H46200
A276 (310 S)	S31008	A296	N10001	A304 (4621 H)	H46210
A276 (314)	S31400	A296 (CF-3)	J92700	A304 (4626H)	H46260
A276 (316)	S31600	A297	N06006	A304 (4718 H)	H47180
A276 (316 L)	S31603	A297	N08001	A304 (4720 H)	H47200
A276 (316 N)	S31651	A297	N08002	A304 (4815 H)	H48150
A276 (317)	S31700	A297	N08004	A304 (4817 H)	H48170
A276 (321)	S32100	A297 (HC)	J92605	A304 (4820 H)	H48200
A276 (347)	S34700	A297 (HD)	J93005	A304 (5046 H)	H50460
A276 (348)	S34800	A297 (HE)	J93403	A304 (5120 H)	H51200
A276 (403)	S40300	A297 (HF)	J92603	A304 (5130 H)	H51300
A276 (405)	S40500	A297 (HH)	J93503	A304 (5132 H)	H51320
A276 (410)	S41000	A297 (HI)	J94003	A304 (5135 H)	H51350
A276 (414)	S41400	A297 (HK)	J94224	A304 (5140 H)	H51400
A276 (420)	S42000	A297 (HL)	J94604	A304 (5145 H)	H51450
A276 (429)	S42900	A297 (HN)	J94213	A304 (5147 H)	H51470
A276 (430)	S43000	A297 (HP)	J95705	A304 (5150 H)	H51500
A276 (431)	S43100	A297 (HT)	J94605	A304 (5155 H)	H51550
A276 (440 A)	S44002	A297 (HU)	J95405	A304 (5160 H)	H51600
A276 (440 B)	S44003	A299/A299M	K02803	A304 (6118 H)	H61180
A276 (440 C)	S44004	A302/A302M (A)	K12021	A304 (6150 H)	H61500
A276 (446)	S44600	A302/A302M (C)	K12039	A304 (8617 H)	H86170
A276 (XM-10)	S21900	A302/A302M (D)	K12054	A304 (8620 H)	H86200
A276 (XM-11)	S21904	A302 (B)	K12022	A304 (8622 H)	H86220
A276 (XM-21)	S30452	A304 (15B35H)	H15351	A304 (8625 H)	H86250
A276 (XM-26)	S31100	A304 (15B37H)	H15371	A304 (8627 H)	H86270
A276 (XM-27)	S44625 ob	A304 (15B41H)	H15411	A304 (8630 H)	H86300
A278 (20)	F11401	A304 (15B48H)	H15481	A304 (8637 H)	H86370
A278 (25)	F11701	A304 (15B62H)	H15621	A304 (8640 H)	H86400
A278 (30)	F12101	A304 (50B40 H)	H50401	A304 (8642 H)	H86420
A278 (35)	F12401	A304 (50B44 H)	H50441	A304 (8645 H)	H86450
A278 (40)	F12803	A304 (50B46 H)	H50461	A304 (8650 H)	H86500
A278 (45)	F13102	A304 (50B50 H)	H50501	A304 (8655 H)	H86550
A278 (50)	F13502	A304 (50B60 H)	H50601	A304 (8660 H)	H86600
A278 (55)	F13802	A304 (51B60 H)	H51601	A304 (8720 H)	H87200
A278 (60)	F14102	A304 (81B45 H)	H81451	A304 (8740 H)	H87400
A278 (70)	F14801	A304 (86B30 H)	H86301	A304 (8822 H)	H88220
A278 (80)	F15501	A304 (86B45 H)	H86451	A304 (9260 H)	H92600
A284	K01804	A304 (94B15 H)	H94151	A304 (9262)	G92620
A284	K02001	A304 (94B17 H)	H94171	A304 (9310 H)	H93100
A284 (C)	K02401	A304 (94B30 H)	H94301	A304 (E4340 H)	H43406
A284 (D)	K02702	A304 (1038H)	H10380	A311 (1137)	G11370
A285/A285M (A)	K01700	A304 (1045H)	H10450	A311 (1141)	G11410
A285/A285M (B)	K02200	A304 (1330 H)	H13300	A311 (1144)	G11440
A285/A285M (C)	K02801	A304 (1335 H)	H13350	A311 (1151)	G11510
A288 (1)	K05002	A304 (1340 H)	H13400	A312	S30600
A288 (2, 3)	K14542	A304 (1345 H)	H13450	A312	S31725
A288 (4, 5, 6, 7, 8)	K24562	A304 (1522H)	H15220	A312	S31726
A289 (A)	K91555	A304 (1526H)	H15240	A312 (304)	S30400
A289 (B)	K91955	A304 (1526H)	H15260	A312 (304 H)	S30409
A290 (A, B)	K04000	A304 (1541H)	H15410	A312 (304 L)	S30403
A290 (C, D)	K04500	A304 (4027 H)	H40270	A312 (304 N)	S30451
A290 (E, F)	K14048	A304 (4028 H)	H40280	A312 (309)	S30900
A290 (G, H, I, J, K, L)	K24045	A304 (4032 H)	H40320	A312 (310)	S31000
A291 (1)	K05500	A304 (4037 H)	H40370	A312 (316)	S31600
A291 (2)	K05000	A304 (4042 H)	H40420	A312 (316 H)	S31609
A291 (3)	K14507	A304 (4047 H)	H40470	A312 (316 L)	S31603
A291 (3A)	K14557	A304 (4118 H)	H41180	A312 (316 N)	S31651
A291 (4 to 7)	K24245	A304 (4130 H)	H41300	A312 (317)	S31700
A293 (1)	K14501	A304 (4135 H)	H41350	A312 (321)	S32100
A293 (2, 3)	K23028	A304 (4137 H)	H41370	A312 (321 H)	S32109
A293 (4)	K23578	A304 (4140 H)	H41400	A312 (347)	S34700
A293 (5)	K23579	A304 (4142 H)	H41420	A312 (347 H)	S34709
A293 (6)	K23705	A304 (4145 H)	H41450	A312 (348)	S34800

Cross Reference Specifications	Unified Number	Cross Reference Specifications	Unified Number	Cross Reference Specifications	Unified Number
A312 (348 H)	S34809	A322 (50B46)	G50461	A325 (2)	K11900
A312 (XM-15)	S38100	A322 (50B50)	G50501	A325 (A)	K13643
A312 (XM-19)	S20910	A322 (50B60)	G50601	A325 (B)	K14358
A312 (XM-29)	S24000	A322 (51B60)	G51601	A325 (C)	K12033
A313 (302)	S30200	A322 (94B30)	G94301	A325 (D)	K12059
A313 (304)	S30400	A322 (1330)	G13300	A325 (E)	K12254
A313 (305)	S30500	A322 (1340)	G13400	A325 (F)	K12238
A313 (309-Cb)	S30940	A322 (1345)	G13450	A331 (50B44)	G50441
A313 (310Cb)	S31040	A322 (4023)	G40230	A331 (50B46)	G50461
A313 (316)	S31600	A322 (4024)	G40240	A331 (50B50)	G50501
A313 (316 Ti)	S31635	A322 (4027)	G40270	A331 (50B60)	G50601
A313 (316Cb)	S31640	A322 (4028)	G40280	A331 (51B60)	G51601
A313 (631)	S17700	A322 (4037)	G40370	A331 (81B45)	G81451
A313 (XM-16)	S45500	A322 (4047)	G40470	A331 (94B30)	G94301
A314	S28200	A322 (4118)	G41180	A331 (1330)	G13300
A314	S42010	A322 (4130)	G41300	A331 (1335)	G13350
A314	S44627	A322 (4137)	G41370	A331 (1340)	G13400
A314 (202)	S20200	A322 (4140)	G41400	A331 (1345)	G13450
A314 (302)	S30200	A322 (4142)	G41420	A331 (3140)	G31400
A314 (302 B)	S30215	A322 (4145)	G41450	A331 (4012)	G40120
A314 (303)	S30300	A322 (4147)	G41470	A331 (4023)	G40230
A314 (303 Se)	S30323	A322 (4150)	G41500	A331 (4024)	G40240
A314 (304)	S30400	A322 (4161)	G41610	A331 (4027)	G40270
A314 (304 L)	S30403	A322 (4320)	G43200	A331 (4028)	G40280
A314 (305)	S30500	A322 (4340)	G43400	A331 (4037)	G40370
A314 (308)	S30800	A322 (4615)	G46150	A331 (4042)	G40420
A314 (309)	S30900	A322 (4620)	G46200	A331 (4047)	G40470
A314 (309 S)	S30908	A322 (4621)	G46210	A331 (4063)	G40630
A314 (310)	S31000	A322 (4626)	G46260	A331 (4118)	G41180
A314 (310 S)	S31008	A322 (4718)	G47180	A331 (4130)	G41300
A314 (314)	S31400	A322 (4720)	G47200	A331 (4135)	G41350
A314 (316)	S31600	A322 (4815)	G48150	A331 (4137)	G41370
A314 (316 L)	S31603	A322 (4817)	G48170	A331 (4140)	G41400
A314 (317)	S31700	A322 (4820)	G48200	A331 (4142)	G41420
A314 (321)	S32100	A322 (5015)	G50150	A331 (4145)	G41450
A314 (347)	S34700	A322 (5117)	G51170	A331 (4147)	G41470
A314 (348)	S34800	A322 (5120)	G51200	A331 (4150)	G41500
A314 (403)	S40300	A322 (5130)	G51300	A331 (4161)	G41610
A314 (405)	S40500	A322 (5132)	G51320	A331 (4320)	G43200
A314 (410)	S41000	A322 (5135)	G51350	A331 (4340)	G43400
A314 (414)	S41400	A322 (5140)	G51400	A331 (4419)	G44190
A314 (416)	S41600	A322 (5145)	G51450	A331 (4422)	G44220
A314 (416 Se)	S41623	A322 (5147)	G51470	A331 (4427)	G44270
A314 (420)	S42000	A322 (5150)	G51500	A331 (4615)	G46150
A314 (429)	S42900	A322 (5155)	G51550	A331 (4620)	G46200
A314 (430)	S43000	A322 (5160)	G51600	A331 (4621)	G46210
A314 (430 F)	S43020	A322 (6118)	G61180	A331 (4626)	G46260
A314 (430F Se)	S43023	A322 (6150)	G61500	A331 (4718)	G47180
A314 (431)	S43100	A322 (8615)	G86150	A331 (4720)	G47200
A314 (440 A)	S44002	A322 (8617)	G86170	A331 (4815)	G48150
A314 (440 B)	S44003	A322 (8620)	G86200	A331 (4817)	G48170
A314 (440 C)	S44004	A322 (8622)	G86220	A331 (4820)	G48200
A314 (446)	S44600	A322 (8622)	G88220	A331 (5015)	G50150
A314 (501)	S50100	A322 (8625)	G86250	A331 (5117)	G51170
A314 (502)	S50200	A322 (8627)	G86270	A331 (5120)	G51200
A314 (XM-10)	S21900	A322 (8630)	G86300	A331 (5130)	G51300
A314 (XM-11)	S21904	A322 (8637)	G86370	A331 (5132)	G51320
A314 (XM-27)	S44625 ob	A322 (8640)	G86400	A331 (5135)	G51350
A319 (I)	F10001	A322 (8642)	G86420	A331 (5140)	G51400
A319 (II)	F10002	A322 (8645)	G86450	A331 (5145)	G51450
A319 (III)	F10003	A322 (8650)	G86500	A331 (5147)	G51470
A320 (303)	S30300	A322 (8655)	G86550	A331 (5150)	G51500
A320 (303 Se)	S30323	A322 (8660)	G86600	A331 (5155)	G51550
A320 (304)	S30400	A322 (8720)	G87200	A331 (5160)	G51600
A320 (316)	S31600	A322 (8740)	G87400	A331 (6118)	G61180
A320 (321)	S32100	A322 (9260)	G92600	A331 (6120)	G61200
A320 (347)	S34700	A322 (E9310)	G93106	A331 (6150)	G61500
A321	K05501	A322 (E51100)	G51986	A331 (8615)	G86150
A322	G81451	A322 (E52100)	G52986	A331 (8617)	G86170
A322 (50B44)	G50441	A325 (1)	K02706	A331 (8620)	G86200

ASTM (American Society for Testing and Materials) Numbers

Cross Reference Specifications	Unified Number	Cross Reference Specifications	Unified Number	Cross Reference Specifications	Unified Number
A331 (8622)	G86220	A351 (CG6MMN)	J93790	A372 (2)	K04001
A331 (8622)	G88220	A351 (CH-8)	J93400	A372 (3)	K04801
A331 (8625)	G86250	A351 (CH-10)	J93401	A372 (4)	K14508
A331 (8627)	G86270	A351 (CH-20)	J93402	A372 (6)	K31820
A331 (8630)	G86300	A351 (CK-20)	J94202	A372 (7)	K24055
A331 (8637)	G86370	A351 (CN-7M)	J95150	A372 (V-A)	K13047
A331 (8640)	G86400	A351 (HK-30)	J94203	A376	S31725
A331 (8642)	G86420	A351 (HK-40)	J94204	A376	S31726
A331 (8645)	G86450	A351 (HT-30)	J94603	A376 (16-8-2-H)	S16800
A331 (8650)	G86500	A352 (LC1)	J12522	A376 (304)	S30400
A331 (8655)	G86550	A352 (LC2)	J22500	A376 (304 H)	S30409
A331 (8720)	G87200	A352 (LC2-1)	J42215	A376 (304 N)	S30451
A331 (8740)	G87400	A352 (LC3)	J31550	A376 (316)	S31600
A331 (8742)	G87420	A352 (LC4)	J41500	A376 (316 H)	S31609
A331 (9260)	G92600	A352 (LCA)	J02504	A376 (316 N)	S31651
A331 (E3310)	G33106	A352 (LCB)	J03003	A376 (321)	S32100
A331 (E4340)	G43406	A352 (LCC)	J02505	A376 (321 H)	S32109
A331 (E9310)	G93106	A353M	K81340	A376 (347)	S34700
A331 (E52100)	G52986	A354	K04100	A376 (347 H)	S34709
A333 (1)	K03008	A355 (A)	K24065	A376 (348)	S34800
A333 (3)	K31918	A355 (A)	K24728	A381	K03013
A333 (4)	K11267	A355 (B)	K23745	A383	K04700
A333 (6)	K03006	A355 (C)	K52440	A387	S50300
A333 (7)	K21903	A355 (D)	K23510	A387	S50400
A333 (8)	K81340	A356 (1)	J03502	A387 (5)	K41545
A333 (9)	K22035	A356 (2)	J12523	A387 (5)	S50100
A334 (1)	K03008	A356 (5)	J12540	A387 (5)	S50200
A334 (3)	K31918	A356 (6)	J12073	A387 (21)	K31545
A334 (6)	K03006	A356 (8)	J11697	A387 (22)	K21590
A334 (7)	K21903	A356 (9)	J21610	A387/A387M (2)	K12143
A334 (8)	K81340	A356 (10)	J22090	A387/A387M (11)	K11789
A334 (9)	K22035	A358	S30600	A387/A387M (12)	K11757
A335	S50200	A358	S31725	A389 (C23)	J12080
A335 (P1)	K11522	A358	S31726	A389 (C24)	J12092
A335 (P2)	K11547	A358 (304)	S30400	A395 (60-40-18)	F32800
A335 (P5)	K41545	A358 (304 L)	S30409	A401	G92540
A335 (P5b)	K51545	A358 (304 N)	S30451	A401	K15590
A335 (P5c)	K41245	A358 (309)	S30900	A403 (304 L)	S30403
A335 (P7)	S50300	A358 (310)	S31000	A403 (304 L)	S30409
A335 (P9)	S50400	A358 (316)	S31600	A403 (304 N)	S30451
A335 P11)	K11597	A358 (316 H)	S31609	A403 (309)	S30900
A335 (P12)	K11562	A358 (316 N)	S31651	A403 (310)	S31000
A335 (P15)	K11578	A358 (321)	S32100	A403 (316)	S31600
A335 (P21)	K31545	A358 (347)	S34700	A403 (316 H)	S31609
A335 (P22)	K21590	A358 (348)	S34800	A403 (316 L)	S31603
A336 (F1)	K12520	A368 (302)	S30200	A403 (316 N)	S31651
A336 (F5)	K41545	A368 (304)	S30400	A403 (317)	S31700
A336 (F5a)	K42544	A368 (305)	S30500	A403 (321)	S32100
A336 (F6)	S41000	A368 (316)	S31600	A403 (321 H)	S32109
A336 (F12)	K11564	A369 (FP1)	K11522	A403 (347)	S34700
A336 (F21, F21a)	K31545	A369 (FP2)	K11547	A403 (347 H)	S34709
A336 (F22, F22a)	K21590	A369 (FP3b)	K21509	A403 (348)	S34800
A336 (F30)	K14520	A369 (FP5)	K41545	A403 (XM-19)	S20910
A336 (F31)	K23545	A369 (FP7)	S50300	A405 (P24)	K11591
A336 (F32)	K33585	A369 (FP9)	K90941	A409	S31725
A336 (FN, F11A)	K11572	A369 (FP11)	K11597	A409	S31726
A350 (LF1)	K03009	A369 (FP12)	K11562	A409 (304)	S30400
A350 (LF2)	K03011	A369 (FP21)	K31545	A409 (309)	S30900
A350 (LF3)	K32025	A369 (FP22)	K21590	A409 (310)	S31000
A350 (LF5)	K13050	A369 (FPA)	K02501	A409 (316)	S31600
A350 (LF9)	K22036	A369 (FPB)	K03006	A409 (317)	S31700
A351	J95151	A372	G13450	A409 (321)	S32100
A351	N08007	A372	G41350	A409 (347)	S34700
A351	N08030	A372	G41370	A409 (348)	S34800
A351 (CF-3)	J92500	A372	G41420	A412 (201)	S20100
A351 (CF3M, CF3MA)	J92800	A372	G41470	A412 (202)	S20200
A351 (CF-8)	J92600	A372	K13049	A412 (XM-10)	S21900
A351 (CF-8C)	J92710	A372	K13547	A412 (XM-11)	S21904
A351 (CF-8M)	J92900	A372	K13548	A412 (XM-14)	S21460
A351 (CF-10MC)	J92971	A372 (1)	K03002	A412 (XM-19)	S20910

Cross Reference Specifications	Unified Number
A412 (XM-29)	S24000
A413	K03700
A414 (A)	K01501
A414 (B)	K02201
A414 (C)	K02503
A414 (D)	K02505
A414 (E)	K02704
A414 (F)	K03102
A414 (G)	K03103
A417	K06201
A423 (1)	K11535
A423 (2)	K11540
A424 (I)	K00100
A424 (IIA)	K00400
A424 (IIB)	K00801
A426 (CP1)	J12521
A426 (CP2)	J11547
A426 (CP5)	J42045
A426 (CP5b)	J51545
A426 (CP7)	J61594
A426 (CP9)	J82090
A426 (CP11)	J12072
A426 (CP12)	J11562
A426 (CP15)	J11522
A426 (CP21)	J31545
A426 (CP22)	J21890
A426 (CPCA-15)	J91150
A426 (CPCA15)	J91171
A429 (201)	S20100
A429 (202)	S20200
A429 (XM-10)	S21900
A429 (XM-11)	S21904
A429 (XM-19)	S20910
A430 (16-8-2-H)	S16800
A430 (304)	S30400
A430 (304 H)	S30409
A430 (304 N)	S30451
A430 (316)	S31600
A430 (316 H)	S31609
A430 (316 N)	S31651
A430 (321)	S32100
A430 (321 H)	S32109
A430 (347)	S34700
A430 (347 H)	S34709
A436 (1)	F41000
A436 (1b)	F41001
A436 (2)	F41002
A436 (2b)	F41003
A436 (3)	F41004
A436 (4)	F41005
A436 (5)	F41006
A436 (6)	F41007
A437 (B4B, B4C)	K91352
A437 (B4D)	K14072
A439 (D-2)	F43000
A439 (D-2B)	F43001
A439 (D-2C)	F43002
A439 (D-3)	F43003
A439 (D-3A)	F43004
A439 (D-4)	F43005
A439 (D-5)	F43006
A439 (D-5B)	F43007
A441	K12211
A442/A442M (55)	K02202
A442/A442M (60)	K02402
A445/A445M	K02802
A447	J93303
A448	J94605
A449	K04200
A451 (CH-2)	J93402

Cross Reference Specifications	Unified Number
A451 (CPF8)	J92600
A451 (CPF8C)	J92710
A451 (CPF8M)	J92900
A451 (CPH-8)	J93400
A451 (CPK20)	J94202
A452 (TP304H)	J92590
A452 (TP316H)	J92920
A452 (TP347H)	J92660
A453 (651)	S63198
A453 (660)	S66286
A453 (662)	S66220
A453 (665)	S66545
A454 (1, 2)	K03700
A454 (3)	K03500
A455 (I)	K03300
A457 (651)	S63198
A458 (651)	S63198
A461	R30816
A469 (1)	K14501
A469 (2)	K22573
A469 (3)	K22773
A469 (4)	K32723
A469 (5)	K33125
A469 (6, 7, 8)	K42885
A470 (1)	K14501
A470 (2)	K22578
A470 (3, 4)	K22878
A470 (5, 6, 7)	K42885
A470 (8)	K23010
A471 (1-9)	K32800
A471 (10)	K23205
A473	S28200
A473	S30815
A473	S31803
A473 (202)	S20200
A473 (302)	S30200
A473 (302 B)	S30215
A473 (303)	S30300
A473 (303 Se)	S30323
A473 (304)	S30400
A473 (304 L)	S30403
A473 (305)	S30500
A473 (308)	S30800
A473 (309)	S30900
A473 (309 S)	S30908
A473 (310)	S31000
A473 (310 S)	S31008
A473 (314)	S31400
A473 (316)	S31600
A473 (316 L)	S31603
A473 (317)	S31700
A473 (321)	S32100
A473 (347)	S34700
A473 (403)	S40300
A473 (405)	S40500
A473 (410)	S41000
A473 (410 S)	S41008
A473 (414)	S41400
A473 (416)	S41600
A473 (416 Se)	S41623
A473 (420)	S42000
A473 (429)	S42900
A473 (430)	S43000
A473 (430 F)	S43020
A473 (430F Se)	S43023
A473 (431)	S43100
A473 (440 A)	S44002
A473 (440 B)	S44003
A473 (440 C)	S44004
A473 (446)	S44600

Cross Reference Specifications	Unified Number
A473 (501)	S50100
A473 (501A)	S50300
A473 (501B)	S50400
A473 (502)	S50200
A473 (XM-10)	S21900
A473 (XM-11)	S21904
A476 (80-60-03)	F34100
A477 (651)	S63198
A478 (302)	S30200
A478 (304)	S30400
A478 (304 L)	S30403
A478 (305)	S30500
A478 (316)	S31600
A478 (316 L)	S31603
A478 (317)	S31700
A479	S21800
A479	S30600
A479	S30815
A479	S31725
A479	S31726
A479	S31803
A479	S32550
A479	S44627
A479	S44700
A479	S44800
A479 (302)	S30200
A479 (304)	S30400
A479 (304 H)	S30409
A479 (304 L)	S30403
A479 (304 N)	S30451
A479 (310 S)	S31008
A479 (316)	S31600
A479 (316 H)	S31609
A479 (316 L)	S31603
A479 (316 N)	S31651
A479 (321)	S32100
A479 (321 H)	S32109
A479 (347)	S34700
A479 (347 H)	S34709
A479 (348)	S34800
A479 (348 H)	S34809
A479 (403)	S40300
A479 (405)	S40500
A479 (410)	S41000
A479 (430)	S43000
A479 (439)	S43035
A479 (XM-17)	S21600
A479 (XM-18)	S21603
A479 (XM-19)	S20910
A479 (XM-27)	S44625 ob
A479 (XM-30)	S41040
A481 (A)	R20990
A481 (B)	R20994
A485 (1)	K19667
A485 (2)	K19195
A485 (3)	K19965
A485 (4)	K19990
A486 (70)	J03503
A487 (1N, 1Q)	J13002
A487 (2N, 2Q)	J13005
A487 (4A, 4N, 4Q, 4QA)	J05003
A487 (4N, 4Q, 4QA)	J13047
A487 (6N, 6Q)	J13855
A487 (7Q)	J12084
A487 (8N, 8Q)	J22091
A487 (9N, 9Q)	J13345
A487 (10N, 10Q)	J23015
A487 (11N, 11Q)	J12082
A487 (12N, 12Q)	J22000
A487 (13N, 13Q)	J13080

Cross Reference Specifications	Unified Number	Cross Reference Specifications	Unified Number	Cross Reference Specifications	Unified Number
A487 (14Q)	J15580	A508 (3 and 3a)	K12042	A511 (329)	S32900
A487 (A, AN, AQ)	J02502	A508 (4, 4a, 4b)	K22375	A511 (347)	S34700
A487 (B, BN, BQ)	J03002	A508 (5, 5a)	K42365	A511 (403)	S40300
A487 (C, CN, CQ)	J02503	A510 (1005)	G10050	A511 (405)	S40500
A487 (CA-6NM)	J91540	A510 (1006)	G10060	A511 (410)	S41000
A487 (CA15)	J91171	A510 (1008)	G10080	A511 (414)	S41400
A487 (CA-15M)	J91151	A510 (1010)	G10100 ob	A511 (416 Se)	S41623
A487 (CA-15a)	J91150	A510 (1012)	G10120	A511 (429)	S42900
A487 (DN)	J04500	A510 (1015)	G10150	A511 (430)	S43000
A489	K04800	A510 (1016)	G10160	A511 (440 A)	S44002
A490	K03900	A510 (1017)	G10170	A511 (443)	S44300
A492 (302)	S30200	A510 (1018)	G10180	A511 (446)	S44600
A492 (304)	S30400	A510 (1019)	G10190	A512 (1025)	G10250
A492 (305)	S30500	A510 (1020)	G10200	A512 (1030)	G10300
A492 (XM-17)	S21600	A510 (1021)	G10210	A513 (1016)	G10160
A492 (XM-18)	S21603	A510 (1022)	G10220	A513 (1017)	G10170
A493	S28200	A510 (1023)	G10230	A513 (1018)	G10180
A493 (302)	S30200	A510 (1025)	G10250	A513 (1019)	G10190
A493 (304)	S30400	A510 (1026)	G10260	A513 (1024)	G15240
A493 (305)	S30500	A510 (1029)	G10290	A513 (1027)	G15270
A493 (321)	S32100	A510 (1030)	G10300	A513 (1033)	G10330
A493 (347)	S34700	A510 (1035)	G10350	A513 (4130)	G41300
A493 (384)	S38400	A510 (1037)	G10370	A513 (8620)	G86200
A493 (385)	S38500 ob	A510 (1038)	G10380	A514 (A)	K11856
A493 (410)	S41000	A510 (1039)	G10390	A514 (B)	K11630
A493 (429)	S42900	A510 (1040)	G10400	A514 (C)	K11511
A493 (430)	S43000	A510 (1042)	G10420	A514 (D)	K11662
A493 (431)	S43100	A510 (1043)	G10430	A514 (E)	K21604
A493 (440 C)	S44004	A510 (1044)	G10440	A514 (F)	K11576
A493 (XM-7)	S30430	A510 (1045)	G10450	A514 (G)	K11872
A493 (XM-27)	S44625 ob	A510 (1046)	G10460	A514 (H)	K11646
A494	N10001	A510 (1049)	G10490	A514 (J)	K11625
A494	N10002	A510 (1050)	G10500	A514 (K)	K11523
A500 (A, B)	K03000	A510 (1053)	G10530	A514 (L)	K11682
A500 (C)	K02705	A510 (1055)	G10550	A514 (M)	K11683
A501	K03000	A510 (1060)	G10600	A514 (N)	K11847
A502 (1)	K01900	A510 (1070)	G10700	A514 (P)	K21650
A502 (2)	K02405	A510 (1078)	G10780	A515 (55)	K02001
A502 (3A)	K11430	A510 (1080)	G10800	A515 (60)	K02401
A502 (3B)	K12244	A510 (1084)	G10840	A515/A515M (65)	K02800
A504 (A)	K05700	A510 (1090)	G10900	A515/A515M (70)	K03101
A504 (B)	K06200	A510 (1095)	G10950	A516 (1116)	G11160
A504 (C)	K07201	A510 (1518)	G15180	A516/A516M (55)	K01800
A504 (U)	K07200	A510 (1522)	G15220	A516/A516M (60)	K02100
A505 (4012)	G40120	A510 (1524)	G15240	A516/A516M (65)	K02403
A505 (4118)	G41180	A510 (1525)	G15250	A516/A516M (70)	K02700
A505 (4130)	G41300	A510 (1526)	G15260	A517 (A)	K11856
A505 (4137)	G41370	A510 (1527)	G15270	A517 (B)	K11630
A505 (4140)	G41400	A510 (1536)	G15360	A517 (C)	K11511
A505 (4142)	G41420	A510 (1541)	G15410	A517 (D)	K11662
A505 (4145)	G41450	A510 (1547)	G15470	A517 (E)	K21604
A505 (4147)	G41470	A510 (1548)	G15480	A517 (F)	K11576
A505 (4150)	G41500	A510 (1551)	G15510	A517 (G)	K11872
A505 (4320)	G43200	A510 (1552)	G15520	A517 (H)	K11646
A505 (4340)	G43400	A510 (1561)	G15610	A517 (J)	K11625
A505 (4615)	G46150	A510 (1566)	G15660	A517 (K)	K11523
A505 (4620)	G46200	A510 (1572)	G15720	A517 (L)	K11682
A505 (4718)	G47180	A511	S44735	A517 (M)	K11683
A505 (4815)	G48150	A511 (302)	S30200	A517 (P)	K21650
A505 (4820)	G48200	A511 (304)	S30400	A518 G Rate 2	F47003
A505 (5015)	G50150	A511 (304 L)	S30403	A519 (50B40)	G50401
A505 (5132)	G51320	A511 (305)	S30500	A519 (50B44)	G50441
A505 (5140)	G51400	A511 (309)	S30900	A519 (50B46)	G50461
A505 (5150)	G51500	A511 (309 S)	S30908	A519 (50B50)	G50501
A505 (5160)	G51600	A511 (310)	S31000	A519 (50B60)	G50601
A505 (E4340)	G43406	A511 (310 S)	S31008	A519 (51B60)	G51601
A505 (E51100)	G51986	A511 (316)	S31600	A519 (81B45)	G81451
A505 (E52100)	G52986	A511 (316 L)	S31603	A519 (86B45)	G86451
A508 (1)	K13502	A511 (317)	S31700	A519 (94B15)	G94151
A508 (2, 2a)	K12766	A511 (321)	S32100	A519 (94B17)	G94171

Cross Reference Specifications	Unified Number	Cross Reference Specifications	Unified Number	Cross Reference Specifications	Unified Number
A519 (94B30)	G94301	A519 (8615)	G86150	A541 (1)	K03506
A519 (94B40)	G94401	A519 (8617)	G86170	A541 (2, 2A)	K12765
A519 (1008)	G10080	A519 (8620)	G86200	A541 (3 and 3a)	K12045
A519 (1010)	G10100 ob	A519 (8622)	G86220	A541 (4)	K11800
A519 (1012)	G10120	A519 (8622)	G88220	A541 (5)	K11598
A519 (1015)	G10150	A519 (8625)	G86250	A541 (6, 6A)	K21590
A519 (1017)	G10170	A519 (8627)	G86270	A541 (7, 7A, 7B)	K42343
A519 (1018)	G10180	A519 (8630)	G86300	A541 (8, 8A)	K42348
A519 (1019)	G10190	A519 (8640)	G86400	A542	K21590
A519 (1020)	G10200	A519 (8642)	G86420	A543/A543M(B)	K42339
A519 (1021)	G10210	A519 (8645)	G86450	A543/A543M (B,C)	K42338
A519 (1022)	G10220	A519 (8650)	G86500	A544 (1017)	G10170
A519 (1025)	G10250	A519 (8655)	G86550	A544 (1018)	G10180
A519 (1026)	G10260	A519 (8660)	G86600	A544 (1020)	G10200
A519 (1030)	G10300	A519 (8720)	G87200	A544 (1022)	G10220
A519 (1035)	G10350	A519 (8735)	G87350	A544 (1030)	G10300
A519 (1040)	G10400	A519 (8740)	G87400	A544 (1035)	G10350
A519 (1045)	G10450	A519 (8742)	G87420	A544 (1038)	G10380
A519 (1050)	G10500	A519 (9262)	G92620	A545 (1006)	G10060
A519 (1330)	G13300	A519 (9840)	G98400	A545 (1008)	G10080
A519 (1335)	G13350	A519 (9850)	G98500	A545 (1010)	G10100 ob
A519 (1340)	G13400	A519 (E3310)	G33106	A545 (1012)	G10120
A519 (1345)	G13450	A519 (E4337)	G43376	A545 (1015)	G10150
A519 (1518)	G15180	A519 (E4340)	G43406	A545 (1016)	G10160
A519 (1524)	G15240	A519 (E7140)	G71406	A545 (1018)	G10180
A519 (1541)	G15410	A519 (E9310)	G93106	A545 (1019)	G10190
A519 (3140)	G31400	A519 (E50100)	G50986	A545 (1021)	G10210
A519 (4012)	G40120	A519 (E51100)	G51986	A545 (1022)	G10220
A519 (4023)	G40230	A519 (E52100)	G52986	A545 (1026)	G10260
A519 (4024)	G40240	A522 (I)	K81340	A545 (1035)	G10300
A519 (4027)	G40270	A522 (II)	K71340	A545 (1035)	G10350
A519 (4028)	G40280	A523 (A)	K02504	A545 (1038)	G10380
A519 (4037)	G40370	A523 (B)	K03005	A545 (1524)	G15240
A519 (4042)	G40420	A524	K02104	A545 (1541)	G15410
A519 (4047)	G40470	A529	K02703	A546 (1030)	G10300
A519 (4063)	G40630	A532 (IA)	F45000	A546 (1035)	G10350
A519 (4118)	G41180	A532 (IB)	F45001	A546 (1038)	G10380
A519 (4130)	G41300	A532 (IC)	F45002	A546 (1039)	G10390
A519 (4135)	G41350	A532 (ID)	F45003	A546 (1040)	G10400
A519 (4137)	G41370	A532 (IIA)	F45004	A546 (1541)	G15410
A519 (4140)	G41400	A532 (IIB)	F45005	A547 (1335)	G13350
A519 (4142)	G41420	A532 (IIC)	F45006	A547 (1340)	G13400
A519 (4145)	G41450	A532 (IID)	F45007	A547 (4037)	G40370
A519 (4147)	G41470	A532 (IIE)	F45008	A547 (4137)	G41370
A519 (4150)	G41500	A532 (IIIA)	F45009	A547 (4140)	G41400
A519 (4320)	G43200	A533/A533M (A)	K12521	A547 (4142)	G41420
A519 (4337)	G43370	A533 (B)	K12539	A547 (4340)	G43400
A519 (4340)	G43400	A533 (C)	K12554	A548 (1016)	G10160
A519 (4422)	G44220	A533 (D)	K12529	A548 (1018)	G10180
A519 (4427)	G44270	A534 (4023)	G40230	A548 (1019)	G10190
A519 (4520)	G45200	A534 (Krupp)	K51210	A548 (1021)	G10210
A519 (4617)	G46170	A535 (4320)	G43200	A548 (1022)	G10220
A519 (4720)	G47200	A535 (4620)	G46200	A549 (1008)	G10080
A519 (4817)	G48170	A535 (4720)	G47200	A549 (1010)	G10100 ob
A519 (4820)	G48200	A535 (4820)	G48200	A549 (1012)	G10120
A519 (5015)	G50150	A535 (E52100)	G52986	A549 (1015)	G10150
A519 (5046)	G50460	A536 (60-40-18)	F32800	A549 (1016)	G10160
A519 (5120)	G51200	A536 (65-45-12)	F33100	A549 (1017)	G10170
A519 (5132)	G51320	A536 (80-55-06)	F33800	A549 (1018)	G10180
A519 (5135)	G51350	A536 (100-70-03)	F34800	A551 (A, AHT)	K05701
A519 (5140)	G51400	A536 (120-90-02)	F36200	A551 (B, BHT)	K06702
A519 (5145)	G51450	A537	K02400	A551 (C, CHT, DHT)	K07701
A519 (5147)	G51470	A537/A537M	K12437	A553 (II)	K71340
A519 (5150)	G51500	A538/A538M (A)	K92810	A553M (I)	K81340
A519 (5155)	G51550	A538/A538M (B)	K92890	A554 (301)	S30100
A519 (5160)	G51600	A538/A538M (C)	K93120	A554 (302)	S30200
A519 (5515)	G51150	A539, A587	K01506	A554 (304)	S30400
A519 (6120)	G61200	A540 (B21)	K14073	A554 (304 L)	S30403
A519 (6150)	G61500	A540 (B24)	K24064	A554 (305)	S30500
A519 (8115)	G81150	A540 (B24V)	K24070	A554 (309)	S30900

313

Cross Reference Specifications	Unified Number	Cross Reference Specifications	Unified Number	Cross Reference Specifications	Unified Number
A554 (309 S)	S30908	A576 (12L14)	G12144	A579 (73)	K93160
A554 (309-S-Cb)	S30940	A576 (1008)	G10080	A579 (74)	K91930
A554 (310)	S31000	A576 (1010)	G10100 ob	A579 (75)	K91940
A554 (310 S)	S31008	A576 (1012)	G10120	A579 (81)	K91122
A554 (316)	S31600	A576 (1015)	G10150	A579 (82)	K91283
A554 (316 L)	S31603	A576 (1016)	G10160	A579 (83)	K91094
A554 (317)	S31700	A576 (1017)	G10170	A579 (632)	S15700
A554 (321)	S32100	A576 (1018)	G10180	A579 (634)	S35500
A554 (347)	S34700	A576 (1019)	G10190	A579 (H-11)	T20811
A554 (429)	S42900	A576 (1020)	G10200	A580	S28200
A554 (430)	S43000	A576 (1021)	G10210	A580 (302 B)	S30215
A554 (430 Ti)	S43036	A576 (1022)	G10220	A580 (304)	S30400
A555	S21800	A576 (1023)	G10230	A580 (304 L)	S30403
A556 (A2)	K01807	A576 (1025)	G10250	A580 (305)	S30500
A556 (B2)	K02707	A576 (1026)	G10260	A580 (308)	S30800
A556 (C2)	K03006	A576 (1029)	G10290	A580 (309)	S30900
A557 (A2)	K01807	A576 (1030)	G10300	A580 (309 S)	S30908
A557 (B2)	K03007	A576 (1035)	G10350	A580 (310 S)	S31008
A557 (C2)	K03505	A576 (1037)	G10370	A580 (314)	S31400
A560 (50 Cr-50 Ni)	R20500	A576 (1038)	G10380	A580 (316 L)	S31603
A560 (60 Cr-40 Ni)	R20600	A576 (1039)	G10390	A580 (317)	S31700
A562/A562M	K11224	A576 (1040)	G10400	A580 (347)	S34700
A563 (C3A)	K13643	A576 (1042)	G10420	A580 (348)	S34800
A563 (C3B)	K14358	A576 (1043)	G10430	A580 (403)	S40300
A563 (C3C)	K12033	A576 (1044)	G10440	A580 (405)	S40500
A563 (C3D)	K12059	A576 (1045)	G10450	A580 (410)	S41000
A563 (C3E)	K12254	A576 (1046)	G10460	A580 (414)	S41400
A563 (C3F)	K12238	A576 (1049)	G10490	A580 (420)	S42000
A563 (D)	K05801	A576 (1050)	G10500	A580 (430)	S43000
A563 (DH)	K03800	A576 (1053)	G10530	A580 (431)	S43100
A563 (DH3)	K13650	A576 (1055)	G10550	A580 (440 A)	S44002
A563 (O, A, B, C)	K05802	A576 (1060)	G10600	A580 (440 B)	S44003
A564 (630)	S17400	A576 (1070)	G10700	A580 (440 C)	S44004
A564 (631)	S17700	A576 (1078)	G10780	A580 (446)	S44600
A564 (632)	S15700	A576 (1080)	G10800	A580 (XM-10)	S21900
A564 (634)	S35500	A576 (1084)	G10840	A580 (XM-11)	S21904
A564 (635)	S17600	A576 (1090)	G10900	A580 (XM-19)	S20910
A564 (XM-9)	S36200	A576 (1095)	G10950	A580 (XM-28)	S24100
A564 (XM-12)	S15500	A576 (1109)	G11090	A581 (303)	S30300
A564 (XM-13)	S13800	A576 (1110)	G11100	A581 (303 Se)	S30323
A564 (XM-16)	S45500	A576 (1117)	G11170	A581 (416)	S41600
A564 (XM-25)	S45000	A576 (1118)	G11180	A581 (416 Se)	S41623
A565 (615)	S41800	A576 (1119)	G11190	A581 (430 F)	S43020
A565 (616)	S42200	A576 (1132)	G11320	A581 (430F Se)	S43023
A565 (619)	S42300	A576 (1137)	G11370	A581 (XM-1)	S20300
A565 (XM-32)	S64152	A576 (1139)	G11390	A581 (XM-2)	S30345
A567	N06002	A576 (1140)	G11400	A581 (XM-3)	S30360
A567	N07500	A576 (1141)	G11410	A581 (XM-5)	S30310
A567	N07713	A576 (1146)	G11460	A581 (XM-6)	S41610
A567	N08008	A576 (1151)	G11510	A581(XM-34)	S18200
A567	N10002	A576 (1211)	G12110	A582 (303)	S30300
A567 (1)	R30021	A576 (1212)	G12120	A582 (303 Se)	S30323
A567 (2)	R30031	A576 (1213)	G12130	A582 (416)	S41600
A567 (661)	R30155	A576 (1215)	G12150	A582 (416 Se)	S41623
A570 (30, 33, 36, 40)	K02502	A579 (11)	K42598	A582 (420 F Se)	S42023
A571 (D-2M)	F43010	A579 (12)	K51255	A582 (430 F)	S43020
A573 (58)	K02301	A579 (13)	K13051	A582 (430F Se)	S43023
A573 (65)	K02404	A579 (21)	K23477	A582 (XM-1)	S20300
A573 (70)	K02701	A579 (22)	K24070	A582 (XM-2)	S30345
A574	K03104	A579 (23)	K24728	A582 (XM-3)	S30360
A575 (1010)	G10100 ob	A579 (31)	K32550	A582 (XM-5)	S30310
A575 (1023)	G10230	A579 (32)	K44220	A582 (XM-6)	S41610
A575 (M1008)	G10080	A579 (33)	K14394	A582(XM-34)	S18200
A575 (M1012)	G10120	A579 (51)	S41001	A583 (A)	K05700
A575 (M1015)	G10150	A579 (52)	S42201	A583 (B)	K06200
A575 (M1017)	G10170	A579 (61)	S35000	A583 (C)	K07201
A575 (M1020)	G10200	A579 (62)	S17700	A583 (U)	K07200
A575 (M1025)	G10250	A579 (63)	S43100	A587	K11500
A575 (M1044)	G10440	A579 (71)	K92820	A588	K12040
A576 (12L13)	G12134	A579 (72)	K92940	A588 (A)	K11430

Cross Reference Specifications	Unified Number
A588 (B)	K12043
A588 (C)	K11538
A588 (D)	K11552
A588 (E)	K11567
A588 (F)	K11541
A588 (H)	K12032
A588 (J)	K12044
A590	K91890
A592 (A)	K11856
A592 (E)	K11695
A592 (F)	K11576
A594 (1)	K01500
A594 (2)	K01000
A594 (3)	K00800
A594 (4)	K00600
A595 (A)	K02004
A595 (B)	K02005
A595 (C)	K11526
A597 (CA-2)	T90102
A597 (CD-2)	T90402
A597 (CD-5)	T90405
A597 (CH-12)	T90812
A597 (CH-13)	T90813
A597 (CO-1)	T91501
A597 (CS-5)	T91905
A600 (M-1)	T11301
A600 (M-2)	T11302
A600 (M-3, Cl.1)	T11313
A600 (M-3, Cl.2)	T11323
A600 (M-4)	T11304
A600 (M-6)	T11306
A600 (M-7)	T11307
A600 (M-10)	T11310
A600 (M-30)	T11330
A600 (M-33)	T11333
A600 (M-34)	T11334
A600 (M-36)	T11336
A600 (M-41)	T11341
A600 (M-42)	T11342
A600 (M-43)	T11343 ob
A600 (M-44)	T11344 ob
A600 (M-46)	T11346
A600 (M-47)	T11347 ob
A600 (T-1)	T12001
A600 (T-2)	T12002 ob
A600 (T-4)	T12004
A600 (T-5)	T12005
A600 (T-6)	T12006
A600 (T-8)	T12008
A600 (T-15)	T12015
A601 (A)	M29951
A601 (B)	M29952
A601 (C)	M29953
A601 (D)	M29450
A601 (E)	M29350
A601 (F)	M29954
A602 (M3210)	F20000
A605/A605M	K91401
A608	N06050
A608	N08005
A608	N08006
A608	N08050
A608 (HC30)	J92613
A608 (HD-50)	J93015
A608 (HE-35)	J93413
A608 (HF30)	J92803
A608 (HH30)	J93513
A608 (HH33)	J93633
A608 (HI35)	J94013
A608 (HK-30)	J94203

Cross Reference Specifications	Unified Number
A608 (HK40)	J94204
A608 (HL-30)	J94613
A608 (HL40)	J94614
A608 (HN40)	J94214
A608 (HT50)	J94805
A608 (HU50)	J95404
A612/A612M	K02900
A618 (I)	K02601
A618 (II)	K12609
A618 (III)	K12700
A631 (A)	K05700
A631 (B)	K06200
A631 (C)	K07201
A631 (U)	K07200
A632 (304)	S30400
A632 (304 L)	S30403
A632 (310)	S31000
A632 (316 L)	S31603
A632 (317)	S31700
A632 (321)	S32100
A632 (347)	S34700
A632 (348)	S34800
A633	K01803
A633 (A)	K01802
A633 (C)	K12000
A633 (D)	K12037
A633 (E)	K12202
A638 (660)	S66286
A638 (662)	S66220
A639 (661)	R30155
A639 (671)	R30816
A643 (A)	J02503
A643 (C)	J22092
A643 (D)	J42065
A645/A645M	K41583
A646 (12)	K14047
A646 (4130)	G41300
A646 (4140)	G41400
A646 (4340)	G43400
A646 (E52100)	G52986
A648 (I)	K06000
A648 (II)	K06700
A648 (III)	K07100
A649 (1A)	K14247
A649 (1B)	K24040
A649 (2)	K05001
A649 (3)	K13047
A651 (304)	S30400
A651 (316)	S31600
A651 (409)	S40900
A651 (430)	S43000
A651 (430 Ti)	S43036
A651 (434)	S43400
A651 (439)	S43035
A656 (1)	K11804
A656 (2)	K11503
A658/A658M	K93601
A659 (1015)	G10150
A659 (1016)	G10160
A659 (1017) (1017)	G10170
A659 (1018)	G10180
A659 (1020)	G10200
A659 (1021)	G10210
A659 (1023)	G10230
A660 (WCA)	J02504
A660 (WCB)	J03003
A660 (WCC)	J02505
A662/A662M (A)	K01701
A662/A662M (B)	K02203
A662/A662M (C)	K02007

Cross Reference Specifications	Unified Number
A666 (201)	S20100
A666 (202)	S20200
A666 (301)	S30100
A666 (302)	S30200
A666 (304)	S30400
A669	S31500
A678 (A)	K01600
A678 (B)	K02002
A678 (C)	K02204
A679	K08200
A681 (0-7)	T31507
A681 (A-2)	T30102
A681 (A-3)	T30103 ob
A681 (A-4)	T30104
A681 (A-5)	T30105 ob
A681 (A-6)	T30106
A681 (A-7)	T30107
A681 (A-8)	T30108
A681 (A-9)	T30109
A681 (A-10)	T30110
A681 (D-2)	T30402
A681 (D-3)	T30403
A681 (D-4)	T30404
A681 (D-5)	T30405
A681 (D-7)	T30407
A681 (F-1)	T60601 ob
A681 (F-2)	T60602 ob
A681 (H-10)	T20810
A681 (H-11)	T20811
A681 (H-12)	T20812
A681 (H-13)	T20813
A681 (H-14)	T20814
A681 (H-19)	T20819
A681 (H-21)	T20821
A681 (H-22)	T20822
A681 (H-23)	T20823
A681 (H-24)	T20824
A681 (H-25)	T20825 ob
A681 (H-26)	T20826
A681 (H-41)	T20841 ob
A681 (H-42)	T20842
A681 (H-43)	T20843 ob
A681 (L-2)	T61202
A681 (L-3)	T61203 ob
A681 (L-6)	T61206
A681 (O-1)	T31501
A681 (O-2)	T31502
A681 (O-6)	T31506
A681 (P-2)	T51602 ob
A681 (P-3)	T51603 ob
A681 (P-4)	T51604 ob
A681 (P-5)	T51605 ob
A681 (P-6)	T51606
A681 (P-20)	T51620
A681 (P-21)	T51621
A681 (S-1)	T41901
A681 (S-2)	T41902
A681 (S-4)	T41904
A681 (S-5)	T41905
A681 (S-6)	T41906
A681 (S-7)	T41907
A682 (1030)	G10300
A682 (1035)	G10350
A682 (1040)	G10400
A682 (1045)	G10450
A682 (1050)	G10500
A682 (1055)	G10550
A682 (1060)	G10600
A682 (1064)	G10640
A682 (1065)	G10650

ASTM (American Society for Testing and Materials) Numbers

Cross Reference Specifications	Unified Number	Cross Reference Specifications	Unified Number	Cross Reference Specifications	Unified Number
A682 (1070)	G10700	A714 (II)	K12609	A744 (CF-3)	J92500
A682 (1074)	G10740	A714 (III)	K12709	A744 (CF-3M)	J92800
A682 (1080)	G10800	A714 (IV)	K11356	A744 (CF-8)	J92600
A682 (1085)	G10850	A714 (V)	K22035	A744 (CF-8C)	J92710
A682 (1086)	G10860	A714 (VI)	K11835	A744 (CF-8M)	J92900
A682 (1095)	G10950	A715 (1)	K11501	A744 (CG-8M)	J93000
A686 (W-1)	T72301	A715 (2)	K11502	A744 (CN-7M)	J95150
A686 (W-2)	T72302	A715 (3)	K11503	A744 (CN-7MS)	J94650
A686 (W-5)	T72305	A715 (4)	K11504	A747 (CB7Cu-1)	J92180
A687 (I)	K13521	A715 (5)	K11505	A747 (CB7Cu-2)	J92110
A687 (II)	K14044	A715 (6)	K11506	A756	S44025
A688 (304)	S30400	A715 (7)	K11507	A757 (A2Q)	J02503
A688 (304 L)	S30403	A715 (8)	K11508	A757 (A7Q)	J03002
A688 (316 L)	S31603	A716	F32900	A757 (B2N, B2Q)	J22501
A688 (XM-29)	S24000	A723 (1)	K23550	A757 (B3N, B3Q)	J31500
A689	G86500	A723 (2)	K34035	A757 (B4N, B4Q)	J41501
A689	G86550	A723 (3)	K44045	A757 (C1Q)	J12582
A689	G86600	A724/A724M (A)	K11831	A757 (D1N, D1Q)	J22092
A689	G92600	A724/A724M (B)	K12031	A757 (E1Q)	J42220
A689 (8637)	G86370	A727	K02506	A757 (E2N, E2Q)	J42065
A690	K12249	A729	K06001	A757 (E3N)	J91550
A692	K12121	A730 (A)	K01502	A758	K02741
A693 (350)	S35000	A730 (B)	K02000	A759	K07500
A693 (630)	S17400	A730 (C, D, E)	K04700	A765 (I)	K03046
A693 (631)	S17700	A730 (F)	K05200	A765 (II)	K03047
A693 (632)	S15700	A731	S44627	A765 (III)	K32026
A693 (634)	S35500	A732 (1A)	J02002	A766	K11711
A693 (635)	S17600	A732 (2A, 2Q)	J03011	A771	S38660
A693 (XM-9)	S36200	A732 (3A, 3Q)	J04002	A774 (A7Q)	J03002
A693 (XM-12)	S15500	A732 (5N)	J13052	A789	S31260
A693 (XM-13)	S13800	A732 (6N)	J13512	A789	S32304
A693 (XM-16)	S45500	A732 (7Q)	J13045	A789	S32950
A693 (XM-25)	S45000	A732 (8Q)	J14049	A790	S31260
A694	K03014	A732 (9Q)	J23055	A790	S32304
A695 (B)	K03504	A732 (10Q)	J24054	A790	S32950
A696 (B, C)	K03200	A732 (11Q)	J12094	B1	C10100
A699	K10614	A732 (12Q)	J15048	B1	C10200
A705 (630)	S17400	A732 (13Q)	J12048	B1	C10400
A705 (631)	S17700	A732 (14Q)	J13051	B1	C10500
A705 (632)	S15700	A732 (15A)	J19966	B1	C10700
A705 (634)	S35500	A734/A734M (A)	K21205	B1	C11000
A705 (635)	S17600	A734/A734M (B)	K11720	B1	C11100
A705 (XM-9)	S36200	A735/A735M	K10623	B1	C11300
A705 (XM-12)	S15500	A736	K20747	B1	C11400
A705 (XM-13)	S13800	A737/A737M (B)	K12001	B1	C11500
A705 (XM-16)	S45500	A737 (C)	K12202	B1	C11600
A705 (XM-25)	S45000	A738/A738M	K12447	B2	C10100
A707 (L1)	K02302	A739 (B11)	K11797	B2	C10200
A707 (L2)	K03301	A739 (B22)	K21390	B2	C10400
A707 (L3)	K12510	A743 (CA-6N)	J91650	B2	C10500
A707 (L4)	K12089	A743 (CA-15)	J91150	B2	C10700
A707 (L5)	K20934	A743 (CA-15M)	J91151	B2	C11000
A707 (L6)	K20902	A743 (CA-40)	J91153	B2	C11100
A707 (L7)	K32218	A743 (CB-30)	J91803	B2	C11300
A707 (L8)	K42247	A743 (CC-50)	J92615	B2	C11400
A710 (A)	K20747	A743 (CE-30)	J93423	B2	C11500
A710 (B)	K20622	A743 (CF-3)	J92500	B2	C11600
A711	G13300	A743 (CF-3F)	J92800	B3	C10100
A711	G41400	A743 (CF-8)	J92600	B3	C10200
A711	G41420	A743 (CF-8C)	J92710	B3	C10400
A711	G41450	A743 (CF-8M)	J92900	B3	C10500
A711	G41500	A743 (CF-16F)	J92701	B3	C10700
A711	G43400	A743 (CF-20)	J92602	B3	C11000
A711	G47200	A743 (CG-8M)	J93000	B3	C11100
A711	G51986	A743 (CG-12)	J93001	B3	C11300
A711	G61200	A743 (CH-20)	J93402	B3	C11400
A711	G86600	A743 (CK-20)	J94202	B3	C11500
A711	G98400	A743 (CN-7M)	J95150	B3	C11600
A711 (1335)	G13350	A743 (CN-7MS)	J94650	B6 (Brass Special)	Z17001 ob
A714 (I)	K12608	A744 (CD-4MCu)	J93370		

Cross Reference Specifications	Unified Number	Cross Reference Specifications	Unified Number	Cross Reference Specifications	Unified Number
B6 (High Grade)	Z15001	B26 (705.0)	A07050	B30 (962)	C96200
B6 (Intermediate)	Z16001 ob	B26 (707.0)	A07070	B30 (964)	C96400
B6 (Prime Western)	Z19001	B26 (713.0)	A07130	B30 (973)	C97300
B6 (Special High Grade)	Z13001	B26 (771.0)	A07710	B30 (976)	C97600
B8	C11000	B26 (771.0)	A08500	B30 (978)	C97800
B8	C11100	B26 (A356.0)	A13560	B32 (1.5S)	L50131
B8	C11300	B26 (A443.0)	A14430	B32 (2.5S)	L50150
B8	C11400	B26 (B443.0)	A24430	B32 (60A)	L13600
B8	C11500	B26 (B850.0)	A08520	B32 (60B)	L13601
B8	C11600	B26 (C355.0)	A33550	B32 (63B)	L13631
B9	C16200	B26 (D712.0)	A07120	B32 (70A)	L13700
B9	C16500	B29 (Chemical Lead)	L51120	B32 (70B)	L13701
B9	C50500	B29 (Common Lead)	L50045	B32 (95TA)	L13950
B11 (110)	C11000	B29 (Copper Bearing Lead)	L51121	B32 (96.5 TS)	L13965
B11 (122)	C12200	B29 (Corroding Lead)	L50042	B32 (Alloy 2A)	L54210
B11 (125)	C12500	B30 (836)	C83600	B32 (Alloy 2B)	L54211
B11 (141)	C14100 ob	B30 (838)	C83800	B32 (Alloy 5A)	L54320
B11 (142)	C14200	B30 (842)	C84200	B32 (Alloy 5B)	L54321
B12 (102)	C10200	B30 (844)	C84400	B32 (Alloy 10B)	L54520
B12 (103)	C10300	B30 (848)	C84800	B32 (Alloy 15B)	L54560
B12 (104)	C10500	B30 (852)	C85200	B32 (Alloy 20B)	L54711
B12 (108)	C10800	B30 (854)	C85400	B32 (Alloy 20C)	L54712
B12 (110)	C11000	B30 (857)	C85700	B32 (Alloy 25A)	L54720
B12 (120)	C12000	B30 (858)	C85800	B32 (Alloy 25B)	L54721
B12 (121)	C12100	B30 (862)	C86200	B32 (Alloy 25C)	L54722
B12 (122)	C12200	B30 (863)	C86300	B32 (Alloy 30A)	L54820
B12 (123)	C12300	B30 (864)	C86400	B32 (Alloy 30B)	L54821
B12 (125)	C12500	B30 (865)	C86500	B32 (Alloy 30C)	L54822
B12 (127)	C12700	B30 (867)	C86700	B32 (Alloy 35A)	L54850
B12 (128)	C12800	B30 (874)	C87400	B32 (Alloy 35B)	L54851
B12 (130)	C13000	B30 (875)	C87500	B32 (Alloy 35C)	L54852
B12 (141)	C14100 ob	B30 (876)	C87600	B32 (Alloy 40A)	L54915
B12 (142)	C14200	B30 (878)	C87800	B32 (Alloy 40B)	L54916
B16 (360)	C36000	B30 (879)	C87900	B32 (Alloy 40C)	L54918
B19 (260)	C26000	B30 (903)	C90300	B32 (Alloy 45A)	L54950
B21	C47940	B30 (905)	C90500	B32 (Alloy 45B)	L54951
B21 (462)	C46200	B30 (907)	C90700	B32 (Alloy 50A)	L55030
B21 (464)	C46400	B30 (908)	C90800	B32 (Alloy 50B)	L55031
B21 (482)	C48200	B30 (910)	C91000	B32 (grade 63A)	L13630
B21 (485)	C48500	B30 (911)	C91100	B33	C10100
B22 (863)	C86300	B30 (913)	C91300	B33	C10200
B22 (905)	C90500	B30 (916)	C91600	B33	C11100
B22 (911)	C91100	B30 (917)	C91700	B36 (210)	C21000
B22 (913)	C91300	B30 (922)	C92200	B36 (220)	C22000
B22 (937)	C93700	B30 (923)	C92300	B36 (230)	C23000
B23 (1)	L13910	B30 (925)	C92500	B36 (240)	C24000
B23 (2)	L13890	B30 (927)	C92700	B36 (260)	C26000
B23 (3)	L13840	B30 (928)	C92800	B36 (268)	C26800
B23 (11)	L13870	B30 (928)	C92900	B36 (272)	C27200
B23 (Alloy 7)	L53585	B30 (932)	C93200	B42 (102)	C10200
B23 (Alloy 8)	L53565	B30 (934)	C93400	B42 (103)	C10300
B23 (Alloy 13)	L53346	B30 (935)	C93500	B42 (108)	C10800
B23 (Alloy 15)	L53620	B30 (937)	C93700	B42 (120)	C12000
B26	A05120 ob	B30 (938)	C93800	B42 (122)	C12200
B26	A07100	B30 (939)	C93900	B43 (230)	C23000
B26	A08510	B30 (940)	C94000	B47	C11000
B26 (201.0)	A02010	B30 (941)	C94100	B47	C11100
B26 (204.0)	A02040	B30 (943)	C94300	B47	C11300
B26 (208.0)	A02080	B30 (944)	C94400	B47	C11400
B26 (222.0)	A02220	B30 (945)	C94500	B47	C11500
B26 (242.0)	A02420	B30 (947)	C94700	B47	C11600
B26 (295.0)	A02950	B30 (948)	C94800	B48	C10100
B26 (319.0)	A03190	B30 (949)	C94900	B48	C10200
B26 (328.0)	A03280	B30 (952)	C95200	B48	C10400
B26 (355.0)	A03550	B30 (953)	C95300	B48	C11000
B26 (356.0)	A03560	B30 (954)	C95400	B48	C11400
B26 (443.0)	A04430	B30 (955)	C95500	B48	C11600
B26 (514.0)	A05140	B30 (956)	C95600	B49	C10200
B26 (520.0)	A05200	B30 (957)	C95700	B49	C10400
B26 (535.0)	A05350	B30 (958)	C95800	B49	C10500

Cross Reference Specifications	Unified Number	Cross Reference Specifications	Unified Number	Cross Reference Specifications	Unified Number
B49	C10700	B90 (HM21A)	M13210	B108	A03360
B49	C11000	B90 (LA141A)	M14141	B108	A05130
B49	C11100	B90 (ZE10A)	M16100	B108	A08510
B49	C11300	B91 (AZ31B)	M11311	B108 (204.0)	A02040
B49	C11400	B91 (AZ61A)	M11610	B108 (208.0)	A02080
B49	C11500	B91 (AZ80A)	M11800	B108 (222.0)	A02220
B49	C11600	B91 (HM21A)	M13210	B108 (238.0)	A02380
B61 (922)	C92200	B91 (MIC)	M15102	B108 (242.0)	A02420
B62 (836)	C83600	B91 (ZK60A)	M16600	B108 (319.0)	A03190
B66 (938)	C93800	B92 (9980A)	M19980	B108 (333.0)	A03330
B66 (943)	C94300	B92 (9990A)	M19990	B108 (354.0)	A03540
B66 (944)	C94400	B92 (9990B)	M19991	B108 (355.0)	A03550
B66 (945)	C94500	B92 (9995A)	M19995	B108 (356.0)	A03560
B67	C94100	B92 (9998A)	M19998	B108 (359.0)	A03590
B68 (102)	C10200	B93 (AM60A)	M10600	B108 (443.0)	A04430
B68 (103)	C10300	B93 (AM100A)	M10100	B108 (535.0)	A05350
B68 (108)	C10800	B93 (AS41A)	M10410	B108 (705.0)	A07050
B68 (120)	C12000	B93 (AZ63A)	M11630	B108 (707.0)	A07070
B68 (122)	C12200	B93 (AZ81A)	M11810	B108 (713.0)	A07130
B69 Rolled Zinc	Z21210	B93 (AZ91A)	M11910	B108 (850.0)	A08500
B69 Rolled Zinc	Z21310	B93 (AZ91B)	M11912	B108 (A356.0)	A13560
B69 (Rolled Zinc)	Z21540	B93 (AZ91C)	M11914	B108 (A357.0)	A13570
B69 (Rolled Zinc)	Z44330	B93 (AZ92A)	M11920	B108 (A443.0)	A14430
B69 (Rolled Zinc)	Z45330	B93 (M1B)	M15101	B108 (A444.0)	A14440
B75 (101)	C10100	B94 (AM60A)	M10600	B108 (B443.0)	A24430
B75 (102)	C10200	B94 (AS41A)	M10410	B108 (B514.0)	A05120 ob
B75 (103)	C10300	B94 (AZ91A)	M11910	B108 (B850.0)	A08520
B75 (108)	C10800	B94 (AZ91B)	M11912	B108 (C112.0)	A07110
B75 (120)	C12000	B94 (AZ91C)	M11914	B108 (C355.0)	A33550
B75 (122)	C12200	B94 (AZ92A)	M11920	B108 (F332.0)	A03320
B75 (142)	C14200	B96 (655)	C65100	B111	C69100
B80 (AM100A)	M10100	B97 (651)	C65100	B111, B395	C71640
B80 (AZ63A)	M11630	B97 (655)	C65500	B111 (102)	C10200
B80 (AZ81A)	M11810	B98 (651)	C65100	B111 (103)	C10300
B80 (AZ91C)	M11914	B98 (655)	C65500	B111 (108)	C10800
B80 (AZ92A)	M11920	B98 (661)	C66100	B111 (120)	C12000
B80 (EK30A)	M12300	B99 (651)	C65100	B111 (122)	C12200
B80 (EK41A)	M12410	B99 (655)	C65500	B111 (142)	C14200
B80 (EZ33A)	M12330	B100 (510)	C51000	B111 (192)	C19200
B80 (HK31A)	M13310	B100 (511)	C51100	B111 (230)	C23000
B80 (HZ32A)	M13320	B100 (655)	C65500	B111 (280)	C28000
B80 (K1A)	M18010	B102 (CY44A)	L13913	B111 (443)	C44300
B80 (QE22A)	M18220	B102 (PY1815A)	L13650	B111 (444)	C44400
B80 (ZE41A)	M16410	B102 (Y10A)	L53340	B111 (445)	C44500
B80 (ZE63A)	M16630	B102 (YC135A)	L13820	B111 (608)	C60800
B80 (ZH62A)	M16620	B102 (YT155A)	L53560	B111 (687)	C68700
B80 (ZK51A)	M16510	B103 (510)	C51000	B111 (704)	C70400
B80 (ZK61A)	M16610	B103 (511)	C51100	B111 (706)	C70600
B85 (68A)	A05180	B103 (521)	C52100	B111 (710)	C71000
B85 (360.0)	A03600	B103 (524)	C52400	B111 (715)	C71500
B85 (380.0)	A03800	B103 (532)	C53200 ob	B116	C11000
B85 (383.0)	A03830	B103 (534)	C53400	B116	C11100
B85 (384.0)	A03840	B103 (544)	C54400	B116	C11300
B85 (413.0)	A04130	B105	C16200	B116	C11400
B85 (A360.0)	A13600	B105	C16400 ob	B116	C11500
B85 (A380.0)	A13800	B105	C16500	B116	C11600
B85 (A413.0)	A14130	B105	C50500	B121	C35330
B85 (C443.0)	A34430	B105 (502)	C50200	B121 (310)	C31000 ob
B86 (AG40A)	Z33520	B105 (507)	C50700	B121 (335)	C33500
B86 (AG40B)	Z33523	B105 (508)	C50800	B121 (340)	C34000
B86 (AG41A)	Z35531	B105 (607)	C60700	B121 (342)	C34200
B88 (102)	C10200	B105 (651)	C65100	B121 (350)	C35000
B88 (103)	C10300	B105 (655)	C65500	B121 (353)	C35300
B88 (108)	C10800	B107 (AZ31B)	M11311	B121 (356)	C35600
B88 (120)	C12000	B107 (AZ31C)	M11312	B122 (114)	C11400
B88 (122)	C12200	B107 (AZ61A)	M11610	B122 (115)	C11500
B90 (AZ31A)	M11310	B107 (AZ80A)	M11800	B122 (706)	C70600
B90 (AZ31B)	M11311	B107 (M1A)	M15100	B122 (710)	C71000
B90 (AZ31C)	M11312	B107 (ZK40A)	M16400	B122 (715)	C71500
B90 (HK31A)	M13310	B107 (ZK60A)	M16600	B122 (732)	C73200

Cross Reference Specifications	Unified Number	Cross Reference Specifications	Unified Number	Cross Reference Specifications	Unified Number
B122 (735)	C73500	B148 (956)	C95600	B171 (630)	C63000
B122 (740)	C74000	B148 (957)	C95700	B171 (706)	C70600
B122 (745)	C74500	B148 (958)	C95800	B171 (715)	C71500
B122 (752)	C75200	B149 (AM100A)	M10100	B172	C11000
B122 (762)	C76200	B150 (614)	C61400	B172	C11100
B122 (770)	C77000	B150 (619)	C61900	B172	C11300
B124	C14510	B150 (623)	C62300	B172	C11400
B124	C14520	B150 (630)	C63000	B172	C11500
B124 (110)	C11000	B150 (642)	C64200	B172	C11600
B124 (145)	C14500	B151 (706)	C70600	B173	C11000
B124 (377)	C37700	B151 (715)	C71500	B173	C11100
B124 (464)	C46400	B151 (745)	C74500	B173	C11300
B124 (485)	C48500	B151 (752)	C75200	B173	C11400
B124 (623)	C62300	B151 (757)	C75700	B173	C11500
B124 (630)	C63000	B151 (764)	C76400	B173	C11600
B124 (642)	C64200	B151 (770)	C77000	B174	C11000
B124 (655)	C65500	B151 (792)	C79200	B174	C11100
B124 (675)	C67500	B152 (101)	C10100	B174	C11300
B124 (774)	C77400	B152 (102)	C10200	B174	C11400
B127	N04400	B152 (103)	C10300	B174	C11500
B129 (260)	C26000	B152 (104)	C10400	B174	C11600
B129 (261)	C26100	B152 (105)	C10500	B176 (858)	C85800
B129 (619)	C61900	B152 (107)	C10700	B176 (878)	C87800
B130 (220)	C22000	B152 (108)	C10800	B176 (879)	C87900
B131 (220)	C22000	B152 (110)	C11000	B179	A03361
B133	C11100	B152 (113)	C11300	B179	A03362
B133 (101)	C10100	B152 (114)	C11400	B179	A05122
B133 (102)	C10200	B152 (116)	C11600	B179	A05132
B133 (104)	C10400	B152 (120)	C12000	B179	A07101
B133 (104)	C10500	B152 (122)	C12200	B179	A08511
B133 (107)	C10700	B152 (123)	C12300	B179 (100.1)	A01001
B133 (110)	C11000	B152 (125)	C12500	B179 (130.1)	A01301
B133 (121)	C12100	B152 (141)	C14100 ob	B179 (150.1)	A01501
B133 (122)	C12200	B159 (510)	C51000	B179 (170.1)	A01701
B133 (123)	C12300	B159 (521)	C52100	B179 (201.2)	A02012
B133 (125)	C12500	B159 (524)	C52400	B179 (204.2)	A02042
B133 (127)	C12700	B160	N02200	B179 (208.1)	A02081
B133 (128)	C12800	B160	N02201	B179 (208.2)	A02082
B133 (130)	C13000	B161	N02200	B179 (222.1)	A02221
B134 (210)	C21000	B161	N02201	B179 (242.1)	A02421
B134 (220)	C22000	B162	N02200	B179 (242.2)	A02422
B134 (230)	C23000	B162	N02201	B179 (295.1)	A02951
B134 (240)	C24000	B163	N02200	B179 (295.2)	A02952
B134 (260)	C26000	B163	N02201	B179 (319.1)	A03191
B134 (270)	C27000	B163	N04400	B179 (319.2)	A03192
B134 (274)	C27400	B163	N06600	B179 (328.1)	A03281
B135 (220)	C22000	B163	N06690	B179 (333.1)	A03331
B135 (230)	C23000	B163	N08800	B179 (354.1)	A03541
B135 (260)	C26000	B163	N08810	B179 (355.1)	A03551
B135 (270)	C27000	B163	N08811	B179 (355.2)	A03552
B135 (272)	C27200	B163	N08825	B179 (356.1)	A03561
B135 (280)	C28000	B164	N04400	B179 (356.2)	A03562
B135 (330)	C33000	B164	N04405	B179 (359.2)	A03592
B135 (332)	C33200	B165	N04400	B179 (360.2)	A03602
B135 (370)	C37000	B166	N06600	B179 (380.2)	A03802
B138 (670)	C67000	B166	N06690	B179 (383.1)	A03831
B138 (675)	C67500	B167	N06600	B179 (383.2)	A03832
B139 (510)	C51000	B167	N06690	B179 (384.1)	A03841
B139 (521)	C52100	B168	N06600	B179 (384.2)	A03842
B139 (524)	C52400	B168	N06690	B179 (443.1)	A04431
B139 (534)	C53400	B169 (606)	C60600	B179 (443.2)	A04432
B139 (544)	C54400	B169 (610)	C61000	B179 (514.1)	A05141
B140 (314)	C31400	B169 (614)	C61400	B179 (514.2)	A05142
B140 (316)	C31600	B170 (102)	C10200	B179 (520.2)	A05202
B140 (320)	C32000	B171 (365)	C36500	B179 (535.2)	A05352
B146	C11500	B171 (443)	C44300	B179 (705.1)	A07051
B148 (952)	C95200	B171 (444)	C44400	B179 (707.1)	A07071
B148 (953)	C95300	B171 (445)	C44500	B179 (712.2)	A07122
B148 (954)	C95400	B171 (464)	C46400	B179 (713.1)	A07131
B148 (955)	C95500	B171 (614)	C61400	B179 (771.2)	A07712

ASTM (American Society for Testing and Materials) Numbers

Cross Reference Specifications	Unified Number	Cross Reference Specifications	Unified Number	Cross Reference Specifications	Unified Number
B179 (850.1)	A08501	B209 (3003)	A93003	B221 (6005)	A96005
B179 (A356.2)	A13562	B209 (3004)	A93004	B221 (6061)	A96061
B179 (A360.1)	A13601	B209 (3005)	A93005	B221 (6063)	A96063
B179 (A380.1)	A13801	B209 (3105)	A93105	B221 (6066)	A96066
B179 (A444.2)	A14442	B209 (5005)	A95005	B221 (6262)	A96262
B179 (B443.1)	A24431	B209 (5050)	A95050	B221 (6351)	A96351
B179 (B850.1)	A08521	B209 (5052)	A95052	B221 (6463)	A96463
B179 (C355.2)	A33552	B209 (5083)	A95083	B221 (7005)	A97005
B179 (C443.1)	A34431	B209 (5086)	A95086	B221 (7072)	A97072
B179 (C443.2)	A34432	B209 (5154)	A95154	B221 (7075)	A97075
B179 (C712.1)	A07111	B209 (5252)	A95252	B221 (7178)	A97178
B179 (F332.1)	A03321	B209 (5254)	A95254	B226	C11000
B179 (F332.2)	A03322	B209 (5454)	A95454	B226	C11100
B187 (101)	C10100	B209 (5456)	A95456	B226	C11300
B187 (102)	C10200	B209 (5457)	A95457	B226	C11400
B187 (103)	C10300	B209 (5652)	A95652	B226	C11500
B187 (104)	C10400	B209 (5657)	A95657	B226	C11600
B187 (105)	C10500	B209 (6061)	A96061	B228	C11000
B187 (107)	C10700	B209 (7008)	A97008	B228	C11100
B187 (110)	C11000	B209 (7011)	A97011 ob	B228	C11300
B187 (113)	C11300	B209 (7072)	A97072	B228	C11400
B187 (114)	C11400	B209 (7075)	A97075	B228	C11500
B187 (116)	C11600	B209 (7178)	A97178	B228	C11600
B187 (120)	C12000	B210 (1060)	A91060	B229	C11000
B188 (102)	C10200	B210 (1100)	A91100	B229	C11100
B188 (103)	C10300	B210 (2011)	A92011	B229	C11300
B188 (104)	C10400	B210 (2014)	A92014	B229	C11400
B188 (105)	C10500	B210 (2024)	A92024	B229	C11500
B188 (107)	C10700	B210 (3003)	A93003	B229	C11600
B188 (110)	C11000	B210 (3102)	A93102	B230	A91350
B188 (113)	C11300	B210 (3303)	A93303	B231	A91350
B188 (114)	C11400	B210 (5005)	A95005	B233	A91350
B188 (116)	C11600	B210 (5050)	A95050	B234 (1060)	A91060
B188 (120)	C12000	B210 (5052)	A95052	B234 (3003)	A93003
B189	C10100	B210 (5083)	A95083	B234 (5052)	A95052
B189	C10200	B210 (5086)	A95086	B234 (5454)	A95454
B189	C10400	B210 (5154)	A95154	B234 (6061)	A96061
B189	C10500	B210 (5456)	A95456	B234 (7072)	A97072
B189	C10700	B210 (6061)	A96061	B236	A91350
B189	C11000	B210 (6063)	A96063	B237 (A)	M00998
B189	C11100	B210 (6262)	A96262	B237 (B)	M00995
B189	C11300	B210 (7075)	A97075	B240 (AC43A)	Z35541
B189	C11400	B211 (1060)	A91060	B240 (AG40A)	Z33520
B189	C11500	B211 (1100)	A91100	B240 (AG40B)	Z33522
B189	C11600	B211 (2011)	A92011	B240 (AG41A)	Z35531
B194 (170)	C17000	B211 (2014)	A92014	B241 (1060)	A91060
B194 (172)	C17200	B211 (2017)	A92017	B241 (1100)	A91100
B196 (170)	C17000	B211 (2024)	A92024	B241 (2014)	A92014
B196 (172)	C17200	B211 (2219)	A92219	B241 (2024)	A92024
B196 (173)	C17300	B211 (3003)	A93003	B241 (2219)	A92219
B197 (172)	C17200	B211 (5052)	A95052	B241 (3003)	A93003
B199 (AZ81A)	M11810	B211 (5056)	A95056	B241 (5052)	A95052
B199 (AZ91C)	M11914	B211 (5154)	A95154	B241 (5083)	A95083
B199 (AZ92A)	M11920	B211 (6061)	A96061	B241 (5086)	A95086
B199 (EZ33A)	M12330	B211 (6253)	A96253	B241 (5254)	A95254
B199 (HK31A)	M13310	B211 (6262)	A96262	B241 (5454)	A95454
B199 (QE22A)	M18220	B211 (7075)	A97075	B241 (5456)	A95456
B206 (710)	C71000	B221 (1060)	A91060	B241 (5652)	A95652
B206 (745)	C74500	B221 (1100)	A91100	B241 (6061)	A96061
B206 (752)	C75200	B221 (2014)	A92014	B241 (6063)	A96063
B206 (757)	C75700	B221 (2024)	A92024	B241 (7072)	A97072
B206 (764)	C76400	B221 (2219)	A92219	B241 (7075)	A97075
B206 (770)	C77000	B221 (3003)	A93003	B241 (7178)	A97178
B209 (1060)	A91060	B221 (3004)	A93004	B246	C10100
B209 (1100)	A91100	B221 (5052)	A95052	B246	C10200
B209 (1230)	A91230	B221 (5083)	A95083	B246	C10400
B209 (2014)	A92014	B221 (5086)	A95086	B246	C10500
B209 (2024)	A92024	B221 (5154)	A95154	B246	C10700
B209 (2124)	A92124	B221 (5454)	A95454	B246	C11000
B209 (2219)	A92219	B221 (5456)	A95456	B246	C11100

Cross Reference Specifications	Unified Number	Cross Reference Specifications	Unified Number	Cross Reference Specifications	Unified Number
B246	C11300	B272	C10500	B283 (464)	C46400
B246	C11400	B272	C10700	B283 (485)	C48500
B246	C11500	B272	C11000	B283 (619)	C61900
B246	C11600	B272	C11100	B283 (623)	C62300
B247	A92025	B272	C11300	B283 (630)	C63000
B247	A96151	B272	C11400	B283 (642)	C64200
B247 (1100)	A91100	B272	C11500	B283 (655)	C65500
B247 (2014)	A92014	B272	C11600	B283 (675)	C67500
B247 (2018)	A92018	B272	C12000	B283 (774)	C77400
B247 (2218)	A92218	B272	C12200	B286	C11000
B247 (2219)	A92219	B275 (9980A)	M19980	B286	C11100
B247 (2618)	A92618	B275 (9990A)	M19990	B286	C11300
B247 (3003)	A93003	B275 (9990B)	M19991	B286	C11400
B247 (4032)	A94032	B275 (9995A)	M19995	B286	C11500
B247 (5083)	A95083	B275 (9998A)	M19998	B286	C11600
B247 (5456)	A95456	B275 (A3A)	M10030	B291 (667)	C66700
B247 (6061)	A96061	B275 (AM60A)	M10600	B293	C11100
B247 (6066)	A96066	B275 (AM80A)	M10800	B298	C10100
B247 (7049)	A97049	B275 (AM90A)	M10900	B298	C10200
B247 (7050)	A97050	B275 (AM100A)	M10100	B298	C10400
B247 (7075)	A97075	B275 (AM100B)	M10102	B298	C10500
B247 (7076)	A97076	B275 (AS41A)	M10410	B298	C10700
B247 (7079)	A97079	B275 (AZ10A)	M11100	B298	C11000
B247 (7175)	A97175	B275 (AZ21A)	M11210	B298	C11300
B248 (2618)	A92618	B275 (AZ31A)	M11310	B298	C11400
B265 (1)	R50250	B275 (AZ31B)	M11311	B298	C11500
B265 (2)	R50400	B275 (AZ31C)	M11312	B298	C11600
B265 (3)	R50550	B275 (AZ61A)	M11610	B301	C14510
B265 (4)	R50700	B275 (AZ63A)	M11630	B301	C14520
B265 (5)	R56400	B275 (AZ80A)	M11800	B301	C14710
B265 (6)	R54520	B275 (AZ81A)	M11810	B301	C14720
B265 (7)	R52400	B275 (AZ90A)	M11900	B301 (145)	C14500
B265 (10)	R58030	B275 (AZ91A)	M11910	B301 (147)	C14700
B265 (11)	R52250	B275 (AZ91B)	M11912	B301 (187)	C18700
B265 (12)	R53400	B275 (AZ91C)	M11914	B302 (103)	C10300
B271	C87200	B275 (AZ92A)	M11920	B302 (108)	C10800
B271 (836)	C83600	B275 (AZ101A)	M11101	B302 (120)	C12000
B271 (838)	C83800	B275 (AZ125A)	M11125	B302 (122)	C12200
B271 (844)	C84400	B275 (EK30A)	M12300	B306 (103)	C10300
B271 (848)	C84800	B275 (EK41A)	M12410	B306 (108)	C10800
B271 (852)	C85200	B275 (EZ33A)	M12330	B306 (120)	C12000
B271 (854)	C85400	B275 (HK31A)	M13310	B306 (122)	C12200
B271 (857)	C85700	B275 (HM21A)	M13210	B308 (6061)	A96061
B271 (862)	C86200	B275 (HM31A)	M13312	B313 (1100)	A91100
B271 (863)	C86300	B275 (HZ32A)	M13320	B313 (3003)	A93003
B271 (864)	C86400	B275 (K1A)	M18010	B313 (3004)	A93004
B271 (865)	C86500	B275 (M1A)	M15100	B313 (5050)	A95050
B271 (867)	C86700	B275 (M1B)	M15101	B313 (5052)	A95052
B271 (874)	C87400	B275 (MIC)	M15102	B313 (5086)	A95086
B271 (875)	C87500	B275 (TA54A)	M18540	B313 (5154)	A95154
B271 (903)	C90300	B275 (ZE10A)	M16100	B313 (6061)	A96061
B271 (905)	C90500	B275 (ZE41A)	M16410	B313 (7072)	A97072
B271 (922)	C92200	B275 (ZE63A)	M16630	B315 (651)	C65100
B271 (923)	C92300	B275 (ZH62A)	M16620	B315 (655)	C65500
B271 (932)	C93200	B275 (ZK21A)	M16210	B315 (658)	C65800
B271 (935)	C93500	B275 (ZK40A)	M16400	B316 (1100)	A91100
B271 (937)	C93700	B275 (ZK51A)	M16510	B316 (2017)	A92017
B271 (938)	C93800	B275 (ZK60A)	M16600	B316 (2024)	A92024
B271 (943)	C94300	B275 (ZK60B)	M16601	B316 (2117)	A92117
B271 (952)	C95200	B275 (ZK61A)	M16610	B316 (2219)	A92219
B271 (953)	C95300	B280 (102)	C10200	B316 (3003)	A93003
B271 (954)	C95400	B280 (103)	C10300	B316 (5005)	A95005
B271 (955)	C95500	B280 (108)	C10800	B316 (5052)	A95052
B271 (958)	C95800	B280 (120)	C12000	B316 (5056)	A95056
B271 (973)	C97300	B280 (122)	C12200	B316 (6053)	A96053
B271 (976)	C97600	B283	C14510	B316 (6061)	A96061
B271 (978)	C97800	B283	C14520	B316 (7075)	A97075
B272	C10100	B283 (110)	C11000	B316 (7178)	A97178
B272	C10200	B283 (145)	C14500	B317 (6101)	A96101
B272	C10400	B283 (377)	C37700	B324	A91350

Cross Reference Specifications	Unified Number	Cross Reference Specifications	Unified Number	Cross Reference Specifications	Unified Number
B333	N10001	B353	R60802	B366	N10665
B333	N10665	B353	R60804	B367 (C-1)	R50250
B334	C10400	B353	R60901	B367 (C-2)	R50400
B334	C10500	B355	C10100	B367 (C-3)	R50550
B334	C10700	B355	C10200	B367 (C-4)	R50700
B334	C11000	B355	C10400	B367 (C-5)	R56400
B334	C11300	B355	C10500	B367 (C-6)	R54520
B334	C11400	B355	C10700	B367 (C-7A)	R52250
B334	C11500	B355	C11000	B367 (C-7B)	R52400
B334	C11600	B355	C11100	B367 (C-8A)	R52550
B335	N10001	B355	C11300	B367 (C-8B)	R52700
B335	N10665	B355	C11400	B369 (962)	C96200
B337 (1)	R50250	B355	C11500	B369 (964)	C96400
B337 (2)	R50400	B355	C11600	B370	C11000
B337 (3)	R50550	B357	L06990	B370	C12200
B337 (7)	R52400	B359 (7l0)	C71000	B370	C12500
B337 (9)	R56320	B359 (102)	C10200	B371 (694)	C69400
B337 (10)	R58030	B359 (103)	C10300	B371 (697)	C69700
B337 (11)	R52250	B359 (108)	C10800	B372 (102)	C10200
B337 (12)	R53400	B359 (120)	C12000	B372 (103)	C10300
B338 (1)	R50250	B359 (122)	C12200	B372 (120)	C12000
B338 (2)	R50400	B359 (142)	C14200	B372 (220)	C22000
B338 (3)	R50550	B359 (192)	C19200	B373 (1145)	A91145
B338 (7)	R52400	B359 (230)	C23000	B373 (1235)	A91235
B338 (9)	R56320	B359 (443)	C44300	B376	N08366
B338 (10)	R58030	B359 (444)	C44400	B379 (103)	C10300
B338 (11)	R52250	B359 (445)	C44500	B379 (108)	C10800
B338 (12)	R53400	B359 (608)	C60800	B381 (F-1)	R50250
B339 (A)	L13008	B359 (687)	C68700	B381 (F-2)	R50400
B339 (AA)	L13004	B359 (704)	C70400	B381 (F-3)	R50550
B339 (AAA)	L13002	B359 (706)	C70600	B381 (F-4)	R50700
B339 (B)	L13006	B359 (715)	C71500	B381 (F-5)	R56400
B339 (C)	L13010	B360	C10800	B381 (F-6)	R54520
B339 (D)	L13012	B360 (122)	C12200	B381 (F-7)	R52400
B339 (E)	L13014	B361	A96061	B381 (F-11)	R52250
B344	N06003	B361	A96063	B381 (F-12)	R53400
B344	N06004	B361 (1060)	A91060	B384 (360)	R03600
B345 (1060)	A91060	B361 (1100)	A91100	B384 (361)	R03610
B345 (3003)	A93003	B361 (5083)	A95083	B384 (362)	R03620
B345 (5083)	A95083	B361 (5086)	A95086	B384 (363)	R03630
B345 (5086)	A95086	B361 (5154)	A95154	B384 (364)	R03640
B345 (6061)	A96061	B364	R05200	B384 (365)	R03650
B345 (6063)	A96063	B364	R05255	B385 (360)	R03600
B345 (6070)	A96070	B364	R05400	B385 (361)	R03610
B345 (6351)	A96351	B365	R05200	B385 (362)	R03620
B345 (7072)	A97072	B365	R05255	B385 (363)	R03630
B348 (1)	R50250	B365	R05400	B385 (364)	R03640
B348 (2)	R50400	B366	N02200	B385 (365)	R03650
B348 (3)	R50550	B366	N02201	B386 (360)	R03600
B348 (4)	R50700	B366	N04400	B386 (361)	R03610
B348 (5)	R56400	B366	N06001	B386 (362)	R03620
B348 (6)	R54520	B366	N06002	B386 (363)	R03630
B348 (7)	R52400	B366	N06007	B386 (364)	R03640
B348 (10)	R58030	B366	N06022	B386 (365)	R03650
B348 (11)	R52250	B366	N06030	B387 (360)	R03600
B348 (12)	R53400	B366	N06600	B387 (361)	R03610
B349	R60001	B366	N06625	B387 (362)	R03620
B350	R60001	B366	N06690	B387 (363)	R03630
B350	R60802	B366	N06975	B387 (364)	R03640
B350	R60804	B366	N08020	B387 (365)	R03650
B350	R60901	B366	N08221	B388	K92100
B351	R60001	B366	N08320	B388	K92350
B351	R60802	B366	N08330	B388	K92510
B351	R60804	B366	N08332	B388	K92850
B351	R60901	B366	N08800	B388	K93600
B352	R60001	B366	N08810	B388	K93800
B352	R60802	B366	N08811	B388	K94000
B352	R60804	B366	N10001	B388	K94200
B352	R60901	B366	N10003	B388	K94500
B353	R60001	B366	N10276	B388	K95000

Cross Reference Specifications	Unified Number	Cross Reference Specifications	Unified Number	Cross Reference Specifications	Unified Number
B388	M27200	B418 (Type III)	Z32121	B459 (4)	R07030
B391	R04200	B418 (Zinc Anodes Type II)	Z13000	B462	N08020
B391	R04210	B418 (Zinc Anodes Type I)	Z32120	B463	N08020
B391	R04251	B422 (647)	C64700	B463	N08024
B391	R04261	B423	N08221	B463	N08026
B392	R04200	B423	N08825	B464	N08020
B392	R04210	B424	N08221	B464	N08024
B392	R04251	B424	N08825	B464	N08026
B393	R04200	B425	N08221	B465	C19400
B393	R04210	B425	N08825	B466 (704)	C70400
B393	R04251	B427 (908)	C90800	B466 (706)	C70600
B393	R04261	B427 (916)	C91600	B466 (710)	C71000
B394	R04200	B427 (917)	C91700	B466 (715)	C71500
B394	R04210	B427 (928)	C92900	B467	A96262
B394	R04251	B429 (6061)	A96061	B467 (706)	C70600
B394	R04261	B429 (6063)	A96063	B467 (710)	C71000
B395 (102)	C10200	B432 (101)	C10100	B467 (715)	C71500
B395 (103)	C10300	B432 (102)	C10200	B468	N08020
B395 (108)	C10800	B432 (103)	C10300	B468	N08024
B395 (120)	C12000	B432 (108)	C10800	B468	N08026
B395 (122)	C12200	B432 (122)	C12200	B469	C19200
B395 (142)	C14200	B432 (365)	C36500	B470	C11100
B395 (192)	C19200	B432 (366)	C36600	B471	N08020
B395 (230)	C23000	B432 (367)	C36700	B472	N08020
B395 (443)	C44300	B432 (368)	C36800	B473	N08020
B395 (444)	C44400	B432 (443)	C44300	B474	N08020
B395 (445)	C44500	B432 (444)	C44400	B474	N08026
B395 (608)	C60800	B432 (445)	C44500	B475	N08020
B395 (687)	C68700	B432 (464)	C46400	B477	P00750
B395 (704)	C70400	B432 (465)	C46500	B483 (1060)	A91060
B395 (706)	C70600	B432 (465)	C46600	B483 (1100)	A91100
B395 (710)	C71000	B432 (467)	C46700	B483 (1435)	A91435
B395 (715)	C71500	B432 (614)	C61400	B483 (3003)	A93003
B396 (5005)	A95005	B432 (651)	C65100	B483 (5005)	A95005
B397 (5005)	A95005	B432 (655)	C65500	B483 (5050)	A95050
B398 (6201)	A96201	B432 (706)	C70600	B483 (5052)	A95052
B399 (6201)	A96201	B432 (715)	C71500	B483 (6061)	A96061
B401	A91350	B433 (962)	C96200	B483 (6063)	A96063
B402 (706)	C70600	B433 (964)	C96400	B483 (6262)	A96262
B402 (715)	C71500	B434	N10003	B486 (60A)	L13600
B403 (AM100A)	M10100	B435	N06002	B486 (60B)	L13601
B403 (AZ81A)	M11810	B436	N06001	B486 (63B)	L13631
B403 (AZ91C)	M11914	B440	L01900	B486 (70A)	L13700
B403 (AZ92A)	M11920	B440	L01950	B486 (70B)	L13701
B403 (EZ33A)	M12330	B441 (175)	C17500	B486 (95TA)	L13950
B403 (HK31A)	M13310	B443	N06625	B486 (96TS)	L13960
B403 (K1A)	M18010	B444	N06625	B486 (grade 63A)	L13630
B403 (QE22A)	M18220	B445	N06102	B491 (1050)	A91050
B403 (ZK61A)	M16610	B446	N06625	B491 (1100)	A91100
B404 (1060)	A91060	B447 (102)	C10200	B491 (1200)	A91200
B404 (3003)	A93003	B447 (103)	C10300	B491 (1235)	A91235
B404 (5052)	A95052	B447 (108)	C10800	B491 (3003)	A93003
B404 (5454)	A95454	B447 (110)	C11000	B491 (6063)	A96063
B404 (6061)	A96061	B447 (120)	C12000	B492	C96300
B404 (7072)	A97072	B447 (122)	C12200	B493	R60701
B407	N08800	B447 (142)	C14200	B493	R60702
B407	N08810	B451	C10100	B493	R60704
B407	N08811	B451	C10200	B493	R60705
B408	N08800	B451	C11000	B494	R60701
B408	N08810	B451	C12000	B494	R60702
B408	N08811	B453 (335)	C33500	B494	R60703
B409	N08800	B453 (340)	C34000	B495	R60701
B409	N08810	B453 (345)	C34500	B495	R60702
B409	N08811	B453 (350)	C35000	B495	R60703
B410	R07004 ob	B453 (353)	C35300	B495	R60704
B411 (647)	C64700	B455 (380)	C38000	B495	R60705
B412 (647)	C64700	B455 (385)	C38500	B495	R60706
B413 (99.90)	P07020	B459 (1)	R07100	B496	C11100
B413 (99.95)	P07015	B459 (2)	R07080	B505	C94100
B413 (Grade 99.99)	P07010	B459 (3)	R07050	B505	C95200

Cross Reference Specifications	Unified Number	Cross Reference Specifications	Unified Number	Cross Reference Specifications	Unified Number
B505 (836)	C83600	B524 (6201)	A96201	B564	N08810
B505 (838)	C83800	B531 (5005)	A95005	B566	C10200
B505 (842)	C84200	B534 (175)	C17500	B566	C11000
B505 (844)	C84400	B535	N08330	B569 (260)	C26000
B505 (848)	C84800	B535	N08332	B570 (170)	C17000
B505 (862)	C86200	B536	N08330	B570 (172)	C17200
B505 (863)	C86300	B536	N08332	B572	N06002
B505 (865)	C86500	B540	P03350	B573	N10003
B505 (903)	C90300	B541	P00710	B574	N06022
B505 (905)	C90500	B543 (108)	C10800	B574	N06455
B505 (907)	C90700	B543 (122)	C12200	B574	N08800
B505 (910)	C91000	B543 (194)	C19400	B574	N10276
B505 (913)	C91300	B543 (230)	C23000	B575	N06022
B505 (922)	C92200	B543 (443)	C44300	B575	N06455
B505 (923)	C92300	B543 (443)	C44400	B575	N10276
B505 (925)	C92500	B543 (445)	C44500	B581	N06007
B505 (927)	C92700	B543 (687)	C68700	B581	N06030
B505 (928)	C92800	B543 (704)	C70400	B581	N06975
B505 (928)	C92900	B543 (706)	C70600	B581	N06985
B505 (932)	C93200	B543 (715)	C71500	B582	N06007
B505 (934)	C93400	B544	A91350	B582	N06030
B505 (935)	C93500	B546	N08330	B582	N06975
B505 (937)	C93700	B546	N08332	B582	N06985
B505 (938)	C93800	B547 (1100)	A91100	B584	C87200
B505 (939)	C93900	B547 (3003)	A93003	B584 (836) (formerly B145)	C83600
B505 (940)	C94000	B547 (3004)	A93004	B584 (838) (formerly B145)	C83800
B505 (943)	C94300	B547 (5050)	A95050	B584 (844) (formerly B145)	C84400
B505 (947)	C94700	B547 (5052)	A95052	B584 (848) (formerly B145)	C84800
B505 (948)	C94800	B547 (5083)	A95083	B584 (852) (formerly B146)	C85200
B505 (953)	C95300	B547 (5086)	A95086	B584 (854) (formerly B146)	C85400
B505 (954)	C95400	B547 (5154)	A95154	B584 (857)	C85700
B505 (955)	C95500	B547 (5454)	A95454	B584 (862) (formerly B147)	C86200
B505 (958)	C95800	B547 (6061)	A96061	B584 (863) (formerly B147)	C86300
B505 (964)	C96400	B547 (7072)	A97072	B584 (864) (formerly B132, B147)	C86400
B505 (973)	C97300	B548	A96061	B584 (865) (formerly B147)	C86500
B505 (976)	C97600	B548 (1060)	A91060	B584 (867) (formerly B132)	C86700
B505 (978)	C97800	B548 (3004)	A93004	B584 (874) (formerly B198)	C87400
B506	C10500	B548 (5050)	A95050	B584 (875) (formerly B198)	C87500
B506 (102)	C10200	B548 (5083)	A95083	B584 (876)	C87600
B506 (104)	C10400	B548 (5086)	A95086	B584 (903) (formerly B143)	C90300
B506 (107)	C10700	B548 (5154)	A95154	B584 (905) (formerly B143)	C90500
B506 (110)	C11000	B548 (5254)	A95254	B584 (922) (formerly B143)	C92200
B506 (113)	C11300	B548 (5454)	A95454	B584 (923) (formerly B143)	C92300
B506 (114)	C11400	B548 (5456)	A95456	B584 (932) (formerly B144)	C93200
B506 (116)	C11600	B548 (5652)	A95652	B584 (935) (formerly B144)	C93500
B506 (120)	C12000	B549	A91350	B584 (937) (formerly B144)	C93700
B506 (122)	C12200	B550	R60701	B584 (938)	C93800
B506 (123)	C12300	B550	R60702	B584 (943) (formerly B144)	C94300
B506 (125)	C12500	B550	R60704	B584 (947) (formerly B292)	C94700
B508 (411)	C41100	B550	R60705	B584 (948) (formerly B292)	C94800
B508 (505)	C50500	B551	R60701	B584 (949) (formerly B292)	C94900
B511	N08330	B551	R60702	B584 (973) (formerly B149)	C97300
B511	N08332	B551	R60704	B584 (976) (formerly B149)	C97600
B512	N08330	B551	R60705	B584 (978) (formerly B149)	C97800
B512	N08332	B551	R60706	B585 (192)	C19200
B514	N08800	B552 (706)	C70600	B586 (194)	C19400
B514	N08810	B552 (715)	C71500	B587 (220)	C22000
B515	N08800	B560 (1)	L13911	B587 (230)	C23000
B515	N08810	B560 (2)	L13912	B587 (260)	C26000
B516	N06600	B560 (3)	L13963	B587 (268)	C26800
B517	N06600	B561 (99.80)	P04980	B587 (270)	C27000
B518	N06102	B561 (99.95)	P04995	B587 (272)	C27200
B519	N06102	B562 (99.5)	P00025	B589 (99.80)	P03980
B522 (I)	P00691	B562 (99.95)	P00020	B589 (99.95)	P03995
B522 (II)	P00692	B562 (99.995)	P00010	B590	N08245 ob
B523	R60701	B562 (Grade 99.99)	P00015	B591 (405)	C40500
B523	R60702	B563	P03440	B591 (408)	C40800
B523	R60704	B564	N04400	B591 (411)	C41100
B523	R60705	B564	N06600	B591 (413)	C41300
B524	A91350	B564	N08800		

Cross Reference Specifications	Unified Number	Cross Reference Specifications	Unified Number	Cross Reference Specifications	Unified Number
B591 (415)	C41500	B671 (99.80)	P06100	F239	N02220
B591 (422)	C42200	B672	N08700	F239	N02225
B591 (425)	C42500	B673	N08904	F239	N02230
B591 (430)	C43000	B673	N08925	F239	N02270
B591 (434)	C43400	B674	N08904	F256 (I)	K91800
B592 (688)	C68800	B674	N08925	F256 (II)	K92801
B596	P00901	B675	N08366	F288 (1A and 1B)	R07005
B599	N08700	B675	N08367	F290	N02211
B603 (I)	K92500	B676	N08366	F290	N03300
B603 (IIA)	K91870	B676	N08367	F290	R07005
B603 (IIB)	K92400	B677	N08904	F364 (1)	R03604
B603 (III)	K91670	B677	N08925	F364 (II)	R03603
B603 (IV)	K91470	B688	N08366	F467	A92024
B609	A91350	B688	N08367	F467	A96061
B616-78 (99.90)	P05981	B690	N08366	F467 (1)	R50250
B616 (99.80)	P05980	B690	N08367	F467 (2)	R50400
B616 (99.90)	P05990	B691	N08366	F467 (4)	R50700
B616 (99.95)	P05982	B691	N08367	F467 (5)	R56401
B616 (99.95)	P05995	B704	N06625	F467 (7)	R52400
B617	P07900	B704	N08825	F468	A92024
B619	N06007	B705	N06625	F468	A96061
B619	N06022	B705	N08825	F468 (1)	R50250
B619	N06030	B709	N08028	F468 (2)	R50400
B619	N06455	B710	N08330	F468 (4)	R50700
B619	N06975	B710	N08332	F468 (5)	R56401
B619	N06985	B718	N06333	F468 (7)	R52400
B619	N10001	B719	N06333	F468 (S)	R56401
B619	N10276	B720	N08310	F563 ISO 5832/8	R30477
B619	N10665	B722	N06333	(LS141A)	M14142
B619, B622	N06002	B723	N06333	(LZ145A)	M14145
B620	N08320	B725	N02200	ZI ZA-8	Z35635
B621	N08320	B725	N02201	ZI ZA-27	Z35840
B622	N06022	B726	N06333		
B622	N06030	B730	N02200		
B622	N06455	B730	N02201		
B622	N06985	B750 (Zn-5Al-MM)	Z38510		
B622	N08320	F1	N02205		
B622	N10276	F1	N02233		
B622	N10665	F1	N02270		
B625	N08904	F2	N02205		
B625	N08925	F2	N02233		
B626	N06002	F2	N02270		
B626	N06022	F3	N02205		
B626	N06030	F3	N02233		
B626	N06455	F3	N02270		
B626	N06985	F4	N02233		
B626	N10276	F9	C10200		
B626	N10665	F9	C18200		
B632	A96061	F9	N02205		
B637	N07001	F15	K94610		
B637	N07080	F29	K94101		
B637	N07252	F30 (42)	K94100		
B637	N07500	F30 (46)	K94600		
B637	N07718	F30 (48)	K94800		
B637	N07750	F30 (52)	N14052		
B649	N08904	F31	K94760		
B649	N08925	F49 (Arc-Casting Grade)	R03601		
B652	R04295	F49 (Powder Grade)	R03602		
B653	R60702	F67 (1)	R50250		
B653	R60704	F67 (2)	R50400		
B653	R60705	F67 (3)	R50550		
B654	R04295	F67 (4)	R50700		
B655	R04295	F68	C10100		
B658	R60702	F72	P00016		
B658	R60704	F73 (Electronic Grade)	R07031		
B668	N08028	F90	R30605		
B669	Z35630	F96	C70690		
B669 (ZA-8)	Z35635	F96	C71590		
B669 (ZA-27)	Z35840	F96	N04404		
B670	N07718	F106 (TB Ag)	P07016		
		F136	R56401		

Cross Index of Commonly Known Documents Which Describe Materials Same as or Similar to Those Covered by UNS Numbers

AWS (American Welding Society) NUMBERS
(The Symbol **ob** Indicates an Obsolete UNS Number.)

Cross Reference Specifications	Unified Number

AWS

Cross Reference Specifications	Unified Number
A5.1 (7016)	W07016
A5.1 (7018)	W07018
A5.1 (E6010)	W06010
A5.1 (E6011)	W06011
A5.1 (E6012)	W06012
A5.1 (E6013)	W06013
A5.1 (E6020)	W06020
A5.1 (E6022)	W06022
A5.1 (E6027)	W06027
A5.1 (E7014)	W07014
A5.1 (E7015)	W07015
A5.1 (E7016-1)	W07116
A5.1 (E7018-1)	W07118
A5.1 (E7024)	W07024
A5.1 (E7024-1)	W07124
A5.1 (E7027)	W07027
A5.1 (E7028)	W07028
A5.1 (E7048)	W07048
A5.2 (R100)	K12048
A5.2 (RG45)	K00045
A5.2 (RG60)	K00060
A5.2 (RG65)	K00065
A5.4 (E7Cr)	W50310
A5.4 (E16-8-2)	W36810
A5.4 (E028L)	W88028
A5.4 (E209)	W32210
A5.4 (E219)	W32310
A5.4 (E240)	W32410
A5.4 (E307)	W30710
A5.4 (E308) (E308H)	W30810
A5.4 (E308L)	W30813
A5.4 (E308Mo)	W30820
A5.4 (E308MoL)	W30823
A5.4 (E309)	W30910
A5.4 (E309Cb)	W30917
A5.4 (E309L)	W30913
A5.4 (E309Mo)	W30920
A5.4 (E310)	W31010
A5.4 (E310Cb)	W31017
A5.4 (E310H)	W31015
A5.4 (E310Mo)	W31020
A5.4 (E312)	W31310
A5.4 (E316) (E316H)	W31610
A5.4 (E316L)	W31613
A5.4 (E317)	W31710
A5.4 (E317L)	W31713
A5.4 (E318)	W31910
A5.4 (E320)	W88021
A5.4 (E320LR)	W88022
A5.4 (E330)	W88331
A5.4 (E330H)	W88335
A5.4 (E347)	W34710
A5.4 (E349)	W34910
A5.4 (E410)	W41010
A5.4 (E410NiMo)	W41016
A5.4 (E430)	W43010
A5.4 (E502)	W50210
A5.4 (E505)	W50410
A5.4 (E630)	W37410
A5.4 (E904L)	W88904
A5.5 (E7010-A1)	W17010
A5.5 (E7011-A1)	W17011
A5.5 (E7015-A1)	W17015
A5.5 (E7015-C1L)	W22115
A5.5 (E7015-C2L)	W23115
A5.5 (E7016-A1)	W17016
A5.5 (E7016-C1L)	W22116
A5.5 (E7016-C2L)	W23116

Cross Reference Specifications	Unified Number
A5.5 (E7018-A1)	W17018
A5.5 (E7018-C1L)	W22118
A5.5 (E7018-C2L)	W23118
A5.5 (E7018-W)	W20018
A5.5 (E7020-A1)	W17020
A5.5 (E7027-A1)	W17027
A5.5 (E8015-B2L)	W52115
A5.5 (E8015-B4L)	W53415
A5.5 (E8016-B1)	W51016
A5.5 (E8016-B2)	W52016
A5.5 (E8016-B5)	W51316
A5.5 (E8016-C1)	W22016
A5.5 (E8016-C2)	W23016
A5.5 (E8016-C3)	W21016
A5.5 (E8016-D3)	W18016
A5.5 (E8018-B1)	W51018
A5.5 (E8018-B2)	W52018
A5.5 (E8018-B2L)	W52118
A5.5 (E8018-C1)	W22018
A5.5 (E8018-C2)	W23018
A5.5 (E8018-C3)	W21018
A5.5 (E8018-D3)	W18018
A5.5 (E8018-NM)	W21118
A5.5 (E8018-W)	W20118
A5.5 (E9015-B3)	W53015
A5.5 (E9015-B3L)	W53115
A5.5 (E9015-D1)	W19015
A5.5 (E9016-B3)	W53016
A5.5 (E9018-B3)	W53018
A5.5 (E9018-B3L)	W53118
A5.5 (E9018-D1)	W19018
A5.5 (E9018-M)	W21218
A5.5 (E10015-D2)	W10015
A5.5 (E10016-D2)	W10016
A5.5 (E10018-D2)	W10018
A5.5 (E10018-M)	W21318
A5.5 (E11018-M)	W21418
A5.5 (E12015-M)	W22218
A5.5 (E12018-M1)	W23218
A5.6 (ECu)	W60189
A5.6 (ECuAl-A2)	W60614
A5.6 (ECuMnNiAl)	W60633
A5.6 (ECuNi)	W60715
A5.6 (ECuNiAl)	W60632
A5.6 (ECuSi)	W60656
A5.6 (ECuSn-A)	W60518
A5.6 (ECuSn-C)	W60521
A5.7 (ERCu)	C18980
A5.7 (ERCuAl-A1)	C61000
A5.7 (ERCuAl-A2)	C61800
A5.7 (ERCuAl-A3)	C62400
A5.7 (ERCuMnNiAl)	C63380
A5.7 (ERCuNi)	C71581
A5.7 (ERCuNiAl)	C63280
A5.7 (ERCuSi-A)	C65600
A5.7 (ERCuSn-A)	C51800
A5.8 (BAg-1)	P07450
A5.8 (BAg-1a)	P07500
A5.8 (BAg-2)	P07350
A5.8 (BAg-2a)	P07300
A5.8 (BAg-3)	P07501
A5.8 (BAg-4)	P07400
A5.8 (BAg-5)	P07453
A5.8 (BAg-6)	P07503
A5.8 (BAg-7)	P07563
A5.8 (BAg-8)	P07720
A5.8 (BAg-8a)	P07723
A5.8 (BAg-9)	P07650
A5.8 (BAg-10)	P07700
A5.8 (BAg-13)	P07540

Cross Reference Specifications	Unified Number
A5.8 (BAg-13a)	P07560
A5.8 (BAg-18)	P07600
A5.8 (BAg-19)	P07925
A5.8 (BAg-20)	P07301
A5.8 (BAg-21)	P07630
A5.8 (BAg-22)	P07490
A5.8 (BAg-23)	P07850
A5.8 (BAg-24)	P07505
A5.8 (BAg-25)	P07200
A5.8 (BAg-26)	P07250
A5.8 (BAg-27)	P07251
A5.8 (BAg-28)	P07401
A5.8 (BAlSi-2)	A94343
A5.8 (BAlSi-3)	A94145
A5.8 (BAlSi-4)	A94047
A5.8 (BAlSi-5)	A94045
A5.8 (BAlSi-7)	A94004
A5.8 (BAlSi-11)	A94104
A5.8 (BAu-1)	P00375
A5.8 (BAu-2)	P00800
A5.8 (BAu-3)	P00350
A5.8 (BAu-4)	P00820
A5.8 (BAu-5)	P00300
A5.8 (BAu-6)	P00700
A5.8 (BCo-1)	R39001
A5.8 (BCu-1)	C14180
A5.8 (BCuP-1)	C55180
A5.8 (BCuP-2)	C55181
A5.8 (BCuP-3)	C55281
A5.8 (BCuP-4)	C55283
A5.8 (BCuP-5)	C55284
A5.8 (BCuP-6)	C55280
A5.8 (BCuP-7)	C55282
A5.8 (BCuZn-E)	C28580
A5.8 (BCuZn-F)	C49080
A5.8 (BCuZn-G)	C26380
A5.8 (BCuZn-H)	C24080
A5.8 (BCuZn-H)	C26000
A5.8 (BMg-1)	M19001
A5.8 (BNi-1)	N99600
A5.8 (BNi-1a)	N99610
A5.8 (BNi-2)	N99620
A5.8 (BNi-3)	N99630
A5.8 (BNi-4)	N99640
A5.8 (BNi-5)	N99650
A5.8 (BNi-6)	N99700
A5.8 (BNi-7)	N99710
A5.8 (BNi-8)	N99800
A5.8 (BNi-9)	N99612
A5.8 (BNi-10)	N99622
A5.8 (BNi-10)	N99624
A5.8 (BVAg-0)	P07017
A5.8 (BVAg-6b)	P07507
A5.8 (BVAg-8)	P07727
A5.8 (BVAg-8b)	P07728
A5.8 (BVAg-18)	P07607
A5.8 (BVAg-29)	P07627
A5.8 (BVAg-30)	P07687
A5.8 (BVAg-31)	P07587
A5.8 (BVAg-32)	P07547
A5.8 (BVAu-2)	P00807
A5.8 (BVAu-4)	P00827
A5.8 (BVAu-7)	P00507
A5.8 (BVAu-8)	P00927
A5.8 (BVCu-1x)	C14181
A5.8 (BVPd-1)	P03657
A5.8 (RBCuZn-A)	C47000
A5.8 (RBCuZn-C)	C68100
A5.8 (RBCuZn-D)	C77300
A5.9 (ER 317L)	S31783

AWS (American Welding Society) Numbers

Cross Reference Specifications	Unified Number	Cross Reference Specifications	Unified Number	Cross Reference Specifications	Unified Number
A5.9 (ER16-8-2)	S16880	A5.9 (ER505)	S50480	A5.13 (RCuAl-D)	C62581
A5.9 (ER16-8-2)	W36840	A5.9 (ER505)	W50440	A5.13 (RCuAl-E)	C62582
A5.9 (ER26-1)	S44687	A5.9 (ER630)	S17480	A5.13 (RCuSn-D)	C52400
A5.9 (ER26-1)	W44647	A5.9 (ER630)	W37440	A5.13 (RFe5-A)	K91308
A5.9 (ER209)	S20980	A5.10 (ER1100)	A91100	A5.13 (RFe5-B)	K90987
A5.9 (ER209)	W32240	A5.10 (ER1188)	A91188	A5.13 (RFeCr-A1)	F45100
A5.9 (ER218)	S21880	A5.10 (ER2319)	A92319	A5.13 (RNiCr-A)	N99644
A5.9 (ER218)	W32540	A5.10 (ER4043)	A94043	A5.13 (RNiCr-B)	N99645
A5.9 (ER219)	S21980	A5.10 (ER4047)	A94047	A5.13 (RNiCr-C)	N99646
A5.9 (ER219)	W32340	A5.10 (ER4145)	A94145	A5.14 (ERNi-1)	N02061
A5.9 (ER240)	S23980	A5.10 (ER4643)	A94643	A5.14 (ERNiCr-3)	N06082
A5.9 (ER240)	W32440	A5.10 (ER5183)	A95183	A5.14 (ERNiCrFe-5)	N06062
A5.9 (ER307)	S30780	A5.10 (ER5356)	A95356	A5.14 (ERNiCrFe-6)	N07092
A5.9 (ER307)	W30740	A5.10 (ER5554)	A95554	A5.14 (ERNiCrFe-7)	N07069
A5.9 (ER308) (ER308H)	S30880	A5.10 (ER5556)	A95556	A5.14 (ERNiCrMo-1)	N06007
A5.9 (ER308) (ER308H)	W30840	A5.10 (ER5654)	A95654	A5.14 (ERNiCrMo-2)	N06002
A5.9 (ER308L)	S30883	A5.10 (R206.0)	A02060	A5.14 (ERNiCrMo-3)	N06625
A5.9 (ER308L)	W30843	A5.10 (R242.0)	A02042	A5.14 (ERNiCrMo-7)	N06455
A5.9 (ER308LSi)	S30888	A5.10 (R242.0)	A02420	A5.14 (ERNiCrMo-8)	N06975
A5.9 (ER308LSi)	W30848	A5.10 (R295.0)	A02950	A5.14 (ERNiCrMo-9)	N06985
A5.9 (ER308Mo)	S30882	A5.10 (R355.0)	A03550	A5.14 (ERNiCu-7)	N04060
A5.9 (ER308Mo)	W30850	A5.10 (R356.0)	A03560	A5.14 (ERNiFeCr-1)	N08065
A5.9 (ER308MoL)	S30886	A5.11 (ENi-1)	W82141	A5.14 (ERNiMo-2)	N10003
A5.9 (ER308MoL)	W30853	A5.11 (ENiCrCoMo-1)	W86117	A5.14 (ERNiMo-3)	N10004
A5.9 (ER308Si)	S30881	A5.11 (ENiCrFe-1)	W86132	A5.14 (ERNiMo-7)	N10665
A5.9 (ER308Si)	W30841	A5.11 (ENiCrFe-2)	W86133	A5.14 (ErNiMo-1)	N10001
A5.9 (ER309)	S30980	A5.11 (ENiCrFe-3)	W86182	A5.14 (NiCrMo-4)	N10276
A5.9 (ER309)	W30940	A5.11 (ENiCrFe-4)	W86134	A5.15 (ENi-CI)	W82001
A5.9 (ER309L)	S30983	A5.11 (ENiCrMo-1)	W86007	A5.15 (ENi-CI-A)	W82003
A5.9 (ER309L)	W30943	A5.11 (ENiCrMo-2)	W86002	A5.15 (ENiCu-A)	W84001
A5.9 (ER309Si)	S30981	A5.11 (ENiCrMo-3)	W86112	A5.15 (ENiCu-B)	W84002
A5.9 (ER309Si)	W30941	A5.11 (ENiCrMo-4)	W80276	A5.15 (ENiFe-CI)	W82002
A5.9 (ER310)	S31080	A5.11 (ENiCrMo-5)	W80002	A5.15 (ENiFe-CI-A)	W82004
A5.9 (ER310)	W31040	A5.11 (ENiCrMo-6)	W86620	A5.15 (ESt)	K01520
A5.9 (ER312)	S31380	A5.11 (ENiCrMo-7)	W86455	A5.15 (RCI)	F10090
A5.9 (ER312)	W31340	A5.11 (ENiCrMo-9)	W86985	A5.15 (RCI-A)	F10091
A5.9 (ER316) (316H)	S31680	A5.11 (ENiCu-7)	W84190	A5.15 (RCI-B)	F10092
A5.9 (ER316) (ER316H)	W31640	A5.11 (ENiMo-1)	W80001	A5.16 (ERTi-0.2Pd)	R52401
A5.9 (ER316L)	S31683	A5.11 (ENiMo-3)	W80004	A5.16 (ERTi-1)	R50100
A5.9 (ER316L)	W31643	A5.11 (ENiMo-7)	W80665	A5.16 (ERTi-2)	R50120
A5.9 (ER316LSi)	S31688	A5.12 (EWP)	R07900	A5.16 (ERTi-3)	R50125
A5.9 (ER316LSi)	W31648	A5.12 (EWTh-1)	R07911	A5.16 (ERTi-3Al-2.5V)	R56320
A5.9 (ER316Si)	S31681	A5.12 (EWTh-2)	R07912	A5.16 (ERTi-3Al-2.5V-1)	R56321
A5.9 (ER316Si)	W31641	A5.12(EWTh-3)	R07913	A5.16 (ERTi-4)	R50130
A5.9 (ER317)	S31780	A5.12 (EWZr)	R07920	A5.16 (ERTi-5Al-2.5Sn)	R54522
A5.9 (ER317)	W31740	A5.13 (ECoCr-A)	W73006	A5.16 (ERTi-5Al-2.5Sn-1)	R54523
A5.9 (ER317L)	W31743	A5.13 (ECoCr-B)	W73012	A5.16 (ERTi-6Al-4V)	R56400
A5.9 (ER318)	S31980	A5.13 (ECoCr-C)	W73001	A5.16 (ERTi-6Al-4V-1)	R56402
A5.9 (ER318)	W31940	A5.13 (ECuAl-B)	W60619	A5.16 (ERTi-8Al-1Mo-1V)	R54810
A5.9 (ER320)	N08021	A5.13 (ECuAl-C)	W60625	A5.16 (ERTi-8Al-2Cb-1Ta-1Mo)	R56210
A5.9 (ER320)	N08321	A5.13 (ECuAl-D)	W61625	A5.16 (ERTi-12)	R53401
A5.9 (ER320LR)	N08022	A5.13 (ECuAl-E)	W62625	A5.16 (ERTi-13V-11Cr-3Al)	R58010
A5.9 (ER321)	S32180	A5.13 (ECuSi)	W60656	A5.17 (EH11K)	K11140
A5.9 (ER321)	W32140	A5.13 (ECuSn-A)	W60518	A5.17 (EH14)	K11585
A5.9 (ER347)	S34780	A5.13 (ECuSn-C)	W60521	A5.17 (EL8)	K01008
A5.9 (ER347)	W34740	A5.13 (EFe5-A)	W77710	A5.17 (EL8K)	K01009
A5.9 (ER347Si)	S34788	A5.13 (EFe5-B)	W77510	A5.17 (EL12)	K01012
A5.9 (ER347Si)	W34748	A5.13 (EFe5-C)	W77310	A5.17 (EM12)	K01112
A5.9 (ER349)	S63197	A5.13 (EFeCr-A1)	W74510	A5.17 (EM12K)	K01113
A5.9 (ER349)	W34940	A5.13 (EFeMn-A)	W79110	A5.17 (EM13K)	K01313
A5.9 (ER410)	S41080	A5.13 (EFeMn-B)	W79310	A5.17 (EM14K)	K01314
A5.9 (ER410)	W41040	A5.13 (ENiCr-A)	W89604	A5.17 (EM15K)	K01515
A5.9 (ER410NiMo)	S41086	A5.13 (ENiCr-B)	W89605	A5.18 (ER70S-2)	K10726
A5.9 (ER410NiMo)	W41046	A5.13 (ENiCr-C)	W89606	A5.18 (ER70S-3)	K11022
A5.9 (ER420)	S42080	A5.13 (ERCuSi-A)	C65600	A5.18 (ER70S-4)	K11132
A5.9 (ER420)	W42040	A5.13 (ERCuSn-A)	C51800	A5.18 (ER70S-5)	K11357
A5.9 (ER430)	S43080	A5.13 (RCoCr-A)	R30006	A5.18 (ER70S-6)	K11140
A5.9 (ER430)	W43040	A5.13 (RCoCr-B)	R30012	A5.18 (ER70S-7)	K11125
A5.9 (ER502)	S50280	A5.13 (RCoCr-C)	R30001	A5.19 ER AZ101A	M11101
A5.9 (ER502)	W50240	A5.13 (RCuAl-C)	C62580	A5.19 (ER AZ61A)	M11611

Cross Reference Specifications	Unified Number
A5.19 (ER AZ92A)	M11922
A5.19 (ER EZ33A)	M12331
A5.20 (E6xT-1)	W06601
A5.20 (E6xT-11)	W06611
A5.20 (E6xT-4)	W06604
A5.20 (E6xT-5)	W06605
A5.20 (E6xT-6)	W06606
A5.20 (E6xT-7)	W06607
A5.20 (E6xT-8)	W06608
A5.20 (E7xT-1)	W07601
A5.20 (E7xT-10)	W07610
A5.20 (E7xT-11)	W07611
A5.20 (E7xT-2)	W07602
A5.20 (E7xT-3)	W07603
A5.20 (E7xT-4)	W07604
A5.20 (E7xT-5)	W07605
A5.20 (E7xT-6)	W07606
A5.20 (E7xT-7)	W07607
A5.20 (E7xT-8)	W07608
A5.21 (EFe5-A)	W77710
A5.21 (EFe5-B)	W77510
A5.21 (EFe5-C)	W77310
A5.21 (EFeCr-A1)	W74510
A5.21 (EFeMn-A)	W79110
A5.21 (EFeMn-B)	W79310
A5.21 (RFe5-A)	W77730
A5.21 (RFe5-B)	W77530
A5.21 (RFeCr-A1)	W74530
A5.22 (E307T-1) (E307T-2)	W30731
A5.22 (E307T-3)	W30733
A5.22 (E308LT-1) (E308LT-2)	W30835
A5.22 (E308LT-3)	W30837
A5.22 (E308MoLT-1) (E308MoLT-2) (E308MoLT-3)	W30838
A5.22 (E308MoT-1) (E308MoT-2) (E308MoT-3)	W30832
A5.22 (E308T-1) (E308T-2)	W30831
A5.22 (E308T-3)	W30833
A5.22 (E309CbLT-1) (E309CbLT-2)	W30932
A5.22 (E309CbLT-3)	W30934
A5.22 (E309LT-1) (E309LT-2)	W30935
A5.22 (E309LT-3)	W30937
A5.22 (E309T-1) (E309T-2)	W30931
A5.22 (E309T-3)	W30933
A5.22 (E310T-1) (E310T-2) (E310T-3)	W31031
A5.22 (E312T-1) (E312T-2) (E312T-3)	W31331
A5.22 (E316LT-1) (E316LT-2)	W31635
A5.22 (E316LT-3)	W31637
A5.22 (E316T-1) (E316T-2)	W31631
A5.22 (E316T-3)	W31633
A5.22 (E317LT-1) (E317LT-2)	W31735
A5.22 (E317LT-3)	W31737
A5.22 (E347T-1) (E347T-2)	W34731
A5.22 (E347T-3)	W34733
A5.22 (E409T-1) (E409T-2) (E409T-3)	W40931
A5.22 (E410NiMoT-1) (E410NiMoT-2) (E410NiMoT-3)	W41036
A5.22 (E410NiTiT-1) (E410NiTiT-2) (E410NiTiT-3)	W41038
A5.22 (E410T-1) (E410T-2) (E410T-3)	W41031
A5.22 (E430T-1) (E430T-2) (E430T-3)	W43031
A5.22 (E502T-1) (E502T-2)	W50231
A5.22 (E505T-1) (E505T-2)	W50431
A5.23 (EA1)	K11222

Cross Reference Specifications	Unified Number
A5.23 (EA1N)	K11122
A5.23 (EA2)	K11223
A5.23 (EA2N)	K11123
A5.23 (EA3)	K11423
A5.23 (EA3N)	K11323
A5.23 (EA4)	K11424
A5.23 (EA4N)	K11324
A5.23 (EB1)	K11043
A5.23 (EB1N)	K10943
A5.23 (EB2)	K11172
A5.23 (EB2H)	K23016
A5.23 (EB2HN)	K23116
A5.23 (EB2N)	K11072
A5.23 (EB3)	K31115
A5.23 (EB3N)	K31015
A5.23 (EB5)	K12187
A5.23 (EB5N)	K12087
A5.23 (EB6)	S50280
A5.23 (EB6H)	S50180
A5.23 (EB6HN)	S50181
A5.23 (EB6N)	S50281
A5.23 (EB8)	S50480
A5.23 (EB8N)	S50481
A5.23 (EC-B8)	W50440
A5.23 (ECA1)	W17041
A5.23 (ECA1N)	W17141
A5.23 (ECA2)	W17042
A5.23 (ECA2N)	W17142
A5.23 (ECA3)	W17043
A5.23 (ECA3N)	W17143
A5.23 (ECA4)	W17044
A5.23 (ECA4N)	W17144
A5.23 (ECB1)	W51040
A5.23 (ECB1N)	W51041
A5.23 (ECB2)	W52040
A5.23 (ECB2H)	W52240
A5.23 (ECB2HN)	W52241
A5.23 (ECB2N)	W52041
A5.23 (ECB3)	W53040
A5.23 (ECB3N)	W53041
A5.23 (ECB4)	W53340
A5.23 (ECB4N)	W53341
A5.23 (ECB5)	W51340
A5.23 (ECB5N)	W51341
A5.23 (ECB6)	W50240
A5.23 (ECB6H)	W50140
A5.23 (ECB6HN)	W50141
A5.23 (ECB6N)	W50241
A5.23 (ECB8N)	W50441
A5.23 (ECF1)	W21150
A5.23 (ECF1N)	W21151
A5.23 (ECF2)	W20240
A5.23 (ECF2N)	W20241
A5.23 (ECF3)	W21140
A5.23 (ECF3N)	W21141
A5.23 (ECF4)	W20440
A5.23 (ECF4N)	W20441
A5.23 (ECF5)	W22540
A5.23 (ECF6)	W22640
A5.23 (ECF6N)	W22641
A5.23 (ECM1)	W21240
A5.23 (ECM1N)	W21241
A5.23 (ECM2)	W21340
A5.23 (ECM2N)	W21341
A5.23 (ECM3)	W22240
A5.23 (ECM3N)	W22241
A5.23 (ECM4)	W22440
A5.23 (ECM4N)	W22441
A5.23 (ECNi1)	W21040
A5.23 (ECNi1N)	W21041

Cross Reference Specifications	Unified Number
A5.23 (ECNi2)	W22040
A5.23 (ECNi2N)	W22041
A5.23 (ECNi3)	W23040
A5.23 (ECNi3N)	W23041
A5.23 (ECNi4)	W21250
A5.23 (ECNi4N)	W21251
A5.23 (ECW)	W20140
A5.23 (EF1)	K11160
A5.23 (EF1N)	K11060
A5.23 (EF2)	K21450
A5.23 (EF2N)	K21350
A5.23 (EF3)	K21485
A5.23 (EF3N)	K21385
A5.23 (EF4)	K12048
A5.23 (EF4N)	K11948
A5.23 (EF5)	K41370
A5.23 (EF6)	K21135
A5.23 (EF6N)	K21035
A5.23 (EL12)	K01012
A5.23 (EL12N)	K00912
A5.23 (EM2)	K10882
A5.23 (EM2N)	K10982
A5.23 (EM3)	K21015
A5.23 (EM3N)	K20915
A5.23 (EM4)	K21030
A5.23 (EM4N)	K20930
A5.23 (EM12K)	K01113
A5.23 (EM12KN)	K01013
A5.23 (ENi1)	K11040
A5.23 (ENi1K)	K11058
A5.23 (ENi1KN)	K10958
A5.23 (ENi1N)	K10940
A5.23 (ENi2)	K21010
A5.23 (ENi2N)	K20910
A5.23 (ENi3)	K31310
A5.23 (ENi3N)	K31210
A5.23 (ENi4)	K11485
A5.23 (ENi4N)	K11385
A5.23 (EW)	K11245
A5.24 (ERZr2)	R60702
A5.24 (ERZr3)	R60704
A5.24 (ERZr4)	R60707
A5.25 (EH10K-EW)	K01010
A5.25 (EH10Mo-EW)	K10945
A5.25 (EH11K-EW)	K11140
A5.25 (EH14-EW)	K11585
A5.25 (EM5K-EW)	K10726
A5.25 (EM12-EW)	K01112
A5.25 (EM12K-EW)	K01113
A5.25 (EM13K-EW)	K01313
A5.25 (EM15K-EW)	K01515
A5.25 (EWS-EW)	K11245
A5.25 (EWT1)	W06040
A5.25 (EWT2) (EWT4)	W20140
A5.25 (EWT3)	W22340
A5.26 (EG6xT1)	W06301
A5.26 (EG6xT2)	W06302
A5.26 (EG7xT1)	W07301
A5.26 (EG7xT2)	W07302
A5.26 (EGxxS-1)	K01313
A5.26 (EGxxS-1B)	K10945
A5.26 (EGxxS-2)	K10726
A5.26 (EGxxS-3)	K11022
A5.26 (EGxxS-5)	K11325
A5.26 (EGxxS-6)	K11140
A5.26 (EGxxT-3)	W21033
A5.26 (EGxxT4)	W22334
A5.26 (EGxxT5)	W20131
A5.27 (ERCu)	C18980
A5.27 (ERCuNi)	C71581

AWS (American Welding Society) Numbers

Cross Reference Specifications	Unified Number
A5.27 (ERCuSi-A)	C65600
A5.27 (RBCuZn-A)	C47000
A5.27 (RBCuZn-D)	C77300
A5.27 (RCuZn-B)	C68000
A5.27 (RCuZn-C)	C68100
A5.28 (E80C-B2)	W52030
A5.28 (E80C-B2L)	W52130
A5.28 (E80C-Ni1)	W21030
A5.28 (E80C-Ni2)	W22030
A5.28 (E80C-Ni3)	W23030
A5.28 (E90C-B3)	W53030
A5.28 (E90C-B3L)	W53130
A5.28 (ER80S-B2)	K20900
A5.28 (ER80S-B2L)	K20500
A5.28 (ER80S-D2)	K10945
A5.28 (ER80S-Ni1)	K11260
A5.28 (ER80S-Ni2)	K21240
A5.28 (ER80S-Ni3)	K31240
A5.28 (ER90S-B3)	K30960
A5.28 (ER90S-B3L)	K30560
A5.28 (ER100S-1)	K10882
A5.28 (ER100S-2)	K11250
A5.28 (ER110S-1)	K21015
A5.28 (ER120S-1)	K21030
A5.29 (E1xxT5-K4)	W22235
A5.29 (E6xT5-B6L)	W50230
A5.29 (E6xT5-B8L)	W50430
A5.29 (E8xT1-A1)	W17031
A5.29 (E8xT1-Ni1)	W21031
A5.29 (E70T4-K2)	W21234
A5.29 (E70T5-A1)	W17035
A5.29 (E71T8-K2)	W21238
A5.29 (E71T8-Ni1)	W21038
A5.29 (E71T8-Ni2)	W22038
A5.29 (E80T1-B2)	W52031
A5.29 (E80T1-B2H)	W52231
A5.29 (E80T1-W)	W20131
A5.29 (E80T5-B2)	W52035
A5.29 (E80T5-B2L)	W52135
A5.29 (E80T5-K1)	W21135
A5.29 (E80T5-K2)	W21235
A5.29 (E80T5-Ni1)	W21035
A5.29 (E80T5-Ni2)	W22035
A5.29 (E81T1-B1)	W51031
A5.29 (E81T1-B2)	W52031
A5.29 (E90T1-B3)	W53031
A5.29 (E90T1-B3H)	W53231
A5.29 (E90T1-B3L)	W53131
A5.29 (E90T1-D3)	W19331
A5.29 (E90T5-B3)	W53035
A5.29 (E90T5-D2)	W19235
A5.29 (E91T1-B3)	W53031
A5.29 (E91T1-D1)	W19131
A5.29 (E100T1-B3)	W53031
A5.29 (E100T5-D2)	W10235
A5.29 (E101T1-K7)	W22051
A5.29 (E111T1-K4)	W22231
A5.29 (E120T1-K5)	W21631
A5.29 (EBxT1-B1L)	W51131
A5.29 (ExxT1-K2)	W21231
A5.29 (ExxT1-Ni2)	W22031
A5.29 (ExxT5-Ni3)	W23035
A5.29 (ExxT8-K6)	W21048
A5.29 (ExxxT1-K3)	W21331
A5.29 (ExxxT5-K3)	W21335
A5.30 (ER316L)	S31683
A5.30 (IN-Ms3)	K11140
A5.30 (IN6A)	N07092
A5.30 (IN60)	N04060
A5.30 (IN61)	N02061
A5.30 (IN67)	C71581
A5.30 (IN82)	N06082
A5.30 (IN308)	S30880
A5.30 (IN308L)	S30883
A5.30 (IN310)	S31080
A5.30 (IN312)	S31380
A5.30 (IN316)	S31680
A5.30 (IN347)	S34780
A5.30 (IN502)	S50280
A5.30 (IN515)	K20900
A5.30 (IN521)	K30960
A5.30 (INMs1)	K10726
A5.30 (INMs2)	K01313
ER330	N08331

The CDA (Copper Development Association) NUMBERS are identical to the UNS designations.

(The Symbol **ob** Indicates an Obsolete UNS Number.)

Cross Index of Commonly Known Documents Which Describe Materials Same as or Similar to Those Covered by UNS Numbers

FEDERAL SPECIFICATION NUMBERS

(The Symbol **ob** Indicates an Obsolete UNS Number.)

Cross Reference Specifications	Unified Number
Cross Reference Specifications	Unified Number

FEDERAL SPECIFICATIONS

Cross Reference Specifications	Unified Number
QQ-A-20/1	A93003
QQ-A-20/2	A92014
QQ-A-20/3	A92024
QQ-A-20/4	A95083
QQ-A-20/5	A95086
QQ-A-20/6	A95454
QQ-A-20/7	A95456
QQ-A-20/8	A96061
QQ-A-20/9	A96063
QQ-A-20/10	A96066
QQ-A-20/11	A97075
QQ-A-20/13	A97178
QQ-A-20/14	A97178
QQ-A-20/15	A97075
QQ-A-20/16	A96061
QQ-A-20/17	A96162
QQ-A-25/1	A91100
QQ-A-25/2	A93003
QQ-A-25/3	A92011
QQ-A-25/4	A92014
QQ-A-25/5	A92017
QQ-A-25/6	A92024
QQ-A-25/7	A95052
QQ-A-25/8	A96061
QQ-A-25/9	A97075
QQ-A-25/10	A96262
QQ-A-20/1	A91100
QQ-A-20/2	A93003
QQ-A-20/4	A92024
QQ-A-20/6	A95083
QQ-A-20/7	A95086
QQ-A-20/8	A95052
QQ-A-20/9	A92124
QQ-A-20/9	A95456
QQ-A-20/10	A95454
QQ-A-20/11	A96061
QQ-A-20/13	A97075
QQ-A-20/14	A97178
QQ-A-20/21	A97178
QQ-A-20/24	A97075
QQ-A-20/28	A97178
QQ-A-20/29	A92124
QQ-A-20/30	A92219
QQ-A-367	A92014
QQ-A-367	A92018
QQ-A-367	A92025
QQ-A-367	A92218
QQ-A-367	A92219
QQ-A-367	A92618
QQ-A-367	A94032
QQ-A-367	A95083
QQ-A-367	A96061
QQ-A-367	A96066
QQ-A-367	A96151
QQ-A-367	A97049
QQ-A-367	A97075
QQ-A-367	A97076
QQ-A-367	A97079
QQ-A-371 (208.1)	A02081
QQ-A-371 (208.2)	A02082
QQ-A-371 (213.1)	A02131
QQ-A-371 (222.1)	A02221
QQ-A-371 (238.1)	A02381
QQ-A-371 (238.2)	A02382
QQ-A-371 (242.1)	A02421
QQ-A-371 (242.2)	A02422
QQ-A-371 (295.1)	A02951

Cross Reference Specifications	Unified Number
QQ-A-371 (295.2)	A02952
QQ-A-371 (308.1)	A03081
QQ-A-371 (308.2)	A03082
QQ-A-371 (319.1)	A03191
QQ-A-371 (319.2)	A03192
QQ-A-371 (324.1)	A03241
QQ-A-371 (328.1)	A03281
QQ-A-371 (333.1)	A03331
QQ-A-371 (354.1)	A03541
QQ-A-371 (355.1)	A03551
QQ-A-371 (355.2)	A03552
QQ-A-371 (356.1)	A03561
QQ-A-371 (356.2)	A03562
QQ-A-371 (357.1)	A03571
QQ-A-371 (360.2)	A03602
QQ-A-371 (364.2)	A03642
QQ-A-371 (380.2)	A03802
QQ-A-371 (384.1)	A03841
QQ-A-371 (384.2)	A03842
QQ-A-371 (390.2)	A03902
QQ-A-371 (413.2)	A04132
QQ-A-371 (443.1)	A04431
QQ-A-371 (443.2)	A04432
QQ-A-371 (514.1)	A05141
QQ-A-371 (514.2)	A05142
QQ-A-371 (518.1)	A05181
QQ-A-371 (518.2)	A05182
QQ-A-371 (520.2)	A05202
QQ-A-371 (535.2)	A05352
QQ-A-371 (705.1)	A07051
QQ-A-371 (707.1)	A07071
QQ-A-371 (713.1)	A07131
QQ-A-371 (771.2)	A07712
QQ-A-371 (850.1)	A08501
QQ-A-371 (A240.1)	A02401
QQ-A-371 (A242.1)	A12421
QQ-A-371 (A242.2)	A12422
QQ-A-371 (A332.1)	A03361
QQ-A-371 (A332.2)	A03362
QQ-A-371 (A356.2)	A13562
QQ-A-371 (A357.2)	A13572
QQ-A-371 (A360.1)	A13601
QQ-A-371 (A360.2)	A13602
QQ-A-371 (A380.1)	A13801
QQ-A-371 (A380.2)	A13802
QQ-A-371 (A390.1)	A13901
QQ-A-371 (A413.1)	A14131
QQ-A-371 (A413.2)	A14132
QQ-A-371 (A444.2)	A14442
QQ-A-371 (A514.2)	A05132
QQ-A-371 (A535.1)	A15351
QQ-A-371 (A712.1)	A07101
QQ-A-371 (A850.1)	A08511
QQ-A-371 (B295.1)	A02961
QQ-A-371 (B295.2)	A02962
QQ-A-371 (B358.2)	A03582
QQ-A-371 (B514.2)	A05122
QQ-A-371 (B850.1)	A08521
QQ-A-371 (C355.2)	A33552
QQ-A-371 (C443.1)	A34431
QQ-A-371 (C443.2)	A34432
QQ-A-371 (C712.1)	A07111
QQ-A-371 (D712.2)	A07122
QQ-A-371 (F332.1)	A03321
QQ-A-371 (F332.2)	A03322
QQ-A-371 (F514.1)	A05111
QQ-A-371 (F514.2)	A05112
QQ-A-371 (L514.2)	A05152
QQ-A-430	A91100
QQ-A-430	A92017

Cross Reference Specifications	Unified Number
QQ-A-430	A92024
QQ-A-430	A92117
QQ-A-430	A92219
QQ-A-430	A93003
QQ-A-430	A95052
QQ-A-430	A95056
QQ-A-430	A96053
QQ-A-430	A96061
QQ-A-430	A97050
QQ-A-430	A97075
QQ-A-430	A97178
QQ-A-430 (5005)	A95005
QQ-A-591 (360.0)	A03600
QQ-A-591 (380)	A03800
QQ-A-591 (383.0)	A03830
QQ-A-591 (384.0)	A03840
QQ-A-591 (413.0)	A04130
QQ-A-591 (443.0)	A04430
QQ-A-591 (518.0)	A05180
QQ-A-591 (A360.0)	A13600
QQ-A-591 (A380.0)	A13800
QQ-A-591 (A413.0)	A14130
QQ-A-596 (213)	A02130
QQ-A-596 (222)	A02220
QQ-A-596 (242)	A02420
QQ-A-596 (296)	A02960
QQ-A-596 (308)	A03080
QQ-A-596 (319)	A03190
QQ-A-596 (332)	A03320
QQ-A-596 (333)	A03330
QQ-A-596 (336)	A03360
QQ-A-596 (355)	A03550
QQ-A-596 (356)	A03560
QQ-A-596 (357)	A03570
QQ-A-596 (513)	A05130
QQ-A-596 (705)	A07050
QQ-A-596 (713-Tenzaloy)	A07130
QQ-A-596 (850)	A08500
QQ-A-596 (851)	A08510
QQ-A-596 (852)	A08520
QQ-A-596 (A356)	A13560
QQ-A-596 (B443.0)	A24430
QQ-A-596 (C355)	A33550
QQ-A-601 (208.0)	A02080
QQ-A-601 (222)	A02220
QQ-A-601 (242.0)	A02420
QQ-A-601 (295.0)	A02950
QQ-A-601 (319.0)	A03190
QQ-A-601 (328.0)	A03280
QQ-A-601 (355.0)	A03550
QQ-A-601 (356.0)	A03560
QQ-A-601 (512.0)	A05120 ob
QQ-A-601 (514.0)	A05140
QQ-A-601 (520.0)	A05200
QQ-A-601 (535.0)	A05350
QQ-A-601 (705.0)	A07050
QQ-A-601 (707.0)	A07070
QQ-A-601 (710.0)	A07100
QQ-A-601 (712.0)	A07120
QQ-A-601 (713.0)	A07130
QQ-A-601 (771.0)	A07710
QQ-A-601 (850.0)	A08500
QQ-A-601 (851)	A08510
QQ-A-601 (852.0)	A08520
QQ-A-601 (A356.0)	A13560
QQ-A-601 (B443.0)	A24430
QQ-A-601 (C355.0)	A33550
QQ-A-671	L01900
QQ-A-1876	A91100

Federal Specification Numbers

Cross Reference Specifications	Unified Number	Cross Reference Specifications	Unified Number	Cross Reference Specifications	Unified Number
QQ-A-1876	A91145	QQ-C-390	C82800	QQ-C-502	C12300
QQ-A-1876	A91235	QQ-C-390	C83600	QQ-C-502	C12500
QQ-B-613	C23000	QQ-C-390	C83800	QQ-C-502	C12700
QQ-B-613	C24000	QQ-C-390	C84200	QQ-C-502	C12800
QQ-B-613	C26000	QQ-C-390	C84400	QQ-C-502	C13000
QQ-B-613	C26800	QQ-C-390	C84800	QQ-C-523	C86100
QQ-B-613	C34200	QQ-C-390	C85200	QQ-C-523	C86200
QQ-B-613	C35300	QQ-C-390	C85400	QQ-C-523	C86300
QQ-B-626	C23000	QQ-C-390	C85500	QQ-C-523	C86400
QQ-B-626	C24000	QQ-C-390	C85700	QQ-C-523	C86500
QQ-B-626	C26000	QQ-C-390	C86100	QQ-C-525	C83600
QQ-B-626	C26800	QQ-C-390	C86200	QQ-C-525	C84200
QQ-B-626	C34200	QQ-C-390	C86300	QQ-C-525	C84400
QQ-B-626	C35300	QQ-C-390	C86400	QQ-C-525	C90300
QQ-B-626	C36000	QQ-C-390	C86500	QQ-C-525	C91000
QQ-B-626	C37700	QQ-C-390	C86800	QQ-C-525	C91300
QQ-B-626	C46700	QQ-C-390	C87200	QQ-C-525	C92200
QQ-B-637	C46200	QQ-C-390	C87400	QQ-C-525	C93200
QQ-B-637	C46400	QQ-C-390	C87500	QQ-C-525	C93400
QQ-B-637	C48200	QQ-C-390	C90300	QQ-C-525	C93900
QQ-B-637	C48500	QQ-C-390	C90500	QQ-C-525	C94000
QQ-B-639	C46200	QQ-C-390	C90700	QQ-C-530	C17200
QQ-B-639	C46400	QQ-C-390	C91000	QQ-C-530	C17300
QQ-B-639	C48200	QQ-C-390	C91300	QQ-C-533	C17000
QQ-B-639	C48500	QQ-C-390	C91600	QQ-C-533	C17200
QQ-B-650	C11010	QQ-C-390	C92200	QQ-C-576	C10100
QQ-B-650	C11020	QQ-C-390	C92500	QQ-C-576	C10200
QQ-B-650	C11030	QQ-C-390	C92700	QQ-C-576	C10400
QQ-B-650	C24000	QQ-C-390	C92900	QQ-C-576	C10500
QQ-B-650	C26000	QQ-C-390	C93200	QQ-C-576	C10700
QQ-B-650	C29800 ob	QQ-C-390	C93400	QQ-C-576	C11000
QQ-B-650	C47200 ob	QQ-C-390	C93500	QQ-C-576	C11300
QQ-B-650 (FS-BCuP-1)	C55180	QQ-C-390	C93700	QQ-C-576	C11400
QQ-B-650 (FS-BCuP-2)	C55181	QQ-C-390	C93800	QQ-C-576	C11600
QQ-B-650 (FS-BCuP-3)	C55281	QQ-C-390	C93900	QQ-C-576	C12000
QQ-B-650 (FS-BCuP-4)	C55283	QQ-C-390	C94000	QQ-C-576	C12100
QQ-B650 (FS-BCuP-5)	C55284	QQ-C-390	C94100	QQ-C-576	C12200
QQ-B-654 (I)	P07453	QQ-C-390	C94300	QQ-C-576	C12300
QQ-B-654 (II)	P07650	QQ-C-390	C94700	QQ-C-576	C12500
QQ-B654 (III)	C55284	QQ-C-390	C94800	QQ-C-576	C12700
QQ-B-654 (IV)	P07500	QQ-C-390	C95200	QQ-C-576	C12800
QQ-B-654 (V)	P07501	QQ-C-390	C95300	QQ-C-576	C13000
QQ-B-654 (VII)	P07450	QQ-C-390	C95400	QQ-C-576	C14180
QQ-B-654 (VIII)	P07350	QQ-C-390	C95500	QQ-C-585	C73500
QQ-B-655	A94043	QQ-C-390	C95700	QQ-C-585	C74500
QQ-B-655	A94045	QQ-C-390	C95800	QQ-C-585	C75200
QQ-B-655	A94047	QQ-C-390	C96200	QQ-C-585	C76200
QQ-B-655	A94145	QQ-C-390	C96400	QQ-C-585	C76600
QQ-B-655	A94343	QQ-C-390 (D3)	C92300	QQ-C-585	C77000
QQ-B-655	M19001	QQ-C-450	C60600	QQ-C-586	C74500
QQ-B-675	C95200	QQ-C-450	C61000	QQ-C-586	C75200
QQ-B-675	C95300	QQ-C-450	C61300	QQ-C-586	C76400
QQ-B-675	C95400	QQ-C-450	C61400	QQ-C-586	C77000
QQ-B-675	C95500	QQ-C-450	C63000	QQ-C-586	C79200
QQ-B-675	C95600	QQ-C-465	C60600	QQ-C-591	C64700
QQ-B-675	C95700	QQ-C-465	C61400	QQ-C-591	C65100
QQ-B-675	C95800	QQ-C-465	C63000	QQ-C-591	C65500
QQ-B-728	C67000	QQ-C-465	C64200	QQ-C-591	C66100
QQ-B-728	C67500	QQ-C-502	C10100	QQ-E-450 (6013)	W06013
QQ-B-750	C51000	QQ-C-502	C10200	QQ-G-540	P00900
QQ-B-750	C52400	QQ-C-502	C10400	QQ-G-00545B	P00015
QQ-B-750	C54400	QQ-C-502	C10500	QQ-L-171	L50050
QQ-C-40 (Grade AA and Grade C)		QQ-C-502	C10700	QQ-L-171	L50080
	L50065	QQ-C-502	C11000	QQ-L-171 (Grade C)	L51121
QQ-C-40 (Grade D)	L51124	QQ-C-502	C11300	QQ-L-201	L50070
QQ-C-390	C82000	QQ-C-502	C11400	QQ-L-201 (Grade C)	L51121
QQ-C-390	C82400	QQ-C-502	C11600	QQ-L-201 (Grade D)	L51123
QQ-C-390	C82500	QQ-C-502	C12000	QQ-M-31	M11100
QQ-C-390	C82600	QQ-C-502	C12100	QQ-M-31	M11311
QQ-C-390	C82700	QQ-C-502	C12200	QQ-M-31	M11610

Cross Reference Specifications	Unified Number	Cross Reference Specifications	Unified Number	Cross Reference Specifications	Unified Number
QQ-M-31	M11800	QQ-S-571 (Sn5)	L54322	QQ-S-766 (347)	S34700
QQ-M-31	M15100	QQ-S-571 (Sn10)	L54525	QQ-S-766 (348)	S34800
QQ-M-31	M16600	QQ-S-571 (Sn20)	L54712	QQ-S-766 (410)	S41000
QQ-M-38	M11910	QQ-S-571 (Sn30)	L54822	QQ-S-766 (420)	S42000
QQ-M-40	M11311	QQ-S-571 (Sn35)	L54852	QQ-S-766 (429)	S42900
QQ-M-40	M11610	QQ-S-571 (Sn40)	L54916	QQ-S-766 (430)	S43000
QQ-M-40	M11800	QQ-S-571 (Sn50)	L55031	QQ-S-766 (446)	S44600
QQ-M-40	M13210	QQ-S-571 (Sn60)	L13600	QQ-T-371	L13007
QQ-M-40	M15100	QQ-S-571 (Sn63)	L13630	QQ-T-390	L05120
QQ-M-40	M16600	QQ-S-571 (Sn70)	L13700	QQ-T-390	L13840
QQ-M-40	M18540	QQ-S-571 (Sn96)	L13961	QQ-T-390	L13890
QQ-M-44	M11311	QQ-S-626	G86200	QQ-T-390	L13910
QQ-M-55	M10100	QQ-S-635	G10400	QQ-T-570 (0-7)	T31507
QQ-M-55	M11630	QQ-S-635 (C1009)	G10090	QQ-T-570 (A-2)	T30102
QQ-M-55	M11810	QQ-S-635 (C1035)	G10350	QQ-T-570 (A-3)	T30103 ob
QQ-M-55	M11914	QQ-S-635 (C1045)	G10450	QQ-T-570 (A-4)	T30104
QQ-M-55	M11920	QQ-S-635 (C1050)	G10500	QQ-T-570 (A-5)	T30105 ob
QQ-M-55	M12330	QQ-S-681	G43400	QQ-T-570 (A-6)	T30106
QQ-M-55	M13310	QQ-S-681	G86300	QQ-T-570 (A-7)	T30107
QQ-M-55	M13320	QQ-S-700 (C1025)	G10250	QQ-T-570 (A-8)	T30108
QQ-M-55	M18220	QQ-S-700 (C1030)	G10300	QQ-T-570 (A-9)	T30109
QQ-M-56	M11630	QQ-S-700 (C1035)	G10350	QQ-T-570 (A-10)	T30110
QQ-M-56	M11810	QQ-S-700 (C1045)	G10450	QQ-T-570 (D-2)	T30402
QQ-M-56	M11914	QQ-S-700 (C1050)	G10500	QQ-T-570 (D-3)	T30403
QQ-M-56	M11920	QQ-S-700 (C1055)	G10550	QQ-T-570 (D-4)	T30404
QQ-M-56	M12330	QQ-S-700 (C1065)	G10650	QQ-T-570 (D-5)	T30405
QQ-M-56	M13310	QQ-S-700 (C1074)	G10740	QQ-T-570 (D-7)	T30407
QQ-M-56	M13320	QQ-S-700 (C1080)	G10800	QQ-T-570 (F-1)	T60601 ob
QQ-M-56	M16510	QQ-S-700 (C1084)	G10840	QQ-T-570 (F-2)	T60602 ob
QQ-M-56	M16610	QQ-S-700 (C1085)	G10850	QQ-T-570 (H-10)	T20810
QQ-M-56	M16620	QQ-S-700 (C1086)	G10860	QQ-T-570 (H-11)	T20811
QQ-M-56	M18220	QQ-S-700 (C1095)	G10950	QQ-T-570 (H-12)	T20812
QQ-N-281	N04400	QQ-S-763	S31008	QQ-T-570 (H-13)	T20813
QQ-N-281	N04405	QQ-S-763	S41623	QQ-T-570 (H-14)	T20814
QQ-N-286	N05500	QQ-S-763	S44300	QQ-T-570 (H-19)	T20819
QQ-N-288	N04019	QQ-S-763 (202)	S20200	QQ-T-570 (H-21)	T20821
QQ-P-00428	P04995	QQ-S-763 (302)	S30200	QQ-T-570 (H-22)	T20822
QQ-R-566	A07120	QQ-S-763 (304)	S30400	QQ-T-570 (H-23)	T20823
QQ-R-566	A91100	QQ-S-763 (304 L)	S30403	QQ-T-570 (H-24)	T20824
QQ-R-566	A92319	QQ-S-763 (305)	S30500	QQ-T-570 (H-25)	T20825 ob
QQ-R-566	A94047	QQ-S-763 (309)	S30900	QQ-T-570 (H-26)	T20826
QQ-R-566	A94145	QQ-S-763 (310)	S31000	QQ-T-570 (H-41)	T20841 ob
QQ-R-566	A95039 ob	QQ-S-763 (316)	S31600	QQ-T-570 (H-42)	T20842
QQ-R-566	A95183	QQ-S-763 (316 L)	S31603	QQ-T-570 (H-43)	T20843 ob
QQ-R-566	A95356	QQ-S-763 (317)	S31700	QQ-T-570 (L-2)	T61202
QQ-R-566	A95554	QQ-S-763 (321)	S32100	QQ-T-570 (L-3)	T61203 ob
QQ-R-566	A95556	QQ-S-763 (347)	S34700	QQ-T-570 (L-6)	T61206
QQ-R-566	A95654	QQ-S-763 (403)	S40300	QQ-T-570 (O-1)	T31501
QQ-R-571	C18900	QQ-S-763 (405)	S40500	QQ-T-570 (O-2)	T31502
QQ-R-571	C51800	QQ-S-763 (410)	S41000	QQ-T-570 (O-6)	T31506
QQ-R-571	C61800	QQ-S-763 (414)	S41400	QQ-T-570 (S-1)	T41901
QQ-R-571	C65600	QQ-S-763 (420)	S42000	QQ-T-570 (S-2)	T41902
QQ-R-571 (RBCuZn-A)	C47000	QQ-S-763 (429)	S42900	QQ-T-570 (S-4)	T41904
QQ-R-571 (RBCuZn-D)	C77300	QQ-S-763 (430)	S43000	QQ-T-570 (S-5)	T41905
QQ-R-571 (RCu-1	C10200	QQ-S-763 (440 A)	S44002	QQ-T-570 (S-6)	T41906
QQ-R-571 (RCuAl-B)	C62200	QQ-S-763 (440 B)	S44003	QQ-T-570 (S-7)	T41907
QQ-R-571 (RCuNi)	C71581	QQ-S-763 (440 C)	S44004	QQ-W-321	C21000
QQ-R-571 (RCuSn-C)	C52100	QQ-S-763 (446)	S44600	QQ-W-321	C22000
QQ-R-571 (RCuZn-B)	C68000	QQ-S-766 (201)	S20100	QQ-W-321	C23000
QQ-R-571 (RCuZn-C)	C68100	QQ-S-766 (202)	S20200	QQ-W-321	C24000
QQ-R-571c	C18980	QQ-S-766 (301)	S30100	QQ-W-321	C26000
QQ-S-498 (C1015)	G10150	QQ-S-766 (302)	S30200	QQ-W-321	C27000
QQ-S-561 (Class 3)	C55284	QQ-S-766 (304)	S30400	QQ-W-321	C27400
QQ-S-571 (Ag1.5)	L50132	QQ-S-766 (304 L)	S30403	QQ-W-321	C51000
QQ-S-571 (Ag2.5)	L50151	QQ-S-766 (305)	S30500	QQ-W-321	C74500
QQ-S-571 (Ag5.5)	L50180	QQ-S-766 (309)	S30900	QQ-W-321	C75200
QQ-S-571 (Pb65)	L54851	QQ-S-766 (310)	S31000	QQ-W-321	C75700
QQ-S-571 (Pb70)	L54821	QQ-S-766 (316)	S31600	QQ-W-321	C76400
QQ-S-571 (Pb80)	L54711	QQ-S-766 (316 L)	S31603	QQ-W-321	C77000
QQ-S-571 (Sb5)	L13940	QQ-S-766 (321)	S32100	QQ-W-343	C10100

Federal Specification Numbers

Cross Index of Commonly Known Documents Which Describe Materials Same as or Similar to Those Covered by UNS Numbers

MIL (Military Specification) NUMBERS
(The Symbol **ob** Indicates an Obsolete UNS Number.)

Cross Reference Specifications	Unified Number
MIL	
MIL-G--S-24149/2	A95083
MIL-C-10375	C26000
MIL-C-10375	C26100
MIL-C-10387	A03560
MIL-C-10387	A96151
MIL-S-10520	G10550
MIL-T-10794	A96061
MIL-A-10841	M00995
MIL-A-10841	M00998
MIL-A-11267 (8280)	A98280
MIL-S-11310 (CS1005)	G10050
MIL-S-11310 (CS1006)	G10060
MIL-S-11310 (CS1008)	G10080
MIL-S-11310 (CS1010)	G10100 ob
MIL-S-11310 (CS1012)	G10120
MIL-S-11310 (CS1017)	G10170
MIL-S-11310 (CS1018)	G10180
MIL-S-11310 (CS1020)	G10200
MIL-S-11310 (CS1022)	G10220
MIL-S-11310 (CS1025)	G10250
MIL-S-11310 (CS1030)	G10300
MIL-S-11310 (CS1040)	G10400
MIL-A-11356	J13025
MIL-S-11595 (ORD4150)	G41500
MIL-S-11713	G10690
MIL-S-11713 (2)	G10700
MIL-C-11866	A03560
MIL-C-11866	A24430
MIL-C-11866 (17)	C82500
MIL-C-11866 (19)	C87200
MIL-C-11866 (195)	A02950
MIL-C-11866 (20)	C86200
MIL-C-11866 (21)	C86300
MIL-C-11866 (214)	A05140
MIL-C-11866 (22)	C95300
MIL-C-11866 (25)	C83600
MIL-C-11866 (26)	C90300
MIL-C-11866 (355)	A03550
Mil-C-11866 (40E)	A07120
MIL-C-11866 (Almag 35)	A05350
MIL-C-12166	C11000
MIL-C-12166	C11100
MIL-R-12221 (7277)	A97277
MIL-S-12504	G10700
MIL-S-12504 (Mn-Cr-Mo)	K15747
MIL-S-12504 (Mn-Mo)	K17145
MIL-A-12545	A91100
MIL-A-12545	A92014
MIL-A-12545	A96061
MIL-A-12545	A96070
MIL-A-12545	A97075
MIL-A-12560	K11918 ob
MIL-S-12875	A93003
MIL-S-12875	A95052
MIL-E-13080 (MIL-307)	W30710
MIL-E-13080 (MIL-308)	W30821
MIL-A-13259	K91209
MIL-S-13282	P07015
MIL-C-13351	C37700
MIL-B-13501	C41100
MIL-B-13501	C51000
MIL-B-13501	C54600 ob
MIL-B-13506	C98200
MIL-B-13506	C98400
MIL-B-13506	C98600
MIL-T-1368	N04400
MIL-T-15005	C70600
MIL-T-15005	C71500

Cross Reference Specifications	Unified Number
MIL-S-15083 (100-70)	J32075
MIL-S-15083 (65-35)	J03009
MIL-S-15083 (70-36)	J03504
MIL-S-15083 (80-40)	J05002
MIL-S-15083 (B)	J03008
MIL-S-15083 (CW)	J03007
MIL-T-15089	A92014
MIL-T-15089	A92024
MIL-C-15345 (1)	C83600
MIL-C-15345 (10)	C92300
MIL-C-15345 (11)	C93400
MIL-C-15345 (12)	C93200
MIL-C-15345 (14)	C95500
MIL-C-15345 (23)	C91600
MIL-C-15345 (24)	C96400
MIL-C-15345 (25)	C96200
MIL-C-15345 (28)	C95800
MIL-C-15345 (3)	C85700
MIL-C-15345 (4)	C86500
MIL-C-15345 (5)	C86100
MIL-C-15345 (6)	C86300
MIL-C-15345 (8)	C90300
MIL-C-15345 (9)	C92200
MIL-S-15464 (1)	J12070
MIL-S-15464 (2)	J21880
MIL-S-15464 (3)	J11875
MIL-E-15597	A91100
MIL-E-15597	A93003
MIL-E-15597	A94043
MIL-C-15726	C70600
MIL-C-15726	C71500
MIL-B-15894 (1)	C85800
MIL-B-15894 (2)	C87900
MIL-B-15894 (3)	C87800
MIL-S-16113 (I)	K11801
MIL-S-16113 (II)	K11802
MIL-B-16166	C62300
MIL-B-16166	C63000
MIL-S-16216	K31820
MIL-S-16216	K32045
MIL-S-16216 (HY100)	J42240
MIL-S-16216 (HY80)	J42015
MIL-T-16286	S50200
MIL-T-16420	C70600
MIL-T-16420	C71500
MIL-Z-16460 (A)	Z35542
MIL-B-16541	C92200
MIL-S-16598	K93602
MIL-S-16788	G10200
MIL-S-16788	G10400
MIL-S-16788 (C10)	G10950
MIL-S-16974	G10200
MIL-S-16974	G10220
MIL-S-16974	G10400
MIL-S-16974	G86220
MIL-S-16974	G86270
MIL-S-16974 (1015)	G10150
MIL-S-16974 (1050)	G10500
MIL-S-16974 (1080)	G10800
MIL-S-16974 (1330)	G13300
MIL-S-16974 (1335)	G13350
MIL-S-16974 (1340)	G13400
MIL-S-16974 (3140)	G31400
MIL-S-16974 (4130)	G41300
MIL-S-16974 (4135)	G41350
MIL-S-16974 (4140)	G41400
MIL-S-16974 (4145)	G41450
MIL-S-16974 (4340)	G43400
MIL-S-16974 (8620)	G86200
MIL-S-16974 (8625)	G86250

Cross Reference Specifications	Unified Number
MIL-S-16974 (8630)	G86300
MIL-S-16974 (8640)	G86400
MIL-S-16974 (8645)	G86450
MIL-S-16974 (Gr.1060)	G10600
MIL-S-16993 (2)	J91261
MIL-S-16993 (Class 1)	J91150
MIL-C-17112	C97600
MIL-R-17131 (MIL-RCoCr-A)	R30006
MIL-R-17131 (MIL-RCoCr-C)	R30001
MIL-R-17131 (MIL-RNiCr-B)	N99645
MIL-R-17131 (MIL-RNiCr-C)	N99646
MIL-S-17249	J91209
MIL-S-17509 (I)	J92502
MIL-S-17509 (II)	J92801
MIL-S-17509 (III)	J92720
MIL-S-17759 (Ships)	S30210
MIL-S-17759 (Ships)	S30260
MIL-A-18001	Z32121
MIL-G-18014	A95052
MIL-G-18014	A95086
MIL-G-18014	A95456
MIL-G-18014	A95554
MIL-G-18014	A96061
MIL-G-18014	A96063
MIL-G-18015	A95052
MIL-G-18015	A96061
MIL-G-18015	A96063
MIL-E-18038 (MIL-8016)	W21016
MIL-F-18280	A92014
MIL-F-18280	A92024
MIL-F-18280	A96061
MIL-F-18280	A97075
MIL-S-18410	K21590
MIL-S-18410 (A)	K11598
MIL-S-18728	G86300
MIL-S-18729	G41300
MIL-S-18732	S43100
MIL-B-18907	C10100
MIL-B-18907	C12000
MIL-B-18907	C12200
MIL-B-18907	C22000
MIL-B-18907	C70200
MIL-E-19141 (MIL-Fe5-B)	W77510
MIL-E-19141 (MIL-FeMn-A)	W79110
MIL-E-19141 (MIL-I-A2C)	W75110
MIL-E-19141 (MIL-I-A2a)	W75010
MIL-B-19231	C10700
MIL-B-19231	C11600
MIL-C-19310	C81500
MIL-C-19311	C18200
MIL-C-19311	C18400
MIL-C-19311	C18500
MIL-S-19434	G10350
MIL-S-19434	G98500
MIL-R-19631 (MIL-RCuNi)	C71581
MIL-E-19933 (MIL-308) (MIL-308Co)	S30880
MIL-E-19933 (MIL-308L) (MIL-308CoL)	S30883
MIL-E-19933 (MIL-309) (MIL-309Co)	S30980
MIL-E-19933 (MIL-310)	S31080
MIL-E-19933 (MIL-312) (MIL-312Co)	S31380
MIL-E-19933 (MIL-316)	S31680
MIL-E-19933 (MIL-316L)	S31683
MIL-E-19933 (MIL-317)	S31780
MIL-E-19933 (MIL-318)	S31980
MIL-E-19933 (MIL-321)	S32180

MIL (Military Specification) Numbers

Cross Reference Specifications	Unified Number
MIL-E-19933 (MIL-347) (MIL-347Co)	
	S34780
MIL-E-19933 (MIL-349)	S63197
MIL-E-19933 (MIL-410)	S41080
MIL-E-19933D, (MIL-308HC)	S30884
MIL-F-20138	S30400
MIL-S-20146	S50200
MIL-B-20148	A94047
MIL-B-20148	A94343
MIL-T-20157	G10260
MIL-C-20159 (I)	C96400
MIL-C-20159 (II)	C96200
MIL-T-20168	C23000
MIL-T-20219	C26000
MIL-T-20219	C26100
MIL-B-20292	C11100
MIL-B-20292	C13000
MIL-B-20292	C22000
MIL-B-20292	C70200
MIL-F-20670	G10260
MIL-A-21180 (354.0)	A03540
MIL-A-21180 (356.0)	A13560
MIL-A-21180 (359.0)	A03590
MIL-A-21180 (A210.0)	A12010
MIL-A-21180 (A357.0)	A13570
MIL-A-21180 (C355.0)	A33550
MIL-W-21425 (I, III)	K06500
MIL-W-21425 (II)	K06000
MIL-E-21562 (EN60, RN60)	N04060
MIL-E-21562 (EN61, RN61)	N02061
MIL-E-21562 (EN62, RN62)	N06062
MIL-E-21562 (EN67, RN67) MIL-I-23413	
(MIL-67)	C71581
MIL-E-21562 (EN6A, RN6A)	N07092
MIL-E-21562 (EN82, RN82)	N06082
MIL-E-21562 (RN625) (EN625)	N06625
MIL-E-21562 (RN65)	N08065
MIL-E-21562 (RN69)	N07069
MIL-C-21657	C17200
MIL-E-21659 (CuAl-A1)	C61000
MIL-E-21659 (MIL-CuAl-A2)	C61800
MIL-E-21659 (MIL-CuAl-B)	C62200
MIL-E-21659 (MIL-CuSi)	C65600
MIL-E-21659 (MIL-CuSn-C)	C52100
MIL-C-21768	C21000
MIL-C-21768	C22000
MIL-S-21952	K31820
MIL-S-21952	K32045
MIL-C-22087	C87200
MIL-C-22087 (1)	C93400
MIL-C-22087 (10)	C82500
MIL-C-22087 (2)	C83600
MIL-C-22087 (3)	C90300
MIL-C-22087 (4)	C87500
MIL-C-22087 (5)	C86500
MIL-C-22087 (7)	C86100
MIL-C-22087 (8)	C95500
MIL-C-22087 (9)	C86200
MIL-C-22087 (9)	C86300
MIL-S-22141 (1020)	J02001
MIL-S-22141 (1040)	J04001
MIL-S-22141 (1050)	J05001
MIL-S-22141 (4130)	J13048
MIL-S-22141 (4140)	J14047
MIL-S-22141 (4335M)	J13432
MIL-S-22141 (4340)	J24055
MIL-S-22141 (4620)	J12093
MIL-S-22141 (52100)	G52986
MIL-S-22141 (52100)	J19965
MIL-S-22141 (6150)	J15047

Cross Reference Specifications	Unified Number
MIL-S-22141 (8620)	J12047
MIL-S-22141 (8630)	J13049
MIL-S-22141 (8640)	J14048
MIL-S-22141 (8735)	J13442
MIL-S-22141 (IC-1030)	J03005
MIL-S-22141 (Nitralloy)	J24056
MIL-E-22200 (MIL-7018)	W07018
MIL-E-22200/1 (MIL-10018-M)	W21318
MIL-E-22200/1 (MIL-11018-M)	W21418
MIL-E-22200/1 (MIL-8018-C3)	
(MIL-8018-C3SR)	W21018
MIL-E-22200/1 (MIL-9018-M)	W21218
MIL-E-22200/10, (MIL-12018-M1)	
	W23218
MIL-E-22200/11 (MIL-14018-HT)	
	W26018
MIL-E-22200/2 (MIL-308) (MIL-308Co)	
	W30810
MIL-E-22200/2 (MIL-308L) (MIL-308CoL)	
	W30813
MIL-E-22200/2 (MIL-309) (MIL-309Co)	
	W30910
MIL-E-22200/2 (MIL-309Cb)	W30917
MIL-E-22200/2 (MIL-310)	W31010
MIL-E-22200/2 (MIL-312)	W31310
MIL-E-22200/2 (MIL-316)	W31610
MIL-E-22200/2 (MIL-316L)	W31613
MIL-E-22200/2 (MIL-317)	W31710
MIL-E-22200/2 (MIL-318)	W31910
MIL-E-22200/2 (MIL-330)	W88331
MIL-E-22200/2 (MIL-347) (MIL-347Co)	
	W34710
MIL-E-22200/2 (MIL-347HC)	W34712
MIL-E-22200/2 (MIL-349)	W34910
MIL-E-22200/2, Type MIL-308HC	
	W30815
MIL-E-22200/2, Types (MIL-309L)	
(MIL-309CoL)	W30913
MIL-E-22200/2C Type MIL-16-8-2	
	W36810
MIL-E-22200/3 (MIL-1N12)	W86112
MIL-E-22200/3 (MIL-3N12)	W86132
MIL-E-22200/3 (MIL-3N1B)	W80001
MIL-E-22200/3 (MIL-3N1C)	W80002
MIL-E-22200/3 (MIL-3N1L)	W73605
MIL-E-22200/3 (MIL-3N1N)	W73155
MIL-E-22200/3 (MIL-4N11)	W82141
MIL-E-22200/3 (MIL-4N1A)	W86133
MIL-E-22200/3 (MIL-4N1W)	W80004
MIL-E-22200/3 (MIL-8N12)	W86182
MIL-E-22200/3 (MIL-9N10)	W84190
MIL-E-22200/4 (MIL-CuNi)	W60715
MIL-E-22200/5 (MIL-10018-N1)	
	W22718
MIL-E-22200/6 (MIL-10015)	W21215
MIL-E-22200/6 (MIL-10016)	W21216
MIL-E-22200/6 (MIL-7015)	W07015
MIL-E-22200/6 (MIL-7016)	W07016
MIL-E-22200/6 (MIL-8015-C3)	W21015
MIL-E-22200/6 (MIL-8016-C3)	W21016
MIL-E-22200/7 (MIL-7010-A1)	W17010
MIL-E-22200/7 (MIL-7011-A1)	W17011
MIL-E-22200/7 (MIL-7018-A1)	W17018
MIL-E-22200/7 (MIL-7020-A1)	W17020
MIL-E-22200/8 (MIL-410)	W41010
MIL-E-22200/8 (MIL-502)	W50210
MIL-E-22200/8 (MIL-502-1x-L)	W50213
MIL-E-22200/8 (MIL-505)	W50410
MIL-E-22200/8 (MIL-8015-B2)	W52015

Cross Reference Specifications	Unified Number
MIL-E-22200/8 (MIL-8015-B2L)	
	W52115
MIL-E-22200/8 (MIL-8016-B2)	W52016
MIL-E-22200/8 (MIL-8018-B2)	W52018
MIL-E-22200/8 (MIL-8018-B2L)	
	W52118
MIL-E-22200/8 (MIL-9015-B3)	W53015
MIL-E-22200/8 (MIL-9015-B3L)	
	W53115
MIL-E-22200/8 (MIL-9016-B3)	W53016
MIL-E-22200/8 (MIL-9018-B3)	W53018
MIL-E-22200/8 (MIL-9018-B3L)	
	W53118
MIL-E-22200/9 (MIL-14018-M1)	
	W23318
MIL-T-22214	C70600
MIL-T-22214	C71500
MIL-C-22229 (1)	C90300
MIL-C-22229 (1)	C93400
MIL-C-22229 (10)	C86100
Mil-C-22229 (2)	C83600
MIL-C-22229 (4)	C87200
MIL-C-22229 (5)	C95200
MIL-C-22229 (6)	C95500
MIL-C-22229 (7)	C86500
MIL-C-22229 (8)	C86300
MIL-C-22229 (9)	C86200
MIL-S-22499	C26000
MIL-S-22499	C26100
MIL-S-22664	K31820
MIL-S-22664	K32045
MIL-S-22698	G10260
MIL-S-22698 (B)	K02102
MIL-S-22698 (C)	K02708
MIL-W-22759	C11100
MIL-A-22771	A92014
MIL-A-22771	A92219
MIL-A-22771	A96061
MIL-A-22771	A96151
MIL-A-22771	A97039
MIL-A-22771	A97075
MIL-A-22771 (2618)	A92618
MIL-S-23008 (HY100)	J42240
MIL-S-23008 (HY80)	J42015
MIL-S-23009	K31820
MIL-S-23009	K32045
MIL-S-23009 (HY100)	J42240
MIL-S-23009 (HY80)	J42015
MIL-W-23068	C10100
MIL-W-23068	C10300
MIL-W-23068	C12000
MIL-W-23068	C22000
MIL-W-23068	C10200
MIL-S-23192	N07750
MIL-S-23195 (304 L)	S30403
MIL-S-23195 (304)	S30400
MIL-S-23195 (347)	S34700
MIL-S-23195 (348)	S34800
MIL-S-23196 (304 L)	S30403
MIL-S-23196 (304)	S30400
MIL-S-23196 (347)	S34700
MIL-S-23196 (348)	S34800
MIL-T-23227	N06600
MIL-N-23228	N06600
MIL-N-23229	N06600
MIL-W-23351	A96061
MIL-I-23413 (60)	N04060
MIL-I-23413 (61)	N02061
MIL-I-23413 (62)	N06062
MIL-I-23413 (6A)	N07092

Cross Reference Specifications	Unified Number
MIL-I-23413 (82)	N06082
MIL-I-23413 (MIL-308) (MIL-308Co)	S30880
MIL-I-23413 (MIL-308L) (MIL-308CoL)	S30883
MIL-I-23413 (MIL-310)	S31080
MIL-I-23413 (MIL-316)	S31680
MIL-I-23413 (MIL-316L)	S31683
MIL-I-23413 (MIL-348) (MIL-348Co)	S34780
MIL-I-23413 (MIL-515)	K20900
MIL-I-23413 (MS2)	K01313
MIL-I-23413B (MIL-521)	K30960
MIL-I-23413B (MIL-Ms-1)	K10726
MIL-I-23431 (MIL-505)	S50280
MIL-S-23495	K03016
MIL-S-23495 (B)	K02102
MIL-S-23495 (C)	K02401
MIL-T-23520	N04400
MIL-E-23765	C65500
MIL-E-23765 (MIL-110S-1)	K21015
MIL-E-23765 (MIL-120S-1)	K21030
MIL-E-23765 (MIL-70S-4)	K11132
MIL-E-23765 (MIL-70S-5)	K11357
MIL-E-23765 (MIL-70S-6)	K11140
MIL-E-23765 (MIL-80S-1)	K21485
MIL-E-23765 (MIL-CuAl)	C61900
MIL-E-23765 (MIL-CuAl-A2)	C61800
MIL-E-23765 (MIL-CuSi)	C65600
MIL-E-23765 (MIL-CuSn-C)	C52100
MIL-E-23765(MIL-70S-7)	K11585
MIL-E-23765/1C (70S-3)	K01313
MIL-E-23765/1C (MIL-705-2) ...	K10726
MIL-E-23765/1C (MIL-70S-1) ..	K01515
MIL-E-23765/1C MIL-70S-8	K11123
MIL-E-23765/2A (MIL-100S-1) .	K10882
MIL-E-23765/3 (CuMnNiAl)	C63380
MIL-E-23765/3, CuNiAl	C63280
MIL-S-24093	G10260
MIL-N-24106	N04400
MIL-T-24107	C12000
MIL-T-24107	C12200
MIL-C-24111	S17400
MIL-N-24114	N07750
MIL-I-24137 (A)	F33101
MIL-I-24137 (B)	F43020
MIL-I-24137 (C)	F43021
MIL-S-24149/2	A95086
MIL-S-24149/2	A95356
MIL-S-24149/2	A95456
MIL-S-24149/5	A91100
MIL-S-24238 (A)	K12042
MIL-S-24371	K51255
MIL-E-24403/1 (MIL-70T-5)	W07605
MIL-S-24412	K11201
MIL-S-24451	K31820
MIL-S-24451	K32045
MIL-N-24469	N07718
MIL-B-24480	C95700
MIL-B-24480	C95800
MIL-S-24502	K14675
MIL-S-24512	K51255
MIL-N-24549	N05500
MIL-S-25043	S17700
MIL-P-25995	A93003
MIL-P-25995	A96061
MIL-P-25995	A96063
MIL-M-26075	M13310
MIL-C-26094	A91100
MIL-C-26094	A95005

Cross Reference Specifications	Unified Number
MIL-C-26094	A95052
MIL-C-26094	A95086
MIL-C-26094	A95154
MIL-M-26696	M16601
MIL-S-27419	S30400
MIL-S-27419	S32100
MIL-E-278 (E-CuAl-A)	W60614
MIL-E-278 (E-CuAl-B)	W60619
MIL-S-3039	K04600
MIL-W-30508	S30560
MIL-T-3235	C10200
MIL-T-3235	C12000
MIL-T-3235	C12200
MIL-S-3289	G10350
MIL-S-3289	K02901
MIL-W-3318	C10100
MIL-W-3318	C10200
MIL-W-3318	C10400
MIL-W-3318	C10500
MIL-W-3318	C10700
MIL-W-3318	C11000
MIL-W-3318	C11100
MIL-W-3318	C11300
MIL-W-3318	C11400
MIL-W-3318	C11600
MIL-W-3318	C12000
MIL-W-3318	C12100
MIL-W-3318	C12200
MIL-W-3318	C12300
MIL-W-3318	C12500
MIL-W-3318	C12700
MIL-W-3318	C12800
MIL-W-3318	C13000
MIL-C-3383	C22000
MIL-T-3520	G10200
MIL-T-3520	G10250
MIL-T-3595	C51000
MIL-F-39000	A02080
MIL-F-39000	A02950
MIL-F-39000	A03550
MIL-F-39000	A03560
MIL-F-39000	A05180
MIL-F-39000	A13600
MIL-F-39000	A96061
MIl-F-39000 (C712.0)	A07110
MIl-F-39000 (D712.0)	A07120
MIL-F-3922	A02080
MIL-F-3922	A02950
MIL-F-3922	A03550
MIL-F-3922	A03560
MIL-F-3922	A05180
MIL-F-3922	A13600
MIL-F-3922	A96061
MIl-F-3922 (C712.0)	A07110
MIl-F-3922 (D712.0)	A07120
MIL-S-4043	S30403
MIL-S-43	K03201
MIL-A-45225	A95083
MIL-A-45225	A95456
MIL-A-45225	A97039
MIL-N-46025	N02205
MIL-A-46027	A95083
MIL-A-46027	A95456
MIL-M-46037	M16100
MIL-M-46039	M16210
MIL-S-46044	S30200
MIL-S-46047	K14185
MIL-S-46049 (1065)	G10650
MIL-S-46049 (1074)	G10740
MIL-S-46059	G10200

Cross Reference Specifications	Unified Number
MIL-S-46059	G10500
MIL-S-46059	G41300
MIL-S-46059	G41400
MIL-S-46059	G43400
MIL-S-46059	G86300
MIL-M-46062	M13310
MIL-M-46062	M16510
MIL-M-46062	M16610
MIL-M-46062	M16620
MIL-M-46062	M16630
MIL-M-46062	M18220
MIL-M-46062B	M10100
MIL-M-46062B	M11914
MIL-M-46062B	M11920
MIL-A-46063	A97039
MIL-B-46066	C83400
MIL-S-46070	G10300
MIL-S-46070	G10350
MIL-S-46070	G10400
MIL-T-46072	C33000
MIL-T-46072	C33100
MIL-T-46072	C33200
MIL-T-46072	C37000
MIL-A-46083	A95083
MIL-A-46083	A95456
MIL-A-46083	A97039
MIL-C-46087	C17500
MIL-A-46104	A96070
MIL-A-46118	A92219
MIL-S-46128	K23080
MIL-M-46130	M14141
MIL-M-46130	M14142
MIL-M-46130	M14145
MIL-M-46143	M14142
MIL-M-46143	M14145
MIL-A-46173	K33370
MIL-S-46409 (1065)	G10650
MIL-A-46808	A92219
MIL-S-46850	K92810
MIL-S-46850	K92820
MIL-S-46850	K92890
MIL-S-46850	K93120
MIL-S-46889	S30115
MIL-S-47036	K24728
MIL-S-47038	K11268
MIL-S-47086	T20811
MIL-S-47139	K92810
MIL-S-47139	K92890
MIL-S-47139	K93120
MIL-A-47182	R30005
MIL-S-47262	T20811
MIL-C-50	C26000
MIL-S-5000	G43400
MIL-S-5000	G43406
MIL-R-5031 (Class 1)	S30880
MIL-R-5031 (Class 10)	N10001
MIL-R-5031 (Class 11)	N10002
MIL-R-5031 (Class 12)	N10004
MIL-R-5031 (Class 13)	R30605
MIL-R-5031 (Class 16)	S30883
MIL-R-5031 (Class 17)	S31683
MIL-R-5031 (Class 2)	S30980
MIL-R-5031 (Class 3)	S31080
MIL-R-5031 (Class 4)	S31680
MIL-R-5031 (Class 6)	S63197
MIL-C-5031 (Class 7)	N06003
MIL-R-5031 (Class 8)	N06600
MIL-R-5031 (Class 8A)	N06082
MIL-R-5031 (Class 9)	R30155

MIL (Military Specification) Numbers

Cross Reference Specifications	Unified Number	Cross Reference Specifications	Unified Number	Cross Reference Specifications	Unified Number
MIL-R-5031 (Classes 5 and 5a)		MIL-C-81519	C96600	MIL-W-85	A96063
	S34780	MIL-T-81556	R50250	MIL-W-85	C10100
MIL-S-5059	S30200	MIL-T-81556	R50400	MIL-W-85	C10200
MIL-S-5059 (301)	S30100	MIL-T-81556	R50550	MIL-W-85	C12000
MIL-S-5059 (304)	S30400	MIL-T-81556	R50700	MIL-W-85	C22000
MIL-S-5059 (316)	S31600	MIL-T-81556	R54520	MIL-S-8503	G33106
MIL-T-5066	G10250	MIL-T-81556	R54521	MIL-S-8503	G61500
MIL-T-50777	A92024	MIL-T-81556	R54810	MIL-T-8504 (304)	S30400
MIL-A-52174	A91100	MIL-T-81556	R56400	MIL-T-8506 (304)	S30400
MIL-A-52174	A93003	MIL-T-81556	R56401	MIL-N-8550	N07750
MIL-A-52177	A91100	MIL-T-81556	R56620	MIL-S-8559	G10950
MIL-A-52242 (7001)	A97001	MIL-T-81556	R56740	MIL-S-861	S40300
MIL-F-5509	A92014	MIL-R-81586 (CP-E)	R50100	MIL-S-861	S40500
MIL-F-5509	A92024	MIL-R-81588 (13V-11Cr-3Al)	R58010	MIL-S-861	S41000
MIL-F-5509	A97075	MIL-R-81588 (5Al-2.5Sn ELI)	R54523	MIL-S-861	S42200
MIL-S-5626	G41400	MIL-R-81588 (6Al-2Cb-1Ta-0.8Mo)		MIL-S-862 (302)	S30200
MIL-R-5674	A91100		R56210	MIL-S-862 (303 Se)	S30323
MIL-R-5674	A92017	MIL-R-81588 (6Al-2Sn-4Zr-2Mo)		MIL-S-862 (303)	S30300
MIL-R-5674	A92024		R54621	MIL-S-862 (304 L)	S30403
MIL-R-5674	A92117	MIL-R-81588 (6Al-4V-ELI)	R56402	MIL-S-862 (304)	S30400
MIL-R-5674	A95056	MIL-R-81588 (8Al-1Mo-1VELI)	R54810	MIL-S-862 (309)	S30900
MIL-T-5695	S30400	MIL-R-81588 (CP-F)	R50125	MIL-S-862 (310)	S31000
MIL-S-6049	G87400	MIL-S-81591	S17400	MIL-S-862 (316 L)	S31603
MIL-S-6050	G86300	MIL-S-81591 (1020)	J02000	MIL-S-862 (316)	S31600
MIL-S-6098	G87350	MIL-S-81591 (1040)	J04000	MIL-S-862 (317)	S31700
MIL-B-63573	N03360	MIL-S-81591 (1050)	J05000	MIL-S-862 (321)	S32100
MIL-S-645	K01602	MIL-S-81591 (17-4)	J92150	MIL-S-862 (347)	S34700
MIL-S-6709	K24065	MIL-S-81591 (302)	J92501	MIL-S-862 (403)	S40300
MIL-W-6712	A91100	MIL-S-81591 (304)	J92610	MIL-S-862 (405)	S40500
MIL-W-6712	A94043	MIL-S-81591 (304L)	J92620	MIL-S-862 (410)	S41000
MIL-W-6712	C11000	MIL-S-81591 (310)	J94302	MIL-S-862 (416 Se)	S41623
MIL-W-6712	C22000	MIL-S-81591 (316)	J92810	MIL-S-862 (416)	S41600
MIL-W-6712	C26800	MIL-S-81591 (321)	J92630	MIL-S-862 (420)	S42000
MIL-W-6712	C51000	MIL-S-81591 (347)	J92640	MIL-S-862 (430 F)	S43020
MIL-T-6733	G87350	MIL-S-81591 (410)	J91152	MIL-S-862 (430)	S42900
MIL-S-6758	G41300	MIL-S-81591 (420)	J91201	MIL-S-862 (430)	S43000
MIL-B-6812	A92024	MIL-S-81591 (431)	J91651	MIL-S-862 (430F Se)	S43023
MIL-E-6843 (10013)	W10013	MIL-S-81591 (440A)	J91606	MIL-S-862 (431)	S43100
MIL-T-6845 (304)	S30400	MIL-S-81591 (440C)	J91639	MIL-S-862 (440 A)	S44002
MIL-Z-7068 (Class I)	Z35543	MIL-S-81591 (IC-1030)	J03006	MIL-S-862 (440 B)	S44003
MIL-Z-7068 (Class II)	Z35542	MIL-S-81591 (IC-303)	J92511	MIL-S-862 (440 C)	S44004
MIL-T-7081	A96061	MIL-A-81596	A92024	MIL-S-862 (440 F Se)	S44023
MIL-S-7108	K32550	MIL-A-81596	A93003	MIL-S-862 (440 F)	S44020
MIL-S-7393	G33106	MIL-A-81596	A95052	MIL-S-862 (446)	S44600
MIL-S-7420	G52986	MIL-A-81596	A95056	MIL-S-866 (1016)	G10160
MIL-S-763	S41623	MIL-T-81915	R50250	MIL-S-866 (8615)	G86150
MIL-S-7720	S30200	MIL-T-81915	R54520	MIL-S-867 (I)	J92650
MIL-S-7720 (316)	S31600	MIL-T-81915	R54620	MIL-S-867 (II)	J92730
MIL-N-7786	N07750	MIL-T-81915	R56400	MIL-S-867 (III)	J92910
MIL-B-7883 (BAg-1)	P07450	MIL-T-8231	C65500	MIL-S-8690	G86200
MIL-B-7883 (BAg-13)	P07540	MIL-W-82598	C16200	MIL-S-8699	K23080
MIL-B-7883 (BAg-19)	P07925	MIL-S-83135	G43406	MIL-S-870	J12520
MIL-B-7883 (BAg-1a)	P07500	MIL-F-83142	R50700	MIL-S-872 (A)	K12220
MIL-B-7883 (BAg-2)	P07350	MIL-F-83142	R54520	MIL-S-872 (B)	K13020
MIL-B-7883 (BAg-23)	P07850	MIL-F-83142	R54521	MIL-W-8777	C11100
MIL-B-7883 (BAg-3)	P07501	MIL-F-83142	R54550	MIL-R-8814	A92117
MIL-B-7883 (BCuP-3)	C55281	MIL-F-83142	R54620	MIL-R-8814	A95056
MIL-B-7883 (BCuP-5)	C55284	MIL-F-83142	R54790	MIL-S-8840	S35000
MIL-B-7883 (BNi-1)	N99600	MIL-F-83142	R54810	MIL-S-8840	S35500
MIL-B-7883 (BNi-1a)	N99610	MIL-F-83142	R56400	MIL-S-8844 (3)	K44220
MIL-B-7883 (BNi-2)	N99620	MIL-F-83142	R56401	MIL-M-8916	M13312
MIL-B-7883 (BNi-4)	N99640	MIL-F-83142	R56620	MIL-M-8917	M13210
MIL-B-7883 (BNi-5)	N99650	MIL-F-83142	R56740	MIL-S-8949	K24728
MIL-B-7883B (BNi-3)	N99630	MIL-F-83142	R58010	MIL-S-8955	S15700
MIL-S-7947	G10950	MIL-F-83142	R58030	MIL-W-8957	K08700
MIL-S-7952	G10200	MIL-F-83142	R58640	MIL-S-8967	S43100
MIL-S-7952	G10250	MIL-F-83142	R58820	MIL-T-9046	R50400
MIL-C-81021	C17500	MIL-S-83311	S45500	MIL-T-9046	R50550
MIL-W-81044	C11100	MIL-W-85	A91100	MIL-T-9046	R50700
MIL-W-81381	C11100	MIL-W-85	A96061	MIL-T-9046	R54520

Cross Reference Specifications	Unified Number
MIL-T-9046	R54521
MIL-T-9046	R54560
MIL-T-9046	R54620
MIL-T-9046	R54810
MIL-T-9046	R56080
MIL-T-9046	R56210
MIL-T-9046	R56320
MIL-T-9046	R56400
MIL-T-9046	R56401
MIL-T-9046	R56430
MIL-T-9046	R56620
MIL-T-9046	R58010
MIL-T-9046	R58030
MIL-T-9046	R58640
MIL-T-9046	R58820
MIL-T-9047	R50700
MIL-T-9047	R54520
MIL-T-9047	R54521
MIL-T-9047	R54560
MIL-T-9047	R54620
MIL-T-9047	R54790
MIL-T-9047	R54810
MIL-T-9047	R56210
MIL-T-9047	R56260
MIL-T-9047	R56320
MIL-T-9047	R56400
MIL-T-9047	R56401
MIL-T-9047	R56620
MIL-T-9047	R56740
MIL-T-9047	R58010
MIL-T-9047	R58030
MIL-T-9047	R58640
MIL-T-9047	R58820
MIL-S-980	K19964
MIL-S-980 (52100)	G52986
MIL-C-I5345 (Alloy 13)	C95400

Cross Index of Commonly Known Documents Which Describe Materials Same as or Similar to Those Covered by UNS Numbers

SAE (Society of Automotive Engineers) "J" NUMBERS
(The Symbol **ob** Indicates an Obsolete UNS Number.)

Cross Reference Specifications	Unified Number
SAE	
J113	K06501
J118 (1009)	G10090
J118 (1011)	G10110
J118 (1033)	G10330
J118 (1034)	G10340
J118 (1059)	G10590
J118 (1109)	G11090
J118 (1119)	G11190
J118 (1132)	G11320
J118 (1145)	G11450
J118 (1518)	G15180
J118 (1525)	G15250
J118 (1547)	G15470
J125 (G4000)	F10008
J132	K15047
J157	K15590
J158 (M3210)	F20000
J158 (M4504)	F20001
J158 (M5003)	F20002
J158 (M5503)	F20003
J158 (M7002)	F20004
J158 (M8501)	F20005
J172	K06701
J178	K08500
J217	S17700
J230	S30200
J316	K06701
J316	K07001
J403 12L13	G12134
J403 (12L14)	G12144
J403 (1005)	G10050
J403 (1006)	G10060
J403 (1008)	G10080
J403 (1010)	G10100 ob
J403 (1012)	G10120
J403 (1013)	G10130
J403 (1015)	G10150
J403 (1016)	G10160
J403 (1017)	G10170
J403 (1018)	G10180
J403 (1019)	G10190
J403 (1020)	G10200
J403 (1021)	G10210
J403 (1022)	G10220
J403 (1023)	G10230
J403 (1025)	G10250
J403 (1026)	G10260
J403 (1029)	G10290
J403 (1030)	G10300
J403 (1035)	G10350
J403 (1037)	G10370
J403 (1038)	G10380
J403 (1039)	G10390
J403 (1040)	G10400
J403 (1042)	G10420
J403 (1043)	G10430
J403 (1044)	G10440
J403 (1045)	G10450
J403 (1046)	G10460
J403 (1049)	G10490
J403 (1050)	G10500
J403 (1053)	G10530
J403 (1055)	G10550
J403 (1060)	G10600
J403 (1064)	G10640
J403 (1065)	G10650
J403 (1069)	G10690
J403 (1070)	G10700
J403 (1074)	G10740
J403 (1075)	G10750
J403 (1078)	G10780
J403 (1080)	G10800
J403 (1084)	G10840
J403 (1085)	G10850
J403 (1086)	G10860
J403 (1090)	G10900
J403 (1095)	G10950
J403 (1108)	G11080
J403 (1110)	G11100
J403 (1117)	G11170
J403 (1118)	G11180
J403 (1123)	G11230
J403 (1137)	G11370
J403 (1139)	G11390
J403 (1140)	G11400
J403 (1141)	G11410
J403 (1144)	G11440
J403 (1146)	G11460
J403 (1151)	G11510
J403 (1211)	G12110
J403 (1212)	G12120
J403 (1213)	G12130
J403 (1215)	G12150
J403 (1513)	G15130
J403 (1522)	G15220
J403 (1524)	G15240
J403 (1526)	G15260
J403 (1527)	G15270
J403 (1533)	G15330
J403 (1534)	G15340
J403 (1536)	G15360
J403 (1541)	G15410
J403 (1547)	G15470
J403 (1548)	G15480
J403 (1551)	G15510
J403 (1552)	G15520
J403 (1553)	G15530
J403 (1561)	G15610
J403 (1566)	G15660
J403 (1570)	G15700
J403 (1572)	G15720
J403 (1580)	G15800
J403 (1590)	G15900
J403 (EX15)	G41200
J403 (EX24)	G41210
J403 (EX30)	G47150
J404 (50B40)	G50401
J404 (50B44)	G50441
J404 (50B46)	G50461
J404 (50B50)	G50501
J404 (50B60)	G50601
J404 (51B60)	G51601
J404 (81B45)	G81451
J404 (86B45)	G86451
J404 (94B15)	G94151
J404 (94B17)	G94171
J404 (94B30)	G94301
J404 (1330)	G13300
J404 (1335)	G13350
J404 (1340)	G13400
J404 (1345)	G13450
J404 (4023)	G40230
J404 (4024)	G40240
J404 (4027)	G40270
J404 (4028)	G40280
J404 (4032)	G40320
J404 (4037)	G40370
J404 (4042)	G40420
J404 (4047)	G40470
J404 (4118)	G41180
J404 (4130)	G41300
J404 (4135)	G41350
J404 (4137)	G41370
J404 (4140)	G41400
J404 (4142)	G41420
J404 (4145)	G41450
J404 (4147)	G41470
J404 (4150)	G41500
J404 (4161)	G41610
J404 (4320)	G43200
J404 (4340)	G43400
J404 (4419)	G44190
J404 (4422)	G44220
J404 (4427)	G44270
J404 (4615)	G46150
J404 (4617)	G46170
J404 (4620)	G46200
J404 (4621)	G46210
J404 (4626)	G46260
J404 (4718)	G47180
J404 (4720)	G47200
J404 (4815)	G48150
J404 (4817)	G48170
J404 (4820)	G48200
J404 (5015)	G50150
J404 (5046)	G50460
J404 (5060)	G50600
J404 (5115)	G51150
J404 (5117)	G51170
J404 (5120)	G51200
J404 (5130)	G51300
J404 (5132)	G51320
J404 (5135)	G51350
J404 (5140)	G51400
J404 (5145)	G51450
J404 (5147)	G51470
J404 (5150)	G51500
J404 (5155)	G51550
J404 (5160)	G51600
J404 (6118)	G61180
J404 (6150)	G61500
J404 (8115)	G81150
J404 (8615)	G86150
J404 (8617)	G86170
J404 (8620)	G86200
J404 (8622)	G86220
J404 (8625)	G86250
J404 (8627)	G86270
J404 (8630)	G86300
J404 (8637)	G86370
J404 (8640)	G86400
J404 (8642)	G86420
J404 (8645)	G86450
J404 (8650)	G86500
J404 (8655)	G86550
J404 (8660)	G86600
J404 (8720)	G87200
J404 (8740)	G87400
J404 (8822)	G88220
J404 (9254)	G92540
J404 (9255)	G92550
J404 (9260)	G92600
J404 (9310)	G93106
J404 (50100)	G50986
J404 (51100)	G51986
J404 (52100)	G52986
J404 (E4340)	G43406
J405 (203 EZ)	S20300

Cross Reference Specifications	Unified Number	Cross Reference Specifications	Unified Number	Cross Reference Specifications	Unified Number
J405 (303 Cu)	S30330	J412 (1012)	G10120	J412 (1526)	G15260
J405 (303 plus X)	S30310	J412 (1013)	G10130	J412 (1527)	G15270
J405 (416 plus X)	S41610	J412 (1015)	G10150	J412 (1536)	G15360
J405 (30201)	S20100	J412 (1016)	G10160	J412 (1541)	G15410
J405 (30202)	S20200	J412 (1017)	G10170	J412 (1547)	G15470
J405 (30301)	S30100	J412 (1018)	G10180	J412 (1548)	G15480
J405 (30302)	S30200	J412 (1019)	G10190	J412 (1551)	G15510
J405 (30302 B)	S30215	J412 (1020)	G10200	J412 (1552)	G15520
J405 (30303)	S30300	J412 (1021)	G10210	J412 (1561)	G15610
J405 (30303 Se)	S30323	J412 (1022)	G10220	J412 (1566)	G15660
J405 (30304)	S30400	J412 (1023)	G10230	J412 (1572)	G15720
J405 (30304 L)	S30403	J412 (1025)	G10250	J412 (4012)	G40120
J405 (30305)	S30500	J412 (1026)	G10260	J412 (4023)	G40230
J405 (30308)	S30800	J412 (1029)	G10290	J412 (4024)	G40240
J405 (30309)	S30900	J412 (1030)	G10300	J412 (4027)	G40270
J405 (30309 S)	S30908	J412 (1034)	G10340	J412 (4028)	G40280
J405 (30310)	S31000	J412 (1035)	G10350	J412 (4032)	G40320
J405 (30310 S)	S31008	J412 (1037)	G10370	J412 (4037)	G40370
J405 (30314)	S31400	J412 (1038)	G10380	J412 (4042)	G40420
J405 (30316)	S31600	J412 (1039)	G10390	J412 (4047)	G40470
J405 (30316 L)	S31603	J412 (1040)	G10400	J412 (4118)	G41180
J405 (30317)	S31700	J412 (1042)	G10420	J412 (4130)	G41300
J405 (30321)	S32100	J412 (1043)	G10430	J412 (4135)	G41350
J405 (30330)	N08330	J412 (1044)	G10440	J412 (4137)	G41370
J405 (30347)	S34700	J412 (1045)	G10450	J412 (4140)	G41400
J405 (30348)	S34800	J412 (1046)	G10460	J412 (4142)	G41420
J405 (30384)	S38400	J412 (1049)	G10490	J412 (4145)	G41450
J405 (30385)	S38500 ob	J412 (1050)	G10500	J412 (4147)	G41470
J405 (51403)	S40300	J412 (1053)	G10530	J412 (4150)	G41500
J405 (51405)	S40500	J412 (1055)	G10550	J412 (4161)	G41610
J405 (51409)	S40900	J412 (1059)	G10590	J412 (4320)	G43200
J405 (51410)	S41000	J412 (1060)	G10600	J412 (4340)	G43400
J405 (51414)	S41400	J412 (1064)	G10640	J412 (4419)	G44190
J405 (51416)	S41600	J412 (1065)	G10650	J412 (4422)	G44220
J405 (51416 Se)	S41623	J412 (1069)	G10690	J412 (4427)	G44270
J405 (51420)	S42000	J412 (1070)	G10700	J412 (4615)	G46150
J405 (51420F Se)	S42023	J412 (1074)	G10740	J412 (4617)	G46170
J405 (51420F)	S42020	J412 (1075)	G10750	J412 (4620)	G46200
J405 (51429)	S42900	J412 (1078)	G10780	J412 (4621)	G46210
J405 (51430)	S43000	J412 (1080)	G10800	J412 (4626)	G46260
J405 (51430 F)	S43020	J412 (1084)	G10840	J412 (4718)	G47180
J405 (51430F Se)	S43023	J412 (1085)	G10850	J412 (4720)	G47200
J405 (51431)	S43100	J412 (1086)	G10860	J412 (4815)	G48150
J405 (51434)	S43400	J412 (1090)	G10900	J412 (4817)	G48170
J405 (51436)	S43600	J412 (1095)	G10950	J412 (4820)	G48200
J405 (51440 A)	S44002	J412 (1108)	G11080	J412 (5015)	G50150
J405 (51440 B)	S44003	J412 (1109)	G11090	J412 (5046)	G50460
J405 (51440 C)	S44004	J412 (1110)	G11100	J412 (5130)	G51300
J405 (51440 F Se)	S44023	J412 (1116)	G11160	J412 (5132)	G51320
J405 (51440 F)	S44020	J412 (1117)	G11170	J412 (5135)	G51350
J405 (51442)	S44200	J412 (1118)	G11180	J412 (5140)	G51400
J405 (51446)	S44600	J412 (1119)	G11190	J412 (5145)	G51450
J405 (51501)	S50100	J412 (1132)	G11320	J412 (5147)	G51470
J405 (51502)	S50200	J412 (1137)	G11370	J412 (5150)	G51500
J412 (12L14)	G12144	J412 (1140)	G11400	J412 (5155)	G51550
J412 (50B40)	G50401	J412 (1141)	G11410	J412 (5160)	G51600
J412 (50B44)	G50441	J412 (1144)	G11440	J412 (6150)	G61500
J412 (50B46)	G50461	J412 (1145)	G11450	J412 (8630)	G86300
J412 (50B50)	G50501	J412 (1146)	G11460	J412 (8637)	G86370
J412 (50B60)	G50601	J412 (1151)	G11510	J412 (8640)	G86400
J412 (51B60)	G51601	J412 (1215)	G12150	J412 (8642)	G86420
J412 (81B45)	G81451	J412 (1330)	G13300	J412 (8645)	G86450
J412 (86B45)	G86451	J412 (1335)	G13350	J412 (8650)	G86500
J412 (94B30)	G94301	J412 (1340)	G13400	J412 (8655)	G86550
J412 (303 plus X)	S30310	J412 (1345)	G13450	J412 (8660)	G86600
J412 (1005)	G10050	J412 (1513)	G15130	J412 (8740)	G87400
J412 (1005)	G10060	J412 (1518)	G15180	J412 (9254)	G92540
J412 (1008)	G10080	J412 (1522)	G15220	J412 (9255)	G92550
J412 (1010)	G10100 ob	J412 (1524)	G15240	J412 (9260)	G92600
J412 (1011)	G10110	J412 (1525)	G15250	J412 (30330)	N08330

Cross Reference Specifications	Unified Number	Cross Reference Specifications	Unified Number	Cross Reference Specifications	Unified Number
J412 (50100)	G50986	J431 (G3500c)	F10011	J452 (34)	A02220
J412 (51100)	G51986	J431 (G4000)	F10008	J452 (35)	A04430
J412 (51410)	S41000	J431 (G4000d)	F10012	J452 (38)	A02950
J412 (52100)	G52986	J434 (D4018)	F32800	J452 (39)	A02420
J414 (12L14)	G12144	J434 (D4512)	F33100	J452 (303)	A03840
J414 (1006)	G10060	J434 (D5506)	F33800	J452 (304)	A34430
J414 (1008)	G10080	J434 (D7003)	F34800	J452 (305)	A14130
J414 (1010)	G10100 ob	J434 (DQ and T)	F30000	J452 (306)	A13800
J414 (1012)	G10120	J435 (0022)	J01700	J452 (308)	A03800
J414 (1015)	G10150	J435 (0025)	J02507	J452 (309)	A13600
J414 (1016)	G10160	J435 (0030)	J03010	J452 (310)	A07120
J414 (1017)	G10170	J435 (0050A, 0050B)	J04501	J452 (311)	A07050
J414 (1018)	G10180	J437	T41901	J452 (312)	A07070
J414 (1019)	G10190	J437 (06)	T31506	J452 (313)	A07100
J414 (1020)	G10200	J437 (A2)	T30102	J452 (314)	A07110
J414 (1021)	G10210	J437 (D2)	T30402	J452 (315)	A07130
J414 (1022)	G10220	J437 (D3)	T30403	J452 (320)	A05140
J414 (1023)	G10230	J437 (D5)	T30405	J452 (321)	A03360
J414 (1025)	G10250	J437 (D7)	T30407	J452 (322)	A03550
J414 (1026)	G10260	J437 (H-11)	T20811	J452 (323)	A03560
J414 (1030)	G10300	J437 (H12)	T20812	J452 (324)	A05200
J414 (1035)	G10350	J437 (H13)	T20813	J452 (326)	A03190
J414 (1037)	G10370	J437 (H21)	T20821	J452 (327)	A03280
J414 (1038)	G10380	J437 (L6)	T61206	J452 (329)	A23190
J414 (1039)	G10390	J437 (M1)	T11301	J452 (331)	A03330
J414 (1040)	G10400	J437 (M2)	T11302	J452 (332)	A03320
J414 (1042)	G10420	J437 (M3)	T11313	J452 (335)	A33550
J414 (1043)	G10430	J437 (M3)	T11323	J452 (336)	A13560
J414 (1044)	G10440	J437 (M4)	T11304	J452 (380)	A02960
J414 (1045)	G10450	J437 (O1)	T31501	J452 (382)	A02010
J414 (1046)	G10460	J437 (O2)	T31502	J452 (383)	A03830
J414 (1049)	G10490	J437 (S2)	T41902	J453 (34)	A02220
J414 (1050)	G10500	J437 (S5)	T41905	J453 (34)	A02221
J414 (1055)	G10550	J437 (T1)	T12001	J453 (35)	A04430
J414 (1060)	G10600	J437 (T2)	T12002 ob	J453 (35)	A04431
J414 (1064)	G10640	J437 (T4)	T12004	J453 (38)	A02950
J414 (1065)	G10650	J437 (T5)	T12005	J453 (38)	A02952
J414 (1070)	G10700	J437 (T8)	T12008	J453 (39)	A02420
J414 (1074)	G10740	J437 (W108), (W109), (W110), (W112)		J453 (303)	A03840
J414 (1078)	G10780		T72301	J453 (303)	A03841
J414 (1080)	G10800	J437 (W209), (W210)	T72302	J453 (304)	A34430
J414 (1084)	G10840	J438 (06)	T31506	J453 (305)	A14130
J414 (1085)	G10850	J438 (A2)	T30102	J453 (305)	A14131
J414 (1086)	G10860	J438 (D2)	T30402	J453 (306)	A13800
J414 (1090)	G10900	J438 (D3)	T30403	J453 (306)	A13801
J414 (1095)	G10950	J438 (D5)	T30405	J453 (308)	A03800
J414 (1109)	G11090	J438 (D7)	T30407	J453 (309)	A13600
J414 (1117)	G11170	J438 (H-11)	T20811	J453 (309)	A13601
J414 (1118)	G11180	J438 (H12)	T20812	J453 (310)	A07120
J414 (1119)	G11190	J438 (H13)	T20813	J453 (311)	A07050
J414 (1132)	G11320	J438 (H21)	T20821	J453 (311)	A07051
J414 (1137)	G11370	J438 (L6)	T61206	J453 (312)	A07070
J414 (1140)	G11400	J438 (M1)	T11301	J453 (312)	A07071
J414 (1141)	G11410	J438 (M2)	T11302	J453 (313)	A07100
J414 (1144)	G11440	J438 (M3)	T11313	J453 (314)	A07110
J414 (1145)	G11450	J438 (M3)	T11323	J453 (315)	A07130
J414 (1146)	G11460	J438 (M4)	T11304	J453 (315)	A07131
J414 (1151)	G11510	J438 (O1)	T31501	J453 (320)	A05140
J414 (1524)	G15240	J438 (O2)	T31502	J453 (320)	A05141
J414 (1536)	G15360	J438 (S-1)	T41901	J453 (321)	A03360
J414 (1541)	G15410	J438 (S2)	T41902	J453 (323)	A03560
J414 (1547)	G15470	J438 (S5)	T41905	J453 (323)	A03561
J414 (1548)	G15480	J438 (T1)	T12001	J453 (324)	A05200
J414 (1552)	G15520	J438 (T2)	T12002 ob	J453 (326)	A03190
J431 (G1800)	F10004	J438 (T4)	T12004	J453 (326)	A03191
J431 (G2500)	F10005	J438 (T5)	T12005	J453 (327)	A03280
J431 (G2500a)	F10009	J438 (T8)	T12008	J453 (327)	A03281
J431 (G3000)	F10006	J438 (W108), (W109), (W110), (W112)		J453 (329)	A23190
J431 (G3500)	F10007		T72301	J453 (329)	A23191
J431 (G3500b)	F10010	J438 (W209), (W210)	T72302	J453 (331)	A03330

Cross Reference Specifications	Unified Number	Cross Reference Specifications	Unified Number	Cross Reference Specifications	Unified Number
J453 (332)	A03320	J461 (CA114)	C11400	J461 (CA922)	C92200
J453 (332)	A03321	J461 (CA115)	C11500	J461 (CA923)	C92300
J453 (335)	A33550	J461 (CA116)	C11600	J461 (CA925)	C92500
J453 (336)	A13560	J461 (CA120)	C12000	J461 (CA927)	C92700
J453 (382)	A02010	J461 (CA122)	C12200	J461 (CA929)	C92900
J453 (382)	A02012	J461 (CA147)	C14700	J461 (CA932)	C93200
J453 (383)	A03830	J461 (CA150)	C15000	J461 (CA935)	C93500
J453 (383)	A03831	J461 (CA162)	C16200	J461 (CA937)	C93700
J454 (1060)	A91060	J461 (CA170)	C17000	J461 (CA938)	C93800
J454 (1100)	A91100	J461 (CA172)	C17200	J461 (CA943)	C94300
J454 (2011)	A92011	J461 (CA175)	C17500	J461 (CA947)	C94700
J454 (2014)	A92014	J461 (CA176)	C17600	J461 (CA948)	C94800
J454 (2017)	A92017	J461 (CA184)	C18400	J461 (CA952)	C95200
J454 (2018)	A92018	J461 (CA187)	C18700	J461 (CA953)	C95300
J454 (2024)	A92024	J461 (CA192)	C19200	J461 (CA954)	C95400
J454 (2025)	A92025	J461 (CA210)	C21000	J461 (CA955)	C95500
J454 (2117)	A92117	J461 (CA220)	C22000	J461 (CA958)	C95800
J454 (2218)	A92218	J461 (CA230)	C23000	J461 (CA962)	C96200
J454 (2219)	A92219	J461 (CA240)	C24000	J462	C87200
J454 (2618)	A92618	J461 (CA260)	C26000	J462 (CA145)	C14500
J454 (3003)	A93003	J461 (CA268)	C26800	J462 (CA836)	C83600
J454 (3004)	A93004	J461 (CA270)	C27000	J462 (CA838)	C83800
J454 (4032)	A94032	J461 (CA330)	C33000	J462 (CA852)	C85200
J454 (4043)	A94043	J461 (CA331)	C33100	J462 (CA854)	C85400
J454 (5005)	A95005	J461 (CA342)	C34200	J462 (CA856)	C85800
J454 (5050)	A95050	J461 (CA345)	C34500	J462 (CA862)	C86200
J454 (5056)	A95056	J461 (CA350)	C35000	J462 (CA863)	C86300
J454 (5083)	A95083	J461 (CA360)	C36000	J462 (CA865)	C86500
J454 (5086)	A95086	J461 (CA377)	C37700	J462 (CA874)	C87400
J454 (5154)	A95154	J461 (CA462)	C64200	J462 (CA875)	C87500
J454 (5252)	A95252	J461 (CA464)	C46400	J462 (CA878)	C87800
J454 (5254)	A95254	J461 (CA465)	C46500	J462 (CA879)	C87900
J454 (5454)	A95454	J461 (CA466)	C46600	J462 (CA903)	C90300
J454 (5456)	A95456	J461 (CA467)	C46700	J462 (CA905)	C90500
J454 (5457)	A95457	J461 (CA510)	C51000	J462 (CA907)	C90700
J454 (5652)	A95652	J461 (CA511)	C51100	J462 (CA922)	C92200
J454 (5657)	A95657	J461 (CA521)	C52100	J462 (CA923)	C92300
J454 (6053)	A96053	J461 (CA524)	C52400	J462 (CA925)	C92500
J454 (6061)	A96061	J461 (CA544)	C54400	J462 (CA927)	C92700
J454 (6063)	A96063	J461 (CA608)	C60800	J462 (CA929)	C92900
J454 (6066)	A96066	J461 (CA614)	C61400	J462 (CA932)	C93200
J454 (6070)	A96070	J461 (CA618)	C61800	J462 (CA935)	C93500
J454 (6101)	A96101	J461 (CA623)	C62300	J462 (CA937)	C93700
J454 (6151)	A96151	J461 (CA624)	C62400	J462 (CA938)	C93800
J454 (6201)	A96201	J461 (CA630)	C63000	J462 (CA939)	C93900
J454 (6262)	A96262	J461 (CA655)	C65500	J462 (CA943)	C94300
J454 (6463)	A96463	J461 (CA670)	C67000	J462 (CA947)	C94700
J454 (7001)	A97001	J461 (CA673)	C67300	J462 (CA948)	C94800
J454 (7075)	A97075	J461 (CA674)	C67400	J462 (CA952)	C95200
J454 (7079)	A97079	J461 (CA675)	C67500	J462 (CA953)	C95300
J454 (7178)	A97178	J461 (CA706)	C70600	J462 (CA954)	C95400
J454 (EC-O)	A91350	J461 (CA710)	C71000	J462 (CA955)	C95500
J455	C83520	J461 (CA715)	C71500	J462 (CA958)	C95800
J459 (791)	C54400	J461 (CA752)	C75200	J462 (CA962)	C96200
J459 (795)	C41100	J461 (CA770)	C77000	J463 (CA102)	C10200
J460	C93720	J461 (CA836)	C83600	J463 (CA110)	C11000
J460	C94320	J461 (CA838)	C83800	J463 (CA113)	C11300
J460	C94330	J461 (CA852)	C85200	J463 (CA114)	C11400
J460	C98820	J461 (CA854)	C85400	J463 (CA115)	C11500
J460	C98840	J461 (CA858)	C85800	J463 (CA116)	C11600
J460 (791)	C54400	J461 (CA862)	C86200	J463 (CA120)	C12000
J460 (795)	C41100	J461 (CA863)	C86300	J463 (CA122)	C12200
J460, Alloy 13	L53345	J461 (CA865)	C86500	J463 (CA145)	C14500
J460, No. 19	L54510	J461 (CA874)	C87400	J463 (CA147)	C14700
J460, No. 190	L54370	J461 (CA875)	C87500	J463 (CA150)	C15000
J460, No.485	L51180	J461 (CA878)	C87800	J463 (CA162)	C16200
J461	C87200	J461 (CA879)	C87900	J463 (CA170)	C17000
J461 (CA102)	C10200	J461 (CA903)	C90300	J463 (CA172)	C17200
J461 (CA110)	C11000	J461 (CA905)	C90500	J463 (CA175)	C17500
J461 (CA113)	C11300	J461 (CA907)	C90700	J463 (CA176)	C17600

Cross Reference Specifications	Unified Number	Cross Reference Specifications	Unified Number	Cross Reference Specifications	Unified Number
J463 (CA184)	C18400	J466 (ZK60A)	M16600	J770 (4718)	G47180
J463 (CA187)	C18700	J467 (17-4PH)	S17400	J770 (4720)	G47200
J463 (CA192)	C19200	J467 (17-7PH)	S17700	J770 (4815)	G48150
J463 (CA210)	C21000	J467 (19-9DL)	S63198	J770 (4817)	G48170
J463 (CA220)	C22000	J467 (19-9DX)	S63199	J770 (4820)	G48200
J463 (CA230)	C23000	J467 (422)	S42200	J770 (5015)	G50150
J463 (CA240)	C24000	J467 (A286)	S66286	J770 (5046)	G50460
J463 (CA260)	C26000	J467 (AM-350)	S35000	J770 (5060)	G50600
J463 (CA268)	C26800	J467 (AM-355)	S35500	J770 (5115)	G51150
J463 (CA270)	C27000	J467 (D-979)	N09979	J770 (5120)	G51200
J463 (CA330)	C33000	J467 (Discaloy)	S66220	J770 (5130)	G51300
J463 (CA331)	C33100	J467 (Greek Ascoloy)	S41800	J770 (5132)	G51320
J463 (CA342)	C34200	J467 (H-11)	T20811	J770 (5135)	G51350
J463 (CA345)	C34500	J467 (H12)	T20812	J770 (5140)	G51400
J463 (CA350)	C35000	J467 (H13)	T20813	J770 (5145)	G51450
J463 (CA360)	C36000	J467 (PH15-7-Mo)	S15700	J770 (5147)	G51470
J463 (CA377)	C37700	J467 (W545)	S66545	J770 (5150)	G51500
J463 (CA464)	C46400	J468 (903) (Castings) (Ingot)	Z33520	J770 (5155)	G51550
J463 (CA465)	C46500	J468 (925) (Castings) (Ingot)	Z35531	J770 (5160)	G51600
J463 (CA466)	C46600	J469 (921)	Z35542	J770 (6118)	G61180
J463 (CA467)	C46700	J470 (4427)	G44270	J770 (6150)	G61500
J463 (CA510)	C51000	J473 (Alloy 9B)	L54250	J770 (8115)	G81150
J463 (CA511)	C51100	J473, Grade 1B	L54940	J770 (8615)	G86150
J463 (CA521)	C52100	J473, No. 2B	L54905	J770 (8617)	G86170
J463 (CA524)	C52400	J473, No. 3B	L54815	J770 (8620)	G86200
J463 (CA544)	C54400	J473, No. 5B	L54610	J770 (8622)	G86220
J463 (CA608)	C60800	J473, No. 6B	L54555	J770 (8625)	G86250
J463 (CA614)	C61400	J770 (50B40)	G50401	J770 (8627)	G86270
J463 (CA618)	C61800	J770 (50B44)	G50441	J770 (8630)	G86300
J463 (CA623)	C62300	J770 (50B46)	G50461	J770 (8637)	G86370
J463 (CA624)	C62400	J770 (50B50)	G50501	J770 (8640)	G86400
J463 (CA630)	C63000	J770 (50B60)	G50601	J770 (8642)	G86420
J463 (CA642)	C64200	J770 (51B60)	G51601	J770 (8645)	G86450
J463 (CA655)	C65500	J770 (81B45)	G81451	J770 (8650)	G86500
J463 (CA670)	C67000	J770 (86B45)	G86451	J770 (8655)	G86550
J463 (CA673)	C67300	J770 (94B15)	G94151	J770 (8660)	G86600
J463 (CA674)	C67400	J770 (94B17)	G94171	J770 (8720)	G87200
J463 (CA675)	C67500	J770 (94B30)	G94301	J770 (8740)	G87400
J463 (CA706)	C70600	J770 (1330)	G13300	J770 (8822)	G88220
J463 (CA710)	C71000	J770 (1335)	G13350	J770 (9254)	G92540
J463 (CA715)	C71500	J770 (1340)	G13400	J770 (9255)	G92550
J463 (CA752)	C75200	J770 (1345)	G13450	J770 (9260)	G92600
J463 (CA770)	C77000	J770 (4012)	G40120	J770 (9310)	G93106
J465 (AM60-A)	M10600	J770 (4023)	G40230	J770 (50100)	G50986
J465 (AM100-A)	M10100	J770 (4024)	G40240	J770 (51100)	G51986
J465 (AS41-A)	M10410	J770 (4027)	G40270	J770 (52100)	G52986
J465 (AZ63A)	M11630	J770 (4028)	G40280	J770 (E4340)	G43406
J465 (AZ91A)	M11910	J770 (4032)	G40320	J775 (EV-11)	S63011
J465 (AZ91B)	M11912	J770 (4037)	G40370	J775 (EV-12)	S63012
J465 (AZ91C)	M11914	J770 (4042)	G40420	J775 (EV-3)	S63016
J465 (AZ92A)	M11920	J770 (4047)	G40470	J775 (EV-4)	S63017
J465 (EZ33A)	M12330	J770 (4118)	G41180	J775 (EV-5)	S63014
J465 (HK31A)	M13310	J770 (4130)	G41300	J775 (EV-6)	S63015
J465 (HZ32A)	M13320	J770 (4135)	G41350	J775 (EV-7)	S63007
J465 (QE22A)	M18220	J770 (4137)	G41370	J775 (EV-8)	S63008
J465 (ZE41A)	M16410	J770 (4140)	G41400	J775 (EV-9)	S66009
J465 (ZE63A)	M16630	J770 (4142)	G41420	J775 (HEV2)	N07002
J465 (ZH62A)	M16620	J770 (4145)	G41450	J775 (HNV-1)	S64005
J465 (ZK51A)	M16510	J770 (4147)	G41470	J775 (HNV-2)	S64006
J465 (ZK61A)	M16610	J770 (4150)	G41500	J775 (HNV-3)	S65007
J466 (AZ31B)	M11311	J770 (4161)	G41610	J775 (HNV-5)	S63005
J466 (AZ61A)	M11610	J770 (4320)	G43200	J775 (HNV-6)	S65006
J466 (AZ80A)	M11800	J770 (4340)	G43400	J775 (VF3)	N06005
J466 (HK31A)	M13310	J770 (4419)	G44190	J775 (VF4)	N06782
J466 (HM21A)	M13210	J770 (4422)	G44220	J775 (VF5)	R30002
J466 (HM31A)	M13312	J770 (4615)	G46150	J775 (VF6)	R30001
J466 (LA141A)	M14141	J770 (4617)	G46170	J775 (VF7)	R30012
J466 (M1A)	M15100	J770 (4620)	G46200	J778 (94B40)	G94401
J466 (ZE10A)	M16100	J770 (4621)	G46210	J778 (3140)	G31400
J466 (ZK40A)	M16400	J770 (4626)	G46260	J778 (3310)	G33106

SAE (Society of Automotive Engineers) "J" Numbers

Cross Reference Specifications	Unified Number	Cross Reference Specifications	Unified Number
J778 (4012)	G40120	J1268 (5135 H)	H51350
J778 (4063)	G40630	J1268 (5140 H)	H51400
J778 (4337)	G43370	J1268 (5147 H)	H51470
J778 (6120)	G61200	J1268 (5150 H)	H51500
J778 (8735)	G87350	J1268 (5155 H)	H51550
J778 (8742)	G87420	J1268 (5160 H)	H51600
J778 (9262)	G92620	J1268 (6118 H)	H61180
J778 (9840)	G98400	J1268 (6150 H)	H61500
J778 (9850)	G98500	J1268 (8617 H)	H86170
J1099 (A538A)	K92820	J1268 (8620 H)	H86200
J1099 (A538B)	K92890	J1268 (8622 H)	H86220
J1099 (A538C)	K93120	J1268 (8625 H)	H86250
J1268 (15B21 H)	H15211	J1268 (8627 H)	H86270
J1268 (15B35 H)	H15351	J1268 (8630 H)	H86300
J1268 (15B37 H)	H15371	J1268 (8637 H)	H86370
J1268 (15B41 H)	H15411	J1268 (8640 H)	H86400
J1268 (15B48 H)	H15481	J1268 (8642 H)	H86420
J1268 (15B62 H)	H15621	J1268 (8645 H)	H86450
J1268 (50B40 H)	H50401	J1268 (8650 H)	H86500
J1268 (50B44 H)	H50441	J1268 (8655 H)	H86550
J1268 (50B46 H)	H50461	J1268 (8660 H)	H86600
J1268 (50B50 H)	H50501	J1268 (8720 H)	H87200
J1268 (50B60 H)	H50601	J1268 (8740 H)	H87400
J1268 (51B60 H)	H51601	J1268 (8822 H)	H88220
J1268 (81B45 H)	H81451	J1268 (9260 H)	H92600
J1268 (86B30 H)	H86301	J1268 (9310 H)	H93100
J1268 (86B45 H)	H86451	J1268 (E4340 H)	H43406
J1268 (94B15 H)	H94151		
J1268 (94B17 H)	H94171		
J1268 (94B30 H)	H94301		
J1268 (1038 H)	H10380		
J1268 (1045 H)	H10450		
J1268 (1330 H)	H13300		
J1268 (1335 H)	H13350		
J1268 (1340 H)	H13400		
J1268 (1345 H)	H13450		
J1268 (1522 H)	H15220		
J1268 (1524 H)	H15240		
J1268 (1526 H)	H15260		
J1268 (1541 H)	H15410		
J1268 (4027 H)	H40270		
J1268 (4028 H)	H40280		
J1268 (4032 H)	H40320		
J1268 (4037 H)	H40370		
J1268 (4042 H)	H40420		
J1268 (4047 H)	H40470		
J1268 (4118 H)	H41180		
J1268 (4130 H)	H41300		
J1268 (4135 H)	H41350		
J1268 (4137 H)	H41370		
J1268 (4140 H)	H41400		
J1268 (4142 H)	H41420		
J1268 (4145 H)	H41450		
J1268 (4147 H)	H41470		
J1268 (4150 H)	H41500		
J1268 (4161 H)	H41610		
J1268 (4320 H)	H43200		
J1268 (4340 H)	H43400		
J1268 (4419 H)	H44190		
J1268 (4620 H)	H46200		
J1268 (4621 H)	H46210		
J1268 (4718 H)	H47180		
J1268 (4720 H)	H47200		
J1268 (4815 H)	H48150		
J1268 (4817 H)	H48170		
J1268 (4820 H)	H48200		
J1268 (5046 H)	H50460		
J1268 (5120 H)	H51200		
J1268 (5130 H)	H51300		
J1268 (5132 H)	H51320		

Index by Common Trade Designations

INDEX BY COMMON TRADE DESIGNATIONS

COMMON TRADE DESIGNATION	UNS NUMBER	COMMON TRADE DESIGNATION	UNS NUMBER	COMMON TRADE DESIGNATION	UNS NUMBER
Electrotype—General	L52830	Hastelloy X	W86002	Lanston Special Case Alloy	L53685
Elgiloy	R30003	Havar	R30004	Lapelloy	S42300
ER 349	S63197	Haynes 36	R30036	Lead Alloys	L5XXXX
Esshete 1250	S21500	Haynes No. 20 Mod.	N08320	Linotype-Special	L53425
ETP Copper	C11000	Hiduminium	A02030	Linotype Alloy	L53420
F-10	S33100	HP9-4-20	K91472	Linotype B	L53455
Ferralium 255	S32550	HP9-4-25	K91122	Lithium Alloys	L06XXX
Frary Metal	L50540	HP9-4-30	K91283	M-1 Tool Steel	T11301
Frary Metal	L50542	HS-188	R30188	M-2 Tool Steel	T11302
Frary Metal	L50543	HS-556	R30556	M-3 Class 1 Tool Steel	T11313
Frary Metal	L50541	HWT	S43035	M-3 Class 2 Tool Steel	T11323
FRSTP Copper	C13000	HY-30	K51255	M-4 Tool Steel	T11304
FRSTP Copper	C12700	HY-80	K31820	M-6 Tool Steel	T11306
FRSTP Copper	C12800	HY-100	K32045	M-7 Tool Steel	T11307
FRSTP Copper	C12900	HY-80, CAST	J42015	M-30 Tool Steel	T11330
FRSTP Copper	C12500	HY-100, CAST	J42240	M-33 Tool Steel	T11333
Gold Alloys	P00XXX	IN 102	N06102	M-34 Tool Steel	T11334
Greek Ascoloy	S41800	IN 713	N07713	M-36 Tool Steel	T11336
Greek Ascoloy	S41880	Incoloy 800	N08800	M-41 Tool Steel	T11341
H-10 Tool Steel	T20810	Incoloy 800H	N08810	M-42 Tool Steel	T11342
H-11 Tool Steel	T20811	Incoloy 800HT	N08811	M-46 Tool Steel	T11346
H-12 Tool Steel	T20812	Incoloy 801	N08801	M-48 Tool Steel	T11348
H-13 Tool Steel	T20813	Incoloy 802	N08802	M-50 Tool Steel	T11350
H-14 Tool Steel	T20814	Incoloy 825	N08825	M-52 Tool Steel	T11352
H-19 Tool Steel	T20819	Incoloy 901	N09901	M-61 Tool Steel	T11361
H-21 Tool Steel	T20821	Incoloy FM65	N08065	M-62 Tool Steel	T11362
H-22 Tool Steel	T20822	Inconel 600	N06600	M152	S64152
H-23 Tool Steel	T20823	Inconel 601	N06601	M220C	N03220
H-24 Tool Steel	T20824	Inconel 625	N06625	M252	N07252
H-26 Tool Steel	T20826	Inconel 690	N06690	MA754	N07754
H-42 Tool Steel	T20842	Inconel 702	N07702	Magnesium Alloys	M1XXXX
Hastelloy B	N10001	Inconel 706	N09706	Manganese Alloys	M2XXXX
Hastelloy B	W80001	Inconel 718	N07718	MF202	C50710
Hastelloy B-2	W80665	Inconel FM62	N06062	MF202	C50715
Hastelloy B-2	N10665	Inconel FM69	N07069	MI88	W21380
Hastelloy C	W80002	Inconel FM82	N06082	Molybdenum Alloys	R03XXX
Hastelloy C	N10002	Inconel FM92	N07092	Monel 400	N04400
Hastelloy C-4	N06455	Inconel MA754	N07754	Monel 401	N04401
Hastelloy C-22	W86022	Inconel X-750	N07750	Monel 404	N04404
Hastelloy C-22	N06022	Incramet 800	C99300	Monel 502	N05502
Hastelloy C-276	N10276	Incramute 1	C99600	Monel FM60	N04060
Hastelloy C-276	W80276	JS 700	N08700	Monel K-500	N05500
Hastelloy F	N06001	KCR-D183	J93183	Monel R-405	N04405
Hastelloy G	N06007	KCR-D283	J93550	MP-35-N	R30035
Hastelloy G	W86007	Kirksite A	Z35543	MP-159	R30159
Hastelloy G-2	N06975	Kirksite B	Z35542	N-155	R30155
Hastelloy G-3	N06985	Kirksite I	Z35543	N100	N13100
Hastelloy G-30	W86030	Kirksite II	Z35542	NAX 9115-AC	K11914
Hastelloy G-30	N06030	KOVAR	K94610	Ni-Span-C 902	N09902
Hastelloy N	N10003	Krupp	K51210	Nichrome	N06004
Hastelloy S	N06635	L-2 Tool Steel	T61202	Nichrome V	N06003
Hastelloy W	N10004	L-6 Tool Steel	T61206	Nickel 200	N02200
Hastelloy W	W80004	L-605	W73605	Nickel 201	N02201
Hastelloy X	N06002	L-605	R30605		

COMMON TRADE DESIGNATION	UNS NUMBER
Nickel 205	N02205
Nickel 211	N02211
Nickel 220	N02220
Nickel 225	N02225
Nickel 230	N02230
Nickel 233	N02233
Nickel 270	N02270
Nickel FM61	N02061
Nimonic 75	N06075
Nimonic 80A	N07080
Nimonic 90	N07090
Nimonic 263	N07263
Niobium (Columbium) Alloys	R04XXX
Nitralloy, Cast	J24056
Nitronic 60	S21800
Nitronic 60W	S21880
Nitronic 60W	W32540
0-1 Tool Steel	T31501
0-2 Tool Steel	T31502
0-6 Tool Steel	T31506
0-7 Tool Steel	T31507
OF Copper	C10200
OFLP Copper	C10800
OFS Copper	C10500
OFS Copper	C10400
OFS Copper	C10700
OFXLP Copper	C10300
P-6 Tool Steel	T51606
P-20 Tool Steel	T51620
P-21 Tool Steel	T51621
Palladium Alloys	P03XXX
Permanickel 300	N03300
Pewter	L13911
Pewter	L13912
PH 13-8 MO	S13800
PH 13-8 MO	S13889
PH 14-8 MO	S14800
PH 15-7 MO	S15789
PH 15-7 MO	S15780
PH 15-7 MO	S15700
Platinum Alloys	P04XXX
Plutonium Alloys	M05XXX
Pyromet 31	N07031
Pyromet X-15	S65150
Pyrowear Alloy 53	K71040
RA 330	N08330
RA 330TX	N08332
RA 333	N06333
Rare Earth Metals	EXXXXX
Rene 41	N07041
Rhodium Alloys	P05XXX
Rules Monotype Alloy . . .	L53580
S-1 Tool Steel	T41901

COMMON TRADE DESIGNATION	UNS NUMBER
S-2 Tool Steel	T41902
S-4 Tool Steel	T41904
S-5 Tool Steel	T41905
S-6 Tool Steel	T41906
S-7 Tool Steel	T41907
S-590	R30590
S-816	R30816
SAF 2304	S32304
Sanicro 28	N08028
SC-1	S44660
Silver Alloys	P07XXX
Stainless W	S17600
Stellite 1	R30001
Stellite 6	W73006
Stellite 6	R30006
Stellite 12	R30012
Stellite 12	W73012
Stellite 21	R30021
Stellite 23	R30023
Stellite 27	R30027
Stellite 30	R30030
Stellite 31	R30031
Stellite F	R30002
Stereotype—Flat	L53530
Stereotype—General	L53510
STP Copper	C11300
STP Copper	C11400
STP Copper	C11500
STP Copper	C11600
T-1 Tool Steel	T12001
T-4 Tool Steel	T12004
T-5 Tool Steel	T12005
T-6 Tool Steel	T12006
T-8 Tool Steel	T12008
T-15 Tool Steel	T12015
Tantalum Alloys	R05XXX
Tenelon	S21400
Tin Alloys	L13XXX
Titanium Alloys	R5XXXX
TPM	N07002
Tungsten (Wolfram) Alloys	R07XXX
Udimet 500	N07500
URANUS 50	S32404
Valve Steel 1	S65007
Valve Steel 2	S64005
Valve Steel 10	S63014
Valve Steel 10N	S63015
Valve Steel 21-2N	S63012
Valve Steel 21-4N	S63008
Valve Steel 21-12	S63016
Valve Steel 21-12N	S63017
Valve Steel 21-55N	S63007
Valve Steel 746	S63011

COMMON TRADE DESIGNATION	UNS NUMBER
Valve Steel CNS	S63005
Valve Steel F	S64006
Valve Steel GAMMAN H .	S63013
Valve Steel TPA	S66009
Valve Steel XB	S65006
W-1 Tool Steel	T72301
W-2 Tool Steel	T72302
W-5 Tool Steel	T72305
W545	S66545
Waspaloy	N07001
White Metal Bearing Alloy	L53332
White Metal Bearing Alloy	L53585
White Metal Bearing Alloy	L53565
Wolfram (Tungsten) Alloys	R07XXX
X-782	N06782
ZA-8	Z35635
ZA-27	Z35840
Zinc Alloys	ZXXXXX
Zirconium Alloys	R6XXXX

APPENDIX

HISTORY OF THE UNIFIED NUMBERING SYSTEM FOR METALS AND ALLOYS

In 1967, the Society of Automotive Engineers (SAE) and the American Society for Testing and Materials (ASTM) began to explore the possibility of developing a unified numbering system for metals and alloys. It had long been recognized that a single orderly system of designating metals and alloys would offer many advantages over the often-confusing array of designation systems which had developed independently during the past 60 years. In addition, many alloys were known only by trade names since there was no central organization to assign numbers. A further complication developed when various companies produced the same alloy under different trade names. Several of the major trade associations were also in the process of considering important changes to their numbering systems since they were no longer adequate to accommodate the needs of the computer age and the influx of new alloys.

The U.S. Army was interested in the idea and in May 1969, a contract was issued by the Army Materials and Mechanics Research Center to the Society of Automotive Engineers to conduct a "Feasibility Study of a Unified Numbering System for Metals and Alloys." This project was jointly sponsored by SAE and ASTM, and a committee was appointed to conduct the study. The committee consisted of the following members under the direction of N. L. Mochel, a past president of ASTM:

> Herbert F. Campbell, Army Materials and Mechanics Research Center
> Harold M. Cobb, American Society for Testing and Materials
> Alvin G. Cook, Allegheny Ludlum Steel Corp.
> Henry B. Fernald, Jr., Technical Consultant
> Muir L. Frey, Engineering Consultant
> John Gadbut, The International Nickel Co.
> S. T. Main, Grumman Aircraft Corp.
> Norman L. Mochel, Engineering Consultant
> R. Thomas Northrup, Society of Automotive Engineers
> Bruce A. Smith, General Motors Engineering Staff
> Harry H. Stout, Phelps Dodge Copper Products Corp.

Many individuals were consulted in the course of the 18-month feasibility study. The major trade associations concerned with numbering systems, including the Aluminum Association, The American Iron and Steel Institute, The Copper Development Association, and The Steel Founders' Society of America, were also consulted since it was recognized at the outset that any new system could be successful only if these organizations were in general agreement with the concept. In January 1971, the study was completed and a report was issued to the Army stating that a unified numbering system for metals was feasible and desirable. This report included a general proposal of how such a system could be established to provide a coherent designation system for all present and future metals and alloys.

In April 1972, SAE and ASTM established an Advisory Board to further develop and refine the proposed unified numbering system. The Advisory Board consisted of the following members:

> Chairman: Bruce A. Smith, General Motors Engineering Staff
> Secretaries: Harold M. Cobb, ASTM Staff
>   R. Thomas Northrup, SAE Staff
> John Artman, Defense Industrial Supply Center
> Lawrence H. Bennett, National Bureau of Standards
> Herbert F. Campbell, Army Materials and Mechanics Research Center
> Alvin G. Cook, Allegheny Ludlum Steel Corp.
> Henry B. Fernald, Jr., Technical Consultant
> John Gadbut, International Nickel Co.
> Joseph M. Engel, Republic Steel Corp.
>   (representing American Iron and Steel Institute)
> W. Stuart Lyman, Copper Development Association
> Robert E. Lyons, Federal Supply Service
> Norman L. Mochel, Consultant-Metallurgy

Edward F. Parker, General Electric Co.
Richard R. Senz, The Aluminum Co. of America
(representing Aluminum Association)
Whitney Snyder, American Motors Corp.
Rudolph Zillman, Steel Founders Society of America

The Advisory Board completed "SAE/ASTM Recommended Practice for Numbering Metals and Alloys" in March 1974 and subsequently coordinated the establishment of specific designations for over 1000 steel, stainless steel, tool steel, super-alloys, aluminum, copper, cobalt, nickel, and rare earth alloys. These were listed in the First Edition of this UNS Handbook, published in January 1975. An expanded Second Edition was published in September 1977. A further expanded Third Edition was published in June 1983.

During the period 1983-1985, further refinements were made in the basic numbering system. UNS numbers were assigned to alloys of each category and many new references were added to the system. The Advisory Board Membership consisted of the following:

Chairman: Alvin G. Cook, Allegheny Ludlum Steel Corp. (Retired)
Secretaries: Earl R. Sullivan, ASTM Staff
             Patricia Couhig, SAE Staff
Lawrence H. Bennett, National Bureau of Standards
Arthur Cohen, Copper Development Association
James A. Conway, U.S. Army Materials Technology Laboratory
Calvin J. Cooley, American Iron & Steel Institute
John R. Cuthill, National Bureau of Standards
Louis S. Falcone, Federal Supply Service
J. G. Gensure, General Electric Co.
Ted Kirk, INCO Alloys International, Inc.
E. J. Kubel, Jr., Steel Founders Society of America
J. W. Tackett, Cabot Corp.
R. David Thomas, R. D. Thomas & Co.
(representing American Welding Society)
J. Douglas Yerger, The Aluminum Co. of America
(representing Aluminum Association)

Work is continuing toward the establishment of designations for other metals, and these will be listed in future editions.

SAE and ASTM gratefully acknowledge the efforts of the many individuals who contributed to this important project.

NUMBERING METALS AND ALLOYS—
SAE J1086 DEC85 and ASTM E527 SAE and ASTM Recommended Practice

Report of the Unified Numbering System Advisory Board, approved August 1974, revised April 1983, corrected December 1985.

UNS designations shall not be used for metals and alloys which are not registered under the system described herein, or for any metal or alloy whose composition differs from those registered.

1. Scope

1.1 This recommended practice describes a unified numbering system (UNS) for metals and alloys which have a "commercial standing" (see Note 1), and covers the procedure by which such numbers are assigned.

φ Section 2 describes the system of alphanumeric designations or "numbers" established for each family of metals and alloys.

Section 3 outlines the organization established for administering the system.

Section 4 describes the procedure for requesting number assignment to metals and alloys for which UNS numbers have not previously been assigned.

1.2 The UNS provides a means of correlating many nationally used numbering systems currently administered by societies, trade associations, and individual users and producers of metals and alloys, thereby avoiding confusion caused by use of more than one identification number for the same material; and by the opposite situation of having the same number assigned to two or more entirely different materials. It provides, also, the uniformity necessary for efficient indexing, record keeping, data storage and retrieval, and cross referencing.

1.3 *A UNS number is not in itself a specification*, since it establishes no requirements for form, condition, quality, etc. It is a unified identification of metals and alloys for which controlling limits have been established in specifications published elsewhere. (See Note 2.)

2. Description of Numbers (or Codes) Established for Metals and Alloys

φ **2.1** The unified numbering system (UNS) establishes 18 series of numbers for metals and alloys, as shown in Table 1. Each UNS number consists of a single letter-prefix followed by five digits. In most cases the letter is suggestive of the family of metals identified, for example, φ A for aluminum, P for precious metals, S for stainless steels. Table 2 shows the secondary division of some primary series of numbers.

2.2 Whereas some of the digits in certain of the UNS number groups have special assigned meaning, each series is independent of the others in such significance; this practice permits greater flexibility and avoids complicated and lengthy UNS numbers. (See Note 3.)

2.3 Wherever feasible, identification "numbers" from existing systems are incorporated into the UNS numbers. For example: The carbon steel which is presently identified by "AISI 1020" (American Iron & Steel Institute), is covered by "UNS G10200" and the free cutting brass, which is presently identified by "CDA (Copper Development Association) φ C36000," is covered by "UNS C36000."

2.4 Welding filler metals fall into two general categories: those whose compositions are determined by the filler metal analysis (e.g., solid bare wire or rods and cast rods) and those whose composition is determined by the weld deposit analysis (e.g., covered electrodes, flux-cored and other composite wire electrodes). The latter are assigned to a new primary series with the letter W as shown in Table 1. The solid bare wire and rods continue to be assigned in the established number series according to their composition.

(Readers are cautioned *not* to make their own "assignments" of numbers from such listings, as this can result in unintended and unexpected duplication and conflict.)

2.5 The ASTM and the SAE periodically publish up-to-date listings of all UNS numbers assigned to specific metals and alloys, with appropriate reference information on each. (See Note 6.) Many trade associations also publish similar listings related to materials of primary interest to their organizations.

3. Organization for Administering Unified Numbering System for Metals and Alloys

3.1 The organization for administering the UNS consists of: (1) an advisory board, (2) several number-assigning offices, (3) a corps of volunteer consultants, and (4) staffs at ASTM and SAE. In addition, SAE and ASTM committees dealing with various groups of materials may be consulted.

φ **TABLE 1—PRIMARY SERIES OF NUMBERS**

UNS Series	Metal
Nonferrous metals and alloys	
A00001–A99999	Aluminum and aluminum alloys
C00001–C99999	Copper and copper alloys
E00001–E99999	Rare earth and rare earth-like metals and alloys (18 Items, see Table 2)
L00001–L99999	Low melting metals and alloys (14 Items, see Table 2)
M00001–M99999	Miscellaneous nonferrous metals and alloys (12 Items, see Table 2)
N00001–N99999	Nickel and nickel alloys
P00001–P99999	Precious metals and alloys (8 Items, see Table 2)
R00001–R99999	Reactive and refractory metals and alloys (14 Items, see Table 2)
Z00001–Z99999	Zinc and zinc alloys
Ferrous metals and alloys	
D00001–D99999	Specified mechanical properties steels
F00001–F99999	Cast irons
G00001–G99999	AISI and SAE carbon and alloy steels (except tool steels)
H00001–H99999	AISI H-steels
J00001–J99999	Cast steels (except tool steels)
K00001–K99999	Miscellaneous steels and ferrous alloys
S00001–S99999	Heat and corrosion resistant (stainless) steels
T00001–T99999	Tool steels
Welding filler metals	
W00001–W99999	Welding filler metals, covered and tubular electrodes, classified by weld deposit composition (see Table 2)

The φ symbol is for the convenience of the user in locating areas where technical revisions have been made to the previous issue of the report. If the symbol is next to the report title, it indicates a complete revision of the report.

UNS Series	Metal	UNS Series	Metal
E00001–E99999 Rare earth and rare earth-like metals and alloys		P00001–P99999 Precious metals and alloys	
E00000–E00999	Actinium	P00001–P00999	Gold
E01000–E20999	Cerium	P01001–P01999	Iridium
E21000–E45999	Mixed rare earths[a]	P02001–P02999	Osmium
E46000–E47999	Dysprosium	P03001–P03999	Palladium
E48000–E49999	Erbium	P04001–P04999	Platinum
E50000–E51999	Europium	P05001–P05999	Rhodium
E52000–E55999	Gadolinium	P06001–P06999	Ruthenium
E56000–E57999	Holmium	P07001–P07999	Silver
E58000–E67999	Lanthanum		
E68000–E68999	Lutetium	R00001–R99999 Reactive and refractory metals and alloys	
E69000–E73999	Neodymium	R01001–R01999	Boron
E74000–E77999	Praseodymium	R02001–R02999	Hafnium
E78000–E78999	Promethium	R03001–R03999	Molybdenum
E79000–E82999	Samarium	R04001–R04999	Niobium (Columbium)
E83000–E84999	Scandium	R05001–R05999	Tantalum
E85000–E86999	Terbium	R06001–R06999	Thorium
E87000–E87999	Thulium	R07001–R07999	Tungsten
E88000–E89999	Ytterbium	R08001–R08999	Vanadium
E90000–E99999	Yttrium	R10001–R19999	Beryllium
		R20001–R29999	Chromium
F00001–F99999 Cast irons		R30001–R39999	Cobalt
	Gray, malleable, pearlitic malleable, and ductile (nodular) cast irons	R40001–R49999	Rhenium
		R50001–R59999	Titanium
K00001–K99999 Miscellaneous steels and ferrous alloys		R60001–R69999	Zirconium
L00001–L99999 Low-melting metals and alloys		W00001–W99999 Welding filler metals, classified by weld deposit composition	
L00001–L00999	Bismuth	W00001–W09999	Carbon steel with no significant alloying elements
L01001–L01999	Cadmium	W10000–W19999	Manganese-molybdenum low alloy steels
L02001–L02999	Cesium		
L03001–L03999	Gallium	W20000–W29999	Nickel low alloy steels
L04001–L04999	Indium	W30000–W39999	Austenitic stainless steels
L50001–L59999	Lead	W40000–W49999	Ferritic stainless steels
L06001–L06999	Lithium	W50000–W59999	Chromium low alloy steels
L07001–L07999	Mercury	W60000–W69999	Copper base alloys
L08001–L08999	Potassium	W70000–W79999	Surfacing alloys
L09001–L09999	Rubidium	W80000–W89999	Nickel base alloys
L10001–L10999	Selenium		
L11001–L11999	Sodium	Z00001–Z99999 Zinc and zinc alloys	Zinc
L12001–L12999	Thallium		
L13001–L13999	Tin		
M00001–M99999 Miscellaneous nonferrous metals and alloys			
M00001–M00999	Antimony		
M01001–M01999	Arsenic		
M02001–M02999	Barium		
M03001–M03999	Calcium		
M04001–M04999	Germanium		
M05001–M05999	Plutonium		
M06001–M06999	Strontium		
M07001–M07999	Tellurium		
M08001–M08999	Uranium		
M10001–M19999	Magnesium		
M20001–M29999	Manganese		
M30001–M39999	Silicon		

[a] Alloys in which the rare earths are used in the ratio of their natural occurrence (that is, unseparated rare earths). In this mixture, cerium is the most abundant of the rare earth elements.

3.1.1 The Advisory Board has approximately 20 volunteer members who are affiliated with major producing and using industries, trade associations, government agencies, and standards societies, and who have extensive experience with identification, classification, and specification of materials. The Board is the administrative arm of SAE and ASTM on all matters pertaining to the UNS. It coordinates thinking on the format of each series of numbers and the administration of each by selected experts. It sets up ground rules for determining eligibility of any material for a UNS number, for requesting such numbers, and for appealing unfavorable rulings. It is the final referee on matters of disagreement between requesters and assigners.

3.1.2 UNS number assigners for certain materials are set up at trade associations which have successfully administered their own numbering systems; for other materials, assigners are located at the offices of SAE and ASTM. Each of these assigners has the responsibility for administering a specific series of numbers, as shown in Table 3. Each considers requests for assignment of new UNS numbers, and informs applicants of the action taken. Trade association UNS number assigners also report immediately to both SAE and ASTM details of each number assignment. ASTM and SAE assigners collaborate with designated consultants when considering requests for assignment of new numbers.

3.1.3 Consultants are selected by the Advisory Board to provide expert knowledge of a specific field of materials. Since they are utilized primarily by the Board and the SAE and ASTM number assigners, they are not listed in this recommended practice. At the request of the ASTM or SAE number assigner, a consultant considers a request for a new number in the light of the ground rules established for the material involved, decides whether a new number is justified, and informs the ASTM or SAE number assigner accordingly.

This utilization of experts (consultants and number assigners) is intended to insure prompt and fair consideration of all requests. It permits each decision to be based on current knowledge of the needs of a specific industry of producers and users.

3.1.4 Staff members at SAE and ASTM maintain duplicate master listings of all UNS numbers assigned.

3.1.5 Established SAE and ASTM committees which normally deal with standards and specifications for the materials covered by the UNS, and other knowledgeable persons, are called upon by the Advisory Board for advice when considering appeals from unfavorable rulings in the matter of UNS number assignments.

4. Procedure for Requesting Number Assignment to Metals and Alloys Not Already Covered by UNS Numbers (or Codes)

4.1 UNS numbers are assigned only to metals and alloys which have a commercial standing (as defined in Note 1).

APPLICATION FOR UNS NUMBER ASSIGNMENT
and
Data Input Sheet for Entering a Specific Material in the
SAE-ASTM Unified Numbering System for Metals and Alloys
(See Reverse Side for Instructions for Completing This Form)

Material Description _____

_____

_____

_____ Suggested UNS No. _____

*UNS Assigned Description _____

_____

_____ *UNS Assigned No. _____

*Chemical Composition (percent by wt.)

Aluminum	Al	_____	Indium	In	_____	Selenium	Se	_____
Antimony	Sb	_____	Iridium	Ir	_____	Silicon	Si	_____
Arsenic	As	_____	Iron	Fe	_____	Silver	Ag	_____
Beryllium	Be	_____	Lead	Pb	_____	Sulfur	S	_____
Bismuth	Bi	_____	Lithium	Li	_____	Tantalum	Ta	_____
Boron	B	_____	Magnesium	Mg	_____	Tellurium	Te	_____
Cadmium	Cd	_____	Manganese	Mn	_____	Thorium	Th	_____
Carbon	C	_____	Mercury	Hg	_____	Tin	Sn	_____
Chromium	Cr	_____	Molybdenum	Mo	_____	Titanium	Ti	_____
Cobalt	Co	_____	Nickel	Ni	_____	Tungsten	W	_____
Columbium	Cb	_____	Nitrogen	N	_____	Uranium	U	_____
Copper	Cu	_____	Oxygen	O	_____	Vanadium	V	_____
Germanium	Ge	_____	Phosphorus	P	_____	Zinc	Zn	_____
Gold	Au	_____	Platinum	Pt	_____	Zirconium	Zr	_____
Hafnium	Hf	_____	Rhenium	Re	_____			_____
Hydrogen	H	_____	Rhodium	Rh	_____			_____

Other _____

_____

_____

*Cross References

AA _____

ACI _____

AISI _____

ANSI _____

AMS _____

ASME _____

ASTM _____

AWS _____

CDA _____

FED _____

MIL SPEC _____

SAE _____

OTHER _____

Requesting Person and Organization (full address) _____

_____ Date of Request _____

*Assigning Org. _____ *Date of UNS Assignment _____

Assigner's Name and Office _____

Applicant: DO NOT write in boxed areas.   * These items for Computer Operator

FIG. 1—APPLICATION FORM FOR UNS NUMBER ASSIGNMENT (FRONT)

GENERAL

Before attempting to complete this form, the applicant should be thoroughly familiar with the objectives of the UNS and the "ground rules" for assigning numbers, as stated in Section 4 of SAE J1086 and ASTM E 527.

MATERIAL DESCRIPTION

Identify the base element; the single alloying element that constitutes 50% or more of the total alloy content; other distinguishing predominant characteristics (such as "casting"); and common or generic names if any (such as "ounce metal" or "Waspalloy"). When no single element makes up 50% or more of the total alloy content, list in decreasing order of abundance, the two alloying elements which together constitute the largest portion of the total alloy contents; except that if no two elements make up at least 50% of the total alloy content, list the three most abundant, and so on. Instead of "iron", use "steel" to identify the base element of those iron-low-carbon alloys commonly known as steels.

When mechanical properties or physical characteristics are the primary defining criteria and chemical composition is secondary or nonsignificant, enter such properties and characteristics with the appropriate values or limits for each.

SUGGESTED UNS NO.

While applicant's suggestion may or may not be the one finally assigned, it will assist proper identification of the material by the UNS Number Assigner.

CHEMICAL COMPOSITION

◊ Enter limits such as 0.13–0.18 (not .13–.18 or 0.13 to 0.18), 1.5 max, 0.040 min, and balance. In space designated "other," enter information such as "Each 0.05 max, Total 0.15 max" and "Sn plus Pb 2.0 min". Additional specific elements not included in the list on this form may be entered in the spaces provided at the end of the list.

CROSS REFERENCES

Letter symbols listed indicate widely known trade associations and standards issuing organizations. Enter after appropriate symbols any known specification numbers or identification numbers issued by such groups to cover material equivalent to, similar to, or closely resembling the subject material.

Examples ; SAE J404 (50B44), AISI 415, ASTM A638 (660)

In space designated "other" enter any pertinent numbers issued by groups not listed above. In these instances, the full name and address of the issuing group shall be included.

SUBMIT COMPLETED FORM TO:
Unified Numbering System
SAE
400 Commonwealth Drive
Warrendale, PA 15096

FIG. 1—APPLICATION FORM FOR UNS NUMBER ASSIGNMENT (BACK)

φ TABLE 3—NUMBER ASSIGNERS AND AREAS OF RESPONSIBILITY

1. The Aluminum Association 750 Third Avenue New York, New York 10017 Attention: Office for Unified Numbering System for Metals Telephone: (212) 972-1800, ext. 32	Aluminum and aluminum alloys UNS Number Series: A00001–A99999
2. American Iron and Steel Institute 1000 16th Street N.W. Washington, D.C. 20036 Attention: Office for Unified Numbering System for Metals Telephone: (202) 452-7100	Carbon and alloy steels UNS Number Series: G00001–G99999 H-steels UNS Number Series: H00001–H99999 Heat and corrosion resistant (stainless) steels UNS Number Series: S00001–S99999 Tool steels UNS Number Series: T00001–T99999
3. American Society for Testing and Materials 1916 Race Street Philadelphia, PA 19103 Attention: Office for Unified Numbering System for Metals Telephone: (215) 299-5521	Miscellaneous steels and ferrous alloys UNS Number Series: K00001–K99999 Rare earth and rare earth-like metals and alloys UNS Number Series: E00001–E99999 Low melting metals and alloys UNS Number Series: L00001–L99999 Miscellaneous nonferrous metals and alloys UNS Number Series: M00001–M99999 Cast steels UNS Number Series: J00001–J99999 Zinc and zinc alloys UNS Number Series: Z00001–Z99999 Precious metals and alloys UNS Number Series: P00001–P99999 Cast irons and cast steels UNS Number Series: F00001–F99999
4. American Welding Society 550 N.W. LeJeune Road P.O. Box 351040 Miami, FL 33135 Attention: Office for Unified Numbering System for Metals Telephone: (305) 443-9353	Welding filler metals UNS Number Series: W00001–W99999
5. Copper Development Association Greenwich Office Park 2 Box 1840 Greenwich, CT 06836-1840 Attention: Office for Unified Numbering System for Metals Telephone: (203) 625-8210	Copper and copper alloys UNS Number Series: C00001–C99999
6. SAE (Society of Automotive Engineers) 400 Commonwealth Drive Warrendale, PA 15096 Attention: Office for Unified Numbering System for Metals Telephone: (412) 776-4841	Nickel and nickel alloys UNS Number Series: N00001–N99999 Steels specified by mechanical properties UNS Number Series: D00001–D99999 Reactive and refractory metals and alloys UNS Number Series: R00001–R99999

4.2 The need for a new number should always be verified by determining from the latest complete listing of already assigned UNS numbers that a usable number is or is not available. (See Note 4.)

4.3 For a new UNS number to be assigned, the composition (or other properties, as applicable) must be significantly different from those of any metal or alloy which has already been assigned a UNS number.

4.3.1 In the case of metals or alloys that are normally identified or specified by chemical composition, the chemical composition limits must be reported.

4.3.2 In the case of metals or alloys which are normally identified or specified by mechanical (or other) properties, such properties and limits thereof must be reported. Only those chemical elements and limits, if any, which are significant in defining such materials need be reported.

4.4 Requests for new numbers shall be submitted on "Application for UNS Number Assignment" forms (Fig. 1). Copies of these are available from any UNS number assigning office (Table 3) or facsimiles may be made of the one herein.

4.5 All instructions on the printed application form should be read carefully and all information provided as indicated. (See Note 5.)

4.6 To further assist in assigning UNS numbers, the requester is encouraged to suggest a possible UNS number in each request, giving appropriate consideration to any existing number presently used by a trade association, standards society, producer, or user.

4.7 Each completed application form shall be sent to the UNS number assigning office having responsibility for the series of numbers which appears to most closely relate to the material described on the form (Table 3).

Note 1. The terms "commercial standing," "production usage," and others, are intended to portray a material in active industrial use, although the actual amount of such use will depend, among other things, upon the type of materials. (Obviously gold will not be used in the same "tonnages" as hot rolled steel.)

Different standardizing groups use different criteria to define the status that a material has to attain before a standard number will be assigned to it. For instance, the American Iron and Steel Institute requires for stainless steels "two or more producers with combined production of 200 tons per year for at least two years;" the Copper Development Association requires that the material be "in commercial use (without tonnage limits);" the Aluminum Association requires that the alloy must be "offered for sale (not necessarily in commercial use);" the SAE Aerospace Materials Division calls for "repetitive procurement by at least two users."

While it is apparent that no hard and fast usage definition can be set up for an all-encompassing system, the UNS numbers are intended to identify metals and alloys that are in more or less regular production and use.

A UNS number will not ordinarily be issued for a material which has just been conceived or which is still in only experimental trial.

Note 2. Organizations that issue specifications should report to appropriate UNS number assigning offices (paragraph 3.1.2), any specification changes which affect descriptions shown in published UNS listings.

Note 3. This arrangement of alpha-numeric six character numbers is a compromise of the thinking that identification numbers should indicate many characteristics of the material, and the thinking that numbers should be short and uncomplicated to be widely accepted and used.

Note 4. In assigning UNS numbers, and consequently in searching complete listings of numbers, the predominant element of the metal or alloy usually determines the prefix letter of the series to which it is assigned. In certain instances where no one element predominates, arbitrary decisions are made as to what prefix letter to use, depending upon the producing industry and other factors.

Note 5. The application form is designed to serve also as a data input sheet to facilitate processing each request through to final printout of the data on electronic data processing equipment and to minimize transcription errors at number-assigning offices and data processing centers.

Note 6. One such listing is ASTM Publication No. DS-56 and SAE Handbook Supplement HS 1086 (a joint ASTM-SAE publication).

This report is published by SAE to advance the state of technical and engineering sciences. The use of this report is entirely voluntary, and its applicability and suitability for any particular use, including any patent infringement arising therefrom, is the sole responsibility of the user.

PRINTED IN U.S.A.